The UK Pesticide Guide 2006

Editor: R. Whitehead BA, MSc

 BCPC

 CABI *Publishing*

CABI Publishing (a division of CAB International) is one of the world's foremost publishers of databases, books and journals in agriculture, and applied life sciences. It has a worldwide reputation for producing high quality, value-added information, drawing on its links with the scientific community. From its Headquarters in Wallingford, UK, CABI Publishing runs a worldwide operation, distributing books, journals and electronic products to customers in over 150 countries, and selling its products through a network of international agents. For further information, please contact CABI Publishing, CAB International, Nosworthy Way, Wallingford, Oxon OX10 8DE, UK.

Telephone: (01491) 832111
Fax: (01491) 833508
e-mail: publishing@cabi.org
Web: www.cabi-publishing.org

BCPC (British Crop Production Council) is a self-supporting limited company with charity status, which was formed in 1968 to promote the knowledge and understanding of the science and practice of crop protection. The corporate members include government departments and research councils; advisory services; associations concerned with the farming industry; agrochemical manufacturers; agricultural engineering, agricultural contracting and distribution services; universities; scientific societies; organisations concerned with the environment; and some experienced independent members. Further details available from BCPC, 7 Omni Business Centre, Omega Park, Alton, Hampshire GU34 2QD, UK.

Telephone: (01420) 593200
Fax: (01420) 593209
e-mail: md@bcpc.org
Web: www.bcpc.org

ISBN 1 84593 045 2

Typeset and printed by Page Bros, Norwich, UK

Contents

Disclaimer

Every effort has been made to ensure that the information in this book is correct at the time of going to press, but the Editor and the publishers do not accept liability for any error or omission in the content, or for any loss, damage or other accident arising from the use of the products listed herein. Omission of a product does not necessarily mean that it is not approved or that it is unavailable.

It is essential to follow the instructions on the approved label when handling, storing or using any crop protection product. Approved 'off-label' extensions of use are undertaken entirely at the risk of the user.

The Dangerous Preparations Directive

The Dangerous Preparations Directive (1999/45/EEC) came into force in UK for pesticide and biocidal products on 30 July 2004. Its aim is to achieve a uniform approach to the classification and labelling of most dangerous preparations, including crop protection products. The Directive is implemented in UK under the Chemicals (Hazard Information and Packaging Supply) Regulations 2002, often referred to under the shorthand acronym 'CHIP3'. In most cases the Regulations have led to additional hazard symbols, and associated risk and safety phrases, relating to environmental and health hazards appearing on the label. All products affected by the Directive entering the supply chain from the implementation date above must be so labelled.

Most of the products in this edition are shown with their new environmental hazard classification, and risk phrases detailed in the pesticide profiles in Section 2. However, no advantage should be inferred for those products that do not show the additional hazards and risks.

Editor's Note

This is the nineteenth edition of *The UK Pesticide Guide* – the only printed reference catalogue of pesticides registered for use in agriculture, forestry, horticulture and amenity in the UK. As usual, the entire contents have been reviewed and updated to ensure that this publication remains as complete and accurate as possible.

Reliable information about pesticides is increasingly demanded by the public, as well as being a requirement of assurance schemes, cross-compliance for the Single Farm Payment Scheme and the Entry Level Stewardship Scheme. This Guide, while not a substitute for the manufacturer's label or literature, nevertheless provides farmers, their advisors, spray operators, and others involved with the use of pesticides with much of the information they need.

In this edition we have replaced several of the crop growth-stage scales in Appendix 3 with those known as the BBCH (Biologische Bundesanstalt, Bundessortenamt und Chemische Industrie) scales, as the latter are now the accepted standard for use on product labels. Elsewhere users of the Guide will notice a significant increase in the number of product uses listed as Specific Off-Label Approvals (SOLAs). This arises as growers, especially of minor crops, seek to replace uses and products lost in the European Review Programme with SOLAs for products that have retained extant approval. This is likely to continue in future editions following termination of the Long-Term Arrangements for Extension of Use (LTAEU) and its replacement with SOLAs for the most frequent uses previously permitted.

There are about 150 new product entries in this edition and five new active ingredients, listed for the first time. Users of the Guide are reminded that, in addition to products notified to the Editor for inclusion in Section 2 (and therefore confirmed by the supplier as available for purchase), many more have extant approval but, for a number of reasons, may not be available. These products are shown in Section 3 to enable comparisons to be made with the main list.

Any important new developments or new products notified to the Editor during the year will again be shown on our website at **www.ukpesticideguide.co.uk**

As always, criticisms and suggestions are welcome and, in particular, notification of any errors or omissions.

R. Whitehead
Editor
ukpg@bcpc.org

The information in this publication has been selected from official sources and from suppliers' product labels and product manuals of pesticides approved for use in the UK under the Control of Pesticides Regulations 1986 or the Plant Protection Products Regulations 1995.

The content is based on information received by the Editor up to October 2005.

For developments since publication of this edition, and new products notified to the Editor during the year, see:

www.ukpesticideguide.co.uk

Changes Since 2005 Edition

Pesticides and adjuvants added or deleted since the 2005 edition are listed below. Every effort has been made to ensure the accuracy of this list, but late notifications to the Editor are not included.

Products Added

Products new to this edition are shown here. In addition, products that were listed in the previous edition whose PSD/HSE registration number *and* supplier have changed are included. Where only the PSD/HSE registration number has changed, the product is not listed.

Product	Reg. No.	Supplier	Product	Reg. No.	Supplier
Advisor S	10948	DuPont	Concert SX	12288	DuPont
Agrichem Hy-Pro Duet	12469	Agrichem	Condor	11791	Doff Portland
Agriguard Chlorothalonil	12201	AgriGuard	Coronet	12267	Bayer CropScience
Agriguard Phenmedipham	12148	AgriGuard	Credo	12368	Syngenta
Agrotech Bentazone	12143	Me2	Cropguard	11835	Nufarm UK
Ally SX	12059	DuPont	CTU Minrinse	12458	Makhteshim
Alpha Pendimethalin 330 EC	12214	Makhteshim	Curzate M WG	11901	DuPont
			Cyflamid	12403	Certis
Alpha Phenmedipham 320 SC	12320	Makhteshim	Danadim Progress	12208	Headland
			Dazide Enhance	11943	Fine
Amaize	12298	AgriGuard	Defiant SC	12302	United Phosphorus
Amistar Opti	12515	Syngenta			
Aristocrat	12343	AgriGuard	Defiant WG	12300	United Phosphorus
Aspect	10516	Syngenta			
Astalavista	12335	Nufarm UK	Delphi	11937	Sipcam
Atlantis WG	12478	Bayer CropScience	Deter	12411	Bayer CropScience
Attenzo	11917	BASF	Difcor 250 EC	12361	Nufarm UK
Ballista	A0524	Interagro	Dithianon WG	12460	BASF
Banco	11191	Makhteshim	Divide	12206	Headland
BAS 500 06F	12338	BASF	Divide 50 SC	12377	Headland
Baythroid	11663	Makhteshim	DP 945	11447	DuPont
Beetaweed	12260	AgriGuard	Emrald Berg	12086	Me2
Bellis	12522	BASF	Emrald Eyetort	12034	Me2
Besiege	08086	DuPont	Emrald Flumet	12000	Me2
Biplay SX	12246	DuPont	Emrald Wotsit	12060	Me2
Blois	12376	Makhteshim	Entice	11096	De Sangosse
Bristol Blues	11798	Doff Portland	Epona	12444	BASF
Bullion	12050	DuPont	ESP	10938	De Sangosse
Calaris	12405	Syngenta	Evict	11673	Bayer CropScience
Calibre SX	12241	DuPont	Excalibur	12047	DuPont
Callisto	12323	Syngenta	Fandango	12276	Bayer CropScience
CDP Minis	12374	De Sangosse			
CleanGuard	12055	Amenity Land	Fenlander 2	12282	Makhteshim
Cleanrun Pro	12083	Scotts	Ferramol	12274	Certis
Colstar	12175	DuPont	Finish SX	12259	DuPont

Product	Reg. No.	Supplier	Product	Reg. No.	Supplier
Firebird	12421	Bayer CropScience	Landgold Deltaland	12114	Teliton
Flight	12534	BASF	Landgold Deputy	12115	Teliton
Fortefog 'P' Fumer	H7187	Agropharm	Landgold Diquat	12116	Teliton
Fury 10 EW	11608	Belchim	Landgold Epoxiconazole	12117	Teliton
Galaxy	12228	AgriGuard	Landgold Fenpropimorph 750	12118	Teliton
GEX 353	11474	DuPont			
Glyfos Proactive	11976	Nomix Enviro	Landgold Fluazinam	12119	Teliton
Goltix Super	12216	Makhteshim	Landgold Glyphosate 360	12120	Teliton
Grasp 400 SC	12280	Syngenta			
Greencrop Duet 603	12171	Greencrop	Landgold Lambda-Z	12383	Teliton
Greencrop Hi-Grass 90	12429	Greencrop	Landgold Mecoprop-P 600	12122	Teliton
Greencrop Martello	12318	Greencrop			
Greencrop Orchid B	12251	Greencrop	Landgold Metamitron	12123	Teliton
Greencrop Reaper	12261	Greencrop	Landgold Metazachlor 50	12124	Teliton
Greencrop Rookie	A0515	Greencrop			
Greencrop Vegex	12439	Greencrop	Landgold Metazachlor 50 SC	12133	Teliton
Greentec Mosskiller	12518	Headland Amenity	Landgold Metribuzin WG	12468	Teliton
Grounded	A0456	Helena	Landgold Minoan	12223	Teliton
Harmony M SX	12258	DuPont	Landgold Minoan Two	A0523	Teliton
Harvesan	12238	DuPont	Landgold Pendimethalin 400	12125	Teliton
Headland Javelin	12451	Headland			
Headland Staff 500	12087	Headland	Landgold Pirimicarb 50	12126	Teliton
Helix	12264	Bayer CropScience	Landgold PQF 100	12127	Teliton
Hermolance	12459	Hermoo	Landgold Strobilurin 250	12128	Teliton
Icon	12202	AgriGuard	Landgold Tepraloxydim	12129	Teliton
Instinct	12317	Headland			
Intracrop Novatex	A0462	Intracrop	Landgold Tralkoxydim	12130	Teliton
Intracrop Sprinter	A0513	Intracrop	Landgold Tribenuron 75	12131	Teliton
Intracrop Stay-Put	A0507	Intracrop			
Intracrop Warrier	A0514	Intracrop	Lanxess Metribuzin 70 WG	12487	Lanxess
IPU Minrinse	12457	Makhteshim			
Jaunt	12350	Bayer CropScience	Lenacil FL	12183	United Phosphorus
Javelin	12367	Bayer CropScience	Lenazar Flo	12286	Hermoo
			Lenazar Flowable	11068	Hermoo
Jubilee SX	12203	DuPont	Lens	12182	Interfarm
Judge	11847	Sipcam	Lexus Class	10809	DuPont
Junction	12493	Rigby Taylor	Lincoln VI	11793	Doff Portland
Koala	12556	Nufarm UK	Low Down	A0459	Helena
Landgold Amidosulfuron	12110	Teliton	Luas	12357	AgriGuard
			Lunar	12366	Sherriff Amenity
Landgold Bedrock	12145	Teliton	Luxan Gro Stop 100	12052	Luxan
Landgold Bentazone SL	12111	Teliton	Luxan Isoproturon 500 Flowable	12426	Luxan
Landgold Clodinafop	12112	Teliton	Luxan Metaldehyde 5	12401	Luxan
Landgold Clopyralid 200	12359	Teliton	Luxan Red 5	12390	Luxan
			Luxan Trigger 5	12389	Luxan
Landgold Cycloxydim	12113	Teliton	Maestro	12307	Bayer CropScience

Product	Reg. No.	Supplier	Product	Reg. No.	Supplier
Magic Tandem	11814	Bayer CropScience	Raxil Pro	12432	Bayer CropScience
Magneto	12339	Nufarm UK	Re-act	12231	Scotts
Masalon	12385	Rigby Taylor	Redigo	12085	Bayer CropScience
Matilda	12006	Nufarm UK	Redigo Deter	12423	Bayer CropScience
Me2 Lambda 2	12437	Me2			
Me2 Pendimethalin	12420	Me2	Redigo Twin	12314	Bayer CropScience
Mediator Sun	A0512	Nufarm UK			
Minuet EW	12304	Belchim	Regalis	12414	BASF
Mitron 70 WG	11516	Hermoo	Reldan 22	12404	Dow
Mitron 90 WG	10888	Hermoo	Renovator Pro	12204	Scotts
Mobius	12324	Bayer CropScience	Rhapsody	11958	DuPont
			Rigid	12319	AgriGuard
Monceren Flowable	11425	Bayer CropScience	Rockett	12344	AgriGuard
Mystique	12348	Nufarm UK	Rubigan	12355	Gowan
Olympus	12407	Syngenta	Safari Lite WSB	12169	DuPont
Oncol 10G	12139	Nufarm UK	Samson	12141	Syngenta
Opera	12167	BASF	Scorpio	12293	Bayer Environ.
Optimol	10939	De Sangosse	SL 567A	12380	Syngenta
Orius	12105	Makhteshim	Snitch	12466	Nufarm UK
Pacifica	12049	Bayer CropScience	Splice	12315	BASF
			Standon Shiva	A0519	Standon
Pan Tepee	12279	Pan Agriculture	Standon Win-Win	12386	Standon
Panarex	12532	Certis	Starion Flo	12455	Belchim
Parador	12372	United Phosphorus	Steam	12473	Chiltern
			Subdue	12358	Fargro
Pathfinder Excel	12501	Barclay	Subdue	12503	Fargro
Penncozeb WDG	12156	Nufarm UK	Super Six	11794	Doff Portland
peroxyacetic acid	PG010	various	Supreme	12233	AgriGuard
Pinnacle	12285	Headland	Talstar 80 Flo	12352	Belchim
Pistol	12173	Bayer Environ.	Teppeki	12402	Belchim
Platoon	12325	BASF	Thrust	12230	Nufarm UK
potassium bicarbonate	PG011	various	Thunder	12464	Makhteshim
Powertwin	12373	Makhteshim	Tonik	12247	AgriGuard
Presite SX	12291	DuPont	Topik	12333	Syngenta
Primo Maxx	11878	Scotts	Toro	12138	Sipcam
Proline	12084	Bayer CropScience	Tracer	12438	Landseer
			Tracker	12295	BASF
Prosaro	12263	Bayer CropScience	Traton SX	12270	DuPont
			Uranus	12506	Makhteshim
Qdos Brefi	A0539	Me2	urea	PG012	various
Qdos Chikara	A0540	Me2	Venture	12316	BASF
Qdos Rockall 2	12488	Me2	Walabi	12265	BASF
Quantum SX	12239	DuPont	Weedazol-TL	11968	Nufarm UK
Rapsan 500 SC	12365	Nufarm UK	Zarado	A0516	De Sangosse

Products Deleted

The appearance of a product name in the following list may not necessarily mean that it is no longer available. In cases of doubt, refer to the supplier.

Product	Reg. No.	Supplier
50/50 Liquid Mosskiller	07191	Vitax
Admire	07481	Bayer
Agate	08826	Bayer
Agate	11235	Bayer CropScience
Agricola Lenacil FL	09481	Agricola
Agricola Lens	10257	Agricola
Agricola Lens Plus	11587	Interfarm
Agriguard 5C Chlormequat 460	09851	AgriGuard
Agriguard Aldicarb	10481	Tronsan
Agriguard Alpha-Cyper	09920	AgriGuard
Agriguard Chlormequat 700	09782	AgriGuard
Agriguard Chlormequat 760	10290	AgriGuard
Agriguard Chlorpyrifos	10626	Tronsan
Agriguard Clopyralid	10625	AgriGuard
Agriguard Cypermethrin EC	12134	AgriGuard
Agriguard Deltamethrin	10770	AgriGuard
Agriguard Duo	10453	AgriGuard
Agriguard Fluroxypyr	09298	AgriGuard
Agriguard Isoproturon	09769	AgriGuard
Agriguard Isoxaben	11652	AgriGuard
Agriguard Metamitron	09859	AgriGuard
Agriguard Metsulfuron	09569	AgriGuard
Agriguard Phenmedipham	10712	AgriGuard
Agriguard Propyzamide Flo	10435	AgriGuard
Agriguard Triclopyr	10679	AgriGuard
Agropen	A0227	Intracrop
Alfacron 10 WP	09439	Novartis A H
Alfadex	H7221	Novartis A H
Aliette 80 WG	09156	Certis
Ally	02977	DuPont
Ally WSB	06588	DuPont
Alpha Protugan Plus	08799	Makhteshim
Alphamouse	H6692	Killgerm
Alphathrin	11163	Nufarm UK
Amazon	10266	Bayer CropScience
Amistar Opti	12433	Syngenta
Apex 5E	08739	Novartis A H

Product	Reg. No.	Supplier
Applaud	10930	Certis
Artist	10658	Bayer
AS Elite	A0406	Monsanto
Ashlade Carbendazim Flowable	06213	Nufarm UK
Atlacide Soluble Powder	00125	Nomix Enviro
Atladox HI	05559	Nomix Enviro
Atlas 5C Quintacel	11130	Nufarm UK
Atlas Adjuvant Oil	A0021	Nufarm UK
Atlas Cropguard	09123	Nufarm UK
Atlas Solan 40	07726	Atlas
Atrazol	07598	Sipcam
Attribut	10671	Bayer
Aurora 50 WG	09807	FMC
Bacara	09976	Bayer CropScience
Bandu	10629	Cheminova
Barclay Gallup Biograde 360	11370	Barclay
Barclay Goalpost	11382	Barclay
Barclay Hurler	11356	Barclay
Barclay Keeper	11375	Barclay
Barclay Keeper 500 FL	11376	Barclay
BAS 493F	11748	BASF
Base 50 W	10202	Interfarm
Base 50 W	10765	Dow
BASF MCPA Amine 50	00209	BASF
Bavistin DF	03848	BASF
Bavistin FL	00218	BASF
Baytan Flowable	02593	Bayer
Baytan Secur	09510	Bayer
Beam	10411	Bayer
Beam	11255	Bayer CropScience
Beetup Flo 160	10382	United Phosphorus
Benlate Fungicide	00229	DuPont
Beret Gold	08390	Syngenta
Besiege WSB	08075	DuPont
Bettix 70 WG	11019	United Phosphorus
Biothene	A0360	Intracrop
Biplay PX	10990	DuPont
Brits	10116	Doff Portland
Brits	11792	Doff Portland

Product	Reg. No.	Supplier	Product	Reg. No.	Supplier
Caddy 240 EC	10631	Bayer	Emerald	A0031	Intracrop
Calibre	07795	DuPont	Enhance	A0147	Greenhill
Calypso	10342	Bayer	Enhance Low Foam	A0148	Greenhill
Casoron G4	09215	Crompton	Ethokem	A0353	Greenhill
CDA Vanquish	09927	Bayer Environ.	Ethos	10736	Bayer
Centium 360 CS	10720	FMC	Euroagkem Pace	A0305	Euroagkem
Challenge	07306	Bayer CropScience	Evade	08071	Dow
			Evict	09150	Bayer
Chum	A0420	Greenhill	Evidence	06934	Aventis
Circium II	11143	Nufarm UK	Exit Wetex	10149	Bayer
Circium II	11801	Nufarm UK	Fernpath Banjo	10182	AgriGuard
Citation 70	09370	United Phosphorus	Fernpath Haptol	10446	AgriGuard
			Fernpath Hatchet	11510	AgriGuard
Colstar	06783	DuPont	Fernpath Tangent	10341	AgriGuard
Compass	10041	Aventis	Fieldgard	11541	Nufarm UK
Conserve	11011	Dow	Fiesta T	10260	BASF
Covershield	10900	BASF	Finish PX	10989	DuPont
Cropsafe 5C Chlormequat	11179	Certis	Flamenco	09913	Aventis
			Flexidor 125	05104	Dow
Croptex Bronze	04087	Certis	Foil	10905	Aventis
Curzate M68 WSB	08073	DuPont	Folicur	08691	Bayer
Cutinol	A0281	Greenhill	Force ST	11085	Bayer
Cutinol Plus	A0395	Greenhill	Force ST	11671	Bayer CropScience
Cyperkill 5	00625	Mitchell Cotts			
Cyren	08358	Cheminova	Fort	11606	Bayer CropScience
Dacthal W-75	10623	AMVAC			
Dagger	10218	BASF	Freeway	09933	Bayer Environ.
Dazide	02691	Fine	Frupica	10944	Certis
Decoy Wetex	09707	Bayer	Frupica	11178	Certis
Deosan Rataway	08803	Johnson Diversey	Fumite Dicloran Smoke	09291	Certis
Deosan Rataway Bait Bags	08805	Johnson Diversey	Fumyl-O-Gas	04833	Brian Jones
			Fungaflor	11326	Certis
DG90	11200	Sipcam	Fungaflor Smoke	11327	Certis
Discovery	10416	United Phosphorus	Fury 10 EW	10718	FMC
			Fusilade 250 EW	10525	Syngenta
Doff Horticultural Slug Killer Blue Mini Pellets	09666	Doff Portland	Galion	A0162	Intracrop
			Gaucho	06590	Bayer
Doxstar	06050	Dow	Genesis	09993	Sipcam
DP 911 PX	10992	DuPont	Gharda Bonanza	11571	Nufarm UK
DP 911 WSB	09867	DuPont	Glydate	10999	Nufarm UK
Druid	08714	Aventis	Glyfos	07109	Cheminova
Du Pont Adjuvant	A0119	DuPont	Glyfos ProActive	07800	Nomix Enviro
Duplosan KV	09431	BASF	Glyper	11095	Interfarm
Easel	11146	Nufarm UK	Glyphosate 360	09151	Applyworld
Eclipse	07361	BASF	Glyphosate 360	09151	Dow
Elan	A0392	Intracrop	Glyphosate 360	11579	Cardel
Electis 75 WG	10565	Interfarm	Goltix 90	08654	Bayer
Elvaron Multi	10080	Bayer	Goltix Flowable	08986	Bayer
Embark Lite	08749	Intracrop	Goltix WG	02430	Bayer

Product	Reg. No.	Supplier
Granit	09995	Bayer CropScience
Greencrop Biscay	12132	Greencrop
Greencrop Cajole	11402	Greencrop
Greencrop Coolfin	09449	Greencrop
Greencrop Reaper	09359	Greencrop
Greencrop Reaper 2	10060	Greencrop
Greenhill Spreader	A0196	Greenhill
Greenmaster Autumn	07508	Scotts
Greenmaster Mosskiller	07509	Scotts
Greenor	07848	Rigby Taylor
Grovex Zinc Phosphide	H6800	Killgerm
GS 800	A0354	Greenhill
Harlequin 500 SC	09779	Makhteshim
Harmony M	03990	DuPont
Headland Link	11091	Headland
Headland Neptune	10230	Headland
Headland Staff	07189	Headland
Helmsman	09934	Bayer Environ.
Herbasan	10734	Nufarm UK
Heritage	11383	Scotts
Hive	10953	Nufarm UK
Huron	10148	Bayer
HY-D	06278	Agrichem
I T Alpha-Cyper	10483	I T Agro
I T Asulam	10186	I T Agro
I T Cyanazine	10179	I T Agro
I T Cyper	10557	I T Agro
I T Dicamba	10976	I T Agro
I T Fosetyl-AL	11717	I T Agro
I T Glyphosate	07212	I T Agro
I T Iprodione	08267	I T Agro
Ibex	10901	BASF
Inter-Asulam	10929	I T Agro
Inter-Metribuzin WG	09801	I T Agro
Inter-Pendimethalin	09841	I T Agro
Intracrop Bla-Tex	A0173	Intracrop
Intracrop Green Oil	A0262	Intracrop
Intracrop Non-Ionic Wetter	A0009	Intracrop
Intracrop Rapeze	A0390	Intracrop
Intracrop Rapide Beta	A0175	Intracrop
Iona	A0232	Intracrop
Iona Low Foam	A0255	Intracrop
Isoguard	10829	Nufarm UK
Isomec	09881	Nufarm UK
Javelin	09998	Bayer CropScience
Jockey	10076	Aventis
Jockey F	10074	Aventis
Jogral	A0226	Intracrop
Jubilee (WSB)	06082	DuPont
Judge	10000	Sipcam
Karan	09637	Bayer
Karate Ready to Use Rat Bait	H6745	Johnson Diversey
Karate Ready to Use Rodenticide Sachets	H6746	Johnson Diversey
Karmex	09475	Headland
Kaspar	10615	Certis
Katamaran	09049	BASF
Kerb 50 W	10200	Interfarm
Kerb 80 EDF	10761	Dow
Kerb Flo	10160	Interfarm
Kerb Pro Flo	10158	SumiAgro Amenity
Kernel	10616	Cheminova
Killgerm Fenitrothion 40 WP	H4858	Killgerm
Killgerm Fenitrothion 50 EC	H4722	Killgerm
Klerat	H6701	Sorex
Klerat Wax Blocks	H6703	Sorex
Knot Out	05163	Vitax
Konker	03988	BASF
Kruga 5EC	09863	Interfarm
Kubist 2	11126	Nufarm UK
Laminator DG	11073	Interfarm
Laminator FL	11072	Interfarm
Laminator WP	11071	Interfarm
Landgold CCC 720	08527	Landgold
Landgold Difenoconazole	09964	Landgold
Landgold Epoxiconazole FM	08806	Landgold
Landgold Ethofumesate 200	08980	Landgold
Landgold Fenpropidin 750	08973	Landgold
Landgold Fluroxypyr	10490	Landgold
Landgold Flusilazole MBC	08528	Landgold
Landgold Isoproturon	06012	Landgold
Landgold Rimsulfuron	08959	Landgold
Landgold Strobilurin KE	09908	Landgold
Landgold Strobilurin KF	09196	Landgold
Landgold Sulfosulfuron	10908	Landgold
Landgold TFS 50	08941	Landgold

Product	Reg. No.	Supplier
Landgold Vinclozolin SC	06459	Landgold
Lexus Class WSB	08637	DuPont
Lexus XPE WSB	08542	DuPont
Leyclene	07263	Interfarm
Lo Dose G	A0349	Reabrook
Loncid	10832	I T Agro
Lotus	09231	BASF
Lupus	09638	Bayer
Luxan 2,4-D	09379	Luxan
Luxan 9363	07359	Luxan
Luxan 9363 Red	11480	Luxan
Luxan Chlorotoluron 500 Flowable	12431	Luxan
Luxan Dichlorvos 600	08297	Luxan
Luxan Dichlorvos Aerosol 15	08298	Luxan
Luxan Gro-Stop Basis	08601	Luxan
Luxan Isoproturon 500 Flowable	07437	Luxan
Luxan MCPA 500	07470	Luxan
Luxan Metaldehyde	06564	Luxan
Luxan Trigger	10419	Luxan
Maneb 80	01276	Rohm & Haas
Manzate 75 WG	11046	Headland
Marnoch Chlorothalonil	09763	Marnoch
Marshal 10G	11682	Belchim
Mascot Mosskiller	02439	Rigby Taylor
Mascot Systemic	08776	Rigby Taylor
Mastiff	11747	BASF
Matador 200 SC	11058	Margarita
Mazide Selective	05753	Vitax
Me2 After	10463	Me2
Me2 Buy	10655	Me2
Me2 New Aldee	10434	Me2
Me2 Notin	11655	Me2
Me2 Snap!	10476	Me2
Me2 Succotash	11815	Me2
Merit	10384	BASF
Metham Sodium 400	08051	United Phosphorus
Micromite	H4480	Killgerm
Minuet EW	10716	FMC
Mircam Plus	11004	Nufarm UK
Mirquat	10604	Nufarm UK
Mitac HF	07358	Bayer CropScience
Mithras 80 EDF	10164	Interfarm
Monceren DS	04160	Bayer
Monceren IM	06259	Bayer

Product	Reg. No.	Supplier
n2n Diuron 80	11821	Nomix Enviro
n2n Diuron Flowable	11822	Nomix Enviro
n2n Garlon 4	11806	Nomix Enviro
n2n Mosskiller	11823	Nomix Enviro
Nectec Paste	08510	Certis
Nemathorin 10G	08915	Syngenta
Nemolt	07012	Cyanamid
Neon	08337	Bayer
Neon	11226	Bayer CropScience
Neosorexa	H6773	Sorex
Neosorexa Ratpacks	H6782	Sorex
Neporex 2SG	H7222	Novartis A H
New Draza	11082	Bayer
Nico Soap	07517	United Phosphorus
Nomix Diuron 80	12079	Nomix Enviro
Novak	08020	Aventis
Novak	11693	BASF
Nu Film 17	A0055	Intracrop
Nufarm Adjuvant Oil	A0474	Nufarm UK
Nuvan 500 EC	08590	Novartis A H
Oncol 10G	08249	Mirfield
Opus Team	07362	BASF
Pal	A0396	Greenhill
Panther	10008	Aventis
Panther WDG	10650	Bayer CropScience
Partna	A0428	Syngenta
Pastor	07440	Dow
Patriot EC	08990	Aventis
Penncozeb WDG	11622	Nufarm UK
Perm-E8	A0304	Intracrop
Pin-o-Film	A0209	Intracrop
Platform S	09706	FMC
Plinth	11094	Interfarm
Plover	08429	Syngenta
Podquat	03003	Mandops
Poncho Beta	11999	Bayer CropScience
Poraz	10474	Aventis
Posse 10G	10725	Bayer
Premiere Granules	07987	Dow
Premis	10177	Aventis
Premis	11742	BASF
Profile	08134	Aventis
Propose	11195	Interfarm
Prospect	09389	Nufarm UK
Quantum	06270	DuPont
Quantum 75 DF	09340	DuPont
Quaver Flo	10162	Interfarm

Product	Reg. No.	Supplier
Racumin Contact Powder	H6755	Bayer Environ.
Ragox	11145	Nufarm UK
Raxil S	06974	Bayer
Raxil Secur	08966	Bayer
RCR Zinc Phosphide	H6801	Killgerm
Redeem Flo	10236	Interfarm
Redeem Flo	10764	Dow
Reldan 22	08191	Dow
Renardine 72-2	06769	Roebuck Eyot
Repulse	07641	Certis
Reward 5EC	09862	Interfarm
Rhythm	09636	Interfarm
Rivet	09512	Bayer
Rizolex Flowable	09358	Certis
Robust	10176	Aventis
Rombus	11301	Bayer CropScience
Ronstar Liquid	08974	Certis
Rouge	11090	Nufarm UK
Rover	09848	Sipcam
Rovral Flo	10013	Aventis
Rovral Green	09938	Bayer Environ.
Rovral Liquid FS	10551	Aventis
Rovral WP	10015	Aventis
Roxam 75 WG	10566	Interfarm
Sage	10413	Bayer
Salute	07660	United Phosphorus
Scala	07806	Aventis
Sceptre	08043	Bayer CropScience
Sectacide 50 EC	H4939	Killgerm
Sencorex WG	03755	Bayer
Seradix 1	10422	Certis
Seradix 2	10423	Certis
Seradix 3	10424	Certis
Shark	11054	FMC
Shark	11616	Belchim
sHYlin	10030	Agrichem
Sibutol	07238	Bayer
Sibutol Secur	09131	Bayer
Sierraron G	09675	Scotts
Silvacur	06387	Bayer
Sinbar	01956	DuPont
SL 567A	08811	Syngenta
Slaymor	H6729	Novartis A H
Slaymor Bait Bags	H6730	Novartis A H
Snooker	10018	Aventis
Sobrom BM 100	04381	Brian Jones
Sobrom BM 98	04189	Brian Jones

Product	Reg. No.	Supplier
Solfa	06959	Nufarm UK
Sorex Fatal	H7087	Sorex
Sorex Super Fly Spray	H6297	Sorex
Sorex Warfarin 250 ppm Rat Bait	H6812	Sorex
Sorex Warfarin 500 ppm Rat Bait	H6813	Sorex
Sorex Warfarin Sewer Bait	H6814	Sorex
Sorex Wasp Nest Destroyer	H6294	Sorex
Sorexa CD Mouse Bait	H6751	Sorex
Speedway 2	09273	Scotts
Spin Out	07610	Fargro
Splice (old)	11918	BASF
Sportak 45 EW	07996	Aventis
Sportak Delta 460 HF	07431	Aventis
Spotlight 24 EC	10702	FMC
Standon Aldicarb 10G	11024	Standon
Standon Chlorothalonil 500	08597	Standon
Standon CIPC 300 HN	09187	Standon
Standon Cymoxanil Extra	09442	Standon
Standon Cyprodinil	09345	Standon
Standon Dichlobenil 6G	08874	Standon
Standon Ethofumesate 200	09360	Standon
Standon Fluazinam 500	08670	Standon
Standon Flupyrsulfuron MM	09098	Standon
Standon Fluroxypyr	08923	Standon
Standon Fluroxypyr 200	11534	Standon
Standon Flusilazole Plus	07403	Standon
Standon Glyphosate 360	05582	Standon
Standon Metamitron	07885	Standon
Standon Metsulfuron	05670	Standon
Standon Pendimethalin 400 SC	10729	Standon
Standon Pyrimethanil	10576	Standon
Standon Triflusulfuron	09487	Standon
Starane 2	05496	Dow
Starion	09795	FMC
Sterilite Tar Oil Winter Wash 60% Stock Emulsion	05061	Coventry Chemicals

Product	Reg. No.	Supplier
Sterilite Tar Oil Winter Wash 80% Miscible Quality	05062	Coventry Chemicals
Stomp Pico	10738	BASF
Storite Clear Liquid	08982	Banks Cargill
Super Selective Plus	11928	Rigby Taylor
Superflor 6% Metaldehyde Slug Killer Mini Pellets	09773	CMI
Surpass 5EC	09861	Interfarm
suSCon Indigo	09902	Fargro
Swift SC	10884	Bayer
Systhane 6 Flo	07334	Landseer
Systol M	08085	Quadrangle
T 25	A0352	Reabrook
Takron	06237	BASF
Talon Rat & Mouse Bait (Cut Wheat)	H6709	Killgerm
Talon Rat & Mouse Bait (Whole Wheat)	H6710	Killgerm
Talstar	10835	Certis
Talzene	A0233	Intracrop
Target SC	10861	Unicrop
Tech E	A0346	Greenhill
Tech G	A0345	Greenhill
Teldor	08955	Bayer
Terbine	11407	Nufarm UK
Thripstick	02134	Aquaspersions
Titus	07908	DuPont
Tolkan Liquid	10023	Aventis
Tolkan Liquid	10895	Aventis
Topas C 50 WP	08459	Novartis
Torch	08336	Bayer
Torch	11258	Bayer CropScience
Tordon 101	05816	Dow
Torpedo	A0336	Loveland
Trimanzone	09584	Intracrop
Tripart Accendo	06110	Tripart
Tripart Acer	A0097	Tripart
Tripart Beta	03111	Tripart

Product	Reg. No.	Supplier
Tripart Brevis	03754	Tripart
Tripart Culmus	06619	Tripart
Tripart Defensor FL	02752	Tripart
Tripart Faber	04549	Tripart
Tripart Gladiator 2	06618	Tripart
Tripart Lentus	A0117	Tripart
Tripart Minax	A0108	Tripart
Tripart Orbis	A0204	Tripart
Tripart Sentinel	03250	Tripart
Tripart Sentinel 2	05140	Tripart
Tripart Trifluralin 48 EC	02215	Tripart
Trireme	11524	Nufarm UK
Triumph	08740	Aventis
Tuberon	10613	Unicrop
Tumbleweed Pro	11083	Scotts
Turf Systemic Fungicide	09349	PBI
Twist	10587	Bayer
Ultrafaber	05627	Tripart
Unix	08764	Syngenta
UPL Diuron 80	07619	United Phosphorus
Uplift	07527	United Phosphorus
Validate	A0500	Loveland
Veto F	08057	Bayer
Vindex	05470	Dow
Vitesse	10042	Bayer Environ.
Vizzler	11409	Interfarm
Wayfarer	A0045	Certis
Weedazol-TL	02979	Bayer
Weedazol-TL	11430	Bayer CropScience
X-Spor SC	08077	United Phosphorus
Zapper	09944	Bayer Environ.
Zenon	09193	Bayer
Zenon	11232	Bayer CropScience
ZP Rodent Pellets	07814	Antec Biosentry

Introduction

Purpose

The primary aim of this book is to provide a practical handbook of pesticides, plant growth regulators and adjuvants that the farmer or grower can realistically and legally obtain in the UK, and to indicate the purposes for which they may be used. It is designed to help in the identification of products appropriate for a particular problem. In addition to uses recommended on product labels, details are provided of those uses that do not appear on labels but which have been granted specific off-label approval (SOLA). As well as identifying the products available, the book provides guidance on how to use them safely and effectively, but without giving details of doses, volumes, spray schedules or approved tank mixtures. Sections 5 and 6 provide essential background information on a wide range of pesticide-related issues including legislation, codes of practice, poisons and treatment of poisoning, products for use in special situations, and weed and crop growth stage keys.

While we have tried to cover all other important factors, this book does **not** provide a full statement of product recommendations. **Before using any pesticide product it is essential that the user should read the label carefully and comply strictly with the instructions it contains. This *Guide* is not a substitute for the product label**.

Scope

The *Guide* is confined to pesticides registered for use in the UK. Details are given of some 350 individual active ingredients, and a further 150 profiles of mixtures containing these active ingredients. For each active ingredient a list is shown of available approved products for use in arable agriculture, horticulture (including amenity horticulture), forestry and areas near water, together with the name of the supplier from whom each may be obtained. Within these fields of use, all types of pesticide covered by the Control of Pesticides Regulations 1986 or the Plant Protection Products Regulations are included. This embraces acaricides, algicides, fungicides, herbicides, insecticides, lumbricides, molluscicides, nematicides and rodenticides, together with plant growth regulators used as straw shorteners, sprout inhibitors and for various horticultural purposes. The total number of pesticides covered in the *Guide* is about 1150.

Products are included in the main section of the *Guide* only if requested by the supplier and supported by evidence of approval. This is intended to ensure that only marketed products are listed. In addition, the *Guide* gives information on more than 100 authorised adjuvants which, although not themselves pesticides, may be added to pesticides to improve their effectiveness in use.

The *Guide* does **not** include products approved solely for amateur, home garden, domestic, food storage, public health or animal health uses.

Sources of Information

The information for this edition has been drawn from these authoritative sources:

- approved labels and product manuals received from suppliers of pesticides up to October 2005
- websites of the Pesticides Safety Directorate (PSD, www.pesticides.gov.uk) and the Health and Safety Executive (HSE, www.hse.gov.uk).

Criteria for Inclusion

To be included in the *Guide*, a product must meet the following conditions:
- it must have extant approval under UK pesticides legislation
- information on the approved uses must have been provided by the supplier
- it must be expected to be on the UK market during the currency of this edition.

When a company changes its name, whether by merger, take-over or joint venture, it is obliged to re-register its products in the name of the new company, and new Ministry registration numbers are normally assigned. Where stocks of the previously registered products remain legally available, both old and new identities are included in the *Guide* and remain until approval for the former lapses, or stocks are notified as exhausted.

Products that have been withdrawn from the market and whose approval will finally lapse during 2005 are identified in each profile. After the indicated date, sale, storage or use of the product bearing that approval number becomes illegal. Where there is a direct replacement product, this is indicated.

The Voluntary Initiative

The Voluntary Initiative (VI) is a five-year programme of measures, agreed by the crop protection industry and related organisations with Government, to minimise the environmental impacts of crop protection products. The programme is providing a framework of practices and principles that will help Government achieve its objective of protection of water quality and enhancement of farmland biodiversity. The programme formally concludes in March 2006 but many of its components are likely to continue, and anyone involved with the products in this *Guide*, whether as a user or advisor, is likely to be affected directly or indirectly.

This edition of *The UK Pesticide Guide* can help in two specific ways.

- It is illegal to use, or even store, products whose approval has expired. While mere storage is unlikely to have any environmental impact, their incorrect or inappropriate disposal could undoubtedly do so. All the products listed in Sections 2 and 3 of this *Guide* have extant approval for all or part of 2006. A date is shown for any whose approval will expire before the end of the year. Therefore any products not listed here are likely to be obsolete, although a check should be made with the manufacturer or supplier before disposal.

- Product labels contain much information that is often too crowded and complicated. Another project in the VI, undertaken by the Crop Protection Association (CPA) together with the Pesticides Safety Directorate and the Health and Safety Executive, is a review of product labels and the production of Best Practice Labelling Guidance. CPA member companies have adapted their labels to be compliant with these guidelines at the same time as making the mandatory changes required by the Dangerous Preparations Directive. The presentation of information in this *Guide* reflects this improved clarity as far as possible. On the sister CD, *The e-UK Pesticide Guide*, we have gone further by using the new symbols for environmental hazard, product activity and LERAP category, as well as providing links to the Voluntary Initiative website (www.voluntaryinitiative.org.uk) to enable access to product Environmental Information Sheets (also being produced as a VI project).

Notes on Contents and Layout

The book consists of six main sections:

1 Crop/Pest Guide
2 Pesticide Profiles
3 Products Also Registered
4 Adjuvants
5 Useful Information
6 Appendices

1 Crop/Pest Guide

This section enables the user to identify which active ingredients are approved for a particular crop/pest combination. The crops are grouped as shown in the Crop Index. For convenience, some crops and targets have been grouped into generic units (for example, 'cereals' are handled as a group, not individually). Therefore indications from the Crop/Pest Guide must always be followed up by reference to the specific entry in the pesticide profiles section because a product may not be approved on all crops in the group. Chemicals indicated as having uses in cereals, for example, may be approved for use only on winter wheat and barley, not for other cereals. Because of differences in the wording of product labels it may sometimes be necessary to refer to broader categories of organism in addition to specific organisms (e.g. annual grasses and annual weeds in addition to annual meadow grass)..

2 Pesticide Profiles

Each active ingredient has a separate, numbered profile entry, as does each mixture of active ingredients. The entries are arranged in alphabetical order using the common names approved by the British Standards Institution, and used in *The Pesticide Manual* (BCPC, 2003). Where an active ingredient is available only in mixtures, this is stated. The ingredients of the mixtures are themselves ordered alphabetically, and the entries appear at the correct point for the first named ingredient.

Within each profile entry, a table lists the products approved and available on the market, in the following style:

Product name	Main supplier	Active ingredient content	Formulation type	Registration number
1 Broadsword	United Phosphorus	200:85:65 g/l	EC	09140
2 Greengard	SumiAgro Amenity	200:85:65 g/l	EC	11715
3 Nu-Shot	Nufarm UK	200:85:65 g/l	EC	11148

Many of the **product names** are registered trademarks, but no special indication of this is given. Individual products are indexed under their entry number in the Index of Proprietary Names at the back of the book. The **main supplier** indicates the marketing outlet through which the product may be purchased. Full addresses and telephone/fax numbers of suppliers are listed in Appendix 1. Some website and e-mail addresses are also given. For mixtures, the **active ingredient contents** are given in the same order as they appear in the profile heading. The **formulation types** are listed in full in the Key to Abbreviations and Acronyms (Appendix 5). The **registration number** normally refers to registration with PSD. In cases where products registered with the Health and Safety Executive are included, HSE numbers are quoted, e.g. H0462.

Below the product table, a **Uses** section lists all approved uses notified by suppliers to the Editor by October 2005, giving both the principal target organisms and the recommended crops or situations (identified in ***bold italics***). Where there is an important condition of use, or the approval is off-label, this is shown in parentheses, e.g. (*off-label*). Numbers in square brackets refer to the numbered products in the table above. Thus, in the example shown above, a use approved for Nu-Shot (product 3) but not for Broadsword or Greengard (products 1 and 2) appears as:

Annual dicotyledons in **rotational grassland** [3]

Any **Specific Off-Label Approvals (SOLAs)** for products in the profile are detailed below the list of approved uses. Each SOLA has a separate entry and shows the crops to which it applies, the Notice of Approval number, the expiry date (if any) and the product reference number in square brackets. SOLAs do not appear on the product label and are undertaken entirely at the risk of the user.

Below the SOLA paragraph, **Notes** are listed under the following headings. Unless otherwise stated, any reference to dose is made in terms of product rather than active ingredient. Where notes refer to particular products, rather than to the entry generally, this is indicated by numbers in square brackets as described above.

Approval information	Information of a general nature about the approval status of the active ingredient or products in the profile is given here. Notes on approval for aerial or ULV application are given.
	Inclusion of the active ingredient in Annex I under EC Directive 91/414 is shown, as well as any acceptance from the British Beer and Pub Association (BBPA) for its use on barley or hops for brewing or malting.
	Where a product approval will finally expire in 2006, the expiry date is shown here together with the registration number of its direct replacement, if any.
Efficacy	Factors important in making the most effective use of the treatment. Certain factors, such as the need to apply chemicals uniformly, the need to spray at appropriate volume rates, and the need to prevent settling out of active ingredient in spray tanks, are assumed to be important for all treatments and are not emphasised in individual profiles.
Restrictions	Notes are included in this section where products are subject to the Poisons Law, and where the label warns about organophosphorus and/or anticholinesterase compounds. Factors important in optimising performance and minimising the risk of crop damage, including statutory conditions relating to the maximum permitted number of applications, are listed. Any restrictions on crop varieties that may not be sprayed are mentioned here.
Environmental safety	Where the label specifies an environmental hazard classification, it is noted, together with the associated risk phrases. Any other special operator precautions, and any conditions concerning withholding livestock from treated areas, are specified. Where any of the products in the profile are subject to Category A, Category B or broadcast air-assisted buffer zone restrictions under the LERAP scheme, the relevant classification is shown.
	Other environmental hazards are also noted here, including potential dangers to livestock, game, wildlife, bees and fish. The need to avoid drift onto neighbouring crops and to wash out equipment thoroughly after use are important with all pesticide treatments, but may receive special mention here if of particular significance.
Crop-specific information	Instructions about timing of application or cultivations that are specific to a particular crop, rather than generally applicable, are listed here. The latest permitted use and harvest intervals (if any) are shown for individual crops.
Following crops guidance	Any specific label instructions about what may be grown after a treated crop, whether harvested normally or after failure, are shown here.

Hazard classification and safety precautions	The label hazard classification(s) and precautions are shown using a series of letter or number codes which are explained in Appendix 4. The codes are listed under a number of sub-headings as follows:

- Hazard
- Risk phrases
- Operator protection
- Environmental protection
- Consumer protection
- Storage and disposal
- Treated seed
- Vertebrate/rodent control products
- Medical advice

This section is given for information only and should not be used for the purpose of making a COSHH assessment without reference to the actual label of the product to be used.

3 Products also Registered

Products with extant approval for all or part of the year of the edition in which they appear are listed in this section if they have not been requested by their supplier or manufacturer for inclusion in Section 2. Details shown are the same (apart from formulation) as those in the product tables in Section 2 and, where relevant, an approval expiry date is shown. Not all the products listed here will be available in the market, but this list, together with the products in Section 2, comprises a comprehensive listing of all approved products in the UK for uses within the scope of the *Guide* for the year in question.

4 Adjuvants

Adjuvants are listed in a table ordered alphabetically by product name. For each product, details are shown of the main supplier, the authorisation number and the mode of action (e.g. wetter, sticker, etc.) as shown on the label. A brief statement of the uses of the adjuvant is given. Protective clothing requirements and label precautions are listed using the codes from Appendix 4.

5 Useful Information

This section summarises legislation covering approval, storage, sale and use of pesticides in the UK. There are brief articles on the broader issues of using crop protection chemicals, including resistance management. Lists are provided of products approved for use in or near water, in forestry, as seed treatments, and for aerial application. Chemicals subject to the Poisons Laws are listed, and there is a summary of first aid measures if pesticide poisoning should be suspected. Finally, this section provides guidance on environmental protection issues and covers the protection of bees and the use of pesticides in or near water.

6 Appendices

Appendix 1	Gives names, addresses, telephone and fax numbers of all companies listed as main suppliers in the pesticide profiles. E-mail and website addresses are also listed where available.
	Where a supplier is no longer trading under that name (usually following a merger or takeover) but products in that name are still listed in the *Guide* (because they are still available in the supply chain), a cross-reference indicates the new 'parent' company from which technical or commercial information can be obtained.
Appendix 2	Gives names, addresses, telephone and fax numbers of useful contacts, including the National Poisons Information Service. Website addresses are included where available.

Appendix 3 | Gives details of the keys to crop and weed growth stages, including the publication reference for each. The numeric codes are used in the descriptive sections of the pesticide profiles (Section 2).

Appendix 4 | Shows the full text for code letters and numbers used in the pesticide profiles (Section 2) to indicate personal protective equipment requirements and label precautions.

Appendix 5 | Shows the full text for the formulation abbreviations used in the pesticide profiles (Section 2). Other abbreviations and acronyms used in the *Guide* are explained here.

Appendix 6 | Provides full definitions of officially agreed descriptive phrases for crops or situations used in the pesticide profiles (Section 2) where misunderstandings can occur.

Appendix 7 | Shows a list of useful reference publications which amplify the summarised information provided in this section.

SECTION 1
CROP/PEST GUIDE

Crop/Pest Guide Index

Important Note: The Crop/Pest Guide Index refers to pages on which the subject can be located.

Crop/Pest Guide

Important note: For convenience, some crops and pests or targets have been brought together into generic groups in this guide, e.g. 'cereals', 'annual grasses'. It is essential to check the profile entry in Section 2 *and* the label to ensure that a product is approved for a specific crop/pest combination, e.g. winter wheat/blackgrass.

Arable and vegetable crops

Brassicas - Brassica seed crops

Diseases	Alternaria	iprodione
	Botrytis	iprodione
Weeds	Broad-leaved weeds	carbetamide, propyzamide, trifluralin (*off-label*)
	Crops as weeds	carbetamide
	Grass weeds	carbetamide, propyzamide, trifluralin (*off-label*)

Brassicas - Brassicas, general

Diseases	Alternaria	iprodione (*seed treatment*)
	Downy mildew	chlorothalonil
	Ring spot	chlorothalonil
Pests	Aphids	chlorpyrifos
	Caterpillars	chlorpyrifos
	Cutworms	chlorpyrifos
	Flies	chlorpyrifos
	Leatherjackets	chlorpyrifos
	Slugs/snails	ferric phosphate

Brassicas - Fodder brassicas

Diseases	Alternaria	azoxystrobin, iprodione, iprodione (*seed treatment*)
	Black rot	copper oxychloride (*off-label*)
	Damping off	fosetyl-aluminium (*off-label*)
	Downy mildew	fosetyl-aluminium (*off-label*), fosetyl-aluminium (*off-label - seedlings*)
	Ring spot	azoxystrobin, difenoconazole (*off-label*), tebuconazole (*off-label*)
	Spear rot	copper oxychloride (*off-label*)
	White blister	azoxystrobin
Pests	Aphids	carbosulfan, imidacloprid (*off-label - seed treatment*), nicotine, pirimicarb, pymetrozine (*off-label*)
	Beetles	alpha-cypermethrin, carbosulfan, deltamethrin
	Caterpillars	alpha-cypermethrin, Bacillus thuringiensis (*off-label*), cypermethrin, deltamethrin, nicotine
	Flies	carbosulfan, chlorpyrifos (*off-label*)
	Leaf miners	nicotine

	Pests, miscellaneous	dimethoate (*off-label*)
	Weevils	carbosulfan
Weeds	Broad-leaved weeds	chlorthal-dimethyl, chlorthal-dimethyl + propachlor, clopyralid, cyanazine (*off-label*), metazachlor (*off-label*), napropamide, propachlor, sodium monochloroacetate, trifluralin
	Crops as weeds	fluazifop-P-butyl (*stockfeed only*)
	Grass weeds	chlorthal-dimethyl + propachlor, fluazifop-P-butyl (*stockfeed only*), napropamide, propachlor, trifluralin
	Mayweeds	clopyralid, metazachlor (*off-label*)

Brassicas - Leaf and flowerhead brassicas

Diseases	Alternaria	azoxystrobin, boscalid + pyraclostrobin, chlorothalonil, chlorothalonil + metalaxyl-M (*moderate control*), difenoconazole, iprodione, tebuconazole
	Black rot	copper oxychloride (*off-label*)
	Botrytis	chlorothalonil
	Damping off	fosetyl-aluminium (*off-label*), thiram (*seed treatment*)
	Damping off and foot rot	etridiazole, etridiazole (*seeds and seedlings*)
	Damping off and wirestem	tolclofos-methyl
	Downy mildew	chlorothalonil, chlorothalonil + metalaxyl-M, fosetyl-aluminium (*off-label*), fosetyl-aluminium (*off-label - seedlings*), propamocarb hydrochloride
	Light leaf spot	tebuconazole
	Phytophthora	etridiazole, etridiazole (*transplants*), propamocarb hydrochloride
	Powdery mildew	tebuconazole
	Pythium	propamocarb hydrochloride
	Ring spot	azoxystrobin, boscalid + pyraclostrobin, chlorothalonil, chlorothalonil + metalaxyl-M (*reduction*), difenoconazole, difenoconazole (*off-label*), tebuconazole, tebuconazole (*off-label*)
	Spear rot	copper oxychloride (*off-label*)
	Storage rots	metalaxyl-M (*off-label*)
	White blister	azoxystrobin, boscalid + pyraclostrobin, boscalid + pyraclostrobin (*qualified minor use*), chlorothalonil + metalaxyl-M, mancozeb + metalaxyl-M (*off-label*)
Pests	Aphids	bifenthrin, carbosulfan, chlorpyrifos, cypermethrin, dimethoate, fatty acids, imidacloprid (*off-label - seed treatment*), lambda-cyhalothrin + pirimicarb, nicotine, pirimicarb, pymetrozine (*off-label*), triazamate
	Beetles	alpha-cypermethrin, carbosulfan, deltamethrin, lambda-cyhalothrin + pirimicarb
	Birds/mammals	aluminium ammonium sulphate
	Caterpillars	alpha-cypermethrin, Bacillus thuringiensis, Bacillus thuringiensis (*off-label*), bifenthrin, chlorpyrifos, cyfluthrin, cypermethrin, deltamethrin, diflubenzuron, lambda-cyhalothrin, lambda-cyhalothrin + pirimicarb, nicotine, spinosad
	Cutworms	chlorpyrifos
	Flies	carbosulfan, chlorpyrifos, chlorpyrifos (*off-label*)

	Leaf miners	nicotine
	Leatherjackets	chlorpyrifos
	Mealybugs	fatty acids
	Pests, miscellaneous	dimethoate (*off-label*)
	Scale insects	fatty acids
	Slugs/snails	methiocarb, thiodicarb
	Spider mites	fatty acids
	Weevils	carbosulfan
	Whiteflies	bifenthrin, chlorpyrifos, cypermethrin, fatty acids, lambda-cyhalothrin, lambda-cyhalothrin + pirimicarb
Weeds	Broad-leaved weeds	carbetamide, chlorthal-dimethyl, chlorthal-dimethyl + propachlor, clopyralid, cyanazine (*off-label*), metazachlor, metazachlor (*off-label*), napropamide, pendimethalin, propachlor, pyridate, sodium monochloroacetate, sodium monochloroacetate (*off-label*), trifluralin
	Crops as weeds	carbetamide, cycloxydim, pendimethalin, tepraloxydim
	Grass weeds	carbetamide, chlorthal-dimethyl + propachlor, cycloxydim, fluazifop-P-butyl (*off-label*), metazachlor, napropamide, pendimethalin, propachlor, tepraloxydim, trifluralin
	Mayweeds	clopyralid, metazachlor (*off-label*)

Brassicas - Mustard

Crop control	Pre-harvest desiccation	diquat (*off-label*), glyphosate
Diseases	Alternaria	iprodione, iprodione (*seed treatment*)
	Botrytis	iprodione
Pests	Beetles	deltamethrin
	Flies	tefluthrin (*off-label - seed treatment*)
	Midges	deltamethrin
	Weevils	deltamethrin
Weeds	Broad-leaved weeds	chlorthal-dimethyl + propachlor, diquat (*off-label*), glyphosate, propachlor, trifluralin
	Crops as weeds	glyphosate
	Grass weeds	chlorthal-dimethyl + propachlor, glyphosate, propachlor, trifluralin
	Weeds, miscellaneous	glyphosate

Brassicas - Root brassicas

Diseases	Alternaria	fenpropimorph (*off-label*), iprodione (*seed treatment*), iprodione + thiophanate-methyl (*off-label*), tebuconazole
	Canker	iprodione (*off-label*)
	Crown rot	fenpropimorph (*off-label*), iprodione + thiophanate-methyl (*off-label*)
	Damping off	thiram (*seed treatment*)
	Downy mildew	propamocarb hydrochloride (*off-label*)
	Fungus diseases	tebuconazole (*off-label*)
	Powdery mildew	fenpropimorph (*off-label*), sulphur, tebuconazole

	Rhizoctonia	azoxystrobin (*off-label*), tolclofos-methyl (*with fleece or mesh covers*), tolclofos-methyl (*without covers*)
	Sclerotinia	boscalid + pyraclostrobin (*off-label*)
	White blister	mancozeb + metalaxyl-M (*off-label*), metalaxyl-M (*off-label*), propamocarb hydrochloride (*off-label*)
Pests	Aphids	nicotine, pirimicarb, pirimicarb (*off-label*)
	Beetles	deltamethrin
	Caterpillars	deltamethrin, nicotine
	Flies	chlorpyrifos (*off-label*)
	Leaf miners	nicotine
	Pests, miscellaneous	dimethoate (*off-label*)
	Slugs/snails	ferric phosphate
	Weevils	lambda-cyhalothrin (*off-label*)
Plant growth regulation	Quality/yield control	sulphur
Weeds	Broad-leaved weeds	chlorthal-dimethyl, chlorthal-dimethyl + propachlor, clopyralid, ethofumesate (*off-label*), glyphosate, metazachlor, propachlor, propachlor (*off-label*), trifluralin, trifluralin (*off-label*)
	Crops as weeds	cycloxydim, fluazifop-P-butyl (*stockfeed only*), glyphosate
	Grass weeds	chlorthal-dimethyl + propachlor, cycloxydim, ethofumesate (*off-label*), fluazifop-P-butyl (*stockfeed only*), glyphosate, metazachlor, propachlor, propachlor (*off-label*), propaquizafop, trifluralin, trifluralin (*off-label*)
	Mayweeds	clopyralid
	Weeds, miscellaneous	ethofumesate (*off-label*), glyphosate

Brassicas - Salad brassicas

Diseases	Alternaria	azoxystrobin (*off-label - for baby leaf production*), iprodione (*seed treatment*)
	Black rot	copper oxychloride (*off-label*)
	Botrytis	fenhexamid (*off-label - for baby leaf production*), iprodione (*off-label*)
	Damping off	fosetyl-aluminium (*off-label*)
	Damping off and wirestem	tolclofos-methyl
	Downy mildew	azoxystrobin (*off-label - for baby leaf production*), copper oxychloride (*off-label - for baby leaf production*), fosetyl-aluminium (*off-label*), fosetyl-aluminium (*off-label - seedlings*), propamocarb hydrochloride
	Phytophthora	propamocarb hydrochloride
	Pythium	propamocarb hydrochloride
	Ring spot	difenoconazole (*off-label*), tebuconazole (*off-label*), tebuconazole (*off-label - for baby leaf production*)
	Spear rot	copper oxychloride (*off-label*)
Pests	Aphids	chlorpyrifos, cypermethrin (*off-label*), deltamethrin (*off-label*), nicotine, pirimicarb, pirimicarb (*off-label - for baby leaf production*), pymetrozine (*off-label - for baby leaf production*)

8

	Beetles	deltamethrin (*off-label*), deltamethrin (*off-label - baby leaf production*), deltamethrin (*off-label - for baby leaf production*)
	Birds/mammals	aluminium ammonium sulphate
	Caterpillars	alpha-cypermethrin (*off-label - for baby leaf production*), Bacillus thuringiensis (*off-label*), chlorpyrifos, cypermethrin (*off-label*), deltamethrin (*off-label*), deltamethrin (*off-label - baby leaf production*), deltamethrin (*off-label - for baby leaf production*), diflubenzuron (*off-label - for baby leaf production*), nicotine
	Cutworms	chlorpyrifos
	Flies	chlorpyrifos, chlorpyrifos (*off-label*), tefluthrin (*off-label - seed treatment*)
	Leaf miners	nicotine
	Leatherjackets	chlorpyrifos
	Pests, miscellaneous	deltamethrin (*off-label*), dimethoate (*off-label*)
	Whiteflies	chlorpyrifos
Weeds	Broad-leaved weeds	metazachlor (*off-label*), propachlor (*off-label*), propachlor (*off-label - for baby leaf production*)
	Grass weeds	fluazifop-P-butyl (*off-label*), propachlor (*off-label*), propachlor (*off-label - for baby leaf production*)
	Mayweeds	metazachlor (*off-label*)

Cereals - Barley

Crop control	Pre-harvest desiccation	diquat (*stockfeed only*), glyphosate
Diseases	Brown foot rot and ear blight	carboxin + thiram (*seed treatment*), clothianidin + prothioconazole (*seed treatment*), fludioxonil (*seed treatment*), fluquinconazole + prochloraz (*seed treatment*), fuberidazole + imidacloprid + triadimenol (*seed treatment*), fuberidazole + imidacloprid + triadimenol (*seed treatment - reduction*), fuberidazole + triadimenol (*seed treatment*), guazatine (*seed treatment - reduction*), guazatine + imazalil (*seed treatment*), guazatine + imazalil (*seed treatment - moderate control*), imazalil + triticonazole (*seed treatment - moderate control*), prochloraz + triticonazole (*seed treatment*), prothioconazole (*seed treatment*), prothioconazole + tebuconazole + triazoxide (*seed treatment*)
	Covered smut	carboxin + thiram (*seed treatment*), clothianidin + prothioconazole (*seed treatment*), fludioxonil (*seed treatment*), fluquinconazole (*seed treatment*), fluquinconazole + prochloraz (*seed treatment*), fuberidazole + imidacloprid + triadimenol (*seed treatment*), fuberidazole + triadimenol (*seed treatment*), guazatine + imazalil (*seed treatment*), imazalil + triticonazole (*seed treatment*), prochloraz + triticonazole (*seed treatment*), prothioconazole (*seed treatment*), prothioconazole + tebuconazole + triazoxide (*seed treatment*)
	Eyespot	azoxystrobin + cyproconazole (*reduction*), boscalid + epoxiconazole (*moderate*), carbendazim, carbendazim + flusilazole, chlorothalonil + cyproconazole, cyproconazole, cyproconazole + cyprodinil, cyprodinil,

cyprodinil + picoxystrobin, epoxiconazole (*reduction*), epoxiconazole + fenpropimorph (*reduction*), epoxiconazole + fenpropimorph + kresoxim-methyl (*reduction*), epoxiconazole + kresoxim-methyl (*reduction*), fluoxastrobin + prothioconazole (*reduction*), fluoxastrobin + prothioconazole + trifloxystrobin (*reduction*), flusilazole, prochloraz, prochloraz + propiconazole, prochloraz + tebuconazole, prothioconazole, prothioconazole + spiroxamine, prothioconazole + tebuconazole (*reduction*), prothioconazole + trifloxystrobin (*reduction*)

Foot rot	guazatine (*seed treatment*)
Late ear diseases	fluoxastrobin + prothioconazole, fluoxastrobin + prothioconazole + trifloxystrobin, prothioconazole, prothioconazole + tebuconazole, prothioconazole + trifloxystrobin
Leaf stripe	carboxin + thiram (*seed treatment*), carboxin + thiram (*seed treatment - reduction*), clothianidin + prothioconazole (*seed treatment*), fludioxonil (*seed treatment - reduction*), fuberidazole + triadimenol (*seed treatment*), guazatine (*seed treatment*), guazatine + imazalil (*seed treatment*), imazalil + triticonazole (*seed treatment*), imidacloprid + tebuconazole + triazoxide (*seed treatment*), prochloraz + triticonazole (*seed treatment*), prothioconazole (*seed treatment*), prothioconazole + tebuconazole + triazoxide (*seed treatment*), tebuconazole + triazoxide (*seed treatment*)
Loose smut	carboxin + thiram (*seed treatment - reduction*), clothianidin + prothioconazole (*seed treatment*), fluquinconazole (*seed treatment*), fluquinconazole + prochloraz (*seed treatment*), fuberidazole + imidacloprid + triadimenol (*seed treatment*), fuberidazole + triadimenol (*seed treatment*), imazalil + triticonazole (*seed treatment*), imidacloprid + tebuconazole + triazoxide (*seed treatment*), prochloraz + triticonazole (*seed treatment*), prothioconazole (*seed treatment*), prothioconazole + tebuconazole + triazoxide (*seed treatment*), tebuconazole + triazoxide (*seed treatment*)
Net blotch	azoxystrobin, azoxystrobin + chlorothalonil, azoxystrobin + cyproconazole, azoxystrobin + fenpropimorph, boscalid + epoxiconazole, carboxin + thiram (*seed treatment*), chlorothalonil + cyproconazole, chlorothalonil + picoxystrobin, chlorothalonil + tetraconazole, chlorothalonil + tetraconazole (*autumn sown*), cyproconazole, cyproconazole + cyprodinil, cyproconazole + propiconazole, cyproconazole + trifloxystrobin, cyprodinil, cyprodinil + picoxystrobin, epoxiconazole, epoxiconazole + fenpropimorph, epoxiconazole + fenpropimorph + kresoxim-methyl, epoxiconazole + fenpropimorph + pyraclostrobin, epoxiconazole + kresoxim-methyl, epoxiconazole + kresoxim-methyl + pyraclostrobin, epoxiconazole + pyraclostrobin, famoxadone + flusilazole, fenpropimorph + flusilazole, fenpropimorph + pyraclostrobin, fluoxastrobin + prothioconazole, fluoxastrobin + prothioconazole + trifloxystrobin, flusilazole, fuberidazole + imidacloprid + triadimenol (*seed treatment - seed-borne only*), guazatine (*seed treatment*), guazatine + imazalil (*seed treatment - moderate control*), imazalil + triticonazole (*seed*

treatment - moderate control), imidacloprid + tebuconazole + triazoxide (*seed treatment*), iprodione, mancozeb, metconazole (*reduction*), picoxystrobin, prochloraz, prochloraz + tebuconazole, prothioconazole, prothioconazole + spiroxamine, prothioconazole + tebuconazole, prothioconazole + tebuconazole + triazoxide (*seed treatment*), prothioconazole + trifloxystrobin, pyraclostrobin, spiroxamine + tebuconazole, tebuconazole, tebuconazole + triadimenol, tebuconazole + triazoxide (*seed treatment*), trifloxystrobin

Powdery mildew   azoxystrobin, azoxystrobin + cyproconazole, azoxystrobin + fenpropimorph, carbendazim + flusilazole, chlorothalonil + cyproconazole, chlorothalonil + picoxystrobin, chlorothalonil + tetraconazole, chlorothalonil + tetraconazole (*autumn sown*), cyflufenamid, cyproconazole, cyproconazole + cyprodinil, cyproconazole + propiconazole, cyproconazole + trifloxystrobin, cyprodinil, cyprodinil + picoxystrobin, epoxiconazole, epoxiconazole + fenpropimorph, epoxiconazole + fenpropimorph + kresoxim-methyl, epoxiconazole + fenpropimorph + pyraclostrobin, epoxiconazole + kresoxim-methyl, fenpropidin, fenpropimorph, fenpropimorph + flusilazole, fenpropimorph + kresoxim-methyl, fenpropimorph + pyraclostrobin, fenpropimorph + quinoxyfen, fluoxastrobin + prothioconazole, fluoxastrobin + prothioconazole + trifloxystrobin, fluquinconazole + prochloraz, flusilazole, flutriafol, fuberidazole + imidacloprid + triadimenol (*seed treatment*), fuberidazole + triadimenol (*seed treatment*), metconazole, metconazole (*moderate control*), metrafenone, picoxystrobin, prochloraz, prochloraz + tebuconazole, propiconazole, prothioconazole, prothioconazole + spiroxamine, prothioconazole + tebuconazole, prothioconazole + trifloxystrobin, quinoxyfen, spiroxamine, spiroxamine + tebuconazole, sulphur, tebuconazole, tebuconazole + triadimenol

Rhynchosporium   azoxystrobin, azoxystrobin + chlorothalonil, azoxystrobin + cyproconazole (*moderate control*), azoxystrobin + fenpropimorph, boscalid + epoxiconazole (*moderate*), carbendazim, chlorothalonil, chlorothalonil + cyproconazole, chlorothalonil + picoxystrobin, chlorothalonil + tetraconazole, chlorothalonil + tetraconazole (*autumn sown*), cyproconazole, cyproconazole + cyprodinil, cyproconazole + propiconazole, cyproconazole + trifloxystrobin, cyprodinil, cyprodinil + picoxystrobin, epoxiconazole, epoxiconazole + fenpropimorph, epoxiconazole + fenpropimorph + kresoxim-methyl, epoxiconazole + fenpropimorph + pyraclostrobin, epoxiconazole + kresoxim-methyl, epoxiconazole + kresoxim-methyl + pyraclostrobin, epoxiconazole + pyraclostrobin, famoxadone + flusilazole, fenpropidin, fenpropimorph, fenpropimorph + flusilazole, fenpropimorph + kresoxim-methyl, fenpropimorph + pyraclostrobin, fluoxastrobin + prothioconazole, fluoxastrobin + prothioconazole + trifloxystrobin, fluquinconazole + prochloraz, flusilazole, flutriafol, mancozeb, metconazole, picoxystrobin, prochloraz, prochloraz + propiconazole, prochloraz + tebuconazole,

		propiconazole, prothioconazole, prothioconazole + spiroxamine, prothioconazole + tebuconazole, prothioconazole + trifloxystrobin, pyraclostrobin (*moderate*), spiroxamine (*reduction*), spiroxamine + tebuconazole, tebuconazole, tebuconazole + triadimenol, trifloxystrobin
	Rust	azoxystrobin, azoxystrobin + chlorothalonil, azoxystrobin + cyproconazole, azoxystrobin + fenpropimorph, boscalid + epoxiconazole, carbendazim + flusilazole, chlorothalonil + cyproconazole, chlorothalonil + picoxystrobin, chlorothalonil + tetraconazole, chlorothalonil + tetraconazole (*autumn sown*), cyproconazole, cyproconazole + cyprodinil, cyproconazole + propiconazole, cyproconazole + trifloxystrobin, cyprodinil + picoxystrobin, epoxiconazole, epoxiconazole + fenpropimorph, epoxiconazole + fenpropimorph + kresoxim-methyl, epoxiconazole + fenpropimorph + pyraclostrobin, epoxiconazole + kresoxim-methyl, epoxiconazole + kresoxim-methyl + pyraclostrobin, epoxiconazole + pyraclostrobin, famoxadone + flusilazole, fenpropidin, fenpropimorph, fenpropimorph + flusilazole, fenpropimorph + pyraclostrobin, fluoxastrobin + prothioconazole, fluoxastrobin + prothioconazole + trifloxystrobin, fluquinconazole (*seed treatment*), fluquinconazole + prochloraz, fluquinconazole + prochloraz (*seed treatment*), flusilazole, flutriafol, fuberidazole + imidacloprid + triadimenol (*seed treatment*), fuberidazole + triadimenol (*seed treatment*), mancozeb, metconazole, picoxystrobin, prochloraz + tebuconazole, propiconazole, prothioconazole, prothioconazole + spiroxamine, prothioconazole + tebuconazole, prothioconazole + trifloxystrobin, pyraclostrobin, spiroxamine, spiroxamine + tebuconazole, tebuconazole, tebuconazole + triadimenol, trifloxystrobin
	Septoria diseases	flusilazole
	Snow mould	fludioxonil (*seed treatment*), guazatine + imazalil (*seed treatment*)
	Snow rot	fuberidazole + imidacloprid + triadimenol (*seed treatment - reduction*)
	Sooty moulds	mancozeb
	Take-all	azoxystrobin (*reduction*), azoxystrobin + chlorothalonil (*reduction*), azoxystrobin + cyproconazole (*reduction*), azoxystrobin + fenpropimorph (*reduction*), fluquinconazole (*seed treatment - reduction*), fluquinconazole + prochloraz (*seed treatment - reduction*), silthiofam (*seed treatment*)
Pests	Aphids	alpha-cypermethrin, bifenthrin, chlorpyrifos, clothianidin (*seed treatment*), clothianidin + prothioconazole (*seed treatment*), cyfluthrin, cypermethrin, cypermethrin (*autumn sown*), deltamethrin, esfenvalerate, fuberidazole + imidacloprid + triadimenol (*seed treatment*), imidacloprid + tebuconazole + triazoxide (*seed treatment*), lambda-cyhalothrin, lambda-cyhalothrin + pirimicarb, pirimicarb, tau-fluvalinate, zeta-cypermethrin
	Birds/mammals	aluminium ammonium sulphate

	Flies	alpha-cypermethrin, chlorpyrifos, cypermethrin, cypermethrin (*autumn sown*), deltamethrin, tefluthrin (*seed treatment*)
	Leafhoppers	clothianidin (*seed treatment*), clothianidin + prothioconazole (*seed treatment*)
	Leatherjackets	chlorpyrifos
	Slugs/snails	thiodicarb
	Thrips	chlorpyrifos
	Wireworms	clothianidin (*seed treatment*), clothianidin + prothioconazole (*seed treatment*), fuberidazole + imidacloprid + triadimenol (*reduction of damage*), imidacloprid + tebuconazole + triazoxide (*reduction of damage*), tefluthrin (*seed treatment*)
Plant growth regulation	Growth control	2-chloroethylphosphonic acid, 2-chloroethylphosphonic acid + mepiquat chloride, chlormequat, chlormequat + 2-chloroethylphosphonic acid, chlormequat + 2-chloroethylphosphonic acid + mepiquat chloride, trinexapac-ethyl
	Quality/yield control	2-chloroethylphosphonic acid + mepiquat chloride (*low lodging situations*), chlormequat, sulphur
Weeds	Broad-leaved weeds	2,4-D, 2,4-D + MCPA, 2,4-DB + MCPA, amidosulfuron, amidosulfuron + iodosulfuron-methyl-sodium, bifenox, bifenox + isoproturon, bromoxynil, bromoxynil + diflufenican + ioxynil, bromoxynil + ioxynil, bromoxynil + ioxynil + mecoprop-P, carfentrazone-ethyl, carfentrazone-ethyl + mecoprop-P, carfentrazone-ethyl + metsulfuron-methyl, chlorotoluron, chlorotoluron + isoproturon, clopyralid, dicamba + MCPA + mecoprop-P, dicamba + mecoprop-P, dichlorprop-P + MCPA + mecoprop-P, diflufenican + flufenacet, diflufenican + flurtamone, diflufenican + flurtamone + isoproturon, diflufenican + isoproturon, diflufenican + trifluralin, florasulam, florasulam + fluroxypyr, flufenacet + pendimethalin, fluroxypyr, fluroxypyr + thifensulfuron-methyl + tribenuron-methyl, glyphosate, iodosulfuron-methyl-sodium, isoproturon, isoproturon (*off-label*), isoproturon + pendimethalin, isoproturon + trifluralin, linuron, linuron + trifluralin, MCPA, MCPA + MCPB, MCPB, mecoprop-P, metsulfuron-methyl, metsulfuron-methyl + thifensulfuron-methyl, metsulfuron-methyl + tribenuron-methyl, pendimethalin, pendimethalin + picolinafen, thifensulfuron-methyl, thifensulfuron-methyl + tribenuron-methyl, tribenuron-methyl, trifluralin
	Crops as weeds	amidosulfuron + iodosulfuron-methyl-sodium, diclofop-methyl + fenoxaprop-P-ethyl, diflufenican + flufenacet, diflufenican + flurtamone, diflufenican + flurtamone + isoproturon, florasulam, fluroxypyr, glyphosate, iodosulfuron-methyl-sodium, pendimethalin, tralkoxydim, tralkoxydim (*autumn sown*)
	General weed control	diquat (*stockfeed only*)
	Grass weeds	bifenox + isoproturon, chlorotoluron, chlorotoluron + isoproturon, diclofop-methyl + fenoxaprop-P-ethyl, diflufenican + flufenacet, diflufenican + flurtamone, diflufenican + flurtamone + isoproturon, diflufenican + isoproturon, diflufenican + trifluralin, flufenacet + pendimethalin, flupyrsulfuron-methyl, glyphosate, isoproturon, isoproturon (*off-label*), isoproturon +

		pendimethalin, isoproturon + trifluralin, linuron, linuron + trifluralin, pendimethalin, pendimethalin + picolinafen, tralkoxydim, tralkoxydim (*autumn sown*), tralkoxydim (*autumn sown - qualified minor use*), tralkoxydim (*qualified minor use*), tri-allate, trifluralin
	Mayweeds	amidosulfuron + iodosulfuron-methyl-sodium, bromoxynil + diflufenican + ioxynil, clopyralid, dicamba + mecoprop-P, dichlorprop-P + MCPA + mecoprop-P, diflufenican + flufenacet, florasulam, florasulam + fluroxypyr, iodosulfuron-methyl-sodium, metsulfuron-methyl, metsulfuron-methyl + thifensulfuron-methyl, metsulfuron-methyl + tribenuron-methyl, thifensulfuron-methyl, thifensulfuron-methyl + tribenuron-methyl, tribenuron-methyl
	Polygonums	dicamba + mecoprop-P, metsulfuron-methyl + thifensulfuron-methyl
	Speedwells	bifenox, bromoxynil + diflufenican + ioxynil, carfentrazone-ethyl + mecoprop-P, diflufenican + flufenacet, flufenacet + pendimethalin, iodosulfuron-methyl-sodium, metsulfuron-methyl + thifensulfuron-methyl, pendimethalin + picolinafen
	Weeds, miscellaneous	glyphosate

Cereals - Cereals, general

Pests	Leatherjackets	methiocarb (*reduction*)
	Slugs/snails	methiocarb
Weeds	Broad-leaved weeds	2,4-DB + MCPA, isoxaben (*off-label - grown for game cover*), MCPA + MCPB
	Weeds, miscellaneous	clopyralid (*off-label - game cover*)

Cereals - Maize/sweetcorn

Diseases	Damping off	thiram (*seed treatment*)
Pests	Aphids	nicotine, pirimicarb, pirimicarb (*off-label*), pymetrozine (*off-label*)
	Caterpillars	nicotine
	Flies	chlorpyrifos, lambda-cyhalothrin (*off-label*)
	Leaf miners	nicotine
	Slugs/snails	methiocarb
Weeds	Broad-leaved weeds	atrazine, bromoxynil, bromoxynil (*off-label*), bromoxynil + prosulfuron, clopyralid, fluroxypyr, isoxaben (*off-label - grown for game cover*), mesotrione, mesotrione + terbuthylazine, nicosulfuron, pendimethalin, pendimethalin (*off-label*), pendimethalin (*off-label - under covers*), pendimethalin (*off-label - under covers and mulches*), pyridate, rimsulfuron
	Crops as weeds	fluroxypyr (*off-label*), mesotrione, nicosulfuron, pendimethalin, rimsulfuron
	Grass weeds	atrazine, mesotrione + terbuthylazine, nicosulfuron, nicosulfuron (*off-label*), pendimethalin, pendimethalin (*off-label - under covers*)
	Mayweeds	bromoxynil + prosulfuron, clopyralid, nicosulfuron

Cereals - Oats

Crop control	Pre-harvest desiccation	diquat (*stockfeed only*), glyphosate
Diseases	Brown foot rot and ear blight	carboxin + thiram (*seed treatment*), fludioxonil (*seed treatment*), fuberidazole + imidacloprid + triadimenol (*seed treatment - reduction*), fuberidazole + triadimenol (*seed treatment*), guazatine (*seed treatment - reduction*)
	Covered smut	carboxin + thiram (*seed treatment*)
	Crown rust	azoxystrobin, azoxystrobin + cyproconazole, chlorothalonil + tetraconazole, cyproconazole, epoxiconazole + fenpropimorph + pyraclostrobin, epoxiconazole + kresoxim-methyl + pyraclostrobin, epoxiconazole + pyraclostrobin, fenpropimorph + pyraclostrobin, fuberidazole + triadimenol (*seed treatment*), picoxystrobin, propiconazole, pyraclostrobin, tebuconazole (*reduction*), tebuconazole + triadimenol, tetraconazole
	Eyespot	epoxiconazole + fenpropimorph + kresoxim-methyl (*reduction*), epoxiconazole + kresoxim-methyl (*reduction*)
	Fusarium root rot	bitertanol + fuberidazole (*seed treatment*), bitertanol + fuberidazole + imidacloprid (*seed treatment*)
	Loose smut	carboxin + thiram (*seed treatment - reduction*), fuberidazole + imidacloprid + triadimenol (*seed treatment*), fuberidazole + triadimenol (*seed treatment*)
	Powdery mildew	azoxystrobin, azoxystrobin + cyproconazole, cyproconazole, epoxiconazole, epoxiconazole + fenpropimorph, epoxiconazole + fenpropimorph + kresoxim-methyl, epoxiconazole + fenpropimorph + pyraclostrobin, epoxiconazole + kresoxim-methyl, fenpropimorph, fenpropimorph + kresoxim-methyl, fenpropimorph + pyraclostrobin, fenpropimorph + quinoxyfen, fuberidazole + imidacloprid + triadimenol (*seed treatment*), picoxystrobin, propiconazole, quinoxyfen, sulphur, tebuconazole, tebuconazole + triadimenol
	Pyrenophora leaf spot	fludioxonil (*seed treatment*), fuberidazole + imidacloprid + triadimenol (*seed treatment*), fuberidazole + triadimenol (*seed treatment*), guazatine (*seed treatment*), guazatine + imazalil (*seed treatment*)
Pests	Aphids	bifenthrin, bitertanol + fuberidazole + imidacloprid (*seed treatment*), chlorpyrifos, cypermethrin, deltamethrin, fuberidazole + imidacloprid + triadimenol (*seed treatment*), lambda-cyhalothrin, lambda-cyhalothrin + pirimicarb, pirimicarb, zeta-cypermethrin
	Birds/mammals	aluminium ammonium sulphate
	Flies	chlorpyrifos, cypermethrin, deltamethrin, tefluthrin (*seed treatment*)
	Leatherjackets	chlorpyrifos
	Midges	chlorpyrifos
	Slugs/snails	thiodicarb
	Thrips	chlorpyrifos
	Wireworms	bitertanol + fuberidazole + imidacloprid (*seed treatment - reduction of damage*), fuberidazole + imidacloprid + triadimenol (*reduction of damage*), tefluthrin (*seed treatment*)

Plant growth regulation	Growth control	chlormequat, trinexapac-ethyl
	Quality/yield control	chlormequat
Weeds	Broad-leaved weeds	2,4-D, 2,4-D + MCPA, 2,4-DB + MCPA, amidosulfuron, bromoxynil, bromoxynil + ioxynil, bromoxynil + ioxynil + mecoprop-P, carfentrazone-ethyl, carfentrazone-ethyl + flupyrsulfuron-methyl, carfentrazone-ethyl + mecoprop-P, carfentrazone-ethyl + metsulfuron-methyl, clopyralid, dicamba + MCPA + mecoprop-P, dicamba + mecoprop-P, dichlorprop-P + MCPA + mecoprop-P, florasulam, florasulam + fluroxypyr, flupyrsulfuron-methyl + thifensulfuron-methyl, flupyrsulfuron-methyl + tribenuron-methyl, fluroxypyr, glyphosate, linuron, MCPA, MCPA + MCPB, MCPB, mecoprop-P, metsulfuron-methyl, metsulfuron-methyl + thifensulfuron-methyl, metsulfuron-methyl + tribenuron-methyl, tribenuron-methyl
	Crops as weeds	florasulam, glyphosate
	General weed control	diquat (*stockfeed only*)
	Grass weeds	carfentrazone-ethyl + flupyrsulfuron-methyl, flupyrsulfuron-methyl + thifensulfuron-methyl, flupyrsulfuron-methyl + tribenuron-methyl, glyphosate, linuron, linuron + trifluralin (*off-label*)
	Mayweeds	clopyralid, dicamba + mecoprop-P, dichlorprop-P + MCPA + mecoprop-P, florasulam, florasulam + fluroxypyr, metsulfuron-methyl, metsulfuron-methyl + thifensulfuron-methyl, metsulfuron-methyl + tribenuron-methyl, tribenuron-methyl
	Polygonums	dicamba + mecoprop-P
	Speedwells	carfentrazone-ethyl + mecoprop-P, metsulfuron-methyl + thifensulfuron-methyl
	Weeds, miscellaneous	glyphosate

Cereals - Rye/triticale

Diseases	Blue mould	fuberidazole + triadimenol (*seed treatment*)
	Brown foot rot and ear blight	carboxin + thiram (*seed treatment*), fuberidazole + triadimenol (*seed treatment*)
	Eyespot	epoxiconazole + fenpropimorph + kresoxim-methyl (*reduction*), epoxiconazole + kresoxim-methyl (*reduction*), fluoxastrobin + prothioconazole (*reduction*), prochloraz, prochloraz + tebuconazole, prothioconazole, prothioconazole + spiroxamine, prothioconazole + tebuconazole (*reduction*)
	Fusarium root rot	bitertanol + fuberidazole (*seed treatment*)
	Late ear diseases	prothioconazole + spiroxamine
	Powdery mildew	azoxystrobin, azoxystrobin + cyproconazole, cyflufenamid, cyproconazole, epoxiconazole, epoxiconazole + fenpropimorph, epoxiconazole + fenpropimorph + kresoxim-methyl, epoxiconazole + kresoxim-methyl, fenpropimorph, fenpropimorph + kresoxim-methyl, fenpropimorph + quinoxyfen, fluoxastrobin + prothioconazole, prochloraz, prochloraz + tebuconazole, propiconazole, prothioconazole, prothioconazole + spiroxamine, prothioconazole +

		tebuconazole, quinoxyfen, spiroxamine, spiroxamine + tebuconazole, sulphur, tebuconazole, tebuconazole + triadimenol
	Rhynchosporium	azoxystrobin, azoxystrobin (*off-label*), azoxystrobin + cyproconazole (*moderate control*), epoxiconazole, epoxiconazole + fenpropimorph, epoxiconazole + fenpropimorph + kresoxim-methyl, epoxiconazole + kresoxim-methyl, fenpropimorph + kresoxim-methyl, fluoxastrobin + prothioconazole, prochloraz, prochloraz + tebuconazole, propiconazole, prothioconazole, prothioconazole + spiroxamine, prothioconazole + tebuconazole, spiroxamine (*reduction*), tebuconazole
	Rust	azoxystrobin, azoxystrobin (*off-label*), azoxystrobin + cyproconazole, cyproconazole, epoxiconazole, epoxiconazole + fenpropimorph, epoxiconazole + fenpropimorph + kresoxim-methyl, epoxiconazole + kresoxim-methyl, fenpropimorph, fluoxastrobin + prothioconazole, fuberidazole + triadimenol (*seed treatment*), prochloraz + tebuconazole, propiconazole, prothioconazole, prothioconazole + spiroxamine, prothioconazole + tebuconazole, spiroxamine, spiroxamine + tebuconazole, tebuconazole, tebuconazole + triadimenol
	Septoria diseases	epoxiconazole, epoxiconazole + fenpropimorph, epoxiconazole + fenpropimorph + kresoxim-methyl, epoxiconazole + kresoxim-methyl, fenpropimorph + kresoxim-methyl (*reduction*), prochloraz, propiconazole
Pests	Aphids	cypermethrin, dimethoate, lambda-cyhalothrin + pirimicarb, pirimicarb
	Flies	cypermethrin, dimethoate
	Slugs/snails	thiodicarb
Plant growth regulation	Growth control	2-chloroethylphosphonic acid, 2-chloroethylphosphonic acid + mepiquat chloride, chlormequat, trinexapac-ethyl
Weeds	Broad-leaved weeds	2,4-D, amidosulfuron, amidosulfuron + iodosulfuron-methyl-sodium, bifenox, bromoxynil + diflufenican + ioxynil, bromoxynil + ioxynil, bromoxynil + ioxynil + mecoprop-P, carfentrazone-ethyl, carfentrazone-ethyl + flupyrsulfuron-methyl, carfentrazone-ethyl + metsulfuron-methyl, chlorotoluron, dicamba + MCPA + mecoprop-P, dicamba + mecoprop-P, diflufenican + isoproturon, diflufenican + trifluralin, fluroxypyr, iodosulfuron-methyl-sodium, isoproturon, isoproturon + pendimethalin, linuron + trifluralin, MCPA, metsulfuron-methyl, metsulfuron-methyl + tribenuron-methyl, pendimethalin, tribenuron-methyl
	Crops as weeds	amidosulfuron + iodosulfuron-methyl-sodium, pendimethalin, tralkoxydim
	Grass weeds	carfentrazone-ethyl + flupyrsulfuron-methyl, chlorotoluron, clodinafop-propargyl, diflufenican + isoproturon, iodosulfuron-methyl-sodium (*from seed*), isoproturon, isoproturon + pendimethalin, linuron + trifluralin, pendimethalin, tralkoxydim, tralkoxydim (*qualified minor use*), tri-allate
	Mayweeds	amidosulfuron + iodosulfuron-methyl-sodium, bromoxynil + diflufenican + ioxynil, dicamba + mecoprop-P, iodosulfuron-methyl-sodium, metsulfuron-methyl, metsulfuron-methyl + tribenuron-methyl, tribenuron-methyl
	Polygonums	dicamba + mecoprop-P

| | Speedwells | bifenox, bromoxynil + diflufenican + ioxynil, iodosulfuron-methyl-sodium |

Cereals - Undersown cereals

Weeds	Broad-leaved weeds	2,4-DB (*undersown with lucerne*), 2,4-DB + linuron + MCPA (*undersown with clover*), 2,4-DB + MCPA, 2,4-DB + MCPA (*red or white clover*), bentazone + MCPA + MCPB, bromoxynil + ioxynil, dicamba + MCPA + mecoprop-P, dicamba + MCPA + mecoprop-P (*grass only*), dicamba + mecoprop-P, MCPA, MCPA (*red clover or grass*), MCPA + MCPB, MCPB, mecoprop-P
	Mayweeds	dicamba + mecoprop-P
	Polygonums	2,4-DB + MCPA, dicamba + mecoprop-P

Cereals - Wheat

Crop control	Pre-harvest desiccation	glyphosate
Diseases	Alternaria	iprodione
	Blue mould	fuberidazole + imidacloprid + triadimenol (*seed treatment*), fuberidazole + triadimenol (*seed treatment*)
	Botrytis	iprodione
	Brown foot rot and ear blight	bitertanol + fuberidazole + imidacloprid (*seed treatment - reduction*), carboxin + thiram (*seed treatment*), clothianidin + prothioconazole (*seed treatment*), dimoxystrobin + epoxiconazole, epoxiconazole, epoxiconazole (*reduction*), epoxiconazole + fenpropimorph, epoxiconazole + fenpropimorph (*reduction*), epoxiconazole + fenpropimorph + kresoxim-methyl (*reduction*), epoxiconazole + fenpropimorph + pyraclostrobin (*reduction*), epoxiconazole + kresoxim-methyl (*reduction*), fludioxonil (*seed treatment*), fluoxastrobin + prothioconazole (*seed treatment*), fuberidazole + imidacloprid + triadimenol (*seed treatment - reduction*), fuberidazole + triadimenol (*seed treatment*), guazatine (*seed treatment - reduction*), metconazole (*reduction*), prochloraz + triticonazole (*seed treatment*), prothioconazole (*seed treatment*), spiroxamine + tebuconazole, tebuconazole, tebuconazole + triadimenol
	Drechslera leaf spot	boscalid + epoxiconazole (*reduction*), chlorothalonil + picoxystrobin, dimoxystrobin + epoxiconazole, fenpropimorph + pyraclostrobin, fluoxastrobin + prothioconazole, fluoxastrobin + prothioconazole + trifloxystrobin, prothioconazole, prothioconazole + spiroxamine, prothioconazole + tebuconazole
	Eyespot	azoxystrobin + cyproconazole (*reduction*), boscalid + epoxiconazole (*moderate*), carbendazim, carbendazim + flusilazole, chlorothalonil + cyproconazole, cyproconazole, cyproconazole + cyprodinil, cyprodinil, cyprodinil + picoxystrobin (*moderate control*), epoxiconazole (*reduction*), epoxiconazole + fenpropimorph (*reduction*), epoxiconazole + fenpropimorph + kresoxim-methyl (*reduction*), epoxiconazole + kresoxim-methyl (*reduction*), fluoxastrobin + prothioconazole (*reduction*),

fluoxastrobin + prothioconazole + trifloxystrobin (*reduction*), flusilazole, metrafenone (*reduction*), picoxystrobin (*reduction*), prochloraz, prochloraz + propiconazole, prochloraz + tebuconazole, prothioconazole, prothioconazole + spiroxamine, prothioconazole + tebuconazole (*reduction*), prothioconazole + trifloxystrobin

Fusarium root rot   bitertanol + fuberidazole (*seed treatment*), bitertanol + fuberidazole + imidacloprid (*seed treatment*), fluquinconazole + prochloraz (*seed treatment*)

Late ear diseases   azoxystrobin, azoxystrobin + fenpropimorph, chlorothalonil + flutriafol, cyproconazole, cyproconazole + propiconazole, fluoxastrobin + prothioconazole, fluoxastrobin + prothioconazole + trifloxystrobin, picoxystrobin, prochloraz + tebuconazole, prothioconazole, prothioconazole + spiroxamine, prothioconazole + tebuconazole, prothioconazole + trifloxystrobin, tebuconazole + trifloxystrobin

Loose smut   bitertanol + fuberidazole (*seed treatment - reduction*), bitertanol + fuberidazole + imidacloprid (*seed treatment - reduction*), clothianidin + prothioconazole (*seed treatment*), fluoxastrobin + prothioconazole (*seed treatment*), fuberidazole + imidacloprid + triadimenol (*seed treatment*), fuberidazole + triadimenol (*seed treatment*), prochloraz + triticonazole (*seed treatment*), prothioconazole (*seed treatment*)

Powdery mildew   azoxystrobin, azoxystrobin + cyproconazole, carbendazim + flusilazole, chlorothalonil + cyproconazole, chlorothalonil + flutriafol, chlorothalonil + tetraconazole, chlorothalonil + tetraconazole (*autumn sown*), cyflufenamid, cyproconazole, cyproconazole + cyprodinil, cyproconazole + propiconazole, cyproconazole + trifloxystrobin, cyprodinil, epoxiconazole, epoxiconazole + fenpropimorph, fenpropidin, fenpropimorph, fenpropimorph + flusilazole, fenpropimorph + quinoxyfen, fluoxastrobin + prothioconazole, fluoxastrobin + prothioconazole + trifloxystrobin, fluquinconazole, fluquinconazole + prochloraz (*moderate control*), flusilazole, flutriafol, fuberidazole + imidacloprid + triadimenol (*seed treatment*), fuberidazole + triadimenol (*seed treatment*), metconazole (*moderate control*), metrafenone, prochloraz, prochloraz + tebuconazole, propiconazole, prothioconazole, prothioconazole + spiroxamine, prothioconazole + tebuconazole, prothioconazole + trifloxystrobin, quinoxyfen, spiroxamine, spiroxamine + tebuconazole, sulphur, tebuconazole, tebuconazole + triadimenol, tebuconazole + trifloxystrobin, tetraconazole

Rust   azoxystrobin, azoxystrobin + chlorothalonil, azoxystrobin + cyproconazole, azoxystrobin + fenpropimorph, boscalid + epoxiconazole, carbendazim + flusilazole, chlorothalonil + cyproconazole, chlorothalonil + flutriafol, chlorothalonil + picoxystrobin, chlorothalonil + tetraconazole, chlorothalonil + tetraconazole (*autumn sown*), cyproconazole, cyproconazole + cyprodinil, cyproconazole + propiconazole, cyproconazole + trifloxystrobin, cyprodinil + picoxystrobin, difenoconazole, dimoxystrobin + epoxiconazole, epoxiconazole,

epoxiconazole + fenpropimorph, epoxiconazole + fenpropimorph + kresoxim-methyl, epoxiconazole + fenpropimorph + pyraclostrobin, epoxiconazole + kresoxim-methyl, epoxiconazole + kresoxim-methyl + pyraclostrobin, epoxiconazole + pyraclostrobin, famoxadone + flusilazole, fenpropidin, fenpropimorph, fenpropimorph + flusilazole, fenpropimorph + pyraclostrobin, fluoxastrobin + prothioconazole, fluoxastrobin + prothioconazole + trifloxystrobin, fluquinconazole, fluquinconazole (*seed treatment*), fluquinconazole + prochloraz, fluquinconazole + prochloraz (*seed treatment*), fluquinconazole + prochloraz (*seed treatment - moderate control*), flusilazole, flutriafol, fuberidazole + imidacloprid + triadimenol (*seed treatment*), fuberidazole + triadimenol (*seed treatment*), mancozeb, metconazole, picoxystrobin, prochloraz + tebuconazole, propiconazole, prothioconazole, prothioconazole + spiroxamine, prothioconazole + tebuconazole, prothioconazole + trifloxystrobin, pyraclostrobin, spiroxamine, spiroxamine + tebuconazole, tebuconazole, tebuconazole + triadimenol, tebuconazole + trifloxystrobin, tetraconazole, trifloxystrobin

Septoria diseases   azoxystrobin, azoxystrobin + chlorothalonil, azoxystrobin + cyproconazole, azoxystrobin + fenpropimorph, bitertanol + fuberidazole (*seed treatment - reduction*), boscalid + epoxiconazole, carbendazim + flusilazole, carboxin + thiram (*seed treatment*), chlorothalonil, chlorothalonil + cyproconazole, chlorothalonil + flutriafol, chlorothalonil + mancozeb, chlorothalonil + picoxystrobin, chlorothalonil + tetraconazole, chlorothalonil + tetraconazole (*autumn sown*), cyproconazole, cyproconazole + cyprodinil, cyproconazole + propiconazole, cyproconazole + trifloxystrobin, cyprodinil + picoxystrobin, difenoconazole, dimoxystrobin + epoxiconazole, epoxiconazole, epoxiconazole + fenpropimorph, epoxiconazole + fenpropimorph + kresoxim-methyl, epoxiconazole + fenpropimorph + pyraclostrobin, epoxiconazole + kresoxim-methyl, epoxiconazole + kresoxim-methyl + pyraclostrobin, epoxiconazole + pyraclostrobin, famoxadone + flusilazole, fenpropidin, fenpropimorph + flusilazole, fenpropimorph + kresoxim-methyl (*reduction*), fenpropimorph + pyraclostrobin, fluoxastrobin + prothioconazole, fluoxastrobin + prothioconazole + trifloxystrobin, fluquinconazole, fluquinconazole (*seed treatment*), fluquinconazole + prochloraz, fluquinconazole + prochloraz (*seed treatment*), fluquinconazole + prochloraz (*seed treatment - moderate control*), flusilazole, flutriafol, fuberidazole + imidacloprid + triadimenol (*seed treatment*), fuberidazole + imidacloprid + triadimenol (*seed treatment - reduction*), fuberidazole + triadimenol (*seed treatment*), guazatine (*seed treatment*), iprodione, mancozeb, metconazole, picoxystrobin, prochloraz, prochloraz + propiconazole, prochloraz + tebuconazole, prochloraz + triticonazole (*seed treatment*), propiconazole, prothioconazole, prothioconazole + spiroxamine, prothioconazole + tebuconazole,

		prothioconazole + trifloxystrobin, pyraclostrobin, spiroxamine + tebuconazole, tebuconazole, tebuconazole + triadimenol, tebuconazole + trifloxystrobin, tetraconazole, trifloxystrobin
	Snow mould	fludioxonil (*seed treatment*)
	Sooty moulds	chlorothalonil + tetraconazole, epoxiconazole (*reduction*), epoxiconazole + fenpropimorph (*reduction*), epoxiconazole + fenpropimorph + kresoxim-methyl (*reduction*), epoxiconazole + kresoxim-methyl (*reduction*), mancozeb, propiconazole, spiroxamine + tebuconazole, tebuconazole, tebuconazole + triadimenol, tetraconazole
	Stinking smut	bitertanol + fuberidazole (*seed treatment*), bitertanol + fuberidazole + imidacloprid (*seed treatment*), carboxin + thiram (*seed treatment*), clothianidin + prothioconazole (*seed treatment*), fludioxonil (*seed treatment*), fluoxastrobin + prothioconazole (*seed treatment*), fluquinconazole (*seed treatment*), fluquinconazole + prochloraz (*seed treatment*), fuberidazole + imidacloprid + triadimenol (*seed treatment*), fuberidazole + triadimenol (*seed treatment*), prochloraz + triticonazole (*seed treatment*), prothioconazole (*seed treatment*)
	Take-all	azoxystrobin (*reduction*), azoxystrobin + chlorothalonil (*reduction*), azoxystrobin + cyproconazole (*reduction*), azoxystrobin + fenpropimorph (*reduction*), fluquinconazole (*seed treatment - reduction*), fluquinconazole + prochloraz (*seed treatment - reduction*), silthiofam (*seed treatment*)
Pests	Aphids	alpha-cypermethrin, bifenthrin, bitertanol + fuberidazole + imidacloprid (*seed treatment*), chlorpyrifos, clothianidin (*seed treatment*), clothianidin + prothioconazole (*seed treatment*), cyfluthrin, cypermethrin, cypermethrin (*autumn sown*), deltamethrin, dimethoate, esfenvalerate, flonicamid, fuberidazole + imidacloprid + triadimenol (*seed treatment*), lambda-cyhalothrin, lambda-cyhalothrin + pirimicarb, pirimicarb, tau-fluvalinate, zeta-cypermethrin
	Birds/mammals	aluminium ammonium sulphate
	Flies	alpha-cypermethrin, chlorpyrifos, cypermethrin, cypermethrin (*autumn sown*), deltamethrin, dimethoate, lambda-cyhalothrin, tefluthrin (*seed treatment*)
	Leafhoppers	clothianidin (*seed treatment*), clothianidin + prothioconazole (*seed treatment*)
	Leatherjackets	chlorpyrifos
	Midges	chlorpyrifos
	Slugs/snails	thiodicarb
	Thrips	chlorpyrifos
	Wireworms	bitertanol + fuberidazole + imidacloprid (*seed treatment - reduction of damage*), clothianidin (*seed treatment*), clothianidin + prothioconazole (*seed treatment*), fuberidazole + imidacloprid + triadimenol (*reduction of damage*), tefluthrin (*seed treatment*)
Plant growth regulation	Growth control	2-chloroethylphosphonic acid, 2-chloroethylphosphonic acid + mepiquat chloride, chlormequat, chlormequat + 2-chloroethylphosphonic acid, chlormequat + 2-chloroethylphosphonic acid + mepiquat chloride, chlormequat + imazaquin, chlormequat + mepiquat

		chloride, fuberidazole + imidacloprid + triadimenol (*seed treatment - reduction*), trinexapac-ethyl, trinexapac-ethyl (*off-label - cv A C Barrie only*)
	Quality/yield control	chlormequat + imazaquin, sulphur
Weeds	Broad-leaved weeds	2,4-D, 2,4-D + MCPA, 2,4-DB + MCPA, amidosulfuron, amidosulfuron + iodosulfuron-methyl-sodium, bifenox, bifenox + isoproturon, bromoxynil, bromoxynil + diflufenican + ioxynil, bromoxynil + ioxynil, bromoxynil + ioxynil + mecoprop-P, carfentrazone-ethyl, carfentrazone-ethyl + flupyrsulfuron-methyl, carfentrazone-ethyl + mecoprop-P, carfentrazone-ethyl + metsulfuron-methyl, chlorotoluron, chlorotoluron + isoproturon, clodinafop-propargyl + trifluralin, clopyralid, dicamba + MCPA + mecoprop-P, dicamba + mecoprop-P, dichlorprop-P + MCPA + mecoprop-P, diflufenican + flufenacet, diflufenican + flurtamone, diflufenican + flurtamone + isoproturon, diflufenican + isoproturon, diflufenican + trifluralin, florasulam, florasulam + fluroxypyr, flufenacet + pendimethalin, flupyrsulfuron-methyl, flupyrsulfuron-methyl + picolinafen, flupyrsulfuron-methyl + thifensulfuron-methyl, flupyrsulfuron-methyl + tribenuron-methyl, fluroxypyr, fluroxypyr + thifensulfuron-methyl + tribenuron-methyl, glyphosate, iodosulfuron-methyl-sodium, iodosulfuron-methyl-sodium + mesosulfuron-methyl, isoproturon, isoproturon (*autumn sown*), isoproturon + pendimethalin, isoproturon + pendimethalin (*autumn sown*), isoproturon + trifluralin, linuron, linuron + trifluralin, MCPA, MCPA + MCPB, MCPB, mecoprop-P, metsulfuron-methyl, metsulfuron-methyl + thifensulfuron-methyl, metsulfuron-methyl + tribenuron-methyl, pendimethalin, pendimethalin + picolinafen, sulfosulfuron, thifensulfuron-methyl, thifensulfuron-methyl + tribenuron-methyl, tribenuron-methyl, trifluralin
	Crops as weeds	amidosulfuron + iodosulfuron-methyl-sodium, diclofop-methyl + fenoxaprop-P-ethyl, diflufenican + flufenacet, diflufenican + flurtamone, diflufenican + flurtamone + isoproturon, florasulam, fluroxypyr, glyphosate, pendimethalin, tralkoxydim, tralkoxydim (*autumn sown*)
	Grass weeds	bifenox + isoproturon, carfentrazone-ethyl + flupyrsulfuron-methyl, chlorotoluron, chlorotoluron + isoproturon, clodinafop-propargyl, clodinafop-propargyl + trifluralin, diclofop-methyl + fenoxaprop-P-ethyl, diflufenican + flufenacet, diflufenican + flurtamone, diflufenican + flurtamone + isoproturon, diflufenican + isoproturon, diflufenican + trifluralin, fenoxaprop-P-ethyl, flufenacet + pendimethalin, flupyrsulfuron-methyl, flupyrsulfuron-methyl + picolinafen, flupyrsulfuron-methyl + thifensulfuron-methyl, flupyrsulfuron-methyl + tribenuron-methyl, glyphosate, iodosulfuron-methyl-sodium (*from seed*), iodosulfuron-methyl-sodium + mesosulfuron-methyl, isoproturon, isoproturon (*autumn sown*), isoproturon + pendimethalin, isoproturon + pendimethalin (*autumn sown*), isoproturon + trifluralin, linuron, linuron + trifluralin, pendimethalin, pendimethalin + picolinafen, propoxycarbazone-sodium, sulfosulfuron, tralkoxydim, tralkoxydim (*autumn sown*), tralkoxydim (*autumn sown -*

		qualified minor use), tralkoxydim (*qualified minor use*), tri-allate, trifluralin
	Mayweeds	amidosulfuron + iodosulfuron-methyl-sodium, bromoxynil + diflufenican + ioxynil, clopyralid, dicamba + mecoprop-P, dichlorprop-P + MCPA + mecoprop-P, diflufenican + flufenacet, florasulam, florasulam + fluroxypyr, iodosulfuron-methyl-sodium, iodosulfuron-methyl-sodium + mesosulfuron-methyl, metsulfuron-methyl, metsulfuron-methyl + thifensulfuron-methyl, metsulfuron-methyl + tribenuron-methyl, sulfosulfuron, thifensulfuron-methyl, thifensulfuron-methyl + tribenuron-methyl, tribenuron-methyl
	Polygonums	2,4-DB + MCPA, dicamba + mecoprop-P, metsulfuron-methyl + thifensulfuron-methyl
	Speedwells	bifenox, bromoxynil + diflufenican + ioxynil, carfentrazone-ethyl + mecoprop-P, diflufenican + flufenacet, flufenacet + pendimethalin, iodosulfuron-methyl-sodium, metsulfuron-methyl + thifensulfuron-methyl, pendimethalin + picolinafen
	Weeds, miscellaneous	glyphosate

Edible fungi - Mushrooms

Diseases	Bacterial blotch	sodium hypochlorite (commodity substance)
	Fungus diseases	formaldehyde (commodity substance) (*spray or fumigant*)
	Trichoderma	carbendazim (*off-label - spawn treatment*)
Pests	Flies	deltamethrin (*off-label*), diflubenzuron

Fruiting vegetables - Aubergines

Diseases	Blight	azoxystrobin (*off-label*)
	Botrytis	azoxystrobin (*off-label*), pyrimethanil (*off-label*)
	Didymella stem rot	azoxystrobin (*off-label*)
	Phytophthora	propamocarb hydrochloride
	Powdery mildew	azoxystrobin (*off-label*), sulphur (*off-label*)
	Pythium	propamocarb hydrochloride
Pests	Aphids	nicotine, Verticillium lecanii
	Leaf miners	deltamethrin (*off-label*), oxamyl (*off-label*)
	Leafhoppers	nicotine
	Thrips	deltamethrin (*off-label*), nicotine
	Whiteflies	buprofezin, nicotine, Verticillium lecanii
Weeds	Broad-leaved weeds	paraquat (*off-label - inter-row directed treatment*)
	Grass weeds	paraquat (*off-label - inter-row directed treatment*)

Fruiting vegetables - Cucurbits

Diseases	Botrytis	mepanipyrim (*off-label*)
	Damping off and foot rot	etridiazole (*seeds and seedlings*)
	Downy mildew	mancozeb (*off-label*)
	Phytophthora	etridiazole (*transplants*), propamocarb hydrochloride

	Powdery mildew	azoxystrobin (*off-label*), bupirimate (*off-label*), bupirimate (*outdoor only*), copper ammonium carbonate, fenarimol (*off-label*)
	Pythium	propamocarb hydrochloride
Pests	Aphids	fatty acids, nicotine, pirimicarb, pirimicarb (*off-label*)
	Caterpillars	nicotine
	Leaf miners	nicotine
	Mealybugs	fatty acids
	Scale insects	fatty acids
	Spider mites	fatty acids
	Whiteflies	fatty acids
Weeds	Broad-leaved weeds	isoxaben (*off-label*)
	Grass weeds	propyzamide (*off-label*)
	Weeds, miscellaneous	chlorthal-dimethyl (*off-label*)

Fruiting vegetables - Peppers

Diseases	Damping off	propamocarb hydrochloride (*off-label*)
	Phytophthora	propamocarb hydrochloride
	Pythium	propamocarb hydrochloride
	Root rot	propamocarb hydrochloride (*off-label*)
Pests	Aphids	fatty acids, nicotine, pirimicarb
	Caterpillars	nicotine
	Leaf miners	nicotine
	Mealybugs	fatty acids
	Scale insects	fatty acids
	Spider mites	fatty acids
	Whiteflies	fatty acids

Fruiting vegetables - Tomatoes

Diseases	Blight	Bordeaux mixture, copper ammonium carbonate, copper oxychloride, propamocarb hydrochloride
	Botrytis	fenhexamid (*off-label*), iprodione, pyrimethanil (*off-label*), thiram
	Damping off	copper oxychloride
	Damping off and foot rot	etridiazole (*seeds and seedlings*)
	Foot rot	copper oxychloride
	Phytophthora	copper oxychloride, etridiazole (*transplants*), propamocarb hydrochloride
	Pythium	propamocarb hydrochloride
Pests	Aphids	fatty acids, nicotine, pirimicarb
	Caterpillars	nicotine
	Leaf miners	nicotine
	Mealybugs	fatty acids
	Scale insects	fatty acids
	Spider mites	fatty acids
	Whiteflies	fatty acids

| Plant growth regulation | Fruiting control | (2-naphthyloxy)acetic acid, 2-chloroethylphosphonic acid |

Herb crops - Herbs

Crop control	Pre-harvest desiccation	diquat (*off-label*)
Diseases	Alternaria	fenpropimorph (*off-label*)
	Botrytis	fenhexamid (*off-label*), iprodione (*off-label*), propamocarb hydrochloride (*off-label*)
	Crown rot	fenpropimorph (*off-label*)
	Downy mildew	copper oxychloride (*off-label*), fosetyl-aluminium (*off-label*), propamocarb hydrochloride (*off-label*)
	Fungus diseases	mancozeb + metalaxyl-M (*off-label*)
	Powdery mildew	azoxystrobin (*off-label*), fenpropimorph (*off-label*), tebuconazole (*off-label*)
	Ring spot	azoxystrobin (*off-label*), prochloraz (*off-label*)
	Rust	azoxystrobin (*off-label*), tebuconazole (*off-label*)
	Seed-borne diseases	thiram (*seed soak*)
Pests	Aphids	nicotine, pirimicarb (*off-label*), pymetrozine (*off-label*)
	Beetles	deltamethrin (*off-label*)
	Caterpillars	Bacillus thuringiensis (*off-label*), deltamethrin (*off-label*), diflubenzuron (*off-label*), nicotine
	Cutworms	lambda-cyhalothrin (*off-label*)
	Flies	lambda-cyhalothrin (*off-label*), tefluthrin (*off-label - seed treatment*)
	Leaf miners	nicotine
	Leafhoppers	deltamethrin (*off-label*)
	Thrips	pirimicarb (*off-label*)
Weeds	Broad-leaved weeds	asulam (*off-label*), chlorpropham, chlorpropham + pentanochlor, chlorthal-dimethyl, chlorthal-dimethyl (*off-label*), clopyralid (*off-label*), diquat (*off-label*), linuron (*off-label*), metamitron (*off-label*), pendimethalin, pendimethalin (*off-label*), pentanochlor, prometryn, prometryn (*off-label*), propachlor, propachlor (*off-label*), propyzamide (*off-label*), trifluralin, trifluralin (*off-label*)
	Crops as weeds	pendimethalin, propaquizafop (*off-label*)
	Grass weeds	chlorpropham, chlorpropham + pentanochlor, linuron (*off-label*), pendimethalin, pendimethalin (*off-label*), pentanochlor, prometryn, prometryn (*off-label*), propachlor, propachlor (*off-label*), propaquizafop (*off-label*), propyzamide (*off-label*), trifluralin, trifluralin (*off-label*)
	Mayweeds	clopyralid (*off-label*)
	Polygonums	chlorpropham
	Weeds, miscellaneous	linuron (*off-label*)

Leafy vegetables - Endives

Diseases	Botrytis	fenhexamid (*off-label*), iprodione (*off-label*)
	Downy mildew	copper oxychloride (*off-label*)
	Sclerotinia	iprodione (*off-label*)

Pests	Aphids	nicotine, pirimicarb (*off-label*)
	Caterpillars	diflubenzuron (*off-label*), nicotine
	Leaf miners	nicotine
	Thrips	pirimicarb (*off-label*)
Weeds	Broad-leaved weeds	pendimethalin (*off-label*), propachlor (*off-label - under crop covers*)
	Grass weeds	propachlor (*off-label - under crop covers*)

Leafy vegetables - Lettuce

Diseases	Big vein	carbendazim (*off-label*)
	Botrytis	fenhexamid (*off-label*), iprodione (*off-label*), iprodione (*outdoor crops*), propamocarb hydrochloride, thiram (*outdoor crops*)
	Damping off	thiram (*seed treatment*)
	Downy mildew	copper oxychloride (*off-label*), fosetyl-aluminium (*off-label*), mancozeb, propamocarb hydrochloride
	Fungus diseases	mancozeb + metalaxyl-M (*off-label*)
	Rhizoctonia	boscalid + pyraclostrobin
	Ring spot	prochloraz (*off-label*)
	Sclerotinia	azoxystrobin (*off-label*), boscalid + pyraclostrobin, iprodione (*off-label*)
Pests	Aphids	cypermethrin, deltamethrin (*off-label*), dimethoate, fatty acids, imidacloprid (*off-label*), lambda-cyhalothrin + pirimicarb, nicotine, pirimicarb (*outdoor crops*), pymetrozine (*off-label*)
	Caterpillars	Bacillus thuringiensis (*off-label*), cypermethrin, deltamethrin (*off-label*), diflubenzuron (*off-label*), nicotine
	Cutworms	cypermethrin, deltamethrin, lambda-cyhalothrin, lambda-cyhalothrin + pirimicarb
	Leaf miners	nicotine
	Mealybugs	fatty acids
	Scale insects	fatty acids
	Slugs/snails	ferric phosphate, methiocarb
	Spider mites	fatty acids
	Whiteflies	fatty acids
Weeds	Broad-leaved weeds	chlorpropham, paraquat (*off-label - inter-row directed treatment*), pendimethalin (*off-label*), propachlor (*off-label*), propachlor (*off-label - under crop covers*), propyzamide, propyzamide (*outdoor crops*), trifluralin
	Grass weeds	chlorpropham, paraquat (*off-label - inter-row directed treatment*), pendimethalin (*off-label*), propachlor (*off-label*), propachlor (*off-label - under crop covers*), propyzamide, propyzamide (*outdoor crops*), trifluralin
	Polygonums	chlorpropham

Leafy vegetables - Spinach

Diseases	Botrytis	fenhexamid (*off-label*)
	Downy mildew	copper oxychloride (*off-label*), fosetyl-aluminium (*off-label*), metalaxyl-M (*off-label*)

Pests	Aphids	cypermethrin (*off-label*), nicotine, pirimicarb (*off-label*), pymetrozine (*off-label*)
	Caterpillars	cypermethrin (*off-label*), diflubenzuron (*off-label*), nicotine
	Flies	tefluthrin (*off-label - seed treatment*)
	Leaf miners	nicotine
	Slugs/snails	methiocarb
Weeds	Broad-leaved weeds	chlorpropham + fenuron, clopyralid (*off-label*), phenmedipham (*off-label*)
	Crops as weeds	propaquizafop (*off-label*)
	Grass weeds	chlorpropham + fenuron, propaquizafop (*off-label*)
	Mayweeds	clopyralid (*off-label*)

Leafy vegetables - Watercress

Diseases	Downy mildew	fosetyl-aluminium (*off-label - during propagation*), propamocarb hydrochloride (*off-label - during propagation*), propamocarb hydrochloride (*off-label - under protection*)
	Phytophthora	etridiazole (*off-label*), fosetyl-aluminium (*off-label - during propagation*), propamocarb hydrochloride (*off-label - during propagation*), propamocarb hydrochloride (*off-label - under protection*)
	Pythium	copper oxychloride (*off-label*), copper oxychloride (*off-label - during propagation*), etridiazole (*off-label*), fosetyl-aluminium (*off-label - during propagation*), metalaxyl-M (*off-label*), propamocarb hydrochloride (*off-label - during propagation*), propamocarb hydrochloride (*off-label - under protection*)
	Rhizoctonia	copper oxychloride (*off-label*), copper oxychloride (*off-label - during propagation*)
Pests	Beetles	deltamethrin (*off-label*)
	Caterpillars	Bacillus thuringiensis (*off-label*)
	Leafhoppers	deltamethrin (*off-label*)

Legumes - Beans (Phaseolus)

Crop control	Pre-harvest desiccation	diquat (*off-label*)
Diseases	Botrytis	azoxystrobin (*off-label*), iprodione (*off-label*), vinclozolin
	Damping off	thiram (*seed treatment*)
	Rust	tebuconazole (*off-label*)
	Sclerotinia	iprodione (*off-label*)
Pests	Aphids	fatty acids, nicotine, pirimicarb
	Caterpillars	lambda-cyhalothrin (*off-label*), nicotine
	Flies	chlorpyrifos (*off-label - harvested as a dry pulse*)
	Leaf miners	nicotine
	Mealybugs	fatty acids
	Pests, miscellaneous	nicotine (*off-label*)
	Scale insects	fatty acids
	Spider mites	fatty acids
	Whiteflies	fatty acids

Weeds	Broad-leaved weeds	bentazone, chlorpropham + fenuron (*off-label*), chlorthal-dimethyl, chlorthal-dimethyl (*off-label*), fomesafen, linuron (*off-label*), pendimethalin (*off-label*), trifluralin
	Crops as weeds	chlorthal-dimethyl (*off-label*), cycloxydim, fomesafen
	Grass weeds	cycloxydim, fluazifop-P-butyl (*off-label*), pendimethalin (*off-label*), trifluralin

Legumes - Beans (Vicia)

Crop control	Pre-harvest desiccation	diquat (*stock or pigeon feed only*), glufosinate-ammonium, glyphosate
Diseases	Ascochyta	azoxystrobin (*off-label*), thiabendazole + thiram (*seed treatment*)
	Botrytis	azoxystrobin (*off-label*)
	Chocolate spot	boscalid + pyraclostrobin (*moderate*), chlorothalonil, chlorothalonil + cyproconazole, cyproconazole (*with chlorothalonil*), iprodione, iprodione + thiophanate-methyl, tebuconazole, vinclozolin
	Damping off	thiabendazole + thiram (*seed treatment*), thiram (*seed treatment*)
	Downy mildew	chlorothalonil + metalaxyl-M, cymoxanil + fludioxonil + metalaxyl-M (*off-label - seed treatment*), fosetyl-aluminium
	Rust	azoxystrobin, boscalid + pyraclostrobin, chlorothalonil + cyproconazole, cyproconazole, tebuconazole, tebuconazole (*off-label*)
Pests	Aphids	fatty acids, nicotine, pirimicarb
	Birds/mammals	aluminium ammonium sulphate
	Caterpillars	nicotine
	Leaf miners	nicotine
	Mealybugs	fatty acids
	Scale insects	fatty acids
	Spider mites	fatty acids
	Weevils	alpha-cypermethrin, cypermethrin, deltamethrin, lambda-cyhalothrin, lambda-cyhalothrin + pirimicarb, zeta-cypermethrin
	Whiteflies	fatty acids
Weeds	Broad-leaved weeds	bentazone, carbetamide, clomazone, cyanazine (*Scotland only*), cyanazine + pendimethalin, fomesafen + terbutryn, fomesafen + terbutryn (*spring sown*), glyphosate, isoxaben + terbuthylazine, linuron + trifluralin, pendimethalin (*off-label*), propyzamide, simazine, terbuthylazine + terbutryn, trifluralin
	Crops as weeds	carbetamide, cycloxydim, fluazifop-P-butyl, glyphosate, propyzamide, quizalofop-P-ethyl, quizalofop-P-tefuryl, tepraloxydim
	Grass weeds	carbetamide, cyanazine (*Scotland only*), cyanazine + pendimethalin, cycloxydim, fluazifop-P-butyl, fluazifop-P-butyl (*off-label*), glyphosate, linuron + trifluralin, propaquizafop, propyzamide, quizalofop-P-ethyl, quizalofop-P-tefuryl, simazine, tepraloxydim, terbuthylazine + terbutryn, tri-allate, trifluralin
	Weeds, miscellaneous	glyphosate

Legumes - Clovers

Crop control	Pre-harvest desiccation	diquat
Weeds	Broad-leaved weeds	asulam (*off-label*), carbetamide, isoxaben (*off-label - grown for game cover*), MCPA + MCPB, MCPB, MCPB (*seed crops*), propyzamide
	Crops as weeds	carbetamide
	Grass weeds	carbetamide, propyzamide, tri-allate
	Weeds, miscellaneous	clopyralid (*off-label - game cover*)

Legumes - Forage legumes, general

Weeds	Broad-leaved weeds	2,4-DB, carbetamide, chlorpropham, MCPA + MCPB, paraquat (*off-label*), propyzamide
	Crops as weeds	carbetamide, fluazifop-P-butyl (*off-label*)
	Grass weeds	carbetamide, chlorpropham, fluazifop-P-butyl (*off-label*), paraquat (*off-label*), propyzamide, tri-allate
	Polygonums	chlorpropham

Legumes - Lupins

Weeds	Broad-leaved weeds	terbuthylazine + terbutryn
	Grass weeds	terbuthylazine + terbutryn

Legumes - Peas

Crop control	Pre-harvest desiccation	diquat, glufosinate-ammonium, glufosinate-ammonium (*not for seed*), glyphosate, sulphuric acid (commodity substance)
Diseases	Ascochyta	azoxystrobin, chlorothalonil, chlorothalonil + pyrimethanil, cymoxanil + fludioxonil + metalaxyl-M (*seed treatment*), thiabendazole + thiram (*seed treatment*), vinclozolin
	Botrytis	chlorothalonil, chlorothalonil + cyproconazole, iprodione (*off-label*), vinclozolin
	Damping off	cymoxanil + fludioxonil + metalaxyl-M (*seed treatment*), thiabendazole + thiram (*seed treatment*), thiram (*seed treatment*)
	Downy mildew	cymoxanil + fludioxonil + metalaxyl-M (*seed treatment*), fosetyl-aluminium (*off-label*), fosetyl-aluminium (*off-label - seed treatment*)
	Mycosphaerella	chlorothalonil, chlorothalonil + pyrimethanil, vinclozolin
	Pythium	cymoxanil + fludioxonil + metalaxyl-M (*seed treatment*)
	Sclerotinia	iprodione (*off-label*)
Pests	Aphids	alpha-cypermethrin, alpha-cypermethrin (*reduction*), bifenthrin, cypermethrin, deltamethrin, fatty acids, lambda-cyhalothrin, lambda-cyhalothrin + pirimicarb, nicotine, pirimicarb, triazamate, zeta-cypermethrin
	Birds/mammals	aluminium ammonium sulphate
	Caterpillars	alpha-cypermethrin, cypermethrin, deltamethrin, lambda-cyhalothrin, lambda-cyhalothrin + pirimicarb, nicotine, zeta-cypermethrin

	Leaf miners	nicotine
	Mealybugs	fatty acids
	Midges	deltamethrin, lambda-cyhalothrin + pirimicarb
	Scale insects	fatty acids
	Spider mites	fatty acids
	Weevils	alpha-cypermethrin, cypermethrin, deltamethrin, lambda-cyhalothrin, lambda-cyhalothrin + pirimicarb, zeta-cypermethrin
	Whiteflies	fatty acids
Weeds	Broad-leaved weeds	bentazone, bentazone + MCPB, bentazone + pendimethalin, clomazone, cyanazine, cyanazine + pendimethalin, fomesafen + terbutryn (*spring sown*), glyphosate, isoxaben + terbuthylazine, linuron + trifluralin, MCPA + MCPB, MCPB, pendimethalin, pendimethalin (*off-label*), terbuthylazine + terbutryn, trifluralin (*off-label*)
	Crops as weeds	cycloxydim, fluazifop-P-butyl, glyphosate, pendimethalin, quizalofop-P-ethyl, quizalofop-P-tefuryl, tepraloxydim
	Grass weeds	cyanazine, cyanazine + pendimethalin, cycloxydim, fluazifop-P-butyl, glyphosate, linuron + trifluralin, pendimethalin, propaquizafop, quizalofop-P-ethyl, quizalofop-P-tefuryl, tepraloxydim, terbuthylazine + terbutryn, tri-allate, trifluralin (*off-label*)
	Mayweeds	bentazone + pendimethalin
	Weeds, miscellaneous	glyphosate

Miscellaneous arable - Miscellaneous arable crops

Crop control	Pre-harvest desiccation	diquat (*off-label*)
Diseases	Botrytis	iprodione + thiophanate-methyl (*off-label*)
Pests	Slugs/snails	metaldehyde
	Whiteflies	alginate/polysaccharide
Weeds	Broad-leaved weeds	carfentrazone-ethyl (*before planting*), diquat, diquat (*between row treatment*), diquat + paraquat, glyphosate, isoxaben (*off-label - grown for game cover*), paraquat, paraquat (*stale seedbed/inter-row*)
	Crops as weeds	carfentrazone-ethyl (*before planting*), diquat + paraquat, glyphosate, paraquat, paraquat (*stale seedbed/inter-row*), quizalofop-P-tefuryl
	Grass weeds	diquat + paraquat, glyphosate, paraquat, paraquat (*stale seedbed/inter-row*), quizalofop-P-tefuryl
	Weeds, miscellaneous	clopyralid (*off-label - game cover*)

Miscellaneous arable - Miscellaneous arable situations

Diseases	Soil-borne diseases	dazomet (*soil fumigation*)
Pests	Birds/mammals	aluminium ammonium sulphate (*seed treatment*), ziram
	Free-living nematodes	dazomet (*soil fumigation*)

	Slugs/snails	metaldehyde
	Soil pests	dazomet (*soil fumigation*)
Weeds	Broad-leaved weeds	amitrole, citronella oil, diquat + paraquat, glufosinate-ammonium, glufosinate-ammonium (*uncropped*), glyphosate, paraquat, paraquat (*minimum cultivation*)
	Crops as weeds	amitrole (*barley stubble*), diquat + paraquat, fluazifop-P-butyl, glyphosate, paraquat, paraquat (*minimum cultivation*), tepraloxydim
	Grass weeds	amitrole, diquat + paraquat, fluazifop-P-butyl, glufosinate-ammonium, glufosinate-ammonium (*uncropped*), glyphosate, paraquat, paraquat (*minimum cultivation*), paraquat (*stubble treatment*), tepraloxydim
	Weeds, miscellaneous	2,4-D + dicamba + triclopyr, amitrole, cycloxydim, dazomet (*soil fumigation*), diquat + paraquat, fluazifop-P-butyl, glufosinate-ammonium, glyphosate, glyphosate (*wiper application*), metsulfuron-methyl, paraquat, tepraloxydim, thifensulfuron-methyl

Miscellaneous field vegetables - All vegetables

Diseases	Soil-borne diseases	dazomet (*soil fumigation*)
Pests	Aphids	natural plant extracts, nicotine, rotenone
	Capsid bugs	nicotine
	Free-living nematodes	dazomet (*soil fumigation*)
	Leafhoppers	natural plant extracts, nicotine
	Mites	natural plant extracts
	Sawflies	nicotine
	Soil pests	dazomet (*soil fumigation*)
	Thrips	natural plant extracts, nicotine
	Whiteflies	natural plant extracts
Weeds	Broad-leaved weeds	glufosinate-ammonium
	Grass weeds	glufosinate-ammonium
	Weeds, miscellaneous	ammonium sulphamate (*pre-planting*), dazomet (*soil fumigation*), glufosinate-ammonium

Oilseed crops - Hemp

Crop control	Pre-harvest desiccation	diquat (*off-label*)

Oilseed crops - Linseed/flax

Crop control	Pre-harvest desiccation	diquat, glufosinate-ammonium, glyphosate
Diseases	Alternaria	iprodione (*seed treatment*)
	Botrytis	tebuconazole (*reduction*)
	Powdery mildew	tebuconazole
	Seed-borne diseases	prochloraz (*seed treatment*)
Pests	Beetles	bifenthrin, zeta-cypermethrin
Weeds	Broad-leaved weeds	amidosulfuron, bentazone, bromoxynil, clopyralid, glyphosate, MCPA, metsulfuron-methyl, trifluralin, trifluralin (*off-label*)

	Crops as weeds	cycloxydim, fluazifop-P-butyl, glyphosate, quizalofop-P-ethyl, quizalofop-P-tefuryl, tepraloxydim
	Grass weeds	cycloxydim, fluazifop-P-butyl, glyphosate, propaquizafop, quizalofop-P-ethyl, quizalofop-P-tefuryl, tepraloxydim, trifluralin, trifluralin (*off-label*)
	Mayweeds	clopyralid, metsulfuron-methyl
	Weeds, miscellaneous	glyphosate

Oilseed crops - Miscellaneous oilseeds

Crop control	Pre-harvest desiccation	diquat (*off-label*), diquat (*off-label - for morphine production*), glyphosate (*off-label - for morphine production*)
Diseases	Alternaria	boscalid (*off-label - for morphine production*), iprodione (*off-label - for morphine production - seed treatment*)
	Botrytis	boscalid (*off-label - for morphine production*)
	Damping off	cymoxanil + fludioxonil + metalaxyl-M (*off-label - for morphine production - seed treatment*), thiram (*off-label - for morphine production - seed treatment*)
	Downy mildew	chlorothalonil + metalaxyl-M (*off-label - for morphine production*), cymoxanil + fludioxonil + metalaxyl M (*off-label - for morphine production - seed treatment*), dimethomorph + mancozeb (*off-label - for morphine production*), mancozeb (*off-label - for morphine production*)
	Powdery mildew	azoxystrobin (*off-label - for morphine production*)
	Pythium	cymoxanil + fludioxonil + metalaxyl-M (*off-label - for morphine production - seed treatment*), thiram (*off-label - for morphine production - seed treatment*)
	Sclerotinia stem rot	boscalid (*off-label - for morphine production*)
	White blister	propiconazole (*off-label*)
Pests	Aphids	deltamethrin (*off-label*), pirimicarb (*off-label*), pirimicarb (*off-label - for morphine production*)
	Beetles	deltamethrin (*off-label*), lambda-cyhalothrin (*off-label - for morphine production*)
Weeds	Broad-leaved weeds	asulam (*off-label - for morphine production*), bentazone (*off-label*), chlorthal-dimethyl (*off-label - for morphine production*), clopyralid (*off-label*), clopyralid (*off-label - for morphine production*), ethofumesate (*off-label - for morphine production*), pendimethalin (*off-label*), trifluralin (*off-label*)
	Crops as weeds	propaquizafop (*off-label - for morphine production*)
	Grass weeds	asulam (*off-label - for morphine production*), ethofumesate (*off-label - for morphine production*), pendimethalin (*off-label*), propaquizafop (*off-label - for morphine production*), propyzamide (*off-label*), tepraloxydim (*off-label - for morphine production*), trifluralin (*off-label*)
	Mayweeds	clopyralid (*off-label*), clopyralid (*off-label - for morphine production*), ethofumesate (*off-label - for morphine production*)
	Weeds, miscellaneous	diquat (*off-label - for morphine production*)

Oilseed crops - Oilseed rape

Crop control	Pre-harvest desiccation	diquat, glufosinate-ammonium, glyphosate
Diseases	Alternaria	azoxystrobin, difenoconazole, iprodione, iprodione (*seed treatment*), iprodione + thiophanate-methyl, metconazole, prochloraz, tebuconazole, vinclozolin
	Black scurf and stem canker	difenoconazole, iprodione + thiophanate-methyl, tebuconazole
	Botrytis	chlorothalonil, iprodione, iprodione + thiophanate-methyl, prochloraz, vinclozolin
	Canker	carbendazim + flusilazole (*reduction*), famoxadone + flusilazole (*suppression*), metconazole (*foliar disease - reduction*), metconazole (*reduction*), prochloraz, prochloraz + propiconazole, prochloraz + thiram (*seed treatment*), tebuconazole
	Damping off	thiram (*seed treatment*)
	Downy mildew	chlorothalonil, mancozeb
	Light leaf spot	carbendazim, carbendazim + flusilazole, cyproconazole, difenoconazole, famoxadone + flusilazole, flusilazole, iprodione + thiophanate-methyl, prochloraz, prochloraz + propiconazole, propiconazole (*reduction*), tebuconazole
	Phoma leaf spot	carbendazim + flusilazole, cyproconazole
	Powdery mildew	sulphur
	Ring spot	tebuconazole (*reduction*)
	Sclerotinia stem rot	azoxystrobin, boscalid, iprodione, iprodione + thiophanate-methyl, metconazole (*reduction*), prochloraz, prochloraz + tebuconazole, prothioconazole, tebuconazole, vinclozolin
	White leaf spot	prochloraz
Pests	Aphids	deltamethrin, lambda-cyhalothrin, lambda-cyhalothrin + pirimicarb, pirimicarb, tau-fluvalinate
	Beetles	alpha-cypermethrin, beta-cyfluthrin + imidacloprid (*seed treatment*), bifenthrin, cyfluthrin, cypermethrin, deltamethrin, lambda-cyhalothrin, lambda-cyhalothrin + pirimicarb, tau-fluvalinate, zeta-cypermethrin
	Birds/mammals	aluminium ammonium sulphate
	Midges	alpha-cypermethrin, bifenthrin, cypermethrin, deltamethrin, lambda-cyhalothrin, lambda-cyhalothrin + pirimicarb, zeta-cypermethrin
	Slugs/snails	methiocarb, thiodicarb
	Weevils	alpha-cypermethrin, bifenthrin, cypermethrin, deltamethrin, lambda-cyhalothrin, lambda-cyhalothrin + pirimicarb, zeta-cypermethrin
Plant growth regulation	Growth control	tebuconazole
	Quality/yield control	sulphur
Weeds	Broad-leaved weeds	bifenox (*off-label*), carbetamide, clomazone, clopyralid, clopyralid + picloram, cyanazine, glyphosate, metazachlor, metazachlor + quinmerac, napropamide, propachlor, propyzamide, pyridate (*off-label*), trifluralin
	Crops as weeds	carbetamide, cycloxydim, fluazifop-P-butyl, glyphosate, propyzamide, quizalofop-P-ethyl, quizalofop-P-tefuryl, tepraloxydim

	Grass weeds	carbetamide, cyanazine, cycloxydim, fluazifop-P-butyl, glyphosate, metazachlor, metazachlor + quinmerac, napropamide, propachlor, propaquizafop, propyzamide, quizalofop-P-ethyl, quizalofop-P-tefuryl, tepraloxydim, trifluralin
	Mayweeds	clopyralid, clopyralid + picloram
	Weeds, miscellaneous	glyphosate

Oilseed crops - Soya

Crop control	Pre-harvest desiccation	diquat (*off-label*)
Weeds	Broad-leaved weeds	fomesafen (*off-label*), trifluralin (*off-label*)
	Crops as weeds	fomesafen (*off-label*)
	Grass weeds	trifluralin (*off-label*)

Oilseed crops - Sunflowers

Crop control	Pre-harvest desiccation	diquat (*off-label*)
Pests	Slugs/snails	methiocarb
Weeds	Broad-leaved weeds	diquat (*off-label*), isoxaben (*off-label - grown for game cover*), pendimethalin, trifluralin (*off-label*)
	Crops as weeds	pendimethalin
	Grass weeds	pendimethalin, trifluralin (*off-label*)

Root and tuber crops - Beet crops

Diseases	Black leg	hymexazol (*seed treatment*)
	Botrytis	iprodione (*off-label*)
	Canker	iprodione (*off-label*)
	Cercospora leaf spot	epoxiconazole + pyraclostrobin
	Damping off	cymoxanil + fludioxonil + metalaxyl-M (*off-label - seed treatment*)
	Downy mildew	mancozeb (*off-label*)
	Powdery mildew	carbendazim + flusilazole, cyproconazole, epoxiconazole + pyraclostrobin, flusilazole, quinoxyfen, sulphur
	Ramularia leaf spots	cyproconazole, epoxiconazole + pyraclostrobin, propiconazole (*reduction*)
	Rhizoctonia	azoxystrobin (*off-label*), mancozeb (*off-label*), metalaxyl-M (*off-label*), propamocarb hydrochloride (*off-label*)
	Rust	carbendazim + flusilazole, cyproconazole, cyproconazole (*off-label*), epoxiconazole + pyraclostrobin, fenpropimorph (*off-label*), flusilazole, propiconazole
	Seed-borne diseases	thiram (*seed soak*)
Pests	Aphids	beta-cyfluthrin + clothianidin (*seed treatment*), carbosulfan, dimethoate, dimethoate (*excluding Myzus persicae*), imidacloprid (*seed treatment*), lambda-cyhalothrin + pirimicarb, nicotine, oxamyl, pirimicarb,

		pirimicarb (*off-label*), triazamate (*including Myzus persicae*)
	Beetles	benfuracarb, carbosulfan, chlorpyrifos, deltamethrin, imidacloprid (*seed treatment*), lambda-cyhalothrin, lambda-cyhalothrin + pirimicarb, oxamyl, tefluthrin (*seed treatment*)
	Birds/mammals	aluminium ammonium sulphate
	Caterpillars	cypermethrin, lambda-cyhalothrin (*off-label*), nicotine
	Cutworms	cypermethrin, lambda-cyhalothrin, lambda-cyhalothrin (*off-label*), lambda-cyhalothrin + pirimicarb, methiocarb (*reduction*), zeta-cypermethrin
	Flies	carbosulfan, dimethoate, imidacloprid (*seed treatment*), oxamyl
	Free-living nematodes	benfuracarb, carbosulfan, oxamyl
	Leaf miners	benfuracarb, dimethoate, lambda-cyhalothrin, lambda-cyhalothrin + pirimicarb, nicotine
	Leatherjackets	chlorpyrifos, methiocarb (*reduction*)
	Millipedes	benfuracarb, carbosulfan, imidacloprid (*seed treatment*), methiocarb (*reduction*), oxamyl, tefluthrin (*seed treatment*)
	Slugs/snails	methiocarb
	Springtails	benfuracarb, carbosulfan, imidacloprid (*seed treatment*), tefluthrin (*seed treatment*)
	Symphylids	benfuracarb, carbosulfan, imidacloprid (*seed treatment*), tefluthrin (*seed treatment*)
	Wireworms	carbosulfan
Plant growth regulation	Quality/yield control	sulphur
Weeds	Broad-leaved weeds	carbetamide, chloridazon, chloridazon + chlorpropham + metamitron, chloridazon + ethofumesate, chloridazon + lenacil, chloridazon + metamitron, chloridazon + quinmerac, chlorpropham + metamitron, clopyralid, desmedipham + ethofumesate + phenmedipham, desmedipham + phenmedipham, diquat, ethofumesate, ethofumesate + metamitron, ethofumesate + phenmedipham, glufosinate-ammonium, glyphosate, lenacil, lenacil + phenmedipham, lenacil + triflusulfuron-methyl, metamitron, paraquat, phenmedipham, propyzamide, trifluralin, triflusulfuron-methyl, triflusulfuron-methyl (*off-label*)
	Crops as weeds	carbetamide, cycloxydim, fluazifop-P-butyl, glyphosate, glyphosate (*wiper application*), lenacil + triflusulfuron-methyl, paraquat, propaquizafop (*off-label*), propyzamide, quizalofop-P-ethyl, quizalofop-P-tefuryl, sodium chloride (commodity substance), tepraloxydim
	Grass weeds	carbetamide, chloridazon, chloridazon + chlorpropham + metamitron, chloridazon + ethofumesate, chloridazon + lenacil, chloridazon + metamitron, chloridazon + quinmerac, chlorpropham + metamitron, cycloxydim, desmedipham + ethofumesate + phenmedipham, ethofumesate, ethofumesate + metamitron, ethofumesate + phenmedipham, fluazifop-P-butyl, fluazifop-P-butyl (*off-label*), glufosinate-ammonium, glyphosate, lenacil, metamitron, paraquat, propaquizafop, propaquizafop (*off-label*), propyzamide,

quizalofop-P-ethyl, quizalofop-P-tefuryl, tepraloxydim, tri-allate, trifluralin

Mayweeds	chloridazon + quinmerac, clopyralid
Polygonums	sodium chloride (commodity substance)
Speedwells	chloridazon + quinmerac
Weeds, miscellaneous	glufosinate-ammonium, glyphosate

Root and tuber crops - Carrots/parsnips/celeriac

Diseases	Alternaria	azoxystrobin, boscalid + pyraclostrobin (*moderate*), fenpropimorph (*off-label*), iprodione + thiophanate-methyl (*off-label*), tebuconazole
	Cavity spot	metalaxyl-M, metalaxyl-M (*off-label*)
	Crown rot	fenpropimorph (*off-label*), iprodione + thiophanate-methyl (*off-label*)
	Damping off	thiram (*seed treatment*)
	Powdery mildew	azoxystrobin, boscalid + pyraclostrobin, fenpropimorph (*off-label*), sulphur (*off-label*), tebuconazole
	Pythium	cymoxanil + fludioxonil + metalaxyl-M (*off-label - seed treatment*)
	Sclerotinia	azoxystrobin (*off-label*), boscalid + pyraclostrobin (*moderate*), boscalid + pyraclostrobin (*off-label*), tebuconazole
	Seed-borne diseases	thiram (*seed soak*)
	Septoria diseases	chlorothalonil (*off-label*), difenoconazole (*off-label*), mancozeb (*off-label*)
Pests	Aphids	aldicarb, carbosulfan, deltamethrin (*off-label*), lambda-cyhalothrin + pirimicarb, nicotine, pirimicarb, pirimicarb (*off-label*), pymetrozine (*off-label*)
	Birds/mammals	aluminium ammonium sulphate
	Caterpillars	deltamethrin (*off-label*), nicotine
	Cutworms	cypermethrin (*off-label*), lambda-cyhalothrin, lambda-cyhalothrin (*off-label*), lambda-cyhalothrin + pirimicarb
	Flies	lambda-cyhalothrin (*off-label*), tefluthrin (*off-label - seed treatment*)
	Free-living nematodes	aldicarb, carbosulfan
	Leaf miners	nicotine
	Pests, miscellaneous	deltamethrin (*off-label*), dimethoate (*off-label*)
	Stem nematodes	oxamyl (*off-label*)
Plant growth regulation	Growth control	maleic hydrazide (*off-label*)
Weeds	Broad-leaved weeds	chlorpropham, chlorpropham + pentanochlor, isoxaben (*off-label*), linuron, linuron (*off-label*), metoxuron, metribuzin (*off-label*), pendimethalin, pentanochlor, prometryn, trifluralin, trifluralin (*off-label*)
	Crops as weeds	cycloxydim, fluazifop-P-butyl, pendimethalin, tepraloxydim
	Grass weeds	chlorpropham, chlorpropham + pentanochlor, cycloxydim, fluazifop-P-butyl, fluazifop-P-butyl (*off-label*), linuron, metribuzin (*off-label*), pendimethalin, pentanochlor, prometryn, propaquizafop, tepraloxydim, trifluralin, trifluralin (*off-label*)
	Mayweeds	metoxuron

	Polygonums	chlorpropham
	Weeds, miscellaneous	linuron (*off-label*)

Root and tuber crops - Potatoes

Crop control	Pre-harvest desiccation	carfentrazone-ethyl, diquat, glufosinate-ammonium, sulphuric acid (commodity substance)
Diseases	Black dot	azoxystrobin
	Black scurf and stem canker	flutolanil (*tuber treatment*), imazalil + pencycuron (*tuber treatment - reduction*), imazalil + thiabendazole (*tuber treatment - reduction*), pencycuron (*tuber treatment*)
	Blight	benalaxyl + mancozeb, Bordeaux mixture, chlorothalonil, chlorothalonil + mancozeb, chlorothalonil + propamocarb hydrochloride, copper oxychloride, cyazofamid, cymoxanil, cymoxanil + famoxadone, cymoxanil + mancozeb, dimethomorph + mancozeb, fenamidone + mancozeb, fenamidone + propamocarb hydrochloride, fluazinam, fluazinam + metalaxyl-M, mancozeb, mancozeb + metalaxyl-M, mancozeb + propamocarb hydrochloride, mancozeb + zoxamide
	Dry rot	imazalil, imazalil + thiabendazole (*tuber treatment - reduction*), thiabendazole (*off-label*), thiabendazole (*tuber treatment - post-harvest*)
	Fungus diseases	peroxyacetic acid (commodity substance) (*tuber treatment*)
	Gangrene	2-aminobutane, imazalil, imazalil + thiabendazole (*tuber treatment - reduction*), thiabendazole (*tuber treatment - post-harvest*)
	Rhizoctonia	azoxystrobin, flutolanil (*off-label - chitted seed treatment*), flutolanil (*tuber treatment*), imazalil + pencycuron (*tuber treatment*), iprodione (*seed treatment*), pencycuron (*tuber treatment*), tolclofos-methyl (*off-label - chitted seed treatment*), tolclofos-methyl (*tuber treatment*)
	Silver scurf	2-aminobutane, imazalil, imazalil + pencycuron (*tuber treatment - reduction*), imazalil + thiabendazole (*tuber treatment - reduction*), thiabendazole (*tuber treatment - post-harvest*)
	Skin spot	2-aminobutane, imazalil, imazalil + thiabendazole (*tuber treatment - reduction*), thiabendazole (*tuber treatment - post-harvest*)
Pests	Aphids	aldicarb, flonicamid, lambda-cyhalothrin, lambda-cyhalothrin + pirimicarb, nicotine, oxamyl, pirimicarb, pymetrozine
	Beetles	deltamethrin (*off-label*)
	Caterpillars	cypermethrin, nicotine
	Cutworms	chlorpyrifos, cypermethrin, lambda-cyhalothrin + pirimicarb, zeta-cypermethrin
	Cyst nematodes	1,3-dichloropropene, ethoprophos, fosthiazate, oxamyl
	Free-living nematodes	1,3-dichloropropene, aldicarb, oxamyl
	Leaf miners	nicotine
	Leatherjackets	methiocarb (*reduction*)
	Slugs/snails	methiocarb, thiodicarb
	Wireworms	ethoprophos, fosthiazate (*reduction*)

Plant growth regulation	Growth control	chlorpropham, chlorpropham (*fog*), chlorpropham (*thermal fog*), ethylene (commodity substance) (*in store*), maleic hydrazide
	Plant growth regulation, miscellaneous	maleic hydrazide
Weeds	Broad-leaved weeds	bentazone, carfentrazone-ethyl, diquat, diquat + paraquat, flufenacet + metribuzin, glufosinate-ammonium, linuron, metribuzin, paraquat, pendimethalin, rimsulfuron
	Crops as weeds	carfentrazone-ethyl, cycloxydim, diquat + paraquat, metribuzin, paraquat, pendimethalin, quizalofop-P-tefuryl, rimsulfuron
	Grass weeds	cycloxydim, diquat + paraquat, flufenacet + metribuzin, glufosinate-ammonium, linuron, metribuzin, paraquat, pendimethalin, propaquizafop, quizalofop-P-tefuryl
	Weeds, miscellaneous	diquat, glufosinate-ammonium (*not for seed*)

Stem and bulb vegetables - Asparagus

Diseases	Phytophthora	metalaxyl-M (*off-label*)
	Rust	azoxystrobin, difenoconazole (*off-label*), mancozeb (*off-label*)
	Stemphylium	azoxystrobin, iprodione (*off-label*), mancozeb (*off-label*)
Pests	Aphids	nicotine
	Beetles	cypermethrin (*off-label*)
	Caterpillars	nicotine
	Leaf miners	nicotine
Weeds	Broad-leaved weeds	clomazone (*off-label*), clopyralid (*off-label*), diuron (*off-label*), isoxaben (*off-label*), metamitron (*off-label*), metribuzin (*off-label*), pendimethalin (*off-label*), simazine, simazine (*off-label*)
	Grass weeds	diuron (*off-label*), metamitron (*off-label*), metribuzin (*off-label*), simazine, simazine (*off-label*)
	Weeds, miscellaneous	diuron (*off-label*), glyphosate (*off-label*), metamitron (*off-label*), metribuzin (*off-label*), triclopyr (*off-label - directed treatment*)

Stem and bulb vegetables - Celery/chicory

Diseases	Botrytis	azoxystrobin (*off-label*), iprodione (*off-label*)
	Damping off and foot rot	etridiazole (*seeds and seedlings*)
	Downy mildew	copper oxychloride (*off-label*), mancozeb (*off-label*)
	Phytophthora	etridiazole (*transplants*), fosetyl-aluminium (*off-label - for forcing*), fosetyl-aluminium (*off-label - in forcing sheds*)
	Rhizoctonia	azoxystrobin (*off-label*)
	Rust	mancozeb (*off-label*)
	Sclerotinia	azoxystrobin (*off-label*), iprodione (*off-label*)
	Seed-borne diseases	thiram (*seed soak*)
	Septoria diseases	Bordeaux mixture, chlorothalonil (*qualified minor use*), copper ammonium carbonate, copper oxychloride, difenoconazole (*off-label*), mancozeb (*off-label*)

Pests	Aphids	deltamethrin (*off-label*), nicotine, pirimicarb, pirimicarb (*off-label*), pirimicarb (*off-label - for forcing*), pymetrozine (*off-label*)
	Beetles	deltamethrin (*off-label*)
	Caterpillars	Bacillus thuringiensis (*off-label*), deltamethrin (*off-label*), diflubenzuron (*off-label*), lambda-cyhalothrin (*off-label*), nicotine
	Cutworms	lambda-cyhalothrin (*off-label - for forcing*)
	Flies	lambda-cyhalothrin (*off-label*)
	Leaf miners	nicotine
	Leafhoppers	deltamethrin (*off-label*)
	Pests, miscellaneous	deltamethrin (*off-label*)
Weeds	Broad-leaved weeds	chlorpropham, chlorpropham + pentanochlor, linuron, pendimethalin (*off-label*), pentanochlor, prometryn, propachlor (*off-label*), propachlor (*off-label - under crop covers*), propyzamide (*off-label*), triflusulfuron-methyl (*off-label*)
	Crops as weeds	propaquizafop (*off-label*)
	Grass weeds	chlorpropham, chlorpropham + pentanochlor, linuron, pentanochlor, prometryn, propachlor (*off-label*), propachlor (*off-label - under crop covers*), propaquizafop (*off-label*), propyzamide (*off-label*)
	Polygonums	chlorpropham

Stem and bulb vegetables - Globe artichokes/cardoons

Diseases	Downy mildew	azoxystrobin (*off-label*)
	Powdery mildew	azoxystrobin (*off-label*)
Pests	Aphids	nicotine
	Caterpillars	nicotine
	Leaf miners	nicotine

Stem and bulb vegetables - Onions/leeks/garlic

Diseases	Bacterial blight	copper oxychloride (*off-label*)
	Botrytis	chlorothalonil, fenhexamid (*off-label*), iprodione, propamocarb hydrochloride (*off-label*), thiabendazole + thiram (*off-label - seed treatment*)
	Collar rot	iprodione
	Damping off	thiram (*seed treatment*)
	Downy mildew	azoxystrobin, azoxystrobin (*off-label*), chlorothalonil + metalaxyl-M (*qualified minor use*), dimethomorph + mancozeb (*off-label*), fosetyl-aluminium (*off-label*), mancozeb (*off-label*), mancozeb + metalaxyl-M (*off-label*), metalaxyl-M (*off-label*), propamocarb hydrochloride (*off-label*)
	Fungus diseases	mancozeb + metalaxyl-M (*off-label*)
	Fusarium diseases	thiabendazole + thiram (*off-label - seed treatment*)
	Phytophthora	propamocarb hydrochloride
	Powdery mildew	azoxystrobin (*off-label*), tebuconazole (*off-label*)
	Purple blotch	azoxystrobin
	Pythium	propamocarb hydrochloride

	Rhynchosporium	propiconazole (*off-label*)
	Ring spot	azoxystrobin (*off-label*), prochloraz (*off-label*)
	Rust	azoxystrobin, azoxystrobin (*off-label*), cyproconazole, propiconazole (*off-label*), tebuconazole, tebuconazole (*off-label*)
	Storage rots	copper oxychloride (*off-label*)
	White rot	tebuconazole (*off-label*)
	White tip	chlorothalonil + metalaxyl-M (*qualified minor use*), mancozeb (*off-label*)
Pests	Aphids	deltamethrin (*off-label*), nicotine, pirimicarb (*off-label*), pymetrozine (*off-label*)
	Caterpillars	Bacillus thuringiensis (*off-label*), deltamethrin (*off-label*), diflubenzuron (*off-label*), nicotine
	Cutworms	chlorpyrifos
	Flies	tefluthrin (*off-label - seed treatment*)
	Leaf miners	nicotine
	Pests, miscellaneous	deltamethrin (*off-label*), dimethoate (*off-label*), dimethoate (*off-label - seedlings*)
	Stem nematodes	aldicarb, oxamyl (*off-label*)
	Thrips	lambda-cyhalothrin (*off-label*), spinosad
Plant growth regulation	Growth control	maleic hydrazide
Weeds	Broad-leaved weeds	bentazone (*off-label*), chloridazon (*off-label*), chloridazon + propachlor (*off-label - post-em. Pre-em on label*), chlorpropham, chlorthal-dimethyl, chlorthal-dimethyl (*off-label*), chlorthal-dimethyl + propachlor, clopyralid, clopyralid (*off-label*), cyanazine, cyanazine (*fen soils only*), ethofumesate (*off-label*), fluroxypyr (*off-label*), glyphosate, ioxynil, ioxynil (*off-label*), linuron (*off-label*), metamitron (*off-label*), pendimethalin, pendimethalin (*off-label*), pendimethalin (*off-label - pre + post emergence treatment*), prometryn, prometryn (*off-label*), propachlor, propachlor (*off-label*), pyridate, sodium monochloroacetate, trifluralin (*off-label*)
	Crops as weeds	cycloxydim, fluazifop-P-butyl, fluroxypyr (*off-label*), glyphosate, pendimethalin, pendimethalin (*off-label*), pendimethalin (*off-label - pre + post emergence treatment*), propaquizafop (*off-label*), tepraloxydim
	Grass weeds	chloridazon (*off-label*), chloridazon + propachlor (*off-label - post-em. Pre-em on label*), chlorpropham, chlorthal-dimethyl + propachlor, cyanazine, cyanazine (*fen soils only*), cycloxydim, ethofumesate (*off-label*), fluazifop-P-butyl, fluazifop-P-butyl (*off-label*), glyphosate, linuron (*off-label*), pendimethalin, pendimethalin (*off-label*), pendimethalin (*off-label - pre + post emergence treatment*), prometryn, prometryn (*off-label*), propachlor, propachlor (*off-label*), propaquizafop, propaquizafop (*off-label*), tepraloxydim, tepraloxydim (*off-label*), trifluralin (*off-label*)
	Mayweeds	clopyralid, clopyralid (*off-label*)
	Weeds, miscellaneous	glyphosate, linuron (*off-label*)

Flowers and ornamentals

Flowers - Bedding plants, general

Diseases	Alternaria	iprodione (*seed treatment*)
	Botrytis	carbendazim (*off-label*), thiram
	Petal blight	mancozeb
	Phytophthora	propamocarb hydrochloride
	Powdery mildew	bupirimate, carbendazim (*off-label*), copper ammonium carbonate, dinocap
	Pythium	propamocarb hydrochloride
	Rust	azoxystrobin (*off-label - in pots*), mancozeb, propiconazole (*off-label*), thiram
Pests	Aphids	imidacloprid, malathion, pirimicarb
	Flies	imidacloprid
	Pests, miscellaneous	aldicarb (*off-label*)
	Stem nematodes	aldicarb (*off-label*)
	Weevils	imidacloprid
	Whiteflies	imidacloprid
Plant growth regulation	Flowering control	paclobutrazol
	Growth control	chlormequat, daminozide, paclobutrazol
Weeds	Broad-leaved weeds	chlorpropham, chlorpropham + pentanochlor, chlorthal-dimethyl, pentanochlor
	Grass weeds	chlorpropham, chlorpropham + pentanochlor, pentanochlor
	Polygonums	chlorpropham

Flowers - Bulbs/corms

Crop control	Pre-harvest desiccation	sulphuric acid (commodity substance)
Diseases	Botrytis	carbendazim (*off-label*), chlorothalonil (*off-label - for galanthamine production*), tebuconazole (*off-label - for galanthamine production*), thiram, vinclozolin (*off-label - for galanthamine production*)
	Crown rot	metalaxyl-M (*off-label*)
	Fire	mancozeb, thiram
	Fungus diseases	formaldehyde (commodity substance) (*dip*)
	Fusarium diseases	carbendazim (*off-label*), thiabendazole (*bulb dip or spray*)
	Ink disease	chlorothalonil (*qualified minor use*)
	Penicillium rot	carbendazim (*off-label*)
	Phytophthora	etridiazole, propamocarb hydrochloride
	Powdery mildew	bupirimate
	Pythium	propamocarb hydrochloride
	Sclerotinia	carbendazim (*off-label*)
	Stagonospora	carbendazim (*off-label*)
	White mould	chlorothalonil (*off-label*), mancozeb (*off-label - for galanthamine production*)
Pests	Aphids	malathion, nicotine

	Birds/mammals	aluminium ammonium sulphate (*bulb treatment*), aluminium ammonium sulphate (*corm treatment*)
	Capsid bugs	nicotine
	Flies	chlorpyrifos (*off-label*)
	Leaf miners	nicotine
	Leafhoppers	nicotine
	Sawflies	nicotine
	Stem nematodes	1,3-dichloropropene
	Thrips	nicotine
Plant growth regulation	Flowering control	paclobutrazol
	Growth control	2-chloroethylphosphonic acid, chlormequat, paclobutrazol
Weeds	Broad-leaved weeds	bentazone, chlorpropham, chlorpropham + pentanochlor, cyanazine, diquat, diquat + paraquat, diuron (*off-label - for galanthamine production*), paraquat, pendimethalin (*off-label - for galanthamine production*), pentanochlor
	Crops as weeds	cycloxydim, pendimethalin (*off-label - for galanthamine production*)
	Grass weeds	chlorpropham, chlorpropham + pentanochlor, cyanazine, cycloxydim, diquat + paraquat, paraquat, pentanochlor
	Speedwells	pendimethalin (*off-label - for galanthamine production*)

Flowers - Miscellaneous flowers

Diseases	Black spot	captan, kresoxim-methyl, mancozeb, myclobutanil
	Powdery mildew	bupirimate, dinocap, dodemorph, fenarimol, kresoxim-methyl, myclobutanil
	Rust	mancozeb, myclobutanil, penconazole
Pests	Aphids	imidacloprid, malathion, pirimicarb, rotenone
	Flies	imidacloprid
	Leaf miners	abamectin
	Sawflies	rotenone
	Spider mites	abamectin
	Thrips	abamectin
	Weevils	imidacloprid
	Whiteflies	imidacloprid
Plant growth regulation	Flowering control	paclobutrazol
	Growth control	paclobutrazol
Weeds	Broad-leaved weeds	chlorthal-dimethyl, dichlobenil, pentanochlor, propyzamide, simazine, simazine (*field grown nursery stock*)
	Grass weeds	dichlobenil, pentanochlor, propyzamide, simazine, simazine (*field grown nursery stock*)
	Weeds, miscellaneous	dichlobenil

Flowers - Pot plants

Diseases	Botrytis	carbendazim (*off-label*), iprodione
	Phytophthora	fosetyl-aluminium, propamocarb hydrochloride
	Powdery mildew	carbendazim (*off-label*)
	Pythium	propamocarb hydrochloride
	Root rot	fosetyl-aluminium
	Rust	mancozeb
Pests	Aphids	deltamethrin, imidacloprid, pirimicarb
	Caterpillars	deltamethrin
	Flies	imidacloprid
	Mealybugs	deltamethrin, petroleum oil
	Pests, miscellaneous	aldicarb (*off-label*)
	Scale insects	deltamethrin, petroleum oil
	Spider mites	petroleum oil
	Stem nematodes	aldicarb (*off-label*)
	Weevils	imidacloprid
	Whiteflies	deltamethrin, imidacloprid
Plant growth regulation	Flowering control	2-chloroethylphosphonic acid, paclobutrazol
	Fruiting control	paclobutrazol
	Growth control	2-chloroethylphosphonic acid, chlormequat, daminozide, paclobutrazol

Flowers - Protected bulbs/corms

Pests	Aphids	malathion, pirimicarb

Flowers - Protected flowers

Diseases	Ring spot	difenoconazole (*off-label*), difenoconazole (*off-label - ground or pot grown*)
	Rust	azoxystrobin (*off-label*), difenoconazole (*off-label*), difenoconazole (*off-label - ground or pot grown*), propiconazole (*off-label*)
Pests	Aphids	malathion, nicotine, pirimicarb, Verticillium lecanii
	Leaf miners	abamectin
	Pests, miscellaneous	aldicarb (*off-label*)
	Spider mites	abamectin, tebufenpyrad
	Stem nematodes	aldicarb (*off-label*)
	Thrips	abamectin
Plant growth regulation	Growth control	2-chloroethylphosphonic acid

Miscellaneous flowers and ornamentals - Miscellaneous uses

Diseases	Fungus diseases	potassium bicarbonate (commodity substance)
Pests	Birds/mammals	aluminium ammonium sulphate
	Caterpillars	diflubenzuron
	Leaf miners	deltamethrin (*off-label*)

	Slugs/snails	metaldehyde, methiocarb (*outdoor only*)
	Thrips	deltamethrin (*off-label*)
	Whiteflies	alginate/polysaccharide
Plant growth regulation	Growth control	maleic hydrazide
Weeds	Broad-leaved weeds	amitrole, carfentrazone-ethyl (*before planting*), diquat + paraquat, glyphosate
	Crops as weeds	carfentrazone-ethyl (*before planting*), glyphosate
	Grass weeds	amitrole, diquat + paraquat, glyphosate
	Weeds, miscellaneous	amitrole, glyphosate, glyphosate (*wiper application*)

Miscellaneous flowers and ornamentals - Soils

Diseases	Soil-borne diseases	metam-sodium
Pests	Free-living nematodes	metam-sodium
	Leaf miners	deltamethrin (*off-label*)
	Soil pests	metam-sodium
	Thrips	deltamethrin (*off-label*)
Weeds	Weeds, miscellaneous	metam-sodium

Ornamentals - Nursery stock

Diseases	Alternaria	iprodione (*seed treatment*)
	Black root rot	carbendazim (*off-label*)
	Black spot	myclobutanil
	Botrytis	chlorothalonil, thiram (*except Hydrangea*)
	Damping off	copper ammonium carbonate, tolclofos-methyl
	Damping off and foot rot	etridiazole (*seeds and seedlings*)
	Foot rot	tolclofos-methyl
	Fungus diseases	dichlorophen, prochloraz
	Phytophthora	etridiazole, etridiazole (*rooted cuttings and transplants*), fosetyl-aluminium, metalaxyl-M, propamocarb hydrochloride
	Powdery mildew	myclobutanil
	Pythium	metalaxyl-M, propamocarb hydrochloride
	Root rot	tolclofos-methyl
	Rust	myclobutanil, oxycarboxin
Pests	Ants	pirimiphos-methyl
	Aphids	cypermethrin, deltamethrin, dimethoate, imidacloprid, natural plant extracts, nicotine, pirimicarb, pirimiphos-methyl, pymetrozine, rotenone, Verticillium lecanii
	Birds/mammals	ziram
	Capsid bugs	cypermethrin, deltamethrin, pirimiphos-methyl
	Caterpillars	Bacillus thuringiensis, cypermethrin, deltamethrin, diflubenzuron, nicotine, teflubenzuron
	Cutworms	cypermethrin
	Earwigs	pirimiphos-methyl
	Flies	imidacloprid

	Leaf miners	abamectin, deltamethrin (*off-label*), dimethoate, nicotine, oxamyl (*off-label*), pirimiphos-methyl
	Leafhoppers	natural plant extracts, nicotine
	Mealybugs	deltamethrin, petroleum oil
	Mites	natural plant extracts
	Pests, miscellaneous	aldicarb (*off-label*)
	Sawflies	pirimiphos-methyl
	Scale insects	deltamethrin, petroleum oil
	Slugs/snails	ferric phosphate, metaldehyde
	Spider mites	abamectin, bifenthrin, dimethoate, fenbutatin oxide, petroleum oil, pirimiphos-methyl, spiromesifen (*off-label - container grown*)
	Stem nematodes	aldicarb (*off-label*)
	Thrips	abamectin, cypermethrin, deltamethrin, deltamethrin (*off-label*), natural plant extracts, nicotine, pirimiphos-methyl, spinosad
	Weevils	chlorpyrifos, chlorpyrifos (*conifers*), fipronil, imidacloprid
	Whiteflies	buprofezin, cypermethrin, deltamethrin, imidacloprid, natural plant extracts, nicotine, pirimiphos-methyl, spiromesifen (*off-label - container grown*), teflubenzuron, Verticillium lecanii
Plant growth regulation	Growth control	2-(1-naphthyl)acetic acid, 4-indol-3-yl-butyric acid, 4-indol-3-yl-butyric acid + 2-(1-naphthyl)acetic acid with dichlorophen, chlormequat, daminozide, indol-3-ylacetic acid
Weeds	Broad-leaved weeds	clopyralid, diquat, diquat + paraquat, diuron (*not container grown*), diuron + paraquat, glufosinate-ammonium, isoxaben, isoxaben (*off-label*), isoxaben + trifluralin (*outdoor stock only*), napropamide, oxadiazon, paraquat (*off-label - inter-row directed treatment*), pentanochlor, propachlor, propyzamide, simazine, simazine (*for resale only*), trifluralin (*off-label*)
	Crops as weeds	cycloxydim
	Grass weeds	cycloxydim, diquat + paraquat, diuron, diuron + paraquat, fluazifop-P-butyl (*off-label*), glufosinate-ammonium, isoxaben + trifluralin (*outdoor stock only*), napropamide, oxadiazon, paraquat (*off-label - inter-row directed treatment*), pentanochlor, propachlor, propyzamide, simazine, simazine (*for resale only*), trifluralin (*off-label*)
	Mayweeds	clopyralid
	Mosses	dichlorophen
	Weeds, miscellaneous	ammonium sulphamate, ammonium sulphamate (*pre-planting*), glyphosate

Ornamentals - Trees and shrubs

Diseases	Fungus diseases	prochloraz
	Powdery mildew	penconazole
	Scab	penconazole
	Verticillium wilt	chloropicrin (*soil fumigation*)
Pests	Aphids	deltamethrin
	Capsid bugs	deltamethrin
	Caterpillars	Bacillus thuringiensis, deltamethrin, diflubenzuron

	Mealybugs	deltamethrin
	Pests, miscellaneous	aldicarb (*off-label*)
	Scale insects	deltamethrin
	Stem nematodes	aldicarb (*off-label*)
	Thrips	deltamethrin
	Whiteflies	deltamethrin
Plant growth regulation	Growth control	maleic hydrazide
Weeds	Broad-leaved weeds	ammonium sulphamate, asulam, chlorthal-dimethyl, dichlobenil, diuron, diuron + glyphosate, glufosinate-ammonium, isoxaben, isoxaben + trifluralin, metazachlor, paraquat, propyzamide, simazine (*field grown nursery stock*)
	Grass weeds	cycloxydim (*off-label*), dichlobenil, diuron, diuron + glyphosate, glufosinate-ammonium, isoxaben + trifluralin, metazachlor, paraquat, propyzamide, simazine (*field grown nursery stock*)
	Weeds, miscellaneous	ammonium sulphamate, dichlobenil, diquat + paraquat, diuron, diuron + glyphosate, glyphosate, glyphosate (*wiper application*)
	Woody weeds/scrub	asulam, glyphosate

Ornamentals - Woody ornamentals

Diseases	Fungus diseases	prochloraz
Pests	Aphids	fatty acids
	Mealybugs	fatty acids
	Scale insects	fatty acids
	Spider mites	fatty acids
	Whiteflies	fatty acids
Weeds	Aquatic weeds	propyzamide
	Bindweeds	oxadiazon
	Broad-leaved weeds	dichlobenil, diuron, diuron + paraquat, glufosinate-ammonium, napropamide, oxadiazon, paraquat, propyzamide
	Grass weeds	dichlobenil, diuron, diuron + paraquat, glufosinate-ammonium, napropamide, oxadiazon, paraquat, propyzamide
	Weeds, miscellaneous	dichlobenil, glyphosate, propyzamide

Forestry

Forest nurseries, general - Forest nurseries

Pests	Aphids	pirimicarb
	Birds/mammals	aluminium ammonium sulphate
	Weevils	carbosulfan (*off-label*)
Weeds	Broad-leaved weeds	cyanazine (*off-label*), diquat + paraquat, napropamide, paraquat, paraquat (*stale seedbed*)

Grass weeds	cyanazine (*off-label*), cycloxydim, diquat + paraquat, napropamide, paraquat (*stale seedbed*), propaquizafop

Forestry plantations, general - Forestry plantations

Crop control	Chemical stripping/ thinning	glyphosate
Pests	Birds/mammals	aluminium ammonium sulphate, warfarin, ziram
	Caterpillars	diflubenzuron
	Weevils	carbosulfan
Weeds	Aquatic weeds	glyphosate, propyzamide
	Broad-leaved weeds	2,4-D, 2,4-D + dicamba + triclopyr, ammonium sulphamate, atrazine, clopyralid (*off-label*), diquat + paraquat, glufosinate-ammonium, glyphosate, glyphosate (*wiper application*), isoxaben, metamitron (*off-label*), metazachlor, paraquat, propyzamide, triclopyr
	Crops as weeds	diquat + paraquat
	Grass weeds	ammonium sulphamate, atrazine, cycloxydim, diquat + paraquat, glufosinate-ammonium, glyphosate, metamitron (*off-label*), metazachlor, paraquat, propaquizafop, propyzamide
	Weeds, miscellaneous	ammonium sulphamate, glufosinate-ammonium, glyphosate, metamitron (*off-label*), paraquat, propyzamide
	Woody weeds/scrub	2,4-D, 2,4-D + dicamba + triclopyr, ammonium sulphamate, asulam, asulam (*off-label*), glyphosate, triclopyr

Miscellaneous forestry situations - Cut logs/timber

Diseases	Fungus diseases	urea (commodity substance)
Pests	Beetles	chlorpyrifos
Weeds	Grass weeds	propaquizafop

Woodland on farms - Woodland

Diseases	Dutch elm disease	thiabendazole
Pests	Aphids	lambda-cyhalothrin (*off-label*)
	Beetles	lambda-cyhalothrin (*off-label*)
	Birds/mammals	aluminium phosphide
	Sawflies	lambda-cyhalothrin (*off-label*)
Weeds	Broad-leaved weeds	2,4-D + dicamba + triclopyr, atrazine, cyanazine (*off-label*), lenacil (*off-label*), metazachlor, pendimethalin (*off-label*)
	Crops as weeds	fluazifop-P-butyl
	Grass weeds	atrazine, cyanazine (*off-label*), cycloxydim, fluazifop-P-butyl, glyphosate, lenacil (*off-label*), metazachlor, pendimethalin (*off-label*), propaquizafop
	Weeds, miscellaneous	amitrole (*off-label*), glyphosate
	Woody weeds/scrub	2,4-D + dicamba + triclopyr, glyphosate

Fruit and hops

Bush fruit - Bilberries/blueberries/cranberries

Diseases	Botrytis	pyrimethanil (*off-label*)
	Powdery mildew	fenpropimorph (*off-label*)
Pests	Caterpillars	Bacillus thuringiensis (*off-label*)
	Gall, rust and leaf & bud mites	tebufenpyrad (*off-label*)
Weeds	Broad-leaved weeds	asulam (*off-label*), propachlor (*off-label*)
	Grass weeds	propachlor (*off-label*)
	Weeds, miscellaneous	glufosinate-ammonium, glyphosate (*off-label*)

Bush fruit - Currants

Diseases	Botrytis	chlorothalonil, chlorothalonil (*qualified minor use*), fenhexamid, fenhexamid + tolylfluanid, pyrimethanil (*off-label*), tolylfluanid
	Leaf spots	Bordeaux mixture, chlorothalonil, chlorothalonil (*qualified minor use*), copper ammonium carbonate, dodine, mancozeb
	Powdery mildew	bupirimate, chlorothalonil, chlorothalonil (*qualified minor use*), fenarimol, fenpropimorph (*off-label*), kresoxim-methyl, myclobutanil, penconazole
	Rust	copper oxychloride, thiram
Pests	Aphids	chlorpyrifos, pirimicarb, pymetrozine (*off-label*)
	Capsid bugs	chlorpyrifos, fenpropathrin
	Caterpillars	Bacillus thuringiensis (*off-label*), chlorpyrifos, diflubenzuron
	Gall, rust and leaf & bud mites	fenpropathrin, sulphur, tebufenpyrad (*off-label*)
	Midges	fenpropathrin
	Sawflies	fenpropathrin
	Slugs/snails	ferric phosphate
	Spider mites	chlorpyrifos, clofentezine (*off-label*), fenpropathrin
Weeds	Bindweeds	oxadiazon
	Broad-leaved weeds	asulam, asulam (*off-label*), chlorpropham, chlorthal-dimethyl, dichlobenil, diuron (*off-label*), glufosinate-ammonium, isoxaben, MCPB, napropamide, oxadiazon, paraquat, pendimethalin, propachlor (*off-label*), propyzamide
	Crops as weeds	fluazifop-P-butyl, pendimethalin
	Grass weeds	chlorpropham, dichlobenil, fluazifop-P-butyl, glufosinate-ammonium, napropamide, oxadiazon, paraquat, pendimethalin, propachlor (*off-label*), propyzamide
	Polygonums	chlorpropham
	Weeds, miscellaneous	dichlobenil, diuron (*off-label*), glufosinate-ammonium, glyphosate (*off-label*)

Bush fruit - Gooseberries

Diseases	Botrytis	chlorothalonil (*qualified minor use*), fenhexamid, fenhexamid + tolylfluanid, pyrimethanil (*off-label*), tolylfluanid
	Leaf spots	chlorothalonil (*qualified minor use*), mancozeb
	Powdery mildew	bupirimate, chlorothalonil (*qualified minor use*), fenarimol, fenpropimorph (*off-label*), myclobutanil, sulphur
Pests	Aphids	chlorpyrifos, pirimicarb, pymetrozine (*off-label*)
	Capsid bugs	chlorpyrifos
	Caterpillars	Bacillus thuringiensis (*off-label*), chlorpyrifos
	Gall, rust and leaf & bud mites	tebufenpyrad (*off-label*)
	Sawflies	nicotine, rotenone
	Spider mites	chlorpyrifos
Weeds	Bindweeds	oxadiazon
	Broad-leaved weeds	asulam (*off-label*), chlorpropham, chlorthal-dimethyl, dichlobenil, isoxaben, MCPB, napropamide, oxadiazon, paraquat, pendimethalin, propachlor (*off-label*), propyzamide
	Crops as weeds	fluazifop-P-butyl, pendimethalin
	Grass weeds	chlorpropham, dichlobenil, fluazifop-P-butyl, napropamide, oxadiazon, paraquat, pendimethalin, propachlor (*off-label*), propyzamide
	Polygonums	chlorpropham
	Weeds, miscellaneous	dichlobenil, glyphosate (*off-label*)

Bush fruit - Miscellaneous bush fruit

Pests	Aphids	nicotine
	Birds/mammals	aluminium ammonium sulphate
	Capsid bugs	nicotine
	Leafhoppers	nicotine
	Mealybugs	petroleum oil
	Sawflies	nicotine
	Scale insects	petroleum oil
	Spider mites	petroleum oil
Weeds	Broad-leaved weeds	diquat + paraquat, glufosinate-ammonium
	Crops as weeds	diquat + paraquat
	Grass weeds	diquat + paraquat, glufosinate-ammonium

Bush fruit - Protected bush fruit

Pests	Aphids	pymetrozine (*off-label*)

Cane fruit - Outdoor cane fruit

Crop control	Sucker/shoot control	sodium monochloroacetate (*off-label*)
Diseases	Botrytis	chlorothalonil, fenhexamid, fenhexamid + tolylfluanid, iprodione, pyrimethanil (*off-label*), thiram, tolylfluanid

	Cane blight	carbendazim (*off-label*), tebuconazole (*off-label*)
	Cane spot	Bordeaux mixture, chlorothalonil, copper ammonium carbonate, copper oxychloride, thiram
	Phytophthora	metalaxyl-M (*off-label*)
	Powdery mildew	azoxystrobin (*off-label*), bupirimate (*outdoor only*), chlorothalonil, fenarimol, fenpropimorph (*off-label*)
	Purple blotch	copper oxychloride
	Spur blight	Bordeaux mixture, thiram
	Verticillium wilt	chloropicrin (*off-label - soil fumigation*)
Pests	Aphids	1,3-dichloropropene, chlorpyrifos, pirimicarb, pymetrozine (*off-label*)
	Beetles	chlorpyrifos, deltamethrin, rotenone
	Birds/mammals	aluminium ammonium sulphate
	Capsid bugs	thiacloprid (*off-label*)
	Caterpillars	Bacillus thuringiensis
	Free-living nematodes	1,3-dichloropropene
	Mealybugs	petroleum oil
	Midges	chlorpyrifos
	Scale insects	petroleum oil
	Spider mites	bifenthrin (*off-label*), chlorpyrifos, clofentezine (*off-label*), petroleum oil, tebufenpyrad (*off-label*)
	Weevils	bifenthrin (*off-label*)
Plant growth regulation	Growth control	sodium monochloroacetate (*off-label*)
Weeds	Bindweeds	oxadiazon
	Broad-leaved weeds	asulam (*off-label*), chlorthal-dimethyl, dichlobenil, diquat + paraquat, glufosinate-ammonium, isoxaben, MCPB, napropamide, oxadiazon, paraquat, pendimethalin, propachlor (*off-label*), propyzamide, propyzamide (*England only*), trifluralin
	Crops as weeds	diquat + paraquat, fluazifop-P-butyl, pendimethalin
	Grass weeds	dichlobenil, diquat + paraquat, fluazifop-P-butyl, glufosinate-ammonium, napropamide, oxadiazon, paraquat, pendimethalin, propachlor (*off-label*), propyzamide, propyzamide (*England only*), trifluralin
	Weeds, miscellaneous	dichlobenil, glufosinate-ammonium

Cane fruit - Protected cane fruit

Diseases	Botrytis	pyrimethanil (*off-label*)
	Cane blight	carbendazim (*off-label*)
	Powdery mildew	azoxystrobin (*off-label*), myclobutanil (*off-label*)
	Root rot	fluazinam (*off-label*)
	Verticillium wilt	chloropicrin (*off-label - soil fumigation*)
Pests	Aphids	pymetrozine (*off-label*)
	Spider mites	bifenthrin (*off-label*), clofentezine (*off-label*)
	Weevils	bifenthrin (*off-label*)

Hops, general - Hops

Crop control	Chemical stripping/ thinning	diquat, diquat + paraquat, paraquat, sodium monochloroacetate (*off-label*)
Diseases	Downy mildew	Bordeaux mixture, chlorothalonil, copper oxychloride, fosetyl-aluminium, metalaxyl-M (*off-label*)
	Powdery mildew	bupirimate, fenpropimorph (*off-label*), myclobutanil (*off-label*), penconazole, sulphur
Pests	Aphids	1,3-dichloropropene, bifenthrin, cyfluthrin (*off-label*), cypermethrin, deltamethrin, imidacloprid, pymetrozine (*off-label*), tebufenpyrad
	Free-living nematodes	1,3-dichloropropene
	Mealybugs	petroleum oil
	Scale insects	petroleum oil
	Spider mites	bifenthrin, petroleum oil, tebufenpyrad
Weeds	Bindweeds	oxadiazon
	Broad-leaved weeds	asulam, diquat + paraquat, isoxaben, oxadiazon, paraquat, pendimethalin, simazine
	Crops as weeds	diquat + paraquat, fluazifop-P-butyl, pendimethalin
	Grass weeds	diquat + paraquat, fluazifop-P-butyl, oxadiazon, paraquat, pendimethalin, propyzamide (*off-label*), simazine

Miscellaneous fruit situations - Fruit crops, general

Pests	Aphids	fatty acids, natural plant extracts
	Leafhoppers	natural plant extracts
	Mealybugs	fatty acids, petroleum oil
	Mites	natural plant extracts
	Scale insects	fatty acids, petroleum oil
	Spider mites	fatty acids, petroleum oil
	Thrips	natural plant extracts
	Whiteflies	fatty acids, natural plant extracts
Plant growth regulation	Fruiting control	ethylene (commodity substance) (*in store*)
Weeds	Broad-leaved weeds	glufosinate-ammonium
	Grass weeds	glufosinate-ammonium

Miscellaneous fruit situations - Fruit nursery stock

Weeds	Broad-leaved weeds	metazachlor, trifluralin (*off-label*)
	Grass weeds	metazachlor, trifluralin (*off-label*)

Miscellaneous fruit situations - Orchards

Weeds	Broad-leaved weeds	paraquat
	Grass weeds	paraquat

Miscellaneous fruit situations - Protected miscellaneous fruit

Diseases	Botrytis	iprodione (*off-label*)
Pests	Aphids	natural plant extracts, nicotine
	Leafhoppers	natural plant extracts
	Mites	natural plant extracts
	Thrips	natural plant extracts
	Whiteflies	natural plant extracts

Miscellaneous nuts - Nuts

Diseases	Bacterial canker	copper oxychloride (*off-label*)
	Blight	copper oxychloride (*off-label*)
	Canker	tebuconazole (*off-label*)
Pests	Aphids	lambda-cyhalothrin (*off-label*)
	Caterpillars	lambda-cyhalothrin (*off-label*)
	Gall, rust and leaf & bud mites	tebufenpyrad (*off-label*), thiacloprid (*off-label*)
Weeds	Broad-leaved weeds	glufosinate-ammonium
	Grass weeds	glufosinate-ammonium
	Weeds, miscellaneous	glufosinate-ammonium, glyphosate (*off-label*)

Other fruit - Rhubarb

Diseases	Downy mildew	mancozeb + metalaxyl-M (*off-label*)
Pests	Aphids	pirimicarb (*off-label*)
Weeds	Broad-leaved weeds	dichlobenil (*off-label - established*), paraquat (*off-label*), propachlor (*off-label*), propyzamide, propyzamide (*outdoor*), simazine, simazine (*off-label*)
	Grass weeds	dichlobenil (*off-label - established*), paraquat (*off-label*), propachlor (*off-label*), propyzamide, propyzamide (*outdoor*), simazine, simazine (*off-label*)
	Weeds, miscellaneous	glyphosate (*off-label*)

Other fruit - Vines

Diseases	Botrytis	fenhexamid, iprodione (*off-label*), pyrimethanil (*off-label*)
	Bunch rots	iprodione (*off-label*)
	Downy mildew	copper oxychloride, fosetyl-aluminium (*off-label*), mancozeb + zoxamide, metalaxyl-M (*off-label*)
	Fungus diseases	mancozeb (*off-label*)
	Powdery mildew	dinocap (*off-label*), fenbuconazole (*off-label*), myclobutanil (*off-label*), sulphur
Pests	Aphids	nicotine
	Mealybugs	petroleum oil
	Scale insects	petroleum oil
	Spider mites	petroleum oil
Weeds	Bindweeds	oxadiazon

	Broad-leaved weeds	dichlobenil (*off-label*), glufosinate-ammonium, isoxaben, oxadiazon, paraquat
	Grass weeds	glufosinate-ammonium, oxadiazon, paraquat, propyzamide (*off-label*)
	Weeds, miscellaneous	glufosinate-ammonium, glyphosate (*off-label*)

Soft fruit - Miscellaneous soft fruit

Pests	Aphids	nicotine, rotenone
	Capsid bugs	nicotine
	Leafhoppers	nicotine
	Sawflies	nicotine

Soft fruit - Strawberries

Crop control	Sucker/shoot control	paraquat
Diseases	Black spot	azoxystrobin (*off-label*)
	Botrytis	boscalid + pyraclostrobin, captan, chlorothalonil, chlorothalonil (*off-label*), chlorothalonil (*outdoor crops only*), fenhexamid, fenhexamid + tolylfluanid, iprodione, mepanipyrim, pyrimethanil, thiram, tolylfluanid
	Crown rot	chloropicrin (*soil fumigation*), fosetyl-aluminium (*off-label*)
	Powdery mildew	bupirimate, dinocap, fenarimol, fenpropimorph (*off-label*), kresoxim-methyl, myclobutanil, quinoxyfen (*off-label*), sulphur
	Red core	chloropicrin (*soil fumigation*), fosetyl-aluminium, fosetyl-aluminium (*off-label*)
	Verticillium wilt	chloropicrin (*soil fumigation*)
Pests	Aphids	1,3-dichloropropene, chlorpyrifos, nicotine, pirimicarb, pymetrozine (*off-label*)
	Beetles	methiocarb
	Birds/mammals	aluminium ammonium sulphate
	Caterpillars	Bacillus thuringiensis, chlorpyrifos
	Free-living nematodes	1,3-dichloropropene, chloropicrin (*soil fumigation*)
	Slugs/snails	ferric phosphate, methiocarb
	Spider mites	abamectin (*off-label*), bifenthrin, chlorpyrifos, clofentezine (*off-label*), fenbutatin oxide, spiromesifen (*off-label - on inert media or by NFT*), tebufenpyrad
	Stem nematodes	1,3-dichloropropene
	Tarsonemid mites	abamectin (*off-label - in propagation*)
	Thrips	spinosad (*off-label*)
	Weevils	chlorpyrifos
	Whiteflies	spiromesifen (*off-label - on inert media or by NFT*)
Weeds	Broad-leaved weeds	asulam (*off-label*), chlorpropham, chlorthal-dimethyl, clopyralid, diquat + paraquat, ethofumesate (*off-label*), glufosinate-ammonium, isoxaben, napropamide, pendimethalin, phenmedipham, propachlor, propachlor (*off-label*), propyzamide, simazine, trifluralin
	Crops as weeds	cycloxydim, diquat + paraquat, ethofumesate (*off-label*), fluazifop-P-butyl, pendimethalin

	Grass weeds	chlorpropham, cycloxydim, diquat + paraquat, ethofumesate (*off-label*), fluazifop-P-butyl, glufosinate-ammonium, napropamide, pendimethalin, propachlor, propachlor (*off-label*), propyzamide, simazine, trifluralin
	Mayweeds	clopyralid
	Polygonums	chlorpropham
	Weeds, miscellaneous	glufosinate-ammonium

Tree fruit - Miscellaneous tree fruit

Diseases	Verticillium wilt	chloropicrin (*soil fumigation*)
Pests	Aphids	nicotine, rotenone
	Birds/mammals	aluminium ammonium sulphate, ziram
	Capsid bugs	nicotine
	Sawflies	nicotine
Weeds	Broad-leaved weeds	paraquat
	Grass weeds	paraquat
	Weeds, miscellaneous	glyphosate

Tree fruit - Pome fruit

Crop control	Sucker/shoot control	glyphosate
Diseases	Blossom wilt	vinclozolin
	Botrytis	iprodione (*off-label*)
	Botrytis fruit rot	thiram
	Canker	Bordeaux mixture, copper oxychloride, tebuconazole (*off-label*)
	Collar rot	copper oxychloride (*off-label*), fosetyl-aluminium
	Crown rot	fosetyl-aluminium
	Phytophthora	mancozeb + metalaxyl-M (*off-label - applied to orchard floor*)
	Powdery mildew	boscalid + pyraclostrobin, bupirimate, dinocap, fenarimol, fenbuconazole (*reduction*), kresoxim-methyl (*reduction*), myclobutanil, penconazole, sulphur
	Scab	Bordeaux mixture, boscalid + pyraclostrobin, captan, carbendazim (*off-label*), dithianon, dodine, fenarimol, fenbuconazole, kresoxim-methyl, mancozeb, myclobutanil, pyrimethanil, sulphur, thiram, tolylfluanid
	Storage rots	captan, carbendazim (*off-label*), metalaxyl-M (*off-label*), thiram
Pests	Aphids	chlorpyrifos, cyfluthrin, cypermethrin, deltamethrin, flonicamid, nicotine, pirimicarb, thiacloprid, triazamate
	Capsid bugs	chlorpyrifos, cypermethrin, deltamethrin, nicotine
	Caterpillars	Bacillus thuringiensis (*off-label*), chlorpyrifos, cyfluthrin, cypermethrin, deltamethrin, diflubenzuron, fenoxycarb, methoxyfenozide, spinosad
	Gall, rust and leaf & bud mites	diflubenzuron, tolylfluanid (*reduction*)
	Sawflies	chlorpyrifos, cypermethrin, deltamethrin, rotenone
	Spider mites	bifenthrin, chlorpyrifos, clofentezine, fenpyroximate, tebufenpyrad, tolylfluanid (*reduction*)

	Suckers	chlorpyrifos, cyfluthrin, cypermethrin, deltamethrin, diflubenzuron, lambda-cyhalothrin
	Weevils	chlorpyrifos
Plant growth regulation	Fruiting control	1-methylcyclopropene (*post-harvest use*), 2-chloroethylphosphonic acid (*off-label - for cider making*), paclobutrazol
	Growth control	gibberellins (*off-label*), paclobutrazol, prohexadione-calcium
	Quality/yield control	gibberellins
Weeds	Bindweeds	oxadiazon
	Broad-leaved weeds	2,4-D, 2,4-D + dichlorprop-P + MCPA + mecoprop-P, asulam, asulam (*off-label*), clopyralid (*off-label*), dicamba + MCPA + mecoprop-P, dichlobenil, diquat + paraquat, diuron, fluroxypyr (*off-label*), glufosinate-ammonium, isoxaben, oxadiazon, pendimethalin, propyzamide
	Crops as weeds	diquat + paraquat, pendimethalin
	Grass weeds	amitrole, dichlobenil, diquat + paraquat, diuron, glufosinate-ammonium, oxadiazon, pendimethalin, propyzamide
	Weeds, miscellaneous	amitrole, amitrole (*off-label*), dichlobenil, glufosinate-ammonium, glyphosate

Tree fruit - Stone fruit

Crop control	Sucker/shoot control	glyphosate
Diseases	Bacterial canker	Bordeaux mixture, copper oxychloride
	Blossom wilt	fenbuconazole (*off-label*), myclobutanil (*off-label*)
	Botrytis	fenhexamid (*off-label*)
	Leaf curl	Bordeaux mixture, copper ammonium carbonate, copper oxychloride
	Rust	myclobutanil (*off-label*)
	Sclerotinia	fenbuconazole (*off-label*)
Pests	Aphids	chlorpyrifos, cypermethrin, deltamethrin, nicotine, pirimicarb, pirimicarb (*off-label*), tebufenpyrad (*off-label*)
	Caterpillars	Bacillus thuringiensis (*off-label*), chlorpyrifos, cypermethrin, deltamethrin, diflubenzuron
	Gall, rust and leaf & bud mites	diflubenzuron
	Sawflies	deltamethrin
	Spider mites	chlorpyrifos, clofentezine, fenpyroximate (*off-label*)
Plant growth regulation	Growth control	paclobutrazol (*off-label*)
Weeds	Broad-leaved weeds	asulam, asulam (*off-label*), diquat + paraquat, glufosinate-ammonium, isoxaben, pendimethalin, propyzamide
	Crops as weeds	diquat + paraquat, pendimethalin
	Grass weeds	diquat + paraquat, glufosinate-ammonium, pendimethalin, propyzamide, propyzamide (*off-label*)
	Weeds, miscellaneous	amitrole (*off-label*), glufosinate-ammonium, glyphosate

Grain/crop store uses

Stored produce - Food/produce storage

Diseases	Alternaria	iprodione
	Botrytis	iprodione

Stored seed - Stored grain/rape/linseed

Pests	Birds/mammals	aluminium ammonium sulphate
	Food/grain storage pests	bifenthrin + malathion, chlorpyrifos-methyl, d-phenothrin + tetramethrin, permethrin, pirimiphos-methyl
	Pests, miscellaneous	aluminium phosphide, magnesium phosphide

Grass

Grass seed - Grass seed crops

Diseases	Crown rust	propiconazole
	Damping off	thiram (*seed treatment*)
	Drechslera leaf spot	propiconazole
	Powdery mildew	propiconazole
	Rhynchosporium	propiconazole
Pests	Aphids	deltamethrin (*off-label*), dimethoate
Plant growth regulation	Growth control	trinexapac-ethyl
Weeds	Broad-leaved weeds	2,4-D, 2,4-D + mecoprop-P, bromoxynil + ethofumesate + ioxynil, clopyralid (*off-label*), dicamba + MCPA + mecoprop-P, ethofumesate, MCPA, mecoprop-P
	Grass weeds	bromoxynil + ethofumesate + ioxynil, ethofumesate

Grassland - Leys

Diseases	Crown rust	propiconazole
	Drechslera leaf spot	propiconazole
	Powdery mildew	propiconazole
	Rhynchosporium	propiconazole
Pests	Flies	chlorpyrifos, cypermethrin
	Leatherjackets	chlorpyrifos, methiocarb (*seed admixture*)
	Slugs/snails	methiocarb (*seed admixture*)
Weeds	Broad-leaved weeds	2,4-D, 2,4-D + dicamba + triclopyr, 2,4-D + MCPA, 2,4-DB + linuron + MCPA, 2,4-DB + MCPA, amidosulfuron, bentazone + MCPA + MCPB, bromoxynil + ethofumesate + ioxynil, bromoxynil + ioxynil + mecoprop-P, clopyralid, clopyralid + fluroxypyr + triclopyr, clopyralid + triclopyr, dicamba + MCPA + mecoprop-P, dicamba + mecoprop-P, ethofumesate, fluroxypyr, fluroxypyr + triclopyr, MCPA, MCPA + MCPB, MCPB, mecoprop-P, thifensulfuron-methyl

	Grass weeds	bromoxynil + ethofumesate + ioxynil, ethofumesate
	Mayweeds	clopyralid, dicamba + mecoprop-P
	Polygonums	2,4-DB + MCPA, dicamba + mecoprop-P
	Weeds, miscellaneous	diquat + paraquat, glyphosate

Grassland - Permanent pasture

Pests	Aphids	pirimicarb
	Birds/mammals	aluminium ammonium sulphate, aluminium phosphide, strychnine hydrochloride (commodity substance) (*areas of restricted public access*)
	Flies	chlorpyrifos
	Leatherjackets	chlorpyrifos
Plant growth regulation	Quality/yield control	sulphur
Weeds	Aquatic weeds	glyphosate, MCPA, triclopyr
	Broad-leaved weeds	2,4-D, 2,4-D + dicamba, 2,4-D + dicamba + triclopyr, 2,4-D + MCPA, 2,4-D + mecoprop-P, amidosulfuron, asulam, citronella oil, clopyralid, clopyralid (*off-label - spot treatment*), clopyralid + 2,4-D + MCPA, clopyralid + fluroxypyr + triclopyr, clopyralid + triclopyr, dicamba + MCPA + mecoprop-P, dicamba + mecoprop-P, ethofumesate, fluroxypyr, fluroxypyr + triclopyr, glyphosate (*wiper application*), MCPA, MCPA + MCPB, MCPB, mecoprop-P, paraquat (*sward destruction/direct drilling*), thifensulfuron-methyl, triclopyr
	Grass weeds	ethofumesate, glyphosate, paraquat (*sward destruction/ direct drilling*)
	Mayweeds	clopyralid, dicamba + mecoprop-P
	Polygonums	dicamba + mecoprop-P
	Weeds, miscellaneous	diquat + paraquat, glufosinate-ammonium, glyphosate
	Woody weeds/scrub	asulam, clopyralid + triclopyr, glyphosate, triclopyr

Turf/amenity grass - Amenity grassland

Diseases	Brown patch	iprodione
	Dollar spot	iprodione
	Fusarium diseases	azoxystrobin, iprodione, trifloxystrobin
	Melting out	iprodione
	Red thread	iprodione, trifloxystrobin
	Snow mould	iprodione
Pests	Flies	chlorpyrifos
	Leatherjackets	chlorpyrifos
	Slugs/snails	metaldehyde
Plant growth regulation	Growth control	maleic hydrazide, trinexapac-ethyl
Weeds	Broad-leaved weeds	2,4-D, 2,4-D + dicamba, 2,4-D + dicamba + triclopyr, asulam, asulam (*not fine turf*), bromoxynil + ioxynil + mecoprop-P, clopyralid + triclopyr, dicamba + MCPA + mecoprop-P, MCPA
	Mosses	ferrous sulphate

Weeds, miscellaneous		glyphosate, glyphosate (*wiper application*)
Woody weeds/scrub		clopyralid + triclopyr

Turf/amenity grass - Managed amenity turf

Crop control	Miscellaneous non-selective situations	glufosinate-ammonium
Diseases	Anthracnose	carbendazim + iprodione, chlorothalonil
	Brown patch	iprodione
	Cladosporium diseases	carbendazim + iprodione
	Dollar spot	carbendazim, chlorothalonil, fenarimol, iprodione, pyraclostrobin (*reduction*), thiophanate-methyl
	Fusarium diseases	azoxystrobin, carbendazim, carbendazim + epoxiconazole, carbendazim + iprodione, chlorothalonil, fenarimol, iprodione, myclobutanil, prochloraz + tebuconazole, pyraclostrobin, thiophanate-methyl, trifloxystrobin
	Melting out	iprodione
	Red thread	carbendazim + iprodione, chlorothalonil, dichlorophen, fenarimol, iprodione, pyraclostrobin, thiophanate-methyl, trifloxystrobin
	Snow mould	iprodione
Pests	Birds/mammals	aluminium phosphide
	Earthworms	carbendazim, thiophanate-methyl
	Flies	chlorpyrifos
	Leatherjackets	chlorpyrifos
	Slugs/snails	metaldehyde
Plant growth regulation	Growth control	maleic hydrazide, trinexapac-ethyl
Weeds	Broad-leaved weeds	2,4-D, 2,4-D + dicamba, 2,4-D + dicamba + fluroxypyr, 2,4-D + florasulam, 2,4-D + mecoprop-P, bromoxynil + ioxynil + mecoprop-P, chlorthal-dimethyl, clopyralid + diflufenican + MCPA, clopyralid + fluroxypyr + MCPA, dicamba + dichlorprop-P + ferrous sulphate + MCPA, dicamba + dichlorprop-P + MCPA, dicamba + MCPA + mecoprop-P, dichlorprop-P + ferrous sulphate + MCPA, dichlorprop-P + MCPA, fluroxypyr + mecoprop-P, MCPA, MCPA + mecoprop-P, MCPA + mecoprop-P + ferrous sulphate, mecoprop-P
	Crops as weeds	2,4-D + dicamba + fluroxypyr, 2,4-D + florasulam, dichlorprop-P + MCPA, mecoprop-P
	Grass weeds	amitrole (*off-label - on amitrole resistant turf*), tepraloxydim (*off-label*)
	Mosses	dicamba + dichlorprop-P + ferrous sulphate + MCPA, dichlorophen, dichlorophen + ferrous sulphate, dichlorprop-P + ferrous sulphate + MCPA, ferrous sulphate, MCPA + mecoprop-P + ferrous sulphate
	Weeds, miscellaneous	diquat + paraquat, glyphosate (*pre-establishment*)

Non-crop pest control

Farm buildings/yards - Farm buildings

Diseases	Fungus diseases	formaldehyde (commodity substance)
Pests	Birds/mammals	aluminium ammonium sulphate, brodifacoum (*indoor use only*), bromadiolone, chlorophacinone, difenacoum, warfarin
	Flies	diflubenzuron, d-phenothrin + tetramethrin, permethrin, pyrethrins, tetramethrin
	Food/grain storage pests	alpha-cypermethrin
	Pests, miscellaneous	alpha-cypermethrin
	Wasps	d-phenothrin + tetramethrin
Weeds	Weeds, miscellaneous	glyphosate

Farm buildings/yards - Farmyards

Pests	Birds/mammals	bromadiolone, difenacoum

Farmland pest control - Farmland situations

Pests	Birds/mammals	aluminium phosphide
Weeds	Crops as weeds	dichlobenil (*blight prevention*)

Miscellaneous non-crop pest control - Manure/rubbish

Pests	Flies	diflubenzuron

Miscellaneous non-crop pest control - Miscellaneous pest control situations

Pests	Birds/mammals	carbon dioxide (commodity substance), paraffin oil (commodity substance) (*egg treatment*), strychnine hydrochloride (commodity substance) (*areas of restricted public access*)

Protected salad and vegetable crops

Protected brassicas - Protected brassica vegetables

Diseases	Black rot	copper oxychloride (*off-label*)
	Damping off	fosetyl-aluminium (*off-label*)
	Downy mildew	fosetyl-aluminium (*off-label*), fosetyl-aluminium (*off-label - seedlings*)
	Spear rot	copper oxychloride (*off-label*)
Pests	Caterpillars	Bacillus thuringiensis (*off-label*)
	Pests, miscellaneous	dimethoate (*off-label*)

Protected brassicas - Protected salad brassicas

Diseases	Black rot	copper oxychloride (*off-label*)
	Botrytis	boscalid + pyraclostrobin (*off-label*)
	Damping off	fosetyl-aluminium (*off-label*)
	Downy mildew	fosetyl-aluminium (*off-label*), fosetyl-aluminium (*off-label - seedlings*)
	Ring spot	boscalid + pyraclostrobin (*off-label*)
	Spear rot	copper oxychloride (*off-label*)
Pests	Aphids	cypermethrin (*off-label*), deltamethrin (*off-label*), pymetrozine (*off-label*), pymetrozine (*off-label - for baby leaf production*)
	Beetles	deltamethrin (*off-label - baby leaf production*), deltamethrin (*off-label - for baby leaf production*)
	Caterpillars	Bacillus thuringiensis (*off-label*), cypermethrin (*off-label*), deltamethrin (*off-label*), deltamethrin (*off-label - baby leaf production*), deltamethrin (*off-label - for baby leaf production*)
	Pests, miscellaneous	deltamethrin (*off-label*), dimethoate (*off-label*)

Protected crops, general - All protected crops

Diseases	Fungus diseases	potassium bicarbonate (commodity substance)
	Soil-borne diseases	dazomet (*soil fumigation*)
Pests	Aphids	nicotine, rotenone
	Capsid bugs	nicotine
	Free-living nematodes	dazomet (*soil fumigation*)
	Leaf miners	nicotine
	Leafhoppers	nicotine
	Sawflies	nicotine
	Slugs/snails	metaldehyde
	Soil pests	dazomet (*soil fumigation*)
	Thrips	nicotine
	Whiteflies	alginate/polysaccaride
Weeds	Weeds, miscellaneous	dazomet (*soil fumigation*)

Protected crops, general - Glasshouses

Diseases	Fungus diseases	dichlorophen, formaldehyde (commodity substance) (*spray, dip or fumigant*)
	Phytophthora	fosetyl-aluminium
	Root rot	fosetyl-aluminium
	Soil-borne diseases	metam-sodium
Pests	Free-living nematodes	metam-sodium
	Soil pests	metam-sodium
Weeds	Mosses	dichlorophen
	Weeds, miscellaneous	metam-sodium

Protected crops, general - Protected vegetables, general

Diseases	White blister	propamocarb hydrochloride (*off-label*)
Pests	Aphids	natural plant extracts, pirimicarb (*off-label*)
	Leafhoppers	natural plant extracts
	Mites	natural plant extracts
	Thrips	natural plant extracts
	Whiteflies	natural plant extracts

Protected crops, general - Soils and compost

Diseases	Soil-borne diseases	formaldehyde (commodity substance) (*drench*), metam-sodium
Pests	Free-living nematodes	metam-sodium
	Pests, miscellaneous	dimethoate (*off-label - for watercress propagation*)
	Soil pests	metam-sodium
Weeds	Weeds, miscellaneous	metam-sodium

Protected fruiting vegetables - Protected aubergines

Diseases	Botrytis	fenhexamid (*off-label*)
Pests	Aphids	pymetrozine (*off-label*)
	Leaf miners	deltamethrin (*off-label*)
	Spider mites	spiromesifen (*off-label - on inert media or by NFT*)
	Thrips	deltamethrin (*off-label*)
	Whiteflies	spiromesifen (*off-label - on inert media or by NFT*)

Protected fruiting vegetables - Protected cucurbits

Diseases	Botrytis	chlorothalonil, fenhexamid (*off-label*), iprodione (*off-label*)
	Damping off	propamocarb hydrochloride (*off-label*)
	Downy mildew	azoxystrobin (*off-label*), metalaxyl-M (*off-label*)
	Phytophthora	propamocarb hydrochloride
	Powdery mildew	azoxystrobin (*off-label*), bupirimate, chlorothalonil, fenarimol (*off-label*), sulphur (*off-label*)
	Pythium	propamocarb hydrochloride
	Root diseases	carbendazim (*off-label*)
	Root rot	propamocarb hydrochloride (*off-label*)
Pests	Ants	pirimiphos-methyl
	Aphids	deltamethrin, nicotine, pirimicarb (*off-label*), pirimiphos-methyl, pymetrozine, pymetrozine (*off-label*), Verticillium lecanii
	Capsid bugs	pirimiphos-methyl
	Caterpillars	Bacillus thuringiensis, deltamethrin
	Earwigs	pirimiphos-methyl
	Leaf miners	deltamethrin (*off-label*), oxamyl (*off-label*), pirimiphos-methyl
	Leafhoppers	nicotine
	Mealybugs	deltamethrin, petroleum oil

	Sawflies	pirimiphos-methyl
	Scale insects	deltamethrin, petroleum oil
	Spider mites	abamectin, abamectin (*off-label*), clofentezine (*off-label*), fenbutatin oxide, fenbutatin oxide (*off-label*), petroleum oil, pirimiphos-methyl, spiromesifen (*off-label - on inert media or by NFT*)
	Thrips	abamectin, deltamethrin (*off-label*), nicotine, pirimiphos-methyl, spinosad
	Whiteflies	buprofezin, cypermethrin, deltamethrin, nicotine, pirimiphos-methyl, spiromesifen (*off-label - on inert media or by NFT*), Verticillium lecanii
Weeds	Broad-leaved weeds	isoxaben (*off-label*), paraquat (*off-label - inter-row directed treatment*)
	Grass weeds	paraquat (*off-label - inter-row directed treatment*), propyzamide (*off-label*)

Protected fruiting vegetables - Protected tomatoes

Diseases	Blight	azoxystrobin (*off-label*), chlorothalonil, copper oxychloride, propamocarb hydrochloride
	Botrytis	azoxystrobin (*off-label*), chlorothalonil, fenhexamid (*off-label*), iprodione, pyrimethanil (*off-label*), thiram
	Damping off	copper oxychloride, propamocarb hydrochloride (*off-label*)
	Damping off and foot rot	etridiazole (*seeds and seedlings*)
	Didymella stem rot	azoxystrobin (*off-label*), carbendazim (*off-label*)
	Foot rot	copper oxychloride
	Leaf mould	chlorothalonil, copper ammonium carbonate
	Phytophthora	copper oxychloride, etridiazole (*transplants*), propamocarb hydrochloride
	Powdery mildew	azoxystrobin (*off-label*), bupirimate (*off-label*), fenarimol (*off-label*), sulphur (*off-label*)
	Pythium	propamocarb hydrochloride
	Root diseases	etridiazole (*off-label*)
	Root rot	propamocarb hydrochloride (*off-label*)
Pests	Ants	pirimiphos-methyl
	Aphids	deltamethrin, fatty acids, nicotine, pirimicarb, pirimiphos-methyl, Verticillium lecanii
	Capsid bugs	pirimiphos-methyl
	Caterpillars	Bacillus thuringiensis, deltamethrin, nicotine
	Earwigs	pirimiphos-methyl
	Leaf miners	abamectin, abamectin (*off-label*), deltamethrin (*off-label*), nicotine, oxamyl (*off-label*), pirimiphos-methyl
	Leafhoppers	nicotine
	Mealybugs	deltamethrin, fatty acids, petroleum oil
	Sawflies	pirimiphos-methyl
	Scale insects	deltamethrin, fatty acids, petroleum oil
	Spider mites	abamectin, fatty acids, fenbutatin oxide, petroleum oil, pirimiphos-methyl
	Thrips	abamectin, deltamethrin (*off-label*), nicotine, pirimiphos-methyl

	Whiteflies	buprofezin, deltamethrin, fatty acids, nicotine, pirimiphos-methyl, pymetrozine (*off-label*), spiromesifen, Verticillium lecanii
Plant growth regulation	Fruiting control	(2-naphthyloxy)acetic acid, 2-chloroethylphosphonic acid
Weeds	Broad-leaved weeds	paraquat (*off-label - inter-row directed treatment*)
	Grass weeds	paraquat (*off-label - inter-row directed treatment*)

Protected herb crops - Protected herbs

Diseases	Botrytis	propamocarb hydrochloride (*off-label*)
	Downy mildew	copper oxychloride (*off-label*), fosetyl-aluminium (*off-label*), propamocarb hydrochloride (*off-label*)
	Fungus diseases	mancozeb + metalaxyl-M (*off-label*)
	Powdery mildew	sulphur (*off-label*)
	Rhizoctonia	azoxystrobin (*off-label*)
	Ring spot	prochloraz (*off-label*)
Pests	Aphids	nicotine, pirimicarb (*off-label*), pymetrozine (*off-label*)
	Beetles	deltamethrin, deltamethrin (*off-label*)
	Caterpillars	Bacillus thuringiensis (*off-label*)
	Leafhoppers	deltamethrin, deltamethrin (*off-label*), nicotine
	Thrips	nicotine, pirimicarb (*off-label*)
	Whiteflies	nicotine
Weeds	Broad-leaved weeds	prometryn (*off-label*)
	Grass weeds	prometryn (*off-label*), propyzamide (*off-label*)

Protected leafy vegetables - Mustard and cress

Diseases	Botrytis	fenhexamid (*off-label*), iprodione (*off-label*), propamocarb hydrochloride (*off-label*)
	Damping off and foot rot	etridiazole
	Downy mildew	copper oxychloride (*off-label*), propamocarb hydrochloride (*off-label*)
	Sclerotinia	iprodione (*off-label*)
Pests	Aphids	nicotine
	Beetles	deltamethrin (*off-label*)
	Caterpillars	nicotine
	Leaf miners	nicotine
	Leafhoppers	deltamethrin (*off-label*)
Weeds	Broad-leaved weeds	pendimethalin (*off-label*), propachlor (*off-label*)
	Grass weeds	propachlor (*off-label*)

Protected leafy vegetables - Protected leafy vegetables

Diseases	Big vein	carbendazim (*off-label*)
	Botrytis	iprodione, iprodione (*off-label*), propamocarb hydrochloride, pyrimethanil (*off-label*), thiram
	Downy mildew	fosetyl-aluminium, mancozeb, propamocarb hydrochloride, thiram
	Fungus diseases	mancozeb + metalaxyl-M (*off-label*)

	Rhizoctonia	azoxystrobin (*off-label*), boscalid + pyraclostrobin, tolclofos-methyl
	Ring spot	prochloraz (*off-label*)
	Sclerotinia	boscalid + pyraclostrobin
Pests	Aphids	nicotine, pirimicarb, pirimicarb (*off-label*), pymetrozine (*off-label*), Verticillium lecanii
	Beetles	deltamethrin (*off-label*)
	Caterpillars	Bacillus thuringiensis (*off-label*)
	Leaf miners	abamectin
	Leafhoppers	deltamethrin (*off-label*), nicotine
	Thrips	abamectin, nicotine, pirimicarb (*off-label*)
	Whiteflies	cypermethrin, nicotine, Verticillium lecanii
Weeds	Broad-leaved weeds	chlorpropham with cetrimide, paraquat (*off-label - inter-row directed treatment*)
	Grass weeds	chlorpropham with cetrimide, paraquat (*off-label - inter-row directed treatment*), propyzamide (*off-label*)
	Polygonums	chlorpropham with cetrimide

Protected leafy vegetables - Protected spinach

Diseases	Downy mildew	metalaxyl-M (*off-label*)
Pests	Aphids	cypermethrin (*off-label*), pirimicarb (*off-label*)
	Caterpillars	cypermethrin (*off-label*)
	Pests, miscellaneous	dimethoate (*off-label*)

Protected legumes - Protected peas and beans

Pests	Aphids	Verticillium lecanii
	Leaf miners	oxamyl (*off-label*)
	Whiteflies	Verticillium lecanii

Protected root and tuber vegetables - Protected carrots/parsnips/celeriac

Pests	Pests, miscellaneous	dimethoate (*off-label*)
Weeds	Broad-leaved weeds	isoxaben (*off-label - temporary protection*)

Protected root and tuber vegetables - Protected potatoes

Diseases	Blight	benalaxyl + mancozeb
Pests	Aphids	nicotine
	Leafhoppers	nicotine
	Thrips	nicotine
	Whiteflies	nicotine

Protected root and tuber vegetables - Protected root brassicas

Diseases	Botrytis	propamocarb hydrochloride (*off-label*)
	Downy mildew	propamocarb hydrochloride (*off-label*)

	Rhizoctonia	tolclofos-methyl (*off-label*)
	White blister	mancozeb + metalaxyl-M (*off-label*)
Pests	Aphids	pirimicarb (*off-label*)

Protected stem and bulb vegetables - Protected asparagus

Pests	Aphids	nicotine

Protected stem and bulb vegetables - Protected celery/chicory

Diseases	Botrytis	azoxystrobin (*off-label*), carbendazim (*off-label*), iprodione (*off-label*)
	Phytophthora	azoxystrobin (*off-label - for forcing*)
	Rhizoctonia	azoxystrobin (*off-label*), tolclofos-methyl (*off-label*)
	Sclerotinia	azoxystrobin (*off-label*)
Pests	Aphids	deltamethrin (*off-label*), nicotine, pymetrozine (*off-label*)
	Beetles	deltamethrin (*off-label*)
	Caterpillars	Bacillus thuringiensis (*off-label*), deltamethrin (*off-label*)
	Leafhoppers	deltamethrin (*off-label*)
	Pests, miscellaneous	deltamethrin (*off-label*)
Weeds	Broad-leaved weeds	paraquat (*off-label - inter-row directed treatment*)
	Grass weeds	paraquat (*off-label - inter-row directed treatment*), propyzamide (*off-label*)

Protected stem and bulb vegetables - Protected onions/leeks/garlic

Pests	Aphids	deltamethrin (*off-label*), nicotine
	Caterpillars	deltamethrin (*off-label*)
	Leafhoppers	nicotine
	Pests, miscellaneous	deltamethrin (*off-label*), dimethoate (*off-label*), dimethoate (*off-label - seedlings*)
	Stem nematodes	oxamyl (*off-label*)
	Thrips	nicotine
	Whiteflies	nicotine
Weeds	Broad-leaved weeds	paraquat (*off-label - inter-row directed treatment*)
	Grass weeds	paraquat (*off-label - inter-row directed treatment*)

Total vegetation control

Aquatic situations, general - Aquatic situations

Plant growth regulation	Growth control	maleic hydrazide
Weeds	Aquatic weeds	2,4-D, dichlobenil, glyphosate
	Broad-leaved weeds	2,4-D
	Grass weeds	glyphosate
	Weeds, miscellaneous	glyphosate, terbutryn

Non-crop areas, general - Farm buildings/yards

Plant growth regulation	Growth control	maleic hydrazide

Non-crop areas, general - Miscellaneous non-crop situations

Pests	Birds/mammals	warfarin
Plant growth regulation	Growth control	maleic hydrazide
Weeds	Broad-leaved weeds	amitrole, asulam, MCPA, triclopyr
	Grass weeds	amitrole
	Weeds, miscellaneous	amitrole, diuron, glyphosate
	Woody weeds/scrub	glyphosate, triclopyr

Non-crop areas, general - Non-crop farm areas

Weeds	Aquatic weeds	triclopyr
	Broad-leaved weeds	2,4-D + dicamba + triclopyr, amitrole, diquat + paraquat, diuron, diuron + glyphosate, glufosinate-ammonium, glyphosate, MCPA, paraquat, picloram, triclopyr
	Crops as weeds	diquat + paraquat
	Grass weeds	amitrole, diquat + paraquat, diuron, diuron + glyphosate, glufosinate-ammonium, glyphosate, paraquat
	Weeds, miscellaneous	amitrole, amitrole + 2,4-D + diuron, dichlobenil, diuron, diuron + paraquat, glufosinate-ammonium, glyphosate, sodium chlorate
	Woody weeds/scrub	asulam, glyphosate, picloram, triclopyr, triclopyr (*directed treatment*)

Non-crop areas, general - Paths/roads etc

Diseases	Fungus diseases	dichlorophen
Pests	Slugs/snails	metaldehyde
Weeds	Broad-leaved weeds	2,4-D + dicamba + triclopyr, diquat + paraquat, diuron, glyphosate, MCPA
	Grass weeds	diquat + paraquat, diuron, glyphosate
	Mosses	dichlorophen
	Weeds, miscellaneous	dichlobenil, dichlorophen, diflufenican + glyphosate, diquat + paraquat, diuron, diuron + glyphosate, diuron + paraquat, glyphosate, sodium chlorate
	Woody weeds/scrub	2,4-D + dicamba + triclopyr, glyphosate

SECTION 2
PESTICIDE PROFILES

1   abamectin

A selective acaricide and insecticide for use in ornamentals

Products

Dynamec	Syngenta Bioline	18 g/l	EC	08701

Uses
- Leaf miner in *miscellaneous flowers, ornamental specimens, protected cherry tomatoes* *(off-label)*, *protected flowers, protected lettuce, protected ornamentals, protected tomatoes*
- Red spider mites in *protected peppers* *(off-label)*, *protected strawberries* *(off-label)*
- Tarsonemid mites in *strawberries* *(off-label - in propagation)*
- Two-spotted spider mite in *miscellaneous flowers, ornamental specimens, protected cucumbers, protected flowers, protected ornamentals, protected tomatoes*
- Western flower thrips in *miscellaneous flowers, ornamental specimens, protected cucumbers, protected flowers, protected lettuce, protected ornamentals, protected tomatoes*

Specific Off-Label Approvals (SOLAs)
- *protected cherry tomatoes, protected peppers, protected strawberries* *(OLA 031128) Dec 2008* [1]
- *strawberries* *(in propagation) (OLA 040165) Dec 2008* [1]

Efficacy guidance
- Treat at first sign of infestation. Repeat sprays may be required
- For effective control total cover of all plant surfaces is essential, but avoid run-off
- Target pests quickly become immobilised but 3-5 d may be required for maximum mortality

Restrictions
- Number of treatments 6 on protected tomatoes and cucumbers (only 4 of which can be made when flowers or fruit present); 5 on protected peppers; 4 on protected lettuce; 3 on strawberries; not restricted on flowers but rotation with other products advised
- Maximum concentration must not exceed 50 ml per 100 l water
- Do not mix with wetters, stickers or other adjuvants
- Do not use on ferns (*Adiantum* spp) or Shasta daisies
- Consult manufacturer for list of plant varieties tested for safety
- There is insufficient evidence to support product compatibility with integrated and biological pest control programmes
- Unprotected persons must be kept out of treated areas until the spray has dried

Crop-specific information
- On tomato or cucumber crops that are in flower, or have started to set fruit, treat only between 1 Mar and 31 Oct. Seedling tomatoes or cucumbers that have not started to flower or set fruit may be treated at any time. Do not treat cherry tomatoes
- Apply to lettuce only between 1 Mar and 31 Oct
- Some spotting or staining may occur on carnation, kalanchoe and begonia foliage
- HI 14 d for protected lettuce; 3 d for other protected edible crops

Environmental safety
- High risk to bees. Do not apply to crops in flower or to those in which bees are actively foraging. Do not apply when flowering weeds are present
- Extremely dangerous to fish or other aquatic life. Do not contaminate surface waters or ditches with chemical or used container
- Keep in original container, tightly closed, in a safe place, under lock and key
- Where bumble bees are used in tomatoes as pollinators, keep them out for 24 h after treatment
- Permissible in organic systems

Hazard classification and safety precautions

Hazard H03, H04

Risk phrases R20, R21, R22a, R36

Operator protection A, C, D, H, K, M; U02a, U05a, U09a, U19a, U20a

Environmental protection E12a, E13a, E15a, E34

Storage and disposal D01, D02, D05, D09b, D10b

Medical advice M04

SEE SECTION 3 FOR PRODUCTS ALSO REGISTERED

SECTION 2

2   aldicarb

A soil-applied, systemic carbamate insecticide and nematicide

Products

Temik 10G	Bayer CropScience	10% w/w	GR	10021

Uses

- Aphids in *carrots, parsnips, potatoes*
- Free-living nematodes in *carrots, parsnips, potatoes*
- Insect control in *bedding plants* (off-label), *chrysanthemums* (off-label), *dahlias* (off-label), *nursery stock* (off-label), *pot plants* (off-label), *protected ornamentals* (off-label), *protected roses* (off-label), *protected stool beds* (off-label), *trees and shrubs* (off-label)
- Spraing vectors in *potatoes*
- Stem and bulb nematodes in *bulb onions*
- Stem nematodes in *bedding plants* (off-label), *bulb onions, chrysanthemums* (off-label), *dahlias* (off-label), *nursery stock* (off-label), *pot plants* (off-label), *protected ornamentals* (off-label), *protected roses* (off-label), *protected stool beds* (off-label), *trees and shrubs* (off-label)

Specific Off-Label Approvals (SOLAs)

- *bedding plants, chrysanthemums, dahlias, nursery stock, pot plants, protected ornamentals, protected roses, protected stool beds, trees and shrubs* (OLA 001932) Dec 2007 [1]

Approval information

- In March 2003 the E U Council decided not to include aldicarb in Annex I under Directive 91/414. This meant that all approved products containing aldicarb have to be withdrawn. However for certain uses it was decided that this withdrawal can take place over a longer period than is usually provided, in order to allow time for the development of alternatives.
- In the UK, for uses on sugar beet, approvals for sale and supply were revoked in September 2003. Off-label approval for use on outdoor and protected sweetcorn seedlings and outdoor leeks was revoked on 18 September 2004.
- Products containing this active ingredient have been granted derogations for specified 'Essential Uses' for use until 31 December 2007. Sale and supply must cease by 30 June 2007 but growers have no guarantee that the products will continue to be available until then.
  For more information see 'The Review Programme' under 'Pesticide Legislation' in Section 5

Efficacy guidance

- Persistence and activity may be reduced in very wet soils or where pH exceeds 8.0. Do not use within 14 d of liming
- Use in potatoes reduces incidence of spraing disease

Restrictions

- Aldicarb is subject to the Poisons Rules 1982 and the Poisons Act 1972. See notes in Section 5
- This product contains an anticholinesterase carbamate compound. Do not use if under medical advice not to work with such compounds
- Maximum number of treatments 1 per crop
- Must be incorporated into soil by physical means. See label for details of application rates, timing, suitable applicators and techniques of incorporation
- No edible crops other than those listed (see label) should be planted into treated soil for at least 8 wk after application
- Keep unprotected persons out of treated glasshouses for at least 1 d

Crop-specific information

- Latest use: at planting, sowing, drilling or transplanting
- HI potatoes 8 wk; carrots, parsnips 12 wk
- Do not harvest spring sown bulb onions until mature bulb stage

Environmental safety

- Dangerous for the environment
- Toxic to aquatic organisms
- Dangerous to fish or other aquatic life. Do not contaminate surface waters or ditches with chemical or used container

---

FOR FULL CONDITIONS OF USE ALWAYS READ THE PRODUCT LABEL

- Dangerous to livestock. Keep all livestock out of treated areas/away from treated water for at least 13 wk. Bury or remove spillages
- Dangerous to game, wild birds and animals
- Keep in original container, tightly closed, in a safe place, under lock and key

Hazard classification and safety precautions

> **Hazard** H01, H11
> **Risk phrases** R21, R28, R51, R53a
> **Operator protection** A, B, H, K, M [1]; C [1] (or D+E); U02a, U04a, U05a, U07, U09a, U12, U13, U19a, U20a
> **Environmental protection** E06b (13 wk); E10a, E13b, E34, E36, E38
> **Storage and disposal** D01, D02, D09b, D11a, D12a, D14
> **Medical advice** M02, M04

3   alginate/polysaccaride

A contact insecticide that works by physical action

Products

Agri-50E	Fargro	-	SC	-

Uses

- Whitefly in *all edible crops*, *all non-edible crops*, *protected crops*

Approval information

- Product not controlled by Control of Pesticides Regulations/Plant Protection Products Regulations because it acts by physical means only

Efficacy guidance

- Product acts by blocking insect spiracles and inhibiting respiration by physical action
- For maximum effectiveness direct spray contact with target insects is essential. Ensure thorough and uniform spray coverage using conventional high volume eqipment
- Treat at first signs of pest infestation or damage
- Best results obtained from treatments in early morning (outdoor crops only) or late afternoon when pests less active
- Apply at first sign of infestation and repeat every 7-21 d as required
- Treatment also gives suppression of leafhoppers, psylids, scale insects and spider mites
- Product most effective when used in water between pH 5.0 and 8.5. Water should be buffered if necessary, especially in areas of very hard water

Restrictions

- No restrictions on number of treatments. Minimum spray interval 7 d
- Do use any tank mixture without first consulting supplier or distributor
- Do not mix with heavy metal products, sulphur, mineral oils or strongly anionic products
- Do not mix with anti-foaming agents

Crop-specific information

- Harvest interval: 0 days for all edible crops
- Carry out plant safety check before large scale treatment of flowering ornamentals

Environmental safety

- Product may be used in conjunction with biological control agents and within integrated pest and resistance management systems
- If using bees for pollination close hives before application and reopen when crop is dry

Hazard classification and safety precautions

> **Operator protection** U05a
> **Storage and disposal** D01, D02, D05

4   alpha-cypermethrin

A contact and ingested pyrethroid insecticide for use in arable crops

Products

1	Antec Durakil 1.5 SC	Antec Biosentry	15 g/l	SC	H7560
2	Antec Durakil 6SC	Antec Biosentry	60 g/l	SC	H7559
3	Contest	BASF	15% w/w	WG	10216
4	Fernpath Dart	AgriGuard	100 g/l	EC	11970

Uses

- Brassica pod midge in *winter oilseed rape* [3, 4]
- Cabbage seed weevil in *spring oilseed rape, winter oilseed rape* [3, 4]
- Cabbage stem flea beetle in *winter oilseed rape* [3, 4]
- Cabbage white butterfly in *salad brassicas* (off-label - for baby leaf production) [3]
- Caterpillars in *broccoli, brussels sprouts, cabbages, calabrese, cauliflowers, kale* [3, 4]
- Cereal aphid in *spring barley, spring wheat, winter barley, winter wheat* [3, 4]
- Diamond-back moth in *salad brassicas* (off-label - for baby leaf production) [3]
- Flea beetle in *broccoli, brussels sprouts, cabbages, calabrese, cauliflowers, kale* [3, 4]
- Lesser mealworm in *poultry houses* [1, 2]
- Pea and bean weevil in *broad beans, combining peas, spring field beans, vining peas, winter field beans* [3, 4]
- Pea aphid in *combining peas, vining peas* [4]; *combining peas* (reduction), *vining peas* (reduction) [3]
- Pea moth in *combining peas, vining peas* [3, 4]
- Pollen beetle in *spring oilseed rape, winter oilseed rape* [3, 4]
- Poultry red mite in *poultry houses* [1, 2]
- Rape winter stem weevil in *winter oilseed rape* [3, 4]
- Small white butterfly in *salad brassicas* (off-label - for baby leaf production) [3]
- Yellow cereal fly in *winter barley, winter wheat* [3]

Specific Off-Label Approvals (SOLAs)

- *salad brassicas* (for baby leaf production) (OLA 011221) Dec 2008 [3]

Approval information

- Accepted by BBPA for use on malting barley
- Approval expiry 31 Aug 2006 [4]

Efficacy guidance

- For cabbage stem flea beetle control spray oilseed rape when adult or larval damage first seen and about 1 mth later
- For flowering pests on oilseed rape apply at any time during flowering, on pollen beetle best results achieved at green to yellow bud stage (GS 3,3-3,7), on seed weevil between 20 pods set stage and 80% petal fall (GS 4,7-5,8)
- Spray cereals in autumn for control of cereal aphids, in spring/summer for grain aphids. (See label for details)
- For flea beetle, caterpillar and cabbage aphid control on brassicas apply when the pest or damage first seen or as a preventive spray. Repeat if necessary
- For pea and bean weevil control in peas and beans apply when pest attack first seen and repeat as necessary
- For lesser mealworm control in poultry houses apply a coarse, low-pressure spray as routine treatment after clean-out and before each new crop. Spray vertical surfaces and ensure an overlap onto ceilings. It is not necessary to treat the floor [1, 2]
- Use the higher recommended concentration in poultry houses where extreme residual action is required or surfaces are dirty or highly absorbent [1, 2]

Restrictions

- Maximum number of treatments 4 per crop on edible brassicas, 3 per crop on winter oilseed rape, peas, 2 on beans, spring oilseed rape (only 1 after yellow bud stage - GS 3,7)
- Maximum number of applications in animal husbandry use 2 per crop when used in premises that are occupied by poultry [1, 2]
- Apply up to 2 sprays on cereals in autumn and spring, 1 in summer between 1 Apr and 31 Aug. See label for details of rates

FOR FULL CONDITIONS OF USE ALWAYS READ THE PRODUCT LABEL

- Only 1 aphicide treatment may be applied in cereals between 1 Apr and 31 Aug in any one year and spray volume must not be reduced in this period
- Do not apply to a cereal crop if any product containing a pyrethroid or dimethoate has been applied after the start of ear emergence (GS 51)

Crop-specific information
- Latest use: before the end of flowering for oilseed rape; before 31 Mar in year of harvest for cereals (autumn and spring application); before early dough stage (GS 83) for cereals (summer application).
- HI vining peas 1 d; brassicas 7 d; combining peas, broad beans, field beans 11 d
- For summer cereal application do not spray within 6 m from edge of crop and do not reduce volume when used after 31 Mar

Environmental safety
- Dangerous for the environment [3, 4]
- Very toxic to aquatic organisms [3, 4]
- Extremely dangerous to fish or other aquatic life. Do not contaminate surface waters or ditches with chemical or used container [1, 2]
- Dangerous to bees. Do not apply to crops in flower, or to those in which bees are actively foraging, except as directed on oilseed rape, wheat and barley. Do not apply when flowering weeds are present
- Risk to non-target insects or other arthropods [4]
- LERAP Category A [3, 4]
- Where possible spray oilseed rape crops in the late evening or early morning or in dull weather
- Do not spray within 6 m of the edge of a cereal crop after 31 Mar in yr of harvest
- Do not apply directly to poultry; collect eggs before application [1, 2]

Hazard classification and safety precautions
Hazard H03, H11 [3, 4]; H08 [4]
Risk phrases R21, R37, R38, R41, R51, R67 [4]; R22a, R53a [3, 4]; R50 [3]
Operator protection A, H [1-4]; C [4]; U02a, U19a [1, 2, 4]; U04a, U11, U14 [4]; U05a, U10, U20b [3, 4]
Environmental protection E02a [1, 2] (until dry); E05a, E12e, E13a [1, 2]; E12e [4] (oilseed rape, wheat, barley); E15a [1-4]; E16c, E16d [3, 4]; E22c, E34 [4]; E38 [3]
Consumer protection C09, C11 [1, 2]
Storage and disposal D01, D02, D05, D09a, D10b [3, 4]; D11a [1, 2]; D12a [3]
Medical advice M03, M05b [4]; M05a [3]

---

5 aluminium ammonium sulphate

An inorganic bird and animal repellent

Products

1	Curb Crop Spray Powder	Sphere	88% w/w	WP	02480
2	Guardsman STP Seed Dressing Powder	Sphere	88% w/w	DS	03606
3	Liquid Curb Crop Spray	Sphere	83 g/l	SC	03164

Uses
- Animal repellent in *agricultural premises, all top fruit, broad beans, bush fruit, cane fruit, carrots, flowerhead brassicas, forest nursery beds, forestry plantations, grain stores, grassland, leaf brassicas, peas, spring barley, spring field beans, spring oats, spring oilseed rape, spring wheat, strawberries, sugar beet, winter barley, winter field beans, winter oats, winter oilseed rape, winter wheat* [1, 3]; *amenity areas* [1]
- Bird repellent in *agricultural premises, all top fruit, broad beans, bush fruit, cane fruit, carrots, flowerhead brassicas, forest nursery beds, forestry plantations, grain stores, grassland, leaf brassicas, peas, spring barley, spring field beans, spring oats, spring oilseed rape, spring wheat, strawberries, sugar beet, winter barley, winter field beans, winter oats, winter oilseed rape, winter wheat* [1, 3]; *amenity areas* [1]
- Birds in *corms* (corm treatment), *flower bulbs* (bulb treatment), *seeds* (seed treatment) [2]
- Damaging mammals in *corms* (corm treatment), *flower bulbs* (bulb treatment), *seeds* (seed treatment) [2]

SEE SECTION 3 FOR PRODUCTS ALSO REGISTERED

Efficacy guidance
- Apply as overall spray to growing crops before damage starts or mix powder with seed depending on type of protection required
- Spray deposit protects growth present at spraying but gives little protection to new growth
- Product must be sprayed onto dry foliage to be effective and must dry completely before dew or frost forms. In winter this may require some wind

Restrictions
- Maximum number of treatments 1 per batch for corms, flower bulbs, seeds [2]

Crop-specific information
- Latest use: no restriction

Hazard classification and safety precautions
  Operator protection U05a, U20a [1, 3]; U20b [2]
  Environmental protection E15a [1-3]; E19b [1, 3]
  Storage and disposal D01, D02, D05, D12a [1, 3]; D09a [2]; D11a [1-3]
  Treated seed S02, S05 [2]
  Medical advice M03 [1, 3]

6   aluminium phosphide

A phosphine generating compound used against vertebrates and grain store pests

Products

1 Degesch Fumigation Tablets	Rentokil	56% w/w	GE	09313
2 Luxan Talunex	Luxan	57% w/w	GE	06563
3 Phostoxin	Rentokil	56% w/w	GE	09315

Uses
- Insect pests in **stored grain** [1]
- Moles in **farm woodland, grassland, lawns, managed amenity turf** [2]; **farmland** [2, 3]
- Rabbits in **farm woodland, grassland, lawns, managed amenity turf** [2]; **farmland** [2, 3]
- Rats in **farm woodland, grassland, lawns, managed amenity turf** [2]; **farmland** [2, 3]

Approval information
- Accepted by BBPA for use in stores for malting barley

Efficacy guidance
- Product releases poisonous hydrogen phosphide gas in contact with moisture
- Place fumigation tablets in grain stores as directed [1]
- Place pellets in burrows or runs and seal hole by heeling in or covering with turf. Do not cover pellets with soil. Inspect daily and treat any new or re-opened holes [2, 3]
- Apply pellets by means of Luxan Topex Applicator [2]

Restrictions
- Aluminium phosphide is subject to the Poisons Rules 1982 and the Poisons Act 1972. See notes in Section 5
- Only to be used by operators instructed or trained in the use of aluminium phosphide and familiar with the precautionary measures to be taken. See label and HSE Guidance Notes for full precautions
- Only open container outdoors [2, 3] and for immediate use. Keep away from liquid or water as this causes immediate release of gas. Do not use in wet weather
- Do not use within 3 m of human or animal habitation. Before application ensure that no humans or domestic animals are in adjacent buildings or structures. Allow a minimum airing-off period of 4 h before re-admission

Environmental safety
- Product liberates very toxic, highly flammable gas
- Dangerous for the environment [2]
- Very toxic to aquatic organisms [2]
- Dangerous to fish or other aquatic life. Do not contaminate surface waters or ditches with chemical or used container [1, 3]
- Prevent access by livestock, pets and other non-target mammals and birds to buildings under fumigation and ventilation [1]

FOR FULL CONDITIONS OF USE ALWAYS READ THE PRODUCT LABEL

- Pellets must never be placed or allowed to remain on ground surface
- Do not use adjacent to watercourses
- Take particular care to avoid gassing non-target animals, especially those protected under the Wildlife and Countryside Act (e.g. badgers, polecat, reptiles, natterjack toads, most birds). Do not use in burrows where there is evidence of badger or fox activity, or when burrows might be occupied by birds
- Dust remaining after decomposition is harmless and of no environmental hazard
- Keep in original container, tightly closed, in a safe place, under lock and key
- Dispose of empty containers as directed on label

Hazard classification and safety precautions
Hazard H01, H07 [1-3]; H03 [3]; H11 [2]
Risk phrases R21, R26, R28 [1-3]; R36, R37, R50 [2]
Operator protection A, H [1-3]; D [1, 3]; G [2]; U01, U13, U19a, U20a [1-3]; U05a [2, 3]; U05b [1]; U07 [1, 3]; U14, U15 [2]; U18 [1, 2]
Environmental protection E02a [1] (4 h min); E02a [3] (4 h); E02b [1]; E13b [1, 3]; E15a [2]; E34 [1-3]
Storage and disposal D01, D02, D09b, D11b [1-3]; D07 [1, 3]
Vertebrate/rodent control products V04a [2, 3]
Medical advice M04

7   amidosulfuron

A post-emergence sulfonylurea herbicide for cleavers and other broad-leaved weed control in cereals

Products

1 Eagle	Bayer CropScience	75% w/w	WG	07318
2 Landgold Amidosulfuron	Landgold	75% w/w	WG	09021
3 Landgold Amidosulfuron	Teliton	75% w/w	WG	12110
4 Pursuit	Bayer CropScience	75% w/w	WG	07333
5 Squire	Bayer CropScience	50% w/w	WG	08715

Uses
- Annual dicotyledons in *durum wheat, spring barley, spring oats, spring rye, spring wheat, triticale, winter barley, winter oats, winter rye, winter wheat* [1-4]; *linseed* [1, 4]
- Charlock in *permanent pasture, rotational grassland* [5]
- Cleavers in *durum wheat, spring barley, spring oats, spring rye, spring wheat, triticale, winter barley, winter oats, winter rye, winter wheat* [1-4]; *linseed* [1, 4]; *permanent pasture, rotational grassland* [5]
- Docks in *permanent pasture, rotational grassland* [5]
- Forget-me-not in *permanent pasture, rotational grassland* [5]
- Shepherd's purse in *permanent pasture, rotational grassland* [5]

Approval information
- Accepted by BBPA for use on malting barley

Efficacy guidance
- For best results apply in spring (from 1 Feb) in warm weather when soil moist and weeds growing actively. When used in grassland following cutting or grazing, docks should be allowed to regrow before treatment
- Weed kill is slow, especially under cool, dry conditions. Weeds may sometimes only be stunted but will have little or no competitive effect on crop
- May be used on all soil types unless certain sequences are used on linseed. See label
- Spray is rainfast after 1 h
- Cleavers controlled from emergence to flower bud stage. If present at application charlock (up to flower bud), shepherds purse (up to flower bud) and field forget-me-not (up to 6 leaves) will also be controlled
- Amidosulfuron is a member of the ALS-inhibitor group of herbicides and products should be used in a planned Resistance Management strategy. See Section 5 for more information

SEE SECTION 3 FOR PRODUCTS ALSO REGISTERED

Restrictions

- Maximum number of treatments 1 per crop, or 1 per yr when used on grass
- Use after 1 Feb and do not apply to rotational grass after 30 Jun, or to permanent grassland after 15 Oct
- Do not apply to crops undersown or due to be undersown with clover or alfalfa
- Do not use on swards containing red clover but may be used on swards with white clover. Where clover has been newly established apply from 3-leaf stage of grass and 1-2 trifoliate leaves of clover [5]
- Do not spray crops under stress, suffering drought, waterlogged, grazed, lacking nutrients or if soil compacted
- Do not spray if frost expected
- Do not roll or harrow within 1 wk of spraying
- Specific restrictions apply to use in sequence or tank mixture with other sulfonylurea or ALS-inhibiting herbicides. See label for details. There are no recommendations for mixtures with metsulfuron-methyl products on linseed
- Certain mixtures with fungicides are expressly forbidden. See label for details

Crop-specific information

- Latest use: before first spikelets just visible (GS 51) for cereals; before flower buds visible for linseed; 15 Oct for grassland
- Broadcast crops should be sprayed post-emergence after plants have a well established root system
- Hay or silage from treated grass crops must not be cut for at least 21 d after treatment [5]

Following crops guidance

- If a treated crop fails cereals may be sown after 15 d and thorough cultivation
- After normal harvest of a treated crop only cereals, winter oilseed rape, mustard, turnips, winter field beans or vetches may be sown in the same year as treatment and these must be preceded by ploughing or thorough cultivation
- Only cereals may be sown within 12 mth of application to grassland [5]
- Cereals or potatoes must be sown as the following crop after use of permitted mixtures or sequences with other sulfonyl urea herbicides in cereals. Only cereals may be sown after the use of such sequences in linseed

Environmental safety

- Dangerous for the environment [5]
- Very toxic to aquatic organisms [5]
- Keep livestock out of treated areas for at least 7d following treatment and until poisonous weeds, such as ragwort, have died down and become unpalatable [5]
- Dangerous to fish or other aquatic life. Do not contaminate surface waters or ditches with chemical or used container
- Take care to wash out sprayers thoroughly. See label for details
- Avoid drift onto neighbouring broad-leaved plants or onto surface waters or ditches

Hazard classification and safety precautions

Hazard H04, H11 [5]
Risk phrases R36, R50, R53a [5]
Operator protection C [5]; U05a [5]; U20a [1-3]; U20b [4, 5]
Environmental protection E07a [5] (7 d); E13b [1-5]; E38 [5]
Storage and disposal D01, D02, D12a [5]; D10a [1-5]

8   amidosulfuron + iodosulfuron-methyl-sodium

A post-emergence sulfonylurea herbicide mixture for cereals

Products

Chekker	Bayer CropScience	12.5:1.25% w/w	WG	10955

Uses

- Annual dicotyledons in **spring barley**, **spring rye**, **spring wheat**, **triticale**, **winter barley**, **winter rye**, **winter wheat**
- Chickweed in **spring barley**, **spring rye**, **spring wheat**, **triticale**, **winter barley**, **winter rye**, **winter wheat**

FOR FULL CONDITIONS OF USE ALWAYS READ THE PRODUCT LABEL

- Cleavers in **spring barley**, **spring rye**, **spring wheat**, **triticale**, **winter barley**, **winter rye**, **winter wheat**
- Mayweeds in **spring barley**, **spring rye**, **spring wheat**, **triticale**, **winter barley**, **winter rye**, **winter wheat**
- Volunteer oilseed rape in **spring barley**, **spring rye**, **spring wheat**, **triticale**, **winter barley**, **winter rye**, **winter wheat**

Approval information
- Accepted by BBPA for use on malting barley

Efficacy guidance
- Best results obtained from treatment in warm weather when soil is moist and the weeds are growing actively
- Weeds must be present at application to be controlled
- Dry conditions resulting in moisture stress may reduce effectiveness
- Weed control is slow especially under cool dry conditions
- Occasionally weeds may only be stunted but they will normally have little or no competitive effect on the crop
- Amidosulfuron and iodosulfuron are members of the ALS-inhibitor group of herbicides and products should be used in a planned Resistance Management strategy. See Section 5 for more information

Restrictions
- Maximum number of treatments 1 per crop
- Must only be applied between 1 Feb in yr of harvest and specified latest time of application
- Do not apply to crops undersown or to be undersown with grass, clover or alfalfa
- Do not roll or harrow within 1 wk of spraying
- Do not spray crops under stress from any cause or if the soil is compacted
- Do not spray if frost expected
- Do not apply in mixture or in sequence with any other ALS inhibitor

Crop-specific information
- Treat drilled crops after the 2-leaf stage; treat broadcast crops after the plants have a well-established root system
- Latest use: before first spikelet of inflorescence just visible (GS 51)

Following crops guidance
- Cereals, winter oilseed rape and winter field beans may be sown in the same yr as treatment provided they are preceded by ploughing or thorough cultivation. Any crop may be sown in the spring of the yr following treatment
- A minimum of 3 mth must elapse between treatment and sowing winter oilseed rape

Environmental safety
- Dangerous for the environment
- Toxic to aquatic organisms
- LERAP Category B
- Take extreme care to avoid damage by drift onto broad-leaved plants outside the target area or onto ponds, waterways and ditches
- Observe carefully label instructions for sprayer cleaning

Hazard classification and safety precautions
Hazard H04, H11
Risk phrases R36, R51, R53a
Operator protection A, C, H; U05a, U08, U11, U14, U15, U20b
Environmental protection E15a, E16a, E16b, E38
Storage and disposal D01, D02, D10a, D12a

9 2-aminobutane

A fumigant alkylamine fungicide permitted for use only on stored seed potatoes

Products

Certis 2-Aminobutane	Certis	720 g/l	VP	11182

SEE SECTION 3 FOR PRODUCTS ALSO REGISTERED

Uses
- Gangrene in **seed potatoes**
- Silver scurf in **seed potatoes**
- Skin spot in **seed potatoes**

Approval information
- Products containing this active ingredient have been granted derogations for specified 'Essential Uses' for use until 31 December 2007. Sale and supply must cease by 30 June 2007 but growers have no guarantee that the products will continue to be available until then.
  For more information see 'The Review Programme' under 'Pesticide Legislation' in Section 5

Efficacy guidance
- Product used for treatment of seed potato tubers by fumigation in appropriate premises (see label)

Restrictions
- Maximum number of treatments 1 per batch. Maximum quantity to be fumigated in a single stack must not exceed 2000 tonnes
- Do not treat immature tubers. Allow period of healing before treating damaged tubers
- Treatment must only be carried out by trained operators in suitable fumigation chambers under licence from the British Technology Group
- Fumigate within 21 d of lifting
- This product must not be used on any crops other than those listed, including any extrapolations that would normally be permissible under the Long Term Arrangements for Extension of Use (see Section 5)

Environmental safety
- Dangerous for the environment
- Very toxic to aquatic organisms
- Keep in original container, tightly closed, in a safe place, under lock and key
- Do not empty into drains
- Do not supply treated potatoes for consumption by humans or lactating dairy cows
- Use must be in accordance with approved Code of Practice for the Control of Substances Hazardous to Health: Fumigation Operations

Hazard classification and safety precautions
Hazard H03, H05, H07, H11
Risk phrases R20, R22a, R35, R50
Operator protection A, C; U02a, U04a, U05b, U10, U11, U13, U19a, U20b
Environmental protection E15a, E19b, E34
Storage and disposal D01, D02, D08, D09b, D10a
Medical advice M04

10 amitrole

A translocated, foliar-acting, non-selective triazole herbicide

Products
1 Aminotriazole Technical	Nufarm UK	92.5% w/w	TC	11137
2 Weedazol-TL	Nufarm UK	225 g/l	SL	11968

Uses
- Annual and perennial weeds in **apricots** *(off-label)*, **cherries** *(off-label)*, **peaches** *(off-label)*, **plums** *(off-label)*, **quinces** *(off-label)* [2]
- Annual weeds in **fallows, headlands, stubbles** [2]
- Barren brome in **apple orchards, pear orchards** [2]
- Couch in **amenity areas, industrial sites, non-crop areas** [1]; **apple orchards, fallows, headlands, pear orchards, stubbles** [2]
- Creeping bent in **amenity areas, industrial sites, non-crop areas** [1]
- Creeping thistle in **amenity areas, industrial sites, non-crop areas** [1]; **fallows, stubbles** [2]
- Docks in **amenity areas, industrial sites, non-crop areas** [1]; **fallows, headlands, stubbles** [2]
- General weed control in **apple orchards, farm woodland** *(off-label)*, **pear orchards** [2]
- Grass weeds in **amenity turf** *(off-label - on amitrole resistant turf)* [2]
- Perennial weeds in **apple orchards, fallows, headlands, pear orchards** [2]

FOR FULL CONDITIONS OF USE ALWAYS READ THE PRODUCT LABEL

- Total vegetation control in **amenity areas, industrial sites, non-crop areas** [1]
- Volunteer potatoes in **stubbles** *(barley stubble)* [2]

Specific Off-Label Approvals (SOLAs)
- **amenity turf** *(on amitrole resistant turf) (OLA 051979) Jan 2007* [2]
- **apricots, cherries, peaches, plums, quinces** *(OLA 051981) Jan 2007* [2]
- **farm woodland** *(OLA 051980) Jan 2007* [2]

Approval information
- Amitrole included in Annex I under EC Directive 91/414

Efficacy guidance
- In non-crop land may be applied at any time from Apr to Oct. Best results achieved in spring or early summer when weeds growing actively. For coltsfoot, hogweed and horsetail summer and autumn applications are preferred
- Uptake is via foliage and heavy rain immediately after application will reduce efficacy. Amitrole is less affected by drought than some residual herbicides and remains effective for up to 2 mth
- Applications made in summer may not give complete control of couch if past the shooting stage or not actively growing
- Effective crop competition and efficient ploughing are essential for good couch control

Restrictions
- Maximum number of treatments 1 per yr for edible crops
- Keep off suckers or foliage of desirable trees or shrubs
- Do not spray areas into which the roots of adjacent trees or shrubs extend
- Application to land intended for spring barley should be in the preceding autumn, not the spring
- Do not spray on sloping ground when rain imminent and run-off may occur
- Do not spray if foliage is wet or rain imminent
- Do not use in low temperatures or in drought
- Do not mix product with acids

Crop-specific information
- Latest use: after harvest but before 30 Jun for apricots, cherries, peaches, plums, quinces; before end Jun or after harvest for apple and pear orchards; end Oct for headlands; end Oct and at least 2 wk before cultivation and drilling for stubbles and fallows [2]

Following crops guidance
- Amitrole breaks down fairly quickly in medium and heavy soils and 3 wk should be allowed between application and sowing or planting. On sandy soils the interval should be 6 wk

Environmental safety
- Harmful to aquatic organisms
- Keep livestock out of treated areas for at least two weeks following treatment and until poisonous weeds, such as ragwort, have died down and become unpalatable
- Harmful to fish or other aquatic life. Do not contaminate surface waters or ditches with chemical or used container

Hazard classification and safety precautions
Hazard H03 [2]
Risk phrases R22a, R40, R48, R52, R53a [2]
Operator protection A [1, 2]; C [2]; U05a, U20b [2]; U08, U19a [1, 2]; U20a [1]
Environmental protection E07a, E15a [2]; E13c [1]
Storage and disposal D01, D02, D05 [2]; D09a, D10b [1, 2]

11   amitrole + 2,4-D + diuron

A total herbicide mixture of translocated and residual chemicals

Products
Trik   Nufarm UK   26.6:11.2:46.4% w/w   WP   11441

Uses
- Total vegetation control in **land not intended to bear vegetation**

Approval information
- Amitrole and 2,4-D included in Annex I under EC Directive 91/414

SEE SECTION 3 FOR PRODUCTS ALSO REGISTERED

Efficacy guidance
- Apply in spring or late summer/early autumn when weeds are growing actively and have sufficient leaf area to absorb chemical
- Apply maintenance treatment if necessary at lower rate when weeds 7-10 cm high
- Increase dose rate on areas of peat or high carbon content

Restrictions
- Maximum number of treatments 1 per yr for land not intended to bear vegetation
- Do not use on ground under which roots of valuable trees or shrubs are growing

Environmental safety
- Dangerous for the environment
- Very toxic to aquatic organisms
- Keep livestock out of treated areas for at least two weeks following treatment and until poisonous weeds, such as ragwort, have died down and become unpalatable

Hazard classification and safety precautions
 Hazard H03, H11
 Risk phrases R22a, R37, R48, R50, R53a, R68
 Operator protection A, C, D, H, M; U05a, U08, U14, U15, U19a, U20a
 Environmental protection E07a, E15a
 Storage and disposal D01, D02, D09a, D11a, D12b
 Medical advice M03, M05a

12   ammonium sulphamate

A non-selective, inorganic, general purpose herbicide and tree-killer

Products

1 Amcide	B H & B	99.5% w/w	CR	04246
2 Root-Out	Dax	98.5% w/w	CR	03510

Uses
- Annual weeds in *forest*, *ornamental specimens* (pre-planting), *vegetables* (pre-planting) [1]; *ornamental specimens*, *trees and shrubs* [2]
- Perennial dicotyledons in *forest* [1]; *trees and shrubs* [2]
- Perennial grasses in *forest* [1, 2]
- Perennial weeds in *ornamental specimens* [2]; *ornamental specimens* (pre-planting), *vegetables* (pre-planting) [1]
- Rhododendrons in *forest* [1, 2]
- Woody weeds in *forest* [1, 2]

Efficacy guidance
- Apply as spray to low scrub and herbaceous weeds from Apr to Sep in dry weather when rain unlikely and cultivate after 3-8 wk
- Apply as crystals in frills or notches in trunks of standing trees at any time of year
- Apply as concentrated solution or crystals to stump surfaces within 48 h of cutting. Rhododendrons must be cut level with ground and sprayed to cover cut surface, bark and immediate root area
- Stainless steel or plastic sprayers are recommended. Solutions are corrosive to mild steel, galvanised iron, brass and copper

Restrictions
- Keep spray at least 30 cm from growing plants. Low doses may be used under mature trees with undamaged bark

Following crops guidance
- Allow 8-12 wk after treatment before replanting

Environmental safety
- Harmful to fish or other aquatic life. Do not contaminate surface waters or ditches with chemical or used container

Hazard classification and safety precautions
 Operator protection U09b, U20b [1]; U11, U14, U19a [1, 2]; U15, U20a [2]

FOR FULL CONDITIONS OF USE ALWAYS READ THE PRODUCT LABEL

Environmental protection E13c
Storage and disposal D01, D09a [1, 2]; D11a [1]

13   asulam

A translocated carbamate herbicide for control of docks and bracken

Products

1 Asulox	Bayer CropScience	400 g/l	SL	09969
2 Greencrop Frond	Greencrop	400 g/l	SL	11912
3 Spitfire	AgriGuard	400 g/l	SL	12002

Uses

- Bracken in *amenity vegetation* [2, 3]; *forest, permanent pasture, rough grazing* [1-3]; *forest* (off-label), *non-crop areas* [1]
- Brome grasses in *poppies* (off-label - for morphine production) [1]
- Docks in *amenity grass* (not fine turf), *apple orchards, blackcurrants, cherries, hops, pear orchards, permanent pasture, plums* [1-3]; *amenity vegetation, blackberries* (off-label), *blueberries* (off-label), *clover seed crops* (off-label), *cranberries* (off-label), *damsons* (off-label), *gooseberries* (off-label), *loganberries* (off-label), *mint* (off-label), *nectarines* (off-label), *parsley* (off-label), *poppies* (off-label - for morphine production), *quinces* (off-label), *raspberries* (off-label), *redcurrants* (off-label), *road verges, strawberries* (off-label), *tarragon* (off-label), *waste ground, whitecurrants* (off-label) [1]
- Meadow grasses in *poppies* (off-label - for morphine production) [1]

Specific Off-Label Approvals (SOLAs)

- *blackberries, blueberries, cranberries, damsons, gooseberries, loganberries, nectarines, quinces, raspberries, redcurrants, whitecurrants* (OLA 040508) Dec 2008 [1]
- *clover seed crops* (OLA 001894) Dec 2008 [1]
- *forest* (OLA 001898) Dec 2008 [1]
- *mint, parsley, tarragon* (OLA 001900) Dec 2008 [1]
- *poppies* (for morphine production) (OLA 040471) Dec 2008 [1]
- *strawberries* (OLA 001896) Dec 2008 [1]

Approval information

- May be applied through CDA equipment
- Approved for aerial application on bracken in agricultural grassland, amenity grassland, forestry and rough upland intended for grazing [1-3]. See notes in Section 5
- Approved for use near surface waters. See notes in Section 5
- Accepted by BBPA for use on hops

Efficacy guidance

- Spray bracken when fronds fully expanded but not senescent, usually Jul-Aug; docks in full leaf before flower stem emergence
- Bracken fronds must not be damaged by stock, frost or cutting before treatment
- Uptake and reliability of bracken control may be improved by use of specified additives - see label. Additives not recommended on forestry land
- To allow adequate translocation do not cut or admit stock for 14 d after spraying bracken or 7 d after spraying docks. Preferably leave undisturbed until late autumn
- Complete bracken control rarely achieved by one treatment. Survivors should be sprayed when they recover to full green frond, which may be in the ensuing year but more likely in the second year following initial application

Restrictions

- Maximum number of treatments 1 per crop or 1 per yr
- Do not apply in drought or hot, dry conditions
- Do not use in pasture before mowing for hay

Crop-specific information

- Harvest interval: 10 wk for tarragon; 6 wk for mint, parsley
- Latest Use: 1st treatment between Mar and Apr; 2nd treatment between Sep and Nov for cane and soft fruit; Aug for amenity vegetation, forest, non-crop areas, grass; before 31 Mar in yr of harvest for strawberries; flower buds emerging for poppies
- In forestry areas some young trees may be checked if sprayed directly (see label)

SEE SECTION 3 FOR PRODUCTS ALSO REGISTERED

- In fruit crops apply as a directed spray
- Do not treat blackcurrant cuttings, hop sets or weak hills
- Some grasses and herbs will be damaged by full dose. Most sensitive are cocksfoot, Yorkshire fog, timothy, bents, annual meadow-grass, daisies, docks, plantains, saxifrage
- Apply as spot treatment in parsley, mint and tarragon, not directly to crop

Following crops guidance
- Allow at least 6 wk between spraying and planting any crop

Environmental safety
- Dangerous for the environment
- Very toxic to aquatic organisms
- Keep livestock out of treated areas for at least two weeks following treatment and until poisonous weeds, such as ragwort, have died down and become unpalatable
- The use of asulam near surface waters has been considered by PSD. Whilst every care should be taken to avoid contamination, any that does occur during the normal course of spraying should offer no harm to operators, to users and consumers of the water, to domestic and farm animals and to wildlife. Before spraying such areas the appropriate regulatory authority should be notified

Hazard classification and safety precautions
Hazard H04, H11
Risk phrases R43 [1, 2]; R50 [1-3]; R53a [2, 3]
Operator protection A, C, D, H [1-3]; M [1] (for ULV application); M [2, 3]; U05a, U20a [3]; U08, U20b [2]; U14 [1, 2]; U19a, U20b [1] (ULV); U19a [2, 3]; U20c [1]
Environmental protection E07a, E15a [1-3]; E38 [1]
Storage and disposal D01, D02 [3]; D05 [2, 3]; D09a, D10a [1-3]; D12a [1]

14   atrazine

A triazine herbicide with residual and foliar activity, with restricted permitted uses

Products

1	Aconite 50	DAPT	500 g/l	SC	10642
2	Alpha Atrazine 50 SC	Makhteshim	500 g/l	SC	04877
3	Gesaprim	Syngenta	500 g/l	SC	08411
4	Greencrop Amaize	Greencrop	500 g/l	SC	11973

Uses
- Annual dicotyledons in **farm forestry**, **forest** [2, 4]; **sweetcorn** [1-4]
- Annual grasses in **farm forestry**, **forest** [2, 4]; **sweetcorn** [1-4]
- Perennial grasses in **farm forestry**, **forest** [2, 4]

Approval information
- Atrazine was reviewed in 1992 and approvals for aerial use, and use on non-crop land (excluding home garden use) revoked. Restrictions were also placed on the number of applications that could be made to crops
- Use on maize was permitted until Sep 2005. Products containing this active ingredient have been granted derogations for other specified 'Essential Uses' for use until 31 December 2007. Sale and supply must cease by 30 June 2007 but growers have no guarantee that the products will continue to be available until then.
  For more information see 'The Review Programme' under 'Pesticide Legislation' in Section 5

Efficacy guidance
- Root activity enhanced by rainfall soon after application and reduced on high organic soils. Foliar activity effective on weeds up to 3 cm high
- Resistant weed strains may develop with repeated use of atrazine or other triazines

Restrictions
- Maximum number of applications (including other atrazine/simazine products) 1 per crop (or lower doses to 3.0 l/ha total) for sweetcorn; 1 per season for conifer plantations, raspberries, roses
- Maximum total dose equivalent to one full dose treatment
- Not recommended for use on soils with more than 10% organic matter
- Do not use on Christmas trees

Crop-specific information
- Harvest interval: 7 mth for sweetcorn

FOR FULL CONDITIONS OF USE ALWAYS READ THE PRODUCT LABEL

- May be used pre- or early post-weed emergence in sweetcorn
- In conifers apply as overall spray in Feb-Apr. May be used in first spring after planting

Following crops guidance
- Latest use: 7 mth before a succeeding crop for maize, sweetcorn; Apr for conifer plantations; before cane emergence for raspberries; before weeds at 3 cm for roses

Environmental safety
- Dangerous for the environment
- Very toxic to aquatic organisms. May cause long-term adverse effects in the aquatic environment
- LERAP Category B
- Use must be restricted to one product containing atrazine or simazine, and either to a single application at the maximum approved rate or (subject to any existing maximum permitted number of treatments) to several applications at lower doses up to the maximum approved rate for a single application
- On slopes heavy rainfall soon after application may cause surface run-off
- To reduce soil run-off, especially from forest plantations, users are advised to plant grass strips 6 m wide between treated areas and surface waters

Hazard classification and safety precautions
Hazard H03, H11
Risk phrases R22a, R43 [2, 4]; R48, R50, R53a [1-4]
Operator protection A, C, D, H, M [1-4]; B [3]; U05a, U15, U20b [2]; U08, U19a [4]; U14 [2, 4]; U20c [1, 3, 4]
Environmental protection E15a, E16a, E16b [1-4]; E38 [1, 3]
Storage and disposal D01, D02 [1-3]; D05, D10b [2, 4]; D09a [1-4]; D10c, D12a [1, 3]; D12b [2]
Medical advice M05a [4]

15   azoxystrobin

A systemic translaminar and protectant strobilurin fungicide for a wide range of crops

Products

1 Amistar	Syngenta	250 g/l	SC	10443
2 Heritage	Scotts	50% w/w	WG	11383
3 Landgold Strobilurin 250	Landgold	250 g/l	SC	09595
4 Landgold Strobilurin 250	Teliton	250 g/l	SC	12128
5 Me2 Azoxystrobin	Me2	250 g/l	SC	09654
6 Standon Azoxystrobin	Standon	250 g/l	SC	09515

Uses
- Alternaria in *broccoli*, *brussels sprouts*, *cabbages*, *calabrese*, *carrots*, *cauliflowers*, *collards*, *kale*, *spring oilseed rape*, *winter oilseed rape* [1]
- Ascochyta in *broad beans* (off-label), *combining peas*, *vining peas* [1]
- Black dot in *potatoes* [1]
- Black scurf and stem canker in *potatoes* [1]
- Black spot in *protected strawberries* (off-label), *strawberries* (off-label) [1]
- Botrytis in *broad beans* (off-label), *celery* (off-label), *dwarf beans* (off-label), *navy beans* (off-label), *protected celery* (off-label), *runner beans* (off-label) [1]
- Brown rust in *spring barley*, *spring wheat*, *winter barley*, *winter wheat* [1, 3-6]; *spring rye* (off-label), *triticale*, *winter rye* [1]
- Crown rust in *spring oats*, *winter oats* [1]
- Dark leaf spot in *salad brassicas* (off-label - for baby leaf production) [1]
- Didymella in *inert substrate aubergines* (off-label), *inert substrate tomatoes* (off-label) [1]
- Didymella stem rot in *aubergines* (off-label), *protected tomatoes* (off-label) [1]
- Downy mildew in *artichokes* (off-label), *bulb onions*, *inert substrate courgettes* (off-label), *inert substrate cucumbers* (off-label), *inert substrate gherkins* (off-label), *protected courgettes* (off-label), *protected cucumbers* (off-label), *protected gherkins* (off-label), *salad brassicas* (off-label - for baby leaf production), *salad onions* (off-label) [1]
- Fusarium patch in *amenity grass*, *managed amenity turf* [2]
- Glume blotch in *spring wheat*, *winter wheat* [1, 3-6]
- Grey mould in *aubergines* (off-label), *inert substrate aubergines* (off-label), *inert substrate tomatoes* (off-label), *protected tomatoes* (off-label) [1]

SEE SECTION 3 FOR PRODUCTS ALSO REGISTERED

- Late blight in *aubergines* (off-label), *inert substrate aubergines* (off-label), *inert substrate tomatoes* (off-label), *protected tomatoes* (off-label) [1]
- Late ear diseases in *spring wheat, winter wheat* [1, 3-6]
- Net blotch in *spring barley, winter barley* [1, 3-6]
- Phytophthora in *protected chicory* (off-label - for forcing) [1]
- Powdery mildew in *artichokes* (off-label), *aubergines* (off-label), *carrots*, *chives* (off-label), *courgettes* (off-label), *herbs (see appendix 6)* (off-label), *inert substrate aubergines* (off-label), *inert substrate courgettes* (off-label), *inert substrate cucumbers* (off-label), *inert substrate gherkins* (off-label), *inert substrate tomatoes* (off-label), *parsley* (off-label), *poppies* (off-label - for morphine production), *protected blackberries* (off-label), *protected courgettes* (off-label), *protected cucumbers* (off-label), *protected gherkins* (off-label), *protected peppers* (off-label), *protected raspberries* (off-label), *protected tomatoes* (off-label), *raspberries* (off-label), *spring oats, triticale, winter oats, winter rye* [1]; *spring barley, winter barley* [1, 3-6]; *spring wheat, winter wheat* [6]
- Purple blotch in *leeks* [1]
- Rhizoctonia in *celery* (off-label), *protected celery* (off-label), *protected chives* (off-label), *protected herbs (see appendix 6)* (off-label), *protected lettuce* (off-label), *protected parsley* (off-label), *swedes* (off-label), *turnips* (off-label) [1]
- Rhynchosporium in *spring barley, winter barley* [1, 3-6]; *spring rye* (off-label), *triticale, winter rye* [1]
- Ring spot in *broccoli, brussels sprouts, cabbages, calabrese, cauliflowers, chives* (off-label), *collards, herbs (see appendix 6)* (off-label), *kale, parsley* (off-label) [1]
- Root malformation disorder in *red beet* (off-label) [1]
- Rust in *asparagus, chives* (off-label), *herbs (see appendix 6)* (off-label), *leeks, parsley* (off-label), *spring field beans, winter field beans* [1]
- Sclerotinia in *celeriac* (off-label), *celery* (off-label), *lettuce* (off-label), *protected celery* (off-label) [1]
- Sclerotinia stem rot in *spring oilseed rape, winter oilseed rape* [1]
- Septoria leaf spot in *spring wheat, winter wheat* [1, 3-6]
- Stemphylium in *asparagus* [1]
- Take-all in *spring barley* (reduction), *spring wheat* (reduction), *winter barley* (reduction), *winter wheat* (reduction) [1]
- White blister in *broccoli, brussels sprouts, cabbages, calabrese, cauliflowers, collards, kale* [1]
- White rust in *chrysanthemums* (off-label - in pots), *protected chrysanthemums* (off-label) [1]
- Yellow rust in *spring wheat, winter wheat* [1, 3-6]

Specific Off-Label Approvals (SOLAs)
- *artichokes* (OLA 051814) Jul 2008 [1]
- *aubergines, protected courgettes, protected cucumbers, protected gherkins, protected tomatoes* (OLA 021533) Jul 2008 [1]
- *broad beans, dwarf beans, navy beans, runner beans* (OLA 032311) Jul 2008 [1]
- *celeriac* (OLA 031862) Jul 2008 [1]
- *celery, protected celery* (OLA 001041) Jul 2008 [1]
- *chives, herbs (see appendix 6), parsley* (OLA 021293) Jul 2008 [1]
- *chrysanthemums* (in pots) (OLA 011684) Jul 2008 [1]
- *courgettes* (OLA 051985) Jul 2008 [1]
- *inert substrate aubergines, inert substrate courgettes, inert substrate cucumbers, inert substrate gherkins, inert substrate tomatoes* (OLA 011685) Jul 2008 [1]
- *lettuce* (OLA 011465) Jul 2008 [1]
- *poppies* (for morphine production) (OLA 031137) Jul 2008 [1]
- *protected blackberries, protected raspberries* (OLA 051194) Jul 2008 [1]
- *protected chicory* (for forcing) (OLA 051814) Jul 2008 [1]
- *protected chives, protected herbs (see appendix 6), protected lettuce, protected parsley* (OLA 030659) Jul 2008 [1]
- *protected chrysanthemums* (OLA 001536) May 2008 [1]
- *protected chrysanthemums* (OLA 011684) Jul 2008 [1]
- *protected peppers* (OLA 021295) Jul 2008 [1]
- *protected strawberries, strawberries* (OLA 021294) Jul 2008 [1]
- *raspberries* (OLA 030365) Jul 2008 [1]

FOR FULL CONDITIONS OF USE ALWAYS READ THE PRODUCT LABEL

- **red beet, swedes, turnips** *(OLA 040614) Dec 2008* [1]
- **salad brassicas** *(for baby leaf production) (OLA 011465) Jul 2008* [1]
- **salad onions** *(OLA 021687) Jul 2008* [1]

Approval information
- Azoxystrobin included in Annex I under EC Directive 91/414
- Accepted by BBPA for use on malting barley

Efficacy guidance
- Best results obtained from use as a protectant or during early stages of disease establishment or when a predictive assessment indicates a risk of disease development
- Azoxystrobin inhibits fungal respiration and should always be used in mixture with fungicides with other modes of action
- Treatment under poor growing conditions may give less reliable results
- For good control of *Fusarium* patch in amenity turf and grass repeat treamtment at minimum intervals of 2 wk
- Azoxystrobin is a member of the QoI cross resistance group. Product should be used preventatively and not relied on for its curative potential
- Use product in cereals as part of an Integrated Crop Management strategy incorporating other methods of control, including where appropriate other fungicides with a different mode of action. Do not apply more than two foliar applications of QoI containing products to any cereal crop
- There is a significant risk of widespread resistance occurring in *Septoria tritici* populations in UK. Failure to follow resistance management action may result in reduced levels of disease control
- On cereal crops product must always be used in mixture with another product, recommended for control of the same target disease, that contains a fungicide from a different cross resistance group and is applied at a dose that will give robust control
- Strains of barley powdery mildew resistant to QoIs are common in the UK

Restrictions
- Maximum total dose ranges from 2-4 times the single full dose depending on crop and product. See labels for details
- On turf the maximum number of treatments is 4 per yr but they must not exceed one third of the total number of fungicide treatments applied
- Do not use where there is risk of spray drift onto neighbouring apple crops
- The same spray equipment should not be used to treat apples

Crop-specific information
- Latest use: grain watery ripe (GS 71) for cereals; before senescence for asparagus
- HI 3 d for inert substrate vegetables, aubergines, protected cucurbits, protected and outdoor strawberries, protected tomatoes; 7 d for artichokes, raspberries salad onions; 10 d for carrots, protected blackberries, protected raspberries; 12 d for dwarf, navy and runner beans; 14 d for onions, celery, celeriac, herbs, lettuce, brassicas, salad brassicas, vining peas, poppies; 21 d for leeks, oilseed rape, protected chicory; 35 d for field beans, red beet, swedes, turnips; 36 d for combining peas
- In cereals control of established infections can be improved by appropriate tank mixtures or application as part of a programme. Always use in mixture with another product from a different cross-resistance group
- In turf use product at full dose rate in a disease control programme, alternating with fungicides of different modes of action
- In potatoes when used as incorporated treatment apply overall to the entire area to be planted, incorporate to 15 cm and plant on the same day. In-furrow spray should be directed at the furrow and not the seed tubers
- Applications to brassica crops must only be made to a developed leaf canopy and not before growth stages specified on the label
- Heavy disease pressure in brassicae and oilseed rape may require a second treatment
- All crops should be treated when not under stress. Check leaf wax on peas if necessary
- Consult processor before treating any crops for processing
- Treat asparagus after the harvest season. Where a new bed is established do not treat within 3 wk of transplanting out the crowns
- Do not apply to turf when ground is frozen or during drought

SEE SECTION 3 FOR PRODUCTS ALSO REGISTERED

Environmental safety
- Dangerous for the environment [1, 3-6]
- Very toxic to aquatic organisms [1, 3-6]
- Extremely dangerous to fish or other aquatic life. Do not contaminate surface waters or ditches with chemical or used container [2]
- Avoid spray drift onto surrounding areas or crops, especially apples, plums or privet
- LERAP Category B (potatoes only) [1]

Hazard classification and safety precautions
> **Hazard** H11 [1, 3-6]
> **Risk phrases** R50, R53a [1, 3-6]
> **Operator protection** A [1]; U05a [1, 3, 4]; U09a, U19a, U20b [1, 3-6]; U20c [2]
> **Environmental protection** E13a [2]; E15a [1, 3-6]; E16a, E16b [1] (potatoes only); E38 [1]
> **Storage and disposal** D01, D02 [1, 3, 4]; D03 [2]; D05, D09a [1-6]; D10b [2-6]; D10c, D12a [1]
> **Medical advice** M05a [2]

16   azoxystrobin + chlorothalonil

A preventative and systemic fungicide mixture for cereals

Products

1 Amistar Opti	Syngenta	80:400 g/l	SC	12515
2 Olympus	Syngenta	100:500 g/l	SC	12407

Uses
- Brown rust in **spring barley**, **winter barley**, **winter wheat** [1, 2]; **spring wheat** [1]
- Glume blotch in **spring wheat** [1]; **winter wheat** [1, 2]
- Net blotch in **spring barley**, **winter barley** [1, 2]
- Rhynchosporium in **spring barley**, **winter barley** [1, 2]
- Septoria leaf spot in **spring wheat** [1]; **winter wheat** [1, 2]
- Take-all in **spring barley** (reduction), **winter barley** (reduction), **winter wheat** (reduction) [1, 2]; **spring wheat** (reduction) [1]
- Yellow rust in **spring wheat** [1]; **winter wheat** [1, 2]

Approval information
- Azoxystrobin and chlorothalonil included in Annex I under EC Directive 91/414
- Accepted by BBPA for use on malting barley

Efficacy guidance
- Best results obtained from applications made as a protectant treatment or in earliest stages of disease development. Further applications may be needed if disease attack is prolonged
- Reduction of barley spotting occurs when used as part of a programme with other fungicides [1]
- Control of *Septoria* and rust diseases may be improved by mixture with a triazole fungicide
- Azoxystrobin is a member of the QoI cross resistance group. Product should be used preventatively and not relied on for its curative potential
- Use product in cereals as part of an Integrated Crop Management strategy incorporating other methods of control, including where appropriate other fungicides with a different mode of action. Do not apply more than two foliar applications of QoI containing products to any cereal crop
- There is a significant risk of widespread resistance occurring in *Septoria tritici* populations in UK. Failure to follow resistance management action may result in reduced levels of disease control

Restrictions
- Maximum number of treatments 2 per crop
- Maximum total dose on barley equivalent to one full dose treatment
- Do not use where there is risk of spray drift onto neighbouring apple crops
- The same spray equipment should not be used to treat apples

Crop-specific information
- Latest Use: before beginning of heading (GS 51) for barley; before caryopsis watery ripe (GS 71) for wheat

Environmental safety
- Dangerous for the environment
- Very toxic to aquatic organisms

FOR FULL CONDITIONS OF USE ALWAYS READ THE PRODUCT LABEL

- LERAP Category B
- Product supplied in returnable refillable container. Follow label instructions [2]

Hazard classification and safety precautions
    Hazard H02 [2]; H03 [1]; H11 [1, 2]
    Risk phrases R20 [1]; R22a, R23 [2]; R37, R40, R41, R43, R50, R53a [1, 2]
    Operator protection A, C, H; U02a, U05a, U09a, U11, U14, U15, U19a, U20b [1, 2]; U07 [2]
    Environmental protection E15a, E16a, E34, E38 [1, 2]; E36 [2]
    Storage and disposal D01, D02, D09a, D12a [1, 2]; D05, D14 [2]; D10c [1]
    Medical advice M04

17   azoxystrobin + cyproconazole

A contact and systemic broad spectrum fungicide mixture for cereals

Products

Priori Xtra	Syngenta	200:80 g/l	SC	11518

Uses
- Brown rust in **spring barley**, **spring rye**, **spring wheat**, **winter barley**, **winter rye**, **winter wheat**
- Crown rust in **spring oats**, **winter oats**
- Eyespot in **spring barley** (reduction), **spring wheat** (reduction), **winter barley** (reduction), **winter wheat** (reduction)
- Net blotch in **spring barley**, **winter barley**
- Powdery mildew in **spring barley**, **spring oats**, **spring rye**, **spring wheat**, **winter barley**, **winter oats**, **winter rye**, **winter wheat**
- Rhynchosporium in **spring barley** (moderate control), **spring rye** (moderate control), **winter barley** (moderate control), **winter rye** (moderate control)
- Septoria diseases in **spring wheat**, **winter wheat**
- Take-all in **spring barley** (reduction), **spring wheat** (reduction), **winter barley** (reduction), **winter wheat** (reduction)
- Yellow rust in **spring wheat**, **winter wheat**

Approval information
- Azoxystrobin included in Annex I under EC Directive 91/414
- Accepted by BBPA for use on malting barley

Efficacy guidance
- Best results obtained from treatment during the early stages of disease development
- A second application may be needed if disease attack is prolonged
- Azoxystrobin is a member of the QoI cross resistance group. Product should be used preventatively and not relied on for its curative potential
- Use product as part of an Integrated Crop Management strategy incorporating other methods of control, including where appropriate other fungicides with a different mode of action. Do not apply more than two foliar applications of QoI containing products to any cereal crop
- There is a significant risk of widespread resistance occurring in *Septoria tritici* populations in UK. Failure to follow resistance management action may result in reduced levels of disease control
- Strains of wheat and barley powdery mildew resistant to QoIs are common in the UK. Control of wheat powdery mildew can only be relied upon from the triazole component
- Where specific control of wheat mildew is required this should be achieved through a programme of measures including products recommended for the control of mildew that contain a fungicide from a different cross-resistance group and applied at a dose that will give robust control

Restrictions
- Maximum total dose equivalent to two full dose treatments
- Do not use where there is risk of spray drift onto neighbouring apple crops
- The same spray equipment should not be used to treat apples

Crop-specific information
- Latest use: up to and including anthesis complete (GS 69) for rye and wheat; up to and including emergence of ear complete (GS 59) for barley and oats

SECTION 2

SEE SECTION 3 FOR PRODUCTS ALSO REGISTERED

Environmental safety
- Dangerous for the environment
- Very toxic to aquatic organisms

Hazard classification and safety precautions
  Hazard H03, H11
  Risk phrases R22a, R50, R53a, R63
  Operator protection A; U05a, U09a, U19a, U20b
  Environmental protection E15a, E34, E38
  Storage and disposal D01, D02, D05, D09a, D10c, D12a
  Medical advice M03

18 azoxystrobin + fenpropimorph

A protectant and eradicant fungicide mixture for cereals

Products

1 Amistar Pro	Syngenta	100:280 g/l	SE	10513
2 Aspect	Syngenta	100:280 g/l	SE	10516

Uses
- Brown rust in *spring barley*, *spring wheat*, *winter barley*, *winter wheat*
- Late ear diseases in *spring wheat*, *winter wheat*
- Net blotch in *spring barley*, *winter barley*
- Powdery mildew in *spring barley*, *winter barley*
- Rhynchosporium in *spring barley*, *winter barley*
- Septoria diseases in *spring wheat*, *winter wheat*
- Take-all in *spring barley* (reduction), *spring wheat* (reduction), *winter barley* (reduction), *winter wheat* (reduction)
- Yellow rust in *spring wheat*, *winter wheat*

Approval information
- Azoxystrobin included in Annex I under EC Directive 91/414
- Accepted by BBPA for use on malting barley

Efficacy guidance
- Best results obtained from application before infection following a disease risk assessment, or when disease first seen in crop
- Results may be less reliable when used on crops under stress
- Treatments for protection against ear disease should be made at ear emergence
- Azoxystrobin is a member of the QoI cross resistance group. Product should be used preventatively and not relied on for its curative potential
- Use product as part of an Integrated Crop Management strategy incorporating other methods of control, including where appropriate other fungicides with a different mode of action. Do not apply more than two foliar applications of QoI containing products to any cereal crop
- There is a significant risk of widespread resistance occurring in *Septoria tritici* populations in UK. Failure to follow resistance management action may result in reduced levels of disease control
- Strains of barley powdery mildew resistant to QoIs are common in the UK
- In wheat product must always be used in mixture with another product, recommended for control of the same target disease, that contains a fungicide from a different cross resistance group and is applied at a dose that will give robust control

Restrictions
- Maximum number of treatments 2 per crop
- Do not use where there is risk of spray drift onto neighbouring apple crops
- The same spray equipment should not be used to treat apples

Crop-specific information
- Latest use: before early milk stage (GS 73)
- HI 5 wk

Environmental safety
- Dangerous for the environment
- Very toxic to aquatic organisms

FOR FULL CONDITIONS OF USE ALWAYS READ THE PRODUCT LABEL

Hazard classification and safety precautions
    Hazard H03, H11
    Risk phrases R20, R38, R43, R50, R53a, R63
    Operator protection A, H; U02a, U05a, U09a, U14, U15, U19a, U20b
    Environmental protection E15a, E38
    Storage and disposal D01, D02, D05, D09a, D10c, D12a
    Medical advice M03

19   Bacillus thuringiensis

A bacterial insecticide for control of caterpillars

Products

Dipel DF	Fargro	32000 IU/mg	WG	11184

Uses
- Caterpillars in **amenity vegetation**, **apples** *(off-label)*, **broccoli**, **brussels sprouts**, **cabbages**, **calabrese** *(off-label)*, **cauliflowers**, **celery** *(off-label)*, **cherries** *(off-label)*, **chinese cabbage** *(off-label)*, **chives** *(off-label)*, **collards** *(off-label)*, **herbs (see appendix 6)** *(off-label)*, **kale** *(off-label)*, **lettuce** *(off-label)*, **ornamental specimens**, **parsley** *(off-label)*, **pears** *(off-label)*, **protected broccoli** *(off-label)*, **protected brussels sprouts** *(off-label)*, **protected cabbages** *(off-label)*, **protected calabrese** *(off-label)*, **protected cauliflowers** *(off-label)*, **protected celery** *(off-label)*, **protected chinese cabbage** *(off-label)*, **protected chives** *(off-label)*, **protected cucumbers**, **protected herbs (see appendix 6)** *(off-label)*, **protected kale** *(off-label)*, **protected lettuce** *(off-label)*, **protected ornamentals**, **protected parsley** *(off-label)*, **protected peppers**, **protected tomatoes**, **protected watercress** *(off-label)*, **raspberries**, **strawberries**, **watercress** *(off-label)*
- Winter moth in **bilberries** *(off-label)*, **blackcurrants** *(off-label)*, **blueberries** *(off-label)*, **cranberries** *(off-label)*, **gooseberries** *(off-label)*, **redcurrants** *(off-label)*, **vaccinium spp.** *(off-label)*, **whitecurrants** *(off-label)*

Specific Off-Label Approvals (SOLAs)
- **apples, cherries, chives, herbs (see appendix 6), lettuce, parsley, pears, protected chives, protected herbs (see appendix 6), protected lettuce, protected parsley** *(OLA 022577) Dec 2008* [1]
- **bilberries, blackcurrants, blueberries, cranberries, gooseberries, redcurrants, vaccinium spp., whitecurrants** *(OLA 040358) Dec 2008* [1]
- **calabrese, celery, chinese cabbage, collards, kale, protected broccoli, protected brussels sprouts, protected cabbages, protected calabrese, protected cauliflowers, protected celery, protected chinese cabbage, protected kale** *(OLA 040739) Dec 2008* [1]
- **protected watercress, watercress** *(OLA 031071) Dec 2008* [1]

Efficacy guidance
- Pest control achieved by ingestion by caterpillars of the treated plant vegetation. Caterpillars cease feeding and die in 1-3 d
- Apply as soon as larvae appear on crop and repeat every 7-10 d until the end of the hatching period
- Good coverage is essential, especially of undersides of leaves. Spray onto dry foliage and do not apply if rain expected within 6 h

Restrictions
- No restriction on number of treatments on edible crops
- Apply spray mixture as soon as possible after preparation

Crop-specific information
- HI zero

Environmental safety
- Store out of direct sunlight

Hazard classification and safety precautions
    Operator protection A, C; U05a, U15, U20c
    Environmental protection E15a
    Storage and disposal D01, D02, D05, D09a, D11a

SEE SECTION 3 FOR PRODUCTS ALSO REGISTERED

SECTION 2

20   benalaxyl

A phenylamide (acylalanine) fungicide available only in mixtures

21   benalaxyl + mancozeb

A systemic and protectant fungicide mixture

Products

1	Galben M	Sipcam	8:65% w/w	WP	07220
2	Intro Plus	Interfarm	8:65% w/w	WP	11630
3	Tairel	Sipcam	8:65% w/w	WB	07767

Uses
- Blight in *potatoes, protected potatoes*

Approval information
- Mancozeb included in Annex I under EC Directive 91/414
- Approved for aerial application on potatoes [1, 3]. See notes in Section 5

Efficacy guidance
- Apply to potatoes at blight warning prior to crop becoming infected and repeat at 10-21 d intervals depending on risk of infection
- Spray irrigated potatoes after irrigation and at 14 d intervals, crops in polythene tunnels at 10 d intervals
- When active potato growth ceases use a non-systemic fungicide to end of season, starting not more than 10 d after last application
- For reduction of downy mildew in oilseed rape apply at seedling to 7-leaf stage, before end Nov as soon as infection seen

Restrictions
- Maximum number of treatments 5 per crop (including other phenylamide-based fungicides) for potatoes
- Do not treat potatoes showing active blight infection

Crop-specific information
- HI 7 d

Environmental safety
- Dangerous for the environment
- Very toxic to aquatic organisms

Hazard classification and safety precautions
  Hazard H04, H11
  Risk phrases R37, R43, R50, R53a
  Operator protection A; U05a, U14, U15, U19a
  Environmental protection E15a
  Storage and disposal D01, D02, D12b
  Medical advice M04

22   benfuracarb

A soil-applied carbamate insecticide and nematicide for beet crops

Products

Oncol 10G	Nufarm UK	10% w/w	GR	12139

Uses
- Docking disorder vectors in *fodder beet, mangels, sugar beet*
- Leaf miner in *fodder beet, mangels, sugar beet*
- Millipedes in *fodder beet, mangels, sugar beet*
- Pygmy beetle in *fodder beet, mangels, sugar beet*
- Springtails in *fodder beet, mangels, sugar beet*
- Symphylids in *fodder beet, mangels, sugar beet*

FOR FULL CONDITIONS OF USE ALWAYS READ THE PRODUCT LABEL

Efficacy guidance
- Apply at sowing with suitable applicator so that the granules are mixed with the moving soil closing the seed furrow. See label for recommended applicators and calibration

Restrictions
- Maximum number of treatments 1 per crop

Crop-specific information
- Latest use: at planting

Environmental safety
- Dangerous for the environment
- Toxic to aquatic organisms
- Do not empty into drains

Hazard classification and safety precautions
Hazard H03, H11
Risk phrases R20, R51, R53a
Operator protection A, B, C, D, E, H, K, M; U02a, U05a, U19a, U20a
Environmental protection E15a, E19b
Storage and disposal D01, D02, D09a, D11a, D12a
Medical advice M02, M05a

23 bentazone

A post-emergence contact diazinone herbicide

Products

1	Agriguard Bentazone 480	AgriGuard	480 g/l	SL	11569
2	Agrotech Bentazone	Me2	480 g/l	SL	12143
3	Basagran SG	BASF	87% w/w	SG	08360
4	Landgold Bentazone SL	Landgold	480 g/l	SL	10841
5	Landgold Bentazone SL	Teliton	480 g/l	SC	12111
6	Standon Bentazone S	Standon	87% w/w	SG	10124

Uses
- Annual dicotyledons in *broad beans, french beans, linseed, peas, potatoes, runner beans, spring field beans, winter field beans* [1-6]; *bulb onions (off-label), evening primrose (off-label), leeks (off-label), salad onions (off-label)* [3]; *narcissi* [1-3, 6]; *navy beans* [1, 3-6]
- Common storksbill in *leeks (off-label)* [3]
- Fool's parsley in *leeks (off-label)* [3]

Specific Off-Label Approvals (SOLAs)
- *bulb onions, salad onions (OLA 971234) Aug 2006* [3]
- *evening primrose (OLA 971232) Aug 2006* [3]
- *leeks (OLA 021551) Aug 2006* [3]

Approval information
- Bentazone included in Annex I under EC Directive 91/414
- Approval expiry 1 Aug 2006 [1, 4-6]

Efficacy guidance
- Most effective control obtained when weeds are growing actively and less than 5 cm high or across. Good spray cover is essential
- Various recommendations are made for use in spray programmes with other herbicides or in tank mixes. See label for details
- The addition of specified adjuvant oils is recommended for use on some crops to improve fat hen control. Do not use under hot or humid conditions. See label for details
- Split dose application may be made in all recommended crops except peas and generally gives better weed control. See label for details

Restrictions
- Maximum number of treatments varies with crop and product - see labels
- Crops must be treated at correct stage of growth to avoid danger of scorch. See labels for details
- Not all varieties of recommended crops are fully tolerant. Use only on tolerant varieties named in labels. Do not use on forage or mange-tout varieties of peas

SEE SECTION 3 FOR PRODUCTS ALSO REGISTERED

- Do not use on crops which have been affected by drought, waterlogging, frost or other stress conditions
- Do not apply insecticides within 7 d of treatment
- Leave 7 d after using a post-emergence grass herbicide or 14 d where treatment precedes the grass herbicide and carry out a leaf wax test where relevant
- Do not spray at temperatures above 21°C. Delay spraying until evening if necessary
- Do not apply if rain or frost expected, if unseasonably cold, if foliage wet or in drought
- A minimum of 6 h (preferably 12 h) free from rain is required after application
- May be used on selected varieties of maincrop and second early potatoes (see label for details), not on seed crops or first earlies

Crop-specific information
- Latest use: before shoots exceed 15 cm high for potatoes and spring field beans (or 6-7 leaf pairs), 4 leaf pairs (6 pairs or 15 cm high with split dose) for broad beans, before flower buds visible for French, navy, runner, soya and winter field beans and linseed, before flower buds can be found enclosed in terminal shoot for peas
- HI onions 3 wk
- Best results in narcissi obtained by using a suitable pre-emergence herbicide first
- Consult processor before using on crops for processing
- A satisfactory wax test must be carried out before use on peas
- Do not treat narcissi during flower bud initiation

Environmental safety
- Dangerous for the environment [1, 2, 4, 5]
- Harmful to aquatic organisms [3, 6]

Hazard classification and safety precautions
Hazard H03 [1-6]; H11 [1, 2, 4, 5]
Risk phrases R22a, R43 [1-6]; R36 [1, 2, 4, 5]; R41, R52, R53a [3, 6]
Operator protection A [1-6]; C [1-3, 6]; U05a, U08, U19a, U20b [1-6]; U11, U14 [3, 6]
Environmental protection E15a [1-6]; E34 [1, 2, 4, 5]
Storage and disposal D01, D02, D09a, D10b [1-6]; D05, D07 [1, 2, 4, 5]
Medical advice M05a [3, 6]

---

24   bentazone + MCPA + MCPB

A post-emergence herbicide for undersown spring cereals and grass

Products

Acumen	BASF	200:80:200 g/l	SL	00028

Uses
- Annual dicotyledons in *newly sown grass, undersown spring cereals*

Approval information
- Bentazone, MCPA and MCPB included in Annex I under EC Directive 91/414

Efficacy guidance
- Best results when weeds small and actively growing provided crop is at correct growth stage
- A minimum of 6 h free from rain is required after treatment

Restrictions
- Maximum number of treatments 1 per crop on undersown cereals; 1 per yr on grassland
- Do not treat red clover after 3-trifoliate leaf stage
- Do not use on crops suffering from herbicide damage or physical stress or in temperatures above 21°C
- Do not use on seed crops or on cereals undersown with lucerne or seed mixtures containing lucerne
- Do not roll or harrow for 7 d before or after spraying
- Clovers may be scorched and undersown crop checked but effects likely to be outgrown. Clover stand in grassland may be reduced
- Do not apply if frost expected, if crop wet or when temperatures at or above 21°C

Crop-specific information
- Latest use: before 1st node detectable for undersown cereals (GS 31), 2 wk before grazing for grassland
- Apply to cereals from 2-fully expanded leaf stage but before first node detectable (GS 12-30) provided clover has reached 1-trifoliate leaf stage
- On first year grass leys use on seedling stage weeds before end of Sep
- Apply to newly sown leys after grass has reached 2-leaf stage provided clovers have at least 1 trifoliate leaf and red clover has not passed 3-trifoliate leaf stage

Environmental safety
- Dangerous for the environment
- Toxic to aquatic organisms
- Keep livestock out of treated areas for at least 2 wk and until foliage of poisonous weeds such as ragwort has died and become unpalatable

Hazard classification and safety precautions
  Hazard H03, H11
  Risk phrases R22a, R36, R38, R43, R51, R53a
  Operator protection A, C; U05a, U08, U14, U19a, U20b
  Environmental protection E07a, E15a, E34, E38
  Storage and disposal D01, D02, D08, D09a, D10c, D12a
  Medical advice M03, M05a

25   bentazone + MCPB

A post-emergence herbicide mixture for use in peas

Products

Pulsar	BASF	200:200 g/l	SL	04002

Uses
- Annual dicotyledons in *peas*

Approval information
- Bentazone and MCPB included in Annex I under EC Directive 91/414

Efficacy guidance
- To be used in tank mix with cyanazine
- Best results achieved when weeds small and actively growing provided crop at correct stage. Good spray cover is essential
- May be applied as single treatment or as split dose treatment applying first spray when susceptible weeds are not beyond the 2 leaf stage
- Do not apply if rain or frost expected or if foliage wet. A minimum of 24 h free from rain required after treatment
- Do not apply during drought or unseasonably cold weather

Restrictions
- Maximum number of treatments 1 per crop or 2 per crop with split dose
- Do not treat forage pea cultivars or mange-tout peas or other varieties specifically listed in the label
- Apply after a satisfactory wax test. Early drilled crops or crops affected by frost or abrasion may not have sufficiently waxy cuticle
- Allow 7 d after spraying before using a grass weed herbicide, or wait 14 d afterwards and test leaf wax before treating
- Do not use as tank mix with any other product than cyanazine, nor after use of TCA
- Do not apply insecticides within 7 d of treatment
- Do not apply to any crop that may have been subjected to stress conditions, where foliage damaged or under hot, sunny conditions when temperature exceeds 21°C
- Do not spray if foliage is wet or if frost or rain expected within 24 h
- Consult processors before use on crops for processing

SEE SECTION 3 FOR PRODUCTS ALSO REGISTERED

SECTION 2

Crop-specific information
- Latest use: before enclosed bud stage of crop (GS 10x)
- Apply only to listed pea cultivars (see label) from 3 fully expanded leaf (for full dose) or from 2 node stage (for split dose) to before flower buds can be found enclosed in terminal shoot (GS 102 or 103 to before GS 201)

Environmental safety
- Dangerous for the environment
- Toxic to aquatic organisms
- Keep livestock out of treated areas for 14 d and until foliage of any poisonous weeds such as ragwort has died and become unpalatable

Hazard classification and safety precautions
Hazard H04, H11
Risk phrases R36, R38, R43, R51, R53a
Operator protection A, C; U05a, U08, U14, U19a, U20a
Environmental protection E07a, E15a, E34, E38
Storage and disposal D01, D02, D09a, D10c, D12a
Medical advice M05a

26   bentazone + pendimethalin

A contact and residual herbicide mixture for combining peas

Products
Impuls	BASF	480:400 g/l	KL	11780

Uses
- Annual dicotyledons in **combining peas**
- Chickweed in **combining peas**
- Cleavers in **combining peas**
- Knotgrass in **combining peas**
- Mayweeds in **combining peas**

Approval information
- Bentazone and pendimethalin included in Annex I under EC Directive 91/414

Efficacy guidance
- Best results achieved by treating weeds when they are small and ensuring maximum spray coverage
- Moist soil required for maximum efficacy. Loose or cloddy seedbeds must be consolidated before spraying
- Residual control reduced on soils with more than 6% OM
- Minimum 6 hr free from rain required for optimum foliar activity. Do not spray when foliage wet

Restrictions
- Maximum total dose equivalent to one full dose treatment
- Severe crop damage may occur if mixed with any other product. Leave minimum 7 d before, or 14 d after, using a post-emergence grass herbicide. Do not apply insecticides within 7 d of use. Do not apply if TCA has been used
- Do not use on soils prone to waterlogging
- If day time temperatures likely to exceed 21°C spray in the evening
- Do not treat crops under stress from any cause
- Equipment must be washed out thoroughly after use. Traces of product may damage susceptible crops sprayed later

Crop-specific information
- Latest use: before 3rd node stage (GS 103)
- Crystal violet leaf wax test must be carried out before use. Transient leaf margin scorch may occur after treatment
- Crop damage may occur if heavy rain follows application on stony or gravelly soils

Following crops guidance
- Land must be ploughed to 150 mm before any succeeding crop except cereals
- In the event of crop failure spring barley may be sown after ploughing to 150 mm

FOR FULL CONDITIONS OF USE ALWAYS READ THE PRODUCT LABEL

Environmental safety
- Dangerous for the environment
- Very toxic to aquatic organisms

Hazard classification and safety precautions
 Hazard H03, H11
 Risk phrases R22a, R43, R50, R53a
 Operator protection A; U05a, U08, U13, U19a, U20b
 Environmental protection E15a, E34, E38
 Storage and disposal D01, D02, D08, D09a, D10c, D12a
 Medical advice M03, M05a

27   beta-cyfluthrin

A non-systemic pyrethroid insecticide available only in mixtures

28   beta-cyfluthrin + clothianidin

An insecticidal seed treatment mixture for beet

Products

Poncho Beta	Bayer CropScience	53:400 g/l	FS	12076

Uses
- Beet virus yellows vectors in **fodder beet** *(seed treatment)*, **sugar beet** *(seed treatment)*

Approval information
- Beta-cyfluthrin included in Annex I under EC Directive 91/414

Efficacy guidance
- In addition to control of aphid virus vectors, product improves crop establishment by reducing damage caused by symphylids, springtails, millipedes, wireworms and leatherjackets
- Additional control measures should be taken where very high populations of soil pests are present
- Product reduces direct feeding damage by foliar pests such as pygmy mangold beetle, beet flea beetle, mangold fly
- Product is not active against nematodes
- Treatment does not alter physical characteristics of pelleted seed and no change to standard drill settings should be necessary

Restrictions
- Maximum number of treatments: 1 per seed batch
- Product must be co-applied with a colouring dye
- Product must only be applied as a coating to pelleted seed using special treatment machinery

Crop-specific information
- Latest use: pre-drilling

Environmental safety
- Extremely dangerous to fish or other aquatic life. Do not contaminate surface waters or ditches with chemical or used container

Hazard classification and safety precautions
 Hazard H03
 Risk phrases R22a
 Operator protection A, H; U04a, U05a, U07, U13, U14, U20b, U24
 Environmental protection E13a, E34, E36
 Storage and disposal D01, D02, D05, D09a, D14
 Treated seed S02, S03, S04a, S04b, S05, S06a, S06b, S07
 Medical advice M03

SECTION 2

SEE SECTION 3 FOR PRODUCTS ALSO REGISTERED

29   beta-cyfluthrin + imidacloprid

An insecticide mixture for seed treatment of oilseed rape

Products

1 Chinook Blue	Bayer CropScience	100:100 g/l	LS	11262
2 Chinook Colourless	Bayer CropScience	100:100 g/l	LS	11206

Uses
- Cabbage stem flea beetle in **spring oilseed rape** *(seed treatment)* [1]; **winter oilseed rape** *(seed treatment)* [1, 2]
- Flea beetle in **spring oilseed rape** *(seed treatment)*, **winter oilseed rape** *(seed treatment)* [1]

Approval information
- Beta-cyfluthrin included in Annex I under EC Directive 91/414

Efficacy guidance
- Drill treated seed as soon as possible after treatment
- Product will reduce damage by early attacks of flea beetle and may also reduce subsequent larval damage. However a follow-up spray treatment may be required if pest activity is heavy and prolonged

Restrictions
- Maximum number of treatments 1 per batch of seed
- Do not use on seed with more than 9% moisture content, on sprouted seed or on cracked, split or otherwise damaged seed
- Product must be co-applied with a colouring agent [2]
- Must be applied using seed treatment machine recommended by manufacturer
- Allow treated seed to dry before packaging. Drying can be assisted by subsequent application of talc
- Product can cause a transient tingling or numbing sensation to exposed skin. Avoid skin contact with the product, treated seed or dust throughout all operations in the treatment plant and during drilling

Crop-specific information
- Latest use: before drilling

Environmental safety
- Dangerous for the environment
- Toxic to aquatic organisms
- Extremely dangerous to fish or other aquatic life. Do not contaminate surface waters or ditches with chemical or used container
- Do not use treated seed as food or feed
- Dangerous to birds, game and other wildlife. Treated seed should not be left on the soil surface. Bury spillages
- Do not broadcast treated seed

Hazard classification and safety precautions
Hazard H03, H11
Risk phrases R22a, R51, R53a
Operator protection A, H; U04a, U05a, U07, U20b
Environmental protection E13a, E34, E36, E38
Storage and disposal D01, D02, D05, D09a, D12a, D14
Treated seed S01, S02, S03, S04c, S05, S06a, S06b, S07, S08
Medical advice M03

30   bifenox

A diphenyl ether herbicide for cereals

Products

Fox	Makhteshim	480 g/l	SC	11981

Uses
- Cleavers in **winter barley**, **winter rye**, **winter wheat**

FOR FULL CONDITIONS OF USE ALWAYS READ THE PRODUCT LABEL

- Field pansy in *winter barley, winter rye, winter wheat*
- Forget-me-not in *winter barley, winter rye, winter wheat*
- Geranium species in *spring oilseed rape* (off-label), *winter oilseed rape* (off-label)
- Poppies in *winter barley, winter rye, winter wheat*
- Red dead-nettle in *winter barley, winter rye, winter wheat*
- Speedwells in *winter barley, winter rye, winter wheat*

Specific Off-Label Approvals (SOLAs)
- *spring oilseed rape, winter oilseed rape* (OLA 051155) Dec 2008 [1]

Approval information
- Accepted by BBPA for use on malting barley

Efficacy guidance
- Bifenox is absorbed by foliage and emerging roots of susceptible species
- Best results obtained when weeds are growing actively with adequate soil moisture

Restrictions
- Maximum number of treatments: one per crop
- Do not apply to crops suffering from stress from whatever cause
- Do not apply if the crop is wet or if rain or frost is expected
- Avoid drift onto broad-leaved plants outside the target area

Crop-specific information
- Latest use: before 2nd node detectable (GS 32) for all crops

Environmental safety
- Dangerous for the environment
- Very toxic to aquatic organisms
- Do not empty into drains

Hazard classification and safety precautions
Hazard H11
Risk phrases R50, R53a
Operator protection A; U05a, U08, U20b
Environmental protection E15a, E19b, E34, E38
Storage and disposal D01, D02, D09a, D12a
Medical advice M03

31   bifenox + isoproturon

A contact and residual herbicide mixture for cereals

Products
| Banco | Makhteshim | 160:400 g/l | SC | 11191 |

Uses
- Annual dicotyledons in *winter barley, winter wheat*
- Annual grasses in *winter barley, winter wheat*
- Blackgrass in *winter barley, winter wheat*
- Field pansy in *winter barley, winter wheat*
- Wild oats in *winter barley, winter wheat*

Approval information
- Accepted by BBPA for use on malting barley

Efficacy guidance
- Best results obtained from treatment when crop and weeds are growing actively in moist soil. Weeds are controlled by leaf and root uptake
- Weed control may be reduced on loose or cloddy seedbeds and where trash or straw is not dispersed and buried before treatment
- Rolling or harrowing after treatment will reduce control
- Grass weeds should not be treated before 1 Jan
- Heavy infestations of ryegrass (e.g. after a seed crop) may not be satisfactorily controlled
- Always follow WRAG guidelines for preventing and managing herbicide resistant weeds. See Section 5 for more information

SEE SECTION 3 FOR PRODUCTS ALSO REGISTERED

Restrictions
- Maximum number of treatments 1 per crop for wheat and barley
- Do not treat durum wheat, oats, triticale or rye
- Do not treat undersown crops, or those to be undersown
- Do not spray crops under stress from water logging or drought, or suffering from pest or disease attack
- Do not roll or harrow during the 7 d prior to post-emergence treatment
- Do not spray if frost imminent or after prolonged frosty weather

Crop-specific information
- Latest use: before 2nd node detectable (GS 32) for barley and wheat
- Early sown winter cereals may be damaged if application takes place during rapid growth in the autumn

Environmental safety
- Dangerous for the environment
- Very toxic to aquatic organisms
- Do not empty into drains
- Do not apply to dry, cracked or waterlogged soils where heavy rain may lead to contamination of drains by isoproturon

Hazard classification and safety precautions
Hazard H03, H11
Risk phrases R40, R50, R53a
Operator protection A, C, H; U05a, U08, U13, U19a, U20b
Environmental protection E15a, E19b
Storage and disposal D01, D02, D05, D09a, D10b, D12a
Medical advice M05a

32   bifenthrin

A contact and residual pyrethroid acaricide/insecticide for use in agricultural and horticultural crops

Products

1 Starion	Belchim	100 g/l	EC	11641
2 Starion Flo	Belchim	80 g/l	SC	12455
3 Talstar	Certis	100 g/l	EC	11325
4 Talstar 80 Flo	Belchim	80 g/l	SC	12352

Uses
- Aphids in *broccoli, brussels sprouts, cabbages, calabrese, cauliflowers* [1-4]; *combining peas, spring barley, spring oats, spring wheat, vining peas* [2]; *winter barley, winter oats, winter wheat* [1, 2]
- Cabbage seed weevil in *spring oilseed rape, winter oilseed rape* [2]
- Cabbage stem flea beetle in *winter oilseed rape* [1, 2]
- Caterpillars in *broccoli, brussels sprouts, cabbages, calabrese, cauliflowers* [1-4]
- Clay-coloured weevil in *blackberries* (off-label), *protected blackberries* (off-label), *protected raspberries* (off-label), *raspberries* (off-label) [3]
- Damson-hop aphid in *hops* [2-4]
- Flea beetle in *linseed, spring oilseed rape, winter oilseed rape* [2]
- Fruit tree red spider mite in *apples, pears* [2-4]
- Pod midge in *spring oilseed rape, winter oilseed rape* [2]
- Pollen beetle in *spring oilseed rape, winter oilseed rape* [2]
- Rape winter stem weevil in *winter oilseed rape* [1, 2]
- Two-spotted spider mite in *blackberries* (off-label), *ornamental specimens, protected blackberries* (off-label), *protected raspberries* (off-label), *raspberries* (off-label) [3]; *hops, strawberries* [2-4]; *ornamental plant production* [2, 4]
- Virus vectors in *winter barley, winter oats, winter wheat* [1, 2]
- Whitefly in *broccoli, brussels sprouts, cabbages, calabrese, cauliflowers* [1-4]

Specific Off-Label Approvals (SOLAs)
- *blackberries, protected blackberries, protected raspberries, raspberries* (OLA 050573) Dec 2008 [3]

FOR FULL CONDITIONS OF USE ALWAYS READ THE PRODUCT LABEL

Approval information
- Accepted by BBPA for use on malting barley and hops

Efficacy guidance
- Timing of application varies with crop and pest. See label for details
- Good spray cover of upper and lower plant surfaces essential to achieve effective pest control. Spray volumes should be increased where crop is dense

Restrictions
- Maximum number of treatments 5 per yr for hops; 2 per yr for apples, pears, strawberries. Other crops either 2 per yr or a maximum total dose is stipulated. See label
- Consult processor before use on crops for processing
- Do not tank mix with triazole fungicides

Crop-specific information
- Latest use: before 31 Mar in yr of harvest for cereals; before 30 Nov in yr of sowing for winter oilseed rape
- HI 2 d for brassicas, raspberries; zero for all other crops
- Do not apply to a cereal crop if any product containing a pyrethroid or dimethoate has been applied after the start of ear emergence (GS 51)

Environmental safety
- Dangerous for the environment
- Very toxic to aquatic organisms
- High risk to bees. Do not apply to crops in flower, or to those in which bees are actively foraging, except as directed on named crops. Do not apply when flowering weeds are present [2, 4]
- Extremely dangerous to bees. Do not apply to crops in flower or to those in which bees are actively foraging. Do not apply when flowering weeds are present [1, 3]
- Extremely dangerous to fish or other aquatic life. Do not contaminate surface waters or ditches with chemical or used container [2, 4]
- High risk to non-target insects or other arthropods [2, 4]
- Do not spray cereal crops within 6 m of the field boundary after 31 Mar in the yr of harvest [2]
- Keep in original container, tightly closed, in a safe place, under lock and key
- LERAP Category A and broadcast air-assisted LERAP (30 m [2, 4]) (50 m [1, 3])

Hazard classification and safety precautions
Hazard H03, H11 [1-4]; H08 [1, 3]
Risk phrases R20, R22a, R50, R53a [1-4]; R22b, R36, R37 [1, 3]; R43 [2, 4]
Operator protection A, H [1-4]; C [1, 3]; U02a, U04a, U05a, U08, U19a, U20b [1-4]; U11 [1, 3]
Environmental protection E12b [2, 4] (named crops); E12c [1, 3]; E13a, E15b [2, 4]; E16c, E16d, E34, E38 [1-4]; E17b [1] (50 m); E17b [2] (30 m); E17b [3] (50 m); E17b [4] (30 m); E22a [2, 4] (after 31 Mar)
Storage and disposal D01, D02, D05, D09b, D10b [1-4]; D12a [2, 4]
Medical advice M03, M05b

33   bifenthrin + malathion

An insecticide mixture for pest control in stored grain and grain stores

Products

1	Prostore 157 UL	Nickerson	7.5:150 g/l	UL	12017
2	Prostore 420 EC	Nickerson	20:400 g/l	EC	12036

Uses
- Grain storage pests in *grain stores* [2]; *stored grain* [1, 2]

Approval information
- Accepted by BBPA for use in stores for malting barley

Efficacy guidance
- Best results in stored grain obtained from treatment at the start of storage after high-temperature drying. Cool warm grain as soon as possible to prevent moisture migration and ensure optimum protection from reinfestation
- Grain stores should be cleaned before treatment and treated 3-4 wk before filling. Treat all surfaces, especially any cracks and crevices [2]

SEE SECTION 3 FOR PRODUCTS ALSO REGISTERED

SECTION 2

- Controls a wide range of stored grain insect pests and cosmopolitan food mites
- Under normal storage conditions a single treatment will give 3-6 mth protection
- Knock-down is complete within 24 h at normal harvest grain teperatures
- Control may not be satisfactory if pest strains resistant to bifenthrin or malathion are present

Restrictions

- Maximum number of treatments 1 per store [2] or batch of grain
- Do not treat wheat (including durum), oats, rye or triticale intended for use as seed
- Must only be applied with appropriate equipment designed to treat grain
- Do not apply with hand-held equipment

Environmental safety

- Dangerous for the environment
- Very toxic to aquatic organisms

Hazard classification and safety precautions

Hazard H03, H11
Risk phrases R22a, R38, R41, R43 [2]; R22b, R50, R53a, R66, R67 [1, 2]
Operator protection A, C, H [1, 2]; D, J, M [2]; U05a, U23a [1, 2]; U14 [2]
Environmental protection E15a, E34, E38
Storage and disposal D01, D02, D11a
Medical advice M03, M05b

34   bitertanol

A conazole fungicide available only in mixtures

35   bitertanol + fuberidazole

A broad spectrum fungicide mixture for seed treatment in cereals

Products

Sibutol	Bayer CropScience	375:23 g/l	FS	11305

Uses

- Bunt in **spring wheat** (seed treatment), **winter wheat** (seed treatment)
- Fusarium root rot in **spring oats** (seed treatment), **spring rye** (seed treatment), **spring wheat** (seed treatment), **triticale** (seed treatment), **winter oats** (seed treatment), **winter rye** (seed treatment), **winter wheat** (seed treatment)
- Loose smut in **spring wheat** (seed treatment - reduction), **winter wheat** (seed treatment - reduction)
- Septoria seedling blight in **spring wheat** (seed treatment - reduction), **winter wheat** (seed treatment - reduction)

Efficacy guidance

- Treated seed should preferably be drilled in the same season
- Control of loose smut may be inadequate for use on seed for multiplication

Restrictions

- Maximum number of treatments 1 per batch of seed
- Do not use on seed with more than 16% moisture content, or on sprouted, cracked or skinned seed
- Must be applied simultaneously with water in the ratio 1 part product to 2 parts water in a recommended seed treatment machine

Crop-specific information

- Before drilling

Environmental safety

- Dangerous for the environment
- Toxic to aquatic organisms
- Do not use treated seed as food or feed
- Treated seed harmful to game and wildlife

FOR FULL CONDITIONS OF USE ALWAYS READ THE PRODUCT LABEL

Hazard classification and safety precautions
> **Hazard** H11
> **Risk phrases** R51, R53a
> **Operator protection** A, H; U07, U20a
> **Environmental protection** E03, E15a, E34, E36, E38
> **Storage and disposal** D05, D09a, D12a, D14
> **Treated seed** S01, S02, S03, S04a, S05, S06a, S07

36 bitertanol + fuberidazole + imidacloprid

A broad spectrum fungicide and insecticide seed treatment for winter wheat and winter oats

Products

Sibutol Secur	Bayer CropScience	140:8.6:87.5 g/l	LS	11308

Uses
- Bunt in **winter wheat** *(seed treatment)*
- Fusarium root rot in **winter oats** *(seed treatment)*, **winter wheat** *(seed treatment)*
- Loose smut in **winter wheat** *(seed treatment - reduction)*
- Seedling blight and foot rot in **winter wheat** *(seed treatment - reduction)*
- Virus vectors in **winter oats** *(seed treatment)*, **winter wheat** *(seed treatment)*
- Wireworm in **winter oats** *(seed treatment - reduction of damage)*, **winter wheat** *(seed treatment - reduction of damage)*

Efficacy guidance
- Best applied through recommended seed treatment machines
- Evenness of seed cover improved by simultaneous application of equal volumes of product and water or dilution of product with an equal volume of water
- Drill treated seed in the same season. Use minimum 125 kg treated seed per ha. Lower drilling rates and/or early drilling affect duration of BYDV protection needed and may require follow-up aphicide treatment
- In high risk areas where aphid activity is heavy and prolonged a follow-up aphicide treatment may be required
- Protection against foliar air-borne and splash-borne diseases later in the season will require appropriate fungicide follow-up sprays

Restrictions
- Maximum number of treatments 1 per batch of seed
- Do not use on seed with more than 16% moisture content, or on sprouted, cracked or skinned seed

Crop-specific information
- Latest use: before drilling
- Slightly delayed and reduced emergence may occur but this is normally outgrown
- Field emergence which is delayed for any reason may be accentuated by treatment

Environmental safety
- Dangerous for the environment
- Toxic to aquatic organisms
- Do not use treated seed as food or feed
- Dangerous to birds, game and other wildlife. Treated seed should not be left on the soil surface. Bury spillages
- Treated seed should not be broadcast, but drilled to a depth of 4 cm in a well prepared seedbed
- If seed is left on the soil surface the field should be harrowed and rolled to ensure good incorporation

Hazard classification and safety precautions
> **Hazard** H11
> **Risk phrases** R51, R53a
> **Operator protection** A, H; U04a, U05a, U13, U20b
> **Environmental protection** E03, E15a, E34, E38
> **Storage and disposal** D01, D02, D05, D09a, D12a
> **Treated seed** S01, S02, S03, S04c, S05, S06a, S06b, S07
> **Medical advice** M03

SEE SECTION 3 FOR PRODUCTS ALSO REGISTERED

37 Bordeaux mixture

A protectant copper sulphate/lime complex fungicide

Products

Wetcol 3	Ford Smith	30 g/l (copper)	SC	02360

Uses
- Bacterial canker in *cherries*
- Blight in *outdoor tomatoes, potatoes*
- Cane spot in *loganberries, raspberries*
- Canker in *apples, pears*
- Celery leaf spot in *celery*
- Currant leaf spot in *blackcurrants*
- Downy mildew in *hops*
- Leaf curl in *apricots, nectarines, peaches*
- Scab in *apples, pears*
- Spur blight in *raspberries*

Efficacy guidance
- Spray interval normally 7-14 d but varies with disease and crop. See label for details
- Commence spraying potatoes before crop meets in row or immediately first blight period occurs
- For canker control spray monthly from Aug to Oct
- For peach leaf curl control spray at leaf fall in autumn and again in Feb
- Spray when crop foliage dry. Do not spray if rain imminent

Restrictions
- Do not use on copper sensitive cultivars, including Doyenne du Comice pears

Environmental safety
- Harmful to fish or other aquatic life. Do not contaminate surface waters or ditches with chemical or used container
- Harmful to livestock. Keep all livestock out of treated areas for at least 3 wk

Hazard classification and safety precautions
Hazard H03, H04
Risk phrases R22a, R36, R37, R38
Operator protection U13, U15, U20a
Environmental protection E06c (3 wk); E13c
Storage and disposal D09a, D11a

38 boscalid

A translocated and translaminar anilide fungicide

Products

Filan	BASF	50% w/w	WG	11449

Uses
- Alternaria in *poppies* (off-label - for morphine production)
- Botrytis in *poppies* (off-label - for morphine production)
- Sclerotinia stem rot in *poppies* (off-label - for morphine production), *spring oilseed rape, winter oilseed rape*

Specific Off-Label Approvals (SOLAs)
- *poppies* (for morphine production) (OLA 040163) Dec 2008 [1]

Efficacy guidance
- For best results apply in high disease risk situations at flowering stage of crop but before symptoms are visible
- Ensure adequate spray penetration and good coverage

Restrictions
- Maximum total dose equivalent to two full dose treatments
- Avoid spray drift onto neighbouring crops

FOR FULL CONDITIONS OF USE ALWAYS READ THE PRODUCT LABEL

Crop-specific information
- Latest use: up to and including full flower

Environmental safety
- Dangerous for the environment
- Toxic to aquatic organisms
- Product represents minimal hazard to bees when used as directed. However local bee-keepers should be notified if crops are to be sprayed when in flower

Hazard classification and safety precautions
    Hazard H11
    Risk phrases R51, R53a
    Operator protection U05a, U20c
    Environmental protection E15a, E38
    Storage and disposal D01, D02, D05, D09a, D10c, D12a

39   boscalid + epoxiconazole

A broad spectrum fungicide mixture for cereals

Products

1	Splice	BASF	233:67 g/l	SC	12315
2	Tracker	BASF	233:67 g/l	SC	12295
3	Venture	BASF	233:67 g/l	SC	12316

Uses
- Brown rust in *spring barley*, *spring wheat*, *winter barley*, *winter wheat*
- Eyespot in *spring barley* (moderate), *spring wheat* (moderate), *winter barley* (moderate), *winter wheat* (moderate)
- Net blotch in *spring barley*, *winter barley*
- Rhynchosporium in *spring barley* (moderate), *winter barley* (moderate)
- Septoria in *spring wheat*, *winter wheat*
- Tan spot in *spring wheat* (reduction), *winter wheat* (reduction)

Approval information
- Accepted by BBPA for use on malting barley (before ear emergence only)

Efficacy guidance
- Apply at the start of foliar or stem based disease attack
- Optimum effect against eyespot achieved by spraying between leaf-sheath erect and second node detectable stages (GS 30-32)

Restrictions
- Maximum number of treatments one per crop

Crop-specific information
- Latest use: flag leaf stage

Environmental safety
- Dangerous for the environment
- Toxic to aquatic organisms
- LERAP Category B
- Avoid drift onto neighbouring crops. May cause damage to broad-leaved plant species

Hazard classification and safety precautions
    Hazard H03, H11
    Risk phrases R36, R40, R51, R53a, R62, R63
    Operator protection A, C, H; U05a, U20b
    Environmental protection E15a, E16a, E38
    Storage and disposal D08, D10c, D12a

SEE SECTION 3 FOR PRODUCTS ALSO REGISTERED

40   boscalid + pyraclostrobin

A protectant and systemic fungicide mixture

Products

1	Bellis	BASF	25.2:12.8% w/w	WG	12522
2	Signum	BASF	26.7:6.7% w/w	WG	11450

Uses

- Alternaria in *carrots* *(moderate)* [2]
- Botrytis in *protected chinese cabbage* *(off-label)*, *protected choi sum* *(off-label)*, *protected pak choi* *(off-label)* [2]
- Chocolate spot in *spring field beans* *(moderate)*, *winter field beans* *(moderate)* [2]
- Dark leaf spot in *cabbages*, *cauliflowers* [2]
- Grey mould in *strawberries* [2]
- Powdery mildew in *apples* [1]; *carrots* [2]
- Rhizoctonia rot in *lettuce*, *protected lettuce* [2]
- Ring spot in *brussels sprouts*, *cabbages*, *cauliflowers*, *protected chinese cabbage* *(off-label)*, *protected choi sum* *(off-label)*, *protected pak choi* *(off-label)* [2]
- Rust in *spring field beans*, *winter field beans* [2]
- Scab in *apples*, *pears* [1]
- Sclerotinia in *carrots* *(moderate)*, *horseradish* *(off-label)*, *lettuce*, *parsnips* *(off-label)*, *protected lettuce* [2]
- White blister in *brussels sprouts*, *cabbages* *(qualified minor use)* [2]

Specific Off-Label Approvals (SOLAs)

- *horseradish*, *parsnips* *(OLA 051317) Nov 2006* [2]
- *protected chinese cabbage*, *protected choi sum*, *protected pak choi* *(OLA 051349) Nov 2006* [2]

Approval information

- Pyraclostrobin included in Annex I under EC Directive 91/414
- Accepted by BBPA for use on malting barley (before ear emergence only)
- Approval expiry 25 Nov 2006 [2]

Efficacy guidance

- On brassicas apply as a protectant spray or at the first sign of disease and repeat at 3-4 wk intervals depending on disease pressure [2]
- Ensure adequate spray penetration and coverage by increasing water volume in dense crops [2]
- For best results on strawberries apply as a protectant spray at the white bud stage. Applications should be made in sequence with other products as part of a fungicide spray programme during flowering at 7-10 day intervals [2]
- On carrots and field beans apply as a protectant spray or at the first sign of disease with a repeat treatment if needed, as directed on the label [2]
- On lettuce apply as a protectant spray 1-2 wk after planting [2]
- Optimum results on apples and pears obtained from a protectant treatment from bud burst [1]
- Application as the final 2 sprays on apples and pears gives reduction in storage rots [1]
- Pyraclostrobin is a member of the QoI cross-resistance group of fungicides and should be used in programmes with fungicides with a different mode of action

Restrictions

- Maximum total dose equivalent to three full dose treatments on brassicas; two full dose treatments on all other field crops [2]
- Maximum number of treatments (including other QoI treatments) on apples and pears 4 per yr if total number of applications is 12 or more, or 3 per yr if the total number is fewer than 12 [1]
- Do not use more than 2 consecutive treatments on apples and pears, and these must be separated by a minimum of 2 applications of a fungicide with a different mode of action [1]
- Use a maximum of three applications per yr on brassicas and no more than two per yr on other field crops [2]
- Do not use consecutive treatments; apply in alternation with fungicides from a different cross resistance group and effective against the target diseases [2]
- Do not apply more than 6kg product to the same area of land per yr [2]

FOR FULL CONDITIONS OF USE ALWAYS READ THE PRODUCT LABEL

- Consult processor before use on crops for processing
- Applications to lettuce may only be made between 1 Apr and 31 Oct [2]

Crop-specific information
- HI 21 d for field beans, horseradish, parsnips; 14 d for brassica crops, carrots, herbs, lettuce; 7d for apples, pears; 3 d for strawberries

Environmental safety
- Dangerous for the environment
- Very toxic to aquatic organisms
- LERAP Category B [2]
- Broadcast air-assisted LERAP (40 m) [1]

Hazard classification and safety precautions
    Hazard H03, H11
    Risk phrases R22a, R50, R53a
    Operator protection A; U05a, U20c [1, 2]; U13 [1]
    Environmental protection E15a, E38 [1, 2]; E16a [2]; E17b [1] (40 m); E34 [1]
    Storage and disposal D01, D02, D10c, D12a [1, 2]; D05, D09a [2]; D08 [1]
    Medical advice M03 [1]; M05a [1, 2]

41   brodifacoum

An anticoagulant coumarin rodenticide

Products

Sorexa Checkatube	Sorex	0.2% w/w	XX	H6702

Uses
- Mice in **farm buildings** *(indoor use only)*

Efficacy guidance
- Product is a self-contained control device which requires no bait handling
- Best results obtained by placing tubes where mice are active and will readily enter the device
- To acquire a lethal dose mice must pass through the tube a number of times
- Where a continuous source of infestation is present, or where sustained protection is needed, tubes can be installed permanently and replaced at monthly intervals
- For complete eradication other control methods should also be used bearing in mind the resistance status of the target population

Restrictions
- For use only by professional pest contractors
- Do not use outdoors. Products must be used in situations where baits are placed within a building or other enclosed structure, and the target is living or feeding predominantly within that building or structure
- Do not attempt to open or recharge the tubes

Environmental safety
- Prevent access to baits by children, birds and other animals
- Do not place tubes where food, feed or water could become contaminated
- Remove all remains of tubes after use and burn or bury
- Search for and burn or bury all rodent bodies. Do not place in refuse bins or on open rubbish tips
- Keep in original container, tightly closed, in a safe place, under lock and key

Hazard classification and safety precautions
    Operator protection U13, U20b
    Storage and disposal D02, D09b
    Vertebrate/rodent control products V01b, V02, V03a, V04a
    Medical advice M03

SEE SECTION 3 FOR PRODUCTS ALSO REGISTERED

42   bromadiolone

An anti-coagulant coumarin-derivative rodenticide

Products

1 Endorats Premium Rat Killer	Irish Drugs	0.005% w/w	GB	H6725
2 Sakarat Bromabait	Killgerm	0.005% w/w	RB	H7902
3 Tomcat 2	Antec Biosentry	0.005% w/w	PT	H6736
4 Tomcat 2 Blox	Antec Biosentry	0.005% w/w	BB	H6731

Uses
- Mice in **farm buildings** [2-4]; **farmyards** [2]
- Rats in **farm buildings** [1-4]; **farmyards** [2]

Efficacy guidance
- Ready-to-use baits are formulated on a mould-resistant, whole-wheat base
- Use in baiting programme. Place baits in protected situations, sufficient for continuous feeding between treatments
- Chemical is effective against warfarin- and coumatetralyl-resistant rats and mice and does not induce bait shyness
- Use bait bags where loose baiting inconvenient (eg behind ricks, silage clamps etc)
- The resiatance status of the rodent population should be assessed when considering the choice of product to use

Restrictions
- For use only by professional operators

Environmental safety
- Access to baits by children, birds and animals, particularly cats, dogs, pigs and poultry, must be prevented
- Baits must not be placed where food, feed or water could become contaminated
- Remains of bait and bait containers must be removed after treatment and burned or buried
- Rodent bodies must be searched for and burned or buried. They must not be placed in refuse bins or on rubbish tips
- Take extreme care to prevent domestic animals having access to the bait

Hazard classification and safety precautions
> **Operator protection** U13 [1-4]; U20b [2-4]
> **Environmental protection** E15a [3, 4]
> **Storage and disposal** D05, D07 [2]; D09a [1-4]; D11a [2-4]
> **Treated seed** S06a [1]
> **Vertebrate/rodent control products** V01a, V03a, V04a [1, 3, 4]; V01b, V03b, V04b [2]; V02 [1-4]
> **Medical advice** M03 [2-4]

43   bromoxynil

A contact acting HBN herbicide

Products

1 Alpha Bromolin 225 EC	Makhteshim	225 g/l	EC	08255
2 Alpha Bromotril P	Makhteshim	240 g/l	SC	07099
3 Barclay Mutiny	Barclay	250 g/l	SC	11379
4 Bravado	DAPT	225 g/l	LI	10487
5 Emblem	Nufarm UK	20% w/w	WB	11559
6 Flagon 400 EC	Makhteshim	400 g/l	EC	08875
7 Greencrop Tassle	Greencrop	240 g/l	SC	09659

Uses
- Annual dicotyledons in **forage maize** [3, 5, 7]; **linseed**, **spring barley**, **spring oats**, **spring wheat**, **winter barley**, **winter oats**, **winter wheat** [1, 4, 6]; **maize**, **sweetcorn** *(off-label)* [2]

Specific Off-Label Approvals (SOLAs)
- **sweetcorn** *(OLA 972230) Dec 2008* [2]

SECTION 2

Approval information
- Bromoxynil was reviewed in 1995 and approvals for home garden use, and most hand held applications revoked
- Accepted by BBPA for use on malting barley
- Approval expiry 31 Aug 2006 [4]

Efficacy guidance
- Spray when main weed flush has germinated and the largest are at the 4 leaf stage
- Weed control can be enhanced by using a split treatment spraying each application when the weeds are seedling to 2 true leaves. Apply the second treatment before the crop canopy covers the ground [2, 3]

Restrictions
- Maximum number of treatments 1 per crop or yr or maximum total dose equivalent to one full dose treatment
- Do not apply with oils or other adjuvants
- Do not apply using hand-held equipment or at concentrations higher than those recommended
- Do not apply during frosty weather, drought, when soil is waterlogged, when rain expected within 4 h or to crops under any stress
- Take particular care to avoid drift onto neighbouring susceptible crops or open water surfaces

Crop-specific information
- Latest use: before 10 fully expanded leaf stage of crop for maize, sweetcorn; before crop 20 cm tall and before flower buds visible for linseed; before 2nd node detectable (GS 32) for cereals
- Apply to spring sown maize from 2 fully expanded leaves up to 9 fully expanded leaves [2, 3]
- Apply to spring sown linseed from 1 fully expanded leaf to 20 cm tall [1, 4]
- Foliar scorch, which rapidly disappears without affecting growth, will occur if treatment made in hot weather or during rapid growth

Environmental safety
- Dangerous for the environment
- Very toxic to aquatic organisms
- Keep livestock out of treated areas for at least 6 wk after treatment
- High risk to bees. Do not apply to crops in flower or to those in which bees are actively foraging
- Dangerous to fish or other aquatic life. Do not contaminate surface waters or ditches with chemical or used container [2, 4]
- LERAP Category B [5]

Hazard classification and safety precautions
Hazard H03 [1-7]; H04 [4]; H08 [1, 4, 6]; H11 [1, 3, 5-7]
Risk phrases R20 [1, 6]; R22a [1-7]; R22b [1, 4, 6]; R36 [1-4, 6, 7]; R38 [6]; R43 [4, 6]; R50 [1, 5, 6]; R51 [7]; R52 [3]; R53a [1, 3, 5-7]; R63 [1, 2, 5-7]
Operator protection A, C [1-7]; H [1, 4, 6]; U02a [1-4, 7]; U05a, U19a [1-7]; U08, U13, U20a [1-4, 6, 7]; U10 [6]; U11 [2, 6]; U14, U15 [1, 4-6]; U22a [5]; U23a [1-3, 5-7]
Environmental protection E06a [1, 4, 6] (6 wk); E12a, E16a, E38 [5]; E12f, E34 [1-4, 6, 7]; E13b [2, 4]; E15a [1, 3, 5-7]
Consumer protection C02 [1, 4, 6] (6 wk)
Storage and disposal D01, D02, D09a [1-7]; D05 [1, 3, 4, 6, 7]; D10b [1-4, 6, 7]; D12a [5]; D12b [1, 2, 6]
Medical advice M03 [2-4, 7]; M04 [1, 2, 6]; M05b [6]

44   bromoxynil + diflufenican + ioxynil

A selective contact and translocated herbicide for cereals

Products

Capture	Bayer CropScience	300:50:200 g/l	SC	09982

Uses
- Annual dicotyledons in **spring barley, spring wheat, triticale, winter barley, winter rye, winter wheat**
- Chickweed in **spring barley, spring wheat, triticale, winter barley, winter rye, winter wheat**
- Knotgrass in **spring barley, spring wheat, triticale, winter barley, winter rye, winter wheat**

SEE SECTION 3 FOR PRODUCTS ALSO REGISTERED

- Mayweeds in **spring barley**, **spring wheat**, **triticale**, **winter barley**, **winter rye**, **winter wheat**
- Speedwells in **spring barley**, **spring wheat**, **triticale**, **winter barley**, **winter rye**, **winter wheat**

Approval information
- Accepted by BBPA for use on malting barley

Efficacy guidance
- Best results obtained on small weeds and when competition removed early
- Good spray coverage essential for good activity

Restrictions
- Maximum number of treatments 1 per crop
- Do not treat crops undersown or to be undersown
- Do not treat frosted crops or those that are under stress from any cause
- Do not apply by hand-held equipment or at concentrations higher than those recommended

Crop-specific information
- Latest use: before 2nd node detectable (GS 32)

Following crops guidance
- In the event of crop failure winter wheat may be drilled immediately after normal cultivations. Winter barley may be re-drilled after ploughing
- Land must be ploughed and an interval of 12 wk elapse after treatment before planting spring crops of wheat, barley, oilseed rape, peas, field beans, sugar beet, potatoes, carrots, edible brassicas or onions
- Successive treatments of any products containing diflufenican can lead to soil build-up and inversion ploughing to a depth of 15 cm must precede sowing any following non-cereal crop. Even where ploughing occurs some crops may be damaged

Environmental safety
- Dangerous for the environment
- Toxic to aquatic organisms
- Keep livestock out of treated areas for at least 2 wk after treatment
- Harmful to bees. Do not apply to crops in flower or to those in which bees are actively foraging
- LERAP Category B

Hazard classification and safety precautions
Hazard H03, H11
Risk phrases R22a, R51, R53a, R63
Operator protection A, C, H; U05a, U08, U13, U19a, U20b
Environmental protection E06a (2 wk); E12f, E15a, E16a, E34, E38
Consumer protection C02 (2 wk)
Storage and disposal D01, D02, D05, D09a, D10b, D12a
Medical advice M03

45   bromoxynil + ethofumesate + ioxynil

A post-emergence herbicide for new grass leys

Products
Leyclene	Interfarm	50:200:25 g/l	EC	07263

Uses
- Annual dicotyledons in **grass seed crops**, **seedling leys**
- Annual meadow grass in **grass seed crops**, **seedling leys**

Approval information
- Ethofumesate included in Annex I under EC Directive 91/414

Efficacy guidance
- Best results when weeds small and growing actively in a vigorous crop, soil moist and further rain within 10 d. Mid Oct to end Dec normally suitable
- Annual meadow grass controlled during early crop establishment. Spray weed grasses before fully tillered
- Ash or trash should be burned, buried or removed before spraying

Restrictions
- Maximum number of treatments 2 per yr

FOR FULL CONDITIONS OF USE ALWAYS READ THE PRODUCT LABEL

- Do not apply by hand-held equipment or at concentrations higher than those recommended
- Do not use on crops under stress, during periods of very dry weather, prolonged frost or waterlogging
- Do not use where clovers or other legumes are valued components of ley
- Do not roll for 7 d before or after spraying
- Do not spray in cold conditions or when heavy rain or frost imminent
- Do not use on soils with more than 10% organic matter
- Do not cut for 14 d after spraying or graze in Jan-Feb after spraying in Oct-Dec

Crop-specific information
- Latest use: 6 wk before cutting or grazing
- Apply to healthy ryegrasses or tall fescue after 2-3 leaf stage, to cocksfoot, timothy and meadow fescue at least 60 d after emergence and after 2-3 leaf stage

Following crops guidance
- Any crop may be sown 5 mth after application following mould board ploughing to at least 15 cm

Environmental safety
- Dangerous for the environment
- Toxic to aquatic organisms
- Keep livestock out of treated areas for at least 6 wk after treatment
- Harmful to bees. Do not apply to crops in flower or to those in which bees are actively foraging

Hazard classification and safety precautions
Hazard H04, H08, H11
Risk phrases R20, R21, R22a, R38, R51, R53a, R63
Operator protection A, C; U05a, U08, U13, U19a, U20b
Environmental protection E06a (6 wk); E12f, E15a, E34, E38
Consumer protection C02 (6 wk)
Storage and disposal D01, D02, D09a, D10b, D12a
Medical advice M03

46   bromoxynil + ioxynil

A contact acting post-emergence HBN herbicide for cereals

Products

1	Alpha Briotril 24/16	Makhteshim	240:160 g/l	EC	04876
2	Alpha Briotril Plus 19/19	Makhteshim	190:190 g/l	EC	04740
3	Biotite 380	DAPT	190:190 g/l	EC	10647
4	Mextrol Biox	Nufarm UK	200:200 g/l	EC	09470
5	Oxytril CM	Bayer CropScience	200:200 g/l	EC	10005

Uses
- Annual dicotyledons in *spring barley, spring oats, spring wheat, winter barley, winter oats, winter wheat* [1-5]; *spring rye, undersown spring cereals, undersown winter cereals* [4, 5]; *triticale, winter rye* [1, 2, 4, 5]; *undersown barley, undersown oats, undersown rye, undersown triticale, undersown wheat* [1, 2]

Approval information
- Accepted by BBPA for use on malting barley
- Approval expiry 31 Aug 2006 [3]

Efficacy guidance
- Best results achieved on young weeds growing actively in a highly competitive crop
- Do not apply during periods of drought or when rain imminent (some labels say 'if likely within 4 or 6 h')
- Recommended for tank mixture with hormone herbicides to extend weed spectrum. See labels for details

Restrictions
- Maximum number of treatments 1 per crop
- Do not spray crops stressed by drought, waterlogging or other factors
- Do not roll or harrow for several days before or after spraying. Number of days specified varies with product. See label for details
- Do not apply by hand-held equipment or at concentrations higher than those recommended

SEE SECTION 3 FOR PRODUCTS ALSO REGISTERED

Crop-specific information
- Latest use: before 2nd node detectable stage (GS 31)
- HI (animal consumption) 6 wk
- Apply to winter or spring cereals from 1-2 fully expanded leaf stage, but before second node detectable (GS 32)
- Spray oats in spring when danger of frost past. Do not spray winter oats in autumn
- Apply to undersown cereals pre-sowing or pre-emergence of legume provided cover crop is at correct stage. Only spray trefoil pre-sowing
- On crops undersown with grasses alone or on direct re-seeds apply from 2-leaf stage of grass

Environmental safety
- Dangerous for the environment [1, 2, 4, 5]
- Very toxic to aquatic organisms
- Dangerous to fish or other aquatic life. Do not contaminate surface waters or ditches with chemical or used container [3, 5]
- Keep livestock out of treated areas for at least 6 wk after treatment
- Harmful to bees. Do not apply to crops in flower or to those in which bees are actively foraging. Do not apply when flowering weeds are present [1-5]
- LERAP Category B [1-3]

Hazard classification and safety precautions
  Hazard H03 [1-5]; H04 [3]; H08 [1-3]; H11 [1, 2, 4, 5]
  Risk phrases R20, R38 [1, 2]; R21 [1]; R22a, R22b [1-5]; R36 [1-3]; R43, R66, R67 [4, 5]; R50 [1, 4, 5]; R51 [2]; R53a, R63 [1, 2, 4, 5]
  Operator protection A, C; U02a, U14, U15, U20a [1-3]; U05a, U08, U13, U19a [1-5]; U10, U11 [1, 2]; U20b [4, 5]; U23a [1, 2, 4, 5]
  Environmental protection E06a [1, 2] (6 wk); E07a, E38 [4, 5]; E07b [3] (6 wk); E12f, E34 [1-5]; E13b [3, 5]; E15a [1, 2, 4]; E16a [1-3]; E16b [3]
  Consumer protection C02 [1-3] (6 wk); C02 [4, 5] (14 d)
  Storage and disposal D01, D02, D09a [1-5]; D05, D10b [1-3]; D11a, D12a [4, 5]; D12b [1, 2]
  Medical advice M03, M05b [1-5]; M04 [1, 2]

---

47   bromoxynil + ioxynil + mecoprop-P

A post-emergence contact and translocated herbicide

Products

1 Astalavista	Nufarm UK	56:56:224	EW	12335
2 Swipe P	Nufarm UK	56:56:224 g/l	EW	11955

Uses
- Annual dicotyledons in *amenity grass, managed amenity turf* [1]; *durum wheat, ryegrass, spring barley, spring oats, spring wheat, triticale, winter barley, winter oats, winter wheat* [2]

Approval information
- Mecoprop-P included in Annex I under EC Directive 91/414
- Accepted by BBPA for use on malting barley

Efficacy guidance
- Best results when weeds small and growing actively in a strongly competitive crop
- Do not spray in rain or when rain imminent. Control may be reduced by rain within 6 h
- Application to wet crops or weeds may reduce control. Rainfall in the 6 hour period following spraying may reduce weed control
- To achieve optimum control of large over-wintered weeds or in advanced crops increase water volume to aid spray penetration and cover
- Recommended for tank-mixing with approved MCPA-amine for hemp nettle control

Restrictions
- Maximum number of treatments 1 per crop. The total amount of mecoprop-P applied in a single yr must not exceed the maximum total dose approved for any single product for the crop/situation
- Do not treat durum wheat or winter oats in autumn
- Do not spray crops undersown with legumes or use on winter oats or durum wheat in autumn
- Winter and spring cereal crops should not be sprayed in the spring until the risk of frost is over

---

FOR FULL CONDITIONS OF USE ALWAYS READ THE PRODUCT LABEL

- Do not treat fine grasses such as those found on bowling or putting greens [1]
- Do not use mowings from the first cut of recently treated turf for mulching and allow compost made from subsequent cuts to rot completely before use. Susceptible crops such as brassicas should not be mulched with such compost at any time [1]
- Do not spray crops under stress from frost, waterlogging, drought or other causes
- Do not roll within 5 d after spraying
- Do not mix with manganese sulphate
- Do not apply by hand-held equipment or at concentrations higher than those recommended

Crop-specific information
- Latest use: before first node detectable (GS31) for spring oats; before second node detectable (GS32) for durum wheat, spring barley, spring wheat, triticale, winter barley, winter oats, winter wheat
- HI (animal consumption) 6 wk
- Spray cereals from 3 leaves unfolded (GS 13) to before second node detectable (GS 32) for winter sown cereals, spring wheat and spring barley and before first node detectable (GS 31) for spring oats
- Apply to direct sown ryegrass or cereals undersown with ryegrass from 2-3 leaf stage of grass
- Yield of barley may be reduced if frost occurs within 3-4 wk of treatment of low vigour crops on light soils or subject to stress
- Some cereal crop yellowing may follow treatment but yield not normally affected
- Amenity grass and turf must be well established (tillered) before treatment

Environmental safety
- Dangerous for the environment
- Toxic to aquatic organisms
- Keep livestock out of treated areas for at least two weeks following treatment and until poisonous weeds, such as ragwort, have died down and become unpalatable
- Harmful to bees. Do not apply to crops in flower or to those in which bees are actively foraging. Do not apply when flowering weeds are present
- LERAP Category B
- Do not harvest crops for animal consumption for at least 6 wk after last application
- Avoid drift onto neighbouring crops, especially beans, beet, brassicas (including oilseed rape), carrots, lettuce, legumes, tomatoes and other horticultural crops, which are all very susceptible

Hazard classification and safety precautions
Hazard H03, H11
Risk phrases R21, R22a, R51, R53a, R63
Operator protection A, C, H, M; U05a, U08, U13, U14, U15, U19a, U20a, U23a
Environmental protection E07a, E12f, E15a, E16a, E34, E38
Consumer protection C02 (6 wk)
Storage and disposal D01, D02, D05, D09a, D10b, D10c, D12a
Medical advice M03

48 bromoxynil + prosulfuron

A contact and residual herbicide mixture for maize

Products

Jester	Syngenta	60:3 % w/w	WG	08681

Uses
- Annual dicotyledons in *maize*
- Black bindweed in *maize*
- Chickweed in *maize*
- Hemp-nettle in *maize*
- Knotgrass in *maize*
- Mayweeds in *maize*

Approval information
- Prosulfuron included in Annex I under EC Directive 91/414

SEE SECTION 3 FOR PRODUCTS ALSO REGISTERED

SECTION 2

Efficacy guidance
- Prosulfuron is a member of the ALS-inhibitor group of herbicides and products should be used in a planned Resistance Management strategy. See Section 5 for more information

Restrictions
- Maximum total dose equivalent to one full dose treatment
- Do not treat crops grown for seed production
- Consult before use on crops intended for processing
- Do not use in frosty weather or on crops under stress
- Do not tank mix with organophosphate insecticides or apply in tank mix or sequence with any other sulfonylurea
- Do not apply by knapsack sprayer or in volumes less than those recommended
- Take care to wash out sprayers thoroughly. See label for details

Crop-specific information
- Latest use: before 5 crop leaves unfolded for maize
- Apply post-emergence up to when the crop has four unfolded leaves
- Product must be used with Agral
- Apply in cool conditions, or during the evening, to avoid scorch

Following crops guidance
- Winter or spring cereals, winter or spring beans, spring sown peas or oilseed rape may be sown following normal harvest of a treated crop. Mould-board ploughing to 20 cm is recommended in some cases

Environmental safety
- Dangerous for the environment
- Toxic to aquatic organisms
- LERAP Category B
- Take special care to avoid drift outside the target area

Hazard classification and safety precautions
Hazard H02, H11
Risk phrases R22a, R23, R36, R51, R53a, R63
Operator protection A, C; U05a, U11, U15, U20c, U23a
Environmental protection E15a, E16a, E38
Storage and disposal D01, D02, D09a, D10c, D12a

49 bupirimate

A systemic pyrimidine fungicide active against powdery mildew

Products

Nimrod	Makhteshim	250 g/l	EC	10563

Uses
- Powdery mildew in **apples**, **begonias**, **blackcurrants**, **chrysanthemums**, **courgettes** (outdoor only), **gooseberries**, **hops**, **marrows** (outdoor only), **pears**, **protected cucumbers**, **protected strawberries**, **protected tomatoes** (off-label), **pumpkins** (off-label), **raspberries** (outdoor only), **roses**, **squashes** (off-label), **strawberries**

Specific Off-Label Approvals (SOLAs)
- **protected tomatoes** (OLA 010910) Dec 2008 [1]
- **pumpkins**, **squashes** (OLA 010909) Dec 2008 [1]

Approval information
- Accepted by BBPA for use on hops

Efficacy guidance
- On apples during periods that favour disease development lower doses applied weekly give better results than higher rates fortnightly
- Not effective in protected crops against strains of mildew resistant to bupirimate

Restrictions
- Maximum number of treatments or maximum total dose depends on crop and variety (see label for details)

FOR FULL CONDITIONS OF USE ALWAYS READ THE PRODUCT LABEL

Crop-specific information
- HI depends on crop and variety (see label for details)
- Apply before or at first signs of disease and repeat at 7-14 d intervals. Timing and maximum dose vary with crop. See label for details
- With apples, hops and ornamentals cultivars may vary in sensitivity to spray. See label for details
- If necessary to spray cucurbits in winter or early spring spray a few plants 10-14 d before spraying whole crop to test for likelihood of leaf spotting problem
- On roses some leaf puckering may occur on young soft growth in early spring or under low light intensity. Avoid use of high rates or wetter on such growth
- Never spray flowering begonias (or buds showing colour) as this can scorch petals
- Do not mix with other chemicals for application to begonias, cucumbers or gerberas

Environmental safety
- Dangerous for the environment
- Toxic to aquatic organisms
- Product has negligible effect on *Phytoseiulus* and *Encarsia* and may be used in conjunction with biological control of red spider mite

Hazard classification and safety precautions
Hazard H03, H08, H11
Risk phrases R20, R22b, R38, R42, R51, R53a
Operator protection A, C; U05a, U20b
Environmental protection E15a
Storage and disposal D01, D02, D09a, D10c
Medical advice M05b

---

50   buprofezin

A moulting inhibitor, thiadiazine insecticide for whitefly control

Products
Applaud	Certis	250 g/l	SC	11532

Uses
- Glasshouse whitefly in **aubergines**, **protected cucumbers**, **protected ornamentals**, **protected peppers**, **protected tomatoes**
- Tobacco whitefly in **aubergines**, **protected cucumbers**, **protected ornamentals**, **protected peppers**, **protected tomatoes**

Efficacy guidance
- Product has contact, residual and some vapour activity
- Whitefly most susceptible at larval stages but residual effect can also kill nymphs emerging from treated eggs and application to pupae reduces emergence
- Adult whitefly not directly affected. Resistant strains of tobacco whitefly are known and where present control likely to be reduced or ineffective

Restrictions
- Maximum number of treatments 8 per crop for tomatoes and cucumbers; 4 per crop on protected ornamentals; 2 per crop for aubergines and peppers
- Do not apply more than 2 sprays within a 65 d period on tomatoes, or within a 45 d period on cucumbers
- Do not treat *Dieffenbachia* or *Closmoplictrum*
- Do not apply to crops under stress
- Do not leave spray liquid in sprayer for long periods
- Do not apply as fog or mist

Crop-specific information
- HI edible crops 3 d
- In IPM programme apply as single application and allow at least 60 d before re-applying
- In All Chemical programme apply twice at 7-14 d interval and allow at least 60 d before re-applying
- See label for list of ornamentals successfully treated but small scale test advised to check varietal tolerance. This is especially important if spraying flowering ornamentals with buds showing colour

SEE SECTION 3 FOR PRODUCTS ALSO REGISTERED

Environmental safety
- Product may be used either in IPM programme in association with *Encarsia formosa* or in All Chemical programme

Hazard classification and safety precautions
    Hazard H04
    Risk phrases R36, R53a
    Operator protection U05a, U14, U15, U20c
    Environmental protection E15a, E34
    Storage and disposal D01, D02, D09a, D11a

51   captan

A protectant dicarboximide fungicide with horticultural uses

Products

1	Alpha Captan 80 WDG	Makhteshim	80% w/w	WG	07096
2	Alpha Captan 83 WP	Makhteshim	83% w/w	WP	04806
3	PP Captan 80-WG	Tomen	80% w/w	WG	11006

Uses
- Black spot in **roses** [1-3]
- Botrytis in **strawberries** [1, 2]
- Gloeosporium in **apples** [3]
- Gloeosporium rot in **apples** [1, 2]
- Scab in **apples**, **pears** [1-3]

Restrictions
- Maximum number of treatments 12 per yr on apples and pears as pre-harvest sprays
- Product must not be used as dip or drench on apples or pears [3]
- Do not use on apple cultivars Bramley, Monarch, Winston, King Edward, Spartan, Kidd's Orange or Red Delicious or on pear cultivar D'Anjou
- Do not mix with alkaline materials or oils
- Do not use on fruit for processing
- Powered visor respirator with hood and neck cape must be used when handling concentrate

Crop-specific information
- HI apples, pears 14 d; strawberries 7 d
- For control of scab apply at bud burst and repeat at 10-14 d intervals until danger of scab infection ceased
- For suppression of fruit storage rots apply from late Jul and repeat at 2-3 wk intervals
- For black spot control in roses apply after pruning with 3 further applications at 14 d intervals or spray when spots appear and repeat at 7-10 d intervals
- For grey mould in strawberries spray at first open flower and repeat every 7-10 d [1]
- Do not leave diluted material for more than 2 h. Agitate well before and during spraying

Environmental safety
- Dangerous for the environment
- Very toxic to aquatic organisms
- Harmful to fish or other aquatic life. Do not contaminate surface waters or ditches with chemical or used container [3]

Hazard classification and safety precautions
    Hazard H02 [2]; H03 [1]; H04 [3]; H11 [1, 2]
    Risk phrases R23, R37, R38 [2]; R36, R43 [1-3]; R40, R50, R53a [1, 2]; R41 [1]
    Operator protection A [1-3]; C, K, M [3]; D, E, H [1, 3]; U05a [1-3]; U09a [1]; U11, U19a [1, 2]; U14, U15, U20b [2]; U20c [1, 3]
    Environmental protection E13c, E34 [3]; E15a [1, 2]
    Storage and disposal D01, D02, D09a [1-3]; D10b [3]; D11a, D12a [1, 2]

FOR FULL CONDITIONS OF USE ALWAYS READ THE PRODUCT LABEL

52   carbendazim

A systemic benzimidazole fungicide with curative and protectant activity

Products

1	AgriGuard Pro-Turf	AgriGuard	500 g/l	SC	11837
2	Delsene 50 Flo	Nufarm UK	500 g/l	SC	11452
3	Mascot Systemic	Rigby Taylor	500 g/l	SC	11335
4	Nuturf Carbendazim	Nufarm UK	500 g/l	SC	11469
5	Ringer	SumiAgro Amenity	500 g/l	SC	10692
6	Turfclear	Scotts	500 g/l	SC	07506

Uses

- Big vein in *lettuce* (off-label), *protected lettuce* (off-label) [2]
- Black root rot in *ornamental plant production* (off-label), *protected ornamentals* (off-label) [2]
- Botrytis in *bedding plants* (off-label), *bulbs/corms* (off-label), *chrysanthemums* (off-label), *pot plants* (off-label), *protected celery* (off-label) [2]
- Cane blight in *blackberries* (off-label), *loganberries* (off-label), *protected blackberries* (off-label), *protected loganberries* (off-label), *protected raspberries* (off-label), *protected rubus hybrids* (off-label), *raspberries* (off-label), *rubus hybrids* (off-label) [2]
- Didymella stem rot in *protected tomatoes* (off-label) [2]
- Dollar spot in *managed amenity turf* [1, 3-6]
- Eyespot in *spring barley*, *winter barley*, *winter wheat* [2]
- Fusarium in *bulbs/corms* (off-label), *freesias* (off-label) [2]
- Fusarium patch in *managed amenity turf* [1, 3-6]
- Light leaf spot in *spring oilseed rape*, *winter oilseed rape* [2]
- Penicillium rot in *bulbs/corms* (off-label) [2]
- Powdery mildew in *bedding plants* (off-label), *chrysanthemums* (off-label), *pot plants* (off-label) [2]
- Rhynchosporium in *spring barley*, *winter barley* [2]
- Root diseases in *inert substrate cucumbers* (off-label) [2]
- Scab in *pears* (off-label) [2]
- Sclerotinia in *bulbs/corms* (off-label) [2]
- Stagonospora in *bulbs/corms* (off-label) [2]
- Storage rots in *pears* (off-label) [2]
- Trichoderma in *mushroom compost* (off-label - spawn treatment) [2]
- Wormcast formation in *managed amenity turf* [1, 3-6]

Specific Off-Label Approvals (SOLAs)

- *bedding plants, bulbs/corms, chrysanthemums, freesias, ornamental plant production, pot plants, protected ornamentals* (OLA 041004) Dec 2008 [2]
- *blackberries, loganberries, protected blackberries, protected loganberries, protected raspberries, protected rubus hybrids, raspberries, rubus hybrids* (OLA 041009) Jul 2008 [2]
- *inert substrate cucumbers* (OLA 041005) Dec 2008 [2]
- *lettuce, protected lettuce* (OLA 041011) Dec 2008 [2]
- *mushroom compost* (spawn treatment) (OLA 041007) Dec 2008 [2]
- *pears* (OLA 041008) Dec 2008 [2]
- *protected celery* (OLA 041006) Dec 2008 [2]
- *protected tomatoes* (OLA 041010) Dec 2008 [2]

Approval information

- Following implementation of Directive 98/82/EC, approval for use of carbendazim on numerous crops was revoked in 1999. UK approvals for use of carbendazim on strawberries revoked in 2001 as a result of implementation of the MRL Directives
- Accepted by BBPA for use on malting barley

Efficacy guidance

- Products vary in the diseases listed as controlled for several crops. Labels must be consulted for full details and for rates and timings
- Mostly applied as spray or drench. Spray treatments normally applied at first sign of disease and repeated after 1 mth if required
- Apply as drench rather than spray where red spider mite predators are being used
- Do not apply during drought conditions. Rain or irrigation after treatment may improve control

SECTION 2

SEE SECTION 3 FOR PRODUCTS ALSO REGISTERED

- Leave 4 mth intervals where used only for worm cast control [1, 3-6]
- To delay appearance of resistant strains alternate treatment with non-MBC fungicide. Eyespot in cereals and *Botrytis cinerea* in many crops are now widely resistant

Restrictions
- Maximum number of treatments (including applications of any product containing benomyl, carbendazim or thiophanate-methyl) varies with crop treated and product used - see labels for details
- Do not treat crops or turf suffering from drought or other physical or chemical stress
- Consult processors before using on crops for processing
- Not compatible with alkaline products such as lime sulphur

Crop-specific information
- Latest use varies with crop and product used. See labels for details
- HI 2 d for inert substrate cucumbers, protected tomatoes; 7-14 d (depends on dose) for apples, pears; 14 d for mushrooms, protected celery; 21 d for oilseed rape, dwarf beans, lettuce, 6 mth for cane fruit
- On turf apply as preventative treatment in spring or autumn during periods of high disease risk
- Apply as a drench to control soil-borne diseases in cucumbers and tomatoes and as a pre-planting dip treatment for bulbs

Environmental safety
- Dangerous for the environment
- Very toxic to aquatic organisms
- Harmful to fish or other aquatic life. Do not contaminate surface waters or ditches with chemical or used container [2, 5]
- After use dipping suspension must not be discharged directly into ditches or drains

Hazard classification and safety precautions
Hazard H03 [1-6]; H11 [1-4, 6]
Risk phrases R22a [1]; R46, R60, R61 [2]; R50, R53a [2-4, 6]; R68 [3-6]
Operator protection A, C, H, M; U19a [3, 4, 6]; U20a [1]; U20b [2]; U20c [3-6]
Environmental protection E13c [2, 5]; E15a [1, 3, 4, 6]; E34 [5]; E38 [3, 4, 6]
Storage and disposal D01, D02, D11a [2]; D03, D12a [3, 4, 6]; D05, D09a [1-6]; D07 [1]; D08, D12b [5]; D10b [1, 3-6]

53   carbendazim + epoxiconazole

A systemic fungicide mixture for managed amenity turf

Products
Capricorn	Bayer Environ.	125:125 g/l	SC	10035

Uses
- Fusarium patch in **managed amenity turf**

Efficacy guidance
- Treat at the first sign of disease and repeat as necessary
- Apply after cutting and delay further mowing for at least 48 h to allow systemic distribution within the plant
- Only moderate control of Fusarium patch claimed

Restrictions
- Maximum number of treatments 2 per yr
- Do not apply during drought conditions or to frozen turf
- Use only on established turf

Environmental safety
- Harmful to fish or other aquatic life. Do not contaminate surface waters or ditches with chemical or used container
- LERAP Category B

Hazard classification and safety precautions
Hazard H04
Risk phrases R43
Operator protection A, C, H, M; U04a, U05a, U09b, U20c

FOR FULL CONDITIONS OF USE ALWAYS READ THE PRODUCT LABEL

Environmental protection E13c, E16a, E16b, E34
Storage and disposal D01, D02, D05, D09a, D10b

54   carbendazim + flusilazole

A broad-spectrum systemic and protectant fungicide for cereals, oilseed rape and sugar beet

Products

1 Contrast	DuPont	125:250 g/l	SC	06150
2 Harvesan	DuPont	125:250 g/l	SC	12238
3 Punch C	DuPont	125:250 g/l	SC	06801

Uses
- Brown rust in *spring barley*, *spring wheat*, *winter barley*, *winter wheat*
- Canker in *spring oilseed rape* (reduction), *winter oilseed rape* (reduction)
- Eyespot in *spring barley*, *spring wheat*, *winter barley*, *winter wheat*
- Light leaf spot in *spring oilseed rape*, *winter oilseed rape*
- Phoma leaf spot in *spring oilseed rape*, *winter oilseed rape*
- Powdery mildew in *spring barley*, *spring wheat*, *sugar beet*, *winter barley*, *winter wheat*
- Rust in *sugar beet*
- Septoria in *spring wheat*, *winter wheat*
- Yellow rust in *spring barley*, *spring wheat*, *winter barley*, *winter wheat*

Approval information
- Accepted by BBPA for use on malting barley

Efficacy guidance
- Apply at early stage of disease development or in routine preventive programme
- Most effective timing of treatment on cereals varies with disease. See label for details
- Higher rate active against both MBC-sensitive and MBC-resistant eyespot
- Rain occurring within 2 h of spraying may reduce effectiveness
- To prevent build-up of resistant strains of cereal mildew tank mix with approved morpholine fungicide

Restrictions
- Maximum number of treatments 1 per crop on sugar beet; 2 per crop on cereals and oilseed rape
- Do not apply to crops under stress or during frosty weather

Crop-specific information
- Latest use on cereals varies with crop and dose used - see label for details; before first flower opened stage for oilseed rape
- HI 7 wk for sugar beet
- Treat oilseed rape in autumn when leaf lesions first appear and spring from the start of stem extension when disease appears

Environmental safety
- Dangerous for the environment
- Very toxic to aquatic organisms
- Dangerous to fish or other aquatic life. Do not contaminate surface waters or ditches with chemical or used container

Hazard classification and safety precautions
Hazard H02, H11
Risk phrases R36, R40, R46, R50, R53a, R60, R61
Operator protection A, C [1-3]; H, M [1, 2]; U05a, U11, U19a, U20b
Environmental protection E13b, E15a, E34, E38
Storage and disposal D01, D02, D05, D09a, D10b, D12a
Medical advice M03, M04

55   carbendazim + iprodione

A systemic and contact fungicide mixture for managed amenity turf

Products

Vitesse	Bayer Environ.	87.5:175 g/l	SC	12037

SEE SECTION 3 FOR PRODUCTS ALSO REGISTERED

SECTION 2

Uses
- Anthracnose in *managed amenity turf*
- Fusarium patch in *managed amenity turf*
- Pink patch in *managed amenity turf*
- Red thread in *managed amenity turf*
- Timothy leaf spot in *managed amenity turf*

Approval information
- Iprodione included in Annex I under EC Directive 91/414

Efficacy guidance
- Maximum efficacy against Anthracnose in turf achieved by treatment at early disease development stage. Curative treatments for well-established Anthracnose are not recommended

Crop-specific information
- May be used on turf all year round, but is best suited for spring or late summer/early autumn application
- Where grass is being mown, apply after mowing. Delay further mowing for at least 48 h after treatment

Environmental safety
- Harmful to fish or other aquatic life. Do not contaminate surface waters or ditches with chemical or used container

Hazard classification and safety precautions
Operator protection A, C, H, M; U20c
Environmental protection E13c
Storage and disposal D05, D09a, D10b

56   carbetamide

A residual pre- and post-emergence carbamate herbicide for a range of field crops

Products

1 Carbetamex	Makhteshim	70% w/w	WP	11150
2 Crawler	Makhteshim	60% w/w	WG	11840

Uses
- Annual grasses in *cabbage seed crops, collards, fodder rape seed crops, kale seed crops, lucerne, red clover, sainfoin, spring cabbage, sugar beet seed crops, swede seed crops, turnip seed crops, white clover, winter field beans, winter oilseed rape*
- Some annual dicotyledons in *cabbage seed crops, collards, fodder rape seed crops, kale seed crops, lucerne, red clover, sainfoin, spring cabbage, sugar beet seed crops, swede seed crops, turnip seed crops, white clover, winter field beans, winter oilseed rape*
- Volunteer cereals in *cabbage seed crops, collards, fodder rape seed crops, kale seed crops, lucerne, red clover, sainfoin, spring cabbage, sugar beet seed crops, swede seed crops, turnip seed crops, white clover, winter field beans, winter oilseed rape*

Efficacy guidance
- Best results pre- or early post-emergence of weeds under cool, moist conditions. Adequate soil moisture is essential
- Dicotyledons controlled include chickweed, cleavers and speedwell
- Weed growth stops rapidly after treatment but full effects may take 6-8 wk to develop
- Various tank mixes are recommended to broaden the weed spectrum. See label for details
- Always follow WRAG guidelines for preventing and managing herbicide resistant weeds. Section 5 for more information

Restrictions
- Maximum number of treatments 1 per crop for all crops
- Do not treat any crop on waterlogged soil
- Do not use on soils with more than 10% organic matter as residual activity is impaired
- Do not apply during prolonged periods of cold weather when weeds are dormant

Crop-specific information
- HI 6 wk for all crops

FOR FULL CONDITIONS OF USE ALWAYS READ THE PRODUCT LABEL

- Apply to brassicas from mid-Oct to end-Feb provided crop has at least 4 true leaves (spring cabbage, spring greens), 3-4 true leaves (seed crops, oilseed rape)
- Apply to established lucerne or sainfoin from Nov to end-Feb
- Apply to established red or white clover from Feb to mid-Mar

Following crops guidance
- Succeeding crops may be sown 2 wk after treatment for brassicas, field beans, 8 wk after treatment for peas, runner beans, 16 wk after treatment for cereals, maize
- Ploughing is not necessary before sowing subsequent crops

Environmental safety
- Do not graze crops for at least 6 wk after treatment

Hazard classification and safety precautions
    Operator protection U19a [1]; U20c [1, 2]
    Environmental protection E15a
    Storage and disposal D09a, D11a

57   carbon dioxide (commodity substance)

A gas for the control of trapped rodents and other vertebrates

Products

carbon dioxide	various	99.9%	GA

Uses
- Birds in *traps*
- Mice in *traps*
- Rats in *traps*

Approval information
- Approval for the use of carbon dioxide as a commodity substance was granted on 8 October 1993 by Ministers under regulation 5 of the Control of Pesticides Regulations 1986
- Only to be used where a licence has been issued in accordance with Section 16(1) of the Wildlife and Countryside Act 1981

Efficacy guidance
- Use to destroy trapped rodent pests
- Use to control birds covered by general licences issued by the Agriculture and Environment Departments under Section 16(1) of the Wildlife and Countryside Act (1981) for the control of opportunistic bird species, where birds have been trapped or stupefied with alphachloralose/seconal

Restrictions
- Operators must wear self-contained breathing apparatus when carbon dioxide levels are greater than 0.5% v/v
- Operators must be suitably trained and competent

Environmental safety
- Unprotected persons and non-target animals must be excluded from the treatment enclosures and surrounding areas unless the carbon dioxide levels are below 0.5% v/v

Hazard classification and safety precautions
    Operator protection G

58   carbosulfan

A systemic carbamate insecticide for control of soil and stem pests

Products

1	Marshal Soil Insecticide suSCon CR granules	Fargro	10% w/w	GR	06978
2	Marshal Soil Insecticide suSCon CR Sachets	Fargro	10% w/w	WB	11754
3	Posse 10G	Belchim	10% w/w	GR	11640

SEE SECTION 3 FOR PRODUCTS ALSO REGISTERED

Uses

- Aphids in *broccoli, brussels sprouts, cabbages, calabrese, carrots, cauliflowers, collards, fodder beet, kale, mangels, parsnips, sugar beet* [3]
- Cabbage root fly in *broccoli, brussels sprouts, cabbages, calabrese, cauliflowers, collards, kale* [3]
- Cabbage stem weevil in *broccoli, brussels sprouts, cabbages, calabrese, cauliflowers, collards, kale* [3]
- Flea beetle in *broccoli, brussels sprouts, cabbages, calabrese, cauliflowers, collards, fodder beet, kale, mangels, sugar beet* [3]
- Free-living nematodes in *carrots, fodder beet, mangels, parsnips, sugar beet* [3]
- Large pine weevil in *forest* [1, 2]; *forest nurseries (off-label)* [1]
- Mangold fly in *fodder beet, mangels, sugar beet* [3]
- Millipedes in *fodder beet, mangels, sugar beet* [3]
- Pygmy mangold beetle in *fodder beet, mangels, sugar beet* [3]
- Springtails in *fodder beet, mangels, sugar beet* [3]
- Symphylids in *fodder beet, mangels, sugar beet* [3]
- Wireworm in *fodder beet, mangels, sugar beet* [3]

Specific Off-Label Approvals (SOLAs)

- *forest nurseries (OLA 992383) Dec 2008* [1]

Efficacy guidance

- At recommended rates seed of drilled arable crops is not damaged by contact with product
- Products formulated as controlled release granules to give pine weevil control throughout the 2-year establishment phase [1, 2]

Restrictions

- This product contains an anticholinesterase carbamate compound. Do not use if under medical advice not to work with such compounds
- Maximum number of treatments 1 per crop or tree

Crop-specific information

- Latest use: before or at planting for forestry uses
- HI 100 d for beet crops, carrots, parsnips; 12 wk for broccoli, cabbages, calabrese, cauliflowers; 14 wk for Brussels sprouts; 23 wk for collards
- Apply to drilled crops with suitable granule applicator feeding directly into seed furrow or immediately behind drill coulter (behind seed drill boot for brassicas) or use bow-wave technique [3]
- See label for details of suitable applicators and settings. Correct calibration is essential [3]
- Where used in forestry, granules should be placed in the planting hole by hand or metered applicator before or after placing the tree. Sachets should be placed in the planting hole by hand before placement of the tree
- Safe to Douglas Fir and Sitka Spruce. Other conifers may vary in their sensitivity [1, 2]
- Forest trees may take 10-15 d to achieve full protection. An additional pre-planting insecticide dip or spray is advised, but post-planting sprays during the 2-yr establishment should not be necessary [1, 2]

Environmental safety

- Dangerous for the environment
- Very toxic to aquatic organisms
- Dangerous to fish or other aquatic life. Do not contaminate surface waters or ditches with chemical or used container [1, 3]
- Dangerous to game, wild birds and animals
- Do not empty into drains [3]
- Keep in original container, tightly closed, in a safe place, under lock and key

Hazard classification and safety precautions

Hazard H03, H11

Risk phrases R20, R51 [1, 2]; R22a, R50 [3]; R43 [1, 3]; R53a [1-3]

Operator protection A [1, 2]; B [3]; C, D, E [1]; C [3] (or D+E); H [1-3]; K, M [1, 3]; U02a, U04a, U05a, U13, U20a [1-3]; U09a, U19a [3]; U10, U12 [1, 2]; U14 [1, 3]; U22b [2]

Environmental protection E10a [1-3]; E13b [1, 3]; E15a [2]; E19b, E34 [3]; E38 [1, 2]

FOR FULL CONDITIONS OF USE ALWAYS READ THE PRODUCT LABEL

Storage and disposal D01 [1-3]; D02 [1, 3]; D09a, D12a [1, 2]; D09b, D11a [3]
Treated seed S04a [1]
Medical advice M02 [1-3]; M03, M05a [3]

59   carboxin

A carboxamide fungicide available only in mixtures

60   carboxin + thiram

A fungicide seed dressing for cereals

Products

Anchor	Crompton	200:200 g/l	FS	08684

Uses
- Bunt in **spring wheat** *(seed treatment)*, **winter wheat** *(seed treatment)*
- Covered smut in **spring barley** *(seed treatment)*, **spring oats** *(seed treatment)*, **winter barley** *(seed treatment)*, **winter oats** *(seed treatment)*
- Fusarium foot rot and seedling blight in **spring barley** *(seed treatment)*, **spring oats** *(seed treatment)*, **spring rye** *(seed treatment)*, **spring wheat** *(seed treatment)*, **triticale** *(seed treatment)*, **winter barley** *(seed treatment)*, **winter oats** *(seed treatment)*, **winter rye** *(seed treatment)*, **winter wheat** *(seed treatment)*
- Leaf stripe in **spring barley** *(seed treatment)*, **winter barley** *(seed treatment - reduction)*
- Loose smut in **spring barley** *(seed treatment - reduction)*, **spring oats** *(seed treatment - reduction)*, **winter barley** *(seed treatment - reduction)*, **winter oats** *(seed treatment - reduction)*
- Net blotch in **spring barley** *(seed treatment)*
- Septoria seedling blight in **spring wheat** *(seed treatment)*, **winter wheat** *(seed treatment)*

Approval information
- Accepted by BBPA for use on malting barley

Efficacy guidance
- Apply through suitable liquid flowable seed treating equipment of the batch treatment or continuous flow type where a secondary mixing auger is fitted
- Drill flow may be affected by treatment. Always re-calibrate seed drill before use

Restrictions
- Maximum number of treatments 1 per batch of seed
- Do not treat seed with moisture content above 16%
- Do not apply to cracked, split or sprouted seed
- Do not store treated seed from one season to the next

Crop-specific information
- Latest use: pre-drilling

Environmental safety
- Dangerous for the environment
- Very toxic to aquatic organisms
- Do not use treated seed as food or feed
- Treated seed harmful to game and wildlife

Hazard classification and safety precautions
Hazard H03, H11
Risk phrases R22a, R48, R50, R53a
Operator protection A, D, H; U05a, U09a, U20c
Environmental protection E15a, E34, E38
Storage and disposal D01, D02, D05, D06a, D09a, D10a
Treated seed S01, S02, S04b, S05, S06a, S07
Medical advice M03

61   carfentrazone-ethyl

A triazolinone contact herbicide

Products

1 Aurora 50 WG	Belchim	50% w/w	WG	11613
2 Shark	Belchim	60 g/l	ME	11616
3 Spotlight 24 EC	Belchim	240 g/l	EC	11617
4 Spotlight Plus	Belchim	60 g/l	ME	12100

Uses
- Annual dicotyledons in *all edible crops* *(before planting)*, *all non-edible crops* *(before planting)*, *potatoes* [2]
- Black bindweed in *potatoes* [2]
- Cleavers in *durum wheat*, *spring barley*, *spring oats*, *spring wheat*, *triticale*, *winter barley*, *winter oats*, *winter wheat* [1]; *potatoes* [2]
- Fat hen in *potatoes* [2]
- Haulm destruction in *seed potatoes*, *ware potatoes* [3, 4]
- Ivy-leaved speedwell in *durum wheat*, *spring barley*, *spring oats*, *spring wheat*, *triticale*, *winter barley*, *winter oats*, *winter wheat* [1]; *potatoes* [2]
- Knotgrass in *potatoes* [2]
- Redshank in *potatoes* [2]
- Volunteer oilseed rape in *all edible crops* *(before planting)*, *all non-edible crops* *(before planting)*, *potatoes* [2]

Approval information
- Carfentrazone-ethyl included in Annex I under EC Directive 91/414
- Accepted by BBPA for use on malting barley and hops

Efficacy guidance
- Best weed control results achieved from good spray cover applied to small actively growing weeds [1]
- Carfentrazone-ethyl acts by contact only; see label for optimum timing on specified weeds. Weeds emerging after application will not be controlled [1]
- For weed control use as two spray programme with one application in autumn and one in spring [1]
- Efficacy of haulm destruction will be reduced where flailed haulm covers the stems at application [3, 4]
- For potato crops with very dense vigorous haulm or where regrowth occurs following a single application a second application may be necessary to achieve satisfactory desiccation. A minimum interval between applications of 7 d should be observed to achieve optimum performance [3, 4]

Restrictions
- Maximum number of treatments 2 per crop for cereals (1 in Autumn and 1 in Spring) [1]; 2 per crop for ware potato haulm destruction [3]; 1 per crop for seed potato haulm destruction [3]
- Do not treat cereal crops under stress from drought, waterlogging, cold, pests, diseases, nutrient or lime deficiency or any factors reducing plant growth [1]
- Do not treat cereals undersown with clover or other legumes [1]
- Allow at least 2-3 wk between application and lifting potatoes to allow skins to set if potatoes are to be stored [3, 4]
- Follow label instructions for sprayer cleaning
- Do not apply through knapsack sprayers [2]
- Contact processor before using a split dose on potatoes for processing

Crop-specific information
- Latest use: before 3rd node detectable (GS 33) on cereals [1]
- HI 7 d for potatoes [3]
- For weed control in cereals treat from 2 leaf stage [1]

FOR FULL CONDITIONS OF USE ALWAYS READ THE PRODUCT LABEL

- For potato haulm destruction ware crops should be treated at the onset of senescence; seed crops should be flailed when tubers have reached the desired size and then treated once the flailed haulm is clear of the stems that remain [3]
- When used as a treatment prior to planting a subsequent crop apply before weeds exceed maximum sizes indicated in the label [2]

Following crops guidance

- No restrictions apply on the planting of succeeding crops 1 mth after application to potatoes for haulm destruction or as a pre-planting treatment, or 3 mth after application to cereals for weed control
- In the event of failure of a treated cereal crop, all cereals, ryegrass, maize, oilseed rape, peas, sunflowers, *Phacelia*, vetches, carrots or onions may be planted within 1 mth of treatment

Environmental safety

- Dangerous for the environment
- Very toxic to aquatic organisms
- Some non-target crops are sensitive. Avoid drift onto broad-leaved plants outside the treated area, or onto ponds waterways or ditches

Hazard classification and safety precautions

Hazard H03 [3]; H04 [1, 2, 4]; H11 [1-4]
Risk phrases R22b [3]; R43 [1, 2, 4]; R50, R53a [1-4]
Operator protection A, H [1, 2, 4]; U05a, U14 [1, 2, 4]; U08, U13 [1]; U20a [2, 4]
Environmental protection E15a, E38 [1-4]; E34 [1, 3]
Storage and disposal D01, D02, D09a, D12a [1-4]; D10b [1, 3]; D10c [2, 4]
Medical advice M05a [1]; M05b [3]

---

62   carfentrazone-ethyl + flupyrsulfuron-methyl

A foliar and residual acting herbicide for cereals

Products

Lexus Class	DuPont	33.3:16.7% w/w	WG	10809

Uses

- Annual dicotyledons in *triticale, winter oats, winter rye, winter wheat*
- Blackgrass in *triticale, winter oats, winter rye, winter wheat*

Approval information

- Carfentrazone-ethyl and flupyrsulfuron-methyl included in Annex I under EC Directive 91/414

Efficacy guidance

- Best results obtained when applied to small actively growing weeds
- Good spray cover of weeds must be obtained
- Increased degradation of active ingredient in high soil temperatures reduces residual activity
- Weed control may be reduced in dry soil conditions but susceptible weeds germinating soon after treatment will be controlled if adequate soil moisture present
- Symptoms may not be apparent on some weeds for up to 4 wk, depending on weather conditons
- Blackgrass should be treated from 1 leaf but before first node
- Flupyrsulfuron-methyl is a member of the ALS-inhibitor group of herbicides and products should be used in a planned Resistance Management strategy. See Section 5 for more information

Restrictions

- Maximum number of treatments 1 per crop
- Do not use on barley, on crops undersown with grasses or legumes, or any other broad-leaved crop
- Do not apply within 7 d of rolling
- Do not treat any crop suffering from drought, waterlogging, pest or disease attack, nutrient deficiency, or any other stress factors
- Specific restrictions apply to use in sequence or tank mixture with other sulfonylurea or ALS-inhibiting herbicides. See label for details

SEE SECTION 3 FOR PRODUCTS ALSO REGISTERED

Crop-specific information
- Latest use: before 31 Dec in yr of sowing for winter oats, winter rye, triticale; before 1st node detectable (GS 31) for winter wheat
- Slight chlorosis and stunting may occur in certain conditions. Recovery is rapid and yield not affected

Following crops guidance
- Only cereals, oilseed rape, field beans, clover or grass may be sown in the yr of harvest of a treated crop
- In the event of crop failure only winter or spring wheat may be sown 1-3 mth after treatment. Land should be ploughed and cultivated to 15 cm minimum before resowing

Environmental safety
- Dangerous for the environment
- Very toxic to aquatic organisms
- Take extreme care to avoid damage by drift outside the target area or onto surface waters or ditches
- Spraying equipment should not be drained or flushed onto land planted, or to be planted, with trees or crops other than cereals and should be thoroughly cleansed after use - see label for instructions

Hazard classification and safety precautions
 Hazard H04, H11
 Risk phrases R43, R50, R53a
 Operator protection A, H; U05a, U08, U14, U19a, U20b
 Environmental protection E15b, E38
 Storage and disposal D01, D02, D09a, D10b, D11a, D12a

63 carfentrazone-ethyl + mecoprop-P

A foliar applied herbicide for cereals

Products
 Platform S   Belchim   1.5:60% w/w   WG 11611

Uses
- Charlock in *spring barley, spring oats, spring wheat, winter barley, winter oats, winter wheat*
- Chickweed in *spring barley, spring oats, spring wheat, winter barley, winter oats, winter wheat*
- Cleavers in *spring barley, spring oats, spring wheat, winter barley, winter oats, winter wheat*
- Red dead-nettle in *spring barley, spring oats, spring wheat, winter barley, winter oats, winter wheat*
- Speedwells in *spring barley, spring oats, spring wheat, winter barley, winter oats, winter wheat*

Approval information
- Carfentrazone-ethyl and mecoprop-P included in Annex I under EC Directive 91/414
- Accepted by BBPA for use on malting barley

Efficacy guidance
- Best results obtained when weeds have germinated and growing vigorously in warm moist conditions
- Treatment of large weeds and poor spray coverage may result in reduced weed control

Restrictions
- Maximum number of treatments 2 per crop. The total amount of mecoprop-P applied in a single yr must not exceed the maximum total dose approved for any single product for the crop/situation
- Do not treat crops suffering from stress from any cause
- Do not treat crops undersown or to be undersown

Crop-specific information
- Latest use: before 3rd node detectable (GS 33)

FOR FULL CONDITIONS OF USE ALWAYS READ THE PRODUCT LABEL

- Can be used on all varieties of wheat and barley in autumn or spring from the beginning of tillering
- Early sown crops may be prone to damage if treated after period of rapid growth in autumn

Following crops guidance
- In the event of crop failure, any cereal, maize, oilseed rape, peas, vetches or sunflowers may be sown 1 mth after a spring treatment. Any crop may be planted 3 mth after treatment

Environmental safety
- Dangerous for the environment
- Very toxic to aquatic organisms
- Keep livestock out of treated areas for at least two weeks following treatment and until poisonous weeds, such as ragwort, have died down and become unpalatable

Hazard classification and safety precautions
Hazard H03, H11
Risk phrases R22a, R41, R43, R50, R53a
Operator protection A, C, H, M; U05a, U08, U11, U13, U14, U20b
Environmental protection E07a, E15a, E34, E38
Storage and disposal D01, D02, D09a, D10b, D12a
Medical advice M05a

64   carfentrazone-ethyl + metsulfuron-methyl

A foliar applied herbicide mixture for cereals

Products

Ally Express	DuPont	40:10% w/w	WG	08640

Uses
- Annual dicotyledons in *durum wheat*, *spring barley*, *spring oats*, *spring wheat*, *triticale*, *winter barley*, *winter oats*, *winter wheat*

Approval information
- Carfentrazone-ethyl and metsulfuron-methyl included in Annex I under EC Directive 91/414
- Accepted by BBPA for use on malting barley

Efficacy guidance
- Best results achieved from applications made in good growing conditions
- Good spray cover of weeds must be obtained
- Growth of weeds is inhibited within hours of treatment but the time taken for visible colour changes to appear will vary according to species and weather
- Product has short residual life in soil. Under normal moisture conditions susceptible weeds germinating soon after treatment will be controlled
- Metsulfuron-methyl is a member of the ALS-inhibitor group of herbicides and products should be used in a planned Resistance Management strategy. See Section 5 for more information

Restrictions
- Maximum number of treatments 1 per crop
- Product must only be used after 1 Feb
- Do not use on any crop suffering stress from drought, waterlogging, cold, pest or disease attack, or nutrient deficiency
- Do not use on crops undersown with grass or legumes or on any broad-leaved crop
- Do not apply within 7 d of rolling
- Specific restrictions apply to use in sequence or tank mixture with other sulfonylurea or ALS-inhibiting herbicides. See label for details
- Do not tank mix with products containing chlorpyrifos, diclofop-methyl, difenzoquat, flamprop-methyl, flutriafol, propiconazole or tralkoxydim

Crop-specific information
- Latest use: before 3rd node detectable (GS 33)
- Can be used on all soil types
- Apply in the spring from the 3-leaf stage on all crops
- Slight necrotic spotting of crops can occur under certain crop and soil conditions. Recovery is rapid and there is no effect on grain yield or quality

SEE SECTION 3 FOR PRODUCTS ALSO REGISTERED

Following crops guidance
- Only cereals, oilseed rape, field beans or grass may be sown as a following crop in the same calendar yr. Any crop may follow in the next spring
- In the event of failure of a treated crop, only winter wheat may be sown 1-3 mth later after ploughing and cultivating to at least 15 cm

Environmental safety
- Dangerous for the environment
- Very toxic to aquatic organisms
- Take extreme care to avoid drift onto broad-leaved plants outside the target area or onto ponds, waterways or ditches, or onto land intended for cropping
- Spraying equipment should not be drained or flushed onto land planted, or to be planted, with trees or crops other than cereals and should be thoroughly cleansed after use - see label for instructions

Hazard classification and safety precautions
    Hazard H04, H11
    Risk phrases R43, R50, R53a
    Operator protection A, H; U05a, U08, U14, U19a, U20b
    Environmental protection E15b, E38
    Storage and disposal D01, D02, D09a, D11a, D12a

65   chloridazon

A residual pyridazinone herbicide for beet crops

Products

1 Better DF	Sipcam	65% w/w	SG	06250
2 Better Flowable	Sipcam	430 g/l	SC	04924
3 Burex 430 SC	Interfarm	430 g/l	SC	09494
4 Parador	United Phosphorus	430 g/l	SC	12372
5 Pyramin DF	BASF	65% w/w	WG	03438
6 Sculptor	Sipcam	430 g/l	SC	08836
7 Takron	BASF	430 g/l	SC	11627

Uses
- Annual dicotyledons in **bulb onions** *(off-label)* [5]; **fodder beet**, **mangels**, **sugar beet** [1-7]; **leeks** *(off-label)*, **salad onions** *(off-label)* [1, 5]; **onions** *(off-label)* [1]
- Annual meadow grass in **bulb onions** *(off-label)* [5]; **fodder beet**, **mangels**, **sugar beet** [1, 2, 4-7]; **leeks** *(off-label)*, **salad onions** *(off-label)* [1, 5]; **onions** *(off-label)* [1]

Specific Off-Label Approvals (SOLAs)
- **bulb onions**, **leeks**, **salad onions** *(OLA 970732) Dec 2008* [5]
- **leeks**, **onions**, **salad onions** *(OLA 012570) Dec 2008* [1]

Efficacy guidance
- Absorbed by roots of germinating weeds and best results achieved pre-emergence of weeds or crop when soil moist and adequate rain falls after application
- Application rate depends on soil type. See label for details

Restrictions
- Maximum number of treatments generally 1 per crop (pre-emergence) for fodder beet and mangels; 1 (pre-emergence) + 3 (post-emergence) per crop for sugar beet, but labels vary slightly. Check labels for maximum total dose for the crop to be treated
- Maximum total dose for onions and leeks equivalent to one full dose recommended for these crops [5]
- Do not use on Coarse Sands, Sands or Fine Sands or where organic matter exceeds 5%

Crop-specific information
- Latest use: pre-emergence for fodder beet and mangels; normally before leaves of crop meet in row for sugar beet, but labels vary slightly; up to and including second true leaf stage for onions and leeks [5]
- Where used pre-emergence spray as soon as possible after drilling in mid-Mar to mid-Apr on fine, firm, clod-free seedbed

FOR FULL CONDITIONS OF USE ALWAYS READ THE PRODUCT LABEL

- Where crop drilled after mid-Apr or soil dry apply pre-drilling and incorporate to 2.5 cm immediately afterwards
- Various tank mixes recommended on sugar beet for pre- and post-emergence use and as repeated low dose treatments. See label for details
- Crop vigour may be reduced by treatment of crops growing under unfavourable conditions including poor tilth, drilling at incorrect depth, soil capping, physical damage, pest or disease damage, excess seed dressing, trace-element deficiency or a sudden rise in temperature after a cold spell

Following crops guidance
- Winter cereals or any spring sown crop may follow a treated beet crop harvested at the normal time and after ploughing
- In the event of failure of a treated crop only a beet crop or maize may be drilled, after cultivation

Environmental safety
- Dangerous for the environment
- Very toxic to aquatic organisms

Hazard classification and safety precautions
Hazard H03 [1, 5]; H04 [2-4, 6, 7]; H11 [1-7]
Risk phrases R22a [1, 5]; R43 [2-4, 6, 7]; R50, R53a [1-7]
Operator protection A [3, 4, 6, 7]; U05a, U20b [1, 2, 4-7]; U08 [1-7]; U14 [2-4, 6, 7]; U19a [2, 4, 6, 7]; U20a [3]
Environmental protection E15a [1-7]; E34 [3]; E38 [1, 2, 4-7]
Storage and disposal D01, D02, D09a [1-7]; D05, D12b [3]; D10b [2-4, 6, 7]; D11a [1, 5]; D12a [1, 2, 4-7]
Medical advice M05a [1, 2, 4-7]

66   chloridazon + chlorpropham + metamitron

A residual herbicide mixture for beet crops

Products
Newtron	Nufarm UK	83:42:333 g/l	SE	11208

Uses
- Annual dicotyledons in **fodder beet**, **mangels**, **sugar beet**
- Annual meadow grass in **fodder beet**, **mangels**, **sugar beet**

Approval information
- Chlorpropham included in Annex I under EC Directive 91/414

Efficacy guidance
- Apply as part of a spray programme. For best results a full programme of pre- and post-emergence treatments is required
- Pre-emergence weed control will be reduced under very dry conditions but subsequent rain may activate the chemical
- Best post-emergence performance requires tank mxture with phenmedipham. See label for details
- Correct post-emergence timing important for good control. First treatment should be made when weeds at early cotyledon stage and subsequent applications made when new weed flushes appear

Restrictions
- Subject to the maximum total dose, maximum number of treatments 1 pre-emergence followed by 3 post-emergence, or 5 post-emergence treatments unless another product has been used pre-emergence, in which case 4 post-emergence sprays may be applied
- Must be tank mixed when used post-emergence. See label
- Do not roll or harrow treated soil. Spray after rolling

Crop-specific information
- Latest use: before crop leaves meet between rows

SECTION 2

SEE SECTION 3 FOR PRODUCTS ALSO REGISTERED

Following crops guidance
- Only sugar beet, fodder beet or mangels should be sown within 4 mth of last application
- After harvesting treated crops land should be mould-board ploughed to 15 cm and any spring crop may then be drilled

Environmental safety
- Dangerous for the environment
- Very toxic to aquatic organisms

Hazard classification and safety precautions
   Hazard H03, H11
   Risk phrases R22a, R36, R38, R43, R50, R53a
   Operator protection A, C; U05a, U08, U11, U14, U15, U19a, U20a
   Environmental protection E15a, E34
   Storage and disposal D01, D02, D09a, D10c, D12b

67   chloridazon + ethofumesate

A contact and residual herbicide for beet crops

Products

1 Gremlin	Sipcam	285:176 g/l	SC	09468
2 Magnum	BASF	275:170 g/l	SC	11727

Uses
- Annual dicotyledons in *fodder beet* [2]; *sugar beet* [1, 2]
- Annual meadow grass in *fodder beet* [2]; *sugar beet* [1, 2]

Approval information
- Ethofumesate included in Annex I under EC Directive 91/414

Efficacy guidance
- Best results achieved on a fine firm clod-free seedbed when soil moist and adequate rain falls after spraying. Efficacy and crop safety may be reduced if heavy rain falls just after incorporation
- Effectiveness may be reduced under conditions of low pH
- May be applied by conventional or repeat low dose method. See label for details

Restrictions
- Maximum number of treatments 1 pre-emergence for fodder beet, mangels; 1 pre-emergence plus 3 post-emergence for sugar beet [2]; 1 pre-emergence or 3 post-emergence on sugar beet [1]
- May be used on soil classes Loamy Sand - Silty Clay Loam [2]. Additional restrictions apply for some tank mixtures
- Do not treat beet post-emergence with recommended tank mixtures when temperature is, or is likely to be, above 21°C on day of spraying or under conditions of high light intensity
- If a mixture with phenmedipham is applied to a crop previously treated with a pre-emergence herbicide and the crop is suffering from stress from whatever cause, no further such post-emergence applications may be made [2]

Crop-specific information
- Latest use: pre-crop emergence for fodder beet, mangels; before crop leaves meet between rows for sugar beet
- Apply up to cotyledon stage of weeds
- Crop vigour may be reduced by treatment of crops growing under unfavourable conditions including poor tilth, drilling at incorrect depth, soil capping, physical damage, excess nitrogen, excess seed dressing, trace element deficiency or a sudden rise in temperature after a cold spell. Frost after pre-emergence treatment may check crop growth

Following crops guidance
- In the event of crop failure only sugar beet, fodder beet or mangels may be re-drilled
- Any crop may be sown 3 mth after spraying following ploughing to 15 cm

Environmental safety
- Dangerous for the environment
- Very toxic to aquatic organisms

FOR FULL CONDITIONS OF USE ALWAYS READ THE PRODUCT LABEL

Hazard classification and safety precautions
Hazard H03, H11
Risk phrases R22a, R50, R53a
Operator protection A, H [1, 2]; C [1]; U05a, U08, U19a, U20a
Environmental protection E15a, E38
Storage and disposal D01, D02, D09a, D10b, D12a
Medical advice M05a

68   chloridazon + lenacil

A residual pre-emergence herbicide for beet crops

Products
Advisor S	DuPont	200:133 g/l	SC	10948

Uses
- Annual dicotyledons in *fodder beet*, *mangels*, *sugar beet*
- Annual meadow grass in *fodder beet*, *mangels*, *sugar beet*

Efficacy guidance
- Best pre-emergence results achieved from application to firm, moist, weed-free seedbed when adequate rain falls afterwards. Weed spectrum can be increased by tank mixture with other herbicides. See label
- If dry conditions prevail after application results can be improved by irrigation
- Season long weed control achieved by a programme of pre- and post-emergence treatments
- After pre-drilling incorporated treatment sugar beet should be drilled within 3 wk

Restrictions
- Maximum number of treatments 1 per crop pre-drilling (sugar beet) or 1 pre-emergence (sugar beet, fodder beet, mangels) followed in either case by 1 post-emergence (or 3 post-emergence at low dose) per crop
- Crops should be drilled to at least 15 mm and the seed well covered
- Do not use on Sands, stony or gravelly soils, Heavy soils or those with more than 10% organic matter
- Only use in post-emergence mixtures on crops growing vigorously and not stressed by drought, pest attack, deficiency or other factors

Crop-specific information
- Latest use: pre-emergence or before crop meets across rows post-emergence
- Heavy rainfall after spraying may reduce crop stand especially when such rain is followed by very hot weather

Following crops guidance
- In the event of crop failure only beet crops may be re-drilled on treated land and no further application should be made for at least 4 mth.
- Any succeeding crop may be sown 4 mth after treatment following ploughing to 15 cm

Environmental safety
- Dangerous for the environment
- Very toxic to aquatic organisms
- Harmful to fish or other aquatic life. Do not contaminate surface waters or ditches with chemical or used container

Hazard classification and safety precautions
Hazard H04, H11
Risk phrases R43, R50, R53a
Operator protection U05a, U08, U14, U19a, U20b
Environmental protection E13c, E38
Storage and disposal D01, D02, D05, D09a, D10a, D12a

69   chloridazon + metamitron

A contact and residual herbicide mixture for sugar beet

Products

Volcan Combi	Sipcam	300:280 g/l	SC	10256

Uses
- Annual dicotyledons in **sugar beet**
- Annual meadow grass in **sugar beet**

Efficacy guidance
- Best results achieved from a sequential programmme of sprays
- Pre-emergence use improves efficacy of post-emergence programme. Best results obtained from application to a moist seed bed
- First post-emergence treatment should be made when weeds at early cotyledon stage and subsequent applications made when new weed flushes reach this stage
- Weeds surviving an earlier treatment should be treated again after 7-10 d even if no new weeds have appeared

Restrictions
- Maximum number of treaments 1 pre-emergence followed by 3 post-emergence
- Take advice if light soils have a high proportion of stones

Crop-specific information
- Latest use: when leaves of crop meet in rows
- Tolerance of crops growing under stress from any cause may be reduced
- Crops treated pre-emergence and subsequently subjected to frost may be checked and recovery may not be complete

Following crops guidance
- After the last application only sugar beet or mangels may be sown within 4 mth; cereals may be sown after 16 wk. Land should be mouldboard ploughed to 15 cm and thoroughly cultivated before any succeeding crop

Environmental safety
- Harmful to fish or other aquatic life. Do not contaminate surface waters or ditches with chemical or used container

Hazard classification and safety precautions
Hazard H04
Risk phrases R36, R38
Operator protection U08, U14, U15, U20a
Environmental protection E13c, E34
Storage and disposal D09a, D10b

70   chloridazon + propachlor

A residual pre-emergence herbicide for use in onions and leeks

Products

Ashlade CP	Nufarm UK	86:400 g/l	SC	06481

Uses
- Annual dicotyledons in **bulb onions** *(off-label - post-em. Pre-em on label)*, **chives** *(off-label - post-em. Pre-em on label)*, **leeks** *(off-label - post-em. Pre-em on label)*, **salad onions** *(off-label - post-em. Pre-em on label)*
- Annual grasses in **bulb onions** *(off-label - post-em. Pre-em on label)*, **chives** *(off-label - post-em. Pre-em on label)*, **leeks** *(off-label - post-em. Pre-em on label)*, **salad onions** *(off-label - post-em. Pre-em on label)*

Specific Off-Label Approvals (SOLAs)
- **bulb onions**, **leeks** *(post-em. Pre-em on label) (OLA 981362) Dec 2008* [1]
- **chives**, **salad onions** *(post-em. Pre-em on label) (OLA 951233) Dec 2008* [1]

FOR FULL CONDITIONS OF USE ALWAYS READ THE PRODUCT LABEL

Efficacy guidance
- Best results achieved from application to firm, moist, weed-free seedbed when adequate rain falls afterwards

Restrictions
- Maximum number of treatments 1 per crop
- Do not use on soils with more than 10% organic matter
- Ensure crops are drilled to 20 mm depth

Crop-specific information
- Latest use: pre-emergence of crop; before 2 true leaf stage for onions and leeks
- HI 12 wk for chives, salad onions
- Apply pre-emergence of sown crops, preferably soon after drilling, before weeds emerge. Loose or fluffy seedbeds must be consolidated before application
- Apply to transplanted crops when soil has settled after planting
- Crops stressed by nutrient deficiency, pests or diseases, poor growing conditions or pesticide damage may be checked by treatment, especially on sandy or gravelly soils

Following crops guidance
- In the event of crop failure only onions, leeks or maize should be planted
- Any crop can follow a treated onion or leek crop harvested normally as long as the ground is cultivated thoroughly before drilling

Environmental safety
- Harmful to fish or other aquatic life. Do not contaminate surface waters or ditches with chemical or used container

Hazard classification and safety precautions
   Hazard H03
   Risk phrases R22a, R36, R43
   Operator protection A, C; U02a, U04a, U05a, U08, U11, U13, U14, U19a, U20a
   Environmental protection E13c
   Storage and disposal D01, D02, D09a, D10b
   Medical advice M03

71   chloridazon + quinmerac

A herbicide mixture for use in beet crops

Products

Fiesta T	BASF	360:60 g/l	SC	11734

Uses
- Annual dicotyledons in *fodder beet, mangels, sugar beet*
- Annual meadow grass in *fodder beet, mangels, sugar beet*
- Chickweed in *fodder beet, mangels, sugar beet*
- Cleavers in *fodder beet, mangels, sugar beet*
- Mayweeds in *fodder beet, mangels, sugar beet*
- Poppies in *fodder beet, mangels, sugar beet*
- Speedwells in *fodder beet, mangels, sugar beet*

Efficacy guidance
- Best results obtained when adequate soil moisture is present at application and afterwards to form an active herbicidal layer in the soil
- A programme of pre-emergence treatment followed by post-emergence application(s) in mixture with a contact herbicide optimises weed control and is essential for some difficult weed species such as cleavers
- Treatment pre-emergence only may not provide sufficient residual activity to give season-long weed control
- Effectiveness may be reduced under conditions of low pH
- Always follow WRAG guidelines for preventing and managing herbicide resistant weeds. Section 5 for more information

SEE SECTION 3 FOR PRODUCTS ALSO REGISTERED

Restrictions

- Maximum total dose on sugar beet equivalent to one treatment pre-emergence plus two treatments post-emergence, all at maximum individual dose; maximum total dose on fodder beet and mangels equivalent to one full dose pre-emergence treatment
- Heavy rain shortly after treatment may check crop growth particularly when it leaves water standing in surface depressions
- Treatment of stressed crops or those growing in unfavourable conditions may depress crop vigour and possibly reduce stand
- Do not use on Sands or soils of high organic matter content
- Where rates of nitrogen higher than those generally recommended are considered necessary, apply at least 3 wk before drilling

Crop-specific information

- Latest use: before plants meet between the rows for sugar beet; pre-emergence for fodder beet, mangels

Following crops guidance

- Winter cereals or any spring sown crop may follow a treated beet crop harvested at the normal time and after ploughing
- In the event of failure of a treated crop only a beet crop may be drilled, after cultivation

Environmental safety

- Dangerous for the environment
- Toxic to aquatic organisms
- To reduce movement to groundwater do not apply to dry soil or when heavy rain is forecast

Hazard classification and safety precautions

Hazard H04, H11
Risk phrases R43, R51, R53a
Operator protection A; U05a, U08, U14, U19a, U20b
Environmental protection E15a, E38
Storage and disposal D01, D02, D08, D09a, D10c, D12a
Medical advice M05a

72   chlormequat

A plant-growth regulator for reducing stem growth and lodging

Products

1	Adjust	Mandops	620 g/l	SL	05589
2	Agriguard Chlormequat 700	AgriGuard	700 g/l	SL	09782
3	Agriguard Chlormequat 720	AgriGuard	720 g/l	SL	09919
4	Barclay Take 5	Barclay	645 g/l	SL	11368
5	Barleyquat B	Mandops	620 g/l	SL	07051
6	BASF 3C Chlormequat 720	BASF	720 g/l	SL	06514
7	Bettaquat B	Mandops	620 g/l	SL	07050
8	Fargro Chlormequat	Fargro	460 g/l	SL	02600
9	Fernpath Tangent	AgriGuard	460 g/l	SL	10341
10	Greencrop Carna	Greencrop	600 g/l	SL	09403
11	Hive	Nufarm UK	730 g/l	SL	11392
12	K2	Mandops	620 g/l	SL	10370
13	Mandops Chlormequat 700	Mandops	700 g/l	SL	06002
14	Manipulator	Mandops	620 g/l	SL	05871
15	Mirquat	Nufarm UK	730 g/l	SL	11406
16	New 5C Cycocel	BASF	645 g/l	SL	01482
17	New 5C Quintacel	Nufarm UK	645 g/l	SL	12074
18	Sigma PCT	Nufarm UK	460 g/l	SL	11209
19	Stabilan 700	Nufarm UK	700 g/l	SL	11393

Uses

- Increasing yield in *spring barley* [1, 5, 12-14]; *spring oats*, *winter oats* [5]; *winter barley* [1, 5, 6, 9, 10, 12-14, 16, 17]

- Lodging control in **spring barley** [1, 5, 12, 14]; **spring oats, winter oats** [1-6, 9-11, 13-17, 19]; **spring rye** [6, 16, 17]; **spring wheat, winter wheat** [1-4, 6, 7, 9-19]; **triticale** [4, 6, 9, 16, 17]; **winter barley** [1-6, 11, 12, 14, 15, 18, 19]; **winter rye** [2-4, 6, 16, 17]
- Stem shortening in **bedding plants, camellias, hibiscus trionum, lilies** [8, 11, 15, 19]; **geraniums, poinsettias** [6, 8, 11, 15-17, 19]

Approval information
- Approved for aerial application on wheat and oats [2, 6, 10, 13, 15, 18]; on winter barley [6, 10]; on triticale, rye [6]. See notes in Section 5
- Accepted by BBPA for use on malting barley

Efficacy guidance
- Most effective results on cereals normally achieved from application from Apr onwards, on wheat and rye from leaf sheath erect to first node detectable (GS 30-31), on oats at second node detectable (GS 32), on winter barley from mid-tillering to leaf sheath erect (GS 25-30). However, recommendations vary with product. See label for details
- Influence on growth varies with crop and growth stage. Risk of lodging reduced by application at early stem extension. Root development and yield can be improved by earlier treatment
- Results on barley can be variable
- In tank mixes with other pesticides on cereals optimum timing for herbicide action may differ from that for growth reduction. See label for details of tank mix recommendations
- Most products recommended for use on oats require addition of approved non-ionic wetter. Check label

Restrictions
- Maximum number of treatments or maximum total dose varies with crop and product and whether split dose treatments are recommended. Check labels
- Do not use on very late sown spring wheat or oats or on crops under stress
- Consult processors before use on crops for processing
- Do not use straw from treated cereals as horticultural growth medium or mulch
- Mixtures with liquid nitrogen fertilizers may cause scorch and are specifically excluded on some labels
- Do not use on soils of low fertility unless such crops regularly receive adequate dressings of nitrogen
- At least 6 h, preferably 24 h, required before rain for maximum effectiveness. Do not apply to wet crops
- Check labels for tank mixtures known to be incompatible

Crop-specific information
- Latest use varies with product. See label for details
- May be used on cereals undersown with grass or clovers
- Ornamentals to be treated must be well established and growing vigorously. Do not treat in strong sunlight or when temperatures are likely to fall below 10°C
- Temporary yellow spotting may occur on poinsettias. It can be minimised by use of a non-ionic wetting agent - see label

Environmental safety
- Dangerous for the environment [2, 3, 19]
- Harmful to aquatic organisms [1, 2, 5, 8, 11, 14, 19]
- Wash equipment thoroughly with water and wetting agent immediately after use and spray out. Spray out again before storing or using for another product. Traces can cause harm to susceptible crops sprayed later

Hazard classification and safety precautions
Hazard H03 [1-19]; H11 [2, 3, 19]
Risk phrases R21 [2, 3]; R22a [1-19]; R52, R53a [1, 2, 5, 8, 11, 14, 19]
Operator protection A; U05a [1-14, 16-19]; U05b [15]; U08, U19a [1-19]; U13 [8, 13]; U20a [13]; U20b [1-10, 12, 14-19]
Environmental protection E15a, E34
Consumer protection C01 [13]
Storage and disposal D01, D02, D09a [1-19]; D05 [2-4, 8, 10, 13, 19]; D10a [2, 13, 19]; D10b [1, 3-5, 7-12, 14, 15, 18]; D10c [6, 16, 17]
Medical advice M03 [1-10, 12-19]; M05a [2, 4, 6, 8, 11, 16, 17, 19]

SEE SECTION 3 FOR PRODUCTS ALSO REGISTERED

73   chlormequat + 2-chloroethylphosphonic acid

A plant growth regulator for use in cereals

Products

1 Greencrop Tycoon	Greencrop	305:155 g/l	SL	09571
2 Strate	Bayer CropScience	360:180 g/l	SL	10020
3 Sypex	BASF	305:155 g/l	SL	04650
4 Upgrade	Bayer CropScience	360:180 g/l	SL	10029

Uses

• Lodging control in **spring barley**, **winter barley**, **winter wheat**

Approval information

• Accepted by BBPA for use on malting barley

Efficacy guidance

• Best results obtained when crops growing vigorously
• Recommended dose varies with growth stage. See labels for details and recommendations for use of sequential treatments

Restrictions

• 2-chloroethylphosphonic acid is an anticholinesterase organophosphorus compound. Do not use if under medical advice not to work with such compounds
• Maximum number of treatments 1 per crop; maximum total dose equivalent to one full dose treatment
• Product must always be used with specified non-ionic wetter - see labels
• Do not use on any crop in sequence with any other product containing 2-chloroethylphosphonic acid
• Do not spray when crop wet or rain imminent
• Do not spray during cold weather or periods of night frost, when soil is very dry, when crop diseased or suffering pest damage, nutrient deficiency or herbicide stress
• If used on seed crops grown for certification inform seed merchant beforehand
• Do not use on wheat variety Moulin or on any winter varieties sown in spring [1, 3]
• Do not use on spring barley Triumph [1]
• Do not treat barley on soils with more than 10% organic matter [1, 3]
• Do not use straw from treated cereals as a horticultural growth medium or as a mulch [1, 3]
• Do not use in programme with any other product containing 2-chloroethylphosphonic acid [1]
• Only crops growing under conditions of high fertility should be treated

Crop-specific information

• Latest use: before flag leaf ligule/collar just visible (GS 39) or 1st spikelet visible (GS 51) for wheat or barley at top dose; or before flag leaf sheath opening (GS 47) for winter wheat at reduced dose
• Apply before lodging has started

Environmental safety

• Dangerous for the environment [2, 4]
• Harmful to aquatic organisms [1, 3]
• Harmful to fish or other aquatic life. Do not contaminate surface waters or ditches with chemical or used container [2, 4]

Hazard classification and safety precautions

Hazard H03 [1-4]; H11 [2, 4]
Risk phrases R22a, R37 [1-4]; R41 [2, 4]; R52 [1, 3]
Operator protection A [1-4]; C [2, 4]; U05a, U08, U19a, U20b [1-4]; U11, U13 [2, 4]
Environmental protection E13c [2, 4]; E15a [1, 3]; E34 [1-4]
Storage and disposal D01, D02, D09a, D10b [1-4]; D05 [2, 4]
Medical advice M01, M03 [1-4]; M05a [1, 3]

74   chlormequat + 2-chloroethylphosphonic acid + mepiquat chloride

A plant growth regulator for reducing lodging in cereals

Products

Cyclade	BASF	230:155:75 g/l	SL	08958

Uses
- Lodging control in **spring barley**, **winter barley**, **winter wheat**

Approval information
- Accepted by BBPA for use on malting barley

Efficacy guidance
- Best results achieved in a vigorous, actively growing crop with adequate fertility and moisture
- Optimum timing on all crops is from second node detectable stage (GS 32)
- Recommended for use as part of an intensive growing system which includes provision for optimum fertilizer treatment and disease control

Restrictions
- 2-chloroethylphosphonic acid is an anticholinesterase organophosphorus compound. Do not use if under medical advice not to work with such compounds
- Maximum number of treatments 1 per crop
- Maximum total dose depends on spraying regime adopted. See label
- Must be used with a non-ionic wetting agent
- Do not apply to stressed crops or those on soils of low fertility unless receiving adequate dressings of fertilizer
- Do not apply in temperatures above 21°C or if crop is wet or if rain expected
- Do not treat variety Moulin nor any winter varieties sown in spring
- Do not use in a programme with any other product containing 2-chloroethylphosphonic acid
- Do not apply to barley on soils with more than 10% organic matter (winter wheat may be treated)
- Notify seed merchant in advance if use on a seed crop is proposed

Crop-specific information
- Latest use: before first spikelet of ear visible (GS 51) using reduced dose on winter barley; before flag leaf sheath opening (GS 47) using reduced dose on winter wheat; before flag leaf just visible on spring barley
- May be applied to crops undersown with grasses or clovers
- Treatment may cause some delay in ear emergence

Environmental safety
- Harmful to aquatic organisms
- Do not use straw from treated crops as a horticultural growth medium

Hazard classification and safety precautions
Hazard H03
Risk phrases R22a, R37, R52
Operator protection A; U05a, U08, U19a, U20b
Environmental protection E15a, E34
Storage and disposal D01, D02, D09a, D10c
Medical advice M01, M03, M05a

75   chlormequat + imazaquin

A plant growth regulator mixture for winter wheat

Products

1	Meteor	BASF	368:0.8 g/l	SL	10403
2	Standon Imazaquin 5C	Standon	368:0.8 g/l	SL	08813

Uses
- Increasing yield in **winter wheat**
- Lodging control in **winter wheat**

Efficacy guidance
- Apply to crops during good growing conditions or to those at risk from lodging
- On soils of low fertility, best results obtained where adequate nitrogen fertilizer used

Restrictions
- Maximum number of applications 1 per crop (2 per crop at split dose)
- Do not treat durum wheat
- Do not apply to undersown crops
- Do not use straw from treated cereals as horticultural growth medium or mulch
- Do not apply when crop wet or rain imminent

Crop-specific information
- Latest use: before second node detectable (GS 31)
- Apply as single dose from leaf sheath lengthening up to and including 1st node detectable or as split dose, the first from tillers formed to leaf sheath lengthening, the second from leaf sheath erect up to and including 1st node detectable

Environmental safety
- Dangerous for the environment
- Toxic to aquatic organisms

Hazard classification and safety precautions
> **Hazard** H03, H11
> **Risk phrases** R22a, R36, R51, R53a
> **Operator protection** A, C; U05a, U08, U13, U19a, U20b
> **Environmental protection** E15a, E34
> **Storage and disposal** D01, D02, D05, D09a, D10b
> **Medical advice** M03

76   chlormequat + mepiquat chloride

A plant growth regulator for reducing lodging in wheat

Products

Stronghold	BASF	345:115 g/l	SL	09134

Uses
- Lodging control in *winter wheat*

Efficacy guidance
- Optimum timing is when leaf sheaths erect (GS 30)
- Benefit will vary according to crop and stage of growth at application

Restrictions
- Maximum total dose equivalent to one full dose treatment
- Do not apply to stressed crops or those on soils of low fertility unless receiving adequate dressings of fertilizer
- Do not treat crops where significant foot diseases, especially take-all, are expected
- Do not treat crops on soils of low fertility
- Do not apply in temperatures above 21°C or if crop is wet or if rain expected
- Do not treat any winter varieties sown in spring
- Notify seed merchant in advance if use on a seed crop is proposed

Crop-specific information
- Latest use: before 3rd node detectable (GS 33)
- Apply during good growing conditions at the correct timings - see label
- May be applied to crops undersown with grasses or clovers
- Treatment may cause some delay in ear emergence
- Mixtures with liquid fertilizers may cause scorching in some circumstances

Environmental safety
- Do not use straw from treated crops as a horticultural growth medium or mulch

Hazard classification and safety precautions
> **Hazard** H03
> **Risk phrases** R22a

FOR FULL CONDITIONS OF USE ALWAYS READ THE PRODUCT LABEL

Operator protection A; U05a, U08, U19a, U20b
Environmental protection E15a, E34
Storage and disposal D01, D02, D09a, D10c
Medical advice M03, M05a

77 2-chloroethylphosphonic acid

A plant growth regulator for cereals and various horticultural crops

See also chlormequat + 2-chloroethylphosphonic acid
chlormequat + 2-chloroethylphosphonic acid + mepiquat chloride

Products

1	Agriguard Cerusite	AgriGuard	480 g/l	SL	11494
2	Cerone	Bayer CropScience	480 g/l	SL	09985
3	Ethrel C	Certis	480 g/l	SL	11387
4	Pan Ethephon	Pan Agriculture	480 g/l	SL	11020
5	Pan Stiffen	Pan Agriculture	480 g/l	SL	11644

Uses
- Basal bud stimulation in **protected roses** [3]
- Fruit ripening in **apples** *(off-label - for cider making)*, **outdoor tomatoes**, **protected tomatoes** [3]
- Increasing branching in **geraniums** [3]
- Inducing flowering in **bromeliads** [3]
- Lodging control in **spring barley** [1, 2]; **triticale, winter barley, winter rye, winter wheat** [1, 2, 4, 5]
- Stem shortening in **narcissi** [3]

Specific Off-Label Approvals (SOLAs)
- **apples** *(for cider making) (OLA 030875) Dec 2008* [3]

Approval information
- Approved for aerial application on winter barley [1, 2]. See notes in Section 5

Efficacy guidance
- Best results achieved on crops growing vigorously under conditions of high fertility
- Optimum timing varies between crops and products. See labels for details
- Do not spray crops when wet or if rain imminent
- Best results on horticultural crops when temperature does not fall below 10°C [3]
- Use on tomatoes 17 d before planned pulling date [3]
- Apply as drench to daffodils when stems average 15 cm [3]
- Apply to glasshouse roses when new growth started after pruning [3]

Restrictions
- 2-chloroethylphosphonic acid is an anticholinesterase organophosphorus compound. Do not use if under medical advice not to work with such compounds
- Maximum number of treatments 1 per crop or yr
- Do not spray crops suffering from stress caused by any factor, during cold weather or period of night frost nor when soil very dry
- Do not apply to cereals within 10 d of herbicide or liquid fertilizer application
- Do not spray wheat or triticale where the leaf sheaths have split and the ear is visible

Crop-specific information
- Latest use: before 1st spikelet visible (GS 51) for spring barley, winter barley, winter rye; before flag leaf sheath opening (GS 47) for triticale, winter wheat
- HI cider apples, tomatoes 5 d

Environmental safety
- Dangerous for the environment [1, 4, 5]
- Harmful to aquatic organisms
- Avoid accidental deposits on painted objects such as cars, trucks, aircraft

Hazard classification and safety precautions
Hazard H04 [1-5]; H11 [1, 4, 5]
Risk phrases R36 [1]; R37, R41, R52, R53a [2-5]; R38 [1-5]

Operator protection A, C; U02a, U09a, U20a [3]; U05a, U13 [1-5]; U08, U20b [1, 2, 4, 5]; U11 [2, 3]
Environmental protection E13c, E38 [2, 3]; E15a [1, 4, 5]; E34 [1, 3]
Storage and disposal D01, D02, D09a, D10b [1-5]; D12a [2, 3]
Medical advice M01 [2-5]

78 2-chloroethylphosphonic acid + mepiquat chloride

A plant growth regulator for reducing lodging in cereals

Products

1 Barclay Banshee	Barclay	155:305 g/l	SL	11343
2 Guilder	Nufarm UK	155:305 g/l	SL	11894
3 Me2 Terpitz	Me2	155:305 g/l	SL	09634
4 Standon Mepiquat Plus	Standon	155:305 g/l	SL	09373
5 Terpal	BASF	155:305 g/l	SL	02103

Uses
- Increasing yield in **winter barley** *(low lodging situations)* [1-3, 5]
- Lodging control in **spring barley** [1-3, 5]; **triticale**, **winter barley**, **winter rye**, **winter wheat** [1-5]

Efficacy guidance
- Best results achieved on crops growing vigorously under conditions of high fertility
- Recommended dose and timing vary with crop, cultivar, growing conditions, previous treatment and desired degree of lodging control. See label for details
- May be applied to crops undersown with grass or clovers
- Do not apply to crops if wet or rain expected as efficacy will be impaired

Restrictions
- 2-chloroethylphosphonic acid is an anticholinesterase organophosphorus compound. Do not use if under medical advice not to work with such compounds
- Maximum number of treatments 2 per crop
- Add an authorised non-ionic wetter to spray solution. See label for recommended product and rate
- Do not treat crops damaged by herbicides or stressed by drought, waterlogging etc
- Do not treat crops on soils of low fertility unless adequately fertilized
- Do not use in a programme with any other product containing 2-chloroethylphosphonic acid
- Do not apply to winter cultivars sown in spring or treat winter barley, triticale or winter rye on soils with more than 10% organic matter (winter wheat may be treated)
- Do not apply at temperatures above 21°C

Crop-specific information
- Latest use: before ear visible (GS 49) for winter barley, spring barley, winter wheat and triticale; flag leaf just visible (GS 37) for winter rye
- Late tillering may be increased with crops subject to moisture stress and may reduce quality of malting barley

Environmental safety
- Do not use straw from treated cereals as a mulch or growing medium

Hazard classification and safety precautions
Hazard H03
Risk phrases R22a, R36, R37, R53a
Operator protection A, C [2, 5]; U20b
Environmental protection E15a
Storage and disposal D01, D02, D08, D09a, D10c
Medical advice M01, M05a

79 chlorophacinone

An anticoagulant rodenticide

Products

1 Drat Rat Bait	B H & B	0.005% w/w	RB	H6743
2 Endorats	Irish Drugs	0.005% w/w	RB	H6744

FOR FULL CONDITIONS OF USE ALWAYS READ THE PRODUCT LABEL

Uses
- Rats in **farm buildings** [1, 2]
- Voles in **farm buildings** [1]

Efficacy guidance
- Chemical formulated with oil, thus improving weather resistance of bait
- Use in baiting programme
- Bait stations should be sited where rats active, by rat holes, along runs or in harbourages. Place bait in suitable containers
- Replenish baits every few days and remove unused bait when take ceases or after 7-10 d

Restrictions
- For use only by professional operators
- Resistance status of target population should be taken into account when considering choice of rodenticide
- Bait stations may be sited conveniently but bait should be inaccessible to non-target animals and protected from prevailing weather

Environmental safety
- Prevent access to baits by children, domestic animals and birds; see label for other precautions required
- Harmful to game, wild birds and animals

Hazard classification and safety precautions
  Operator protection U13 [1, 2]; U20a [2]; U20b [1]
  Storage and disposal D09a [1, 2]; D10a [1]; D11a [2]
  Vertebrate/rodent control products V01a, V02, V03a, V04a

80   chloropicrin

A highly toxic horticultural soil fumigant

Products

1	Chloropicrin Fumigant	Dewco-Lloyd	99.5% w/w	VP	04216
2	K & S Chlorofume	K & S Fumigation	99.3% w/w	VP	08722

Uses
- Crown rot in **strawberries** *(soil fumigation)*
- Nematodes in **strawberries** *(soil fumigation)*
- Red core in **strawberries** *(soil fumigation)*
- Replant disease in **all top fruit** *(soil fumigation)*, **hardy ornamentals** *(soil fumigation)*
- Verticillium wilt in **blackberries** *(off-label - soil fumigation)*, **protected blackberries** *(off-label - soil fumigation)*, **protected raspberries** *(off-label - soil fumigation)*, **protected rubus hybrids** *(off-label - soil fumigation)*, **raspberries** *(off-label - soil fumigation)*, **rubus hybrids** *(off-label - soil fumigation)*, **strawberries** *(soil fumigation)*

Specific Off-Label Approvals (SOLAs)
- **blackberries**, **protected blackberries**, **protected raspberries**, **protected rubus hybrids**, **raspberries**, **rubus hybrids** *(soil fumigation) (OLA 051948) Dec 2008* [1]
- **blackberries**, **protected blackberries**, **protected raspberries**, **protected rubus hybrids**, **raspberries**, **rubus hybrids** *(soil fumigation) (OLA 051579) Dec 2008* [2]

Efficacy guidance
- Treat pre-planting
- Apply with specialised injection equipment
- For treating small areas or re-planting a single tree, a hand-operated injector may be used. Mark the area to be treated and inject to 22 cm at intervals of 22 cm
- Polythene sheeting (150 gauge) should be progressively laid over soil as treatment proceeds. The margin of the sheeting around the treated area must be embedded or covered with treated soil. Remove progressively after at least 4 d provided good air movement conditions prevail. Aerate soil for 15 d before planting

Restrictions
- Chloropicrin is subject to the Poisons Rules 1982 and the Poisons Act 1972. See notes in Section 5

SEE SECTION 3 FOR PRODUCTS ALSO REGISTERED

- Before use, consult the code of practice for the fumigation of soil with chloropicrin. 2 fumigators must be present
- Remove contaminated gloves, boots or other clothing immediately and ventilate them in the open air until all odour is eliminated
- Keep unprotected persons out of treated areas and adjacent premises until advised othrerwise by professional operator

Crop-specific information
- Latest use: at least 20 d before planting

Following crops guidance
- Carry out a cress test before replanting treated soil

Environmental safety
- Avoid treatment or vapour release when persistent still air conditions prevail
- Dangerous to game, wild birds and animals
- Dangerous to bees
- Dangerous to fish or other aquatic life. Do not contaminate surface waters or ditches with chemical or used container
- Dangerous to livestock. Keep all livestock out of treated areas until advised otherwise by fumigator
- Keep in original container, tightly closed, in a safe place, under lock and key

Hazard classification and safety precautions
    Hazard H01, H04
    Risk phrases R26, R27, R28, R36, R37, R38
    Operator protection A, G, H, K, M; U04a, U05a, U10, U13, U14, U15, U19a, U20a
    Environmental protection E02a, E06b (until advised); E10a, E12d, E13b, E15a, E34
    Storage and disposal D01, D02, D09b, D11a
    Medical advice M04

81   chlorothalonil

A protectant chlorophenyl fungicide for use in many crops and turf

See also azoxystrobin + chlorothalonil

Products

1	Agriguard Chlorothalonil	AgriGuard	500 g/l	SC	12201
2	Bombardier FL	Unicrop	500 g/l	SC	07910
3	Bravo 500	Syngenta	500 g/l	SC	10518
4	Clortosip 500	Sipcam	500 g/l	SC	09320
5	Cropguard	Nufarm UK	500 g/l	SC	11835
6	Daconil Turf	Scotts	500 g/l	SC	09265
7	Delphi	Sipcam	720 g/l	SC	11937
8	Flute	Sipcam	720 g/l	SC	08953
9	Fusonil Turf	Rigby Taylor	500 g/l	SC	09695
10	Greencrop Orchid	Greencrop	500 g/l	SC	09566
11	Greencrop Orchid B	Greencrop	500 g/l	SC	12251
12	Inzacur	Nufarm UK	500 g/l	SC	12014
13	Joules	Nufarm UK	500 g/l	SC	11047
14	Repulse	Certis	500 g/l	SC	11328
15	Scorpio 500 SC	DAPT	500 g/l	SC	10927
16	Sipcam Echo 75	Sipcam	75% w/w	WG	08302
17	Supreme	AgriGuard	500 g/l	SC	12233

Uses
- Anthracnose in *managed amenity turf* [6, 9, 12]
- Ascochyta in *combining peas* [1-5, 7, 8, 10, 11, 13, 14, 16, 17]
- Blight in *potatoes* [1-5, 7, 8, 10, 11, 13-17]; *protected tomatoes* [1-5, 11, 13, 14, 17]
- Botrytis in *blackberries*, *raspberries* [1-5, 11, 13, 16, 17]; *blackcurrants*, *gooseberries* (qualified minor use), *redcurrants* (qualified minor use), *strawberries* (outdoor crops only) [1, 4, 5, 13, 17]; *broccoli*, *brussels sprouts*, *cabbages*, *cauliflowers* [1-5, 7, 8, 11, 13, 14, 16, 17]; *calabrese* [7, 8, 16]; *combining peas* [1, 4, 5, 13, 14, 16, 17]; *daffodils* (off-label - for galanthamine production) [3]; *onions* [1, 2, 4, 5, 13, 14, 17]; *protected cucumbers*, *protected ornamentals* [1-5, 11, 13, 14, 16,

17]; *protected tomatoes* [2, 3, 11, 14, 16]; *spring oilseed rape, winter oilseed rape* [1, 3, 5, 11, 13, 16, 17]; *strawberries* [3, 11, 14, 16]; *strawberries* *(off-label)* [3, 4]
- Cane spot in *blackberries, raspberries* [3, 11, 14]
- Celery leaf spot in *celery (qualified minor use)* [1-5, 11, 13, 14, 16, 17]
- Chocolate spot in *spring field beans, winter field beans* [1-5, 7, 8, 10, 11, 13, 15-17]
- Currant leaf spot in *blackcurrants* [1-5, 7, 8, 10, 11, 13, 14, 16, 17]; *gooseberries (qualified minor use)*, *redcurrants (qualified minor use)* [1-5, 11, 13, 14, 16, 17]
- Dark leaf spot in *broccoli, brussels sprouts, cabbages, calabrese, cauliflowers* [10, 16]
- Dollar spot in *managed amenity turf* [6, 9, 12]
- Downy mildew in *brassica seed beds* [3, 11]; *broccoli, brussels sprouts, cabbages, cauliflowers* [1-5, 7, 8, 11, 13, 14, 16, 17]; *calabrese* [7, 8, 16]; *edible brassicas* [14]; *hops* [1-5, 11, 13, 14, 16, 17]; *spring oilseed rape, winter oilseed rape* [1, 3-5, 11, 13, 16, 17]
- Fusarium patch in *managed amenity turf* [6, 9, 12]
- Glume blotch in *spring wheat* [1, 3-5, 7, 8, 10, 13, 15-17]; *winter wheat* [1-5, 7, 8, 10, 11, 13, 15-17]
- Ink disease in *irises (qualified minor use)* [1-5, 11, 13, 16, 17]
- Leaf mould in *protected tomatoes* [1, 2, 4, 5, 13, 14, 16, 17]
- Leaf rot in *onions* [1-5, 11, 13, 14, 16, 17]
- Mycosphaerella in *combining peas* [1-5, 11, 13-17]
- Neck rot in *onions* [1-5, 11, 13, 14, 16, 17]
- Powdery mildew in *blackberries* [3, 11]; *blackcurrants, gooseberries (qualified minor use)*, *redcurrants (qualified minor use)* [1, 4, 5, 13, 17]; *protected cucumbers* [16]; *raspberries* [3, 11, 16]
- Red thread in *managed amenity turf* [6, 9, 12]
- Rhynchosporium in *spring barley, winter barley* [3-5, 7, 8, 11, 13, 15]
- Ring spot in *broccoli, brussels sprouts, cabbages, cauliflowers* [3, 7, 8, 11, 14, 16]; *calabrese* [7, 8, 16]; *edible brassicas* [14]
- Septoria leaf spot in *celeriac (off-label)* [3]; *spring wheat* [1, 3-5, 7, 8, 10, 13, 15-17]; *winter wheat* [1-5, 7, 8, 10, 11, 13, 15-17]
- White mould in *daffodils (off-label)* [3]

Specific Off-Label Approvals (SOLAs)
- *celeriac (OLA 032083) Dec 2008* [3]
- *daffodils (for galanthamine production) (OLA 041518) Dec 2008* [3]
- *daffodils (OLA 041518) Dec 2008* [3]
- *strawberries (OLA 011635) Dec 2008* [3]
- *strawberries (OLA 021253) Dec 2008* [4]

Approval information
- Chlorothalonil included in Annex I under EC Directive 91/414
- Following implementation of Directive 98/82/EC, approval for use of chlorothalonil on numerous crops was revoked in 1999, although uses on blackcurrants, redcurrants and gooseberries were subsequently reinstated for some products. Uses on loganberries revoked in 2001 following implementation of MRL Directives
- Approval for aerial spraying on potatoes [1-4, 8, 10, 14, 16]. See notes in Section 5
- Accepted by BBPA for use on malting barley and hops

Efficacy guidance
- For some crops products differ in diseases listed as controlled. See label for details and for application rates, timing and number of sprays
- Apply as protective spray or as soon as disease appears and repeat as directed
- Activity against Septoria may be reduced where serious mildew or rust present. In such conditions mix with suitable mildew or rust fungicide
- May be used for preventive and curative treatment of turf [6, 9]

Restrictions
- Maximum number of treatments 2 per crop for brassicas; 3 per crop for celery, peas, cereals
- Maximum total dose for other crops varies with crop and product - see labels for details
- Operators must use a vehicle fitted with a cab and a forced air filtration unit with a pesticide filter complying with HSE Guidance Note PM74 or to an equivalent or higher standard when making broadcast or air-assisted applications

SEE SECTION 3 FOR PRODUCTS ALSO REGISTERED

- Must only be applied by a pedestrian controlled sprayer or vehicle mounted/drawn equipment [6, 9]
- Do not mow or water turf for 24 h after treatment. Do not add surfactant or mix with liquid fertilizer [6, 9]

Crop-specific information

- Latest time of application to cereals varies with product. Consult label
- Latest use for winter oilseed rape before flowering; end Aug in yr of harvest for blackcurrants, gooseberries, redcurrants, blueberries, loganberries, raspberries
- HI 8 wk for field beans; 6 wk for combining peas; 28 d or before 31 Aug for post-harvest treatment for blackcurrants, gooseberries, redcurrants; 14 d for onions, strawberries; 10 d for hops; 7-14 d for broccoli, Brussels sprouts, cabbages, cauliflowers, celery, onions, potatoes; 3 d for hops, cane fruit; 12-48 h for protected cucumbers, protected tomatoes; 24 h for mushrooms
- For Botrytis control in strawberries important to start spraying early in flowering period and repeat at least 3 times at 10 d intervals
- On strawberries some scorching of calyx may occur with protected crops

Environmental safety

- Dangerous for the environment
- Very toxic to aquatic organisms
- Dangerous to fish or other aquatic life. Do not contaminate surface waters or ditches with chemical or used container
- Do not spray from the air within 250 m horizontal distance of surface waters or ditches
- LERAP Category B
- Broadcast air-assisted LERAP (18 m) [1-3, 5, 9-14, 16, 17]

Hazard classification and safety precautions

Hazard H03 [1-3, 5, 6, 9-14, 17]; H04 [4, 7, 8, 15, 16]; H11 [1-14, 17]

Risk phrases R20 [1, 3, 5, 6, 9, 11, 13, 14, 17]; R36 [1, 3-13, 15, 17]; R37 [1-3, 5, 6, 10, 13, 14, 16, 17]; R38 [2, 4, 7-11, 14, 15]; R40 [1, 3-6, 9-14, 17]; R41 [2, 10, 14-16]; R43 [1, 3-11, 13, 14, 17]; R50, R53a [1, 3-14, 17]

Operator protection A, C, H [1-17]; D [16]; J [1-5, 7, 8, 10, 11, 17]; M [1-12, 14, 16, 17]; U02a [3, 4, 7-11, 15, 16]; U05a [1-6, 9-17]; U09a, U19a, U20a [1-17]; U11 [4, 7, 8]; U13, U15 [4, 7, 8, 10, 12, 15, 16]; U14 [4, 7, 8, 10, 12, 14-16]

Environmental protection E13b [15, 16]; E15a [1-14, 17]; E16a [1-17]; E16b [1-11, 13-17]; E17a [4, 7, 8] (18 m); E17b [1-3, 5, 9-14, 16, 17] (18 m); E18 [1-4, 7-11, 16, 17]; E34 [1, 2, 5, 6, 12, 13, 15, 17]; E38 [1, 3, 5, 6, 9, 11, 13, 14, 17]

Consumer protection C02 [15] (field beans 14 d; potatoes 7 d)

Storage and disposal D01, D02, D09a [1-17]; D05 [1-3, 5, 6, 9-13, 15, 17]; D10a [6, 12, 14]; D10b [2, 4, 7, 8, 10, 15, 16]; D10c [1, 3, 5, 9, 11, 13, 17]; D12a [1, 3-9, 11, 13, 14, 17]

Medical advice M05a [10]

82   chlorothalonil + cyproconazole

A systemic protectant and curative fungicide mixture for cereals

Products

Alto Elite	Syngenta	375:40 g/l	SC	08467

Uses

- Brown rust in **spring barley, winter barley, winter wheat**
- Chocolate spot in **spring field beans, winter field beans**
- Eyespot in **winter barley, winter wheat**
- Grey mould in **combining peas**
- Net blotch in **spring barley, winter barley**
- Powdery mildew in **spring barley, winter barley, winter wheat**
- Rhynchosporium in **spring barley, winter barley**
- Rust in **spring field beans, winter field beans**
- Septoria diseases in **winter wheat**
- Yellow rust in **spring barley, winter barley, winter wheat**

FOR FULL CONDITIONS OF USE ALWAYS READ THE PRODUCT LABEL

Approval information
- Chlorothalonil included in Annex I under EC Directive 91/414
- Accepted by BBPA for use on malting barley

Efficacy guidance
- Apply at first signs of infection or as soon as disease becomes active
- A repeat application may be made if re-infection occurs
- For established mildew tank-mix with an approved mildewicide
- When applied prior to third node detectable (GS 33) a useful reduction of eyespot will be obtained

Restrictions
- Maximum total dose equivalent to 2 full dose treatments
- Do not apply at concentrations higher than recommended

Crop-specific information
- Latest use: completion of ear emergence in wheat and barley
- HI peas, field beans 6 wk
- If applied to winter wheat in spring at GS 30-33 straw shortening may occur but yield is not reduced

Environmental safety
- Dangerous for the environment
- Very toxic to aquatic organisms
- LERAP Category B

Hazard classification and safety precautions
Hazard H03, H11
Risk phrases R20, R37, R40, R41, R43, R50, R53a
Operator protection A, C, H, M; U05a, U11, U19a, U20a
Environmental protection E15a, E16a, E16b, E38
Storage and disposal D01, D02, D09a, D10c, D12a
Medical advice M04

83   chlorothalonil + flutriafol

A systemic eradicant and protectant fungicide for winter wheat

Products

1 Halo	Headland	375:47 g/l	SC	11546
2 Impact Excel	Headland	300:47 g/l	SC	11547

Uses
- Brown rust in *winter wheat*
- Late ear diseases in *winter wheat*
- Powdery mildew in *winter wheat*
- Septoria in *winter wheat*
- Yellow rust in *winter wheat*

Approval information
- Chlorothalonil included in Annex I under EC Directive 91/414

Efficacy guidance
- Generally disease control and yield benefit will be optimised when application is made at an early stage of disease development
- Apply as soon as disease is seen establishing in the crop

Restrictions
- Maximum number of treatments 2 per crop (including other products containing flutriafol)

Crop-specific information
- Latest use: before early milk stage (GS 73)
- On certain cultivars with erect leaves high transpiration can result in flag leaf tip scorch. This may be increased by treatment but does not affect yield

Environmental safety
- Dangerous for the environment [2]
- Very toxic to aquatic organisms [2]

SECTION 2

SEE SECTION 3 FOR PRODUCTS ALSO REGISTERED

- Dangerous to fish or other aquatic life. Do not contaminate surface waters or ditches with chemical or used container [1]
- LERAP Category B

Hazard classification and safety precautions
    Hazard H02, H11 [2]; H04 [1]
    Risk phrases R23, R40, R50, R53a [2]; R38 [1]; R41, R43 [1, 2]
    Operator protection A, C, H; U02a, U05a, U09a, U19a, U20b [1, 2]; U11 [2]
    Environmental protection E13b [1]; E15a, E38 [2]; E16a [1, 2]
    Storage and disposal D01, D02, D09a, D10b

84   chlorothalonil + mancozeb

A multi-site protectant fungicide mixture

Products

1 Adagio	Interfarm	201:274 g/l	SC	10796
2 Guru	Interfarm	286: 194 g/l	SC	10801

Uses
- Blight in **potatoes** [1]
- Septoria diseases in **winter wheat** [2]

Approval information
- Chlorothalonil and mancozeb included in Annex I under EC Directive 91/414

Efficacy guidance
- Start spray treatments on potatoes immediately after a blight warning or just before the haulm meets in the row [1]
- It is essential to start the blight spray programme before the disease appears in the crop [1]
- Repeat treatments for blight at 7, 10 or 14 d intervals depending on blight risk and continue until haulm is to be destroyed [1]
- Best results on winter wheat achieved from protective applications to the flag leaf. If disease already present on lower leaves treat as soon as flag leaf is just visible (GS 37) [2]
- Activity on Septoria may be reduced in presence of severe mildew infection [2]

Restrictions
- Maximum number of treatments 5 per crop for potatoes [1]
- Maximum total dose on winter wheat equivalent to 1.5 full dose treatments [2]
- Do not treat crops under stress from frost, drought, waterlogging, trace element deficiency or pest attack [2]
- Broadcast air assisted applications must only be made by equipment fitted with a cab with a forced air filtration unit plus a pesticide filter complying with HSE Guidance Note PM 74 or an equivalent or higher standard

Crop-specific information
- Before grain watery ripe (GS 71) for winter wheat
- HI potatoes 7 d
- Irrigated potato crops should be sprayed immediately after irrigation [1]

Environmental safety
- Dangerous for the environment
- Very toxic to aquatic organisms
- LERAP Category B

Hazard classification and safety precautions
    Hazard H04, H11
    Risk phrases R37, R43, R50, R53a [1, 2]; R40 [2]
    Operator protection A, C, H, M; U02a, U05a, U08, U10, U11, U13, U14, U19a [1, 2]; U20a [1]; U20b [2]
    Environmental protection E15a [1]; E16a, E16b, E38 [1, 2]
    Storage and disposal D01, D02, D05, D09a, D10b, D12a
    Medical advice M04 [1]

FOR FULL CONDITIONS OF USE ALWAYS READ THE PRODUCT LABEL

85   chlorothalonil + metalaxyl-M

A systemic and protectant fungicide for various crops

Products
Folio Gold   Syngenta   500:37.5 g/l   SC   10704

Uses
- Alternaria in **brussels sprouts** *(moderate control)*, **calabrese** *(moderate control)*, **cauliflowers** *(moderate control)*
- Downy mildew in **broad beans**, **brussels sprouts**, **bulb onions** *(qualified minor use)*, **calabrese**, **cauliflowers**, **poppies** *(off-label - for morphine production)*, **shallots** *(qualified minor use)*, **spring field beans**, **winter field beans**
- Ring spot in **brussels sprouts** *(reduction)*, **calabrese** *(reduction)*, **cauliflowers** *(reduction)*
- White blister in **brussels sprouts**, **calabrese**, **cauliflowers**
- White tip in **leeks** *(qualified minor use)*

Specific Off-Label Approvals (SOLAs)
- **poppies** *(for morphine production) (OLA 040469) Dec 2008* [1]

Approval information
- Chlorothalonil and metalaxyl-M included in Annex I under EC Directive 91/414

Efficacy guidance
- Apply at first signs of disease or when weather conditions favourable for disease pressure
- Best results obtained when used in a full and well-timed programme. Repeat treatment at 14-21 d intervals if necessary
- Treatment of established disease will be less effective
- Evidence of effectiveness in bulb onions, shallots and leeks is limited

Restrictions
- Maximum total dose equivalent to 2 full doses on broad beans, field beans, cauliflowers, calabrese; 3 full doses on Brussels sprouts, leeks, onions and shallots

Crop-specific information
- HI 14 d for all crops

Environmental safety
- Dangerous for the environment
- Very toxic to aquatic organisms
- LERAP Category B

Hazard classification and safety precautions
Hazard H03, H11
Risk phrases R20, R36, R37, R38, R40, R43, R50, R53a
Operator protection A, C, H; U05a, U09a, U15, U20a
Environmental protection E15a, E16a, E38
Consumer protection C02 (14 d)
Storage and disposal D01, D02, D09a, D10c, D12a

86   chlorothalonil + picoxystrobin

A broad spectrum fungicide mixture for cereals

Products
Credo   Syngenta   500:100 g/l   SC   12368

Uses
- Brown rust in **spring barley**, **spring wheat**, **winter barley**, **winter wheat**
- Glume blotch in **spring wheat**, **winter wheat**
- Net blotch in **spring barley**, **winter barley**
- Powdery mildew in **spring barley**, **winter barley**
- Rhynchosporium in **spring barley**, **winter barley**
- Septoria leaf spot in **spring wheat**, **winter wheat**
- Tan spot in **spring wheat**, **winter wheat**
- Yellow rust in **spring wheat**, **winter wheat**

SEE SECTION 3 FOR PRODUCTS ALSO REGISTERED

SECTION 2

Approval information
- Chlorothalonil included in Annex I under EC Directive 91/414
- Accepted by BBPA for use on malting barley

Efficacy guidance
- Best results obtained as a protectant treatment, or when disease first seen in crop, applied in good growing conditions with adequate soil moisture
- Disease control for 4-6 wk normally achieved during stem elongation
- Results may be less reliable when used on crops under stress
- Treatments for protection against ear disease should be made at ear emergence
- Picoxystrobin is a member of the QoI cross resistance group. Product should be used preventatively and not relied on for its curative potential
- Use product as part of an Integrated Crop Management strategy incorporating other methods of control, including where appropriate other fungicides with a different mode of action. Do not apply more than two foliar applications of QoI containing products to any cereal crop
- Control of many diseases may be improved by use in mixture with an appropriate triazole
- There is a significant risk of widespread resistance occurring in *Septoria tritici* populations in UK. Failure to follow resistance management action may result in reduced levels of disease control
- Strains of barley powdery mildew resistant to QoIs are common in the UK

Restrictions
- Maximum number of treatments 2 per crop of wheat or barley

Crop-specific information
- Latest use: before first spikelet just visible (GS 51) for barley; before grain watery ripe (GS 71) for wheat

Environmental safety
- Dangerous for the environment
- Very toxic to aquatic organisms
- LERAP Category B

Hazard classification and safety precautions
    Hazard H03, H11
    Risk phrases R20, R37, R40, R43, R50, R53a
    Operator protection A, H; U05a, U09a, U14, U15, U19a, U20a
    Environmental protection E15a, E16a, E16b, E38
    Storage and disposal D01, D02, D05, D09a, D10c, D12a

87   chlorothalonil + propamocarb hydrochloride

A contact and systemic fungicide mixture for blight control in potatoes

Products

1 Merlin	Bayer CropScience	375:375 g/l	SC	07943
2 Pan Magician	Pan Agriculture	375:375 g/l	SC	11992

Uses
- Blight in **potatoes**

Approval information
- Chlorothalonil included in Annex I under EC Directive 91/414

Efficacy guidance
- Commence treatment early in the season as soon as there is risk of infection
- In the absence of a blight warning treatment should start just before potatoes meet along the row
- Use only as a protectant. Stop use when blight readily visible (1% leaf area destroyed)
- Repeat sprays at 10-14 d intervals depending on blight infection risk. See label for details
- Complete blight spray programme after end Aug up to haulm destruction with protectant fungicides preferably fentin based

Restrictions
- Maximum number of treatments 5 per crop
- Apply to dry foliage. Do not apply if rainfall or irrigation imminent

Crop-specific information
- HI 7 d

FOR FULL CONDITIONS OF USE ALWAYS READ THE PRODUCT LABEL

Environmental safety
- Dangerous for the environment
- Very toxic to aquatic organisms
- LERAP Category B

Hazard classification and safety precautions
    Hazard H03, H11
    Risk phrases R20, R40, R41, R43, R50, R53a
    Operator protection A, C, H; U05a, U08, U11, U14, U19a, U20a
    Environmental protection E15a, E16a, E16b, E34, E38
    Storage and disposal D01, D02, D05, D09a, D10b, D12a

88   chlorothalonil + pyrimethanil

A fungicide mixture for use in combining peas

Products

Walabi	BASF	375:150 g/l	SC	12265

Uses
- Ascochyta in **combining peas**
- Mycosphaerella in **combining peas**

Approval information
- Chlorothalonil included in Annex I under EC Directive 91/414

Efficacy guidance
- Use only as a protectant spray in a two spary programme
- For best results ensure good foliar cover

Restrictions
- Maximum total dose equivalent to two full dose treatments
- Consult processors before use on crops for processing

Crop-specific information
- HI 6 wk for combining peas

Environmental safety
- Dangerous for the environment
- Very toxic to aquatic organisms
- LERAP Category B

Hazard classification and safety precautions
    Hazard H03, H11
    Risk phrases R36, R40, R43, R50, R53a
    Operator protection A, C, H; U02a, U11
    Environmental protection E15a, E16a
    Storage and disposal D01, D02, D05, D09a, D10c, D12a
    Medical advice M03

89   chlorothalonil + tetraconazole

A systemic protectant and curative fungicide mixture for cereals

Products

Voodoo	Sipcam	250:62.5 g/l	SE	09414

Uses
- Brown rust in **spring barley**, **spring barley** *(autumn sown)*, **spring wheat**, **spring wheat** *(autumn sown)*, **winter barley**, **winter wheat**
- Crown rust in **spring oats**, **winter oats**
- Net blotch in **spring barley**, **spring barley** *(autumn sown)*, **winter barley**
- Powdery mildew in **spring barley**, **spring barley** *(autumn sown)*, **spring wheat**, **spring wheat** *(autumn sown)*, **winter barley**, **winter wheat**
- Rhynchosporium in **spring barley**, **spring barley** *(autumn sown)*, **winter barley**
- Septoria diseases in **spring wheat**, **spring wheat** *(autumn sown)*, **winter wheat**

SEE SECTION 3 FOR PRODUCTS ALSO REGISTERED

SECTION 2

- Sooty moulds in **spring wheat**, **winter wheat**
- Yellow rust in **spring barley**, **spring barley** *(autumn sown)*, **spring wheat**, **spring wheat** *(autumn sown)*, **winter barley**, **winter wheat**

Approval information
- Chlorothalonil included in Annex I under EC Directive 91/414

Efficacy guidance
- For best results apply before onset of disease attack, or at an early stage of disease development. Further treatments may be necessary if disease pressure remains high
- May be used in a programme in combination with a number of other fungicides to improve overall control and reduce potential for development of resistance

Restrictions
- Maximum total dose equivalent to three full dose treatments on wheat; two full dose treatments on barley

Crop-specific information
- Latest use: before early flowering (GS 63) in wheat; before end of ear emergence (GS 59) in barley

Environmental safety
- Harmful to fish or other aquatic life. Do not contaminate surface waters or ditches with chemical or used container
- LERAP Category B

Hazard classification and safety precautions
Hazard H04
Risk phrases R36, R38
Operator protection A, C; U05a, U19a, U20a
Environmental protection E13c, E16a, E16b, E34
Storage and disposal D01, D02, D09a, D10b
Medical advice M03

90   chlorotoluron

A contact and residual urea herbicide for cereals

Products

1	Alpha Chlortoluron 500	Makhteshim	500 g/l	SC	04848
2	Atol	Nufarm UK	700 g/l	SC	11138
3	Copal 500	DAPT	500 g/l	SC	10633
4	CTU Minrinse	Makhteshim	90% w/w	WG	12458
5	Headland Tolerate	Headland	500 g/l	SC	10774
6	Lentipur CL 500	Nufarm UK	500 g/l	SC	08743
7	Luxan Chlorotoluron 500 Flowable	Luxan	500 g/l	SC	12431
8	Tolugan 700	Makhteshim	700 g/l	SC	08064
9	Tolurex 90 WDG	Makhteshim	90% w/w	WG	11403

Uses
- Annual dicotyledons in **durum wheat** [4, 6, 7, 9]; **triticale, winter barley, winter wheat** [1-9]
- Annual grasses in **durum wheat** [4, 6, 7, 9]; **triticale, winter barley, winter wheat** [1-9]
- Blackgrass in **durum wheat** [4, 6, 7, 9]; **triticale, winter barley, winter wheat** [1-9]
- Rough meadow grass in **durum wheat** [4, 6, 7, 9]; **triticale, winter barley, winter wheat** [1-9]
- Wild oats in **durum wheat** [4, 6, 7, 9]; **triticale, winter barley, winter wheat** [1-9]

Approval information
- Chlorotoluron included in Annex I under EC Directive 91/414
- May be applied through CDA equipment. See labels for details [2-4, 6, 7]
- Approved for aerial application on winter wheat, winter barley, durum wheat, triticale [6]. See notes in Section 5
- Accepted by BBPA for use on malting barley

Efficacy guidance
- Best results achieved by application soon after drilling. Application in autumn controls most weeds germinating in early spring
- For wild oat control apply within 1 wk of drilling, not after 2-leaf stage. Blackgrass and meadow grasses controlled to 5 leaf, ryegrasses to 3 leaf stage

FOR FULL CONDITIONS OF USE ALWAYS READ THE PRODUCT LABEL

- Any trash or burnt straw should be buried and dispersed during seedbed preparation
- Control may be reduced if prolonged dry conditions follow application
- Harrowing after treatment may reduce weed control
- Always follow WRAG guidelines for preventing and managing herbicide resistant weeds. Section 5 for more information

Restrictions
- Maximum number of treatments 1 per crop
- Use only on listed crop varieties. See label. Ensure seed well covered at drilling
- Apply only as pre-emergence spray in durum wheat, pre- or post-emergence in wheat, barley or triticale
- Do not apply pre-emergence to crops sown after 30 Nov
- Do not apply to crops severely checked by waterlogging, pests, frost or other factors
- Do not use on undersown crops or those due to be undersown
- Do not apply post-emergence in mixture with liquid fertilizers
- Do not roll for 7 d before or after application to an emerged crop
- Do not use on soils with more than 10% organic matter
- Crops on stony or gravelly soils may be damaged, especially after heavy rain

Crop-specific information
- Latest use: pre-emergence for durum wheat. Post emergence timings on other crops vary - see labels
- Early sown crops may be damaged if applied before or during a period of rapid autumn growth
- Autumn treated crops must be drilled by 1 Dec and spring treated crops must be drilled by 1 Feb

Environmental safety
- Dangerous for the environment
- Very toxic to aquatic organisms. May cause long-term adverse effects in the aquatic environment
- Harmful to fish or other aquatic life. Do not contaminate surface waters or ditches with chemical or used container [3]

Hazard classification and safety precautions
Hazard H03 [1-3, 5, 6]; H04 [1, 3]; H11 [1, 2, 4-9]
Risk phrases R22a, R36, R38 [1, 3, 5]; R50, R53a [1, 2, 4-9]; R63, R68 [2, 6]
Operator protection A [1-9]; C [1-6, 8, 9]; H [7]; U05a [1-5, 7-9]; U08, U19a [1-9]; U10, U11 [5]; U20a [2, 8]; U20b [1, 3-7, 9]
Environmental protection E13c [3]; E15a [1, 2, 4-9]; E34 [1-5, 7-9]
Storage and disposal D01, D02, D09a, D10b [1-9]; D05 [5, 7]; D12a [2, 5-7]; D12b [1, 4, 8, 9]
Medical advice M03 [1-5, 7-9]

91   chlorotoluron + isoproturon

A urea herbicide mixture for cereals

Products
Tolugan Extra	Makhteshim	300:300 g/l	SC	09393

Uses
- Annual dicotyledons in **winter barley, winter wheat**
- Annual grasses in **winter barley, winter wheat**
- Blackgrass in **winter barley, winter wheat**
- Rough meadow grass in **winter barley, winter wheat**
- Wild oats in **winter barley, winter wheat**

Approval information
- Chlorotoluron and isoproturon included in Annex I under EC Directive 91/414
- Accepted by BBPA for use on malting barley

Efficacy guidance
- Best results achieved by application soon after drilling. Application in autumn controls most weeds germinating in early spring
- For wild oat control apply within 1 wk of drilling, not after 2-leaf stage. Blackgrass and meadow grasses controlled to 5 leaf, ryegrasses to 3 leaf stage
- Any trash or burnt straw should be buried and dispersed during seedbed preparation

SEE SECTION 3 FOR PRODUCTS ALSO REGISTERED

- Control may be reduced if prolonged dry conditions follow application
- Harrowing after treatment may reduce weed control
- Where strains of herbicide-resistant blackgrass occur control may be unsatisfactory
- Always follow WRAG guidelines for preventing and managing herbicide resistant weeds. Section 5 for more information

Restrictions

- Maximum number of treatments 1 per crop
- Use only on listed crop varieties. See label. Ensure seed well covered at drilling
- Apply only as post-emergence treatment
- Do not apply to crops severely checked by waterlogging, pests, frost or other factors
- Do not use on undersown crops or those due to be undersown
- Do not apply post-emergence in mixture with liquid fertilizers
- Do not roll for 7 d before or after application to an emerged crop
- Do not use on soils with more than 10% organic matter
- Crops on stony or gravelly soils may be damaged, especially after heavy rain

Crop-specific information

- Latest use: before end of tillering (GS 29)

Environmental safety

- Dangerous for the environment
- Very toxic to aquatic organisms
- Do not apply to dry, cracked or waterlogged soils where heavy rain may lead to contamination of drains by isoproturon

Hazard classification and safety precautions

Hazard H03, H11
Risk phrases R40, R43, R50, R53a
Operator protection A, C, H; U05a, U08, U19a, U20b
Environmental protection E15a, E34
Storage and disposal D01, D02, D05, D09a, D10c, D12b

92   chlorpropham

A residual carbamate herbicide and potato sprout suppressant

See also chloridazon + chlorpropham + metamitron

Products

1	BL 500	Whyte Agrochemicals	500 g/l	HN	00279
2	Comrade	United Phosphorus	400 g/l	EC	10181
3	Luxan Gro Stop 100	Luxan	300 g/l	HN	12052
4	Luxan Gro-Stop Fog	Luxan	300 g/l	HN	09388
5	Luxan Gro-Stop HN	Luxan	300 g/l	HN	07689
6	MSS CIPC 5 G	Whyte Agrochemicals	5% w/w	GR	01402
7	MSS CIPC 50 M	Whyte Agrochemicals	500 g/l	HN	11165
8	MSS Sprout Nip	Whyte Agrochemicals	100% w/w	HN	11804
9	Warefog 25	Whyte Agrochemicals	600 g/l	HN	06776

Uses

- Annual dicotyledons in *acroclinium, african marigolds, blackcurrants, calendula, carrots, celery, china asters, chrysanthemums, coreopsis, flower bulbs, french marigolds, gooseberries, lettuce, lucerne, onions, parsley, straw flower, strawberries, sweet sultan* [2]
- Annual grasses in *acroclinium, african marigolds, blackcurrants, calendula, carrots, celery, china asters, chrysanthemums, coreopsis, flower bulbs, french marigolds, gooseberries, lettuce, lucerne, onions, parsley, straw flower, strawberries, sweet sultan* [2]
- Chickweed in *acroclinium, african marigolds, blackcurrants, calendula, carrots, celery, china asters, chrysanthemums, coreopsis, french marigolds, gooseberries, lettuce, lucerne, parsley, straw flower, strawberries, sweet sultan* [2]

- Polygonums in *acroclinium*, *african marigolds*, *blackcurrants*, *calendula*, *carrots*, *celery*, *china asters*, *chrysanthemums*, *coreopsis*, *french marigolds*, *gooseberries*, *lettuce*, *lucerne*, *parsley*, *straw flower*, *strawberries*, *sweet sultan* [2]
- Sprout suppression in *potatoes* [6, 7]; *potatoes* (thermal fog) [8]; *ware potatoes* (fog) [9]; *ware potatoes* (thermal fog) [1, 3-5]

Approval information

- Chlorpropham included in Annex I under EC Directive 91/414
- Some products are formulated for application by thermal fogging. See labels for details [1, 3-5, 8, 9]

Efficacy guidance

- Apply weed control sprays to freshly cultivated soil. Adequate rainfall must occur after spraying. Activity is greater in cold, wet than warm, dry conditions
- For sprout suppression apply with suitable fogging or rotary atomiser equipment or sprinkle granules over dry tubers before sprouting commences. Repeat applications may be needed. See labels for details
- Best results on potatoes obtained in purpose-built box stores with suitable forced draft ventilation. Potatoes in bulk stores should not be stacked more than 3 m high
- Blockage of air spaces between tubers prevents circulation of vapour and consequent loss of efficacy
- It is important to treat potatoes before the eyes open to obtain best results
- Effectiveness of fogging reduced in non-dedicated stores without proper insulation and temperature controls. Best results obtained at 5-10°C and 75-80% humidity

Restrictions

- Maximum number of treatments normally 1 per batch for sprout suppression of potatoes; 2 per yr for flower bulbs; 1 per yr for grass seed crops [2]; 1 per crop for other named crops
- Not to be used on grass seed crops if grass to be grazed or cut for fodder before 31 May following treatment [2]
- Excess rainfall after application may result in crop damage
- Do not use on Sands, Very Light soils or soils low in organic matter
- Poor conditions at drilling or planting, soil compaction, surface capping, waterlogging or attack by pests may result in crop damage
- On crops under glass high temperatures and poor ventilation may cause crop damage
- Only clean, mature, disease-free potatoes should be treated for sprout suppression. Use of chlorpropham can inhibit tuber wound healing and the severity of skin spot infection in store may be increased if damaged tubers are treated
- Do not fog potatoes with a high level of skin spot
- Do not use on potatoes for seed. Do not handle, dry or store seed potatoes or any other seed or bulbs in boxes or buildings in which potatoes are being or have been treated

Crop-specific information

- Latest use: 3 d before drilling carrots; 7d before planting out celery; 2 wk before planting or at least 10 d after planting for chrysanthemums [2]; 21 d before removal for sale or processing for potatoes [1, 6, 7, 9]; before tulip leaves unfurl or flower bulbs 5 cm high
- Cure potatoes according to label instructions before treatment and allow 3 wk between completion of loading into store and first treatment
- Apply weed control sprays to seeded crops pre-emergence of crop or weeds, to onions as soon as first crop seedlings visible, to planted crops a few days before planting, to bulbs immediately after planting, to fruit crops in late autumn-early winter. See label for further details

Following crops guidance

- There is a risk of damage to seed potatoes which are handled or stored in boxes or buildings previously treated with chlorpropham

Environmental safety

- Dangerous for the environment
- Toxic to aquatic organisms
- Keep unprotected persons out of treated stores for at least 24 h after application
- Keep in original container, tightly closed, in a safe place, under lock and key

SEE SECTION 3 FOR PRODUCTS ALSO REGISTERED

Hazard classification and safety precautions

Hazard H02, H07 [1, 7, 8]; H03 [2-5, 9]; H08 [2]; H11 [1, 3-9]

Risk phrases R20, R21, R22a [2]; R23, R24, R39 [1, 7, 8]; R25 [1]; R36 [1, 7-9]; R38 [2, 9]; R40, R42, R43 [3-5]; R48 [9]; R51 [1, 3-5, 7, 9]; R52 [6]; R53a [1, 3-7, 9]

Operator protection A [1-5, 7, 8]; C [1, 2, 7, 8]; D, E, H, M [1, 3-5, 7-9]; J [1, 7-9]; U02a, U04a, U20a [3-5]; U05a [1-9]; U08 [1-5, 7-9]; U09a [2]; U14, U15 [3-5, 9]; U19a [1, 3-5, 7-9]; U20b [1, 2, 6-8]; U20c [9]

Environmental protection E02a [1, 7, 8] (24 h); E02a [9] (12 h); E15a, E34 [1-9]

Consumer protection C12 [3-5]

Storage and disposal D01, D02 [1-9]; D05 [1-5, 7, 8]; D06b [1, 2, 7, 8]; D07, D12a [3-5]; D08, D10b [2]; D09a [2, 6, 9]; D09b [1, 3-5, 7, 8]; D10a [1, 7-9]; D11a [3-6]; D12b [1, 2, 6-9]

Medical advice M03 [2]; M04 [1, 3-5, 7, 8]

93   chlorpropham with cetrimide

A soil-acting herbicide for lettuce under cold glass

Products

Croptex Pewter	Certis	80:80 g/l	SC	11181

Uses

- Annual dicotyledons in **protected lettuce**
- Annual grasses in **protected lettuce**
- Chickweed in **protected lettuce**
- Polygonums in **protected lettuce**

Approval information

- Chlorpropham included in Annex I under EC Directive 91/414

Efficacy guidance

- Adequate irrigation must be applied before or after treatment
- Best results achieved on firm soil of fine tilth, free from clods and weeds
- Control of susceptible weeds is achieved at, or shortly after, germination and residual control lasts for 6-8 wk depending on weather conditions

Restrictions

- Maximum number of treatments 1 per crop
- Apply within 24 h of drilling
- Do not apply to crop foliage or use where seed has germinated
- Excess irrigation may cause temporary check to crop under certain circumstances
- Do not apply where tomatoes, brassicas or other sensitive crops are growing in the same house

Crop-specific information

- Latest use: before crop emergence or pre-planting
- Apply to drilled lettuce under cold glass within 24 h post-drilling, to transplanted crops pre-planting and treat up to 7 d later

Following crops guidance

- In the event of crop failure only lettuce should be grown within 2 mth
- After cutting a normally harvested crop the treated area should be cultivated to a minimum depth of 15 cm before sowing any succeeding crop

Environmental safety

- Dangerous for the environment
- Toxic to aquatic organisms
- Highly flammable

Hazard classification and safety precautions

Hazard H03, H07, H11

Risk phrases R21, R22a, R36, R37, R38, R51, R53a

Operator protection A, C; U05a, U08, U14, U15, U19a, U20a

Environmental protection E15a, E34

Storage and disposal D01, D02, D12a

Medical advice M03

FOR FULL CONDITIONS OF USE ALWAYS READ THE PRODUCT LABEL

94   chlorpropham + fenuron

A residual herbicide for runner beans and spinach

Products

Croptex Chrome	Certis	80:15 g/l	EC	11180

Uses
- Annual dicotyledons in **leaf spinach**, **runner beans** *(off-label)*
- Annual grasses in **leaf spinach**
- Chickweed in **leaf spinach**

Specific Off-Label Approvals (SOLAs)
- **runner beans** *(OLA 030871) Dec 2008* [1]

Approval information
- Chlorpropham included in Annex I under EC Directive 91/414
- Products containing fenuron have been granted derogations for specified 'Essential Uses' for use until 31 December 2007. Sale and supply must cease by 30 June 2007 but growers have no guarantee that the products will continue to be available until then.
  For more information see 'The Review Programme' under 'Pesticide Legislation' in Section 5

Efficacy guidance
- Apply to soil freshly cultivated and free of established weeds. Adequate rainfall must occur after spraying. Activity is greater in cold, wet than warm, dry conditions

Restrictions
- Maximum number of treatments 1 per crop
- Do not use on Sands, Very Light soils or soils low in organic matter
- Poor conditions at drilling or planting, soil compaction, surface capping, waterlogging or attack by pests may result in crop damage
- This product must not be used on any crops other than those listed, including any extrapolations that would normally be permissible under the Long Term Arrangements for Extension of Use (see Section 5)

Crop-specific information
- Latest use: pre crop emergence

Following crops guidance
- In the event of crop failure only recommended crops should be replanted in treated soil
- Plough or cultivate to 15 cm after harvest to dissipate any residues

Environmental safety
- Dangerous for the environment
- Very toxic to aquatic organisms
- Do not empty into drains

Hazard classification and safety precautions
  Hazard H03, H08, H11
  Risk phrases R21, R22a, R36, R37, R38, R50, R53a
  Operator protection A, C; U05a, U08, U11, U13, U14, U19a, U20b
  Environmental protection E15a, E19b, E34
  Storage and disposal D01, D02, D09a, D10a, D12b
  Medical advice M03, M05a

95   chlorpropham + metamitron

A contact and residual herbicide mixture for beet crops

Products

Golmet	Nufarm UK	24.5:280 g/l	SC	10684

Uses
- Annual dicotyledons in **fodder beet**, **mangels**, **sugar beet**
- Annual meadow grass in **fodder beet**, **mangels**, **sugar beet**

SECTION 2

SEE SECTION 3 FOR PRODUCTS ALSO REGISTERED

Approval information
- Chlorpropham included in Annex I under EC Directive 91/414

Efficacy guidance
- May be used pre-emergence alone or post-emergence in tank mixture or with an authorised adjuvant oil
- Use as part of a spray programme. For optimum efficacy a full programme of pre- and post-emergence sprays is recommended
- Ideally one pre-emergence application should be followed by a repeat overall low dose programme on mineral soils. On organic soils a programme of up to five post-emergence sprays is likely to be most effective
- Product combines residual and contact activity. Adequate soil moisture and a fine firm seedbed are essential for optimum residual activity

Restrictions
- Maximum number of treatments one pre-emergence plus three post-emergence or, where no pre-emergence treatment has been applied, up to five post-emergence treatments with an adjuvant oil

Crop-specific information
- Latest use: before crop foliage meets across the rows

Following crops guidance
- Only sugar beet, fodder beet or mangels should be sown as following crops within 4 mth of treatment. After normal harvest of a treated crop land should be mouldboard ploughed to 15 cm after which any spring crops may be sown or planted

Environmental safety
- Dangerous for the environment
- Very toxic to aquatic organisms

Hazard classification and safety precautions
Hazard H03, H11
Risk phrases R22a, R36, R38, R50, R53a
Operator protection A, C; U05a, U08, U11, U14, U15, U19a, U20a
Environmental protection E15a, E34
Storage and disposal D01, D02, D05, D09a, D10c

96   chlorpropham + pentanochlor

A contact and residual herbicide for horticultural crops

Products

Atlas Brown	Nufarm UK	150:300 g/l	EC	07703

Uses
- Annual dicotyledons in *carrots, celeriac, celery, chrysanthemums, narcissi, parsley, parsnips, tulips*
- Annual meadow grass in *carrots, celeriac, celery, chrysanthemums, narcissi, parsley, parsnips, tulips*

Approval information
- Chlorpropham included in Annex I under EC Directive 91/414
- Products containing pentanochlor have been granted derogations for specified 'Essential Uses' for use until 31 December 2007. Sale and supply must cease by 30 June 2007 but growers have no guarantee that the products will continue to be available until then.
  For more information see 'The Review Programme' under 'Pesticide Legislation' in Section 5

Efficacy guidance
- Apply as pre- or post-weed emergence spray
- Best results by application to weeds up to 2-leaf stage on fine, firm, moist seedbed
- Greatest contact action achieved under warm, moist conditions, the short residual action greatest in earlier part of year

Restrictions
- Maximum number of treatments 2 per crop
- This product must not be used on any crops other than those listed, including any extrapolations that would normally be permissible under the Long Term Arrangements for Extension of Use (see Section 5), except herbs and ornamentals

Crop-specific information
- Latest use: pre-emergence for narcissi, tulips
- HI 28 d for carrots, parsnips
- Apply to carrots and related crops pre- or post-emergence after fully expanded cotyledon stage, to narcissi and tulips at any time before emergence
- Apply to chrysanthemums either pre-planting or after planting as carefully directed spray, avoiding foliage

Following crops guidance
- Any crop may be sown or planted after 4 wk following ploughing and cultivation

Environmental safety
- Dangerous for the environment
- Toxic to aquatic organisms

Hazard classification and safety precautions
Hazard H03, H11
Risk phrases R21, R22a, R22b, R36, R48, R51, R53a
Operator protection C; U05a, U09a, U14, U15, U19a, U20a
Environmental protection E15a, E38
Consumer protection C02 (carrots, parsnips 28 d)
Storage and disposal D01, D02, D09a, D10b, D12b
Medical advice M04, M05b

97   chlorpyrifos

A contact and ingested organophosphorus insecticide and acaricide

Products

1	Alpha Chlorpyrifos 48 EC	Makhteshim	480 g/l	EC	04821
2	Ballad	Headland	480 g/l	EC	11659
3	Chevron 48	DAPT	480 g/l	EC	10645
4	Crossfire 480	Bayer Environ.	480 g/l	EC	09929
5	Cyren	Headland	480 g/l	EC	11028
6	Dursban WG	Dow	75% w/w	WG	09153
7	Equity	Dow	480 g/l	EC	11520
8	Greencrop Pontoon	Greencrop	480 g/l	EC	09667
9	Lorsban T	Rigby Taylor	480 g/l	EC	07813
10	Lorsban WG	Landseer	75% w/w	WG	11962
11	Maraud	Scotts	480 g/l	EC	09274
12	Pyrinex 48EC	Makhteshim	480 g/l	EC	08644
13	Spannit	SumiAgro Amenity	480 g/l	EC	08744
14	Spannit Granules	SumiAgro	6% w/w	GR	10935
15	Standon Chlorpyrifos 48	Standon	480 g/l	EC	08286
16	suSCon Green Soil Insecticide	Fargro	10% w/w	CG	06312

Uses
- Ambrosia beetle in *cut logs* [1, 3, 5, 6, 8, 9, 12]
- Aphids in *apples, gooseberries, pears, plums, raspberries, strawberries* [1, 3, 5, 7, 8, 10, 12, 13, 15]; *blackcurrants* [1, 10, 12]; *broccoli, cabbages, calabrese, cauliflowers, chinese cabbage* [1, 3, 5-8, 12, 15]; *brussels sprouts* [1, 12, 15]; *currants* [3, 5, 7, 8, 13, 15]; *edible brassicas* [13]; *redcurrants, whitecurrants* [8, 10]; *spring barley* [12]; *spring oats, spring wheat* [5]
- Apple blossom weevil in *apples* [1, 3, 5, 7, 8, 10, 12, 13, 15]
- Apple sucker in *apples* [15]
- Bean seed fly in *green beans* (off-label - harvested as a dry pulse) [1, 5, 6]
- Cabbage root fly in *broccoli, cabbages, cauliflowers* [1, 3, 5-8, 12, 14, 15]; *brussels sprouts* [1, 6, 7, 12, 15]; *calabrese, chinese cabbage* [1, 3, 5-8, 12, 15]; *choi sum* (off-label), *pak choi*

(off-label), **tatsoi** *(off-label)* [5]; **collards** *(off-label)*, **kale** *(off-label)*, **kohlrabi** *(off-label)*, **mooli** *(off-label)*, **radishes** *(off-label)* [1, 5, 6]; **edible brassicas** [13]

- Capsids in **apples, gooseberries, pears** [1, 3, 5, 7, 8, 10, 12, 13, 15]; **blackcurrants** [1, 10, 12]; **currants** [3, 5, 7, 8, 13, 15]; **redcurrants, whitecurrants** [8, 10]
- Caterpillars in **apples** [1, 3, 5, 7, 8, 12, 13]; **blackcurrants** [1, 10, 12]; **broccoli, cabbages, calabrese, cauliflowers, chinese cabbage** [1, 3, 5-8, 12, 15]; **brussels sprouts** [1, 12, 15]; **currants** [3, 5, 7, 8, 13, 15]; **edible brassicas** [13]; **gooseberries** [1, 3, 5, 7, 8, 10, 12, 13, 15]; **pears** [1, 3, 5, 7, 8, 10, 12, 13]; **redcurrants, whitecurrants** [8, 10]
- Codling moth in **apples, pears** [1, 3, 5, 7, 8, 10, 12, 13, 15]
- Cutworms in **broccoli, cabbages, calabrese, cauliflowers, chinese cabbage** [1, 3, 5-8, 12, 15]; **brussels sprouts** [1, 12, 15]; **edible brassicas** [13]; **onions** [1, 5-8, 12, 15]; **potatoes** [1, 5, 8, 12, 13, 15]; **seed potatoes** [6, 7]
- Damson-hop aphid in **plums** [1, 8, 10, 12, 15]
- Elm bark beetle in **cut logs** [3, 5, 9]
- Frit fly in **amenity grass** [2, 4, 5, 11, 15]; **amenity turf** [8]; **golf courses** [5, 13]; **grassland, maize, spring oats, spring wheat, winter oats, winter wheat** [1, 3, 5-8, 12, 13, 15]; **managed amenity turf** [1-5, 9, 11, 12, 15]; **rotational grassland** [1, 12]; **spring barley** [7, 13]; **winter barley** [7]
- Larch shoot beetle in **cut logs** [1, 3, 5-9, 12]
- Leatherjackets in **amenity grass** [2, 4, 5, 11, 15]; **amenity turf** [8]; **broccoli, cabbages, calabrese, cauliflowers, chinese cabbage** [3, 5-7, 15]; **brussels sprouts** [15]; **edible brassicas** [13]; **golf courses** [5, 13]; **grassland, spring oats, spring wheat, sugar beet, winter oats, winter wheat** [1, 3, 5-8, 12, 13, 15]; **managed amenity turf** [1-5, 9, 11, 12, 15]; **rotational grassland** [1, 12]; **spring barley** [1, 3, 5-7, 12, 13]; **winter barley** [1, 3, 5-7, 12, 13]
- Mealy aphid in **plums** [1, 8, 10, 12]
- Narcissus fly in **flower bulbs** *(off-label)* [1, 5]
- Pear sucker in **pears** [15]
- Pine shoot beetle in **cut logs** [1, 3, 5-9, 12]
- Pygmy mangold beetle in **sugar beet** [3, 5-7, 13]
- Raspberry beetle in **raspberries** [1, 3, 5, 7, 8, 10, 12, 13, 15]
- Raspberry cane midge in **raspberries** [1, 3, 5, 7, 8, 10, 12, 13, 15]
- Red spider mites in **apples, gooseberries, pears, plums, raspberries, strawberries** [1, 3, 5, 7, 8, 10, 12, 13, 15]; **blackcurrants** [1, 10, 12]; **currants** [3, 5, 7, 8, 13, 15]; **redcurrants, whitecurrants** [8, 10]
- Sawflies in **apples** [1, 3, 5, 7, 8, 10, 12, 13, 15]; **pears** [7, 13]
- Strawberry blossom weevil in **strawberries** [7, 10]
- Suckers in **apples, pears** [1, 3, 5, 7, 8, 10, 12, 13]
- Summer-fruit tortrix moth in **apples** [15]
- Thrips in **spring oats, spring wheat, winter oats, winter wheat** [3, 5, 13]; **winter barley** [13]
- Tortrix moths in **apples, pears, plums, strawberries** [1, 3, 5, 7, 8, 10, 12, 13, 15]
- Vine weevil in **container-grown ornamentals, ornamental plant production** [16]; **ornamental plant production** *(conifers)* [1]; **strawberries** [1, 3, 5, 7, 8, 10, 12, 15]
- Wheat bulb fly in **spring barley, winter barley** [6, 7]; **spring oats, spring wheat, winter oats, winter wheat** [1, 3, 5-8, 12, 13, 15]
- Wheat-blossom midge in **spring oats, spring wheat** [5, 7]; **winter oats, winter wheat** [7]
- Whitefly in **broccoli, cabbages, calabrese, cauliflowers, chinese cabbage** [3, 5-7]
- Winter moth in **apples, pears, plums** [1, 3, 5, 7, 8, 10, 12, 13, 15]
- Woolly aphid in **apples** [1, 3, 5, 7, 8, 10, 12, 13, 15]; **pears** [3, 5, 7, 13]
- Yellow cereal fly in **spring wheat, winter wheat** [6]

Specific Off-Label Approvals (SOLAs)

- **choi sum, collards, kale, pak choi, tatsoi** *(OLA 050237) Dec 2008* [5]
- **collards, kale** *(OLA 021688) Dec 2008* [1]
- **collards, kale** *(OLA 031390) Dec 2008* [6]
- **flower bulbs** *(OLA 021458) Dec 2008* [1]
- **flower bulbs** *(OLA 050236) Dec 2008* [5]
- **green beans** *(harvested as a dry pulse) (OLA 041977) Dec 2008* [1]
- **green beans** *(harvested as a dry pulse) (OLA 050235) Dec 2008* [5]
- **green beans** *(harvested as a dry pulse) (OLA 031391) Dec 2008* [6]
- **kohlrabi, mooli, radishes** *(OLA 041977) Dec 2008* [1]

FOR FULL CONDITIONS OF USE ALWAYS READ THE PRODUCT LABEL

- **kohlrabi**, **mooli**, **radishes** *(OLA 050235) Dec 2008* [5]
- **kohlrabi**, **mooli**, **radishes** *(OLA 031391) Dec 2008* [6]

Approval information
- Chlorpyrifos included in Annex I under EC Directive 91/414
- Accepted by BBPA for use on malting barley
- Following implementation of Directive 98/82/EC, approval for use of chlorpyrifos on numerous crops was revoked in 1999
- Following the environmental review of chlorpyrifos uses on ware potatoes, carrots, conifers (drench), forestry trees (dip) and cereals (for aphids and orange blossom midge) were revoked in September 2005. In addition the maximum permitted number of treatments, or the maximum total dose, on several other crops was reduced. These changes have been included in the 2006 edition but use of existing stocks as previously labelled is permitted until 30 September 2006.

Efficacy guidance
- Brassicas raised in plant-raising beds may require retreatment at transplanting
- Activity on field crops may be reduced when soil temperature below 5°C or on organic soils
- In dry conditions the effect of granules applied as a surface band may be reduced [14]
- Where pear suckers resistant to chlorpyrifos occur control is unlikely to be satisfactory
- Efficacy against frit fly and leatherjackets in grass may be reduced if applied during periods of frost
- Thorough incorporation by hand or mixing machine is essential when used in potting media for ornamental plant production
- When used for control of vine weevil in ornamental plant production the recently emerged and young larvae are controlled by incorporation into the growing media. A single application will control successive generations for the growing cycle of plants kept in the same containers for 2 yr, with partial control for a third yr. Where plants are potted-on into larger containers the fresh growing media must also be treated to obtain control [16]

Restrictions
- Maximum number of treatments and timing vary with crop, product and pest. See label for details
- Do not apply to sugar beet under stress or within 4 d of applying a herbicide
- Once incorporated, treated growing media must be used within 30 d
- Do not use product in growing media used for aquatic or semi-aquatic plants, and not for propagation of any edible crop
- Do not graze lactating cows on treated pasture for 14 d after treatment

Crop-specific information
- Latest use 4d after transplanting edible brassica crops; end Jul for sugar beet; varies for other crops. See individual labels
- HI range from 7 d to 6 wk depending on crop. See labels
- For vine weevil control in ornamental nursery stock incorporate in growing medium when plants first potted from rooted cutting stage. Treat the fresh growing medium when plants are potted on into larger containers [16]
- For turf pests apply from Nov where high larval populations detected or damage seen
- On lettuce apply only to strong well developed plants when damage first seen, or on professional advice
- In apples use pre-blossom up to pink/white bud and post-blossom after petal fall
- Test tolerance of glasshouse ornamentals before widescale use for propagating unrooted cuttings, or potting unusual plants and new species, and when using any media with a high content of non-peat ingredients

Environmental safety
- Dangerous for the environment
- Very toxic to aquatic organisms [1, 2, 4-12, 15]
- Toxic to aquatic organisms [14, 16]
- Extremely dangerous to fish or other aquatic life. Do not contaminate surface waters or ditches with chemical or used container [1-3, 5, 12, 13]
- Keep livestock out of treated areas for at least 14 d after treatment [1-3, 5-7, 9-12]
- High risk to bees. Do not apply to crops in flower or to those in which bees are actively foraging. Do not apply when flowering weeds are present [2, 4, 5, 10]
- Dangerous to bees. Do not apply to crops in flower or to those in which bees are actively foraging. Do not apply when flowering weeds are present [1, 3, 6-9, 11-13, 15]

SECTION 2

SEE SECTION 3 FOR PRODUCTS ALSO REGISTERED

- Broadcast air-assisted LERAP (18 m) and LERAP Category A (all except [16])
- Chlorpyrifos is dangerous to some beneficial arthropods, especially parasitoid wasps, ground beetles, rove beetles and hoverflies. Manufacturer makes advisory recommendation not to spray summer applications within 12m of the edge of the growing crop [6]

Hazard classification and safety precautions

Hazard H03 [1-16]; H04 [3, 13]; H08 [1, 3, 4, 7, 8, 11-13, 15]; H11 [1, 2, 4-12, 14-16]

Risk phrases R20 [1, 2, 5, 12]; R21 [4, 8, 13-15]; R22a [1-16]; R22b [1-5, 7, 8, 11, 12, 15]; R36 [1, 12, 13]; R37, R67 [4, 7-9, 11, 15]; R38 [1-5, 7-9, 11-13, 15]; R40 [1, 12]; R41, R43 [1, 3, 12]; R50 [1, 2, 4-12, 15]; R51 [14, 16]; R53a [1, 2, 4-12, 14-16]

Operator protection A [1-16]; C [1-3, 5, 12, 13]; H [1-3, 5, 8-10, 12, 13, 15]; P [7]; U02a [1, 3, 12-14, 16]; U05a [1-5, 7-9, 11-13, 15]; U08, U19a [1-13, 15]; U14 [1, 2, 5, 12]; U15 [1, 12]; U20a [13, 14]; U20b [1-12, 15, 16]

Environmental protection E06a [1-3, 5-7, 9-12] (14 d); E12a [2, 4, 5, 10]; E12d [1, 3, 6-9, 11-13, 15]; E13a [1-3, 5, 12]; E13b [13]; E15a [4, 6-11, 14, 15]; E16c, E16d, E34 [1-13, 15]; E17a [2, 5] (18 m); E17b [1, 3, 4, 6-12, 15] (18 m); E38 [4, 6, 7, 9-11, 14, 16]

Consumer protection C01 [16]; C02 [13, 14] (6 wk)

Storage and disposal D01 [1-9, 11-16]; D02 [1, 3, 4, 6-9, 11-13, 15]; D05 [1-3, 5, 8, 15, 16]; D09a [1-16]; D10b [1-3, 5-13, 15]; D10c [4]; D11a [13, 14, 16]; D12a [4, 6, 7, 9-11, 14, 16]; D12b [2, 5]

Medical advice M01 [1, 3, 4, 6, 8-16]; M03 [1-13, 15]; M04 [16]; M05b [1-5, 7-9, 11, 12, 15]

98   chlorpyrifos-methyl

An organophosphorus insecticide and acaricide for grain store use

Products

1 Greencrop Storeclean 225	Greencrop	225 g/l	EC	11098
2 Reldan 22	Dow	225 g/l	EC	08191
3 Reldan 22	Dow	225 g/l	EC	12404

Uses

- Grain storage pests in **stored grain**
- Pre-harvest hygiene in **grain stores**

Approval information

- Chlorpyrifos-methyl included in Annex I under EC Directive 91/414
- Accepted by BBPA for use in stores for malting barley

Efficacy guidance

- Apply to grain after drying to moisture content below 14%, cooling and cleaning
- Insecticide may become depleted at grain surface if grain is being cooled by continuous extraction of air from the base leading to reduced control of grain store pests especially mites
- Resistance to organophosphorus compounds sometimes occurs in insect and mite pests of stored products

Restrictions

- This product contains an anticholinesterase organophosphorus compound. Do not use if under medical advice not to work with such compounds
- Maximum number of treatments 1 per batch or 1 per store, prior to storage
- Only treat grain in good condition
- Do not treat grain intended for sowing
- Minimum of 90 d must elapse between treatment of grain and removal from store for consumption or processing

Crop-specific information

- May be applied pre-harvest to surfaces of empty store and grain handling machinery and as admixture with grain
- May be used in wheat, barley, oats, rye or triticale stores

Environmental safety

- Dangerous for the environment
- Very toxic to aquatic organisms

FOR FULL CONDITIONS OF USE ALWAYS READ THE PRODUCT LABEL

Hazard classification and safety precautions
  Hazard H03, H11
  Risk phrases R22b, R41, R50, R53a
  Operator protection A, C, H, M; U05a, U08, U19a, U20b
  Environmental protection E15a, E34
  Storage and disposal D01, D02, D05, D09a, D10b
  Medical advice M01, M03, M05b

99 chlorthal-dimethyl

A residual benzoic acid herbicide for use in horticulture

Products

Dacthal W-75	Certis	75% w/w	WP	11323

Uses

- Annual dicotyledons in **bedding plants, blackcurrants, broccoli, brussels sprouts, bulb onions, cabbages, calabrese, cauliflowers, chives** (off-label), **fodder rape, gooseberries, herbs (see appendix 6)** (off-label), **kale, leeks, parsley** (off-label), **raspberries, roses, runner beans, sage, shrubs, strawberries, swedes, trees, turnips**
- Black nightshade in **dwarf beans** (off-label), **navy beans** (off-label), **poppies** (off-label - for morphine production)
- Fat hen in **poppies** (off-label - for morphine production)
- General weed control in **courgettes** (off-label), **marrows** (off-label)
- Slender speedwell in **managed amenity turf**
- Volunteer potatoes in **dwarf beans** (off-label), **navy beans** (off-label)

Specific Off-Label Approvals (SOLAs)

- **chives, herbs (see appendix 6), parsley** (OLA 030873) Dec 2008 [1]
- **courgettes, marrows** (OLA 041561) Dec 2008 [1]
- **dwarf beans, navy beans** (OLA 030874) Dec 2008 [1]
- **poppies** (for morphine production) (OLA 032407) Dec 2008 [1]

Efficacy guidance

- Best results on fine firm weed-free soil when adequate rain or irrigation follows
- Apply after drilling or planting prior to weed emergence. Rates and timing vary with crop and soil type. See label for details
- For control of slender speedwell in turf apply when weeds growing actively. Do not mow for at least 3 d after treatment

Restrictions

- Maximum number of treatments 1 per crop or 1 per season for all crops
- Do not apply mixture with propachlor to newly planted strawberries after rolling or application of other herbicides
- Do not use on strawberries between flowering and harvest
- Many types of ornamental have been treated successfully. See label for details. For species of unknown susceptibility treat a small number of plants first
- Do not use on turf where bent grasses form a major constituent of sward
- Do not plant lettuce within 6 mth of application, seeded turf within 2 mth, other crops within 3 mth. In the event of crop failure deep plough before re-drilling or planting
- Do not use on dwarf French beans
- Do not use on organic soils

Crop-specific information

- Latest use: before crop emergence on runner beans, sage, other herbs and poppies; end of flowering for strawberries. Some restrictions apply to other crops. See label for details
- Recommended alone on roses, runner beans, various ornamentals, strawberries and turf, in tank-mix with propachlor on brassicas, onions, leeks, sage, ornamentals, established strawberries and newly planted soft fruit

SECTION 2

Following crops guidance
- The interval between trreatment and planting a following crop should be 6 mth for lettuce, 2 mth for seeded turf, 3 mth for other crops
- In the event of failure of a treated crop the land must be deep mould-board ploughed before re-drilling or re-planting

Hazard classification and safety precautions
    Hazard H03
    Risk phrases R40, R53a
    Operator protection U19a, U20a
    Environmental protection E15a
    Storage and disposal D01, D02, D09a, D11a
    Medical advice M04

100   chlorthal-dimethyl + propachlor

A residual herbicide mixture for a range of vegetable crops

Products

1 Decimate	Certis	225:216 g/l	SC	11008
2 Greencrop Vegex	Greencrop	225:216 g/l	SC	12439

Uses
- Annual dicotyledons in *broccoli, brussels sprouts, bulb onions, cabbages, calabrese, cauliflowers, fodder rape, kale, leeks, mustard, salad onions, swedes, turnips*
- Annual meadow grass in *broccoli, brussels sprouts, bulb onions, cabbages, calabrese, cauliflowers, fodder rape, kale, leeks, mustard, salad onions, swedes, turnips*

Efficacy guidance
- Product may be applied pre- or post-emergence of the crop but pre-emergence of the weeds. Emerged weeds are not controlled
- Soils should be fine, firm and free of weed growth
- For best results apply to moist soil and before rain. In dry weather irrigate lightly after application and before seedling weeds appear
- Weed control may be reduced under very wet, dry or windy conditions
- Do not use on soils with more than 10% organic matter as weed control will be impaired

Restrictions
- Maximum number of treatments 1 per crop for all crops

Crop-specific information
- Use on brassica crops pre-crop emergence may result in a check to growth but this is usually outgrown within 8 wk without effect on yield or quality
- Post-emergence treatment on brassica crops should not be made until they have 3-4 true leaves. Post-emergence treatment on onions should be made post-crook stage
- Young transplanted plants, particularly cabbages and cauliflowers, raised under glass should be hardened off before treatment
- Care should be taken with modular transplants to ensure that that roots are not exposed to spray

Following crops guidance
- The interval between treatment and planting succeeding crops should be 2 mth for seeded turf, 6 mth for lettuce, 3 mth for all other crops
- In the event of failure of a treated crop the land must be deep mouldboard ploughed before redrilling or replanting

Environmental safety
- Dangerous for the environment
- Toxic to aquatic organisms

Hazard classification and safety precautions
    Hazard H03, H11
    Risk phrases R36 [1]; R38 [2]; R40, R43, R51, R53a [1, 2]
    Operator protection A, C; U02a, U04a, U05a, U09b, U11, U14, U15, U19a, U20a
    Environmental protection E15a

FOR FULL CONDITIONS OF USE ALWAYS READ THE PRODUCT LABEL

Storage and disposal D01, D02, D09a, D10a
Medical advice M04

101   citronella oil

A natural plant extract herbicide

Products

Barrier H	Barrier	22.9% w/w	EW	10136

Uses
- Ragwort in *grassland, land not being used for crop production*

Efficacy guidance
- Best results obtained from spot treatment of ragwort in the rosette stage, during dry still conditions
- Aerial growth of ragwort is rapidly destroyed. Longer term control depends on overall management strategy
- Check for regrowth after 28 d and re-apply as necessary

Crop-specific information
- Contact with grasses will result in transient scorch which is outgrown in good growing conditions

Environmental safety
- Apply away from bees
- Harmful to fish or other aquatic life. Do not contaminate surface waters or ditches with chemical or used container
- Keep livestock out of treated areas for at least 2 wk and until foliage of any poisonous weeds such as ragwort has died and become unpalatable

Hazard classification and safety precautions
   Hazard H04
   Risk phrases R36, R38
   Operator protection A, C; U05a, U20c
   Environmental protection E07b (2 wk); E12g, E13c
   Storage and disposal D01, D02, D05, D09a, D11a

102   clodinafop-propargyl

A contact acting herbicide for annual grass weed control in wheat, triticale and rye

Products

1	Greencrop Boulevard	Greencrop	240 g/l	EC	09960
2	Landgold Clodinafop	Landgold	240 g/l	EC	10172
3	Landgold Clodinafop	Teliton	240 g/l	EC	12112
4	Marathon	Me2	240 g/l	EC	09959
5	Standon Clodinafop 240	Standon	240 g/l	EC	10174
6	Topik	Syngenta	240 g/l	EC	08461
7	Topik	Syngenta	240 g/l	EC	12333

Uses
- Blackgrass in *durum wheat, spring rye, spring wheat, triticale, winter rye, winter wheat*
- Rough meadow grass in *durum wheat, spring rye, spring wheat, triticale, winter rye, winter wheat*
- Wild oats in *durum wheat, spring rye, spring wheat, triticale, winter rye, winter wheat*

Approval information
- Approval expiry 31 Oct 2006 [1-3, 5]

Efficacy guidance
- Spray when majority of weeds have germinated but before competition reduces yield
- Products contain a herbicide safener (cloquintocet-mexyl) that improves crop tolerance to clodinafop-propargyl
- Optimum control achieved when all grass weeds emerged. Wait for delayed germination on dry or cloddy seedbed

SECTION 2

- A mineral oil additive is recommended to give more consistent control of very high blackgrass populations or for late season treatments. See label for details
- Weed control not affected by soil type, organic matter or straw residues
- Control may be reduced if rain falls within 1 h of treatment
- Always follow WRAG guidelines for preventing and managing herbicide resistant weeds. Section 5 for more information

Restrictions
- Maximum total dose equivalent to one full dose treatment
- Do not use on barley or oats
- Do not treat crops under stress or suffering from waterlogging, pest attack, disease or frost
- Do not treat crops undersown with grass mixtures
- Do not mix with products containing MCPA, mecoprop-P, 2,4-D or 2,4-DB
- MCPA, mecoprop, 2,4-D or 2,4-DB should not be applied within 21 d before, or 7 d after, treatment

Crop-specific information
- Latest use: before second node detectable stage (GS 32) for durum wheat, triticale, rye; before flag leaf sheath extending (GS 41) for wheat
- Spray in autumn, winter or spring from 1 true leaf stage (GS 11) to before second node detectable (GS 32) on durum, rye, triticale; before flag leaf sheath extends (GS 41) on wheat

Following crops guidance
- Only a broad leaved crop may be sown after failure of a treated crop. After normal harvest any broad leaved crop or wheat, durum wheat, rye, triticale or barley should be sown

Environmental safety
- Dangerous for the environment
- Very toxic to aquatic organisms

Hazard classification and safety precautions
    Hazard H03 [2-6]; H04 [1]; H11 [1-7]
    Risk phrases R36, R38, R43 [1-6]; R50, R53a [1-7]
    Operator protection A, C, H, K; U02a, U05a, U09a [1-7]; U19a, U20b [1]; U20a [2-7]
    Environmental protection E15a [1-7]; E38 [5-7]
    Storage and disposal D01, D02, D09a [1-7]; D05 [1-5, 7]; D10c, D12a [5-7]; D11a [1-4]

103   clodinafop-propargyl + trifluralin

A contact and residual herbicide for winter wheat

Products

Hawk	Syngenta	12:383 g/l	EC	08417

Uses
- Annual dicotyledons in **winter wheat**
- Blackgrass in **winter wheat**
- Rough meadow grass in **winter wheat**
- Wild oats in **winter wheat**

Efficacy guidance
- Product contains a herbicide safener (cloquintocet-mexyl) that improves crop tolerance to clodinafop-propargyl
- Optimum weed control obtained when most grass weeds emerged and broad leaved weeds at susceptible stage before they compete with the crop. Pre-emergence control of annual meadow-grass and broad leaved weeds may not be satisfactory on medium and heavy soils
- Delay treatment if dry or cloddy seedbeds favour late weed germination
- Optimum weed control obtained in crops growing in a fine, firm seedbed cleared of trash and straw
- Wild oats and other weeds germinating from beneath treated zone not controlled
- Speed of action depends on temperature and growing conditions and may appear slow in dry or cold weather

- Rain within 1 h of application may reduce grass weed control
- Always follow WRAG guidelines for preventing and managing herbicide resistant weeds. Section 5 for more information

Restrictions
- Maximum number of treatments 1 per crop
- Do not use on barley or oats
- Do not treat crops under stress or suffering from waterlogging, pest attack, disease or frost
- Do not treat crops undersown with grass mixtures
- Do not mix with products containing MCPA, mecoprop-P, 2,4-D or 2,4-DB
- MCPA, mecoprop, 2,4-D or 2,4-DB should not be applied within 21 d before, or 7 d after, treatment
- Do not use on Sands, stony or gravelly soils or soils with over 10% organic matter
- Product must be applied with specified adjuvant - see label

Crop-specific information
- Latest use: before ear at 1 cm stage (GS 30)
- Spray in autumn, winter or spring from first to third leaf unfolded stage (GS 11-13) but before ear at 1 cm stage (GS 30)

Following crops guidance
- After normal harvest any broad leaved crop (except sugar beet), barley, wheat, durum wheat, rye or triticale may be sown. Before drilling or planting subsequent crops soil must be mouldboard ploughed to 15 cm
- See label for details of crops that may be safely drilled or planted within 5 mth of treatment. If a treated crop should fail after 5 mth but before normal harvest, sow only a broad-leaved crop (not sugar beet)
- 12 mth must elapse after treatment before sugar beet is drilled

Environmental safety
- Dangerous for the environment
- Very toxic to aquatic organisms
- Crops must not be treated if there is risk of run-off (eg from frozen foliage)

Hazard classification and safety precautions
Hazard H04, H11
Risk phrases R36, R43, R50, R53a
Operator protection A, C, H, K; U02a, U05a, U09a, U20a
Environmental protection E15a, E38
Storage and disposal D01, D02, D09a, D10c, D12a

104 clofentezine

A selective ovicidal tetrazine acaricide for use in top fruit

Products
Apollo 50 SC	Makhteshim	500 g/l	SC	10590

Uses
- Red spider mites in **apples, cherries, pears, plums**
- Spider mites in **blackcurrants** *(off-label)*, **protected blackberries** *(off-label)*, **protected peppers** *(off-label)*, **protected raspberries** *(off-label)*, **protected strawberries** *(off-label)*, **raspberries** *(off-label)*, **strawberries** *(off-label)*

Specific Off-Label Approvals (SOLAs)
- **blackcurrants, raspberries, strawberries** *(OLA 010912) Dec 2008* [1]
- **blackcurrants, raspberries, strawberries** *(OLA 012269) Dec 2008* [1]
- **protected blackberries, protected raspberries** *(OLA 010911) Dec 2008* [1]
- **protected blackberries, protected raspberries** *(OLA 012268) Dec 2008* [1]
- **protected peppers** *(OLA 011431) Jul 2007* [1]
- **protected strawberries** *(OLA 012271) Dec 2008* [1]

Efficacy guidance
- Acts on eggs and early motile stages of mites. For effective control total cover of plants is essential, particular care being needed to cover undersides of leaves

SEE SECTION 3 FOR PRODUCTS ALSO REGISTERED

Restrictions
- Maximum number of treatments 1 per yr for top fruit; 2 per crop on protected peppers, 3 per yr for soft and cane fruit

Crop-specific information
- HI protected cane fruit 7 d; blackcurrants, raspberries, strawberries 14 d; apples, pears 28 d, plums, cherries 8 wk
- For red spider mite control spray apples and pears between bud burst and pink bud, plums and cherries between white bud and first flower. Rust mite is also suppressed
- On established infestations apply in conjunction with an adult acaricide

Environmental safety
- Harmful to aquatic organisms
- Product safe on predatory mites, bees and other predatory insects

Hazard classification and safety precautions
Risk phrases R52
Operator protection U05a, U08, U20b
Environmental protection E15a
Storage and disposal D01, D02, D05, D09a, D11a

105   clomazone

An isoxazolidinone residual herbicide for oilseed rape and field beans

Products

1 Centium 360 CS	Belchim	360 g/l	CS	11607
2 Cirrus CS	Belchim	360 g/l	CS	11751

Uses
- Chickweed in *asparagus (off-label)* [1]; *combining peas, spring field beans, spring oilseed rape, vining peas, winter field beans, winter oilseed rape* [1, 2]
- Cleavers in *asparagus (off-label)* [1]; *combining peas, spring field beans, spring oilseed rape, vining peas, winter field beans, winter oilseed rape* [1, 2]
- Fool's parsley in *asparagus (off-label)* [1]; *combining peas, spring field beans, spring oilseed rape, vining peas* [1, 2]
- Red dead-nettle in *asparagus (off-label)* [1]; *combining peas, spring field beans, spring oilseed rape, vining peas, winter field beans, winter oilseed rape* [1, 2]
- Shepherd's purse in *asparagus (off-label)* [1]; *combining peas, spring field beans, spring oilseed rape, vining peas, winter field beans, winter oilseed rape* [1, 2]

Specific Off-Label Approvals (SOLAs)
- *asparagus (OLA 052012) Dec 2008* [1]

Efficacy guidance
- Best results obtained from application as soon as possible after sowing crop and before emergence of crop or weeds
- Uptake is via roots and shoots. Seedbeds should be firm, level and free from clods. Loose puffy seedbeds should be consolidated before spraying
- Efficacy is reduced on organic soils, on dry cloddy seedbeds and if prolonged dry weather follows application
- Clomazone acts by inhibiting synthesis of chlorophyll pigments. Susceptible weeds emerge but are chlorotic and die shortly afterwards
- Season-long control of weeds may not be achieved

Restrictions
- Maximum number of treatments one per crop or yr
- Crops must be covered by a minimum of 20 mm settled soil. Do not apply to broadcast crops. Direct-drilled crops should be harrowed across the slits to cover seed before spraying
- Do not use on compacted or soils of poor structure that may be liable to waterlogging
- Do not use on Sands or Very Light soils or those with more than 10% organic matter

Crop-specific information
- Latest use: pre-emergence of crop ot at sprouting (asparagus)
- Crop plants emerged at time of treatment may be severely damaged

FOR FULL CONDITIONS OF USE ALWAYS READ THE PRODUCT LABEL

- Some transient crop bleaching may occur under certain climatic conditions and can be severe where heavy rain follows application. This is normally rapidly outgrown and has no effect on final crop yield. Overlapping spray swaths may cause severe damage to field beans

Following crops guidance

- Following normal harvest of a spring or autumn treated crop, cereals, oilseed rape, field beans, combining peas, potatoes, maize, turnips, linseed or sugar beet may be sown
- In the event of failure of an autumn treated crop, winter cereals or winter beans may be sown in the autumn if 6 wk have elapsed since treatment. In the spring following crop failure combining peas, field beans or potatoes may be sown if 6 wk have elapsed since treatment, and spring cereals, maize, turnips, onions, carrots or linseed may be sown if 7 mth have elapsed since treatment
- In the event of a failure of a spring treated crop a wide range of crops may be sown provided intervals of 6-9 wk have elapsed since treatment. See label for details
- Prior to resowing any listed replacement crop the soil should be ploughed and cultivated to 15 cm

Environmental safety

- Take extreme care to avoid drift outside the target area, or on to ponds, waterways or ditches as considerable damage may occur. Apply using a coarse quality spray

Hazard classification and safety precautions

Hazard H04
Risk phrases R43
Operator protection A, H; U05a, U14, U20b
Environmental protection E15a [1, 2]; E34 [1]
Storage and disposal D01, D02, D09a, D10b [1, 2]; D12a [1]

106   clopyralid

A foliar translocated picolinic herbicide for a wide range of crops

Products

	1	Dow Shield	Dow	200 g/l	SL	10988
	2	Fernpath Torate	AgriGuard	200 g/l	SL	11033
	3	Greencrop Champion	Greencrop	200 g/l	SL	11755
	4	Landgold Clopyralid 200	Teliton	200 g/l	SL	12359
	5	Pirlid	AgriGuard	200 g/l	SL	11946

Uses

- Annual dicotyledons in **broccoli, brussels sprouts, cabbages, calabrese, fodder beet, fodder rape, kale, linseed, maize, mangels, onions, permanent pasture, red beet, spring oilseed rape, strawberries, sugar beet, swedes, sweetcorn, turnips, winter barley, winter oats, winter oilseed rape** [1-5]; **cauliflowers, conifers and broadleaved trees** (off-label) [1]; **ornamental plant production** [1-3, 5]; **rotational grassland** [3, 4]; **spring barley, spring oats, spring wheat, winter wheat** [1, 3, 4]
- Corn marigold in **broccoli, brussels sprouts, cabbages, calabrese, cauliflowers, fodder beet, fodder rape, kale, linseed, maize, mangels, onions, permanent pasture, red beet, spring oilseed rape, strawberries, sugar beet, swedes, sweetcorn, turnips, winter barley, winter oats, winter oilseed rape** [1-5]; **ornamental plant production** [1-3, 5]; **rotational grassland** [3, 4]; **spring barley, spring oats, spring wheat, winter wheat** [1, 3, 4]
- Creeping thistle in **broccoli, brussels sprouts, cabbages, calabrese, cauliflowers, fodder beet, fodder rape, kale, linseed, maize, mangels, onions, permanent pasture, red beet, spring oilseed rape, strawberries, sugar beet, swedes, sweetcorn, turnips, winter barley, winter oats, winter oilseed rape** [1-5]; **ornamental plant production** [1-3, 5]; **rotational grassland** [3, 4]; **spring barley, spring oats, spring wheat, winter wheat** [1, 3, 4]
- General weed control in **canary grass** (off-label - game cover), **quinoa** (off-label - game cover), **sweet clover** (off-label - game cover), **tanka millet** (off-label - game cover), **white millet** (off-label - game cover) [1]
- Groundsel in **chives** (off-label), **herbs (see appendix 6)** (off-label), **leaf spinach** (off-label), **parsley** (off-label), **spinach beet** (off-label) [1]
- Mayweeds in **broccoli, brussels sprouts, cabbages, calabrese, cauliflowers, fodder beet, fodder rape, kale, linseed, maize, mangels, onions, permanent pasture, red beet, spring oilseed rape, strawberries, sugar beet, swedes, sweetcorn, turnips, winter barley, winter**

oats, *winter oilseed rape* [1-5]; *chives* *(off-label)*, *herbs (see appendix 6)* *(off-label)*, *honesty* *(off-label)*, *leaf spinach* *(off-label)*, *parsley* *(off-label)*, *poppies* *(off-label - for morphine production)*, *spinach beet* *(off-label)* [1]; *ornamental plant production* [1-3, 5]; *rotational grassland* [3, 4]; *spring barley*, *spring oats*, *spring wheat*, *winter wheat* [1, 3, 4]
- Thistles in *apple orchards* *(off-label)*, *asparagus* *(off-label)*, *established grassland* *(off-label - spot treatment)*, *grass seed crops* *(off-label)*, *honesty* *(off-label)*, *pear orchards* *(off-label)*, *poppies* *(off-label - for morphine production)* [1]

Specific Off-Label Approvals (SOLAs)
- *apple orchards*, *pear orchards* *(OLA 050471) Dec 2008* [1]
- *asparagus* *(OLA 050470) Dec 2008* [1]
- *canary grass*, *quinoa*, *sweet clover*, *tanka millet*, *white millet* *(game cover) (OLA 050474) Dec 2008* [1]
- *chives*, *herbs (see appendix 6)*, *parsley* *(OLA 050473) Dec 2008* [1]
- *conifers and broadleaved trees* *(OLA 050468) Dec 2008* [1]
- *established grassland* *(spot treatment) (OLA 050467) Dec 2008* [1]
- *grass seed crops* *(OLA 050467) Dec 2008* [1]
- *honesty* *(OLA 050469) Dec 2008* [1]
- *leaf spinach*, *spinach beet* *(OLA 050472) Dec 2008* [1]
- *poppies* *(for morphine production) (OLA 040161) Dec 2008* [1]

Approval information
- Accepted by BBPA for use on malting barley

Efficacy guidance
- Best results achieved by application to young actively growing weed seedlings. Treat creeping thistle at rosette stage and repeat 3-4 wk later as directed
- High activity on weeds of Compositae family. For most crops recommended for use in tank mixes. See label for details

Restrictions
- Maximum total dose varies between the equivalent of one and two full dose treatments, depending on the crop treated. See labels for details
- Do not apply to cereals later than the second node detectable stage (GS 32)
- Do not apply when crop damp or when rain expected within 6 h
- Do not use straw from treated cereals in compost or any other form for glasshouse crops. Straw may be used for strawing down strawberries
- Straw from treated grass seed crops or linseed should be baled and carted away. If incorporated do not plant winter beans in same year
- Do not use on onions at temperatures above 20°C or when under stress
- Do not treat maiden strawberries or runner beds or apply to early leaf growth during blossom period or within 4 wk of picking. Aug or early Sep sprays may reduce yield

Crop-specific information
- Latest use: before 3rd node detectable (GS 33) for cereals; before flower buds visible from above for oilseed rape, linseed and honesty; post final harvest of spears and pre-emergence of fern for asparagus
- HI grassland 7 d; apples, pears, strawberries 4 wk; maize, sweetcorn, onions, Brussels sprouts, broccoli, cabbage, cauliflowers, calabrese, kale, fodder rape, oilseed rape, swedes, turnips, sugar beet, red beet, fodder beet, mangels, sage, honesty 6 wk
- Timing of application varies with weed problem, crop and other ingredients of tank mixes. See labels for details
- Apply as directed spray in woody ornamentals, avoiding leaves, buds and green stems. Do not apply in root zone of families Compositae or Leguminosae

Following crops guidance
- Do not plant susceptible autumn-sown crops in same year as treatment. Do not apply later than Jul where susceptible crops are to be planted in spring. See label for details

Environmental safety
- Dangerous for the environment
- Harmful to aquatic organisms

FOR FULL CONDITIONS OF USE ALWAYS READ THE PRODUCT LABEL

- Wash spray equipment thoroughly with water and detergent immediately after use. Traces of product can damage susceptible plants sprayed later
- Keep livestock out of treated areas for at least 7 d and until foliage of any poisonous weeds such as ragwort has died and become unpalatable

Hazard classification and safety precautions
Hazard H11 [2, 5]
Risk phrases R52, R53a [1]
Operator protection A, C; U08, U19a [1-5]; U20a [3, 4]; U20b [1, 2, 5]
Environmental protection E07a (7 d); E15a
Storage and disposal D01, D12a [1]; D05, D09a, D10b [1-5]

107   clopyralid + 2,4-D + MCPA

A translocated herbicide mixture for grassland

Products

Lonpar	Dow	35:150:175 g/l	SL	08686

Uses
- Creeping thistle in **established grassland**

Approval information
- 2,4-D and MCPA included in Annex I under EC Directive 91/414

Efficacy guidance
- Treatment must be made when weeds and grass are actively growing
- Important to ensure sufficient leaf area for uptake especially on established thistles with extensive root system. Treat at rosette stage
- On large well established thistles and where there is a large soil seed reservoir further treatment in the following year may be needed
- To allow maximum translocation do not cut grass for 28 d after treatment

Restrictions
- Maximum number of treatments 1 or 2 per yr depending on dose used - see label
- Do not spray in drought, very hot or very cold weather or if rain expected within 6 h
- Do not treat grass less than 1 yr old or sports or amenity turf
- Product kills or severely checks clover and should not be used where clover is an important constituent of the sward
- Do not drill grass, grass mixtures, kale, swedes or turnips into the sward within 6 wk of spraying

Crop-specific information
- Very occasionally some yellowing of sward may occur after treatment which is quickly outgrown

Following crops guidance
- Susceptible crops (see label) must not be planted in the same calendar yr as treatment. Ensure that remains of treated crop have completely decayed before planting a subsequent susceptible crop. Where a susceptible crop is to be planted in the spring, product must not be sprayed later than end Jul in previous yr

Environmental safety
- Harmful to fish or other aquatic life. Do not contaminate surface waters or ditches with chemical or used container
- Keep livestock out of treated areas for at least 1 wk and until foliage of any poisonous weeds, such as ragwort, has died and become unpalatable
- Wash spray equipment thoroughly with water and detergent immediately after use. Traces of product can damage susceptible plants sprayed later

Hazard classification and safety precautions
Hazard H03, H04
Risk phrases R22a, R41
Operator protection A, C, H, M; U05a, U15
Environmental protection E07b (1 wk); E13c, E34
Storage and disposal D01, D02, D05

SEE SECTION 3 FOR PRODUCTS ALSO REGISTERED

108   clopyralid + diflufenican + MCPA

A selective herbicide for use in managed amenity turf

Products

| Spearhead | Bayer Environ. | 20:15:300 g/l | SL | 09941 |

Uses
- Annual dicotyledons in *managed amenity turf*
- Perennial dicotyledons in *managed amenity turf*

Approval information
- MCPA included in Annex I under EC Directive 91/414

Efficacy guidance
- Best results achieved by application when grass and weeds are actively growing
- Treatment during early part of the season is recommended, but not during drought

Restrictions
- Maximum number of treatments 1 per yr
- Only use when sward is satisfactorily established and regular mowing has begun
- Turf sown in spring or early summer may be ready for treatment after 2 mth. Later sown turf should not be sprayed until growth is resumed in the following spring
- Avoid mowing within 3-4 d before or after treatment
- Do not use cuttings from treated area as a mulch for any crop

Environmental safety
- Extremely dangerous to fish or other aquatic life. Do not contaminate surface waters or ditches with chemical or used container
- Keep livestock out of treated areas/away from treated water for at least 1 wk and until foliage of any poisonous weeds, such as ragwort, has died and become unpalatable
- LERAP Category B
- Avoid drift. Small amounts of spray can cause serious injury to herbaceous plants, vegetables, fruit and glasshouse crops
- Keep livestock out of treated areas

Hazard classification and safety precautions
    Hazard H03, H04
    Risk phrases R20, R21, R22a, R36, R38
    Operator protection A, C; U02a, U05a, U08, U19a, U20b
    Environmental protection E06a (unstipulated period); E13a, E16a, E16b, E34
    Storage and disposal D01, D02, D05, D09a, D10b
    Medical advice M03

109   clopyralid + fluroxypyr + MCPA

A translocated herbicide mixture for use in sports and amenity turf

Products

| Greenor | Rigby Taylor | 20:40:200 g/l | ME | 10909 |

Uses
- Annual dicotyledons in *managed amenity turf*

Approval information
- Fluroxypyr and MCPA included in Annex I under EC Directive 91/414

Efficacy guidance
- Best results achieved when weeds actively growing and turf grass competitive
- Treatment should normally be between Apr-Sep when the soil is moist
- Do not apply during drought unless irrigation is applied
- Allow 3 d before or after mowing established turf to ensure sufficient weed leaf surface present to allow uptake and movement

FOR FULL CONDITIONS OF USE ALWAYS READ THE PRODUCT LABEL

Restrictions
- Maximum number of treatments 2 per yr
- Do not treat grass under stress from frost, drought, waterlogging, trace element deficiency, disease or pest attack
- Do not treat if night temperatures are low, when frost is imminent or during prolonged cold weather

Crop-specific information
- Treat young turf only in spring when at least 2 mth have elapsed since sowing
- Allow 5 d after mowing young turf before treatment
- Product selective on a number of turf grass species (see label) but consultation or testing recommended before treatment of any cultivar

Environmental safety
- Dangerous for the environment
- Very toxic to aquatic organisms
- Wash spray equipment thoroughly with water and detergent immediately after use. Traces of product can damage susceptible plants sprayed later

Hazard classification and safety precautions
   Hazard H04, H11
   Risk phrases R36, R43, R50, R53a
   Operator protection A, C; U05a, U08, U14, U19a, U20b
   Environmental protection E15a, E38
   Storage and disposal D01, D02, D05, D09a, D10b, D12a

110   clopyralid + fluroxypyr + triclopyr

A foliar acting herbicide mixture for grassland

Products
Pastor	Dow	50:75:100 g/l	EC	11168

Uses
- Annual dicotyledons in **newly sown grass**
- Docks in **established grassland**, **newly sown grass**
- Stinging nettle in **established grassland**
- Thistles in **established grassland**

Approval information
- Fluroxypyr included in Annex I under EC Directive 91/414

Efficacy guidance
- Apply in spring or autumn depending on weeds present or as a split treatment in spring followed by autumn
- Treatment must be made when weeds and grass are actively growing
- It is important to ensure sufficient leaf area is present for uptake especially on established docks and thistles
- On large well established docks and where there is a large soil seed reservoir further treatment in the following year may be needed
- To allow maximum translocation in the weeds do not cut grass for 4 wk after treatment

Restrictions
- Maximum number of treatments 1 per yr at full dose or 2 per yr at half dose
- Do not spray in drought, very hot or very cold weather
- Do not treat sports or amenity turf
- Product kills or severely checks clover and should not be used where clover is an important constituent of the sward
- Do not roll or harrow 10 d before or 7 d after treatment
- Do not allow spray or drift to reach other crops, amenity plantings, gardens, ponds, lakes or water courses

Crop-specific information
- HI 7 d
- Application during active growth ensures minimal check to grass. Newly sown grass may be treated from the third leaf visible stage
- Product may be used in established grassland which is under non-rotational setaside arrangements
- Very occasionally some yellowing of the sward may occur after treatment but is quickly outgrown

Following crops guidance
- Residues in incompletely decayed plant tissue may affect succeeding susceptible crops such as peas, beans and other legumes, carrots and related crops, potatoes, tomatoes, lettuce
- Do not plant susceptible autumn-sown crops in the same yr as treatment with product. Spring sown crops may follow if treatment was before end Jul in the previous yr

Environmental safety
- Dangerous for the environment
- Toxic to aquatic organisms
- Keep livestock out of treated areas for at least 7 d following treatment and until foliage of poisonous weeds such as ragwort has died and become unpalatable
- Wash spray equipment thoroughly with water and detergent immediately after use. Traces of product can damage susceptible plants sprayed later

Hazard classification and safety precautions
Hazard H03, H08, H11
Risk phrases R22b, R37, R38, R41, R43, R51, R53a, R67
Operator protection A, C; U02a, U05a, U08, U11, U14, U19a, U20b
Environmental protection E07c (7 d); E15a, E34, E38
Consumer protection C01
Storage and disposal D01, D02, D05, D09a, D10b, D12a
Medical advice M05b

111   clopyralid + picloram

A post-emergence herbicide mixture for oilseed rape

Products

1 Galaxy	AgriGuard	267:67 g/l	SL	12228
2 Galera	Dow	267:67 g/l	SL	11961

Uses
- Cleavers in *winter oilseed rape*
- Mayweeds in *winter oilseed rape*

Efficacy guidance
- Best results obtained from treatment when weeds are small and actively growing
- Cleavers that germinate after treatment will not be controlled

Restrictions
- Maximum total dose equivalent to one full dose treatment
- Do not treat crops under stress from cold, drought, pest damage, nutrient deficiency or any other cause
- Do not roll or harrow for 7 d before or after spraying
- Do not apply through CDA applicators
- Do not use any treated plant material for composting or mulching
- Do not use manure from animals fed on treated crops for composting
- Chop and incorporate all treated plant remains in early autumn, or as soon as possible after harvest, to release any residues into the soil. Ensure that all treated plant remains have completely decayed before planting susceptible crops

Following crops guidance
- Any crop may be sown in the calendar yr following treatment
- Ploughing or thorough cultivation should be carried out before planting leguminous crops

FOR FULL CONDITIONS OF USE ALWAYS READ THE PRODUCT LABEL

- Do not attempt to plant peas, beans, other legumes, carrots, other umbelliferous crops, potatoes, lettuce, other Compositae, or any glasshouse or protected crops if treated crop remains have not fully decayed by the time of planting
- In the event of failure of an autumn treated crop only oilseed rape, wheat, barley, oats, maize or ryegrass may be sown in the spring and only after ploughing or thorough cultivation

Environmental safety
- Dangerous for the environment
- Toxic to aquatic organisms
- Take extreme care to avoid drift onto crops and non-target plants outside the target area

Hazard classification and safety precautions
>**Hazard** H11
>**Risk phrases** R51, R53a
>**Operator protection** A, C; U05a
>**Environmental protection** E15a, E34
>**Storage and disposal** D01, D02, D05, D07, D09a

112   clopyralid + triclopyr

A perennial and woody weed herbicide for use in grassland

Products

1	Blaster	Headland Amenity	60:240 g/l	EC	10571
2	Grazon 90	Dow	60:240 g/l	EC	05456
3	Greencrop Hi-Grass 90	Greencrop	60:240 g/l	EC	12429
4	Thistlex	Dow	200:200 g/l	EC	11533

Uses
- Brambles in *amenity grass* [1]; *established grassland* [2, 3]
- Broom in *amenity grass* [1]; *established grassland* [2, 3]
- Creeping thistle in *permanent pasture*, *rotational grassland* [4]
- Docks in *amenity grass* [1]; *established grassland* [2, 3]
- Gorse in *amenity grass* [1]; *established grassland* [2, 3]
- Perennial dicotyledons in *amenity grass* [1]; *established grassland* [2, 3]
- Stinging nettle in *amenity grass* [1]; *established grassland* [2, 3]
- Thistles in *amenity grass* [1]; *established grassland* [2, 3]

Efficacy guidance
- Must be applied to actively growing weeds
- Correct timing crucial for good control. Spray stinging nettle before flowering, docks in rosette stage in spring, creeping thistle before flower stems 15-20 cm high, brambles, broom and gorse in Jun-Aug
- Allow 2-3 wk regrowth after grazing or mowing before spraying perennial weeds
- Where there is a large reservoir of weed seed in the soil further treatment in the following yr may be needed

Restrictions
- Maximum number of treatments 1 per yr
- Only use on permanent pasture or rotational grassland established for at least 1 yr
- Do not apply where clover is an important constituent of sward
- Do not roll or harrow within 10 d before or 7 d after spraying
- Do not cut grass for 21 d before or 28 d after spraying
- Do not use any treated plant material for composting or mulching, and do not use manure for composting from animals fed on treated crops
- Do not apply by hand-held rotary atomiser equipment
- Do not allow drift onto other crops, amenity plantings or gardens, ponds, lakes or water courses. All conifers, especially pine and larch, are very sensitive

Crop-specific information
- Latest use: 7 d before grazing or cutting
- Some transient yellowing of treated swards may occur but is quickly outgrown

SEE SECTION 3 FOR PRODUCTS ALSO REGISTERED

Following crops guidance
- Residues in plant tissues which have not completely decayed may affect succeeding susceptible crops such as peas, beans, other legumes, carrots, parsnips, potatoes, tomatoes, lettuce, glasshouse and protected crops
- Do not plant susceptible autumn-sown crops (eg winter beans) in same year as treatment and allow at least 9 mth from treatment before planting a susceptible crop in the following yr
- Do not direct drill kale, swedes, turnips, grass or grass mixtures within 6 wk of spraying
- Do not spray after end Jul where susceptible crops are to be planted in the next spring

Environmental safety
- Dangerous for the environment
- Very toxic to aquatic organisms
- Keep livestock out of treated areas for at least 7 d after spraying and until foliage of any poisonous weeds such as ragwort or buttercup has died down and become unpalatable

Hazard classification and safety precautions
Hazard H03 [1-3]; H04 [4]; H11 [1-4]
Risk phrases R22a, R38, R43, R50 [1-3]; R41, R53a [1-4]; R52 [4]
Operator protection A, C [1-4]; H, M [1-3]; U02a, U05a, U11, U20b [1-4]; U08, U19a [1-3]; U14, U23b [1, 2]; U15, U23a [4]
Environmental protection E07a [1-4] (7 d); E15a [1-4]; E23, E34 [1-3]; E38 [1, 2]
Consumer protection C01 [1, 2]
Storage and disposal D01, D02, D05, D09a, D10b, D12a
Medical advice M03 [4]; M05a [3]

113   clothianidin

A nitromethylene neonicotinoid insecticide

See also beta-cyfluthrin + clothianidin

Products

Deter	Bayer CropScience	250 g/l	FS	12411

Uses
- Leafhoppers in **winter barley** *(seed treatment)*, **winter wheat** *(seed treatment)*
- Virus vectors in **winter barley** *(seed treatment)*, **winter wheat** *(seed treatment)*
- Wireworm in **winter barley** *(seed treatment)*, **winter wheat** *(seed treatment)*

Approval information
- Accepted by BBPA for use on malting barley

Efficacy guidance
- May only be used in conjunction with manufacturer's approved seed treatment application equipment or by following procedures given in the operating instructions
- Treated seed should preferably be drilled in the same season
- Evenness of seed cover improved by simultaneous application with a small volume (1.5-3 litres/tonne) of water
- Calibrate drill for treated seed and drill at 2.5-4 cm into firm, well prepared seedbed
- Use minimum 125 kg treated seed per ha
- When aphid activity unusually late, or is heavy and prolonged in areas of high risk, and mild weather predominates, follow-up foliar aphicide may be required
- Incidental suppression of leafhoppers in early spring (March/April) may be achieved but if a specific attack develops an additional foliar insecticide may be required

Restrictions
- Maximum number of seed treatments one per batch
- Product must be fully re-dispersed and homogeneous before use
- Do not use on seed with more than 16% moisture content, or on sprouted, cracked or skinned seed
- All seed batches should be tested to ensure they are suitable for treatment

Crop-specific information
- Latest use: pre-drilling

FOR FULL CONDITIONS OF USE ALWAYS READ THE PRODUCT LABEL

Hazard classification and safety precautions
    Hazard H04
    Risk phrases R43
    Operator protection A, H; U04a, U05a, U07, U13, U14, U20b
    Environmental protection E03, E15b, E34, E36
    Storage and disposal D01, D02, D05, D09a, D14
    Treated seed S01, S02, S03, S04b, S05, S06a, S06b, S07, S08
    Medical advice M03

114   clothianidin + prothioconazole

A combined fungicide and insecticide seed dressing for cereals

Products
    Redigo Deter   Bayer CropScience   250:50 g/l   FS   12423

Uses
- Bunt in **winter wheat** *(seed treatment)*
- Covered smut in **winter barley** *(seed treatment)*
- Fusarium foot rot and seedling blight in **winter barley** *(seed treatment)*, **winter wheat** *(seed treatment)*
- Leaf stripe in **winter barley** *(seed treatment)*
- Leafhoppers in **winter barley** *(seed treatment)*, **winter wheat** *(seed treatment)*
- Loose smut in **winter barley** *(seed treatment)*, **winter wheat** *(seed treatment)*
- Virus vectors in **winter barley** *(seed treatment)*, **winter wheat** *(seed treatment)*
- Wireworm in **winter barley** *(seed treatment)*, **winter wheat** *(seed treatment)*

Approval information
- Accepted by BBPA for use on malting barley

Efficacy guidance
- May only be used in conjunction with manufacturer's approved seed treatment application equipment or by following procedures given in the operating instructions
- Treated wheat seed should preferably be drilled in the same season. Treated barley seed must be used in the same season
- Evenness of seed cover improved by simultaneous application with a small volume (1.5-3 litres/tonne) of water
- Calibrate drill for treated seed and drill at 2.5-4 cm into firm, well prepared seedbed
- Use minimum 125 kg treated seed per ha
- When aphid activity unusually late, or is heavy and prolonged in areas of high risk, and mild weather predominates, follow-up foliar aphicide may be required
- Incidental suppression of leafhoppers in early spring (March/April) may be achieved but if a specific attack develops an additional foliar insecticide may be required

Restrictions
- Maximum number of seed treatments one per batch
- Product must be fully re-dispersed and homogeneous before use
- Do not use on seed with more than 16% moisture content, or on sprouted, cracked or skinned seed
- All seed batches should be tested to ensure they are suitable for treatment

Crop-specific information
- Latest use: pre-drilling

Environmental safety
- Harmful to aquatic organisms

Hazard classification and safety precautions
    Hazard H04
    Risk phrases R43, R52, R53a
    Operator protection A, H; U14, U19a, U20b
    Environmental protection E15a, E34, E38
    Storage and disposal D01, D02, D09a, D11a, D12a
    Treated seed S01, S02, S03, S04b, S05, S06a, S07, S08

SEE SECTION 3 FOR PRODUCTS ALSO REGISTERED

SECTION 2

115   copper ammonium carbonate

A protectant copper fungicide

Products

Croptex Fungex	Certis	8% w/w (copper)	SL	11049

Uses
- Blight in *outdoor tomatoes*
- Cane spot in *loganberries, raspberries*
- Celery leaf spot in *celery*
- Currant leaf spot in *blackcurrants*
- Damping off in *seedlings of ornamentals*
- Leaf curl in *peaches*
- Leaf mould in *protected tomatoes*
- Powdery mildew in *chrysanthemums, cucumbers*

Approval information
- Accepted by BBPA for use on hops

Efficacy guidance
- Apply spray to both sides of foliage
- With protected crops keep foliage dry before and after spraying
- Ventilate glasshouse immediately after spraying

Restrictions
- Maximum number of treatments 5 per crop for celery; 3 per yr for blackcurrants and loganberries; 2 per yr for peaches and raspberries; 1 per tray when used as a drench for ornamental seedlings
- Do not spray plants which are dry at the roots

Environmental safety
- Harmful to fish or other aquatic life. Do not contaminate surface waters or ditches with chemical or used container
- Keep all livestock out of treated areas for at least 3 wk. Bury or remove spillages

Hazard classification and safety precautions
> **Hazard** H03
> **Risk phrases** R22a, R41
> **Operator protection** U05a, U11, U15, U20c
> **Environmental protection** E06c (3 wk); E13c, E34
> **Storage and disposal** D01, D02, D09a, D10a, D12a
> **Medical advice** M03

116   copper oxychloride

A protectant copper fungicide and bactericide

Products

1	Cuprokylt	Unicrop	50% w/w (copper)	WP	00604
2	Cuprokylt FL	Unicrop	270 g/l (copper)	SC	08299
3	Headland Copper	Headland	256 g/l (copper)	SC	07799

Uses
- Bacterial canker in *cherries* [1, 3]; *cob nuts* (off-label), *hazel nuts* (off-label), *walnuts* (off-label) [2, 3]; *plums* [1, 2]
- Bacterial rot in *bulb onions* (off-label), *garlic* (off-label), *leeks* (off-label), *salad onions* (off-label), *shallots* (off-label) [1, 3]
- Black rot in *broccoli* (off-label), *brussels sprouts* (off-label), *cabbages* (off-label), *calabrese* (off-label), *cauliflowers* (off-label), *chinese cabbage* (off-label), *collards* (off-label), *kale* (off-label) [1, 3]; *protected brassica seedlings* (off-label) [2]; *protected broccoli* (off-label), *protected brussels sprouts* (off-label), *protected cabbages* (off-label), *protected calabrese* (off-label), *protected cauliflowers* (off-label), *protected chinese cabbage* (off-label), *protected collards* (off-label), *protected kale* (off-label) [3]
- Blight in *cob nuts* (off-label), *hazel nuts* (off-label), *walnuts* (off-label) [2, 3]; *outdoor tomatoes, potatoes* [1-3]; *protected tomatoes* [1, 2]

FOR FULL CONDITIONS OF USE ALWAYS READ THE PRODUCT LABEL

- Buck-eye rot in **outdoor tomatoes, protected tomatoes** [1-3]
- Cane spot in **loganberries, raspberries** [1, 2]
- Canker in **apples, pears** [1-3]
- Celery leaf spot in **celery** [1-3]
- Collar rot in **apples** *(off-label)* [2]
- Damping off in **outdoor tomatoes, protected tomatoes** [1, 2]
- Downy mildew in **chard** *(off-label)*, **cress** *(off-label)*, **frise** *(off-label)*, **grapevines, herbs (see appendix 6)** *(off-label)*, **leaf spinach** *(off-label)*, **lettuce** *(off-label)*, **protected herbs (see appendix 6)** *(off-label)*, **radicchio** *(off-label)*, **salad brassicas** *(off-label - for baby leaf production)*, **spinach beet** *(off-label)* [3]; **hops** [1-3]
- Foot rot in **outdoor tomatoes, protected tomatoes** [1, 2]
- Leaf curl in **peaches** [1, 2]
- Pseudomonas storage rots in **bulb onions** *(off-label)*, **garlic** *(off-label)*, **leeks** *(off-label)*, **salad onions** *(off-label)*, **shallots** *(off-label)* [1, 3]
- Purple blotch in **blackberries** [3]
- Pythium in **watercress** *(off-label)* [3]; **watercress** *(off-label - during propagation)* [2]
- Rhizoctonia in **watercress** *(off-label)* [3]; **watercress** *(off-label - during propagation)* [2]
- Rust in **blackcurrants** [1-3]
- Spear rot in **broccoli** *(off-label)*, **brussels sprouts** *(off-label)*, **cabbages** *(off-label)*, **calabrese** *(off-label)*, **cauliflowers** *(off-label)*, **chinese cabbage** *(off-label)*, **collards** *(off-label)*, **kale** *(off-label)* [1, 3]; **protected brassica seedlings** *(off-label)* [2]; **protected broccoli** *(off-label)*, **protected brussels sprouts** *(off-label)*, **protected cabbages** *(off-label)*, **protected calabrese** *(off-label)*, **protected cauliflowers** *(off-label)*, **protected chinese cabbage** *(off-label)*, **protected collards** *(off-label)*, **protected kale** *(off-label)* [3]

Specific Off-Label Approvals (SOLAs)
- **apples** *(OLA 982336) Dec 2008* [2]
- **broccoli, brussels sprouts, cabbages, calabrese, cauliflowers, chinese cabbage, collards, kale** *(OLA 010115) Dec 2008* [1]
- **broccoli, brussels sprouts, bulb onions, cabbages, calabrese, cauliflowers, chinese cabbage, cob nuts, collards, garlic, hazel nuts, kale, leeks, protected broccoli, protected brussels sprouts, protected cabbages, protected calabrese, protected cauliflowers, protected chinese cabbage, protected collards, protected kale, salad onions, shallots, walnuts, watercress** *(OLA 020415) Dec 2008* [3]
- **bulb onions, garlic, leeks, salad onions, shallots** *(OLA 991127) Dec 2008* [1]
- **chard, cress, frise, herbs (see appendix 6), leaf spinach, lettuce, protected herbs (see appendix 6), radicchio, spinach beet** *(OLA 051057) Dec 2008* [3]
- **cob nuts, hazel nuts, walnuts** *(OLA 990385) Dec 2008* [2]
- **protected brassica seedlings** *(OLA 010117) Dec 2008* [2]
- **salad brassicas** *(for baby leaf production) (OLA 051057) Dec 2008* [3]
- **watercress** *(during propagation) (OLA 001538) Dec 2008* [2]

Approval information
- Approved for aerial application on potatoes [1]. See notes in Section 5
- Accepted by BBPA for use on hops

Efficacy guidance
- Spray crops at high volume when foliage dry but avoid run off. Do not spray if rain expected soon
- Spray interval commonly 10-14 d but varies with crop, see label for details
- If buck-eye rot occurs, spray soil surface and lower parts of tomato plants to protect unaffected fruit [1, 2]
- A follow-up spray in the following spring should be made to top fruit severely infected with bacterial canker

Restrictions
- Maximum number of treatments 3 per crop for apples, blackberries, grapevines, pears; 2 per crop for watercress. Not specified for other crops
- Some peach cultivars are sensitive to copper. Treat non-sensitive varieties only [1, 2]

Crop-specific information
- Latest use: before bud burst for apples and pears; before planting out protected brassica seedlings
- HI calabrese 3 d; hops 7 d; onions, garlic, leeks, shallots 14 d; nuts 3 mth
- Slight damage may occur to leaves of cherries and plums [1, 2]

SEE SECTION 3 FOR PRODUCTS ALSO REGISTERED

Environmental safety
- Dangerous for the environment
- Very toxic to aquatic organisms
- Keep all livestock out of treated areas for at least 3 wk

Hazard classification and safety precautions
    Hazard H03, H11 [2]
    Risk phrases R22a, R50, R53a [2]
    Operator protection A [3]; U20a [1, 3]; U20c [2]
    Environmental protection E06a [1-3] (3 wk); E13c [1, 3]; E15a [2]; E34 [3]
    Storage and disposal D05, D10c, D12a [2]; D09a [1-3]; D10a [1]; D10b [3]

117   copper sulphate

See Bordeaux Mixture

118   cyanazine

A contact and residual triazine herbicide

Products
Fortrol	Makhteshim	500 g/l	SC	11174

Uses
- Annual dicotyledons in **broad beans** *(Scotland only)*, **broccoli** *(off-label)*, **bulb onions**, **cabbages** *(off-label)*, **calabrese** *(off-label)*, **cauliflowers** *(off-label)*, **collards** *(off-label)*, **combining peas**, **farm forestry** *(off-label)*, **forest nurseries** *(off-label)*, **kale** *(off-label)*, **narcissi**, **salad onions** *(fen soils only)*, **vining peas**, **winter oilseed rape**
- Annual grasses in **broad beans** *(Scotland only)*, **bulb onions**, **combining peas**, **farm forestry** *(off-label)*, **forest nurseries** *(off-label)*, **narcissi**, **salad onions** *(fen soils only)*, **vining peas**, **winter oilseed rape**

Specific Off-Label Approvals (SOLAs)
- **broccoli, cabbages, calabrese, cauliflowers, collards, kale** *(OLA 031074) Dec 2007* [1]
- **farm forestry** *(OLA 031454) Dec 2007* [1]
- **forest nurseries** *(OLA 031455) Dec 2007* [1]

Approval information
- Products containing this active ingredient have been granted derogations for specified 'Essential Uses' for use until 31 December 2007. Sale and supply must cease by 30 June 2007 but growers have no guarantee that the products will continue to be available until then.
  For more information see 'The Review Programme' under 'Pesticide Legislation' in Section 5

Efficacy guidance
- Weeds controlled before emergence or at young seedling stage. Soil should be moist and some rain should follow application
- Best results achieved when applied during mild, bright weather. Avoid applications in dull, cold or wet conditions
- Numerous tank mixtures recommended to broaden weed spectrum. See label for details of tank mix partners and timings
- Some strains of blackgrass have developed resistance to many blackgrass herbicides which may lead to poor control

Restrictions
- Maximum number of treatments 1 per crop for edible crops
- Maximum total dose equivalent to one full dose treatment on onions
- Variety restrictions apply in peas. See label for details of tolerant varieties
- Do not treat crops suffering from physical damage or stress from pest attack
- Treatment directly after conditions that reduce wax formation may lead to damage
- Do not use as pre-emergence treatment on soils with more than 10% organic matter
- Do not use on Sands, Very Light or stony soils

FOR FULL CONDITIONS OF USE ALWAYS READ THE PRODUCT LABEL

• These products must not be used on any crops other than those listed, including any extrapolations that would normally be permissible under the Long Term Arrangements for Extension of Use (see Section 5)

Crop-specific information
• Latest use: pre-emergence for broad beans; before 31 Jan for winter oilseed rape; pre-emergence (high dose) or before flower buds visible for peas; before flower buds visible for field beans
• HI 8 wk for onions, salad onions; 11 wk for brassica crops
• In early drillings of peas in late Feb or Mar application may be delayed until immediately prior to emergence to increase effective duration of control
• Peas should be drilled to give 20-25 mm of settled soil above the seed. On light soils crops may be damaged if heavy rain falls soon after spraying
• Apply to oilseed rape after 1 Nov from 5-leaf stage when winter hardened. Do not apply after 31 Jan
• Apply post-emergence to onions after 2 true leaf stage only on fen soils with more than 10% organic matter

Environmental safety
• Dangerous for the environment
• Very toxic to aquatic organisms
• Do not empty into drains

Hazard classification and safety precautions
Hazard H03, H11
Risk phrases R22a, R50, R53a
Operator protection A, C; U05a, U08, U19a, U20b
Environmental protection E15a, E19b, E34
Storage and disposal D01, D02, D05, D09a, D10c, D12a
Medical advice M03, M05a

119   cyanazine + pendimethalin

A contact and residual herbicide mixture

Products

Bullet	Makhteshim	150:264 g/l	SC	11204

Uses
• Annual dicotyledons in *combining peas, spring field beans*
• Annual meadow grass in *combining peas, spring field beans*
• Rough meadow grass in *combining peas, spring field beans*

Approval information
• Pendimethalin included in Annex I under EC Directive 91/414
• Products containing cyanazine have been granted derogations for specified 'Essential Uses' for use until 31 December 2007. Sale and supply must cease by 30 June 2007 but growers have no guarantee that the products will continue to be available until then.
For more information see 'The Review Programme' under 'Pesticide Legislation' in Section 5

Efficacy guidance
• Best results achieved in crops growing on firm, fine seedbeds, where light rain falls soon after treatment. Do not disturb soil after application. Weed control may be reduced if prolonged dry conditions follow application
• Weeds present before crop emergence should be controlled with a contact herbicide as a tank-mix or in sequence
• Weed control may be reduced on soils with a high Kd factor, where OM exceeds 6% or ash content is high. Surface organic crop residues should be dispersed
• Weeds germinating more than 2 mth after spraying may not be controlled

Restrictions
• Maximum number of treatments 1 per crop
• Do not use on very light, very stony or gravelly soils or those with more than 10% organic matter
• Do not apply where crop root growth is likely to be restricted
• Seed should be covered by a minimum of 25 mm of settled soil after seedbed consolidation

- Do not apply later than when the growing point of the crop is within 13 mm of the soil surface
- Do not treat pea variety Vedette

Crop-specific information
- Latest use: pre-emergence of crop

Following crops guidance
- After a dry season land must be ploughed to 150 mm before drilling ryegrass
- Before drilling a winter crop plough or cultivate to 150 mm
- In the event of crop failure land must be ploughed or cultivated to 150 mm and then an interval of 8 wk must elapse before drilling any crop

Environmental safety
- Dangerous for the environment
- Very toxic to aquatic organisms
- Do not empty into drains

Hazard classification and safety precautions
> **Hazard** H03, H11
> **Risk phrases** R22a, R50, R53a
> **Operator protection** A; U05a, U08, U13, U19a, U20b
> **Environmental protection** E15a, E19b, E34
> **Storage and disposal** D01, D02, D05, D09a, D10b, D12a
> **Medical advice** M03

120   cyazofamid

A cyanoimidazolesulfonamide protectant fungicide for potatoes

Products

Ranman Twinpack	Belchim	400 g/l	KL	11851

Uses
- Blight in *potatoes*

Approval information
- Cyazofamid included in Annex I under EC Directive 91/414

Efficacy guidance
- Apply as a protectant treatment before blight enters the crop and repeat every 7-10 d depending on severity of disease pressure
- Commence spray programme immediately the risk of blight in the locality occurs, usually when the crop meets along the rows
- Product must always be used with organosilicone adjuvant provided in the twin pack
- To minimise the chance of development of resistance no more than three applications should be made consecutively (out of a permissible total of six) in the blight control programme. For more information on Resistance Management see Section 5

Restrictions
- Maximum number of treatments 6 per crop (no more than three of which should be consecutive)
- Mixed product must not be allowed to stand overnight
- Consult processor before using on crops intended for processing

Crop-specific information
- HI 7 d

Environmental safety
- Dangerous for the environment
- Very toxic to aquatic organisms
- Do not empty into drains

Hazard classification and safety precautions
> **Hazard** H03 (adjuvant); H11
> **Risk phrases** R20, R36, R48 (adjuvant); R41, R50, R53a
> **Operator protection** A, C; U02a, U05a, U11, U15, U20a; U19a (adjuvant)
> **Environmental protection** E15a, E19b, E34, E38

FOR FULL CONDITIONS OF USE ALWAYS READ THE PRODUCT LABEL

Storage and disposal D01, D02, D09a, D10c, D12a
Medical advice M03

121 cycloxydim

A translocated post-emergence oxime herbicide for grass weed control

Products

1	Greencrop Pomeroy	Greencrop	200 g/l	EC	11769
2	Greencrop Valentia	Greencrop	200 g/l	EC	10197
3	Landgold Cycloxydim	Landgold	200 g/l	EC	06269
4	Landgold Cycloxydim	Teliton	200 g/l	EC	12113
5	Laser	BASF	200 g/l	EC	05251
6	Marnoch Clodim	Me2	200 g/l	EC	11460
7	Standon Cycloxydim	Standon	200 g/l	EC	08830

Uses

- Annual grasses in *amenity vegetation (off-label)* [5]; *brussels sprouts, cabbages, cauliflowers, fodder beet, mangels, spring field beans, sugar beet, swedes, winter field beans, winter oilseed rape* [1-7]; *bulb onions, carrots, dwarf beans, farm forestry, forest, leeks, linseed, parsnips, salad onions, spring oilseed rape, strawberries* [1, 2, 5-7]; *calabrese, peas* [2]; *combining peas, vining peas* [3-6]; *early potatoes* [1, 3-5]; *flower bulbs* [2, 5-7]; *forest nurseries, ornamental plant production* [1]; *maincrop potatoes* [2-6]
- Annual meadow grass in *combining peas, maincrop potatoes, vining peas* [1, 7]
- Black bent in *brussels sprouts, cabbages, cauliflowers, fodder beet, maincrop potatoes, mangels, spring field beans, sugar beet, swedes, winter field beans, winter oilseed rape* [1-7]; *bulb onions, carrots, dwarf beans, flower bulbs, leeks, linseed, parsnips, salad onions, spring oilseed rape, strawberries* [1, 2, 5-7]; *calabrese, peas* [2]; *combining peas, vining peas* [1, 3-7]; *early potatoes* [1, 3-5]; *ornamental plant production* [1]
- Blackgrass in *brussels sprouts, cabbages, cauliflowers, fodder beet, maincrop potatoes, mangels, spring field beans, sugar beet, swedes, winter field beans, winter oilseed rape* [1-7]; *bulb onions, carrots, dwarf beans, flower bulbs, leeks, linseed, parsnips, salad onions, spring oilseed rape, strawberries* [1, 2, 5-7]; *calabrese, peas* [2]; *combining peas, vining peas* [1, 3-7]; *early potatoes* [1, 3-5]; *ornamental plant production* [1]
- Couch in *brussels sprouts, cabbages, cauliflowers, fodder beet, maincrop potatoes, mangels, spring field beans, sugar beet, swedes, winter field beans, winter oilseed rape* [1-7]; *bulb onions, carrots, dwarf beans, flower bulbs, leeks, linseed, parsnips, salad onions, spring oilseed rape, strawberries* [1, 2, 5-7]; *calabrese, peas* [2]; *combining peas, vining peas* [1, 3-7]; *early potatoes* [1, 3-5]; *ornamental plant production* [1]
- Creeping bent in *brussels sprouts, cabbages, cauliflowers, fodder beet, maincrop potatoes, mangels, spring field beans, sugar beet, swedes, winter field beans, winter oilseed rape* [1-7]; *bulb onions, carrots, dwarf beans, flower bulbs, leeks, linseed, parsnips, salad onions, spring oilseed rape, strawberries* [1, 2, 5-7]; *calabrese, peas* [2]; *combining peas, vining peas* [1, 3-7]; *early potatoes* [1, 3-5]; *ornamental plant production* [1]
- Green cover in *land not being used for crop production* [1, 2, 5, 6]
- Onion couch in *brussels sprouts, bulb onions, cabbages, carrots, cauliflowers, dwarf beans, flower bulbs, fodder beet, leeks, linseed, maincrop potatoes, mangels, parsnips, salad onions, spring field beans, spring oilseed rape, strawberries, sugar beet, swedes, winter field beans, winter oilseed rape* [1, 2, 5-7]; *calabrese, peas* [2]; *combining peas, vining peas* [1, 5-7]; *early potatoes* [1, 5]; *ornamental plant production* [1]
- Perennial grasses in *amenity vegetation (off-label)* [5]; *farm forestry, forest* [1, 2, 5-7]; *forest nurseries* [1]
- Volunteer cereals in *brussels sprouts, cabbages, cauliflowers, fodder beet, maincrop potatoes, mangels, spring field beans, sugar beet, swedes, winter field beans, winter oilseed rape* [1-7]; *bulb onions, carrots, dwarf beans, flower bulbs, leeks, linseed, parsnips, salad onions, spring oilseed rape, strawberries* [1, 2, 5-7]; *calabrese, peas* [2]; *combining peas, vining peas* [1, 3-7]; *early potatoes* [1, 3-5]; *ornamental plant production* [1]
- Wild oats in *brussels sprouts, cabbages, cauliflowers, fodder beet, maincrop potatoes, mangels, spring field beans, sugar beet, swedes, winter field beans, winter oilseed rape* [1-7]; *bulb onions, carrots, dwarf beans, flower bulbs, leeks, linseed, parsnips, salad onions,*

SEE SECTION 3 FOR PRODUCTS ALSO REGISTERED

spring oilseed rape, strawberries [1, 2, 5-7]; ***calabrese, peas*** [2]; ***combining peas, vining peas*** [1, 3-7]; ***early potatoes*** [1, 3-5]; ***ornamental plant production*** [1]

Specific Off-Label Approvals (SOLAs)
- ***amenity vegetation*** *(OLA 962585) Dec 2008* [5]

Efficacy guidance
- Best results achieved when weeds small and have not begun to compete with crop. Effectiveness reduced by drought, cool conditions or stress. Weeds emerging after application are not controlled
- Foliage death usually complete after 3-4 wk but longer under cool conditions, especially late treatments to winter oilseed rape
- Perennial grasses should have sufficient foliage to absorb spray and should not be cultivated for at least 14 d after treatment
- On established couch pre-planting cultivation recommended to fragment rhizomes and encourage uniform emergence
- Split applications to volunteer wheat and barley at GS 12-14 will often give adequate control in winter oilseed rape. See label for details
- Apply to dry foliage when rain not expected for at least 2 h

Restrictions
- Maximum number of treatments 1 per crop for spring oilseed rape, early potatoes; 2 per yr on amenity vegetation; 2 per crop (the second at reduced dose) for other crops. See label for details
- Must be used with Actipron (see label)
- Do not apply to crops damaged or stressed by adverse weather, pest or disease attack or other pesticide treatment
- Prevent drift onto other crops, especially cereals and grass

Crop-specific information
- HI cabbage, cauliflower, calabrese, salad onions 4 wk; peas, dwarf beans 5 wk; bulb onions, carrots, parsnips, strawberries 6 wk; sugar and fodder beet, leeks, mangels, potatoes, field beans, swedes, Brussels sprouts, winter field beans 8 wk; oilseed rape, soya beans, linseed 12 wk
- Recommended time of application varies with crop. See label for details
- On peas a crystal violet wax test should be done if leaf wax likely to have been affected by weather conditions or other chemical treatment. The wax test is essential if other products are to be sprayed before or after treatment
- May be used on ornamental bulbs when crop 5-10 cm tall. Product has been used on tulips, narcissi, hyacinths and irises but some subjects may be more sensitive and growers advised to check tolerance on small number of plants before treating the rest of the crop
- May be applied to land temporarily removed from production where the green cover is made up predominantly of tolerant crops listed on label. Use on industrial crops of linseed and oilseed rape on land temporarily removed from production also permitted.

Following crops guidance
- Guideline intervals for sowing succeeding crops after failed treated crop: field beans, peas, sugar beet, rape, kale, swedes, radish, white clover, lucerne 1 wk; dwarf French beans 4 wk; wheat, barley, maize 8 wk
- Oats should not be sown after failure of a treated crop

Environmental safety
- Dangerous for the environment
- Toxic to aquatic organisms
- Harmful to fish or other aquatic life. Do not contaminate surface waters or ditches with chemical or used container

Hazard classification and safety precautions
    Hazard H03, H11 [1-5, 7]; H04 [6]
    Risk phrases R22b, R51, R53a [1-5, 7]; R36 [1-4, 6, 7]; R38 [1-7]
    Operator protection A, C; U05a, U08, U20b [1-7]; U19a [1, 2]
    Environmental protection E13c [6]; E15a [1-5, 7]; E38 [5]
    Storage and disposal D01, D02, D09a [1-7]; D05, D10b [1-4, 7]; D10c [5, 6]; D12a [5]
    Medical advice M05b [1-5, 7]

FOR FULL CONDITIONS OF USE ALWAYS READ THE PRODUCT LABEL

122   cyflufenamid

An acetamide fungicide for cereals

Products

Cyflamid	Certis	50 g/l	EW	12403

Uses
- Powdery mildew in **durum wheat**, **spring barley**, **spring wheat**, **triticale**, **winter barley**, **winter rye**, **winter wheat**

Efficacy guidance
- Best results obtained from treatment at the first visible signs of infection
- Sustained disease presssure may require a second treatment
- Disease spectrum may broadened by appropriate tank mixtures. See label
- Product is rainfast within 1 h
- Must be used as part of an Integrated Crop Management programme that includes alternating use, or mixture, with fungicides with a different mode of action effective against powdery mildew

Restrictions
- Maximum number of treatments 2 per crop on all recommended cereals
- Apply only in the spring
- Do not apply to crops under stress from drought, waterlogging, cold, pests or diseases, lime or nutrient deficiency or other factors affecting crop growth

Crop-specific information
- Latest use: before start of flowering of cereal crops

Following crops guidance
- No restrictions

Environmental safety
- Dangerous for the environment
- Very toxic to aquatic organisms
- Avoid drift onto ponds, waterways or ditches

Hazard classification and safety precautions
  Hazard H11
  Risk phrases R50, R53a
  Operator protection A; U05a
  Environmental protection E15b, E34
  Storage and disposal D01, D02, D05, D10c, D12b

123   cyfluthrin

A non-systemic pyrethroid insecticide

Products

Baythroid	Makhteshim	50 g/l	EC	11663

Uses
- Aphids in **apples**
- Apple sucker in **apples**
- Barley yellow dwarf vectors in **winter barley**, **winter wheat**
- Cabbage stem flea beetle in **winter oilseed rape**
- Caterpillars in **broccoli**, **brussels sprouts**, **cabbages**, **cauliflowers**
- Damson-hop aphid in **hops** *(off-label)*
- Winter moth in **apples**

Specific Off-Label Approvals (SOLAs)
- **hops** *(OLA 031867) Dec 2008* [1]

Efficacy guidance
- Best control of virus vectors in winter cereals and cabbage stem flea beetle in oilseed rape obtained from treatment normally in Oct or early Nov. Follow specialist or ADAS advice
- Complete control of cabbage stem flea beetle in oilseed rape may require a second treatment to control later hatches

SECTION 2

SEE SECTION 3 FOR PRODUCTS ALSO REGISTERED

- Brassica pests should be treated during first stages of attack. A second treatment may be needed if reinfestation occurs
- Apple pests should be treated at green cluster

Restrictions
- Maximum number of treatments 1 per crop or yr for barley, wheat, apples; 2 per crop for brassicas, oilseed rape; 3 per yr for hops

Crop-specific information
- Latest use: before 31 Mar in yr of harvest for barley, winter oilseed rape, wheat; before flowerhead formation for broccoli, cauliflowers; before pink bud for apples
- HI 7 d for hops

Environmental safety
- Dangerous for the environment
- Toxic to aquatic organisms
- Extremely dangerous to bees. Do not apply to crops in flower or to those in which bees are actively foraging. Do not apply when flowering weeds are present
- Extremely dangerous to fish or other aquatic life. Do not contaminate surface waters or ditches with chemical or used container
- LERAP Category A
- Broadcast air-assisted LERAP (18 m)

Hazard classification and safety precautions
Hazard H03, H08, H11
Risk phrases R20, R22a, R22b, R38, R41, R51, R53a
Operator protection A, C, H; U04a, U05a, U10, U11, U14, U15, U19a, U20b
Environmental protection E12c, E13a, E16c, E16d, E34, E38; E17b (18 m)
Storage and disposal D01, D02, D05, D09a, D10b, D12a
Medical advice M03, M05b

124   cymoxanil

A urea fungicide for potatoes

Products
1 Option	DuPont	60% w/w	WG	11834
2 Sipcam C 50	Sipcam	50% w/w	WP	10610

Uses
- Blight in **potatoes**

Efficacy guidance
- Product to be used in mixture with recommended partner containing mancozeb or fluazinam in order to combine systemic and protective activity
- Commence spray programme before infection appears as soon as weather conditions favourable for disease development occur. At latest the first treatment should be made as the foliage meets along the rows
- Repeat treatments at 7-10 day intervals according to disease incidence and weather conditions

Restrictions
- Do not allow packs to become wet during storage
- Do not apply in tank mixture with other fungicides [1]

Crop-specific information
- HI for potatoes 14 d [1], 7 d [2]

Environmental safety
- Dangerous for the environment
- Very toxic to aquatic organisms
- Extremely dangerous [1] or harmful [2] to fish or other aquatic life. Do not contaminate surface waters or ditches with chemical or used container

Hazard classification and safety precautions
Hazard H03 [1, 2]; H11 [1]
Risk phrases R22a [1, 2]; R50, R53a [1]
Operator protection A; U05a, U08, U20b [1, 2]; U19a [1]

FOR FULL CONDITIONS OF USE ALWAYS READ THE PRODUCT LABEL

Environmental protection E13a, E34, E38 [1]; E13c [2]
Storage and disposal D01, D02, D09a, D11a [1, 2]; D12a [1]
Medical advice M03

125 cymoxanil + famoxadone

A preventative and curative fungicide mixture for potatoes

Products
Tanos	DuPont	25:25% w/w	WG	10677

Uses
- Blight in **potatoes**

Approval information
- Famoxadone included in Annex I under EC Directive 91/414

Efficacy guidance
- Commence spray programme before infection appears as soon as weather conditions favourable for disease development occur. At latest the first treatment should be made as the foliage meets along the rows
- Repeat treatments at 7-10 day intervals according to disease incidence and weather conditions
- Reduce spray interval if conditions are conducive to the spread of blight
- Spray as soon as possible after irrigation
- Increase water volume in dense crops
- Product combines active substances with different modes of action and is effective against strains of potato blight that are insensitive to phenylamide fungicides
- To minimise the likelihood of development of resistance to QoI fungicides these products should be used in a planned Resistance Management strategy. See Section 5 for more information

Restrictions
- Maximum number of treatments 6 per crop
- Consult processors before using on crops for processing
- Do not apply more than 3 consecutive treatments containing QoI active fungicides

Crop-specific information
- HI 14 d for potatoes

Environmental safety
- Dangerous for the environment
- Very toxic to aquatic organisms
- Dangerous to fish or other aquatic life. Do not contaminate surface waters or ditches with chemical or used container
- LERAP Category B

Hazard classification and safety precautions
Hazard H03, H11
Risk phrases R22a, R50, R53a
Operator protection A, D; U05a, U20b
Environmental protection E13b, E16a, E16b, E34, E38
Storage and disposal D01, D02, D09a, D10b, D12a
Medical advice M03

126 cymoxanil + fludioxonil + metalaxyl-M

A fungicide seed dressing for peas

Products
Wakil XL	Syngenta	10:5:17.5% w/w	WS	10562

Uses
- Ascochyta in **combining peas** *(seed treatment)*, **vining peas** *(seed treatment)*
- Damping off in **combining peas** *(seed treatment)*, **poppies** *(off-label - for morphine production - seed treatment)*, **red beet** *(off-label - seed treatment)*, **vining peas** *(seed treatment)*

SEE SECTION 3 FOR PRODUCTS ALSO REGISTERED

- Downy mildew in **broad beans** *(off-label - seed treatment)*, **combining peas** *(seed treatment)*, **poppies** *(off-label - for morphine production - seed treatment)*, **spring field beans** *(off-label - seed treatment)*, **vining peas** *(seed treatment)*, **winter field beans** *(off-label - seed treatment)*
- Pythium in **carrots** *(off-label - seed treatment)*, **combining peas** *(seed treatment)*, **parsnips** *(off-label - seed treatment)*, **poppies** *(off-label - for morphine production - seed treatment)*, **vining peas** *(seed treatment)*

Specific Off-Label Approvals (SOLAs)

- **broad beans**, **spring field beans**, **winter field beans** *(seed treatment) (OLA 023932) Dec 2008* [1]
- **carrots**, **parsnips** *(seed treatment) (OLA 021191) Dec 2008* [1]
- **poppies** *(for morphine production - seed treatment) (OLA 040683) Dec 2008* [1]
- **red beet** *(seed treatment) (OLA 032313) Dec 2008* [1]

Approval information

- Metalaxyl-M included in Annex I under EC Directive 91/414

Efficacy guidance

- Apply through continuous flow seed treaters which should be calibrated before use

Restrictions

- Max no of treatments 1 per seed batch
- Ensure moisture content of treated seed satisfactory and store in a dry place
- Check calibration of seed drill with treated seed before drilling and sow as soon as possible after treatment
- Consult before using on crops for processing

Crop-specific information

- Latest use: pre-drilling

Environmental safety

- Harmful to aquatic organisms
- Do not use treated seed as food or feed

Hazard classification and safety precautions

Risk phrases R52, R53a
Operator protection A, D, H; U02a, U05a, U20b
Environmental protection E03, E15a
Storage and disposal D01, D02, D05, D07, D09a, D11a, D12a
Treated seed S02, S04b, S05, S07

127   cymoxanil + mancozeb

A protectant and systemic fungicide for potato blight control

Products

1	Agriguard Cymoxanil Plus	AgriGuard	4.5:68% w/w	WP	10893
2	Besiege	DuPont	4.5:68 % w/w	WP	08086
3	Curzate M WG	DuPont	4.5:68% w/w	WG	11901
4	Curzate M68	DuPont	4.5:68% w/w	WP	08072
5	Globe	Sipcam	6:70 % w/w	WP	10339
6	Matilda	Nufarm UK	4.5:68% w/w	WP	12006
7	Me2 Cymoxeb	Me2	4.5:68% w/w	WP	09486
8	Rhapsody	DuPont	4.5:68% w/w	WG	11958
9	Rhythm	Interfarm	4.5:68% w/w	WP	11669
10	Solace	Nufarm UK	4.5:68% w/w	WP	11936

Uses

- Blight in **potatoes**

Approval information

- Mancozeb included in Annex I under EC Directive 91/414
- Approval expiry 30 Apr 2006 [9]

Efficacy guidance

- Apply immediately after blight warning or as soon as local conditions dictate and repeat at 7-14 d intervals until haulm dies down or is burnt off

FOR FULL CONDITIONS OF USE ALWAYS READ THE PRODUCT LABEL

- Spray interval should not be more than 7-10 d in irrigated crops (see product label). Apply treatment after irrigation
- To minimise the likelihood of development of resistance these products should be used in a planned Resistance Management strategy. See Section 5 for more information

Restrictions
- Maximum number of treatments 6 per crop [5]
- Do not allow packs to become wet during storage

Crop-specific information
- HI 7 d [5]
- Destroy and remove any haulm that remains after harvest of early varieties to reduce blight pressure on neighbouring maincrop potatoes

Environmental safety
- Dangerous for the environment
- Very toxic to aquatic organisms
- Harmful (dangerous [5]) to fish or other aquatic life. Do not contaminate surface waters or ditches with chemical or used container
- Keep product away from fire or sparks

Hazard classification and safety precautions
  Hazard H04 [1-10]; H11 [1-4, 7-10]
  Risk phrases R36, R37, R42 [1-4, 6-10]; R38 [1, 6, 7, 10]; R43 [1-10]; R50, R53a [2-4, 8, 9]
  Operator protection A, C [1-10]; H [5]; U02a [5]; U05a, U08 [1-10]; U11 [2, 4, 6, 9]; U14 [2-4, 8, 9]; U19a [1-4, 6-10]; U20a [1, 7, 10]; U20b [2-6, 8, 9]
  Environmental protection E13b [5]; E13c [2-4, 6, 8]; E15a [1-5, 7-10]; E38 [2-4, 8, 9]
  Storage and disposal D01, D02, D09a, D11a [1-10]; D12a [2, 4, 9]

128   cypermethrin

A contact and stomach acting pyrethroid insecticide

Products

1 Agriguard Cypermethrin EC	AgriGuard	100 g/l	EC	12134
2 Jundi 100 EC	DAPT	100 g/l	EC	10848
3 Permasect C	Nufarm UK	100 g/l	EC	11121
4 Toppel 10	United Phosphorus	100 g/l	EC	08772

Uses
- Aphids in **apples, vining peas** [1-4]; **broccoli, brussels sprouts, cabbages, calabrese, cauliflowers, cherries, durum wheat, hops, lettuce, ornamental plant production, pears, plums, triticale, winter barley, winter oats, winter rye, winter wheat** [1, 2, 4]; **chinese cabbage** (off-label), **leaf spinach** (off-label), **protected chinese cabbage** (off-label), **protected spinach** (off-label) [4]
- Asparagus beetle in **asparagus** (off-label) [4]
- Barley yellow dwarf vectors in **triticale, winter oats, winter rye** [3]
- Barley yellow dwarf virus vectors in **durum wheat, triticale, winter oats, winter rye** [1, 2, 4]; **spring barley** (autumn sown), **spring wheat** (autumn sown) [3]; **winter barley, winter wheat** [1-4]
- Bladder pod midge in **spring oilseed rape, winter oilseed rape** [1, 2, 4]
- Cabbage seed weevil in **spring oilseed rape, winter oilseed rape** [2-4]
- Cabbage stem flea beetle in **winter oilseed rape** [1-4]
- Capsids in **apples, ornamental plant production** [1, 2, 4]; **pears** [1-4]
- Caterpillars in **apples, cherries, lettuce, ornamental plant production, pears, plums** [1, 2, 4]; **broccoli, brussels sprouts, cabbages, calabrese, cauliflowers** [1-4]; **chinese cabbage** (off-label), **leaf spinach** (off-label), **protected chinese cabbage** (off-label), **protected spinach** (off-label) [4]; **fodder beet, kale, mangels, potatoes, red beet, sugar beet** [3]
- Codling moth in **apples** [1-4]
- Cutworms in **carrots** (off-label), **parsnips** (off-label) [4]; **fodder beet, mangels, red beet** [3]; **lettuce, ornamental plant production** [1, 2, 4]; **potatoes, sugar beet** [1-4]
- Frit fly in **grass re-seeds** [1-4]
- Pea and bean weevil in **spring field beans, vining peas, winter field beans** [1-4]

SEE SECTION 3 FOR PRODUCTS ALSO REGISTERED

- Pea moth in **vining peas** [1-4]
- Pod midge in **spring oilseed rape, winter oilseed rape** [3]
- Pollen beetle in **spring oilseed rape, winter oilseed rape** [1-4]
- Rape winter stem weevil in **winter oilseed rape** [1, 2, 4]
- Sawflies in **apples** [1, 2, 4]
- Seed weevil in **spring oilseed rape, winter oilseed rape** [1]
- Suckers in **apples** [3]; **pears** [1, 2, 4]
- Thrips in **ornamental plant production** [1, 2, 4]
- Tortrix moths in **apples** [1-4]; **pears** [3]
- Whitefly in **broccoli, brussels sprouts, cabbages, calabrese, cauliflowers, ornamental plant production, protected cucumbers, protected lettuce, protected ornamentals** [1, 2, 4]
- Winter moth in **apples, pears** [3]
- Yellow cereal fly in **durum wheat** [1, 2, 4]; **spring barley** *(autumn sown)*, **spring wheat** *(autumn sown)* [3]; **triticale, winter barley, winter oats, winter rye, winter wheat** [1-4]

Specific Off-Label Approvals (SOLAs)
- **asparagus** *(OLA 032223) Dec 2008* [4]
- **carrots, parsnips** *(OLA 982184) Dec 2008* [4]
- **chinese cabbage, leaf spinach, protected chinese cabbage, protected spinach** *(OLA 983133) Dec 2008* [4]

Approval information
- Cypermethrin included in Annex I under EC Directive 91/414
- Following implementation of Directive 98/82/EC, approval for use of cypermethrin on numerous crops was revoked in 1999
- Accepted by BBPA for use on malting barley and hops

Efficacy guidance
- Products combine rapid action, good persistence, and high activity on Lepidoptera.
- As effect is mainly via contact good coverage is essential for effective action. Spray volume should be increased on dense crops
- A repeat spray after 10-14 d is needed for some pests of outdoor crops, several sprays at shorter intervals for whitefly and other glasshouse pests
- Rates and timing of sprays vary with crop and pest. See label for details
- Add a non-ionic wetter to improve results on leaf brassicas. In Brussels sprouts use of a drop-leg sprayer may be beneficial
- Where aphids in hops, pear suckers or glasshouse whitefly resistant to cypermethrin occur control is unlikely to be satisfactory

Restrictions
- Maximum number of treatments varies with crop and product. See label or approval notice for details

Crop-specific information
- Latest use: normally before end Mar in yr of harvest, but varies with product and crop. See labels for details
- HI apples, pears 14 d; vining peas 7 d; chicory, lettuce, courgettes, cucumbers, gherkins, endives, asparagus 1 d; other crops 0 d
- Test spray sample of new or unusual ornamentals before committing whole batches

Environmental safety
- Dangerous for the environment
- Very toxic to aquatic organisms
- High risk to bees. Do not apply to crops in flower, or to those in which bees are actively foraging, except as directed. Do not apply when flowering weeds are present [1]
- Extremely dangerous to fish or other aquatic life. Do not contaminate surface waters or ditches with chemical or used container
- Flammable
- Do not empty into drains
- LERAP Category A
- Broadcast air-assisted LERAP (18 m) [2, 3]
- Do not spray cereals after 31 Mar within 6 m of the edge of the growing crop

FOR FULL CONDITIONS OF USE ALWAYS READ THE PRODUCT LABEL

Hazard classification and safety precautions

Hazard H03, H08 [1-4]; H04 [2]; H11 [1, 3, 4]

Risk phrases R21 [4]; R22a, R50, R66, R67 [3, 4]; R22b [1-4]; R36 [1, 2]; R37 [1, 3]; R38 [1, 2, 4]; R51 [1]; R53a [1, 3, 4]

Operator protection A, C; U02a, U04a, U11 [3]; U05a, U10, U19a, U20b [1-4]

Environmental protection E12b [1]; E13a [2]; E15a [1, 3, 4]; E16c, E16d [1-4]; E17b [2, 3] (18 m); E19b [4]; E34 [3]

Storage and disposal D01, D02, D05, D09a, D10b [1-4]; D12a [3]

Medical advice M03 [4]; M04 [3]; M05b [1-4]

129   cyproconazole

A contact and systemic conazole fungicide for cereals and other field crops

See also azoxystrobin + cyproconazole
chlorothalonil + cyproconazole

Products

1	Cabaret	Bayer CropScience	240 g/l	EC	11915
2	Caddy 240 EC	Bayer CropScience	240 g/l	EC	11256

Uses

- Brown rust in **red beet** *(off-label)* [2]; **spring barley, spring rye, winter barley, winter rye, winter wheat** [1, 2]
- Chocolate spot in **spring field beans** *(with chlorothalonil)*, **winter field beans** *(with chlorothalonil)* [1, 2]
- Crown rust in **spring oats, winter oats** [1, 2]
- Eyespot in **winter barley, winter wheat** [1, 2]
- Late ear diseases in **winter wheat** [1, 2]
- Light leaf spot in **winter oilseed rape** [1, 2]
- Net blotch in **spring barley, winter barley** [1, 2]
- Phoma leaf spot in **winter oilseed rape** [1, 2]
- Powdery mildew in **spring barley, spring oats, spring rye, sugar beet, winter barley, winter oats, winter rye, winter wheat** [1, 2]
- Ramularia leaf spots in **sugar beet** [1, 2]
- Rhynchosporium in **spring barley, winter barley** [1, 2]
- Rust in **leeks, spring field beans, sugar beet, winter field beans** [1, 2]; **red beet** *(off-label)* [1]
- Septoria diseases in **winter wheat** [1, 2]
- Yellow rust in **spring barley, winter barley, winter wheat** [1, 2]

Specific Off-Label Approvals (SOLAs)

- **red beet** *(OLA 041024) Dec 2008* [1]
- **red beet** *(OLA 031883) Dec 2008* [2]

Approval information

- Accepted by BBPA for use on malting barley

Efficacy guidance

- Apply at start of disease development or as preventive treatment and repeat as necessary
- Most effective time of treatment varies with disease and use of tank mixes may be desirable. See label for details
- Product alone gives useful reduction of cereal eyespot. Where high infections probable a tank mix with prochloraz recommended
- On oilseed rape a two spray autumn/spring programme recommended for high risk situations and on susceptible varieties

Restrictions

- Maximum total dose equivalent to 1-3 full dose treatments depending on crop treated. See labels for details

Crop-specific information

- HI 6 wk for field beans; 14 d for sugar beet, red beet, leeks
- Latest use: completion of flowering for rye, wheat; emergence of ear complete (GS 59) for barley, oats; before lowest pods more than 2 cm long (GS 5,1) for winter oilseed rape
- Application to winter wheat in spring between the start of stem elongation and the third node detectable stage (GS 30-33) may cause straw shortening, but does not cause loss of yield

Environmental safety

- Dangerous for the environment
- Toxic to aquatic organisms
- Dangerous to fish or other aquatic life. Do not contaminate surface waters or ditches with chemical or used container

Hazard classification and safety precautions

Hazard H03, H11
Risk phrases R22b, R36, R38, R51, R53a, R63
Operator protection A, C, H; U05a, U08, U20b
Environmental protection E13b [1]; E15a [2]; E38 [1, 2]
Storage and disposal D01, D02, D05, D09a, D10b, D12a [1, 2]; D07 [1]
Medical advice M05b

130   cyproconazole + cyprodinil

A broad spectrum fungicide mixture for cereals

Products

Radius	Syngenta	5.33:40 % w/w	WG	09387

Uses

- Brown rust in **spring barley, winter barley, winter wheat**
- Eyespot in **spring barley, winter barley, winter wheat**
- Net blotch in **spring barley, winter barley**
- Powdery mildew in **spring barley, winter barley, winter wheat**
- Rhynchosporium in **spring barley, winter barley**
- Septoria diseases in **winter wheat**
- Yellow rust in **spring barley, winter barley, winter wheat**

Approval information

- Accepted by BBPA for use on malting barley

Efficacy guidance

- Best results obtained from treatment at early stages of development. Later treatments may be required if disease pressure remains high
- Eyespot is fully controlled by treatment in spring during stem extension. Control may be reduced when very dry conditions follow application

Restrictions

- Maximum total dose equivalent to two full dose treatments

Crop-specific information

- Latest use: before first spikelet of inflorescence visible (GS 51) for barley; before caryopsis watery ripe (GS 71) for winter wheat

Environmental safety

- Dangerous for the environment
- Very toxic to aquatic organisms
- Risk to certain non-target insects or other arthropods. For advice on risk management and use in Integrated Pest Management (IPM) see directions for use

Hazard classification and safety precautions

Hazard H03, H11
Risk phrases R38, R43, R50, R53a, R63
Operator protection A, H; U05a, U20a
Environmental protection E15a, E22b, E38
Storage and disposal D01, D02, D05, D07, D09a, D12a

FOR FULL CONDITIONS OF USE ALWAYS READ THE PRODUCT LABEL

131   cyproconazole + propiconazole

A broad-spectrum mixture of conazole fungicides for wheat and barley

Products

1 Alto Xtra	Syngenta	160:250 g/l	EC	11839
2 Menara	Syngenta	160:250 g/l	EC	09321

Uses

- Brown rust in *spring barley, winter barley, winter wheat* [1, 2]
- Glume blotch in *winter wheat* [1]
- Late ear diseases in *winter wheat* [1, 2]
- Net blotch in *spring barley, winter barley* [1, 2]
- Powdery mildew in *spring barley, winter barley* [1, 2]; *winter wheat* [1]
- Rhynchosporium in *spring barley, winter barley* [1, 2]
- Septoria diseases in *winter wheat* [2]
- Septoria leaf spot in *winter wheat* [1]
- Yellow rust in *spring barley, winter barley* [1]; *winter wheat* [1, 2]

Approval information

- Propiconazole included in Annex I under EC Directive 91/414
- Accepted by BBPA for use on malting barley

Efficacy guidance

- Treatment should be made at first sign of disease
- Useful reduction of eyespot when applied in spring during stem extension (GS 30-32)

Restrictions

- Maximum total dose equivalent to one full dose treatment

Crop-specific information

- Latest use: before beginning of anthesis (GS 60) for barley; before grain watery ripe (GS 71) for wheat
- Application to winter wheat in spring may cause straw shortening but does not cause loss of yield

Environmental safety

- Dangerous for the environment
- Very toxic to aquatic organisms

Hazard classification and safety precautions

Hazard H03, H11
Risk phrases R36, R38, R50, R53a, R63, R67
Operator protection A, C, H; U05a, U08, U20b
Environmental protection E15a, E38
Storage and disposal D01, D02, D05, D09a, D10c, D12a

132   cyproconazole + trifloxystrobin

A conazole and strobilurin fungicide mixture for wheat and barley

Products

Sphere	Bayer CropScience	80:187.5 g/l	EC	11429

Uses

- Brown rust in *spring barley, winter barley, winter wheat*
- Net blotch in *spring barley, winter barley*
- Powdery mildew in *spring barley, winter barley, winter wheat*
- Rhynchosporium in *spring barley, winter barley*
- Septoria diseases in *winter wheat*
- Yellow rust in *spring barley, winter barley, winter wheat*

Approval information

- Trifloxystrobin included in Annex I under EC Directive 91/414
- Accepted by BBPA for use on malting barley

SEE SECTION 3 FOR PRODUCTS ALSO REGISTERED

Efficacy guidance
- Best results obtained from treatment at early stages of disease development. Further treatment may be needed if disease attack is prolonged
- Trifloxystrobin is a member of the QoI cross resistance group. Product should be used preventatively and not relied on for its curative potential
- Use product as part of an Integrated Crop Management strategy incorporating other methods of control, including where appropriate other fungicides with a different mode of action. Do not apply more than two foliar applications of QoI containing products to any cereal crop
- There is a significant risk of widespread resistance occurring in *Septoria tritici* populations in UK. Failure to follow resistance management action may result in reduced levels of disease control
- Strains of barley powdery mildew resistant to QoIs are common in the UK

Restrictions
- Maximum total dose per crop equivalent to two full dose treatments

Crop-specific information
- HI 35 d

Environmental safety
- Dangerous for the environment
- Very toxic to aquatic organisms

Hazard classification and safety precautions
   Hazard H03, H11
   Risk phrases R36, R50, R53a, R63
   Operator protection A, C, H; U05a, U09a, U19a, U20b
   Environmental protection E15a, E38
   Storage and disposal D01, D02, D09a, D10b, D12a

133   cyprodinil

An anilinopyrimidine systemic broad spectrum fungicide for cereals

See also cyproconazole + cyprodinil

Products

Unix	Syngenta	75% w/w	WG	11512

Uses
- Eyespot in **winter barley, winter wheat**
- Net blotch in **spring barley, winter barley**
- Powdery mildew in **spring barley, spring wheat, winter barley, winter wheat**
- Rhynchosporium in **spring barley, winter barley**

Approval information
- Accepted by BBPA for use on malting barley

Efficacy guidance
- Best results obtained from treatment at early stages of disease development
- For best control of eyespot spray before or during the period of stem extension in spring. Control may be reduced if very dry conditions follow treatment

Restrictions
- Maximum number of treatments equal to one and two thirds full dose on wheat and twice full dose on barley

Crop-specific information
- Latest use: up to and including first awns visible (GS 49) for spring barley, winter barley; up to and including grain watery ripe (GS 71) for spring wheat, winter wheat

Environmental safety
- Dangerous for the environment
- Very toxic to aquatic organisms

Hazard classification and safety precautions
   Hazard H11
   Risk phrases R50, R53a
   Operator protection A, H; U05a, U20a

FOR FULL CONDITIONS OF USE ALWAYS READ THE PRODUCT LABEL

Environmental protection E15a, E38
Storage and disposal D01, D02, D05, D07, D09a, D12a

134   cyprodinil + picoxystrobin

A broad spectrum fungicide mixture for cereals

Products

Acanto Prima	Syngenta	30:8 % w/w	WG	11864

Uses
- Brown rust in **spring barley**, **spring wheat**, **winter barley**, **winter wheat**
- Eyespot in **spring barley**, **spring wheat** *(moderate control)*, **winter barley**, **winter wheat** *(moderate control)*
- Glume blotch in **spring wheat**, **winter wheat**
- Net blotch in **spring barley**, **winter barley**
- Powdery mildew in **spring barley**, **winter barley**
- Rhynchosporium in **spring barley**, **winter barley**
- Septoria leaf spot in **spring wheat**, **winter wheat**
- Yellow rust in **spring wheat**, **winter wheat**

Approval information
- Accepted by BBPA for use on malting barley

Efficacy guidance
- Best results obtained from use as a protectant treatment or in the earliest stages of disease development. Further applications may be needed if disease attack is prolonged
- Picoxystrobin is a member of the QoI cross resistance group. Product should be used preventatively and not relied on for its curative potential
- Use product as part of an Integrated Crop Management strategy incorporating other methods of control, including where appropriate other fungicides with a different mode of action. Do not apply more than two foliar applications of QoI containing products to any cereal crop
- There is a significant risk of widespread resistance occurring in *Septoria tritici* populations in UK. Failure to follow resistance management action may result in reduced levels of disease control
- In wheat product must always be used in mixture with another product, recommended for control of the same target disease, that contains a fungicide from a different cross resistance group and is applied at a dose that will give robust control
- Strains of barley powdery mildew resistant to QoIs are common in the UK

Restrictions
- Maximum number of treatments 2 per crop

Crop-specific information
- Latest use: before early milk stage (GS 73) wheat; before ear just visible (GS 51) for barley

Environmental safety
- Dangerous for the environment
- Very toxic to aquatic organisms

Hazard classification and safety precautions
Hazard H11
Risk phrases R50, R53a
Operator protection A, C, H; U05a, U09a, U11, U14, U15, U20a
Environmental protection E15a, E34, E38
Storage and disposal D01, D02, D09a, D10c, D12a

135   2,4-D

A translocated phenoxy herbicide for cereals, grass and amenity use

See also amitrole + 2,4-D + diuron
clopyralid + 2,4-D + MCPA

Products

1 Agricorn D II	FCC	490 g/l	SL	09415

SEE SECTION 3 FOR PRODUCTS ALSO REGISTERED

SECTION 2

Products – continued

2	Depitox	Nufarm UK	490 g/l	SL	11149
3	Dicotox Extra	Bayer Environ.	400 g/l	EC	09930
4	Dioweed 50	United Phosphorus	500 g/l	SL	08050
5	Dormone	Bayer Environ.	465 g/l	SL	09932
6	Headland Staff 500	Headland	500 g/l	SL	12087
7	Herboxone	Headland	500 g/l	SL	10032
8	HY-D Super	Agrichem	500 g/l	SL	11618
9	Syford	Vitax	500 g/l	SL	02062

Uses

- Annual dicotyledons in **amenity grass**, **apple orchards**, **pear orchards** [1, 2]; **amenity turf** [3, 5, 9]; **conifer plantations**, **forest** [3]; **established grassland** [1, 2, 4, 6-9]; **grass seed crops** [9]; **managed amenity turf** [1, 2, 4, 5]; **rotational grassland** [4]; **spring barley**, **spring wheat**, **winter barley**, **winter rye**, **winter wheat** [1, 2, 4, 6-8]; **spring rye** [1, 2, 4]; **winter oats** [1, 2, 6-8]
- Aquatic weeds in **aquatic situations** [2]; **water or waterside areas** [5]
- Heather in **conifer plantations**, **forest** [3]
- Perennial dicotyledons in **amenity grass**, **apple orchards**, **pear orchards** [1, 2]; **amenity turf** [3, 5, 9]; **conifer plantations**, **forest** [3]; **established grassland** [1, 2, 4, 6-9]; **grass seed crops** [9]; **managed amenity turf** [1, 2, 4, 5]; **rotational grassland** [4]; **spring barley**, **spring wheat**, **winter barley**, **winter rye**, **winter wheat** [1, 2, 4, 6-8]; **spring rye** [1, 2, 4]; **water or waterside areas** [5]; **winter oats** [1, 2, 6-8]
- Willows in **conifer plantations**, **forest** [3]
- Woody weeds in **conifer plantations**, **forest** [3]

Approval information

- 2,4-D included in Annex I under EC Directive 91/414
- May be applied through CDA equipment (see label for details) [9]. See notes in Section 5 on ULV application
- Approved for aquatic weed control [2, 5]. See notes in Section 5 on use of herbicides in or near water
- Accepted by BBPA for use on malting barley

Efficacy guidance

- Best results achieved by spraying weeds in seedling to young plant stage when growing actively in a strongly competing crop
- Most effective stage for spraying perennials varies with species. See label for details
- Spray aquatic weeds when in active growth between May and Sep

Restrictions

- Maximum number of treatments normally 1 per crop and in forestry, 2 per yr in grassland and 2 or 3 per yr in amenity turf. Check individual labels
- Do not use on newly sown leys containing clover
- Do not spray grass seed crops after ear emergence
- Do not spray within 6 mth of laying turf or sowing fine grass
- Do not dump surplus herbicide in water or ditch bottoms
- Do not plant conifers until at least 1 mth after treatment
- Do not spray crops stressed by cold weather or drought or if frost expected
- Do not roll or harrow within 7 d before or after spraying
- Do not spray if rain falling or imminent
- Do not mow amenity grass 2 d before or 1 d after spraying. Do not mow grassland or graze for at least 10 d after spraying

Crop-specific information

- Latest use: before 1st node detectable in cereals; end Aug for conifer plantations; before established grassland 25 cm high
- Spray winter cereals in spring when leaf-sheath erect but before first node detectable (GS 31), spring cereals from 5-leaf stage to before first node detectable (GS 15-31)
- Cereals undersown with grass and/or clover, but not with lucerne, may be treated
- Selective treatment of resistant conifers can be made in Aug when growth ceased and plants hardened off, spray must be directed if applied earlier. See label for details

Following crops guidance

- Do not use shortly before or after sowing any crop
- Do not direct drill brassicas or grass/clover mixtures within 3 wk of application

FOR FULL CONDITIONS OF USE ALWAYS READ THE PRODUCT LABEL

Environmental safety
- Dangerous for the environment
- Toxic (very toxic [3]) to aquatic organisms
- Harmful to fish or other aquatic life. Do not contaminate surface waters or ditches with chemical or used container [5, 8]
- Dangerous to aquatic higher plants. Do not contaminate surface waters or ditches with chemical or used container
- 2,4-D is active at low concentrations. Take extreme care to avoid drift onto neighbouring crops, especially beet crops, brassicas, most market garden crops including lettuce and tomatoes under glass, pears and vines
- May be used to control aquatic weeds in presence of fish if used in strict accordance with directions for waterweed control and precautions needed for aquatic use [2, 5]
- Keep livestock out of treated areas for at least 2 wk following treatment and until poisonous weeds, such as ragwort, have died down and become unpalatable
- Water containing the herbicide must not be used for irrigation purposes within 3 wk of treatment or until the concentration in water is below 0.05 ppm

Hazard classification and safety precautions
Hazard H03 [1-9]; H11 [3-9]
Risk phrases R21 [3, 5]; R22a, R53a [1-9]; R22b, R36, R50, R66, R67 [3]; R38 [1-3]; R41 [1, 2, 4-9]; R43 [3-9]; R51 [4-9]
Operator protection A, C, H, M [1-9]; D [4]; U05a, U08 [1-9]; U11 [1, 2, 4-9]; U14 [3-9]; U15 [4, 6-8]; U20a [1, 4]; U20b [2, 3, 5-9]
Environmental protection E07a, E34 [1-9]; E13b [1]; E13c [5, 8]; E14b [4, 6, 7]; E15a [2-7, 9]; E21 [5] (3 wk); E38 [3, 5-9]
Storage and disposal D01, D02, D09a [1-9]; D05 [2-4, 6-8]; D10a [1, 6-8]; D10b [2, 4, 9]; D10c [3]; D12a [3, 5, 9]
Medical advice M03 [1-9]; M05a [4, 6-8]; M05b [3]

136   2,4-D + dicamba

A translocated herbicide for use on turf

Products

1	Magneto	Nufarm UK	344:120 g/l	SL	12339
2	New Estermone	Vitax	200:35 g/l	EC	06336
3	Thrust	Nufarm UK	344:120 g/l	SL	12230

Uses
- Annual dicotyledons in *amenity grass*, *established grassland* [1, 3]; *managed amenity turf* [2]
- Perennial dicotyledons in *amenity grass*, *established grassland* [1, 3]; *managed amenity turf* [2]

Approval information
- 2,4-D included in Annex I under EC Directive 91/414

Efficacy guidance
- Best results achieved by application when weeds growing actively in spring or early summer (later with irrigation and feeding)
- More resistant weeds may need repeat treatment after 3 wk
- Improved control of some weeds can be obtained by use of specifed oil adjuvant [1, 3]
- Do not use during drought conditions or mow for 3 d before or after treatment

Restrictions
- Maximum number of treatments 2 per yr on amenity grass and established grassland [1]; 3 per yr on managed amenity turf [2]
- Do not treat newly sown or turfed areas or grass less than 1 yr old
- Do not treat grass crops intended for seed production [1, 3]
- Do not treat grass suffering from drought, disease or other adverse factors [1, 3]
- Do not roll or harrow for 7 d before or after treatment [1, 3]
- Do not apply when grassland is flowering [1, 3]
- Do not re-seed for 6 wk after application [2]
- Avoid spray drift onto cultivated crops or ornamentals

SEE SECTION 3 FOR PRODUCTS ALSO REGISTERED

Crop-specific information
- Latest use: before grass 25 cm high for amenity grass, established grassland [1, 3]
- The first four mowings after treatment must be composted for at least 6 mth before use

Environmental safety
- Dangerous for the environment
- Toxic to aquatic organisms
- Keep livestock out of treated areas for at least two weeks following treatment and until poisonous weeds, such as ragwort, have died down and become unpalatable

Hazard classification and safety precautions
Hazard H03 [1, 3]; H04 [2]; H11 [1-3]
Risk phrases R22a, R41, R43 [1, 3]; R51, R53a [1-3]
Operator protection A, C, H, M; U05a [1, 3]; U08, U20b [2]; U11 [1-3]
Environmental protection E07a, E15a, E38 [1-3]; E34 [1, 3]
Storage and disposal D01, D02, D05 [1, 3]; D09a, D10a, D12a [1-3]
Medical advice M03 [1, 3]

137   2,4-D + dicamba + fluroxypyr

A translocated and contact herbicide mixture for amenity turf

Products
Holster	SumiAgro Amenity	285:52.2:105 g/l	SL	10593

Uses
- Annual dicotyledons in *managed amenity turf*
- Buttercups in *managed amenity turf*
- Chickweed in *managed amenity turf*
- Clover in *managed amenity turf*
- Dandelions in *managed amenity turf*

Approval information
- 2,4-D and fluroxypyr included in Annex I under EC Directive 91/414

Efficacy guidance
- Apply when weeds actively growing (normally between Apr and Sep) and when soil is moist
- Best results obtained from treatment in spring or early summer before weeds begin to flower
- Do not apply if turf is wet or if rainfall expected within 4 h of treatment. Both circumstances will reduce weed control

Restrictions
- Maximum number of treatments 2 per yr
- Do not mow for 3 d before or after treatment
- Avoid overlapping or overdosing, especially on newly sown turf
- Do not spray in drought conditions or if turf under stress from frost, waterlogging, trace element deficiency, pest or disease attack

Crop-specific information
- Latest Use: normally Sep for managed amenity turf
- New turf may be treated in spring provided at least 2 mth have elapsed since sowing

Environmental safety
- Dangerous for the environment
- Toxic to aquatic organisms
- Keep livestock out of treated areas for at least two weeks following treatment and until poisonous weeds, such as ragwort, have died down and become unpalatable

Hazard classification and safety precautions
Hazard H03, H11
Risk phrases R22a, R22b, R36, R38, R51, R53a
Operator protection A, C, H, M; U05a, U08, U19a, U20b
Environmental protection E07a, E15a, E38
Storage and disposal D01, D02, D05, D09a, D10b, D12a
Medical advice M03

FOR FULL CONDITIONS OF USE ALWAYS READ THE PRODUCT LABEL

138   2,4-D + dicamba + triclopyr

A translocated herbicide for perennial and woody weed control

Products

1	Broadsword	United Phosphorus	200:85:65 g/l	EC	09140
2	Greengard	SumiAgro Amenity	200:85:65 g/l	EC	11715
3	Nu-Shot	Nufarm UK	200:85:65 g/l	EC	11148

Uses

- Annual dicotyledons in *amenity grass* [2, 3]; *land not intended to bear vegetation* [1]; *permanent pasture, rotational grassland* [1, 3]
- Brambles in *farm forestry, forest, natural surfaces not intended to bear vegetation* [1-3]
- Docks in *amenity grass* [2, 3]; *land not intended to bear vegetation* [1]; *permanent pasture, rotational grassland* [1, 3]
- Gorse in *farm forestry, forest, natural surfaces not intended to bear vegetation* [1-3]
- Green cover in *land not being used for crop production* [1]
- Japanese knotweed in *farm forestry, forest, natural surfaces not intended to bear vegetation* [1-3]
- Perennial dicotyledons in *amenity grass* [2, 3]; *farm forestry, forest, natural surfaces not intended to bear vegetation* [1-3]; *land not intended to bear vegetation* [1]; *permanent pasture, rotational grassland* [1, 3]
- Plantains in *amenity grass* [2, 3]; *land not intended to bear vegetation* [1]; *permanent pasture, rotational grassland* [1, 3]
- Rhododendrons in *farm forestry, forest, natural surfaces not intended to bear vegetation* [1-3]
- Stinging nettle in *amenity grass* [2, 3]; *permanent pasture, rotational grassland* [1, 3]
- Thistles in *amenity grass* [2, 3]; *land not intended to bear vegetation* [1]; *permanent pasture, rotational grassland* [1, 3]
- Woody weeds in *farm forestry, forest, natural surfaces not intended to bear vegetation* [1-3]

Approval information

- 2,4-D included in Annex I under EC Directive 91/414

Efficacy guidance

- Apply as foliar spray to herbaceous or woody weeds. Timing and growth stage for best results vary with species. See label for details
- Dilute with water for stump treatment and apply after felling up to the start of regrowth. Treat any regrowth with a spray to the growing foliage
- May be applied at 1/3 dilution in weed wipers or 1/8 dilution with ropewick applicators

Restrictions

- Maximum total dose per yr equivalent to two full dose treatments
- Do not use on pasture established less than 1 yr or on grass grown for seed
- Where clover a valued constituent of sward only use as a spot treatment
- Do not graze for 7 d or mow for 14 d after treatment
- Do not direct drill grass, clover or brassicas for at least 6 wk after grassland treatment
- Do not plant trees for 1-3 mth after spraying depending on dose applied. See label
- Avoid spray drift into greenhouses or onto crops or ornamentals. Vapour drift may occur in hot conditions

Crop-specific information

- Latest Use: before weed flower buds open
- Sprays may be applied in pines, spruce and fir providing drift is avoided. Optimum time is mid-autumn when tree growth ceased but weeds not yet senescent

Environmental safety

- Dangerous for the environment
- Toxic (very toxic [1]) to aquatic organisms [2, 3]
- Docks and other weeds may become increasingly palatable after treatment and may be preferentially grazed
- Keep livestock out of treated areas for at least two weeks following treatment and until poisonous weeds, such as ragwort, have died down and become unpalatable

- Take extreme care to avoid drift onto neighbouring crops, especially beet crops, brassicas, most market garden crops including lettuce and tomatoes under glass, pears and vines

Hazard classification and safety precautions
Hazard H03, H08, H11
Risk phrases R22a, R22b, R36, R53a [1-3]; R37 [1, 3]; R38 [1, 2]; R43 [3]; R50, R67 [1]; R51 [2, 3]
Operator protection A, C, H, M [1-3]; D [3]; U02a, U04a, U13, U20a [1, 2]; U05a, U08, U11, U19a [1-3]; U14, U20b [3]
Environmental protection E07a, E34 [1-3]; E15a [2, 3]; E15b [1]; E38 [3]
Consumer protection C01 [1, 2]
Storage and disposal D01, D02, D05, D09a, D10b [1-3]; D12a [1, 3]
Medical advice M03, M05b

139   2,4-D + dichlorprop-P + MCPA + mecoprop-P

A translocated herbicide for use in apple and pear orchards

Products

UPL Camppex	United Phosphorus	34.5:66.6:52.8:82.1 g/l	SL	11661

Uses
- Annual dicotyledons in *apple orchards*, *pear orchards*
- Chickweed in *apple orchards*, *pear orchards*
- Cleavers in *apple orchards*, *pear orchards*
- Perennial dicotyledons in *apple orchards*, *pear orchards*

Approval information
- 2,4-D, MCPA and mecoprop-P included in Annex I under EC Directive 91/414

Efficacy guidance
- Use on emerged weeds in established apple and pear orchards (from 1 yr after planting) as directed application
- Spray when weeds in active growth and at growth stage recommended on label
- Effectiveness may be reduced by rain within 12 h

Restrictions
- Maximum number of treatments 2 per yr
- Applications must be made around, and not directly to, trees. Do not allow drift onto trees
- Do not spray during blossom period. Do not spray to run-off
- Do not roll, harrow or cut grass crops on orchard floor within at least 3 d before or after spraying

Environmental safety
- Harmful to aquatic organisms
- Harmful to fish or other aquatic life. Do not contaminate surface waters or ditches with chemical or used container
- Keep livestock out of treated areas for at least two weeks following treatment and until poisonous weeds, such as ragwort, have died down and become unpalatable
- Take extreme care to avoid drift onto neighbouring sensitive crops

Hazard classification and safety precautions
Hazard H03
Risk phrases R20, R21, R22a, R43, R52, R53a
Operator protection A, C, H, M; U05a, U08, U20b
Environmental protection E07a, E13c, E34
Storage and disposal D01, D02, D05, D09a, D10b
Medical advice M03, M05a

140   2,4-D + florasulam

A post-emergence herbicide mixture for control of broad leaved weeds in managed amenity turf

Products

Junction	Rigby Taylor	452:6.25 g/l	ME	12493

FOR FULL CONDITIONS OF USE ALWAYS READ THE PRODUCT LABEL

Uses

- Clover in **managed amenity turf**
- Daisies in **managed amenity turf**
- Dandelions in **managed amenity turf**
- Plantains in **managed amenity turf**
- Sticky mouse-ear in **managed amenity turf**

Efficacy guidance

- Best results obtained from treatment of actively growing weeds between Mar and Oct when soil is moist
- Do not apply when rain is imminent or during periods of drought unless irrigation is applied

Restrictions

- Maximum number of treatments on managed amenity turf: 1 per yr

Crop-specific information

- Avoid mowing turf 3 d before and after spraying
- Ensure newly sown turf has established before treating. Turf sown in late summer or autumn should not be sprayed until growth is resumed in the following spring

Following crops guidance

- An interval of 4 wk must elapse bewteen application and re-seeding turf

Environmental safety

- Dangerous for the environment
- Toxic to aquatic organisms

Hazard classification and safety precautions

Hazard H03, H11
Risk phrases R22a, R43, R51, R53a
Operator protection A, H; U02a, U05a, U08, U14, U20a
Environmental protection E15a, E34, E38
Storage and disposal D01, D02, D09a, D10b, D12a

---

141 2,4-D + MCPA

A translocated herbicide mixture for cereals and grass

Products

Headland Polo	Headland	360:315 g/l	SL	10283

Uses

- Annual dicotyledons in **established grassland, rotational grassland, spring barley, spring wheat, winter barley, winter oats, winter wheat**
- Perennial dicotyledons in **established grassland, rotational grassland, spring barley, spring wheat, winter barley, winter oats, winter wheat**

Approval information

- 2,4-D and MCPA included in Annex I under EC Directive 91/414
- Accepted by BBPA for use on malting barley

Efficacy guidance

- Best results achieved by spraying weeds in seedling to young plant stage when growing actively in a strongly competing crop
- Most effective stage for spraying perennials varies with species. See label for details

Restrictions

- Maximum number of treatments 1 per crop in cereals and established grassland; 2 per yr in rotational grassland
- Do not spray if rain falling or imminent
- Do not cut grass or graze for at least 10 d after spraying
- Do not use on newly sown leys containing clover or other legumes
- Do not spray crops stressed by cold weather or drought or if frost expected
- Do not use shortly before or after sowing any crop
- Do not roll or harrow within 7 d before or after spraying

SECTION 2

SEE SECTION 3 FOR PRODUCTS ALSO REGISTERED

Crop-specific information
- Spray winter cereals in spring when leaf-sheath erect but before first node detectable (GS 31), spring cereals from 5-leaf stage to before first node detectable (GS 15-31)
- Latest use: before first node detectable (GS 31) in cereals

Environmental safety
- Dangerous for the environment
- Toxic to aquatic organisms
- Keep livestock out of treated areas for at least two weeks following treatment and until poisonous weeds, such as ragwort, have died down and become unpalatable
- 2,4-D and MCPA are active at low concentrations. Take extreme care to avoid drift onto neighbouring crops, especially beet crops, brassicas, most market garden crops including lettuce and tomatoes under glass, pears and vines

Hazard classification and safety precautions
Hazard H03, H11
Risk phrases R20, R21, R22a, R41, R43, R51, R53a
Operator protection A, C, H, M; U05a, U08, U11, U14, U15, U20b
Environmental protection E07a, E15a, E34
Storage and disposal D01, D02, D05, D09a, D10a
Medical advice M03, M05a

142   2,4-D + mecoprop-P

A translocated herbicide for use in amenity turf

Products

1 Supertox 30	Bayer Environ.	93.5:95 g/l	SL	09946
2 Sydex	Vitax	125:125 g/l	SL	06412

Uses
- Annual dicotyledons in *established grassland, grass seed crops* [2]; *managed amenity turf* [1, 2]
- Perennial dicotyledons in *established grassland, grass seed crops* [2]; *managed amenity turf* [1, 2]

Approval information
- 2,4-D and mecoprop-P included in Annex I under EC Directive 91/414

Efficacy guidance
- May be applied from Apr to Sep, best results in May-Jun when weeds in active growth. Less susceptible weeds may require second treatment not less than 6 wk later
- For best results apply fertilizer 1-2 wk before treatment
- Do not spray during drought conditions or when rain imminent as efficacy will be impaired
- Do not close mow infrequently mown turf for 3-4 d before or after treatment. On fine turf do not mow for 24 h before or after treatment

Restrictions
- Maximum number of treatments varies with product. Consult label. The total amount of mecoprop-P applied in a single yr must not exceed the maximum total dose approved for any single product for the crop/situation
- Do not use first 4 mowings for mulching unless composted for at least 6 mth (do not use first 2 mowings for composting on some labels)
- Grass cuttings should not be used for mulching but may be composted and used 6 mth later
- Do not apply during frosty weather
- Do not roll or harrow within 7 d before or after treatment

Crop-specific information
- Newly laid turf or newly sown grass should not be treated for at least 6 mth

Environmental safety
- Dangerous for the environment [2]
- Harmful to aquatic organisms
- Keep livestock out of treated areas for at least two weeks following treatment and until poisonous weeds, such as ragwort, have died down and become unpalatable

FOR FULL CONDITIONS OF USE ALWAYS READ THE PRODUCT LABEL

- Take extreme care to avoid drift onto neighbouring crops, especially beet crops, brassicas, most market garden crops including lettuce and tomatoes under glass, pears and vines

Hazard classification and safety precautions

Hazard H03 [1, 2]; H04 [1]; H11 [2]

Risk phrases R22a [2]; R36, R41 [1]; R43, R52, R53a [1, 2]

Operator protection A, C, H, M; U05a, U08, U11, U14, U20b [1, 2]; U15 [1]

Environmental protection E07a, E15a, E34 [1, 2]; E23 [1]; E38 [2]

Storage and disposal D01, D02, D09a, D12a [1, 2]; D05, D10c [1]; D10b [2]

Medical advice M03 [2]

143   daminozide

A hydrazide plant growth regulator for use in certain ornamentals

Products

1 B-Nine	Certis	85% w/w	SP	11471
2 Dazide Enhance	Fine	85% w/w	SG	11943

Uses

- Internode reduction in *azaleas*, *bedding plants*, *chrysanthemums*, *hydrangeas* [1, 2]; *ornamental specimens*, *poinsettias* [1]

Approval information

- Daminozide included in Annex I under EC Directive 91/414
- Sales of daminozide for use on food crops were halted worldwide by the manufacturer in Oct 1989. Sales for use on flower crops were not affected

Efficacy guidance

- Best results obtained by application in late afternoon when glasshouse has cooled down
- Apply a fine spray to give good coverage of dry foliage without run off
- A reduced rate tank mix with chlormequat + choline chloride is recommended for use on poinsettias. See label for details [1]

Restrictions

- Maximum number of treatments 2 per crop
- Apply only to turgid, well watered plants. Do not water for 24 h after spraying
- Do not use on chrysanthemum Fandango
- Do not mix with other spray chemicals except as recommended above
- Do not store product in metal containers [2]

Crop-specific information

- Latest use: end Sep for Poinsettias [1]

Hazard classification and safety precautions

Hazard H04 [1]

Risk phrases R41 [1]

Operator protection A [1, 2]; C [1]; H [2]; U05a, U19a [1, 2]; U08, U20b [2]; U11 [1]

Environmental protection E15a

Storage and disposal D01, D02, D09a [1, 2]; D08, D10b, D12a [2]

Medical advice M05a [2]

144   dazomet

A methyl isothiocyanate releasing soil fumigant

Products

Basamid	Certis	97% w/w	GR	11324

Uses

- Nematodes in *field crops* (soil fumigation), *protected crops* (soil fumigation), *vegetables* (soil fumigation)
- Soil insects in *field crops* (soil fumigation), *protected crops* (soil fumigation), *vegetables* (soil fumigation)

SEE SECTION 3 FOR PRODUCTS ALSO REGISTERED

- Soil-borne diseases in **field crops** *(soil fumigation)*, **protected crops** *(soil fumigation)*, **vegetables** *(soil fumigation)*
- Weed seeds in **field crops** *(soil fumigation)*, **protected crops** *(soil fumigation)*, **vegetables** *(soil fumigation)*

Efficacy guidance

- Dazomet acts by releasing methyl isothiocyanate in contact with moist soil
- Soil sterilization is carried out after harvesting one crop and before planting the next
- The soil must be of fine tilth, free of clods and evenly moist to the depth of sterilization
- Soil moisture must not be less than 50% of water-holding capacity or oversaturated. If too dry, water at least 7-14 d before treatment
- In order to obtain short treatment times it is recommended to treat soils when soil temperature is above 7°C. Treatment should be used outdoors before winter rains make soil too wet to cultivate - usually early Nov
- For club root control treat only in summer when soil temperature above 10°C
- Where onion white rot a problem unlikely to give effective control where inoculum level high or crop under stress
- Apply granules with suitable applicators, mix into soil immediately to desired depth and seal surface with polythene sheeting, by flooding or heavy rolling. See label for suitable application and incorporation machinery
- With 'planting through' technique polythene seal is left in place to form mulch into which new crop can be planted

Restrictions

- Maximum number of treatments 1 per crop or batch of soil
- Do not treat ground where water table may rise into treated layer

Crop-specific information

- Latest use: pre-planting

Following crops guidance

- With 'planting through' technique no gas release cultivations are made and safety test with cress is particularly important. Conduct cress test on soil samples from centre as well as edges of bed. Observe minimum of 30 d from application to cress test in soils at 10°C or above, at least 50 d in soils below 10°C. See label for details
- In all other situations 14-28 d after treatment cultivate lightly to allow gas to disperse and conduct cress test after a further 14-28 d (timing depends on soil type and temperature). Do not treat structures containing live plants or any ground within 1 m of live plants

Environmental safety

- Dangerous for the environment
- Very toxic to aquatic organisms

Hazard classification and safety precautions

Hazard H03, H11
Risk phrases R22a, R50, R53a
Operator protection A, M; U04a, U05a, U09a, U19a, U20a
Environmental protection E15a, E34, E38
Storage and disposal D01, D02, D07, D09a, D11a, D12a
Medical advice M03, M05a

145   2,4-DB

A translocated phenoxy herbicide for use in lucerne and undersown cereals

Products

DB Straight	United Phosphorus	300 g/l	SL	07523

Uses

- Annual dicotyledons in **lucerne**, **undersown spring cereals** *(undersown with lucerne)*, **undersown winter cereals** *(undersown with lucerne)*
- Thistles in **lucerne**, **undersown spring cereals** *(undersown with lucerne)*, **undersown winter cereals** *(undersown with lucerne)*

FOR FULL CONDITIONS OF USE ALWAYS READ THE PRODUCT LABEL

Approval information
- 2,4-DB included in Annex I under EC Directive 91/414
- Accepted by BBPA for use on malting barley

Efficacy guidance
- Best results achieved on young seedling weeds under good growing conditions. Treatment less effective in cold weather and dry soil conditions
- Rain within 12 h may reduce effectiveness

Restrictions
- Do not allow spray drift onto neighbouring crops

Crop-specific information
- Latest use: before first node detectable (GS 31) for undersown cereals; fourth trifoliate leaf for lucerne
- In direct sown lucerne spray when seedlings have reached first trifoliate leaf stage. Optimum time 3-4 trifoliate leaves
- Do not treat any lucerne after fourth trifoliate leaf
- In spring barley and spring oats undersown with lucerne spray from when cereal has 1 leaf unfolded and lucerne has first trifoliate leaf
- In spring wheat undersown with lucerne spray from when cereal has 3 leaves unfolded and lucerne has first trifoliate leaf

Environmental safety
- Dangerous for the environment
- Toxic to aquatic organisms
- Keep livestock out of treated areas for at least two weeks following treatment and until poisonous weeds, such as ragwort, have died down and become unpalatable

Hazard classification and safety precautions
- **Hazard** H03, H11
- **Risk phrases** R22a, R38, R41, R51, R53a
- **Operator protection** U05a, U08, U20a
- **Environmental protection** E07a, E15a, E19b, E34
- **Storage and disposal** D05, D09a, D10b
- **Medical advice** M03, M05a

146   2,4-DB + linuron + MCPA

A translocated herbicide for undersown cereals and grass

Products

Alistell	United Phosphorus	220:30:30 g/l	EC	11053

Uses
- Annual dicotyledons in **seedling grassland**, **undersown spring cereals** (undersown with clover), **undersown winter cereals** (undersown with clover)

Approval information
- 2,4-DB, linuron and MCPA included in Annex I under EC Directive 91/414
- Accepted by BBPA for use on malting barley

Efficacy guidance
- Best results achieved on young seedling weeds growing actively in warm, moist weather
- May be applied at any time of year provided crop at correct stage and weather suitable
- Avoid spraying if rain falling or imminent

Restrictions
- Maximum number of treatments 1 per crop or yr
- Spray winter cereals when fully tillered but before first node detectable (GS 29-30)
- Spray spring wheat from 5-fully expanded leaf stage (GS 15), barley and oats from 2-fully expanded leaves (GS 12)
- Do not spray cereals undersown with lucerne, peas or beans
- Apply to clovers after 1-trifoliate leaf, to grasses after 2-fully expanded leaf stage
- Do not spray in conditions of drought, waterlogging or extremes of temperature
- In frosty weather clover leaf scorch may occur but damage normally outgrown

SEE SECTION 3 FOR PRODUCTS ALSO REGISTERED

SECTION 2

- Do not use on sand or soils with more than 10% organic matter
- Do not roll or harrow within 7 d before or after spraying
- Avoid drift of spray or vapour onto susceptible crops
- Do not apply by hand-held sprayers

Crop-specific information
- Latest use: before first node detectable (GS 31) for undersown cereals; 2 wk before grazing for grassland

Environmental safety
- Dangerous for the environment
- Toxic to aquatic organisms
- Keep livestock out of treated areas for at least two weeks following treatment and until poisonous weeds, such as ragwort, have died down and become unpalatable
- LERAP Category B

Hazard classification and safety precautions
    Hazard H03, H11
    Risk phrases R22a, R22b, R40, R41, R51, R53a, R66
    Operator protection A, C, H; U05a, U08, U11, U13, U19a, U20b
    Environmental protection E07a, E15a, E16a, E34, E38
    Storage and disposal D01, D02, D09a, D10b, D12a
    Medical advice M03, M05a

147   2,4-DB + MCPA

A translocated herbicide for cereals, clovers and leys

Products

1 Agrichem DB Plus	Agrichem	243:40 g/l	SL	00044
2 Headland Cedar	Headland	240:40 g/l	SL	10489
3 Redlegor	United Phosphorus	244:44 g/l	SL	07519

Uses
- Annual dicotyledons in *all cereals*, *rotational grassland* [1]; *grass re-seeds*, *spring barley*, *spring oats*, *undersown barley* (red or white clover), *undersown oats* (red or white clover), *undersown wheat* (red or white clover), *winter barley*, *winter oats* [2]; *seedling leys* [3]; *spring wheat*, *winter wheat* [2, 3]; *undersown spring cereals*, *undersown winter cereals* [1, 3]
- Perennial dicotyledons in *all cereals*, *rotational grassland* [1]; *grass re-seeds*, *spring barley*, *spring oats*, *undersown barley* (red or white clover), *undersown oats* (red or white clover), *undersown wheat* (red or white clover), *winter barley*, *winter oats* [2]; *seedling leys* [3]; *spring wheat*, *winter wheat* [2, 3]; *undersown spring cereals*, *undersown winter cereals* [1, 3]
- Polygonums in *seedling leys*, *spring wheat*, *undersown spring cereals*, *undersown winter cereals*, *winter wheat* [3]

Approval information
- 2,4-DB and MCPA included in Annex I under EC Directive 91/414
- Accepted by BBPA for use on malting barley
- Approval expiry 30 Apr 2006 [2]

Efficacy guidance
- Best results achieved on young seedling weeds under good growing conditions
- Spray thistles and other perennials when 10-20 cm high provided clover at correct stage
- Effectiveness may be reduced by rain within 12 h, by very cold conditions or drought

Restrictions
- Maximum number of treatments 1 per crop
- Do not spray established clover crops or lucerne
- Do not roll or harrow within 7 d before or after spraying
- Do not spray immediately before or after sowing any crop
- Avoid drift onto neighbouring sensitive crops

Crop-specific information
- Latest use: before first node detectable (GS 31) for cereals; 4th trifoliate leaf stage of clover for grass re-seeds

FOR FULL CONDITIONS OF USE ALWAYS READ THE PRODUCT LABEL

- Apply in spring to winter cereals from leaf sheath erect stage, to spring barley or oats from 2-leaf stage (GS 12), to spring wheat from 5-leaf stage (GS 15)
- Spray clovers as soon as possible after first trifoliate leaf, grasses after 2-3 leaf stage
- Red clover may suffer temporary distortion after treatment

Environmental safety
- Harmful to aquatic organisms
- Harmful to fish or other aquatic life. Do not contaminate surface waters or ditches with chemical or used container
- Keep livestock out of treated areas for at least two weeks following treatment and until poisonous weeds, such as ragwort, have died down and become unpalatable

Hazard classification and safety precautions

Hazard H03

Risk phrases R20, R21 [1]; R22a, R41, R52 [1-3]; R38 [2, 3]; R53a [1, 3]

Operator protection A, C [1, 2]; U05a, U08, U11, U20b [1-3]; U14 [3]; U15 [2, 3]

Environmental protection E07a, E13c [1-3]; E19b [1]; E34 [1, 2]; E38 [2]

Storage and disposal D01, D09a [1, 2]; D02, D10c [1]; D05 [1-3]; D10a [2]; D10b [3]

Medical advice M03 [1-3]; M05a [1, 3]

148   deltamethrin

A pyrethroid insecticide with contact and residual activity

Products

1	Agriguard Deltamethrin	AgriGuard	25 g/l	EC	10770
2	Bandu	Headland	25 g/l	EC	10994
3	Decis	Bayer CropScience	25 g/l	EC	07172
4	Decis Protech	Bayer CropScience	15 g/l	EW	11502
5	Landgold Deltaland	Landgold	25 g/l	EC	09906
6	Landgold Deltaland	Teliton	25 g/l	EC	12114
7	Pearl Micro	Bayer CropScience	6.25% w/w	GR	08620

Uses

- American serpentine leaf miner in *aubergines* (off-label), *non-edible ornamentals* (off-label), *rooting beds* (off-label) [3]; *non-edible glasshouse crops* (off-label), *protected aubergines* (off-label) [4]; *protected courgettes* (off-label), *protected cucumbers* (off-label), *protected gherkins* (off-label), *protected peppers* (off-label), *protected tomatoes* (off-label) [3, 4]
- Aphids in *amenity vegetation* [1-4]; *apples, spring barley, spring oats, spring wheat, winter barley, winter oats, winter wheat* [1-7]; *bulb onions* (off-label), *carrots* (off-label), *celery* (off-label), *garlic* (off-label), *leeks* (off-label), *parsnips* (off-label), *peas, protected celery* (off-label), *protected chinese cabbage* (off-label), *protected salad onions* (off-label), *salad onions* (off-label) [7]; *chinese cabbage* (off-label), *grass seed crops* (off-label), *lettuce* (off-label) [3, 4, 7]; *evening primrose* (off-label), *protected ornamentals* [4]; *nursery stock, radicchio* (off-label) [3, 7]; *ornamental plant production, protected cucumbers, protected peppers, protected pot plants, protected tomatoes* [1-4, 7]
- Apple sawfly in *apples* [7]
- Apple sucker in *apples* [1-7]
- Barley yellow dwarf vectors in *winter barley, winter wheat* [1-4]
- Barley yellow dwarf virus vectors in *spring barley, spring oats, spring wheat, winter oats* [7]; *winter barley, winter wheat* [5-7]
- Beet virus yellows vectors in *winter oilseed rape* [1-7]
- Brassica pod midge in *mustard, spring oilseed rape, winter oilseed rape* [7]
- Cabbage seed weevil in *mustard, spring oilseed rape, winter oilseed rape* [1-7]
- Cabbage stem flea beetle in *winter oilseed rape* [1-7]
- Cabbage stem weevil in *mustard, spring oilseed rape, winter oilseed rape* [1-6]
- Capsids in *amenity vegetation* [1-4]; *apples* [1-7]; *nursery stock* [3, 7]; *ornamental plant production* [1-4, 7]
- Caterpillars in *amenity vegetation* [1-4]; *apples, broccoli, brussels sprouts, cabbages, cauliflowers, kale, plums, swedes, turnips* [1-7]; *bulb onions* (off-label), *carrots* (off-label), *celery* (off-label), *garlic* (off-label), *leeks* (off-label), *parsnips* (off-label), *protected celery* (off-label), *protected chinese cabbage* (off-label), *protected salad onions* (off-label), *salad*

onions (off-label) [7]; *chinese cabbage (off-label)*, *lettuce (off-label)*, *protected salad brassicas (off-label - for baby leaf production)*, *radicchio (off-label)* [3, 4, 7]; *marjoram (off-label)*, *protected ornamentals*, *sorrel (off-label)* [4]; *nursery stock* [3, 7]; *ornamental plant production*, *protected cucumbers*, *protected peppers*, *protected pot plants*, *protected tomatoes* [1-4, 7]; *protected salad brassicas (off-label - baby leaf production)*, *salad brassicas (off-label - baby leaf production)* [2]; *salad brassicas (off-label)* [3]; *salad brassicas (off-label - for baby leaf production)* [4, 7]

- Codling moth in *apples* [1-7]
- Colorado beetle in *potatoes (off-label)* [3]
- Cutworms in *lettuce* [1-4, 7]
- Damson-hop aphid in *hops*, *plums* [1-7]
- Flea beetle in *broccoli*, *brussels sprouts*, *cabbages*, *cauliflowers*, *kale*, *swedes*, *turnips* [3, 7]; *cress (off-label)*, *protected herbs (see appendix 6)*, *salad brassicas (off-label)* [3]; *evening primrose (off-label)*, *protected salad brassicas (off-label - for baby leaf production)*, *sorrel (off-label)* [3, 4, 7]; *marjoram (off-label)*, *protected lamb's lettuce (off-label)*, *protected scarole (off-label)*, *protected watercress (off-label)*, *radicchio (off-label)* [4]; *protected chives (off-label)*, *protected parsley (off-label)* [2, 4]; *protected herbs (see appendix 6) (off-label)*, *protected lettuce (off-label)* [2, 4, 7]; *protected radicchio (off-label)*, *salad brassicas (off-label - for baby leaf production)* [4, 7]; *protected salad brassicas (off-label - baby leaf production)*, *salad brassicas (off-label - baby leaf production)* [2]; *sugar beet* [1-7]
- Fruit tree tortrix moth in *plums* [1-6]
- Insect control in *celery (off-label)*, *protected celery (off-label)*, *protected chinese cabbage (off-label)*, *protected salad onions (off-label)*, *salad onions (off-label)* [2]
- Insect pests in *bulb onions (off-label)*, *carrots (off-label)*, *celery (off-label)*, *garlic (off-label)*, *leeks (off-label)*, *protected celery (off-label)*, *protected spring onions (off-label)*, *salad onions (off-label)* [3, 4]; *chinese cabbage (off-label)* [3]; *protected chinese cabbage (off-label)* [4]
- Leafhoppers in *cress (off-label)*, *protected herbs (see appendix 6)* [3]; *marjoram (off-label)* [3, 4, 7]; *protected chives (off-label)*, *protected parsley (off-label)* [2, 4]; *protected herbs (see appendix 6) (off-label)*, *protected lettuce (off-label)* [2, 4, 7]; *protected lamb's lettuce (off-label)*, *protected scarole (off-label)*, *protected watercress (off-label)*, *radicchio (off-label)*, *sorrel (off-label)* [4]; *protected radicchio (off-label)* [4, 7]
- Mealybugs in *amenity vegetation* [1-4]; *nursery stock* [3, 7]; *ornamental plant production*, *protected cucumbers*, *protected peppers*, *protected pot plants*, *protected tomatoes* [1-4, 7]; *protected ornamentals* [4]
- Pea and bean weevil in *broad beans*, *peas*, *spring field beans*, *winter field beans* [1-7]
- Pea midge in *peas* [3, 4, 7]
- Pea moth in *peas* [1-7]
- Pear sucker in *pears* [1-7]
- Phorid flies in *mushrooms (off-label)* [2-4]
- Plum fruit moth in *plums* [7]
- Plum sawfly in *plums* [7]
- Pollen beetle in *evening primrose (off-label)* [3, 4, 7]; *mustard*, *spring oilseed rape*, *winter oilseed rape* [1-7]
- Rape winter stem weevil in *winter oilseed rape* [7]
- Raspberry beetle in *raspberries* [1-7]
- Sawflies in *apples*, *plums* [1-6]
- Scale insects in *amenity vegetation* [1-4]; *nursery stock* [3, 7]; *ornamental plant production*, *protected cucumbers*, *protected peppers*, *protected pot plants*, *protected tomatoes* [1-4, 7]; *protected ornamentals* [4]
- Sciarid flies in *mushrooms (off-label)* [2-4]
- Thrips in *amenity vegetation* [1-4]; *nursery stock* [3, 7]; *ornamental plant production* [1-4, 7]
- Tortrix moths in *apples* [1-7]
- Western flower thrips in *aubergines (off-label)*, *non-edible ornamentals (off-label)*, *rooting beds (off-label)* [3]; *non-edible glasshouse crops (off-label)*, *protected aubergines (off-label)* [4]; *protected courgettes (off-label)*, *protected cucumbers (off-label)*, *protected gherkins (off-label)*, *protected peppers (off-label)*, *protected tomatoes (off-label)* [3, 4]
- Whitefly in *amenity vegetation* [1-4]; *nursery stock* [3, 7]; *ornamental plant production*, *protected cucumbers*, *protected peppers*, *protected pot plants*, *protected tomatoes* [1-4, 7]; *protected ornamentals* [4]

FOR FULL CONDITIONS OF USE ALWAYS READ THE PRODUCT LABEL

- Yellow cereal fly in **spring barley, spring oats, spring wheat, winter barley, winter oats, winter wheat** [7]

Specific Off-Label Approvals (SOLAs)

- **aubergines, non-edible ornamentals, protected courgettes, protected cucumbers, protected gherkins, protected peppers, protected tomatoes, rooting beds** *(OLA 040511) Dec 2008* [3]
- **bulb onions, garlic** *(OLA 001847) Dec 2008* [3]
- **bulb onions, carrots, celery, chinese cabbage, evening primrose, garlic, grass seed crops, leeks, lettuce, marjoram, protected celery, protected chinese cabbage, protected chives, protected herbs (see appendix 6), protected lamb's lettuce, protected lettuce, protected parsley, protected radicchio, protected scarole, protected spring onions, protected watercress, radicchio, salad onions, sorrel** *(OLA 031140) Dec 2008* [4]
- **bulb onions, carrots, celery, chinese cabbage, evening primrose, garlic, grass seed crops, leeks, lettuce, marjoram, parsnips, radicchio, salad onions, sorrel** *(OLA 040504) Dec 2008* [7]
- **carrots** *(OLA 001845) Dec 2008* [3]
- **celery, protected celery, protected chinese cabbage, protected salad onions, salad onions** *(OLA 041187) Dec 2008* [2]
- **celery, chinese cabbage, protected celery, protected spring onions, salad onions** *(OLA 001849) Dec 2008* [3]
- **chinese cabbage, lettuce** *(OLA 001843) Dec 2008* [3]
- **cress** *(OLA 022908) Dec 2008* [3]
- **evening primrose, grass seed crops, marjoram, radicchio, sorrel** *(OLA 040512) Dec 2008* [3]
- **leeks** *(OLA 001851) Dec 2008* [3]
- **mushrooms** *(OLA 041189) Dec 2008* [2]
- **mushrooms** *(OLA 030039) Dec 2008* [3]
- **mushrooms** *(OLA 030040) Dec 2008* [3]
- **mushrooms** *(OLA 031138) Dec 2008* [4]
- **non-edible glasshouse crops, protected aubergines, protected courgettes, protected cucumbers, protected gherkins, protected peppers, protected tomatoes** *(OLA 031139) Dec 2008* [4]
- **potatoes** *(OLA 001855) Dec 2008* [3]
- **protected celery, protected chinese cabbage, protected salad onions** *(OLA 011796) Dec 2008* [7]
- **protected chives, protected herbs (see appendix 6), protected lettuce, protected parsley** *(OLA 041188) Dec 2008* [2]
- **protected herbs (see appendix 6), protected lettuce, protected radicchio** *(OLA 040740) Dec 2008* [7]
- **protected salad brassicas, salad brassicas** *(baby leaf production) (OLA 041190) Dec 2008* [2]
- **protected salad brassicas** *(for baby leaf production) (OLA 001853) Dec 2008* [3]
- **protected salad brassicas, salad brassicas** *(for baby leaf production) (OLA 031140) Dec 2008* [4]
- **protected salad brassicas** *(for baby leaf production) (OLA 011796) Dec 2008* [7]
- **salad brassicas** *(OLA 001853) Dec 2008* [3]
- **salad brassicas** *(for baby leaf production) (OLA 040504) Dec 2008* [7]

Approval information

- Deltamethrin included in Annex I under EC Directive 91/414
- Accepted by BBPA for use on malting barley and hops

Efficacy guidance

- A contact and stomach poison with 3-4 wk persistence, particularly effective on caterpillars and sucking insects
- Normally applied at first signs of damage with follow-up treatments where necessary at 10-14 d intervals. Rates, timing and recommended combinations with other pesticides vary with crop and pest. See label for details
- Spray is rainfast within 1 h
- May be applied in frosty weather provided foliage not covered in ice
- Temperatures above 35°C may reduce effectiveness or persistence

Restrictions

- Maximum number of treatments varies with crop and pest, 4 per crop for wheat and barley, only 1 application between 1 Apr and 31 Aug. See label or off-label approval notice for other crops

SEE SECTION 3 FOR PRODUCTS ALSO REGISTERED

- Do not apply more than 1 aphicide treatment to cereals in summer
- Do not spray crops suffering from drought or other physical stress
- Consult processer before treating crops for processing
- Do not apply to a cereal crop if any product containing a pyrethroid insecticide or dimethoate has been applied to that crop after the start of ear emergence (GS 51)
- Do not spray cereals after 31 Mar in the year of harvest within 6 m of the outside edge of the crop
- Reduced volume spraying must not be used on cereals after 31 Mar in yr of harvest

Crop-specific information
- Latest use: early dough (GS 83) for barley, oats, wheat; before flowering for mustard, oilseed rape, evening primrose; before 31 Mar for grass seed crops, marjoram, radicchio, sorrel
- HI mushrooms 2 d; leeks, protected spring onions 3 d; chinese cabbage, most herbs, protected celery, protected lettuce, protected salad brassicas, radicchio, sorrel 7 d; marjoram, radicchio, sorrel 14 d; carrots, parsnips 21 d

Environmental safety
- Dangerous for the environment
- Very toxic to aquatic organisms
- Extremely dangerous to fish or other aquatic life. Do not contaminate surface waters or ditches with chemical or used container
- High risk to bees. Do not apply to crops in flower or to those in which bees are actively foraging. Do not apply when flowering weeds are present
- Do not apply in tank-mixture with a triazole-containing fungicide when bees are likely to be actively foraging in the crop
- High risk to non-target insects or other arthropods. Do not spray within 6 m of the field boundary
- Extremely dangerous to fish or other aquatic life. Do not contaminate surface waters or ditches with chemical or used container
- LERAP Category A
- Broadcast air-assisted LERAP (18 m) [3-7]

Hazard classification and safety precautions
> **Hazard** H02 [7]; H03, H08 [1-3, 5, 6]; H11 [1-7]
> **Risk phrases** R20 [2, 3]; R21 [1, 5, 6]; R22a [1-3, 5, 6]; R22b, R41 [2, 3, 5, 6]; R25 [7]; R36 [1, 7]; R38, R51 [1-3, 5-7]; R50 [4]; R53a [1-7]
> **Operator protection** A, C, H; U04a, U05a, U08, U19a [1-3, 5-7]; U10 [2]; U11, U14, U15 [4]; U20b [1-7]
> **Environmental protection** E12b [3, 4, 7] (see label for guidance on cereals, oilseed rape, peas, beans); E12e [1, 2] (cereals, oilseed rape, peas, beans); E12e [5, 6] (oilseed rape, cereals, peas, beans); E13a [1, 7]; E15a [2-6]; E16c, E16d [1-7]; E17b [3-7] (18 m); E22a [1, 4]; E34 [1-3, 5-7]; E38 [2-4, 7]
> **Storage and disposal** D01, D02, D09a [1-7]; D05 [1-3, 5-7]; D08, D11a [7]; D10b [5, 6]; D10c [1-4]; D12a [3, 4, 7]
> **Medical advice** M03 [1-3, 5-7]; M04 [7]; M05b [2, 3, 5, 6]

149   desmedipham

A contact carbamate herbicide available only in mixtures

150   desmedipham + ethofumesate + phenmedipham

A selective contact and residual herbicide for beet

Products

1 Beetaweed	AgriGuard	25:151:75 g/l	EC	12260
2 Betanal Expert	Bayer CropScience	25:151:75 g/l	EC	10592
3 Landgold Deputy	Landgold	25:151:75 g/l	EC	11975
4 Landgold Deputy	Teliton	25:151:75 g/l	EC	12115

Uses
- Annual dicotyledons in **sugar beet**
- Annual meadow grass in **sugar beet**

SECTION 2

Approval information
- Ethofumesate included in Annex I under EC Directive 91/414

Efficacy guidance
- Product recommended for low-volume overall application in a planned spray programme
- Product acts mainly by contact action. A full programme also gives some residual control but this may be reduced on soils with more than 5% organic matter
- Best results achieved from treatments applied at fully expanded cotyledon stage of largest weeds present. Occasional larger weeds will usually be controlled from a full programme of sprays
- Where a pre-emergence band spray has been applied treatment must be timed according to size of the untreated weeds between the rows
- Susceptible weeds may not all be killed by the first spray. Repeat applications as each flush of weeds reaches cotyledon size normally necessary for season long control
- Sequential treatments should be applied when the previous one is still showing an effect on the weeds
- Various mixtures with other beet herbicides are recommended. See label for details

Restrictions
- Maximum total dose equivalent to three full dose treatments
- Do not spray crops stressed by nutrient deficiency, wind damage, pest or disease attack, or previous herbicide treatments. Stressed crops treated under conditions of high light intensity may be checked and not recover fully
- If temperature likely to exceed 21°C spray after 5 pm
- Crystallisation may occur if spray volume exceeds that recommended or spray mixture not used within 2 h, especially if the water temperature is below 5°C
- Before use, wash out sprayer to remove all traces of previous products, especially hormone and sulfonyl urea weedkillers

Crop-specific information
- Latest use: before crop meets between rows
- Apply first treatment when majority of crop plants have reached the fully expanded cotyledon stage
- Frost within 7 d of treatment may cause check from which the crop may not recover

Following crops guidance
- Beet crops may be sown at any time after treatment. Any other crop may be sown 3 mth after treatment following mouldboard ploughing to 15 cm minimum

Environmental safety
- Dangerous for the environment
- Toxic to aquatic organisms

Hazard classification and safety precautions
    Hazard H11
    Risk phrases R51, R53a
    Operator protection A, H; U08, U20a
    Environmental protection E15a, E38 [1, 2]
    Storage and disposal D05, D09a, D10b [1-4]; D07 [3, 4]; D12a [1, 2]
    Medical advice M03

151   desmedipham + phenmedipham

A mixture of contact herbicides for use in sugar beet

Products

Betanal Carrera	Bayer CropScience	40:125 g/l	SE	11029

Uses
- Annual dicotyledons in *sugar beet*

Efficacy guidance
- Product is recommended for low volume overall spraying in a planned programme involving pre and/or post-emergence treatments at doses recommended for low volume programmes
- Best results obtained from treatment when earliest germinating weeds have reached cotyledon stage

SEE SECTION 3 FOR PRODUCTS ALSO REGISTERED

- Further treatments must be applied as each flush of weeds reaches cotyledon stage but allowing a minimum of 7 d between each spray
- Where a pre-emergence band spray has been applied, the first treatment should be timed according to the size of the weeds in the untreated area between the rows
- Product is absorbed by leaves of emerged weeds which are killed by scorching action in 2-10 d. Apply overall as a fine spray to optimise weed cover and spray retention
- Various tank mixtures and sequences recommended to widen weed spectrum and add residual activity - see label for details

Restrictions
- Maximum total dose equivalent to three full dose treatments
- Do not spray crops stressed by nutrient deficiency, frost, wind damage, pest or disease attack, or previous herbicide treatments. Stressed crops may be checked and not recover fully
- If temperature likely to exceed 21°C spray after 5 pm
- Before use, wash out sprayer to remove all traces of previous products, especially hormone and sulfonyl urea weedkillers
- Crystallisation may occur if spray volume exceeds that recommended or spray mixture not used within 2 h, especially if the water temperature is below 5°C
- Product may cause non-reinforced PVC pipes and hoses to soften and swell. Wherever possible, use reinforced PVC or synthetic rubber hoses

Crop-specific information
- Latest use: before crop leaves meet between rows
- Product safe to use on all soil types

Environmental safety
- Dangerous for the environment
- Toxic to aquatic organisms
- Risk to certain non-target insects or other arthropods. Avoid spraying within 6 m of field boundary

Hazard classification and safety precautions
Hazard H04, H11
Risk phrases R43, R51, R53a
Operator protection A, H; U05a, U08, U19a, U20b
Environmental protection E15a, E22b, E38
Storage and disposal D01, D02, D05, D09a, D10b, D12a

152   dicamba + dichlorprop-P + ferrous sulphate + MCPA

A herbicide/fertilizer combination for moss and weed control in turf

Products
Renovator 2	Scotts	0.03:0.22:16.3: 0.22% w/w	GR	11411

Uses
- Annual dicotyledons in **managed amenity turf**
- Moss in **managed amenity turf**
- Perennial dicotyledons in **managed amenity turf**

Approval information
- MCPA included in Annex I under EC Directive 91/414

Efficacy guidance
- Apply from mid-Apr to mid-Aug when weeds are growing
- For best control of moss scarify vigorously after 2 wk to remove dead moss
- Where regrowth of moss or weeds occurs a repeat treatment may be made after 6 wk
- Avoid treatment of wet grass or during drought. If no rain falls within 48 h water in thoroughly

Restrictions
- Do not treat new turf until established for 6-9 mth
- The first 4 mowings after treatment should not be used to mulch cultivated plants unless composted at least 6 mth
- Avoid walking on treated areas until it has rained or they have been watered
- Do not re-seed or turf within 8 wk of last treatment

FOR FULL CONDITIONS OF USE ALWAYS READ THE PRODUCT LABEL

- Do not cut grass for at least 3 d before and at least 4 d after treatment
- Do not apply during freezing conditions or when rain imminent

Crop-specific information
- Apply with a suitable calibrated fertilizer distributor

Hazard classification and safety precautions
Operator protection U09a, U20a
Environmental protection E15a
Storage and disposal D09a, D11a

153   dicamba + dichlorprop-P + MCPA

A translocated herbicide mixture for turf

Products

1	Cleanrun 3	Scotts	0.03:0.22:0.22% w/w	MG	11410
2	Intrepid 2	Scotts	20.8:167:167 g/l	SL	11594

Uses
- Annual dicotyledons in *managed amenity turf*
- Perennial dicotyledons in *managed amenity turf*

Approval information
- MCPA included in Annex I under EC Directive 91/414

Efficacy guidance
- Apply as directed on label between Apr and Sep when weeds growing actively
- If no rain falls within 48 h of treatment irrigate thoroughly [1]
- Do not mow for at least 3 d before and 3-4 d after treatment so that there is sufficient leaf growth for spray uptake and sufficient time for translocation. See label for details
- If re-growth occurs or new weeds germinate re-treatment recommended

Restrictions
- Do not use during drought unless irrigation is carried out before and after treatment
- Do not apply during freezing conditions or when heavy rain is imminent
- Do not treat new turf until established for about 6 mth after seeding or turfing
- The first 4 mowings after treatment should not be used to mulch cultivated plants unless composted for at least 6 mth
- Avoid walking where possible on treated areas until after rain or irrigation [1]
- Do not re-seed turf within 8 wk of last treatment

Crop-specific information
- Latest use: end Sep for managed amenity turf

Hazard classification and safety precautions
Hazard H03 [2]
Risk phrases R22a, R38, R41, R43 [2]
Operator protection A, C [2]; U05a, U08, U11, U19a, U20a [2]; U09a, U20b [1]
Environmental protection E15a [1, 2]; E34 [2]
Storage and disposal D01, D02, D05, D10b [2]; D09a [1, 2]; D11a [1]
Medical advice M03, M05a [2]

154   dicamba + MCPA + mecoprop-P

A translocated herbicide for cereals, grassland, amenity grass and orchards

Products

1	Banlene Super	Bayer CropScience	18:252:42 g/l	SL	10053
2	Field Marshal	United Phosphorus	18:360:80	SL	08956
3	Greencrop Triathlon	Greencrop	25:200:200 g/l	SL	10956
4	Headland Relay P	Headland	25:200:200 g/l	SL	08580
5	Headland Relay Turf	Headland Amenity	25:200:200 g/l	SL	08935
6	Headland Transfer	Headland	18:315:50 g/l	SL	11010
7	Headland Trinity	Headland	18:315:50 g/l	SL	10842
8	Hycamba Plus	Agrichem	20.4:125.9:240.2 g/l	SL	10180
9	Hyprone-P	Agrichem	16:101:92 g/l	SL	09125

SEE SECTION 3 FOR PRODUCTS ALSO REGISTERED

SECTION 2

Products – continued

10 Hysward-P	Agrichem	16:101:92 g/l	SL	09052
11 Mircam Plus	Nufarm UK	19.5:245:43.3 g/l	SL	11525
12 Nocweed	Bayer Environ.	18:252:42 g/l	SL	10609
13 Outrun	SumiAgro Amenity	20.4:125.9:240.2 g/l	SL	10581
14 Pasturol Plus	Headland	25:200:200 g/l	SL	10278
15 Pierce	Nufarm UK	31:256:238 g/l	SL	11924
16 Re-act	Scotts	31:256:237 g/l	SL	12231
17 Super Selective Plus	Rigby Taylor	31.25:256:238 g/l	SL	11928
18 T2 Green	Nufarm UK	31: 256: 237 g/l	SL	11925
19 Tribute	Nomix Enviro	18:252:42 g/l	SL	06921
20 Tribute Plus	Nomix Enviro	18:252:42 g/l	SL	09493
21 Tritox	Scotts	15:178:54 g/l	SL	07764
22 UPL Grassland Herbicide	United Phosphorus	25:200:200	SL	08934

Uses

- Annual dicotyledons in *amenity grass* [3, 7, 8, 10, 13, 14, 22]; *amenity turf* [5, 11, 19-21]; *apple orchards* [1, 7, 11, 12]; *established grassland* [1, 10, 12]; *grass seed crops* [1, 6, 7, 9, 11, 12, 15]; *grassland* [11]; *managed amenity turf* [3, 5, 7, 8, 10, 12-14, 16-21]; *newly sown grass* [9]; *pear orchards*, *spring rye*, *winter rye* [1, 7, 12]; *permanent pasture* [1, 3, 4, 6, 7, 12, 14, 15, 22]; *rotational grassland* [1-4, 6, 7, 10-12, 14, 15, 22]; *spring barley*, *spring oats*, *spring wheat*, *winter barley*, *winter oats*, *winter wheat* [1, 2, 6, 7, 9, 11, 12]; *undersown barley*, *undersown oats*, *undersown wheat* [9, 12]; *undersown rye* [12]; *undersown spring cereals*, *undersown winter cereals* [1, 7]; *undersown spring cereals* (grass only), *undersown winter cereals* (grass only) [2, 6]
- Docks in *amenity grass* [10, 22]; *apple orchards*, *pear orchards* [1, 7, 12]; *established grassland* [1, 10, 12]; *grass seed crops* [1, 7, 9, 12, 15]; *managed amenity turf* [10]; *permanent pasture* [1, 7, 12, 15, 22]; *rotational grassland* [1, 7, 10, 12, 15, 22]
- Perennial dicotyledons in *amenity grass* [3, 7, 8, 10, 13, 14, 22]; *amenity turf* [5, 11, 19-21]; *apple orchards* [1, 7, 11, 12]; *established grassland* [1, 10, 12]; *grass seed crops* [1, 6, 7, 9, 11, 12, 15]; *grassland* [11]; *managed amenity turf* [3, 5, 7, 8, 10, 13, 14, 16-21]; *newly sown grass* [9]; *pear orchards*, *spring rye*, *winter rye* [1, 7, 12]; *permanent pasture* [1, 3, 4, 6, 7, 12, 14, 15, 22]; *rotational grassland* [1-4, 6, 7, 10-12, 14, 15, 22]; *spring barley*, *spring oats*, *spring wheat*, *winter barley*, *winter oats*, *winter wheat* [1, 2, 6, 7, 9, 11, 12]; *undersown barley*, *undersown oats*, *undersown wheat* [9, 12]; *undersown rye* [12]; *undersown spring cereals*, *undersown winter cereals* [1, 7]; *undersown spring cereals* (grass only), *undersown winter cereals* (grass only) [2, 6]

Approval information

- MCPA and mecoprop-P included in Annex I under EC Directive 91/414
- Accepted by BBPA for use on malting barley

Efficacy guidance

- Treatment should be made when weeds growing actively. Weeds hardened by winter weather may be less susceptible
- For best results apply in fine warm weather, preferably when soil is moist. Do not spray if rain expected within 6 h or in drought
- Application of fertilizer 1-2 wk before spraying aids weed control in turf
- Where a second treatment later in the season is needed in amenity situations and on grass allow 4-6 wk between applications to permit sufficient foliage regrowth for uptake

Restrictions

- Maximum number of treatments (including other mecoprop-P products) or maximum total dose varies with crop and product. See label for details. The total amount of mecoprop-P applied in a single yr must not exceed the maximum total dose approved for any single product for the crop/situation
- Do not apply to cereals after the first node is detectable (GS 31), or to grass under stress from drought or cold weather
- Do not spray cereals undersown with clovers or legumes, to be undersown with grass or legumes or grassland where clovers or other legumes are important
- Do not spray leys established less than 18 mth or orchards established less than 3 yr
- Do not roll or harrow within 7 d before or after treatment, or graze for at least 7 d afterwards (longer if poisonous weeds present)

FOR FULL CONDITIONS OF USE ALWAYS READ THE PRODUCT LABEL

- Do not use on turf or grass in year of establishment. Allow 6-8 wk after treatment before seeding bare patches
- The first mowings after use should not be used for mulching unless composted for 6 mth
- Turf should not be mown for 24 h before or after treatment (3-4 d for closely mown turf)
- Avoid drift onto all broad-leaved plants outside the target area

Crop-specific information

- Latest use: before first node detectable (GS 31) for cereals; 5-6 wk before head emergence for grass seed crops; mid-Oct for established grass
- HI 7-14 d before cutting or grazing for leys, permanent pasture
- Apply to winter cereals from the leaf sheath erect stage (GS 30), and to spring cereals from the 5 expanded leaf stage (GS 15)
- Spray grass seed crops 4-6 wk before flower heads begin to emerge (timothy 6 wk)
- Turf containing bulbs may be treated once the foliage has died down completely [4, 10, 20, 21]

Environmental safety

- Harmful to aquatic organisms
- Keep livestock out of treated areas for at least 2 wk following treatment and until poisonous weeds, such as ragwort, have died down and become unpalatable
- Harmful to fish or other aquatic life. Do not contaminate surface waters or ditches with chemical or used container

Hazard classification and safety precautions

Hazard H03 [1-8, 11-20]; H04 [9, 10, 21, 22]

Risk phrases R20 [4, 6, 7, 13, 14]; R21 [1-7, 11, 12, 14, 19, 20]; R22a [1-8, 11-18]; R36 [2-4, 11, 14, 19-22]; R38 [6, 7, 11, 13, 22]; R41 [1, 4-8, 12-18]; R43 [22]; R52 [4-10, 14, 19-22]; R53a [6-10, 21, 22]

Operator protection A, C [1-22]; H, M [2-11, 13-22]; U05a [1-22]; U08 [1-11, 13, 14, 19-22]; U09a [12]; U11 [1, 4, 6-18, 21, 22]; U15 [4, 14]; U19a [3, 8-10, 13, 21]; U20b [1-14, 19-22]

Environmental protection E07a [1-4, 6-21]; E13c [1-10, 12-18, 22]; E15a [11, 15, 16, 18-21]; E23 [19, 20]; E34 [2-7, 11, 12, 14-21]; E38 [4, 14, 19, 20]

Storage and disposal D01, D02, D09a [1-22]; D05 [1-20, 22]; D10a [6, 7, 17, 21]; D10b [1-5, 11, 12, 14-16, 18-20]; D10c [8-10, 13, 22]; D12a [19, 20, 22]

Medical advice M03 [2-7, 11, 14-18, 22]; M05a [7-10, 13, 22]

155   dicamba + mecoprop-P

A translocated post-emergence herbicide for cereals and grassland

Products

1	Camber	Headland	42:319 g/l	SL	09901
2	Di-Farmon R	Headland	42:319 g/l	SL	08472
3	Dockmaster	Nufarm UK	18.7:150 g/l	SL	11651
4	Foundation	Syngenta	84:600 g/l	SL	08475
5	Foundation	Headland	84:600 g/l	SL	11708
6	Headland Saxon	Headland	84:600 g/l	SL	11947
7	High Load Mircam	Nufarm UK	80:600 g/l	SL	11930
8	Hyban-P	Agrichem	18.7:150 g/l	SL	09129
9	Hygrass-P	Agrichem	18.7:150 g/l	SL	09130
10	Mircam	Nufarm UK	18.7:150 g/l	SL	11707
11	Prompt	Headland	84:600 g/l	SL	11948

Uses

- Annual dicotyledons in *established grassland* [1, 2, 4-7, 11]; *permanent pasture* [9, 10]; *rotational grassland* [7, 9, 10]; *spring barley, spring oats, spring wheat, winter barley, winter oats, winter wheat* [1-8, 10, 11]; *spring rye, triticale, undersown rye, winter rye* [8]; *undersown barley, undersown oats, undersown wheat* [3, 8]
- Buttercups in *permanent pasture* [3, 8]; *rotational grassland* [8]
- Chickweed in *established grassland* [1, 2]; *permanent pasture, undersown barley, undersown oats, undersown wheat* [3, 8]; *rotational grassland* [7, 8]; *spring barley, spring oats, spring wheat, winter barley, winter oats, winter wheat* [1-8, 10, 11]; *spring rye, triticale, undersown rye, winter rye* [8]

SEE SECTION 3 FOR PRODUCTS ALSO REGISTERED

- Cleavers in **established grassland** [1, 2]; **rotational grassland** [7]; **spring barley, spring oats, spring wheat, winter barley, winter oats, winter wheat** [1-8, 10, 11]; **spring rye, triticale, undersown rye, winter rye** [8]; **undersown barley, undersown oats, undersown wheat** [3, 8]
- Creeping thistle in **permanent pasture** [3, 8]; **rotational grassland** [8]
- Docks in **permanent pasture** [3, 8-10]; **rotational grassland** [8-10]
- Mayweeds in **established grassland** [1, 2]; **rotational grassland** [7]; **spring barley, spring oats, spring wheat, winter barley, winter oats, winter wheat** [1-8, 10, 11]; **spring rye, triticale, undersown rye, winter rye** [8]; **undersown barley, undersown oats, undersown wheat** [3, 8]
- Perennial dicotyledons in **established grassland** [4-7, 11]; **permanent pasture, rotational grassland** [9, 10]; **spring barley, spring oats, spring wheat, winter barley, winter oats, winter wheat** [3, 8, 10]; **spring rye, triticale, undersown rye, winter rye** [8]; **undersown barley, undersown oats, undersown wheat** [3, 8]
- Polygonums in **established grassland** [1, 2]; **rotational grassland** [7]; **spring barley, spring oats, spring wheat, winter barley, winter oats, winter wheat** [1-8, 10, 11]; **spring rye, triticale, undersown rye, winter rye** [8]; **undersown barley, undersown oats, undersown wheat** [3, 8]
- Scentless mayweed in **permanent pasture** [3, 8]; **rotational grassland** [8]
- Stinging nettle in **permanent pasture, rotational grassland** [8]
- Thistles in **permanent pasture, rotational grassland** [9, 10]

Approval information
- Mecoprop-P included in Annex I under EC Directive 91/414
- Accepted by BBPA for use on malting barley

Efficacy guidance
- Best results by application in warm, moist weather when weeds are actively growing

Restrictions
- Maximum number of treatments 1 per crop for cereals and 1 or 2 per yr on grass depending on label. The total amount of mecoprop-P applied in a single yr must not exceed the maximum total dose approved for any single product for the crop/situation
- Do not spray in cold or frosty conditions
- Do not spray if rain expected within 6 h
- Do not treat undersown grass until tillering begins
- Do not spray cereals undersown with clover or legume mixtures
- Do not roll, harrow within 7 d before or after spraying
- Do not treat crops suffering from stress from any cause
- Use product immediately following dilution; do not allow diluted product to stand before use [6]
- Avoid treatment when drift may damage neighbouring susceptible crops

Crop-specific information
- Latest use: before 1st node detectable for cereals; 7 d before cutting or 14 d before grazing grass
- Apply to winter sown crops from 5 expanded leaf stage (GS 15)
- Apply to spring sown cereals from 5 expanded leaf stage but before first node is detectable (GS 15-31)
- Treat grassland just before perennial weeds flower
- Transient crop prostration may occur after spraying but recovery is rapid

Environmental safety
- Dangerous for the environment [1, 2, 4-6, 11]
- Toxic to aquatic organisms
- Harmful to fish or other aquatic life. Do not contaminate surface waters or ditches with chemical or used container
- Keep livestock out of treated areas for at least 2 wk and until foliage of poisonous weeds such as ragwort has died and become unpalatable

Hazard classification and safety precautions
Hazard H03 [1-7, 10, 11]; H04 [8, 9]; H11 [1, 2, 4-6, 11]
Risk phrases R21 [3, 7]; R22a [1-7, 10, 11]; R36 [3, 10]; R38 [1-6, 10, 11]; R41 [1, 2, 4-9, 11]; R51 [1, 2, 4-6, 11]; R52 [8-10]; R53a [1, 2, 4-6, 8-11]
Operator protection A, C [1-11]; H [1-3, 5, 6, 8-11]; M [1-3, 5, 6, 9-11]; U05a [1-11]; U08 [1-8, 10, 11]; U09a [9]; U11 [1-9, 11]; U19a [8-10]; U20a [2]; U20b [1, 3-11]

FOR FULL CONDITIONS OF USE ALWAYS READ THE PRODUCT LABEL

Environmental protection E07a [1-9, 11]; E13c [8, 9]; E15a [1-7, 10, 11]; E34 [1-11]; E38 [1, 2, 4, 5]
Storage and disposal D01, D02, D05, D09a [1-11]; D10b [1-7, 10, 11]; D10c [8, 9]
Medical advice M03 [1-11]; M05a [8-10]

156   dichlobenil

A residual benzonitrile herbicide for woody crops and non-crop uses

Products

1 Casoron G	Nomix Enviro	6.75% w/w	GR	09023
2 Casoron G	Scotts	6.75% w/w	GR	10709
3 Casoron G	Certis	6.75% w/w	GR	11632
4 Casoron G4	Scotts	4% w/w	GR	10708
5 Embargo G	SumiAgro Amenity	6.75% w/w	GR	10367
6 Luxan Dichlobenil Granules	Luxan	6.75% w/w	GR	09250
7 Midstream GSR	Scotts	20% w/w	GR	11674

Uses
- Annual dicotyledons in **grapevines** *(off-label)*, **rhubarb** *(off-label - established)* [3]
- Annual grasses in **rhubarb** *(off-label - established)* [3]
- Annual weeds in **apple orchards, blackberries, blackcurrants, gooseberries, loganberries, pear orchards, raspberries, redcurrants** [3, 5, 6]; **established woody ornamentals, roses** [1, 2, 4-6]; **ornamental trees** [4, 5]
- Aquatic weeds in **enclosed waters, land immediately adjacent to aquatic areas, open waters** [1-3, 6, 7]
- Perennial dicotyledons in **apple orchards, blackberries, blackcurrants, gooseberries, loganberries, pear orchards, raspberries, redcurrants** [3, 5, 6]; **established woody ornamentals, roses** [1, 2, 4-6]; **grapevines** *(off-label)*, **rhubarb** *(off-label - established)* [3]; **ornamental trees** [4, 5]
- Perennial grasses in **apple orchards, blackberries, blackcurrants, gooseberries, loganberries, pear orchards, raspberries, redcurrants** [3, 5, 6]; **established woody ornamentals, roses** [1, 2, 4-6]; **ornamental trees** [4, 5]; **rhubarb** *(off-label - established)* [3]
- Total vegetation control in **hard surfaces, natural surfaces not intended to bear vegetation, permeable surfaces overlying soil** [1-3, 5, 6]; **non-crop areas** [1, 2]
- Volunteer potatoes in **potato dumps** *(blight prevention)* [3, 5, 6]

Specific Off-Label Approvals (SOLAs)
- **grapevines** *(OLA 041405) Dec 2008* [3]
- **rhubarb** *(established) (OLA 041404) Dec 2008* [3]

Approval information
- Approved for aquatic weed control [1-3, 6, 7]. See notes in Section 5 on use of herbicides in or near water

Efficacy guidance
- Best results achieved by application in winter to moist soil during cool weather, particularly if rain follows soon after. Do not disturb treated soil by hoeing
- In aquatic situations apply as soon as active growth commences to reduce the likelihood of subsequent water deoxygenation caused by die-back of heavy weed stands. Later partial treatment for rooted species may be effective [1, 3, 6, 7]
- Where mulching is practised, apply beforehand
- Lower rates control annuals, higher rates perennials, see label for details
- Residual activity lasts 3-6 mth with selective, up to 12 mth with non-selective rates
- Do not use on fen peat or moss soils as efficacy will be severely impaired
- For control of emergent, floating and submerged aquatics apply to water surface in early spring. Intended for use in still or sluggish flowing water [1, 3, 6, 7]
- Control in water is effective where flow does not exceed 90 m/hr (2.5 cm/second). Treatment of faster flowing water will only be effective if the flow can be checked for a minimum of 7 d after application

Restrictions
- Maximum number of treatments 1 per yr
- Maximum total dose equivalent to one full dose treatment

SEE SECTION 3 FOR PRODUCTS ALSO REGISTERED

SECTION 2

- Do not treat crops established for less than 2 yr. See label for lists of resistant and sensitive species. Do not treat stone fruit or Norway spruce (Christmas trees)
- Do not apply within 300 m of glasshouses or hops or near areas underplanted with bulbs, annuals or herbaceous stock
- Do not apply to frozen, snow-covered or waterlogged ground or when crop foliage wet
- Do not apply to sites less than 18 mth before replanting or sowing
- Store well away from corms, bulbs, tubers and seed

Crop-specific information
- Latest use: before end of spring for aquatic situations; 4 wk before budding for grapevines; before onset of spring growth for potato dumps and all other crops. See labels for precise details
- HI 200 d for grapevines
- Apply to crops in dormant period. See label for details of timing and rates
- Apply evenly by hand applicator or suitable mechanical spreader
- Apply to potato dump sites before emergence of potatoes for blight prevention

Environmental safety
- Dangerous for the environment
- Harmful to aquatic organisms
- Harmful to fish or other aquatic life. Do not contaminate surface waters or ditches with chemical or used container
- If fish are known to be present in a water body intended for treatment for aquatic weed control, partial treatment should be adopted. See label for details [7]
- Do not dump surplus herbicide in water or ditch bottoms [1-3, 6, 7]
- Use not permitted on reservoirs that form part of a water supply system
- Do not use treated water for irrigation purposes within 2 wk of treatment or until concentration of dichlobenil falls below 0.3 ppm [7]
- Use as an aquatic herbicide subject to requirements set out in *Guidelines for the use of herbicides in or near watercourses or lakes* available from Defra

Hazard classification and safety precautions
Hazard H03, H11 [5]; H04 [6]
Risk phrases R21, R51 [5]; R43 [6]; R52 [1, 3, 4, 7]; R53a [1, 3-5, 7]
Operator protection A, H [5, 6]; U05a [5, 6]; U20b [2]; U20c [1, 3-7]
Environmental protection E13c, E19a [1-3, 6, 7]; E15a [4, 5]; E21 [2, 7] (2 wk); E34 [6]; E38 [1, 3, 4]
Storage and disposal D01, D02 [5, 6]; D04 [2]; D07 [2, 4, 7]; D09a, D11a [1-7]
Medical advice M03 [6]; M05b [5]

157   dichlorophen

A chlorophenol moss-killer, fungicide, bactericide and algicide

Products
1	Enforcer	Scotts	360 g/l	SL	09288
2	Fungo	Dax	170 g/l	SL	H4768
3	Mossicide	SumiAgro Amenity	360 g/l	SL	09606
4	Nomix Mosskiller	Nomix Enviro	360 g/l	SL	12080
5	Super Mosstox	Bayer Environ.	360 g/l	SL	09942

Uses
- Algae in *hard surfaces, paths and drives* [2]
- Fungus diseases in *glasshouses, ornamental specimens* [1]; *hard surfaces* [1, 4]
- Moss in *glasshouses, ornamental specimens* [1]; *hard surfaces* [1-5]; *managed amenity turf* [1, 3-5]; *paths and drives* [2, 5]
- Red thread in *managed amenity turf* [5]

Approval information
- Certain products containing this active ingredient have been granted derogations for specified 'Essential Uses' for use until 31 December 2007. Sale and supply must cease by 30 June 2007 but growers have no guarantee that the products will continue to be available until then.
  For more information see 'The Review Programme' under 'Pesticide Legislation' in Section 5

FOR FULL CONDITIONS OF USE ALWAYS READ THE PRODUCT LABEL

Efficacy guidance
- For moss control in turf spray at any time when moss is growing but best results obtained from treatment in spring or early summer. Supplement spraying with other measures to improve fertility and drainage
- Rake out dead moss 2-3 wk after treatment
- Treat hard surfaces when rain not expected and brush away dead material when treatment fully dried
- Efficacy will be impaired if used during freezing, frosty or drought conditions

Restrictions
- To be used only by professional operators
- Keep unprotected persons out of treated areas for 48 h or until surfaces are dry

Environmental safety
- Dangerous for the environment
- Very toxic to aquatic organisms
- Dangerous to fish or other aquatic life. Do not contaminate surface waters or ditches with chemical or used container
- Prevent any surface run-off from entering storm drains

Hazard classification and safety precautions
  Hazard H04 [2, 5]; H05, H11 [1, 3-5]
  Risk phrases R22a, R53a [1, 3-5]; R34, R51 [1]; R35, R50 [3-5]; R36, R38 [1, 2, 4, 5]; R41 [2]
  Operator protection A, C, H [1, 2, 4, 5]; M [1, 4, 5]; U02b, U04a [1, 3, 4]; U05a [1-5]; U08, U15 [4]; U09a [1-3, 5]; U10 [1, 3, 5]; U11 [1, 3]; U14 [1, 4]; U19a [1]; U20a [3]; U20b [1, 2, 4]; U20c [5]
  Environmental protection E02a [1, 3, 4] (48 h); E13b [2]; E15a [1, 2, 4, 5]; E20 [1-4]; E34 [4]; E38 [3]
  Consumer protection C09 [1-4]; C12 [1, 4]
  Storage and disposal D01, D02, D09a [1-5]; D05 [3, 5]; D10a [1, 3-5]; D12a [3, 4]; D12b [1]
  Medical advice M03 [4]; M04 [1, 3, 5]; M05b [2]

158 dichlorophen + ferrous sulphate

A mosskiller/fertilizer mixture for use on turf

Products

SHL Lawn Sand Plus	Sinclair	0.3:5.4% w/w	SA	04439

Uses
- Moss in *amenity turf*

Efficacy guidance
- Apply to established turf from late spring to early autumn when soil moist but not when grass wet or damp with dew
- Heavy infestations may need a repeat treatment

Restrictions
- Maximum number of treatments 2 per yr
- Do not treat newly sown grass or freshly laid turf for the first year
- Do not apply during drought or freezing conditions or when rain imminent
- Avoid walking on treated areas until it has rained or turf has been watered
- Do not mow for 3-4 d before or after application

Environmental safety
- Apply away from fish

Hazard classification and safety precautions
  Operator protection U20c
  Environmental protection E13d
  Storage and disposal D01, D09a, D11a

159 1,3-dichloropropene

A halogenated hydrocarbon soil nematicide

Products

| Telone II | Dow | 94% w/w | VP | 05749 |

Uses
- Free-living nematodes in **potatoes**
- Nematodes in **hops**, **raspberries**, **strawberries**
- Potato cyst nematode in **potatoes**
- Stem and bulb nematodes in **narcissi**
- Stem nematodes in **strawberries**
- Virus vectors in **hops**, **raspberries**, **strawberries**

Approval information
- Accepted by BBPA for use on hops

Efficacy guidance
- Before treatment soil should be in friable seed bed condition above 5°C, with adequate moisture and all crop remains decomposed. Tilling to 30 cm improves results
- Apply with sub-surface A-blade injector combined with soil-sealing roller or, in hops, with a hollow tined injector. See label for details of suitable machinery
- Leave soil undisturbed for 21 d after treatment. Wet or cold soils need longer exposure
- For control of potato-cyst nematode contact firm's representative, check nematode level and use in integrated control programme
- For use in hops apply in May or Jun following the yr of grubbing and leave as long as possible before replanting
- Do not use water to clean out apparatus nor use aluminium or magnesium alloy containers which may corrode

Restrictions
- Maximum number of treatments 2 per yr for hops, 1 per yr for other crops
- Do not drill or plant until odour of fumigant is eliminated
- Do not use fertilizer containing ammonium salts after treatment. Use only nitrate nitrogen fertilizers until crop well established and soil temperature above 18°C
- Do not use on extremely heavy clays or soils with many large stones

Crop-specific information
- Latest use: 6 wk before planting for raspberries, strawberries; pre-planting of crop for hops, narcissi, potatoes

Following crops guidance
- Allow at least 6 wk after treatment before planting raspberries or strawberries

Environmental safety
- Dangerous for the environment
- Toxic to aquatic organisms
- Keep in original container, tightly closed, in a safe place, under lock and key

Hazard classification and safety precautions
   Hazard H02, H08, H11
   Risk phrases R20, R21, R22b, R24, R25, R37, R38, R43, R51, R53a
   Operator protection A, C, H, M; U02a, U04a, U05a, U09a, U14, U15, U19a, U20a
   Environmental protection E15a, E34, E38
   Storage and disposal D01, D02, D03, D06b, D09b, D12a, D14
   Medical advice M04, M05b

160 dichlorprop-P + ferrous sulphate + MCPA

A herbicide/fertilizer combination for moss and weed control in turf

Products

1 SHL Granular Feed, Weed & Mosskiller	Sinclair	0.2:10.9:0.3% w/w	GR	10972
2 SHL Turf Feed, Weed & Mosskiller	Sinclair	0.2:10.9:0.3% w/w	DP	10973

Uses
- Annual dicotyledons in *managed amenity turf*
- Buttercups in *managed amenity turf*
- Moss in *managed amenity turf*
- Perennial dicotyledons in *managed amenity turf*

Approval information
- MCPA included in Annex I under EC Directive 91/414

Efficacy guidance
- Apply between Mar and Sep when grass in active growth and soil moist
- A repeat treatment may be needed after 4-6 wk to control perennial weeds or if moss regrows
- Water in if rainfall does not occur within 48 h [2]

Restrictions
- Do not treat newly sown grass for 6 mth after establishment [1]
- Do not apply during drought or freezing conditions or when rain imminent
- Avoid walking on treated areas until it has rained or turf has been watered
- Do not mow for 3-4 d before or after application
- Do not use first 4 mowings after treatment for mulching. Mowings should be composted for 6 mth before use
- Do not treat areas of fine turf such as golf or bowling greens [1]
- Avoid contact with tarmac surfaces as staining may occur

Environmental safety
- Avoid drift onto nearby plants and borders [1]

Hazard classification and safety precautions
Operator protection U20c
Storage and disposal D01, D09a, D11a

161 dichlorprop-P + MCPA

A translocated herbicide for use in managed amenity turf

Products

1 SHL Granular Feed and Weed	Sinclair	0.2:0.3% w/w	GR	10970
2 SHL Turf Feed and Weed	Sinclair	0.2:0.3% w/w	DP	10963

Uses
- Annual dicotyledons in *managed amenity turf* [1, 2]
- Buttercups in *managed amenity turf* [1, 2]
- Clovers in *managed amenity turf* [2]
- Daisies in *managed amenity turf* [2]
- Dandelions in *managed amenity turf* [2]
- Perennial dicotyledons in *managed amenity turf* [1, 2]

Approval information
- MCPA included in Annex I under EC Directive 91/414

Efficacy guidance
- Best results achieved by application to young seedling weeds in active growth between Mar and Sep
- Water in if rainfall does not occur within 48 h [2]

SEE SECTION 3 FOR PRODUCTS ALSO REGISTERED

Restrictions

- Do not use on newly sown turf for 6 mth after establishment
- Do not cut or mow for 2-3 d before and after treatment
- Do not spray during cold weather, if rain or frost expected, if crop wet or in drought
- The first four mowings should not be used for mulching and should be composted for 6 mth before use
- Avoid close mowing for 3-4 d before or after treatment
- Do not treat areas of fine turf such as golf or bowling greens [1]

Environmental safety

- Avoid drift onto nearby plants and borders [2]
- Avoid contact with tarmac surfaces as staining may occur

Hazard classification and safety precautions

Operator protection U20c
Storage and disposal D01, D09a, D11a

162   dichlorprop-P + MCPA + mecoprop-P

A translocated herbicide mixture for winter and spring cereals

Products

Hymec Triple	Agrichem	310:160:130 g/l	SL	09949

Uses

- Annual dicotyledons in *durum wheat, spring barley, spring oats, spring wheat, winter barley, winter oats, winter wheat*
- Chickweed in *durum wheat, spring barley, spring oats, spring wheat, winter barley, winter oats, winter wheat*
- Cleavers in *durum wheat, spring barley, spring oats, spring wheat, winter barley, winter oats, winter wheat*
- Field pansy in *durum wheat, spring barley, spring oats, spring wheat, winter barley, winter oats, winter wheat*
- Mayweeds in *durum wheat, spring barley, spring oats, spring wheat, winter barley, winter oats, winter wheat*
- Poppies in *durum wheat, spring barley, spring oats, spring wheat, winter barley, winter oats, winter wheat*

Approval information

- MCPA and mecoprop-P included in Annex I under EC Directive 91/414
- Accepted by BBPA for use on malting barley

Efficacy guidance

- Best results obtained if application is made while majority of weeds are at seedling stage but not if temperatures are too low
- Optimum results achieved by spraying when temperature is above 10°C. If temperatures are lower delay spraying until growth becomes more active

Restrictions

- Maximum number of treatments 1 per crop
- Do not spray in windy conditions where spray drift may cause damage to neighbouring crops, especially sugar beet, oilseed rape, peas, turnips and most horticultural crops including lettuce and tomatoes under glass

Crop-specific information

- Latest use: before second node detectable (GS 32) for all crops

Environmental safety

- Dangerous for the environment
- Very toxic to aquatic organisms
- Harmful to fish or other aquatic life. Do not contaminate surface waters or ditches with chemical or used container

Hazard classification and safety precautions

Hazard H03, H11
Risk phrases R20, R21, R22a, R41, R50, R53a

FOR FULL CONDITIONS OF USE ALWAYS READ THE PRODUCT LABEL

Operator protection A, C, H, M; U05a, U08, U11, U20b
Environmental protection E13c, E34
Storage and disposal D01, D02, D05, D09a, D10c, D12a
Medical advice M03, M05a

163   diclofop-methyl

A translocated phenoxypropionic herbicide available only in mixtures

164   diclofop-methyl + fenoxaprop-P-ethyl

A foliar acting herbicide mixture for grass control in wheat and barley

Products

1	Corniche	Bayer CropScience	250:20 g/l	EW	08947
2	Tigress Ultra	Bayer CropScience	250:20 g/l	EW	08946

Uses
- Blackgrass in *spring barley*, *winter barley*, *winter wheat*
- Ryegrass in *spring barley*, *winter barley*, *winter wheat*
- Wild oats in *spring barley*, *winter barley*, *winter wheat*

Approval information
- Accepted by BBPA for use on malting barley

Efficacy guidance
- Apply from emergence of crop up to before second node detectable (GS 32)
- Wild oats and blackgrass controlled from 2 fully expanded leaves to end of tillering but before 1st node detectable
- Ryegrass from seed controlled from 2 fully expanded leaves up to 3 tillers
- Spray is rainfast from 1 h after spraying. Do not apply to wet or icy foliage
- Any conditions resulting in moisture stress may reduce effectiveness especially with later applications to spring barley
- Performance not affected by soils with high OM content or high Kd factor
- Always follow WRAG guidelines for preventing and managing herbicide resistant weeds. Section 5 for more information

Restrictions
- Maximum number of treatments 1 per crop
- Do not apply to undersown crops or those to be undersown
- Broadcast crops should be sprayed post-emergence after root system well established
- Do not roll or harrow within 1 wk
- Do not spray crops under stress from drought, waterlogging etc
- Do not spray immediately before or after a sudden drop in temperature, a period of warm days/ cold nights or when extremely low temperatures forecast
- Interval of 7 d must elapse before or after application of any other product
- Must not be mixed with weedkillers containing hormones or bifenox

Crop-specific information
- Latest use: before 2nd node detectable (GS 32)
- Can be sprayed in frosty weather provided crop hardened off
- Applications may be followed by transient leaf discoloration

Environmental safety
- Dangerous for the environment
- Very toxic to aquatic organisms
- Keep livestock out of treated areas for at least 7 d
- Dangerous to fish or other aquatic life. Do not contaminate surface waters or ditches with chemical or used container

Hazard classification and safety precautions
    Hazard H11
    Risk phrases R50, R53a
    Operator protection A, C, H; U05a, U20b

SEE SECTION 3 FOR PRODUCTS ALSO REGISTERED

Environmental protection E06a (7 d); E13b, E38
Consumer protection C02 (6 wk)
Storage and disposal D01, D02, D05, D09a, D10b, D12a

165 difenacoum

An anticoagulant coumarin rodenticide

Products

1	Neokil	Sorex	0.005% w/w	RB	H6769
2	Neosorexa	Sorex	0.005% w/w	RB	H6773
3	Sakarat D (Whole Wheat)	Killgerm	0.005% w/w	RB	H7109
4	Sakarat D Wax Bait	Killgerm	0.005% w/w	RB	H7489
5	Sorexa Gel	Sorex	0.005% w/w	RB	H6790

Uses
- Mice in *farm buildings*, *farmyards* [1-5]
- Rats in *farm buildings*, *farmyards* [1-3]

Efficacy guidance
- Difenacoum is a chronic poison and rodents need to feed several times before accumulating a lethal dose. Effective against rodents resistant to other commonly used anticoagulants
- Best results achieved by placing baits at points between nesting and feeding places, at entry points, in holes and where droppings are seen
- A minimum of five baiting points normally required for a small infestation; more than 40 for a large infestation
- Inspect bait sites frequently and top up as long as there is evidence of feeding
- Product formulated for application through a skeleton or caulking gun [5]
- Maintain a few baiting points to guard against reinfestation after a successful control campaign
- Products contain an insecticide and fungistat to prevent insect and mould attack [1, 2]

Restrictions
- Only for use by farmers, horticulturists and other professional users
- When working in rodent infested areas wear synthetic rubber/PVC gloves to protect against rodent-borne diseases

Environmental safety
- Cover baits by placing in bait boxes, drain pipes or under boards to prevent access by children, animals or birds
- Products contain human taste deterrent

Hazard classification and safety precautions
Operator protection A [5]; U13, U20b
Environmental protection E15a [5]
Storage and disposal D09a, D11a
Vertebrate/rodent control products V01a, V03a, V04a [1, 2, 5]; V01b, V03b, V04b [4]; V02 [1, 2, 4, 5]
Medical advice M03

166 difenoconazole

A diphenyl-ether triazole protectant and curative fungicide

Products

1	Difcor 250 EC	Nufarm UK	250 g/l	EC	12361
2	Plover	Syngenta	250 g/l	EC	11763

Uses
- Alternaria in *broccoli*, *brussels sprouts*, *cabbages*, *calabrese*, *cauliflowers*, *spring oilseed rape*, *winter oilseed rape*
- Brown rust in *winter wheat*
- Light leaf spot in *spring oilseed rape*, *winter oilseed rape*

FOR FULL CONDITIONS OF USE ALWAYS READ THE PRODUCT LABEL

- Ring spot in *broccoli, brussels sprouts, cabbages, calabrese, cauliflowers, chinese cabbage* *(off-label),* *choi sum* *(off-label),* *collards* *(off-label),* *kale* *(off-label),* *pak choi* *(off-label),* *protected pinks* *(off-label - ground or pot grown),* *protected sweet williams* *(off-label)*
- Rust in *asparagus* *(off-label),* *protected pinks* *(off-label - ground or pot grown),* *protected sweet williams* *(off-label)*
- Septoria in *winter wheat*
- Septoria leaf spot in *celeriac* *(off-label),* *celery* *(off-label)*
- Stem canker in *spring oilseed rape, winter oilseed rape*
- Yellow rust in *winter wheat*

Specific Off-Label Approvals (SOLAs)
- *asparagus (OLA 051493) Dec 2008* [1]
- *asparagus (OLA 050158) Dec 2008* [2]
- *celeriac, celery (OLA 051491) Apr 2006* [1]
- *celeriac, celery (OLA 050157) Apr 2006* [2]
- *chinese cabbage, choi sum, collards, pak choi (OLA 051490) Mar 2008* [1]
- *chinese cabbage, choi sum, collards, pak choi (OLA 050558) Dec 2008* [2]
- *kale (OLA 051492) Dec 2008* [1]
- *kale (OLA 050559) Dec 2008* [2]
- *protected pinks (ground or pot grown) (OLA 051486) Dec 2008* [1]
- *protected pinks (ground or pot grown) (OLA 050156) Dec 2008* [2]
- *protected sweet williams (OLA 051486) Dec 2008* [1]
- *protected sweet williams (OLA 050156) Dec 2008* [2]

Efficacy guidance
- For most effective control of Septoria, apply as part of a programme of sprays which includes a suitable flag leaf treatment
- Adequate control of yellow rust may require an earlier appropriate treatment
- Improved control of established infections on oilseed rape achieved by mixture with carbendazim. See label
- In cabbages a 3-spray programme should be used starting at the first sign of disease and repeated at 14-21 d intervals
- Product is fully rainfast 2 h after application

Restrictions
- Maximum number of treatments 3 per crop for cabbages; 2 per crop for oilseed rape; 1 per crop for wheat
- Apply to wheat any time from ear fully emerged stage but before early milk-ripe stage (GS 59-73)

Crop-specific information
- Latest use: before grain early milk-ripe stage (GS 73) for cereals; end of flowering for oilseed rape
- HI brassicas 21 d
- Recommended for use on all varieties of winter and spring sown oilseed rape and cabbage
- Treat oilseed rape in autumn from 4 expanded true leaf stage (GS 1,4). A repeat spray may be made in spring at the beginning of stem extension (GS 2,0) if visible symptoms develop

Environmental safety
- Dangerous for the environment
- Very toxic to aquatic organisms
- Product supplied in returnable refillable containers. Follow label instructions

Hazard classification and safety precautions
Hazard H11 [2]
Risk phrases R50, R53a [2]
Operator protection A, C, H [2]; U05a, U07, U09a, U20b [2]
Environmental protection E15a, E34, E36, E38 [2]
Storage and disposal D01, D02, D05, D07, D09a, D12a, D14 [2]

SEE SECTION 3 FOR PRODUCTS ALSO REGISTERED

167   diflubenzuron

A selective, persistent, contact and stomach acting insecticide

Products

1 Dimilin Flo	Crompton	480 g/l	SC	08769
2 Dimilin Flo	Certis	480 g/l	SC	11056

Uses

- Browntail moth in **amenity trees and shrubs, hedges, nursery stock, ornamental specimens** [2]
- Bud moth in **apples, pears** [2]
- Carnation tortrix moth in **amenity trees and shrubs, hedges, nursery stock, ornamental specimens** [2]
- Caterpillars in **broccoli, brussels sprouts, cabbages, calabrese, cauliflowers, chives** (off-label), **endives** (off-label), **frise** (off-label), **herbs (see appendix 6)** (off-label), **leaf spinach** (off-label), **lettuce** (off-label), **parsley** (off-label), **radicchio** (off-label), **salad brassicas** (off-label - for baby leaf production), **scarole** (off-label) [2]
- Clouded drab moth in **apples, pears** [2]
- Codling moth in **apples, pears** [2]
- Fruit tree tortrix moth in **apples, pears** [2]
- Houseflies in **livestock houses, manure heaps, refuse tips** [2]
- Lackey moth in **amenity trees and shrubs, hedges, nursery stock, ornamental specimens** [2]
- Oak leaf roller moth in **forest** [2]
- Pear sucker in **pears** [2]
- Pine beauty moth in **forest** [2]
- Pine looper in **forest** [2]
- Plum fruit moth in **plums** [2]
- Rust mite in **apples, pears, plums** [2]
- Sciarid flies in **mushrooms** [1]
- Small ermine moth in **amenity trees and shrubs, hedges, nursery stock, ornamental specimens** [2]
- Tortrix moths in **plums** [2]
- Winter moth in **amenity trees and shrubs, apples, blackcurrants, forest, hedges, nursery stock, ornamental specimens, pears, plums** [2]

Specific Off-Label Approvals (SOLAs)

- **chives, endives, frise, herbs (see appendix 6), leaf spinach, lettuce, parsley, radicchio, scarole** (OLA 051321) Dec 2008 [2]
- **salad brassicas** (for baby leaf production) (OLA 051321) Dec 2008 [2]

Approval information

- Approved for aerial application in forestry when average wind velocity does not exceed 18 knots and gusts do not exceed 20 knots [2]. See notes in Section 5

Efficacy guidance

- Most active on young caterpillars and most effective control achieved by spraying as eggs start to hatch
- Dose and timing of spray treatments vary with pest and crop. See label for details
- Addition of wetter recommended for use on brassicas and for pear sucker control in pears

Restrictions

- Maximum number of treatments 3 per yr for apples, pears; 2 per yr for plums, blackcurrants; 2 per crop for brassicas; 1 per spawning for mushrooms [1]
- Before treating ornamentals check varietal tolerance on a small sample
- Do not use as a compost drench or incorporated treatment on ornamental crops
- Do not spray protected plants in flower or with flower buds showing colour
- For use only on the food crops specified on the label
- Do not apply directly to livestock/poultry

Crop-specific information

- HI apples, pears, plums, blackcurrants, brassicas 14 d
- Apply as casing mixing treatment (use immediately) on mushrooms or as post-casing drench [1]

FOR FULL CONDITIONS OF USE ALWAYS READ THE PRODUCT LABEL

Environmental safety
- Dangerous for the environment
- Very toxic to aquatic organisms
- LERAP Category B
- Broadcast air-assisted LERAP (20 m in orchards, 10 m in blackcurrants, forestry or ornamentals)

Hazard classification and safety precautions
Hazard H11
Risk phrases R50
Operator protection U20c
Environmental protection E05a, E16a, E16b, E18 [2]; E15a, E38 [1, 2]; E17b [2] (20 m when used in orchards; 10 m when used in blackcurrants, forestry or ornamentals)
Storage and disposal D05, D12b [2]; D09a, D11a [1, 2]

168 diflufenican

A shoot absorbed pyridinecarboxamide herbicide available only in mixtures

See also bromoxynil + diflufenican + ioxynil
clopyralid + diflufenican + MCPA

169 diflufenican + flufenacet

A contact and residual herbicide mixture for cereals

Products

1	Firebird	Bayer CropScience	200:400 g/l	SC	12421
2	Liberator	Bayer CropScience	100:400 g/l	SC	12032

Uses
- Annual meadow grass in *winter barley, winter wheat* [1, 2]
- Blackgrass in *winter barley, winter wheat* [1, 2]
- Chickweed in *winter barley, winter wheat* [1, 2]
- Field pansy in *winter barley, winter wheat* [1, 2]
- Field speedwell in *winter barley, winter wheat* [2]
- Mayweeds in *winter barley, winter wheat* [1, 2]
- Red dead-nettle in *winter barley, winter wheat* [2]
- Shepherd's purse in *winter barley, winter wheat* [1]
- Speedwells in *winter barley, winter wheat* [1]
- Volunteer oilseed rape in *winter barley, winter wheat* [1]

Approval information
- Accepted by BBPA for use on malting barley

Efficacy guidance
- Best results obtained when there is moist soil at and after application and rain falls within 7 d
- Residual control may be reduced under prolonged dry conditions
- Activity may be slow under cool conditions and final level of weed control may take some time to appear
- Good weed control depends on burying any trash or straw before or during seedbed preparation
- Established perennial grasses and broad-leaved weeds will not be controlled
- Do not use as a stand-alone treatment for blackgrass control. Always follow WRAG guidelines for preventing and managing herbicide resistant weeds. Section 5 for more information

Restrictions
- Maximum number of treatments 1 per crop
- Do not treat undersown cereals or those to be undersown
- Do not use on waterlogged soils or soils prone to waterlogging
- Do not use on Sands or Very Light soils, or very stony or gravelly soils, or on soils containing more than 10% organic matter
- Do not treat broadcast crops and treat shallow-drilled crops post-emergence only

SECTION 2

SEE SECTION 3 FOR PRODUCTS ALSO REGISTERED

- Do not incorporate into the soil or disturb the soil after application by rolling or harrowing
- Avoid treating crops under stress from whatever cause and avoid treating during periods of prolonged or severe frosts

Crop-specific information
- Latest use: before 31 Dec in yr of sowing and before 3rd tiller stage (GS 23) for wheat or 4th tiller stage (GS 24) for barley
- For pre-emergence treatments the seed should be covered with a minimum of 32 mm settled soil

Following crops guidance
- In the event of crop failure wheat, barley or potatoes may be sown provided the soil is ploughed to 15 cm, and a minimum of 12 weeks elapse between treatment and sowing spring wheat or spring barley
- After normal harvest wheat, barley or potatoes my be sown without special cultivations. Soil must be ploughed or cultivated to 15 cm before sowing oilseed rape, field beans, peas, sugar beet, carrots, onions or edible brassicae
- Successive treatments of any products containing diflufenican can lead to soil build-up and inversion ploughing must precede sowing any following non-cereal crop. Even where ploughing occurs some crops may be damaged

Environmental safety
- Dangerous for the environment
- Very toxic to aquatic organisms
- Risk to non-target insects or other arthropods. Avoid spraying within 6 m of the field boundary to reduce the effects on non-target insects or other arthropods
- LERAP Category B

Hazard classification and safety precautions
    Hazard H03, H11
    Risk phrases R22a, R43, R48, R50, R53a
    Operator protection A, H; U05a, U14
    Environmental protection E15a [1]; E16a, E22c, E34, E38 [1, 2]
    Storage and disposal D01, D02, D09a, D10b, D12a
    Medical advice M03, M05a

170   diflufenican + flurtamone

    A contact and residual herbicide mixture for cereals

Products

1 Bacara	Bayer CropScience	100:250 g/l	SC	10744
2 Graduate	Bayer CropScience	80:400 g/l	SC	10776

Uses
- Annual dicotyledons in *winter barley*, *winter wheat* [1, 2]
- Annual meadow grass in *winter barley*, *winter wheat* [1, 2]
- Blackgrass in *winter barley*, *winter wheat* [1]
- Loose silky bent in *winter barley*, *winter wheat* [1]
- Volunteer oilseed rape in *winter barley*, *winter wheat* [1, 2]

Approval information
- Accepted by BBPA for use on malting barley

Efficacy guidance
- Apply pre-emergence or from when crop has first leaf unfolded before susceptible weeds pass recommended size
- Best results obtained on firm, fine seedbeds with adequate soil moisture present at and after application. Increase water volume where crop or weed foliage is dense
- Good weed control requires ash, trash and burnt straw to be buried during seed bed preparation
- Loose fluffy seedbeds should be rolled before application and the final seed bed should be fine and firm without large clods
- Speed of control depends on weather conditions and activity can be slow under cool conditions
- Always follow WRAG guidelines for preventing and managing herbicide resistant weeds. Section 5 for more information

FOR FULL CONDITIONS OF USE ALWAYS READ THE PRODUCT LABEL

Restrictions
- Maximum number of treatments 1 per crop
- Crops should be drilled to a normal depth of 25 mm and the seed well covered. Do not treat broadcast crops as uncovered seed may be damaged
- Do not treat spring sown cereals, durum wheat, oats, undersown cereals or those to be undersown
- Do not treat frosted crops or when frost is imminent. Severe frost after application, or any other stress, may lead to transient discoloration or scorch
- Do not use on Sands or Very Light soils or those that are very stony or gravelly
- Do not use on waterlogged soils, or on crops subject to temporary waterlogging by heavy rainfall, as there is risk of persistent crop damage which may result in yield loss
- Do not use on soils with more than 10% organic matter
- Do not harrow at any time after application and do not roll autumn treated crops until spring

Crop-specific information
- Latest use: before 2nd node detectable (GS32)
- Take particular care to match spray swaths otherwise crop discoloration and biomass reduction may occur which may lead to yield reduction

Following crops guidance
- In the event of crop failure winter wheat may be redrilled immediately after normal cutlivation, and winter barley may be sown after ploughing. Fields must be ploughed to a depth of 15 cm and 20 wk must elapse before sowing spring crops of wheat, barley, oilseed rape, peas, field beans or potatoes
- After normal harvest autumn cereals can be drilled after ploughing. Thorough mixing of the soil must take place before drilling field beans, leaf brassicae or winter oilseed rape. For sugar beet seed crops and winter onions complete inversion of the furrow slice is essential
- Do not broadcast or direct drill oilseed rape or other brassica crops as a following crop on treated land. See label for detailed advice on preparing land for subsequent autumn cropping in the normal rotation
- Successive treatments of any products containing diflufenican can lead to soil build-up and inversion ploughing must precede sowing any following non-cereal crop. Even where ploughing occurs some crops may be damaged

Environmental safety
- Dangerous for the environment
- Very toxic to aquatic organisms
- Extremely dangerous to fish or other aquatic life. Do not contaminate surface waters or ditches with chemical or used container
- LERAP Category B

Hazard classification and safety precautions
Hazard H11
Risk phrases R50, R53a
Operator protection A, H; U20c
Environmental protection E13a, E16a, E16b, E38
Storage and disposal D05, D09a, D10b, D12a

171   diflufenican + flurtamone + isoproturon

A contact and residual herbicide for winter cereals

Products

Ingot	Bayer CropScience	27:67:400 g/l	SC	09997

Uses
- Annual dicotyledons in *winter barley*, *winter wheat*
- Annual meadow grass in *winter barley*, *winter wheat*
- Blackgrass in *winter barley*, *winter wheat*
- Loose silky bent in *winter barley*, *winter wheat*
- Volunteer oilseed rape in *winter barley*, *winter wheat*

SEE SECTION 3 FOR PRODUCTS ALSO REGISTERED

SECTION 2

Approval information

- Isoproturon included in Annex I under EC Directive 91/414
- Accepted by BBPA for use on malting barley

Efficacy guidance

- Apply from when crop has first leaf unfolded before susceptible weeds pass recommended size
- Best results obtained on firm, fine seedbeds with adequate soil moisture. Good spray coverage of soil and weeds is essential
- Speed of control depends on weather conditions and activity will be reduced during prolonged dry periods especially in spring. Weed control, especially of grasses, may also be reduced in wet seasons or where heavy rain falls shortly after treatment
- Always follow WRAG guidelines for preventing and managing herbicide resistant weeds. Section 5 for more information

Restrictions

- Maximum number of treatments 1 per crop. Maximum total dose of isoproturon 2.5 kg a.i./ha per crop
- Crops should be drilled to a normal depth of 25 mm and the seed well covered. Do not treat broadcast crops
- Do not apply pre-emergence to wheat or barley
- Do not use on crops undersown or to be undersown
- Do not treat frosted crops or when frost is imminent. Severe frost after application may cause transient discoloration or scorch
- Do not use on Sands or Very Light soils or those that are very stony or gravelly as there is risk of crop damage
- Do not treat on soils with more than 10% organic matter
- Do not use on waterlogged soils, or on crops subject to temporary waterlogging by heavy rainfall, as there is risk of persistent crop damage which may result in yield loss
- Do not roll autumn treated crops until the spring and do not harrow at any time after application
- Take particular care to match spray swaths otherwise crop discoloration and biomass reduction may occur which may lead to yield reduction

Crop-specific information

- Latest use: before 2nd node detectable (GS32)

Following crops guidance

- In the event of crop failure winter wheat may be redrilled immediately after normal cutivation, and winter barley may be sown after ploughing. Fields must be ploughed to a depth of 15 cm and 20 wk must elapse before sowing spring crops of wheat, barley, oilseed rape, peas, field beans or potatoes
- After normal harvest autumn cereals can be drilled after ploughing. Thorough mixing of the soil must take place before drilling field beans, leaf brassicae or winter oilseed rape. For sugar beet seed crops and winter onions complete inversion of the furrow slice is essential
- Do not broadcast or direct drill oilseed rape or other brassica crops as a following crop on treated land. See label for detailed advice on preparing land for subsequent autumn cropping in the normal rotation
- Successive treatments of any products containing diflufenican can lead to soil build-up and inversion ploughing must precede sowing any following non-cereal crop. Even where ploughing occurs some crops may be damaged

Environmental safety

- Dangerous for the environment
- Very toxic to aquatic organisms
- Extremely dangerous to fish or other aquatic life. Do not contaminate surface waters or ditches with chemical or used container
- LERAP Category B
- Do not apply to dry, cracked or waterlogged soils where heavy rain may lead to contamination of drains by isoproturon

Hazard classification and safety precautions

Hazard H03, H11
Risk phrases R40, R50, R53a
Operator protection A, C, H; U20c

FOR FULL CONDITIONS OF USE ALWAYS READ THE PRODUCT LABEL

Environmental protection E13a, E16a, E16b, E38
Storage and disposal D05, D09a, D10b, D12a

172 diflufenican + glyphosate

A foliar non-selective herbicide mixture

Products

Pistol	Bayer Environ.	40:250 g/l	SC	12173

Uses
- Annual and perennial weeds in **natural surfaces not intended to bear vegetation, permeable surfaces overlying soil**

Efficacy guidance
- Treat when weeds actively growing from Mar to end Sep and before they begin to senesce
- Performace may be reduced if application is made to plants growing under stress, such as drought or water-logging
- Pre-emergence activity is reduced on soils containing more than 10% organic matter or where organic debris has collected
- Perennial weeds such as docks, perennial sowthistle and willowherb are best treated before flowering or setting seed
- Perennial weeds emerging from established rootstocks after treatment will not be controlled
- A rainfree period of at least 6 h (preferably 24 h) should follow spraying for optimum control
- For oprtimum control do not cultivate or rake after treatment

Restrictions
- Maximum number of treatments 1 per yr
- May only be used on porous surfaces overlying soil. Must not be used if an impermeable membrane lies between the porous surface and the soil, and must not be used on any non-porous man-made surfaces
- Do not add any wetting agent or adjuvant oil
- Do not spray in windy weather

Following crops guidance
- A period of at least 6 mth must be allowed after treatment of sites that are to be cleared and grubbed before sowing and planting. Soil should be ploughed or dug first to ensure thorough mixing and dilution of any herbicide residues

Environmental safety
- Dangerous for the environment
- Very toxic to aquatic organisms
- LERAP Category B
- Avoid drift onto non-target plants
- Heavy rain after application may wash product onto sensitive areas such as newly sown grass or areas about to be planted

Hazard classification and safety precautions
Hazard H11
Risk phrases R50, R53a
Operator protection A, C; U20b
Environmental protection E15a, E16a, E16b, E38
Storage and disposal D01, D02, D09a, D11a, D12a

173 diflufenican + isoproturon

A contact and residual herbicide for use in winter cereals

Products

1	Fernpath Ipex	AgriGuard	50:500 g/l	SC	11391
2	Headland Javelin	Headland	62.5:500 g/l	SC	12451
3	Hermolance	Hermoo	62.5:500 g/l	SC	12459
4	Javelin	Bayer CropScience	62.5:500 g/l	SC	10440
5	Javelin	Bayer CropScience	62.5: 500 g/l	SC	12367
6	Javelin Gold	Bayer CropScience	20:500 g/l	SC	09999

SEE SECTION 3 FOR PRODUCTS ALSO REGISTERED

Products – continued

7	Koala	Nufarm UK	97.5:785 g/l	WG	12556
8	Me2 Sylvester	Me2	50:500 g/l	SC	12021
9	Panther	Bayer CropScience	50:500 g/l	SC	10745
10	Standon Diflufenican 625	Standon	62.5:500 g/l	SC	10753
11	Standon Diflufenican-IPU	Standon	50:500 g/l	SC	11109

Uses

- Annual dicotyledons in *triticale, winter rye* [1-6, 8-11]; *winter barley, winter wheat* [1-11]
- Annual grasses in *triticale, winter barley, winter rye, winter wheat* [1-6, 8-11]
- Annual meadow grass in *winter barley, winter wheat* [7]
- Blackgrass in *triticale, winter barley, winter rye, winter wheat* [1-6, 8-11]
- Rough meadow grass in *winter barley, winter wheat* [7]
- Wild oats in *triticale, winter barley, winter rye, winter wheat* [1-6, 8-11]

Approval information

- Isoproturon included in Annex I under EC Directive 91/414
- Accepted by BBPA for use on malting barley
- Approval expiry 31 Oct 2006 [4, 10]

Efficacy guidance

- May be applied in autumn or spring (but only post-emergence on wheat or barley). Best control normally achieved by early post-emergence treatment
- Best results by application to fine, firm seedbed moist at or after application
- Weeds controlled from before emergence to 6 true leaf stage
- Apply to moist, but not waterlogged, soils
- Any trash or ash should be buried during seedbed preparation
- Always follow WRAG guidelines for preventing and managing herbicide resistant weeds. Section 5 for more information

Restrictions

- Maximum number of treatments 1 per crop. Maximum total dose of isoproturon 2.5 kg a.i./ha per yr
- On triticale and winter rye only treat named varieties and apply as pre-emergence spray
- Do not use on other cereals, broadcast or undersown crops or crops to be undersown
- Do not use on Sands or Very Light soils, or those that are very stony or gravelly or on soils with more than 10% organic matter
- Do not spray when heavy rain is forecast or on crops suffering from stress, frost, deficiency, pest or disease attack
- Do not harrow after application nor roll autumn-treated crops until spring

Crop-specific information

- Latest use: before 2nd node detectable (GS 32) (low dose) or before end Feb (high dose) for wheat and barley; pre-emergence of crop for triticale and rye [2-5, 9, 10];post-emergence and before end Feb for winter wheat and barley [1, 7-9, 11]
- Drill crop to normal depth (25 mm) and ensure seed well covered
- Early sown crops may be damaged if application is made during a period of rapid growth in the autumn

Following crops guidance

- In the event of crop failure winter wheat may be redrilled immediately after normal cutlivation, and winter barley may be sown after ploughing. Fields must be ploughed to a depth of 15 cm and 20 wk must elapse before sowing spring crops of wheat, barley, oilseed rape, peas, field beans, sugar beet, potatoes, carrots, edible brassicas or onions
- After normal harvest autumn cereals can be drilled after ploughing. Thorough mixing of the soil must take place before drilling field beans, leaf brassicae or winter oilseed rape. For sugar beet seed crops and winter onions complete inversion of the furrow slice is essential
- Do not broadcast or direct drill oilseed rape or other brassica crops as a following crop on treated land. See label for detailed advice on preparing land for subsequent autumn cropping in the normal rotation
- Successive treatments of any products containing diflufenican can lead to soil build-up and inversion ploughing must precede sowing any following non-cereal crop. Even where ploughing occurs some crops may be damaged

FOR FULL CONDITIONS OF USE ALWAYS READ THE PRODUCT LABEL

Environmental safety
- Dangerous for the environment
- Very toxic to aquatic organisms
- Extremely dangerous to fish or other aquatic life. Do not contaminate surface waters or ditches with chemical or used container
- Do not apply to dry, cracked or waterlogged soils where heavy rain may lead to contamination of drains by isoproturon
- LERAP Category B

Hazard classification and safety precautions
Hazard H03 [1-6, 8-11]; H04 [7]; H11 [1-11]
Risk phrases R36 [7]; R40, R50, R53a [1-11]
Operator protection A, C, H [1-5, 7-11]; D [7]; U20a [1]; U20b [6]; U20c [2-5, 7-11]
Environmental protection E13a [2, 4-7, 9]; E15a [1, 3, 8, 10, 11]; E16a [1-11]; E16b [1, 2, 4-11]; E38 [2-7, 9]
Storage and disposal D05 [1-7, 10]; D09a [1-11]; D10b [1-6, 8-11]; D11a [7]; D12a [2-7, 9]

174 diflufenican + trifluralin

A contact and residual herbicide for use in winter cereals

Products

Ardent	Bayer CropScience	40:400 g/l	SC	09968

Uses
- Annual dicotyledons in *triticale*, *winter barley*, *winter rye*, *winter wheat*
- Annual meadow grass in *winter barley*, *winter wheat*

Approval information
- Accepted by BBPA for use on malting barley

Efficacy guidance
- Weeds controlled pre-emergence up to 4 leaves (2 leaves for annual meadow grass)
- Good weed control depends on trash being buried before or during seedbed preparation
- Spring germinating weeds are not controlled
- Best results achieved on firm, fine moist seedbeds. Seed should be well covered
- Rolling after autumn treatment will reduce weed control. Roll in spring if necessary
- Speed of control depends on temperature and growing conditions; activity can be slow under cool conditions
- Always follow WRAG guidelines for preventing and managing herbicide resistant weeds. Section 5 for more information

Restrictions
- Maximum number of treatments 1 per crop
- Ensure that crop is evenly drilled and seed well covered. Do not treat broadcast crops or those that are frosted or stressed
- Do not treat undersown crops or those to be undersown
- Do not harrow after application
- Do not use on oats or any spring sown cereals
- Do not use on Sands or Very Light soils, or those that are very stony or gravelly
- Do not use on soils with more than 10% organic matter; this may include newly ploughed grassland for a time

Crop-specific information
- Latest use: before 2nd tiller stage (GS 22), or end Nov, whichever is sooner
- Apply pre- or early post-emergence of crop up to maximum recommended weed size
- Crops occasionally show transient leaf discoloration after treatment. Symptoms are quickly outgrown and yield not affected

Following crops guidance
- In the event of crop failure treated soil cannot be redrilled with wheat or barley until the following autumn; sugar beet should not be drilled in the spring following autumn treatment; 5 mth must elapse between treatment and sowing spring oilseed rape, carrots or edible brassicas

SEE SECTION 3 FOR PRODUCTS ALSO REGISTERED

- After normal harvest of a treated crop soils must be ploughed to 15 cm before drilling field beans, leaf brassicas, winter oilseed rape, autumn sown cereals or winter onions
- Successive treatments of any products containing diflufenican can lead to soil build-up and inversion ploughing must precede sowing any following non-cereal crop. Even where ploughing occurs some crops may be damaged

Environmental safety
- Dangerous for the environment
- Very toxic to aquatic organisms
- Extremely dangerous to fish or other aquatic life. Do not contaminate surface waters or ditches with chemical or used container
- LERAP Category B

Hazard classification and safety precautions
Hazard H04, H11
Risk phrases R36, R43, R50, R66, R67
Operator protection A, C; U05a, U08, U13, U14, U20b
Environmental protection E13a, E16a, E38
Storage and disposal D01, D02, D05, D09a, D10b, D12a

175   dimethoate

A contact and systemic organophosphorus insecticide and acaricide

Products

1	BASF Dimethoate 40	BASF	400 g/l	EC	00199
2	Danadim	Headland	400 g/l	EC	11550
3	Danadim Progress	Headland	400 g/l	EC	12208
4	Rogor L40	Interfarm	400 g/l	EC	07611

Uses

- Aphids in *broccoli, brussels sprouts, calabrese, cauliflowers, ornamental specimens* [2]; *fodder beet, mangels, red beet* [4]; *fodder beet* (excluding Myzus persicae), *lettuce, mangels* (excluding Myzus persicae), *red beet* (excluding Myzus persicae) [1-3]; *grass seed crops* [1]; *mangel seed crops, sugar beet seed crops* (excluding Myzus persicae) [1, 4]; *ornamental plant production* [1, 3]; *spring rye, spring wheat* [1, 3, 4]; *sugar beet* (excluding Myzus persicae), *triticale, winter rye, winter wheat* [1-4]
- Cabbage aphid in *broccoli, brussels sprouts, calabrese, cauliflowers* [1, 3]
- General insect control in *calabrese* (off-label), *celeriac* (off-label), *chinese cabbage* (off-label), *collards* (off-label), *compost* (off-label - for watercress propagation), *kale* (off-label), *kohlrabi* (off-label), *protected broccoli* (off-label), *protected cabbages* (off-label), *protected calabrese* (off-label), *protected cauliflowers* (off-label), *protected celeriac* (off-label), *protected chinese cabbage* (off-label), *protected collards* (off-label), *protected kale* (off-label), *protected kohlrabi* (off-label), *protected roscoff cauliflowers* (off-label), *protected savoy cabbage* (off-label), *protected spinach* (off-label), *roscoff cauliflowers* (off-label), *savoy cabbage* (off-label) [3]
- Insect pests in *bulb onions* (off-label), *calabrese* (off-label), *celeriac* (off-label), *chinese cabbage* (off-label), *collards* (off-label), *garlic* (off-label), *kale* (off-label), *kohlrabi* (off-label), *leeks* (off-label), *protected broccoli* (off-label), *protected brussels sprouts* (off-label), *protected cabbages* (off-label), *protected calabrese* (off-label), *protected cauliflowers* (off-label), *protected celeriac* (off-label), *protected chinese cabbage* (off-label), *protected collards* (off-label), *protected kale* (off-label), *protected kohlrabi* (off-label), *protected spinach* (off-label), *shallots* (off-label) [1, 2]; *bulb onions* (off-label - seedlings), *garlic* (off-label - seedlings), *leeks* (off-label - seedlings), *protected garlic* (off-label - seedlings), *protected leeks* (off-label - seedlings), *protected onions* (off-label - seedlings), *protected roscoff cauliflowers* (off-label), *protected salad onions* (off-label - seedlings), *protected savoy cabbage* (off-label), *protected shallots* (off-label - seedlings), *roscoff cauliflowers* (off-label), *salad onions* (off-label - seedlings), *savoy cabbage* (off-label), *shallots* (off-label - seedlings) [1]; *compost* (off-label - for watercress propagation), *protected garlic* (off-label), *protected leeks* (off-label), *protected onions* (off-label), *protected shallots* (off-label), *salad onions* (off-label) [2]
- Leaf miner in *fodder beet, mangels, red beet, sugar beet* [1-4]; *mangel seed crops, sugar beet seed crops* [4]; *ornamental plant production* [1, 3]; *ornamental specimens* [2]
- Mangold fly in *fodder beet, mangels, red beet, sugar beet* [2]

FOR FULL CONDITIONS OF USE ALWAYS READ THE PRODUCT LABEL

- Red spider mites in **ornamental plant production** [1, 3]
- Wheat bulb fly in **spring rye**, **triticale**, **winter rye** [4]; **spring wheat** [1, 3, 4]; **winter wheat** [1-4]

Specific Off-Label Approvals (SOLAs)

- **bulb onions**, **garlic**, **leeks**, **shallots** *(OLA 001511) May 2006* [1]
- **bulb onions**, **garlic**, **leeks**, **protected garlic**, **protected leeks**, **protected onions**, **protected salad onions**, **protected shallots**, **salad onions**, **shallots** *(seedlings) (OLA 001511) May 2006* [1]
- **bulb onions**, **garlic**, **leeks**, **protected garlic**, **protected leeks**, **protected onions**, **protected shallots**, **salad onions**, **shallots** *(OLA 051697) May 2006* [2]
- **calabrese**, **celeriac**, **chinese cabbage**, **collards**, **kale**, **kohlrabi**, **protected broccoli**, **protected brussels sprouts**, **protected cabbages**, **protected calabrese**, **protected cauliflowers**, **protected celeriac**, **protected chinese cabbage**, **protected collards**, **protected kale**, **protected kohlrabi**, **protected roscoff cauliflowers**, **protected savoy cabbage**, **protected spinach**, **roscoff cauliflowers**, **savoy cabbage** *(OLA 940389) Dec 2008* [1]
- **calabrese**, **celeriac**, **chinese cabbage**, **collards**, **kale**, **kohlrabi**, **protected broccoli**, **protected brussels sprouts**, **protected cabbages**, **protected calabrese**, **protected cauliflowers**, **protected celeriac**, **protected chinese cabbage**, **protected collards**, **protected kale**, **protected kohlrabi**, **protected spinach** *(OLA 041505) Dec 2008* [2]
- **calabrese**, **celeriac**, **chinese cabbage**, **collards**, **kale**, **kohlrabi**, **protected broccoli**, **protected cabbages**, **protected calabrese**, **protected cauliflowers**, **protected celeriac**, **protected chinese cabbage**, **protected collards**, **protected kale**, **protected kohlrabi**, **protected roscoff cauliflowers**, **protected savoy cabbage**, **protected spinach**, **roscoff cauliflowers**, **savoy cabbage** *(OLA 050682) Dec 2008* [3]
- **compost** *(for watercress propagation) (OLA 041502) Dec 2008* [2]
- **compost** *(for watercress propagation) (OLA 050683) Dec 2008* [3]

Approval information

- Approved for aerial application on cereals, sugar beet [1, 4]. See notes in Section 5

Efficacy guidance

- Chemical has quick knock-down effect and systemic activity lasts for up to 14 d
- With some crops, products differ in range of pests listed as controlled. Uses section above provides summary. See labels for details
- For most pests apply when pest first seen and repeat 2-3 wk later or as necessary. Timing and number of sprays varies with crop and pest. See labels for details
- Best results achieved when crop growing vigorously. Systemic activity reduced when crops suffering from drought or other stress
- In hot weather apply in early morning or late evening
- Where aphids or spider mites resistant to organophosphorus compounds occur control is unlikely to be satisfactory and repeat treatments may result in lower levels of control

Restrictions

- Maximum number of treatments 6 per crop for Brussels sprouts, broccoli, cauliflower, calabrese, lettuce; 4 per crop for grass seed crops; 1 per crop for beet crops (see also below), cereals, watercress
- Contains an anticholinesterase organophosphorus compound. Do not use if under medical advice not to work with such compounds
- On beet crops only one treatment per crop may be made for control of leaf miners (max 84 g a.i./ha) and black bean aphid (max 400 g a.i./ha)
- In beet crops resistant strains of peach-potato aphid (*Myzus persicae*) are common and dimethoate products must not be used to control this pest
- Test for varietal susceptibility on all unusual plants or new cultivars
- Do not tank mix with alkaline materials. See label for recommended tank-mixes
- Consult processor before spraying crops grown for processing
- Must not be applied to cereals if any product containing a pyrethroid insecticide or dimethoate has been sprayed after the start of ear emergence (GS 51)

Crop-specific information

- Latest use: before 30 Jun in yr of harvest for beet crops; flowering just complete stage (GS 30) for ground applications to cereals; before 31 Mar in yr of harvest for aerial application to cereals; before 6 true leaves for shallots; before transfer to cropping beds for watercress; before 7 leaf stage for brassicas and protected vegetables
- HI brassicas 21 d; grass seed crops, lettuce, cereals 14 d; shallots 7 d

SEE SECTION 3 FOR PRODUCTS ALSO REGISTERED

SECTION 2

Environmental safety
- Dangerous for the environment
- Very toxic to aquatic organisms
- Harmful to game, wild birds and animals
- Harmful to livestock. Keep all livestock out of treated areas for at least 7 d
- Likely to cause adverse effects on beneficial arthropods
- High risk to bees. Do not apply to crops in flower or to those in which bees are actively foraging. Do not apply when flowering weeds are present
- Dangerous to fish or other aquatic life. Do not contaminate surface waters or ditches with chemical or used container
- Surface residues may also cause bee mortality following spraying
- High risk to non-target insects or other arthropods
- LERAP Category A
- Keep in original container, tightly closed, in a safe place, under lock and key
- Do not treat cereals after 1 Apr within 6 m of edge of crop

Hazard classification and safety precautions
Hazard H03, H08 [1-4]; H11 [4]
Risk phrases R20 [1, 2, 4]; R21 [1, 4]; R22a, R43 [1-4]; R48 [1]; R50, R53a [4]; R52 [2, 3]
Operator protection A, H, M [1-4]; C, J [1, 4]; U02a, U04a, U19a [1-4]; U05a, U10, U13, U20a [1, 4]; U08, U15, U20b [2, 3]; U14 [3]
Environmental protection E06c [1-4] (7 d); E10b, E16c, E16d, E34 [1-4]; E12a [1, 2]; E12b [3]; E12d, E15a [4]; E13b, E22a [1-3]; E17a [2, 4] (18 m); E18 [1, 4]
Storage and disposal D01, D02, D09b [1-4]; D05 [1, 3]; D08 [1, 4]; D10b [4]; D10c [1]; D12b [2, 3]
Treated seed S04b [4]
Medical advice M01, M03 [1-4]; M05a [1]

176   dimethomorph

A cinnamic acid fungicide with translaminar activity available only in mixtures

177   dimethomorph + mancozeb

A systemic and protectant fungicide for potato blight control

Products

Invader	BASF	7.5:66.7% w/w	WG	11978

Uses
- Blight in **potatoes**
- Downy mildew in **bulb onions** (off-label), **garlic** (off-label), **poppies** (off-label - for morphine production), **shallots** (off-label)

Specific Off-Label Approvals (SOLAs)
- **bulb onions**, **garlic**, **shallots** (OLA 042334) Dec 2008 [1]
- **poppies** (for morphine production) (OLA 050909) Dec 2008 [1]

Approval information
- Mancozeb included in Annex I under EC Directive 91/414

Efficacy guidance
- Commence treatment as soon as there is a risk of blight infection
- In the absence of a warning treatment should start before the crop meets along the rows
- Repeat treatments every 7-14 d depending in the degree of infection risk
- Irrigated crops should be regarded as at high risk and treated every 7 d
- For best results good spray coverage of the foliage is essential
- To minimise the likelihood of development of resistance these products should be used in a planned Resistance Management strategy. See Section 5 for more information

Restrictions
- Maximum total dose equivalent to eight full dose treatments

Crop-specific information
- HI 7 d for all crops

FOR FULL CONDITIONS OF USE ALWAYS READ THE PRODUCT LABEL

Environmental safety
- Dangerous for the environment
- Very toxic to aquatic organisms
- LERAP Category B

Hazard classification and safety precautions
    Hazard H04, H11
    Risk phrases R36, R37, R38, R43, R50, R53a
    Operator protection A; U02a, U04a, U05a, U08, U13, U19a, U20b
    Environmental protection E16a, E16b, E38
    Storage and disposal D01, D02, D05, D09a, D12a
    Medical advice M05a

178   dimoxystrobin

A protectant strobilurin fungicide for cereals available only in mixtures

179   dimoxystrobin + epoxiconazole

A protectant and curative fungicide mixture for cereals

Products

Swing Gold	BASF	133:50 g/l	SC	11658

Uses
- Brown rust in **winter wheat**
- Fusarium ear blight in **winter wheat**
- Septoria in **winter wheat**
- Tan spot in **winter wheat**

Efficacy guidance
- For best results apply from the start of ear emergence
- Dimoxystrobin is a member of the QoI cross resistance group. Product should be used preventatively and not relied on for its curative potential
- Use product as part of an Integrated Crop Management strategy incorporating other methods of control, including where appropriate other fungicides with a different mode of action. Do not apply more than two foliar applications of QoI containing products to any cereal crop
- There is a significant risk of widespread resistance occurring in *Septoria tritici* populations in UK. Failure to follow resistance management action may result in reduced levels of disease control

Restrictions
- Maximum number of treatments 1 per crop
- Do not apply before the start of ear emergence

Crop-specific information
- Latest use: up to and including flowering (GS 69)

Environmental safety
- Dangerous for the environment
- Very toxic to aquatic organisms
- LERAP Category B
- Avoid drift onto neighbouring crops

Hazard classification and safety precautions
    Hazard H03, H11
    Risk phrases R20, R22a, R40, R50, R53a
    Operator protection A; U05a, U20b
    Environmental protection E15a, E16a, E34, E38
    Storage and disposal D01, D02, D05, D09a, D10c, D12a
    Medical advice M05a

SECTION 2

180   dinocap

A protectant dinitrophenyl fungicide for powdery mildew control

Products

Karathane Liquid	Landseer	350 g/l	EC	09262

Uses
- Powdery mildew in **apples**, **chrysanthemums**, **grapevines** (off-label), **roses**, **strawberries**

Specific Off-Label Approvals (SOLAs)
- **grapevines** (OLA 021410) Dec 2008 [1]

Efficacy guidance
- Spray at 7-14 d intervals to maintain protective film
- Product must be applied to roses before disease becomes established
- Regular use on apples suppresses red spider mites and rust mites

Restrictions
- Maximum total dose equivalent to 10 full dose treatments on apples; 8 full dose treatments on grapevines
- Maximum number of treatments on strawberries 5 per yr
- When applied to apples during blossom period may cause spotting but no adverse effect on pollination or fruit set. Do not apply to Golden Delicious during blossom period
- Do not apply when temperature is above 24°C. Do not apply with white oils
- Certain chrysanthemum cultivars may be susceptible
- May cause petal spotting on white roses

Crop-specific information
- HI grapevines 21 d, apples 14 d, strawberries 7 d

Environmental safety
- Dangerous for the environment
- Very toxic to aquatic organisms
- LERAP Category B
- Broadcast air-assisted LERAP (18 m)

Hazard classification and safety precautions
> **Hazard** H02, H08, H11
> **Risk phrases** R20, R22b, R36, R38, R43, R48, R50, R53a, R61, R67
> **Operator protection** A, C, H, J, M; U04a, U05a, U10, U11, U19a, U20a
> **Environmental protection** E15a, E16a, E34, E38; E17b (18 m)
> **Storage and disposal** D01, D02, D05, D09a, D10a, D12a
> **Medical advice** M04

181   diquat

A non-residual bipyridyl contact herbicide and crop desiccant

Products

1	Agriguard Diquat	AgriGuard	200 g/l	SL	10836
2	Greencrop Boomerang	Greencrop	200 g/l	SL	10691
3	Landgold Diquat	Landgold	200 g/l	SL	10649
4	Landgold Diquat	Teliton	200 g/l	SL	12116
5	Reglone	Syngenta	200 g/l	SL	10534
6	Standon Diquat SL	Standon	200 g/l	SL	10732
7	Waterloo	Interfarm	200 g/l	SL	11196

Uses
- Annual dicotyledons in **flower bulbs**, **sugar beet** [1, 2, 7]; **mustard** (off-label), **sage** (off-label), **sunflowers** (off-label) [5]; **ornamental plant production**, **row crops** [1]; **potatoes** [1, 2, 6]; **row crops** (between row treatment) [2, 5-7]
- Chemical stripping in **hops** [1, 2, 5, 7]
- General weed control in **poppies** (off-label - for morphine production), **potatoes** [5]
- General weed growth in **laid barley** (stockfeed only), **laid oats** (stockfeed only) [1-4, 6, 7]
- Haulm destruction in **potatoes** [3-6]

FOR FULL CONDITIONS OF USE ALWAYS READ THE PRODUCT LABEL

- Pre-harvest desiccation in **borage** *(off-label)*, **echium plantaginium** *(off-label)*, **hemp for oilseed production** *(off-label)*, **honesty** *(off-label)*, **laid barley** *(stockfeed only)*, **laid oats** *(stockfeed only)*, **mustard** *(off-label)*, **navy beans** *(off-label)*, **peas**, **poppies** *(off-label - for morphine production)*, **sage** *(off-label)*, **soya beans** *(off-label)*, **sunflowers** *(off-label)* [5]; **clover seed crops** [1, 2, 5, 7]; **combining peas** [1-4, 6, 7]; **linseed**, **spring field beans** *(stock or pigeon feed only)*, **spring oilseed rape**, **winter field beans** *(stock or pigeon feed only)*, **winter oilseed rape** [1-7]; **potatoes** [1, 2, 7]

Specific Off-Label Approvals (SOLAs)
- **borage** *(OLA 051630) Jan 2007* [5]
- **echium plantaginium** *(OLA 051629) Jan 2007* [5]
- **hemp for oilseed production** *(OLA 051635) Jan 2007* [5]
- **honesty** *(OLA 051631) Jan 2007* [5]
- **mustard**, **sage**, **sunflowers** *(OLA 051632) Jan 2007* [5]
- **navy beans** *(OLA 051633) Jan 2007* [5]
- **poppies** *(for morphine production) (OLA 051634) Jan 2007* [5]
- **soya beans** *(OLA 011653) Dec 2008* [5]

Approval information
- Diquat included in Annex I under EC Directive 91/414
- All approvals for aquatic weed control expired in 2004
- Accepted by BBPA for use on hops

Efficacy guidance
- Acts rapidly on green parts of plants and rainfast in 15 min
- Best results for potato desiccation achieved by spraying in bright light and low humidity conditions
- For weed control in row crops apply as overall spray before crop emergence or before transplanting or as an inter-row treatment using a spray guard
- Treatments for weed control normally require co-application with paraquat. Use of authorised non-ionic wetter for improved weed control essential for most uses except potato desiccation, treatment of hops and aquatic weed control (except *Lemna*)

Restrictions
- Maximum number of treatments 3 per crop on hops; 1 per crop in most other situations - see labels for details
- Do not apply potato haulm destruction treatment when soil dry. Tubers may be damaged if spray applied during or shortly after dry periods. See label for details of maximum allowable soil-moisture deficit and varietal drought resistance scores
- Do not add wetters to desiccant sprays for potatoes or for water weed control except for *Lemna* control.
- Consult processor before adding Agral or other wetter on peas. Staining may be increased
- Do not use treated straw or haulm as animal feed or bedding within 4 d of spraying
- Do not use on barley, oats and field beans intended for human consumption

Crop-specific information
- Latest use varies with crop and product to ensure best results - see labels
- HI zero when used as a crop desiccant but see label for advisory intervals
- For pre-harvest desiccation, apply to potatoes when tubers the desired size and to other crops when mature or approaching maturity. See label for details of timing and of period to be left before harvesting potatoes and for timing of pea and bean desiccation sprays
- Spray linseed when seed matured evenly over whole field; direct combining can normally begin 10-20 d after spraying
- Apply as hop stripping treatment when shoots have reached top wire
- Apply to floating and submerged aquatic weeds in still or slow moving water [1]
- Treated laid barley and oats may be used only for stock feed; treated peas must be harvested dry; treated field beans may be used for pigeon and animal feed only
- If potato tubers are to be stored leave 14 d after treatment before lifting
- Use on thin or stressed crops of oilseed rape may increase seed losses and decrease oil yield to an unacceptable level
- Treatment of hops under stress or when leaves wet with a heavy dew may cause damage

Environmental safety
- Dangerous for the environment
- Very toxic to aquatic organisms

SEE SECTION 3 FOR PRODUCTS ALSO REGISTERED

- Keep all livestock out of treated areas and away from treated water for at least 24 h
- Do not feed treated straw or haulm to livestock within 4 days of spraying
- Do not use on crops if the straw is to be used as animal feed/bedding
- Do not dump surplus herbicide in water or ditch bottoms

Hazard classification and safety precautions

Hazard H02, H11

Risk phrases R22a, R37, R50, R53a [1-7]; R36, R38 [1-4, 6]; R48 [2-7]

Operator protection A, C; U02a, U04a, U05a, U08, U19a, U20b

Environmental protection E06c [1, 5, 7] (24 h); E08 [5, 7] (4 d); E09 [1] (4 d); E15a, E34 [1-7]; E19a [1]; E38 [5, 7]

Storage and disposal D01, D02, D09a, D10a [1-7]; D05 [1-4, 6]; D12a [5, 7]

Medical advice M03 [1]; M04 [2-7]

182   diquat + paraquat

A non-selective non-residual bipyridyl contact herbicide

Products

1 ASAP	Me2	80:120 g/l	SC	11485
2 Fernpath Pronto	AgriGuard	80:120 g/l	SL	11988
3 PDQ	Syngenta	80:120 g/l	SL	10532
4 Speedway 2	Scotts	2.5:2.5% w/w	WG	10790

Uses

- Annual and perennial weeds in *amenity vegetation, hard surfaces, natural surfaces not intended to bear vegetation, permeable surfaces overlying soil* [4]
- Annual dicotyledons in *all edible crops, all non-edible crops, forest nurseries, natural surfaces not intended to bear vegetation, ornamental plant production* [1, 2]; *apple orchards, bush fruit, cane fruit, cherries, cultivated land/soil, damsons, flower bulbs, non-crop areas, pear orchards, plums, row crops, strawberries, stubbles* [3]; *forest, hops, potatoes* [1-3]
- Annual grasses in *all edible crops, all non-edible crops, forest nurseries, natural surfaces not intended to bear vegetation, ornamental plant production* [1, 2]; *apple orchards, bush fruit, cane fruit, cherries, cultivated land/soil, damsons, flower bulbs, non-crop areas, pear orchards, plums, row crops, strawberries, stubbles* [3]; *forest, hops, potatoes* [1-3]
- Chemical stripping in *hops* [3]
- Grass weeds in *field margins* [3]
- Green cover in *land not being used for crop production* [1-3]
- Perennial non-rhizomatous grasses in *apple orchards, bush fruit, cane fruit, cherries, cultivated land/soil, damsons, flower bulbs, forest, hops, non-crop areas, pear orchards, plums, potatoes, row crops, strawberries, stubbles* [3]
- Sward destruction in *grassland* [3, 4]; *managed amenity turf* [4]; *permanent pasture, rotational grassland* [1, 2]
- Volunteer cereals in *apple orchards, bush fruit, cane fruit, cherries, cultivated land/soil, damsons, forest, hops, non-crop areas, pear orchards, plums, potatoes, row crops, strawberries, stubbles* [3]

Approval information

- Diquat and paraquat included in Annex I under EC Directive 91/414
- Accepted by BBPA for use on hops

Efficacy guidance

- Rapid kill obtained under bright conditions but most effective results on difficult weeds obtained from slower action in winter
- Apply to young emerged weeds less than 15 cm high, annual grasses must have at least 2 leaves when sprayed [3]
- Addition of approved non-ionic wetter recommended for control of certain species and with low dose rates. See label for details [3]
- Interval between spraying and cultivation varies. See label for details [3]
- For chemical stripping apply in Jul or after hops have reached top wire. Do not use on hops under drought conditions [3]
- Chemical rapidly inactivated in moist soil, activity reduced in dirty or muddy water
- Spray is rainfast in 10 min

FOR FULL CONDITIONS OF USE ALWAYS READ THE PRODUCT LABEL

Restrictions
- Paraquat is subject to the Poisons Rules 1982 and the Poisons Act 1972. See notes in Section 5
- Product for use only by local authorities and professional spray operators [4]
- Diquat may cause an allergic reaction
- Maximum number of treatments 2 per yr for grassland destruction
- In non-crop areas do not use around green bark [4]
- Do not put in a food or drinks container

Crop-specific information
- Latest use varies according to situation - see label
- Apply up to just before sown crops emerge or just before planting [3]
- In potatoes spray earlies up to 10% emergence, maincrop to 40% emergence, provided plants are less than 15 cm high. Do not use post-emergence on potatoes from diseased or small tubers or under very hot, dry conditions [3]
- Use guarded no-drift sprayers to kill inter-row weeds and strawberry runners [3]
- Apply to fruit crops as a directed spray, preferably in dormant season [3]
- If spraying bulbs at end of season ensure all crop foliage is detached from bulbs. Do not use on very sandy soils [3]

Following crops guidance
- On sandy or immature peat soils and on forest nursery seedbeds allow 3 d between spraying and planting [3]
- Where trash or dying weeds are left on surface allow at least 3 d before planting

Environmental safety
- Dangerous for the environment
- Very toxic to aquatic organisms
- Paraquat can be harmful to hares; spray stubbles early in the day
- Keep all livestock out of treated areas for at least 24 h
- Keep in original container, tightly closed, in a safe place, under lock and key

Hazard classification and safety precautions
Hazard H02 [1-3]; H03 [4]; H11 [1-4]
Risk phrases R21, R37, R38, R41, R50 [1-3]; R22a, R48, R53a [1-4]; R51 [4]
Operator protection A, C [1-4]; H, M [1-3]; U02a, U04a, U20a [2]; U04b, U04c, U19a, U20b [1, 3, 4]; U05a, U09a [1-4]; U14 [4]
Environmental protection E02a [4] (until dry); E06c [1-4] (24 h); E11 [1-3]; E15a, E34 [1-4]; E38 [1, 3, 4]
Storage and disposal D01, D02, D09b [1-4]; D05, D10a [2, 4]; D10c, D12a [1, 3]; D12b [4]
Medical advice M03 [2]; M04 [1, 3, 4]; M05c [1-3]

183 dithianon

A protectant and eradicant dicarbonitrile fungicide for scab control

Products

1	Dithianon Flowable	BASF	750 g/l	SC	10219
2	Dithianon WG	BASF	70% w/w	WG	12460

Uses
- Scab in *apples*, *pears*

Efficacy guidance
- Apply at bud-burst and repeat every 10-14 d until danger of scab infection ceases
- Application at high rate within 48 h of a Mills period prevents new infection
- Spray programme also reduces summer infection with apple canker

Restrictions
- Maximum number of treatments 8 per crop [1]; 12 per crop [2]
- Do not use on Golden Delicious apples after green cluster
- Do not mix with lime sulphur or highly alkaline products

Crop-specific information
- HI 4 wk

Environmental safety
- Dangerous for the environment
- Very toxic to aquatic organisms

Hazard classification and safety precautions

Hazard H03 [1]; H04 [2]; H11 [1, 2]

Risk phrases R20 [1]; R22a, R50, R53a [1, 2]; R41 [2]

Operator protection A [1, 2]; D [2]; H [1]; U05a [1, 2]; U08, U19a, U20a [1]; U11 [2]

Environmental protection E15a, E34, E38

Storage and disposal D01, D02, D05 [1, 2]; D09a, D10b [1]; D12a [2]

Medical advice M03 [1]; M05a [1, 2]

184   diuron

A residual urea herbicide for non-crop areas and woody crops

See also amitrole + 2,4-D + diuron

Products

1	Diurex 50SC	Makhteshim	500 g/l	SC	11472
2	Diuron 80 WP	Bayer Environ.	80% w/w	WP	09931
3	Divide	Headland	80% w/w	WP	12206
4	Divide 50 SC	Headland	500 g/l	SC	12377
5	Freeway	Bayer Environ.	500 g/l	SC	11129
6	MSS Diuron 500 FL	Nufarm UK	500 g/l	SC	11493
7	MSS Diuron 80 WP	Nufarm UK	80% w/w	WP	11503
8	Nomix Diuron Flowable	Nomix Enviro	500 g/l	SC	12078
9	Unicrop Flowable Diuron	Unicrop	500 g/l	SC	02270
10	UPL Diuron 80	United Phosphorus	80% w/w	WP	07619

Uses
- Annual and perennial weeds in *asparagus (off-label)*, *blackcurrants (off-label)* [1]
- Annual dicotyledons in *amenity trees and shrubs* [1, 5]; *amenity vegetation* [4, 6-8]; *apple orchards*, *pear orchards* [1, 5, 9]; *asparagus (off-label)* [9]; *established woody ornamentals* [2]; *hard surfaces* [1, 4, 6, 8]; *land not intended to bear vegetation* [2-8]; *natural surfaces not intended to bear vegetation* [1]; *permeable surfaces overlying soil* [1, 4, 6-8]; *trees and shrubs*, *woody nursery stock (not container grown)* [5]
- Annual grasses in *amenity trees and shrubs* [1, 5]; *apple orchards*, *pear orchards* [1, 5, 9]; *asparagus (off-label)* [9]; *established woody ornamentals* [2]; *hard surfaces*, *natural surfaces not intended to bear vegetation*, *permeable surfaces overlying soil* [1]; *land not intended to bear vegetation* [2, 5]; *trees and shrubs*, *woody nursery stock* [5]
- Annual meadow grass in *land not intended to bear vegetation* [3]
- Black nightshade in *daffodils (off-label - for galanthamine production)* [9]
- Perennial weeds in *amenity vegetation*, *land not intended to bear vegetation*, *permeable surfaces overlying soil* [4, 6-8]; *hard surfaces* [4, 6, 8]
- Stinging nettle in *daffodils (off-label - for galanthamine production)* [9]
- Total vegetation control in *industrial sites*, *land not intended for cropping* [10]
- Willowherb in *blackcurrants (off-label)* [1, 9]; *daffodils (off-label - for galanthamine production)* [9]

Specific Off-Label Approvals (SOLAs)
- *asparagus*, *blackcurrants* (OLA 032172) Dec 2008 [1]
- *asparagus* (OLA 982352) Dec 2008 [9]
- *blackcurrants* (OLA 951318) Dec 2008 [9]
- *daffodils* (for galanthamine production) (OLA 041521) Dec 2008 [9]

Efficacy guidance
- Best results when applied to moist soil and rain falls soon afterwards
- Length of residual activity may be reduced on heavy or highly organic soils or those with ash substrates
- Selective rates must be applied to weed-free soil and activity persists for 2-3 mth
- Application for total vegetation control may be at any time of year, best results obtained from start of Feb to end Apr (end May in Scotland). See label for list of resistant species
- Application to frozen ground not recommended

FOR FULL CONDITIONS OF USE ALWAYS READ THE PRODUCT LABEL

Restrictions
- Maximum number of treatments 1 per yr on land not intended for cropping (high rate), 2 per yr around trees and shrubs (low rate); one per year for apples, pears; 2 per yr (low rate) for blackcurrants [9]
- Do not treat trees and shrubs less than 5 cm tall or established less than 12 mth or in areas of container grown trees and shrubs. See label for list of sensitive ornamental species that should not be treated
- Do not use around trees and shrubs on Sands or Very Light soils or those that are gravelly or where less than 1% organic matter
- Do not use on lawns, grass tennis courts or similar areas of turf
- Do not apply non-selective rates on or near desirable plants where chemical may be washed into contact with roots
- Application for amenity use should only take place between the beginning of Feb and end of Apr (May in Scotland)

Crop-specific information
- Latest use: 30 April (31 May in Scotland) on non-porous surfaces; at any time on porous surfaces for land not intended to bear vegetation; before bud-burst for blackcurrants [9]; pre-emergence for asparagus
- Around trees and shrubs spray in late winter or early spring
- Treat hard surfaces between Feb and end May
- Apply to weed-free soil in apple and pear orchards established for at least 1 yr during Feb-Mar

Following crops guidance
- Succeeding crops should not be planted for at least 12 mth after treatment. Do not plant susceptible crops such as vegetables for at least 2 yr

Environmental safety
- Dangerous for the environment
- Very toxic to aquatic organisms
- Harmful to fish or other aquatic life. Do not contaminate surface waters or ditches with chemical or used container
- Do not apply over drains or in drainage channels or gullies
- Avoid run-off when using on paved and similar surfaces
- Do not apply, or allow to drift visibly, on to water
- Do not mix or load within 20 m of well waters unless on an impervious pad to prevent ground water infiltration

Hazard classification and safety precautions
Hazard H03 [1, 2, 4-10]; H04 [2, 3]; H11 [1-10]
Risk phrases R22a [1, 10]; R36 [2-4]; R38 [2, 4]; R40 [1-5, 9, 10]; R48, R50, R53a [1-10]; R68 [6-8]
Operator protection A [1, 3-6, 8, 10]; C [3, 4]; D [10]; H [1, 5, 6, 8, 10]; M [1, 5, 6, 8]; N [1, 6, 8]; U05a [1-9]; U08 [2-10]; U14, U15 [1]; U19a [2-5, 7, 9, 10]; U20a [1, 6-8]; U20b [2-5, 9, 10]
Environmental protection E13b [10]; E13c [3, 4]; E15a [1, 2, 5-9]; E34 [2, 4-6, 8, 9]; E38 [2-9]
Storage and disposal D01, D02, D09a [1-10]; D05 [5, 9, 10]; D06a, D08 [10]; D10b [1, 4, 6, 8]; D10c [5, 9]; D11a [2-4, 7, 10]; D12a [2-9]; D12b [1, 10]
Medical advice M03 [2, 5, 9]; M04 [1]; M05a [3, 6-8, 10]

185 diuron + glyphosate

A non-selective residual herbicide mixture for non-crop and amenity use

Products

1 CleanGuard	Amenity Land	125:135 g/l	SC	12055
2 Touché	Nomix Enviro	217.6:109 g/l	RH	07913
3 Xanadu	Bayer Environ.	125:100 g/l	EW	09943

Uses
- Annual dicotyledons in **amenity trees and shrubs**, **land not intended to bear vegetation** [2, 3]
- Annual grasses in **amenity trees and shrubs**, **land not intended to bear vegetation** [2, 3]
- General weed control in **amenity vegetation**, **hard surfaces**, **natural surfaces not intended to bear vegetation**, **permeable surfaces overlying soil** [1]

SEE SECTION 3 FOR PRODUCTS ALSO REGISTERED

- Perennial dicotyledons in *amenity trees and shrubs, land not intended to bear vegetation* [2, 3]
- Perennial grasses in *amenity trees and shrubs, land not intended to bear vegetation* [2, 3]

Approval information
- Glyphosate included in Annex I under EC Directive 91/414
- Approved for use through ULV applicators [2]

Efficacy guidance
- Best results achieved from treatment when weeds are green and actively growing, usually in early spring. Symptoms may be slow to appear in poor growing conditions
- Perennial grasses are susceptible when tillering and making new rhizome growth, normally when plants have 4-5 new leaves
- Perennial dicotyledons most susceptible if treated at or near flowering but will be severely checked if treated at other times when growing actively
- Most species of germinating weeds are controlled. See label for details of more resistant species
- Weed control may be reduced on heavy or highly organic soils and when weeds are suffering stress in any situation
- At least 6 and preferably 24 h rain-free must follow spraying. Rain soon after treatment may reduce initial weed control
- Apply with manufacturer's specialist equipment [2]

Restrictions
- Maximum number of treatments 1 per yr
- Ornamental trees and shrubs should be established for at least 12 mth and be at least 50 mm tall
- Among ornamental trees and shrubs take care to avoid foliage or the stems of young plants
- Do not allow spray to contact desired plants or crops
- Do not use on tree and shrub nurseries or before planting
- Do not treat ornamental trees and shrubs on Sands or Very Light soils or those that are gravelly or with less than 1% organic matter [2]
- Do not decant, connect directly to Nomix applicator [2]

Crop-specific information
- Treat hard surfaces between Feb and end May

Environmental safety
- Dangerous for the environment
- Toxic to aquatic organisms
- Not to be used on food crops
- Do not apply over drains or in drainage channels, gullies or similar structures
- When treating paved or similar surfaces direct spray at joints and cracks using hand held equipment. If this is not possible ensure overall sprays do not lead to run-off

Hazard classification and safety precautions
Hazard H03, H11
Risk phrases R40, R48, R51, R53a
Operator protection A [1, 3]; H [1-3]; M [2, 3]; U02a [2]; U09a, U19a, U20b [1-3]
Environmental protection E15a, E38 [1-3]; E34 [1, 3]
Consumer protection C01 [1, 3]
Storage and disposal D01, D09a, D11a, D12a [1-3]; D02 [1, 3]; D05 [2]
Medical advice M03 [1, 3]

186   diuron + paraquat

A total herbicide with contact and residual activity

Products

Dexuron	Nomix Enviro	300:100 g/l	SC	07169

Uses
- Annual dicotyledons in *established woody ornamentals, woody nursery stock*
- Annual grasses in *established woody ornamentals, woody nursery stock*
- Perennial dicotyledons in *established woody ornamentals, woody nursery stock*

FOR FULL CONDITIONS OF USE ALWAYS READ THE PRODUCT LABEL

- Perennial grasses in **established woody ornamentals**, **woody nursery stock**
- Total vegetation control in **non-crop areas**, **paths**

Approval information
- Paraquat included in Annex I under EC Directive 91/414
- Approval expiry 1 May 2006 [1]

Efficacy guidance
- Apply to emerged weeds at any time of year. Best results achieved by application in spring or early summer
- Effectiveness not reduced by rain soon after treatment

Restrictions
- Paraquat is subject to the Poisons Rules 1982 and the Poisons Act 1972. See notes in Section 5
- Maximum number of treatments 1 per yr
- Avoid contact of spray with green bark, buds or foliage of desirable trees or shrubs

Crop-specific information
- In nurseries use low dose rate as inter-row spray, not more than once per year

Environmental safety
- Dangerous for the environment
- Toxic to aquatic organisms
- Keep in original container, tightly closed, in a safe place, under lock and key
- Keep livestock out of treated areas for at least 2 wk and until foliage of any poisonous weeds such as ragwort has died and become unpalatable

Hazard classification and safety precautions
Hazard H02, H11
Risk phrases R21, R25, R36, R38, R51, R53a, R68
Operator protection A, C; U02a, U04a, U05a, U09a, U11, U19a, U20b
Environmental protection E06c (24 h); E15a, E34, E38
Storage and disposal D01, D02, D09b, D11a, D12a
Medical advice M04

187   dodemorph

A systemic morpholine fungicide for powdery mildew control

Products

F238	BASF	385 g/l	EC	00206

Uses
- Powdery mildew in **roses**

Efficacy guidance
- Spray roses every 10-14 d during mildew period or every 7 d and at increased dose if cleaning up established infection or if disease pressure high
- Add Citowett when treating rose varieties which are difficult to wet
- Product has negligible effect on *Phytoseiulus* spp being used to control red spider mites

Restrictions
- Do not use on seedling roses
- Do not apply to roses under hot, sunny conditions, particularly under glass, but spray early in the morning or during the evening. Increase the humidity some hours before spraying
- Check tolerance of new varieties before treating rest of crop

Environmental safety
- Dangerous for the environment
- Toxic to aquatic organisms

Hazard classification and safety precautions
Hazard H04, H08, H11
Risk phrases R22a, R22b, R38, R41, R51, R53a, R63
Operator protection A, C; U04a, U05a, U09a, U11, U14, U15, U20b
Environmental protection E15a, E38
Storage and disposal D01, D02, D05, D09a, D10c, D12a
Medical advice M05b

SEE SECTION 3 FOR PRODUCTS ALSO REGISTERED

SECTION 2

188   dodine

A protectant and eradicant guanidine fungicide

Products

1 Greencrop Budburst	Greencrop	450 g/l	SC	11042
2 Radspor FL	Truchem	450 g/l	SC	01685

Uses
- Currant leaf spot in **blackcurrants**
- Scab in **apples**, **pears**

Efficacy guidance
- Apply protective spray on apples and pears at bud-burst and at 10-14 d intervals until late Jun to early Jul
- Apply post-infection spray within 36 h of rain responsible for initiating infection. Where scab already present spray prevents production of spores
- On blackcurrants commence spraying at early grape stage and repeat at 2-3 wk intervals, and at least once after picking

Restrictions
- Do not apply in very cold weather (under 5°C) or under slow drying conditions to pears or dessert apples during bloom or immediately after petal fall
- Do not mix with lime sulphur or tetradifon
- Consult processors before use on crops grown for processing

Crop-specific information
- Latest use: early Jul for culinary apples; pre-blossom for dessert apples and pears

Environmental safety
- Dangerous for the environment
- Very toxic to aquatic organisms

Hazard classification and safety precautions
  Hazard H03, H11
  Risk phrases R21 [1]; R22a, R36, R38, R43, R50, R53a [1, 2]; R37, R41 [2]
  Operator protection A, C; U05a, U08, U19a, U20b [1, 2]; U11 [2]
  Environmental protection E15a, E34 [1, 2]; E38 [2]
  Storage and disposal D01, D02, D09a, D10b [1, 2]; D05 [1]
  Medical advice M03 [2]; M04 [1]

189   epoxiconazole

A systemic, protectant and curative triazole fungicide for use in cereals

See also boscalid + epoxiconazole
carbendazim + epoxiconazole
dimoxystrobin + epoxiconazole

Products

1 Agriguard Epoxiconazole	AgriGuard	125 g/l	SC	09407
2 Epic	BASF	125 g/l	SC	12136
3 Greencrop Martello	Greencrop	125 g/l	SC	12318
4 Landgold Epoxiconazole	Landgold	125 g/l	SC	09821
5 Landgold Epoxiconazole	Teliton	125 g/l	SC	12117
6 Me2 Puddy	Me2	125 g/l	SC	12164
7 Opus	BASF	125 g/l	SC	12057
8 Standon Epoxiconazole	Standon	125 g/l	SC	09517

Uses
- Brown rust in **spring barley**, **winter barley**, **winter wheat** [1-8]; **spring rye**, **spring wheat**, **triticale**, **winter rye** [2-7]
- Eyespot in **winter barley** (reduction), **winter wheat** (reduction) [1-8]
- Fusarium ear blight in **spring wheat**, **winter wheat** [2-7]; **winter wheat** (reduction) [1, 8]
- Net blotch in **spring barley**, **winter barley** [1-8]

FOR FULL CONDITIONS OF USE ALWAYS READ THE PRODUCT LABEL

- Powdery mildew in **spring barley, winter barley, winter wheat** [1-8]; **spring oats, spring rye, spring wheat, triticale, winter oats, winter rye** [2-7]
- Rhynchosporium in **spring barley, winter barley** [1-8]; **spring rye, winter rye** [2-7]
- Septoria in **spring wheat, triticale** [2-7]; **winter wheat** [1-8]
- Sooty moulds in **spring wheat** *(reduction)* [2-7]; **winter wheat** *(reduction)* [1-8]
- Yellow rust in **spring barley, winter barley, winter wheat** [1-8]; **spring rye, spring wheat, triticale, winter rye** [2-7]

Approval information
- Accepted by BBPA for use on malting barley (before ear emergence only)

Efficacy guidance
- Apply at the start of foliar disease attack
- Optimum effect against eyespot achieved by spraying between leaf-sheath erect and second node detectable stages (GS 30-32)
- Best control of ear diseases of wheat obtained by treatment during ear emergence
- For Septoria spray after third node detectable stage (GS 33) when weather favouring disease development has occurred
- Mildew control improved by use of tank mixtures. See label for details

Restrictions
- Maximum total dose equivalent to two full dose treatments
- Product may cause damage to broad-leaved plant species
- Avoid spray drift onto neighbouring crops

Crop-specific information
- Latest use: up to and including flowering just complete (GS 69) in wheat, rye, triticale; up to and including emergence of ear just complete (GS 59) in barley, oats

Environmental safety
- Dangerous for the environment
- Very toxic to aquatic organisms
- LERAP Category B

Hazard classification and safety precautions
Hazard H03 [2-8]; H11 [1-8]
Risk phrases R38, R40, R50, R62, R63 [2-8]; R51 [1]; R53a [1-8]
Operator protection A; U05a [1-8]; U19a [3]; U20a [1]; U20b [2-8]
Environmental protection E15a, E16a [1-8]; E16b [1, 8]; E38 [2, 6, 7]
Storage and disposal D01, D02, D05, D09a [1-8]; D08 [4, 5]; D10b [1, 3-5, 8]; D10c, D12a [2, 6, 7]
Medical advice M05a [2, 4-7]; M05b [8]

190   epoxiconazole + fenpropimorph

A systemic, protectant and curative fungicide mixture for cereals

Products

1	Eclipse	BASF	84:250 g/l	SE	11731
2	Greencrop Galore	Greencrop	84:250 g/l	SE	09561
3	Opus Team	BASF	84:250 g/l	SE	11759
4	Standon Epoxifen	Standon	84:250 g/l	SE	08972

Uses
- Brown rust in **spring barley, winter barley, winter wheat** [1-4]; **spring rye, spring wheat, triticale, winter rye** [1, 3]
- Eyespot in **winter barley** *(reduction)*, **winter wheat** *(reduction)* [1-4]
- Fusarium ear blight in **spring wheat, winter wheat** [1, 3]; **winter wheat** *(reduction)* [2, 4]
- Net blotch in **spring barley, winter barley** [1-4]
- Powdery mildew in **spring barley, winter barley, winter wheat** [1-4]; **spring oats, spring rye, spring wheat, triticale, winter oats, winter rye** [1, 3]
- Rhynchosporium in **spring barley, winter barley** [1-4]; **spring rye, winter rye** [1, 3]
- Septoria in **spring wheat, triticale** [1, 3]; **winter wheat** [1-4]
- Sooty moulds in **spring wheat** *(reduction)* [1, 3]; **winter wheat** *(reduction)* [1-3]
- Yellow rust in **spring barley, winter barley, winter wheat** [1-4]; **spring rye, spring wheat, triticale, winter rye** [1, 3]

SEE SECTION 3 FOR PRODUCTS ALSO REGISTERED

Approval information
- Accepted by BBPA for use on malting barley

Efficacy guidance
- Apply at the start of foliar disease attack
- Optimum effect against eyespot achieved by spraying between leaf-sheath erect and second node detectable stages (GS 30-32)
- Best control of ear diseases obtained by treatment during ear emergence
- For Septoria spray after third node detectable stage (GS 33) when weather favouring disease development has occurred

Restrictions
- Maximum total dose equivalent to two full dose treatments
- Product may cause damage to broad-leaved plant species

Crop-specific information
- Latest use: up to and including flowering just complete (GS 69) in wheat, rye, triticale; up to and including emergence of ear just complete (GS 59) in barley, oats

Environmental safety
- Dangerous for the environment
- Toxic to aquatic organisms
- LERAP Category B
- Avoid spray drift onto neighbouring crops

Hazard classification and safety precautions
Hazard H03, H11
Risk phrases R20, R43 [2]; R38 [4]; R40, R62, R63 [1, 3, 4]; R51, R53a [1-4]
Operator protection A [1-4]; H [1, 3]; U05a, U20b [1-4]; U14 [2]; U19a [2, 4]
Environmental protection E15a, E16a [1-4]; E16b [2, 4]; E38 [1, 3]
Storage and disposal D01, D02, D05, D09a [1-4]; D10a [2]; D10b [4]; D10c, D12a [1, 3]
Medical advice M05a [1, 3]; M05b [4]

191   epoxiconazole + fenpropimorph + kresoxim-methyl

A protectant, systemic and curative fungicide mixture for cereals

Products

1	Asana	BASF	125:150:125 g/l	SE	11934
2	Mantra	BASF	125:150:125 g/l	SE	11728
3	Standon Kresoxim Super	Standon	125:150:125 g/l	SE	09794

Uses
- Brown rust in *spring barley*, *winter barley*, *winter wheat* [1-3]; *spring rye*, *spring wheat*, *triticale*, *winter rye* [1, 2]
- Eyespot in *spring rye* (reduction), *triticale* (reduction), *winter oats* (reduction), *winter rye* (reduction) [1, 2]; *winter barley* (reduction), *winter wheat* (reduction) [1-3]
- Fusarium ear blight in *spring wheat* (reduction) [1, 2]; *winter wheat* (reduction) [1-3]
- Net blotch in *spring barley*, *winter barley* [1-3]
- Powdery mildew in *spring barley*, *winter barley* [1-3]; *spring oats*, *spring rye*, *triticale*, *winter oats*, *winter rye* [1, 2]
- Rhynchosporium in *spring barley*, *winter barley* [1-3]; *spring rye*, *winter rye* [1, 2]
- Septoria diseases in *spring wheat*, *triticale* [1, 2]; *winter wheat* [1-3]
- Sooty moulds in *winter wheat* (reduction) [1-3]
- Yellow rust in *spring barley*, *winter barley*, *winter wheat* [1-3]; *spring rye*, *spring wheat*, *triticale*, *winter rye* [1, 2]

Approval information
- Kresoxim-methyl included in Annex I under EC Directive 91/414
- Accepted by BBPA for use on malting barley (before ear emergence only)

Efficacy guidance
- For best results spray at the start of foliar disease attack and repeat if infection conditions persist
- Optimum effect against eyespot obtained by treatment between leaf sheaths erect and first node detectable stages (GS 30-32)
- For protection against ear diseases apply during ear emergence

FOR FULL CONDITIONS OF USE ALWAYS READ THE PRODUCT LABEL

- Kresoxim-methyl is a member of the QoI cross resistance group. Product should be used preventatively and not relied on for its curative potential
- Use product as part of an Integrated Crop Management strategy incorporating other methods of control, including where appropriate other fungicides with a different mode of action. Do not apply more than two foliar applications of QoI containing products to any cereal crop
- There is a significant risk of widespread resistance occurring in *Septoria tritici* populations in UK. Failure to follow resistance management action may result in reduced levels of disease control
- Strains of barley powdery mildew resistant to QoIs are common in the UK

Restrictions
- Maximum total dose equivalent to two full dose treatments

Crop-specific information
- Latest use: mid flowering (GS 65) for wheat, rye, triticale; completion of ear emergence (GS 59) for barley, oats

Environmental safety
- Dangerous for the environment
- Very toxic to aquatic organisms
- LERAP Category B
- Avoid spray drift onto neighbouring crops. Product may damage broad-leaved species

Hazard classification and safety precautions
Hazard H03, H11
Risk phrases R40, R50, R53a, R62, R63 [1-3]; R43 [3]
Operator protection A [1-3]; H [1, 2]; U05a, U14, U20b
Environmental protection E15a, E16a [1-3]; E16b [3]; E38 [1, 2]
Storage and disposal D01, D02, D05, D09a [1-3]; D08, D10b [3]; D10c, D12a [1, 2]
Medical advice M05a [1, 2]; M05b [3]

192   epoxiconazole + fenpropimorph + pyraclostrobin

A protectant and systemic fungicide mixture for cereals

Products

Diamant	BASF	43:214:114 g/l	SE	11557

Uses
- Brown rust in *spring barley*, *spring wheat*, *winter barley*, *winter wheat*
- Crown rust in *spring oats*, *winter oats*
- Fusarium ear blight in *spring wheat* (reduction), *winter wheat* (reduction)
- Net blotch in *spring barley*, *winter barley*
- Powdery mildew in *spring barley*, *spring oats*, *winter barley*, *winter oats*
- Rhynchosporium in *spring barley*, *winter barley*
- Septoria diseases in *spring wheat*, *winter wheat*
- Yellow rust in *spring barley*, *spring wheat*, *winter barley*, *winter wheat*

Approval information
- Pyraclostrobin included in Annex I under EC Directive 91/414
- Accepted by BBPA for use on malting barley (before ear emergence only)

Efficacy guidance
- For best results spray at the start of foliar disease attack and repeat if infection conditions persist
- For protection against ear diseases apply during ear emergence
- Pyraclostrobin is a member of the QoI cross resistance group. Product should be used preventatively and not relied on for its curative potential
- Use product as part of an Integrated Crop Management strategy incorporating other methods of control, including where appropriate other fungicides with a different mode of action. Do not apply more than two foliar applications of QoI containing products to any cereal crop
- There is a significant risk of widespread resistance occurring in *Septoria tritici* populations in UK. Failure to follow resistance management action may result in reduced levels of disease control
- Strains of barley powdery mildew resistant to QoIs are common in the UK

Restrictions
- Maximum number of treatments 2 per crop

SEE SECTION 3 FOR PRODUCTS ALSO REGISTERED

SECTION 2

Crop-specific information
- Latest use: up to and including ear emergence just complete (GS 59) for barley, oats; up to and including flowering just complete (GS 69) for wheat

Environmental safety
- Dangerous for the environment
- Very toxic to aquatic organisms
- Dangerous to fish or other aquatic life. Do not contaminate surface waters or ditches with chemical or used container
- LERAP Category B
- Avoid spray drift onto neighbouring crops. Product may damage broad-leaved species

Hazard classification and safety precautions
    Hazard H03, H11
    Risk phrases R20, R22a, R38, R40, R50, R53a, R63
    Operator protection A; U05a, U14, U20b
    Environmental protection E13b, E16a, E16b, E34, E38
    Storage and disposal D01, D02, D05, D09a, D10c, D12a
    Medical advice M03, M05a

193   epoxiconazole + kresoxim-methyl

A protectant, systemic and curative fungicide mixture for cereals

Products

1	Landmark	BASF	125:125 g/l	SC	11730
2	Me2 KME	Me2	125:125 g/l	SC	09594
3	Standon Kresoxim-Epoxiconazole	Standon	125:125 g/l	SC	09281

Uses
- Brown rust in *spring barley*, *winter barley*, *winter wheat* [1-3]; *spring rye*, *spring wheat*, *triticale*, *winter rye* [1]
- Eyespot in *spring rye* (reduction), *triticale* (reduction), *winter oats* (reduction), *winter rye* (reduction) [1]; *winter barley* (reduction), *winter wheat* (reduction) [1-3]
- Fusarium ear blight in *spring wheat* (reduction) [1]; *winter wheat* (reduction) [1-3]
- Net blotch in *spring barley*, *winter barley* [1-3]
- Powdery mildew in *spring barley*, *winter barley* [1-3]; *spring oats*, *spring rye*, *triticale*, *winter oats*, *winter rye* [1]
- Rhynchosporium in *spring barley*, *winter barley* [1-3]; *spring rye*, *winter rye* [1]
- Septoria diseases in *spring wheat*, *triticale* [1]; *winter wheat* [1-3]
- Sooty moulds in *winter wheat* (reduction) [1-3]
- Yellow rust in *spring barley*, *winter barley*, *winter wheat* [1-3]; *spring rye*, *spring wheat*, *triticale*, *winter rye* [1]

Approval information
- Kresoxim-methyl included in Annex I under EC Directive 91/414
- Accepted by BBPA for use on malting barley (before ear emergence only)

Efficacy guidance
- For best results spray at the start of foliar disease attack and repeat if infection conditions persist
- Optimum effect against eyespot obtained by treatment between leaf sheaths erect and first node detectable stages (GS 30-32)
- For protection against ear diseases apply during ear emergence
- Kresoxim-methyl is a member of the QoI cross resistance group. Product should be used preventatively and not relied on for its curative potential
- Use product as part of an Integrated Crop Management strategy incorporating other methods of control, including where appropriate other fungicides with a different mode of action. Do not apply more than two foliar applications of QoI containing products to any cereal crop
- There is a significant risk of widespread resistance occurring in *Septoria tritici* populations in UK. Failure to follow resistance management action may result in reduced levels of disease control
- Strains of barley powdery mildew resistant to QoIs are common in the UK

FOR FULL CONDITIONS OF USE ALWAYS READ THE PRODUCT LABEL

Restrictions
- Maximum total dose equivalent to two full dose treatments

Crop-specific information
- Latest use: mid flowering (GS 65) for wheat, rye, triticale; completion of ear emergence (GS 59) for barley, oats

Environmental safety
- Dangerous for the environment
- Very toxic to aquatic organisms
- LERAP Category B
- Avoid spray drift onto neighbouring crops. Product may damage broad-leaved species

Hazard classification and safety precautions
>**Hazard** H03, H11
>**Risk phrases** R40, R50, R53a, R62, R63 [1-3]; R43 [2, 3]
>**Operator protection** A; U05a, U14, U20b
>**Environmental protection** E15a, E16a [1-3]; E16b [2, 3]; E38 [1]
>**Storage and disposal** D01, D02, D05, D09a [1-3]; D10b [2, 3]; D10c, D12a [1]
>**Medical advice** M05a [1]; M05b [2, 3]

194   epoxiconazole + kresoxim-methyl + pyraclostrobin

A protectant, systemic and curative fungicide mixture for cereals

Products

Opponent	BASF	50:67:133 g/l	SE	10877

Uses
- Brown rust in *spring barley*, *spring wheat*, *winter barley*, *winter wheat*
- Crown rust in *spring oats*, *winter oats*
- Net blotch in *spring barley*, *winter barley*
- Rhynchosporium in *spring barley*, *winter barley*
- Septoria diseases in *spring wheat*, *winter wheat*
- Yellow rust in *spring barley*, *spring wheat*, *winter barley*, *winter wheat*

Approval information
- Kresoxim-methyl and pyraclostrobin included in Annex I under EC Directive 91/414
- Accepted by BBPA for use on malting barley (before ear emergence only)

Efficacy guidance
- For best results apply at the start of foliar disease attack
- Best results on Septoria leaf blotch when treated in the latent phase
- Good reduction of Fusarium ear blight can be obtained from treatment during flowering
- Yield response may be obtained in the absence of visual disease symptoms
- Kresoxim-methyl and pyraclostrobin are members of the QoI cross resistance group. Product should be used preventatively and not relied on for its curative potential
- Use product as part of an Integrated Crop Management strategy incorporating other methods of control, including where appropriate other fungicides with a different mode of action. Do not apply more than two foliar applications of QoI containing products to any cereal crop
- There is a significant risk of widespread resistance occurring in *Septoria tritici* populations in UK. Failure to follow resistance management action may result in reduced levels of disease control

Restrictions
- Maximum total dose equivalent to two full dose treatments

Crop-specific information
- Latest use: before grain watery ripe (GS 71) for wheat; up to and including emergence of ear just complete (GS 59) for barley and oats

Environmental safety
- Dangerous for the environment
- Very toxic to aquatic organisms
- Dangerous to fish or other aquatic life. Do not contaminate surface waters or ditches with chemical or used container
- LERAP Category B

SEE SECTION 3 FOR PRODUCTS ALSO REGISTERED

SECTION 2

Hazard classification and safety precautions
> **Hazard** H03, H11
> **Risk phrases** R20, R22a, R40, R50, R53a
> **Operator protection** A; U05a, U14, U20b
> **Environmental protection** E13b, E16a, E16b, E34, E38
> **Storage and disposal** D01, D02, D05, D09a, D10c, D12a
> **Medical advice** M05a

195 epoxiconazole + pyraclostrobin

A protectant, systemic and curative fungicide mixture for cereals

Products

1 Euro	Me2	50:133 g/l	SE	11511
2 Opera	BASF	50:133 g/l	SE	12167

Uses
- Brown rust in *spring barley*, *spring wheat*, *winter barley*, *winter wheat* [1, 2]
- Cercospora leaf spot in *sugar beet* [2]
- Crown rust in *spring oats*, *winter oats* [1, 2]
- Net blotch in *spring barley*, *winter barley* [1, 2]
- Powdery mildew in *sugar beet* [2]
- Ramularia leaf spots in *sugar beet* [2]
- Rhynchosporium in *spring barley*, *winter barley* [1, 2]
- Rust in *sugar beet* [2]
- Septoria diseases in *spring wheat*, *winter wheat* [1, 2]
- Yellow rust in *spring barley*, *spring wheat*, *winter barley*, *winter wheat* [1, 2]

Approval information
- Accepted by BBPA for use on malting barley (before ear emergence only)

Efficacy guidance
- For best results apply at the start of foliar disease attack
- Good reduction of Fusarium ear blight can be obtained from treatment during flowering
- Yield response may be obtained in the absence of visual disease symptoms
- Pyraclostrobin is a member of the QoI cross resistance group. Product should be used preventatively and not relied on for its curative potential
- Use product as part of an Integrated Crop Management strategy incorporating other methods of control, including where appropriate other fungicides with a different mode of action. Do not apply more than two foliar applications of QoI containing products to any cereal crop
- There is a significant risk of widespread resistance occurring in *Septoria tritici* populations in UK. Failure to follow resistance management action may result in reduced levels of disease control

Restrictions
- Maximum number of treatments 2 per crop
- Do not use on sugar beet crops grown for seed

Crop-specific information
- Latest use: before grain watery ripe (GS 71) for wheat; up to and including emergence of ear just complete (GS 59) for barley and oats
- HI 6 wk for sugar beet

Environmental safety
- Dangerous for the environment
- Very toxic to aquatic organisms
- LERAP Category B

Hazard classification and safety precautions
> **Hazard** H03, H11
> **Risk phrases** R20, R22a, R38, R40, R50, R53a
> **Operator protection** A; U05a, U14, U20b
> **Environmental protection** E15a, E16a, E16b, E34, E38
> **Storage and disposal** D01, D02, D05, D09a, D10c, D12a
> **Medical advice** M03, M05a

FOR FULL CONDITIONS OF USE ALWAYS READ THE PRODUCT LABEL

196   esfenvalerate

A contact and ingested pyrethroid insecticide

Products

1	Greencrop Cajole	Greencrop	25 g/l	EC	11402
2	Sumi-Alpha	BASF	25 g/l	EC	10401

Uses
- Aphids in *winter barley*, *winter wheat*

Approval information
- Esfenvalerate included in Annex I under EC Directive 91/414
- Accepted by BBPA for use on malting barley
- Approval expiry 1 Aug 2006 [1]

Efficacy guidance
- Crops at high risk (e.g. after grass or in areas with history of BYDV) should be treated when aphids first seen or by mid-Oct. Otherwise treat in late Oct-early Nov
- High risk crops will need a second treatment
- Product also recommended between onset of flowering and milky ripe stages (GS 61-73) for control of summer cereal aphids

Restrictions
- Maximum number of treatments 3 per crop of which 2 may be in autumn
- Do not use if another pyrethroid or dimethoate has been applied to crop after start of ear emergence (GS 51)

Crop-specific information
- Latest use: 31 Mar in yr of harvest (winter use)
- HI 20 d (summer use)

Environmental safety
- Dangerous for the environment
- Very toxic to aquatic organisms
- Extremely dangerous to bees. Do not apply to crops in flower or to those in which bees are actively foraging. Do not apply when flowering weeds are present
- High risk to non-target insects or other arthropods. Do not spray within 6 m of the field boundary
- LERAP Category A
- Store product in dark away from direct sunlight

Hazard classification and safety precautions
Hazard H03, H11
Risk phrases R20, R22a, R38, R41, R43, R50, R53a
Operator protection A, C, H; U04a, U05a, U08, U11, U19a, U20b [1, 2]; U14 [2]
Environmental protection E12c, E15a, E16c, E16d, E22a, E34 [1, 2]; E38 [2]
Storage and disposal D01, D02, D09a [1, 2]; D05, D07, D10b [1]; D08, D10c [2]
Medical advice M03

197   ethofumesate

A benzofuran herbicide for grass weed control in various crops

See also bromoxynil + ethofumesate + ioxynil
chloridazon + ethofumesate
desmedipham + ethofumesate + phenmedipham

Products

1	Agriguard Ethofumesate 200	AgriGuard	200 g/l	EC	10698
2	Agriguard Ethofumesate Flo	AgriGuard	500 g/l	SC	09478
3	Ethosat 500	Makhteshim	500 g/l	SC	11501
4	Kubist Flo	Nufarm UK	500 g/l	SC	11167
5	Nortron Flo	Bayer CropScience	500 g/l	SC	08154

SEE SECTION 3 FOR PRODUCTS ALSO REGISTERED

SECTION 2

Uses

- Annual dicotyledons in **bulb onions** (off-label) [3]; **fodder beet, mangels, red beet, sugar beet** [1-5]; **garlic** (off-label), **horseradish** (off-label), **strawberries** (off-label) [3, 5]; **onions** (off-label) [5]
- Annual grasses in **garlic** (off-label), **horseradish** (off-label), **onions** (off-label), **strawberries** (off-label) [5]; **grass seed crops, permanent pasture, rotational grassland** [1]
- Annual meadow grass in **bulb onions** (off-label), **horseradish** (off-label), **strawberries** (off-label) [3]; **fodder beet, mangels, red beet, sugar beet** [1-5]; **garlic** (off-label) [3, 4]; **onions** (off-label) [4]
- Blackgrass in **bulb onions** (off-label), **garlic** (off-label), **horseradish** (off-label), **strawberries** (off-label) [3]; **fodder beet, mangels, red beet, sugar beet** [1-5]; **grass seed crops, permanent pasture, rotational grassland** [1]
- Chickweed in **garlic** (off-label), **onions** (off-label) [4]; **grass seed crops, permanent pasture, rotational grassland** [1]
- Cleavers in **grass seed crops, permanent pasture, rotational grassland** [1]
- Clover in **strawberries** (off-label) [4]
- Fat hen in **poppies** (off-label - for morphine production) [5]
- General weed control in **horseradish** (off-label) [4]
- Grass weeds in **poppies** (off-label - for morphine production) [5]; **strawberries** (off-label) [4]
- Mayweeds in **poppies** (off-label - for morphine production) [5]

Specific Off-Label Approvals (SOLAs)

- **bulb onions, garlic, horseradish, strawberries** (OLA 040153) Dec 2008 [3]
- **garlic, horseradish, onions, strawberries** (OLA 032082) Dec 2008 [4]
- **garlic, horseradish, onions, strawberries** (OLA 040520) Dec 2008 [5]
- **poppies** (for morphine production) (OLA 040470) Dec 2008 [5]

Approval information

- Ethofumesate included in Annex I under EC Directive 91/414

Efficacy guidance

- Most products may be applied pre- or post-emergence of crop or weeds but some restricted to pre-emergence or post-emergence use only. Check label
- Volunteer cereals not well controlled pre-emergence, weed grasses should be sprayed before fully tillered
- Grass crops may be sprayed during rain or when wet. Not recommended in very dry conditions or prolonged frost

Restrictions

- Maximum number of treatments 1 pre- plus 1 or 2 post-emergence per crop or yr for beet crops and grass seed crops; 1 per crop for horseradish, strawberries; 2 per yr for amenity turf and established grassland
- Do not use on Sands or Heavy soils, Very Light soils containing a high percentage of stones, or soils with more than 5-10% organic matter (percentage varies according to label)
- Do not use on swards reseeded without ploughing
- Clovers will be killed or severely checked
- Do not graze or cut grass for 14 d after, or roll less than 7 d before or after spraying

Crop-specific information

- Latest use: before crops meet across rows for beet crops and mangels; 14 d before cutting or grazing for grass leys; pre-emergence for horseradish
- HI strawberries 11 wk; garlic, onions 3 mth [5]
- Apply in beet crops in tank mixes with other pre- or post-emergence herbicides. Recommendations vary for different mixtures. See label for details
- Safe timing on beet crops varies with other ingredient of tank mix. See label for details
- In grass crops apply to moist soil as soon as possible after sowing or post-emergence when crop in active growth, normally mid-Oct to mid-Dec. See label for details
- May be used in Italian, hybrid and perennial ryegrass, timothy, cocksfoot, meadow fescue and tall fescue. Apply pre-emergence to autumn-sown leys, post-emergence after 2-3 leaf stage. See label for details

Following crops guidance

- Any crop may be sown 3 mth after application of mixtures in beet crops following ploughing, 5 mth after application in grass crops

FOR FULL CONDITIONS OF USE ALWAYS READ THE PRODUCT LABEL

Environmental safety
- Dangerous for the environment
- Toxic to aquatic organisms
- Do not empty into drains

Hazard classification and safety precautions
    Hazard H03, H08 [1]; H11 [1-5]
    Risk phrases R20, R21, R22b, R36, R38 [1]; R51, R53a [3-5]
    Operator protection A, C, H [1]; U05a [1]; U08 [1-3]; U09a [4, 5]; U19a, U20a [1-5]
    Environmental protection E15a [1-4]; E19b [3]; E34 [1, 3]; E38 [4, 5]
    Storage and disposal D01, D02 [1]; D05, D09a [1-5]; D10b [1-4]; D12a [3-5]
    Medical advice M03 [1]

198   ethofumesate + metamitron

A contact and residual herbicide mixture for beet crops

Products

1	Galahad	Bayer CropScience	100:400 g/l	SC	10727
2	Goltix Super	Makhteshim	150:350 g/l	SC	12216
3	Torero	Makhteshim	150:350 g/l	SC	11158

Uses
- Annual dicotyledons in *fodder beet*, *mangels*, *sugar beet*
- Annual meadow grass in *fodder beet*, *mangels*, *sugar beet*

Approval information
- Ethofumesate included in Annex I under EC Directive 91/414

Efficacy guidance
- Best results obtained from a series of treatments applied as an overall fine spray commencing when earliest germinating weeds are no larger than fully expanded cotyledon and the majority of the crop at fully expanded cotyledon
- Apply subsequent sprays as each new flush of weeds reaches early cotyledon and continue until weed emergence ceases (maximum 3 sprays)
- Product may be used on all soil types but residual activity may be reduced on those with more than 5% organic matter

Restrictions
- Maximum total dose equivalent to three full dose treatments

Crop-specific information
- Latest use: before crop leaves meet between rows
- Crop tolerance may be reduced by stress caused by growing conditions, effects of pests, disease or other pesticides, nutrient deficiency etc

Following crops guidance
- Beet crops may be sown at any time after treatment. Any other crop may be sown after mouldboard ploughing to 15 cm and a minimum interval of 3 mth after treatment

Environmental safety
- Dangerous for the environment
- Very toxic to aquatic organisms
- Harmful to fish or other aquatic life. Do not contaminate surface waters or ditches with chemical or used container
- Do not empty into drains

Hazard classification and safety precautions
    Hazard H03 [1]; H11 [1-3]
    Risk phrases R22a, R43, R50 [1]; R51 [2, 3]; R53a [1-3]
    Operator protection A, H [1]; U05a, U08, U19a [1-3]; U14, U15, U20a [2, 3]; U20b [1]
    Environmental protection E13c, E34 [1-3]; E19b [2, 3]; E38 [1]
    Storage and disposal D01, D02, D05, D09a, D10b, D12a
    Medical advice M03 [1]

SECTION 2

SEE SECTION 3 FOR PRODUCTS ALSO REGISTERED

199 ethofumesate + phenmedipham

A contact and residual herbicide for use in beet crops

Products

1	Fenlander 2	Makhteshim	94:97 g/l	EC	12282
2	Magic Tandem	Bayer CropScience	190:200 g/l	SC	11814
3	Powertwin	Makhteshim	200:200 g/l	SC	11187
4	Powertwin	Makhteshim	200:200 g/l	SC	12373
5	Thunder	Makhteshim	200:200 g/l	SC	11186
6	Thunder	Makhteshim	200:200 g/l	SC	12464
7	Twin	Makhteshim	94:97 g/l	EC	11172

Uses

- Annual dicotyledons in *fodder beet*, *mangels*, *sugar beet*
- Annual meadow grass in *fodder beet*, *mangels*, *sugar beet*
- Blackgrass in *fodder beet*, *mangels*, *sugar beet*

Approval information

- Ethofumesate included in Annex I under EC Directive 91/414
- Approval expiry 31 Oct 2006 [3]
- Approval expiry 31 Dec 2006 [5]

Efficacy guidance

- Best results achieved by repeat applications to cotyledon stage weeds. Larger susceptible weeds not killed by first treatment usually checked and controlled by second application
- Apply on all soil types at 7-10 d intervals
- On soils with more than 5-10% organic matter residual activity may be reduced

Restrictions

- Maximum number of treatments normally 3 per crop or maximum total dose equivalent to three full dose treatments, or less - see labels for details
- Product must be used in conjunction with specified adjuvant. See label [2, 3]
- Do not spray wet foliage or if rain imminent
- Spray must be applied low volume. See label for details
- Spray in evening if daytime temperatures above 21°C expected
- Avoid or delay treatment if frost expected within 7 d
- Avoid or delay treating crops under stress from wind damage, manganese or lime deficiency, pest or disease attack etc

Crop-specific information

- Latest use: before crop foliage meets in the rows
- Check from which recovery may not be complete may occur if treatment made during conditions of sharp diurnal temperature fluctuation

Following crops guidance

- Beet crops may be sown at any time after treatment. Any other crop may be sown after mouldboard ploughing to 15 cm and a minimum interval of 3 mth after treatment

Environmental safety

- Dangerous for the environment
- Toxic (harmful [2]) to aquatic organisms
- Harmful to fish or other aquatic life. Do not contaminate surface waters or ditches with chemical or used container [2]
- Do not empty into drains
- Extra care necessary to avoid drift because product is recommended for use as a fine spray

Hazard classification and safety precautions

Hazard H03 [1, 7]; H04 [3-6]; H11 [1, 3-7]

Risk phrases R37, R40 [1, 7]; R43, R51 [1, 3-7]; R52 [2]; R53a [1-7]

Operator protection A [1, 3-7]; H [3-6]; U05a, U08, U19a, U20a [1-7]; U14 [3-6]

Environmental protection E13c [2]; E15a, E19b [1, 3-7]; E34 [1-7]

Storage and disposal D01, D02, D09a, D10b, D12a [1-7]; D05 [2-4]

Medical advice M05a [1, 3-7]

FOR FULL CONDITIONS OF USE ALWAYS READ THE PRODUCT LABEL

200   ethoprophos

An organophosphorus nematicide and insecticide

Products

Mocap 10G	Bayer CropScience	10% w/w	GR	10003

Uses
- Potato cyst nematode in *potatoes*
- Wireworm in *potatoes*

Efficacy guidance
- Broadcast shortly before or during final soil preparation with suitable fertilizer spreader and incorporate immediately to 10-15 cm. Deeper incorporation may reduce efficacy. See label for details
- Treatment can be applied on all soil types. Control of pests reduced on organic soils
- Effectiveness dependent on soil moisture. Drought after application may reduce control
- Where pre-planting nematode populations are high, loss of crop vigour and subsequent loss of yield may occur in spite of treatment

Restrictions
- This product contains an anticholinesterase organophosphorous compound. Do not use if under medical advice not to work with such compounds
- Maximum number of treatments 1 per crop
- Do not harvest crops for human or animal consumption for at least 8 wk after application
- Vehicles with a closed cab must be used when applying and incorporating
- Do not apply by hand or hand held equipment

Crop-specific information
- Latest use: pre-planting of crop
- Field scale application must only be through positive displacement type specialist granule applicators or dual purpose fertilizer applicators with microgranule setting

Environmental safety
- Dangerous for the environment
- Toxic to aquatic organisms
- Dangerous to livestock. Keep all livestock out of treated areas/away from treated water for at least 13 wk. Bury or remove spillages
- Dangerous to game, wild birds and animals
- Product supplied in returnable/refillable containers. Follow instructions on label

Hazard classification and safety precautions
Hazard H02, H11
Risk phrases R22a, R23, R43, R51, R53a
Operator protection A, D, H; U02a, U04a, U05a, U07, U08, U13, U14, U19a, U20a
Environmental protection E06b (13 wk); E10a, E13b, E34, E36, E38
Consumer protection C02 (8 wk)
Storage and disposal D01, D02, D09a, D12a, D14
Medical advice M01, M03, M04

201   ethylene (commodity substance)

A gas used for fruit ripening and potato storage

Products

ethylene	various	-	GA	

Uses
- Fruit ripening in *fruit crops* *(in store)*
- Sprout suppression in *potatoes* *(in store)*

Approval information
- Ethylene is a supported substance in the fourth stage of the EC Review Programme. If ethylene is included in Annex I under Directive 91/414, any commodity chemical approvals will no longer apply

SEE SECTION 3 FOR PRODUCTS ALSO REGISTERED

Efficacy guidance
- For use in stored fruit or potatoes after harvest

Restrictions
- Handling and release of ethylene must only be undertaken by operators suitably trained and competent to carry out the work
- Operators must vacate treated areas immediately after ethylene introduction
- Unprotected persons must be excluded from the treated areas until atmospheres have been thoroughly ventilated for 15 minutes minimum before re-entry
- Ambient atmospheric ethylene concentration must not exceed 1000 ppm. Suitable self-contained breathing apparatus must be worn in atmospheres containing ethylene in excess of 1000 ppm
- A minimum 3 d post treatment period is required before removal of treated crop from storage
- Ethylene treatment must only be undertaken in fully enclosed storage areas that are air tight with appropriate air circulation and venting facilities

202   etridiazole

A protective thiadiazole fungicide for soil or compost incorporation

Products

1	Standon Etridiazole 35	Standon	35% w/w	WP	08778
2	Terrazole 35 WP	Scotts	35% w/w	WP	10468

Uses
- Damping off and foot rot in **cabbages, cauliflowers** *(seeds and seedlings)*, **celery** *(seeds and seedlings)*, **cucumbers** *(seeds and seedlings)*, **mustard and cress, ornamental plant production** *(seeds and seedlings)*, **outdoor tomatoes** *(seeds and seedlings)*, **protected tomatoes** *(seeds and seedlings)* [1, 2]
- Phytophthora in **watercress** *(off-label)* [1]
- Phytophthora root rot in **cabbages, cauliflowers** *(transplants)*, **celery** *(transplants)*, **cucumbers** *(transplants)*, **hardy ornamental nursery stock, ornamental plant production** *(rooted cuttings and transplants)*, **outdoor tomatoes** *(transplants)*, **protected tomatoes** *(transplants)*, **tulips** [1, 2]
- Phytophthora wilt in **cabbages, cauliflowers** *(transplants)*, **celery** *(transplants)*, **cucumbers** *(transplants)*, **hardy ornamental nursery stock, ornamental plant production** *(rooted cuttings and transplants)*, **outdoor tomatoes** *(transplants)*, **protected tomatoes** *(transplants)*, **tulips** [1, 2]
- Pythium in **watercress** *(off-label)* [1]
- Root diseases in **inert substrate tomatoes** *(off-label)* [1]

Specific Off-Label Approvals (SOLAs)
- *inert substrate tomatoes, watercress* (OLA 020409) Dec 2008 [1]

Efficacy guidance
- Best results obtained when used as prophylactic treatment
- Compost/soil incorporation more effective than when applied as drench
- After drenching wash spray residue from crop foliage
- When treating compost for blocking reduce dose by 50% to allow for compaction
- Uniform distribution and thorough incorporation essential for best results. Dilute powder with dry sand to facilitate handling
- Treat compost as soon as possible before use
- Do not apply to wet soil

Restrictions
- Maximum number of treatments 1 per crop or yr for compost incorporation. No restrictions on drenches
- Do not use on *Escallonia, Pyracantha, Gloxinia* spp., pansies or lettuces
- Do not drench seedlings until well established
- When using compost containing more than 20% inert material or a high proportion of young peat, check crop safety before widespread use
- Test on small numbers of plants in advance when treating subjects of unknown susceptibility or using compost with more than 20% inert material

FOR FULL CONDITIONS OF USE ALWAYS READ THE PRODUCT LABEL

Crop-specific information
- Latest use: normally before sowing or transplanting. 24 h after seeding watercress
- HI tomatoes, cucumbers, mustard and cress 3 d; inert substrate tomatoes 24 h
- Germination of lettuce in previously treated soil may be impaired

Environmental safety
- Dangerous for the environment
- Very toxic to aquatic organisms

Hazard classification and safety precautions
 Hazard H03, H11
 Risk phrases R22a, R43 [2]; R36, R40, R50, R53a [1, 2]; R38 [1]
 Operator protection A, C; U02a, U04a, U05a, U08, U20b [1, 2]; U11, U14, U15, U19a [2]
 Environmental protection E15a
 Storage and disposal D01, D02, D09a, D11a
 Medical advice M04 [2]

203 famoxadone

A strobilurin fungicide available only in mixtures

See also cymoxanil + famoxadone

204 famoxadone + flusilazole

A contact, preventative and curative fungicide mixture for cereals

Products

1	Charisma	DuPont	100:106.7 g/l	EC	10415
2	Medley	DuPont	100:106.7 g/l	EC	10933

Uses
- Brown rust in **spring barley, winter barley, winter wheat**
- Light leaf spot in **spring oilseed rape, winter oilseed rape**
- Net blotch in **spring barley, winter barley**
- Phoma in **spring oilseed rape** (suppression), **winter oilseed rape** (suppression)
- Rhynchosporium in **spring barley, winter barley**
- Septoria diseases in **winter wheat**
- Yellow rust in **spring barley, winter barley, winter wheat**

Approval information
- Famoxadone included in Annex I under EC Directive 91/414
- Accepted by BBPA for use on malting barley

Efficacy guidance
- Best results obtained from application at early stage of disease development before infection spreads to new growth
- Famoxadone is a member of the QoI cross resistance group. Product should be used preventatively and not relied on for its curative potential
- Use product as part of an Integrated Crop Management strategy incorporating other methods of control, including where appropriate other fungicides with a different mode of action. Do not apply more than two foliar applications of QoI containing products to any cereal crop
- There is a significant risk of widespread resistance occurring in *Septoria tritici* populations in UK. Failure to follow resistance management action may result in reduced levels of disease control

Restrictions
- Maximum total dose equivalent to two full dose treatments in wheat and barley
- Do not apply to crops under stress
- Do not apply during frosty weather

Crop-specific information
- Latest use: before flowering (GS 60) for winter wheat; before quarter ear emergence (GS 53) for barley

SEE SECTION 3 FOR PRODUCTS ALSO REGISTERED

Environmental safety
- Dangerous for the environment
- Very toxic to aquatic organisms
- Dangerous to fish or other aquatic life. Do not contaminate surface waters or ditches with chemical or used container
- LERAP Category B

Hazard classification and safety precautions
> **Hazard** H02, H11
> **Risk phrases** R36, R40, R51, R53a, R61
> **Operator protection** A, C; U05a, U19a, U20b
> **Environmental protection** E13b, E16a, E16b, E38
> **Storage and disposal** D01, D02, D05, D09a, D10b, D12a
> **Medical advice** M03, M04

205　fatty acids

A soap concentrate insecticide and acaricide

Products

Savona	Koppert	49% w/w	SL	06057

Uses
- Aphids in *broad beans, brussels sprouts, cabbages, cucumbers, fruit trees, lettuce, outdoor tomatoes, peas, peppers, protected tomatoes, pumpkins, runner beans, woody ornamentals*
- Mealybugs in *broad beans, brussels sprouts, cabbages, cucumbers, fruit trees, lettuce, outdoor tomatoes, peas, peppers, protected tomatoes, pumpkins, runner beans, woody ornamentals*
- Scale insects in *broad beans, brussels sprouts, cabbages, cucumbers, fruit trees, lettuce, outdoor tomatoes, peas, peppers, protected tomatoes, pumpkins, runner beans, woody ornamentals*
- Spider mites in *broad beans, brussels sprouts, cabbages, cucumbers, fruit trees, lettuce, outdoor tomatoes, peas, peppers, protected tomatoes, pumpkins, runner beans, woody ornamentals*
- Whitefly in *broad beans, brussels sprouts, cabbages, cucumbers, fruit trees, lettuce, outdoor tomatoes, peas, peppers, protected tomatoes, pumpkins, runner beans, woody ornamentals*

Efficacy guidance
- Use only soft or rain water for diluting spray
- Pests must be sprayed directly to achieve any control. Spray all plant parts thoroughly to run off
- For glasshouse use apply when insects first seen and repeat as necessary. For scale insects apply several applications at weekly intervals after egg hatch
- To control whitefly spray when required and use biological control after 12 h

Restrictions
- Do not use on new transplants, newly rooted cuttings or plants under stress
- Do not use on specified susceptible shrubs. See label for details

Crop-specific information
- HI zero

Environmental safety
- Harmful to fish or other aquatic life. Do not contaminate surface waters or ditches with chemical or used container

Hazard classification and safety precautions
> **Operator protection** U20c
> **Environmental protection** E13c
> **Storage and disposal** D05, D09a, D10a

FOR FULL CONDITIONS OF USE ALWAYS READ THE PRODUCT LABEL

206 fenamidone

A strobilurin fungicide available only in mixtures

207 fenamidone + mancozeb

A fungicide mixture for potatoes

Products

Sonata	Bayer CropScience	10:50% w/w	WG	11570

Uses
- Blight in *potatoes*

Approval information
- Fenamidone and mancozeb included in Annex I under EC Directive 91/414

Efficacy guidance
- Apply as soon as there is risk of blight infection or immediately after a blight warning
- Fenamidone is a member of the QoI cross resistance group. To minimise the likelihood of development of resistance these products should be used in a planned Resistance Management strategy. In addition before application consult and adhere to the latest FRAG-UK resistance guidance on application of QoI fungicides to potatoes. See Section 5 for more information

Restrictions
- Maximum number of treatments (including any other QoI containing fungicide) 6 per yr but no more than three should be applied consecutively. Use in alternation with fungicides from a different cross-resistance group
- Observe a 7 d interval between treatments
- Use only as a protective treatment. Do not use when blight has become readily visible (1% leaf area destroyed)

Crop-specific information
- HI 7 d for potatoes

Environmental safety
- Dangerous for the environment
- Very toxic to aquatic organisms
- Dangerous to fish or other aquatic life. Do not contaminate surface waters or ditches with chemical or used container
- Risk to certain non-target insects or other arthropods. See directions for use
- LERAP Category B

Hazard classification and safety precautions
Hazard H04, H11
Risk phrases R36, R37, R50, R53a
Operator protection D, E, H; U05a, U08, U11, U19a, U20a
Environmental protection E13b, E16a, E22b, E34, E38
Storage and disposal D01, D02, D05, D09a, D10b, D12a

208 fenamidone + propamocarb hydrochloride

A systemic fungicide mixture for potatoes

Products

Consento	Bayer CropScience	75:375 g/l	SC	11889

Uses
- Blight in *potatoes*

Approval information
- Fenamidone included in Annex I under EC Directive 91/414

Efficacy guidance
- Apply as soon as there is risk of blight infection or immediately after a blight warning
- Fenamidone is a member of the QoI cross resistance group. To minimise the likelihood of development of resistance these products should be used in a planned Resistance Management strategy. In addition before application consult and adhere to the latest FRAG-UK resistance guidance on application of QoI fungicides to potatoes. See Section 5 for more information

Restrictions
- Maximum number of treatments (including any other QoI containing fungicide) 6 per yr but no more than three should be applied consecutively. Use in alternation with fungicides from a different cross-resistance group
- Observe a 7 d interval between treatments
- Use only as a protective treatment. Do not use when blight has become readily visible (1% leaf area destroyed)

Crop-specific information
- HI 7 d for potatoes

Environmental safety
- Dangerous for the environment
- Very toxic to aquatic organisms
- Dangerous to fish or other aquatic life. Do not contaminate surface waters or ditches with chemical or used container
- LERAP Category B

Hazard classification and safety precautions
Hazard H04, H11
Risk phrases R36, R50, R53a
Operator protection A, C; U05a, U08, U11, U20a
Environmental protection E13b, E16a, E38
Storage and disposal D01, D02, D05, D09a, D10b, D12a

209 fenarimol

A systemic curative and protective pyrimidine fungicide

Products

1 Rimidin	Rigby Taylor	120 g/l	SC	05907
2 Rubigan	Gowan	120 g/l	SC	12355

Uses
- Dollar spot in *managed amenity turf* [1]
- Fusarium patch in *managed amenity turf* [1]
- Powdery mildew in *apples*, *blackcurrants*, *gooseberries*, *marrows* (off-label), *protected courgettes* (off-label), *protected gherkins* (off-label), *protected peppers* (off-label), *protected tomatoes* (off-label), *pumpkins* (off-label), *raspberries*, *roses*, *squashes* (off-label), *strawberries* [2]
- Red thread in *managed amenity turf* [1]
- Scab in *apples* [2]

Specific Off-Label Approvals (SOLAs)
- *marrows*, *pumpkins*, *squashes* (OLA 050960) Dec 2008 [2]
- *protected courgettes*, *protected gherkins* (OLA 050961) Dec 2008 [2]
- *protected peppers*, *protected tomatoes* (OLA 050962) Dec 2008 [2]

Efficacy guidance
- For turf disease control spray as preventive treatment and repeat as necessary. See label for details

Restrictions
- Maximum number of treatments 15 per yr on apples, 4 per yr on managed amenity turf, 3 per yr on raspberries and protected vegetables
- Do not mow within 24 h after treatment

FOR FULL CONDITIONS OF USE ALWAYS READ THE PRODUCT LABEL

Crop-specific information
- HI 14 d for apples, blackcurrants, gooseberries, raspberries, strawberries; 7 d for marrows, protected courgettes, protected gherkins, pumpkins, squashes; 2 d for protected peppers, protected tomatoes,

Environmental safety
- Harmful to aquatic organisms

Hazard classification and safety precautions
Hazard H03
Risk phrases R52, R62 [1, 2]; R53a, R61 [2]; R63, R64 [1]
Operator protection A, C; U08, U19a [1, 2]; U20c [1]
Environmental protection E15a
Consumer protection C02 [2] (14 d)
Storage and disposal D01, D09a, D10b
Medical advice M05b [2]

210   fenbuconazole

A systemic protectant and curative triazole fungicide for top fruit and grapevines

Products

Indar 5EW	Landseer	50 g/l	EW	09518

Uses
- Blossom wilt in **cherries** *(off-label)*, **mirabelles** *(off-label)*, **plums** *(off-label)*
- Brown rot in **cherries** *(off-label)*, **mirabelles** *(off-label)*, **plums** *(off-label)*
- Powdery mildew in **apples** *(reduction)*, **grapevines** *(off-label)*, **pears** *(reduction)*
- Scab in **apples, pears**

Specific Off-Label Approvals (SOLAs)
- *cherries, mirabelles, plums (OLA 031372) Dec 2008* [1]
- *grapevines (OLA 032081) Dec 2008* [1]

Efficacy guidance
- Most effective when used as part of a routine preventative programme from bud burst to onset of petal fall
- After petal fall, tank mix with other protectant fungicides to enhance scab control
- See label for recommended spray intervals. In periods of rapid growth or high disease pressure, a 7 d interval should be used

Restrictions
- Maximum total dose on top fruit equivalent to ten full doses per yr on apples, pears; six full doses on grapevines; five full doses on plums; four full doses on cherries, mirabelles
- Consult processors before using on pears for processing
- Do not harvest for human or animal consumption for at least 4 wk after last application

Crop-specific information
- HI 28 d for apples, pears; 21 d for grapevines; 3 d for cherries, mirabelles, plums
- Safe to use on all main commercial varieties of apples and pears in UK

Environmental safety
- Dangerous for the environment
- Toxic to aquatic organisms

Hazard classification and safety precautions
Hazard H04, H11
Risk phrases R36, R41, R51, R53a
Operator protection A, C; U05a, U08, U11, U20a
Environmental protection E15a, E38
Consumer protection C02 (4 wk)
Storage and disposal D01, D02, D05, D11a, D12a

SEE SECTION 3 FOR PRODUCTS ALSO REGISTERED

211 fenbutatin oxide

A selective contact and ingested organotin acaricide

Products

Torq	Fargro	50% w/w	WP	08370

Uses
- Spider mites in **protected peppers** *(off-label)*
- Two-spotted spider mite in **protected cucumbers**, **protected ornamentals**, **protected tomatoes**, **tunnel grown strawberries**

Specific Off-Label Approvals (SOLAs)
- **protected peppers** *(OLA 022124) Dec 2008* [1]

Efficacy guidance
- Active on larvae and adult mites. Spray may take 7-10 d to effect complete kill but mites cease feeding and crop damage stops almost immediately
- Apply as soon as mites first appear and repeat as necessary
- On tunnel-grown strawberries apply when mites first appear, usually before flowering starts, and repeat 10-14 d later. A post-harvest spray is also recommended

Restrictions
- Maximum number of treatments 2 per crop pre-harvest + 1 post harvest for tunnel grown strawberries; 2 per crop for protected peppers
- Allow at least 10 d between spray applications and do not apply within 10 d of a previous spray
- Do not add wetters or mix with anything other than water
- Do not use white petroleum oil within 28 d of treatment, or any other pesticide within 7 d
- Do not apply to crops which are under stress for any reason
- On subjects of unknown susceptibility test treat on a small number of plants in advance

Crop-specific information
- HI glasshouse cucumbers, tomatoes, peppers 3 d; tunnel-grown strawberries 7 d

Environmental safety
- Dangerous for the environment
- Very toxic to aquatic organisms
- Suitable for use in IPM programmes with *Encarsia* or *Phytoseiulus* being used for biological control

Hazard classification and safety precautions

> **Hazard** H01, H11
> **Risk phrases** R26, R36, R38, R50, R53a
> **Operator protection** A, C, D, H, M; U05a, U08, U10, U19a, U20a
> **Environmental protection** E15a, E34, E38
> **Storage and disposal** D01, D02, D09a, D11a, D12a
> **Medical advice** M04

212 fenhexamid

A protectant fungicide for soft fruit and a range of horticultural crops

Products

Teldor	Bayer CropScience	50% w/w	WG	11229

Uses
- Botrytis in **blackberries**, **blackcurrants**, **cherries** *(off-label)*, **chives** *(off-label)*, **cress** *(off-label)*, **gooseberries**, **grapevines**, **herbs (see appendix 6)** *(off-label)*, **lamb's lettuce** *(off-label)*, **leaf spinach** *(off-label)*, **lettuce** *(off-label)*, **loganberries**, **outdoor tomatoes** *(off-label)*, **parsley** *(off-label)*, **plums** *(off-label)*, **protected aubergines** *(off-label)*, **protected courgettes** *(off-label)*, **protected cucumbers** *(off-label)*, **protected gherkins** *(off-label)*, **protected peppers** *(off-label)*, **protected squashes** *(off-label)*, **protected tomatoes** *(off-label)*, **raspberries**, **redcurrants**, **rubus hybrids**, **salad brassicas** *(off-label - for baby leaf production)*, **scarole** *(off-label)*, **strawberries**, **whitecurrants**

FOR FULL CONDITIONS OF USE ALWAYS READ THE PRODUCT LABEL

SECTION 2

Specific Off-Label Approvals (SOLAs)
- *cherries, plums* (OLA 031866) May 2011 [1]
- *chives, cress, herbs (see appendix 6), lamb's lettuce, leaf spinach, lettuce, parsley, scarole* (OLA 050026) May 2011 [1]
- *outdoor tomatoes* (OLA 042399) May 2011 [1]
- *protected aubergines, protected tomatoes* (OLA 042087) Nov 2010 [1]
- *protected courgettes, protected cucumbers, protected gherkins, protected squashes* (OLA 042085) Nov 2010 [1]
- *protected peppers* (OLA 042086) Nov 2010 [1]
- *salad brassicas* (for baby leaf production) (OLA 050026) May 2011 [1]

Approval information
- Fenhexamid included in Annex I under EC Directive 91/414

Efficacy guidance
- Use as part of a programme of sprays throughout the flowering period to achieve effective control of Botrytis
- To minimise possibility of development of resistance, no more than two sprays of the product may be applied consecutively. Other fungicides from a different chemical group should then be used for at least two consecutive sprays. If only two applications are made on grapevines, only one may include fenhexamid
- Complete spray cover of all flowers and fruitlets throughout the blossom period is essential for successful control of Botrytis
- Spray programmes should normally start at the start of flowering

Restrictions
- Maximum number of treatments 2 per yr on grapevines; 4 per yr on other listed crops but no more than 2 sprays may be applied consecutively

Crop-specific information
- HI 1 d for strawberries, raspberries, loganberries, blackberries, Rubus hybrids; 3 d for cherries; 7 d for blackcurrants, redcurrants, whitecurrants, gooseberries; 21d for outdoor grapes

Environmental safety
- Dangerous for the environment
- Harmful to fish or other aquatic life. Do not contaminate surface waters or ditches with chemical or used container

Hazard classification and safety precautions
Hazard H11
Risk phrases R53a
Operator protection U08, U19a, U20b
Environmental protection E13c, E38
Storage and disposal D05, D09a, D11a, D12a

213   fenhexamid + tolylfluanid

A protectant fungicide mixture for soft fruit

Products

Talat	Certis	16.7:33.3% w/w	WG	11311

Uses
- Grey mould in *blackberries, blackcurrants, gooseberries, loganberries, raspberries, redcurrants, rubus hybrids, strawberries, whitecurrants*

Approval information
- Fenhexamid and tolyfluanid included in Annex I under EC Directive 91/414

Efficacy guidance
- Effective control of grey mould in soft fruit requires a programme of treatment throughout the blossom period
- Complete even spray coverage of all flowers and developing fruitlets is essential for good results
- To minimise the likelihood of development of resistance use in a planned Resistance Management strategy. See Section 1 for more details

SEE SECTION 3 FOR PRODUCTS ALSO REGISTERED

Restrictions
- Maximum total dose equivalent to four full dose treatments for strawberries, raspberries, loganberries, blackberries and other Rubus hybrids, and to two full dose treatments for currants and gooseberries
- Do not apply more than two treatments consecutively (including other products containing fenhexamid or tolylfluanid)
- Do not treat crops growing under glass or polythene
- Consult processor before using on crops for processing

Crop-specific information
- HI 14 days for blackberries, loganberries, raspberries, Rubus hybrids, strawberries; 21 days for currants, gooseberries

Environmental safety
- Dangerous for the environment
- Very toxic to aquatic organisms
- Dangerous to fish or other aquatic life. Do not contaminate surface waters or ditches with chemical or used container
- Risk to certain non-target insects or other arthropods. See directions for use
- LERAP Category B
- Broadcast air-assisted LERAP (7.5 m)

Hazard classification and safety precautions
Hazard H04, H11
Risk phrases R43, R50, R53a
Operator protection A, H, M; U02a, U05a, U10, U14, U20b
Environmental protection E13b, E16a, E16b, E38; E17b (7.5 m)
Storage and disposal D01, D02, D05, D09a, D11a, D12a

214   fenoxaprop-P-ethyl

A phenoxypropionic acid herbicide for use in wheat

See also diclofop-methyl + fenoxaprop-P-ethyl

Products

1	Cheetah Super	Bayer CropScience	55 g/l	EW	08723
2	Triumph	Bayer CropScience	120 g/l	EC	10902

Uses
- Blackgrass in *spring wheat, winter wheat*
- Canary grass in *spring wheat, winter wheat*
- Rough meadow grass in *spring wheat, winter wheat*
- Wild oats in *spring wheat, winter wheat*

Efficacy guidance
- Treat weeds from 2 fully expanded leaves up to flag leaf ligule just visible; for awned canary-grass and rough meadow-grass from 2 leaves to the end of tillering
- A second application may be made in spring where susceptible weeds emerge after an autumn application
- Spray is rainfast 1 h after application
- Dry conditions resulting in moisture stress may reduce effectiveness
- Always follow WRAG guidelines for preventing and managing herbicide resistant weeds. Section 5 for more information

Restrictions
- Maximum total dose equivalent to one or two full dose treatments depending on product used
- Do not apply to barley, durum wheat, undersown crops or crops to be undersown
- Do not roll or harrow within 1 wk of spraying
- Do not spray crops under stress, suffering from drought, waterlogging or nutrient deficiency or those grazed or if soil compacted
- Avoid spraying immediately before or after a sudden drop in temperature or a period of warm days/cold nights
- Do not mix with hormone weedkillers

FOR FULL CONDITIONS OF USE ALWAYS READ THE PRODUCT LABEL

Crop-specific information

- Latest use: before flag leaf sheath extending (GS 41)
- Treat from crop emergence to flag leaf fully emerged (GS 41).
- Product may be sprayed in frosty weather provided crop hardened off but do not spray wet foliage or leaves covered with ice
- Broadcast crops should be sprayed post-emergence after plants have developed well-established root system

Environmental safety

- Dangerous for the environment [1, 2]
- Very toxic (toxic [1]) to aquatic organisms
- Harmful to fish or other aquatic life. Do not contaminate surface waters or ditches with chemical or used container

Hazard classification and safety precautions

Hazard H04 [2]; H11 [1, 2]
Risk phrases R36, R38, R50 [2]; R51 [1]; R53a [1, 2]
Operator protection A, C, H; U05a, U20b [1, 2]; U08 [2]; U09a [1]
Environmental protection E13c, E38
Storage and disposal D01, D02, D09a, D10b, D12a

SECTION 2

215 fenoxycarb

An insect specific growth regulator for top fruit

Products

Insegar WG	Syngenta	25% w/w	WG	09789

Uses

- Summer-fruit tortrix moth in **apples**, **pears**

Efficacy guidance

- Best results from application at 5th instar stage before pupation. Product prevents transformation from larva to pupa
- Correct timing best identified from pest warnings
- Because of mode of action rapid knock-down of pest is not achieved and larvae continue to feed for a period after treatment
- Adequate water volume necessary to ensure complete coverage of leaves

Restrictions

- Maximum number of treatments 2 per crop
- Consult processors before use

Crop-specific information

- HI 42 d for apples, pears
- Use on all varieties of apples and pears

Environmental safety

- Dangerous for the environment
- Toxic to aquatic organisms
- High risk to bees. Do not apply to crops in flower or to those in which bees are actively foraging. Do not apply when flowering weeds are present
- Risk to certain non-target insects or other arthropods. See directions for use
- Broadcast air-assisted LERAP (8 m)
- Apply to minimise off-target drift to reduce effects on non-target organisms. Some margin of safety to beneficial arthropods is indicated.

Hazard classification and safety precautions

Hazard H11
Risk phrases R51, R53a
Operator protection A, C, H; U05a, U20a, U23a
Environmental protection E12a, E15a, E22b, E38; E17b (8 m)
Storage and disposal D01, D02, D07, D09a, D11a, D12a

SEE SECTION 3 FOR PRODUCTS ALSO REGISTERED

216   fenpropathrin

A contact and ingested pyrethroid acaricide and insecticide

Products

Meothrin	BASF	100 g/l	EC	10400

Uses
- Blackcurrant gall mite in **blackcurrants**
- Blackcurrant leaf midge in **blackcurrants**
- Capsids in **blackcurrants**
- Currant sawfly in **blackcurrants**
- Two-spotted spider mite in **blackcurrants**

Approval information
- Products containing this active ingredient have been granted derogations for specified 'Essential Uses' for use until 31 December 2007. Sale and supply must cease by 30 June 2007 but growers have no guarantee that the products will continue to be available until then.
  For more information see 'The Review Programme' under 'Pesticide Legislation' in Section 5

Efficacy guidance
- Acts on motile stages of mites and gives rapid kill

Restrictions
- Maximum number of treatments 3 per yr for blackcurrants
- This product must not be used on any crops other than those listed, including any extrapolations that would normally be permissible under the Long Term Arrangements for Extension of Use (see Section 5)

Crop-specific information
- Latest use: 14 d after end of flowering for blackcurrants

Environmental safety
- Dangerous for the environment
- Very toxic to aquatic organisms
- High risk to bees. Do not apply to crops in flower or to those in which bees are actively foraging. Do not apply when flowering weeds are present
- Do not operate air-assisted sprayers within 18 m of surface water or ditches. Direct spray away from water
- LERAP Category A
- Broadcast air-assisted LERAP (18 m)

Hazard classification and safety precautions
    Hazard H02, H04, H11
    Risk phrases R25, R36, R38, R50, R53a
    Operator protection A, C, H, J; U02c, U04a, U05a, U08, U19a, U20b
    Environmental protection E12a, E15a, E16c, E16d, E34; E17b (18 m)
    Storage and disposal D01, D02, D08, D09a, D10c, D12b
    Medical advice M04, M05b

217   fenpropidin

A systemic, curative and protective piperidine (morpholine) fungicide

Products

1	Instinct	Headland	750 g/l	EC	12317
2	Tern	Syngenta	750 g/l	EC	08660

Uses
- Brown rust in **spring barley, spring wheat, winter barley, winter wheat**
- Glume blotch in **spring wheat, winter wheat**
- Leaf blotch in **spring barley, winter barley**
- Powdery mildew in **spring barley, spring wheat, winter barley, winter wheat**
- Septoria leaf spot in **spring wheat, winter wheat**
- Yellow rust in **spring barley, spring wheat, winter barley, winter wheat**

FOR FULL CONDITIONS OF USE ALWAYS READ THE PRODUCT LABEL

Approval information
- Accepted by BBPA for use on malting barley

Efficacy guidance
- Best results obtained when applied at early stage of disease development. See label for details of recommended timing alone and in mixtures
- Disease control enhanced by vapour-phase activity. Control can persist for 4-6 wk
- Alternate with triazole fungicides to discourage build-up of resistance

Restrictions
- Maximum number of treatments 3 per crop (up to 2 in yr of harvest) for winter crops; 2 per crop for spring crops

Crop-specific information
- Latest use: up to and including ear emergence complete (GS 59).
- HI 5 wk

Environmental safety
- Dangerous for the environment
- Very toxic to aquatic organisms

Hazard classification and safety precautions
Hazard H03, H11
Risk phrases R22a, R37 [2]; R38, R41 [1]; R50, R53a [1, 2]
Operator protection A, C, H; U02a, U05a [1, 2]; U04a, U10, U20a [2]; U09a, U11, U19a, U20b [1]
Environmental protection E15a, E34, E38
Consumer protection C02 [1] (5 wk)
Storage and disposal D01, D02, D09a, D12a [1, 2]; D05, D10c [1]
Medical advice M03

218   fenpropimorph

A contact and systemic morpholine fungicide

See also azoxystrobin + fenpropimorph
epoxiconazole + fenpropimorph
epoxiconazole + fenpropimorph + kresoxim-methyl
epoxiconazole + fenpropimorph + pyraclostrobin

Products

1	Corbel	BASF	750 g/l	EC	00578
2	Landgold Fenpropimorph 750	Landgold	750 g/l	EC	10472
3	Landgold Fenpropimorph 750	Teliton	750 g/l	EC	12118
4	Marnoch Phorm	Me2	750 g/l	EC	11087
5	Standon Fenpropimorph 750	Standon	750 g/l	EC	08965

Uses
- Alternaria in **carrots** *(off-label)*, **horseradish** *(off-label)*, **parsley root** *(off-label)*, **parsnips** *(off-label)*, **salsify** *(off-label)* [1]
- Brown rust in **spring barley**, **spring wheat**, **triticale**, **winter barley**, **winter wheat** [1-5]
- Crown rot in **carrots** *(off-label)*, **horseradish** *(off-label)*, **parsley root** *(off-label)*, **parsnips** *(off-label)*, **salsify** *(off-label)* [1]
- Powdery mildew in **bilberries** *(off-label)*, **blackcurrants** *(off-label)*, **blueberries** *(off-label)*, **carrots** *(off-label)*, **cranberries** *(off-label)*, **dewberries** *(off-label)*, **gooseberries** *(off-label)*, **hops** *(off-label)*, **horseradish** *(off-label)*, **loganberries** *(off-label)*, **parsley root** *(off-label)*, **parsnips** *(off-label)*, **raspberries** *(off-label)*, **redcurrants** *(off-label)*, **salsify** *(off-label)*, **strawberries** *(off-label)*, **tayberries** *(off-label)*, **whitecurrants** *(off-label)* [1]; **spring barley**, **spring oats**, **spring wheat**, **winter barley**, **winter oats**, **winter rye**, **winter wheat** [1-5]
- Rhynchosporium in **spring barley**, **winter barley** [1-5]
- Rust in **red beet** *(off-label)*, **sugar beet seed crops** *(off-label)* [1]
- Yellow rust in **spring barley**, **spring wheat**, **triticale**, **winter barley**, **winter wheat** [1-5]

Specific Off-Label Approvals (SOLAs)
- **bilberries**, **blackcurrants**, **blueberries**, **cranberries**, **dewberries**, **gooseberries**, **loganberries**, **raspberries**, **redcurrants**, **strawberries**, **tayberries**, **whitecurrants** *(OLA 040804) Dec 2008* [1]

SEE SECTION 3 FOR PRODUCTS ALSO REGISTERED

- *carrots, horseradish, parsley root, parsnips, salsify* (OLA 023753) Dec 2008 [1]
- *carrots, horseradish, parsley root, parsnips, salsify* (OLA 962483) Dec 2008 [1]
- *hops* (OLA 010595) Dec 2008 [1]
- *hops* (OLA 023759) Dec 2008 [1]
- *red beet* (OLA 023751) Dec 2008 [1]
- *red beet* (OLA 941246) Dec 2008 [1]
- *sugar beet seed crops* (OLA 023757) Dec 2008 [1]

Approval information
- Accepted by BBPA for use on malting barley and hops

Efficacy guidance
- On cereals spray at start of disease attack. See labels for recommended tank mixes. Follow-up treatments may be needed if disease pressure remains high
- Product rainfast after 2 h

Restrictions
- Maximum number of treatments 2 per crop for spring cereals, red beet, sugar beet seed crops; 6 per crop for hops; 3 per crop for all other crops
- Consult processors before using on crops for processing

Crop-specific information
- Latest use: before 2 true leaves for sugar beet seed crops
- HI cereals 5 wk; carrots, horseradish, parsley root, parsnips, salsify 4 wk; red beet 3 wk; raspberries, strawberries 2 wk; hops 10 d
- Scorch may occur if applied during frosty weather or in high temperatures

Environmental safety
- Dangerous for the environment
- Very toxic to aquatic organisms
- Dangerous to fish or other aquatic life. Do not contaminate surface waters or ditches with chemical or used container

Hazard classification and safety precautions
Hazard H03 [1-3]; H04 [4, 5]; H11 [1-3, 5]
Risk phrases R36 [2-4]; R38 [1-5]; R50 [1]; R51 [2, 3, 5]; R53a [1-3, 5]; R63 [1-3]
Operator protection A [1-5]; C [2-5]; U05a, U08, U14, U15, U19a [1-5]; U20a [4]; U20b [1-3, 5]
Environmental protection E13b [4]; E15a [1-3, 5]; E38 [1]
Storage and disposal D01, D02, D09a [1-5]; D05 [2-5]; D08, D12a [1]; D10b [4, 5]; D10c [1-3]
Medical advice M05a [1-3, 5]

219 fenpropimorph + flusilazole

A broad-spectrum eradicant and protectant fungicide mixture for cereals

Products

1 Colstar	DuPont	375:160 g/l	EC	12175
2 Pluton	DuPont	375:160 g/l	EC	12200

Uses
- Brown rust in *spring barley, winter barley, winter wheat*
- Net blotch in *spring barley, winter barley*
- Powdery mildew in *spring barley, winter barley, winter wheat*
- Rhynchosporium in *spring barley, winter barley*
- Septoria diseases in *winter wheat*
- Yellow rust in *spring barley, winter barley, winter wheat*

Approval information
- Accepted by BBPA for use on malting barley

Efficacy guidance
- Disease control is more effective if treatment made at an early stage of disease development
- Treat winter cereals in spring or early summer before diseases spread to new growth
- Spring barley should be treated when diseases are first evident
- Treatment may be repeated after 3-4 wk if necessary

Restrictions

- Maximum number of treatments (including other products containing flusilazole) 3 per crop (winter wheat), 2 per crop (winter or spring barley)
- Do not apply to crops under stress
- Do not apply during frosty weather

Crop-specific information

- Latest use: before beginning of anthesis (GS 60) for winter wheat; up to and including completion of ear emergence (GS 59) for barley

Environmental safety

- Dangerous for the environment
- Toxic to aquatic organisms
- Dangerous to fish or other aquatic life. Do not contaminate surface waters or ditches with chemical or used container

Hazard classification and safety precautions

Hazard H02, H11
Risk phrases R36, R40, R51, R53a, R61
Operator protection A, C; U05a, U11, U19a, U20b
Environmental protection E13b, E15a, E38
Storage and disposal D01, D02, D05, D09a, D10b, D12a
Medical advice M04

---

220   fenpropimorph + kresoxim-methyl

A protectant and systemic fungicide mixture for cereals

Products

1	Ensign	BASF	300:150 g/l	SE	11729
2	Greencrop Monsoon	Greencrop	300:150 g/l	SE	09573
3	Standon Kresoxim FM	Standon	300:150 g/l	SE	08922

Uses

- Powdery mildew in *spring barley*, *winter barley* [1-3]; *spring oats*, *spring rye*, *triticale*, *winter oats*, *winter rye* [1]
- Rhynchosporium in *spring barley*, *winter barley* [1-3]; *spring rye*, *triticale*, *winter rye* [1]
- Septoria in *spring wheat* (reduction), *triticale* (reduction) [1]; *winter wheat* (reduction) [1-3]

Approval information

- Kresoxim-methyl included in Annex I under EC Directive 91/414
- Accepted by BBPA for use on malting barley

Efficacy guidance

- For best results spray at the start of foliar disease attack and repeat if infection conditions persist
- Kresoxim-methyl is a member of the QoI cross resistance group. Product should be used preventatively and not relied on for its curative potential
- Use product as part of an Integrated Crop Management strategy incorporating other methods of control, including where appropriate other fungicides with a different mode of action. Do not apply more than two foliar applications of QoI containing products to any cereal crop
- There is a significant risk of widespread resistance occurring in *Septoria tritici* populations in UK. Strains of barley powdery mildew resistant to QoIs are common in UK. Failure to follow resistance management action may result in reduced levels of disease control
- Strains of barley powdery mildew resistant to QoIs are common in the UK

Restrictions

- Maximum total dose equivalent to two full dose treatments for all crops

Crop-specific information

- Latest use: completion of ear emergence (GS 59) for barley and oats; completion of flowering (GS 69) for wheat, rye and triticale

Environmental safety

- Dangerous for the environment
- Very toxic to aquatic organisms

SEE SECTION 3 FOR PRODUCTS ALSO REGISTERED

Hazard classification and safety precautions
    Hazard H03, H11
    Risk phrases R40, R50, R53a [1-3]; R43 [2, 3]; R63 [1, 2]
    Operator protection A [1-3]; H [1]; U05a, U14, U20b [1-3]; U19a [2]
    Environmental protection E15a [1-3]; E38 [1]
    Storage and disposal D01, D02, D05, D09a [1-3]; D10b [2, 3]; D10c, D12a [1]
    Medical advice M05a [1, 3]

221   fenpropimorph + pyraclostrobin

A protectant and curative fungicide mixture for cereals

Products
    Jenton   BASF   375:100 g/l   EC   11898

Uses
- Brown rust in *spring barley*, *spring wheat*, *winter barley*, *winter wheat*
- Crown rust in *spring oats*, *winter oats*
- Drechslera leaf spot in *spring wheat*, *winter wheat*
- Net blotch in *spring barley*, *winter barley*
- Powdery mildew in *spring barley*, *spring oats*, *winter barley*, *winter oats*
- Rhynchosporium in *spring barley*, *winter barley*
- Septoria in *spring wheat*, *winter wheat*
- Yellow rust in *spring barley*, *spring wheat*, *winter barley*, *winter wheat*

Approval information
- Accepted by BBPA for use on malting barley

Efficacy guidance
- Best results obtained from treatment at the start of foliar disease attack
- Yield response may be obtained in the absence of visual disease
- Pyraclostrobin is a member of the QoI cross resistance group. Product should be used preventatively and not relied on for its curative potential
- Use product as part of an Integrated Crop Management strategy incorporating other methods of control, including where appropriate other fungicides with a different mode of action. Do not apply more than two foliar applications of QoI containing products to any cereal crop
- There is a significant risk of widespread resistance occurring in *Septoria tritici* populations in UK. Strains of barley powdery mildew resistant to QoIs are common in UK. Failure to follow resistance management action may result in reduced levels of disease control

Restrictions
- Maximum number of treatments 2 per crop

Crop-specific information
- Latest use: up to and including emergence of ear just complete (GS 59) for barley

Environmental safety
- Dangerous for the environment
- Very toxic to aquatic organisms
- LERAP Category B

Hazard classification and safety precautions
    Hazard H03, H11
    Risk phrases R20, R22a, R36, R38, R50, R53a, R63
    Operator protection A, C; U05a, U14, U20b
    Environmental protection E15a, E16a, E34, E38
    Storage and disposal D01, D02, D05, D09a, D10c, D12a
    Medical advice M03, M05a

222 fenpropimorph + quinoxyfen

A systemic fungicide mixture for cereals

Products

| Orka | Dow | 250:66.7 g/l | EW | 08879 |

Uses
- Powdery mildew in **durum wheat**, **spring barley**, **spring oats**, **spring rye**, **spring wheat**, **triticale**, **winter barley**, **winter oats**, **winter rye**, **winter wheat**

Approval information
- Quinoxyfen included in Annex I under Directive 91/414
- Accepted by BBPA for use on malting barley

Efficacy guidance
- For best results treat at early stage of disease development before infection spreads to new crop growth. Further treatment may be necessary if disease pressure remains high
- For control of established infections and broad spectrum disease control use in tank mixtures. See label
- Product rainfast after 1 h
- Systemic activity may be reduced in severe drought

Restrictions
- Maximum total dose 3.0 l product per ha
- Apply only in the spring from mid-tillering stage (GS 25)

Crop-specific information
- Latest use: when first awns visible (GS 49)
- Crop scorch may occur when treatment made in high temperatures

Environmental safety
- Dangerous for the environment
- Very toxic to aquatic organisms
- LERAP Category B

Hazard classification and safety precautions
   Hazard H03, H11
   Risk phrases R43, R50, R53a, R61
   Operator protection A, H; U05a, U14
   Environmental protection E15a, E16a, E34, E38
   Storage and disposal D01, D02, D12a

223 fenpyroximate

A mitochondrial electron transport inhibitor (METI) acaricide for apples

Products

| Sequel | Certis | 51.3 g/l | SC | 11408 |

Uses
- Fruit tree red spider mite in **apples**, **plums** *(off-label)*

Specific Off-Label Approvals (SOLAs)
- **plums** *(OLA 042327) Dec 2008* [1]

Efficacy guidance
- Kills motile stages of fruit tree red spider mite. Best results achieved if applied in warm weather
- Total spray cover of trees essential. Use higher volumes for large trees
- Apply when majority of winter eggs have hatched

Restrictions
- Maximum number of treatments 1 per yr or total dose equivalent to one full dose treatment
- Other mitochondrial electron transport inhibitor (METI) acaricides should not be applied to the same crop in the same calendar yr either separately or in mixture
- Do not apply when apple crops or pollinator are in flower
- Consult processor before use on crops for processing

SEE SECTION 3 FOR PRODUCTS ALSO REGISTERED

SECTION 2

Crop-specific information
- HI 2 wk

Environmental safety
- Dangerous for the environment
- Very toxic to aquatic organisms
- Risk to non-target insects or other arthropods
- Broadcast air-assisted LERAP (38 m)

Hazard classification and safety precautions
    Hazard H03, H11
    Risk phrases R20, R36, R41, R43, R50
    Operator protection A, C, H; U05a, U08, U14, U15, U20b
    Environmental protection E15a, E22c; E17b (38 m)
    Storage and disposal D01, D02, D05, D09a, D10c, D12b

224   fenuron

A urea herbicide available only in mixtures

See also chlorpropham + fenuron

225   ferric phosphate

A molluscicide bait for controlling slugs and snails

Products

Ferramol	Certis	1.0% w/w	GB	12274

Uses
- Slugs in **blackcurrants**, **lettuce**, **ornamental plant production**, **protected ornamentals**, **root brassicas**, **strawberries**, **vegetable brassicas**
- Snails in **blackcurrants**, **lettuce**, **ornamental plant production**, **protected ornamentals**, **root brassicas**, **strawberries**, **vegetable brassicas**

Efficacy guidance
- Treat as soon as damage first seen preferably in early evening. repeat as necessary to maintain control
- Best results obtained from moist soaked granules. This will occur naturally on moist soils or in humid conditions
- Ferric phosphate does not cause excessive slime secretion and has no requirement to collect moribund slugs from the soil surface
- Active ingredient is degraded by micro-organisms to beneficial plant nutrients

Restrictions
- Maximum number of treatments not specified

Crop-specific information
- Latest use not specified for any crop

Hazard classification and safety precautions
    Operator protection U05a, U20b
    Environmental protection E15a, E34
    Storage and disposal D01, D09a

226   ferrous sulphate

A herbicide/fertilizer combination for moss control in turf

See also dicamba + dichlorprop-P + ferrous sulphate + MCPA
        dichlorophen + ferrous sulphate
        MCPA + mecoprop-P + ferrous sulphate

Products

1	Elliott's Lawn Sand	Elliott	9.3% w/w	SA	04860

FOR FULL CONDITIONS OF USE ALWAYS READ THE PRODUCT LABEL

Products – continued

2	Elliott's Mosskiller	Elliott	24.6% w/w	GR	04909
3	Greenmaster Autumn	Scotts	18.2% w/w	GR	11185
4	Greenmaster Mosskiller	Scotts	24.1% w/w	GR	11576
5	Greentec Mosskiller	Headland Amenity	9% w/w	GR	12518
6	SHL Lawn Sand	Sinclair	5.4% w/w	SA	05254
7	Taylors Lawn Sand	Rigby Taylor	4.1% w/w	SA	04451
8	Vitax Microgran 2	Vitax	10.9% w/w	MG	04541
9	Vitax Turf Tonic	Vitax	15% w/w	SA	04354

Uses
- Moss in *amenity grass* [5]; *managed amenity turf* [1-9]

Approval information
- Approval expiry 30 Apr 2006 [4]

Efficacy guidance
- For best results apply when turf is actively growing and the soil is moist
- Fertilizer component of most products encourages strong root growth and tillering
- Mow 3 d before treatment and do not mow for 3-4 d afterwards
- Water after 2 d if no rain
- Rake out dead moss thoroughly 7-14 d after treatment. Re-treatment may be necessary for heavy infestations

Restrictions
- Maximum number of treatments - see labels
- Do not apply during drought or when heavy rain expected
- Do not apply in frosty weather or when the ground is frozen
- Do not walk on treated areas until well watered

Crop-specific information
- If spilt on paving, concrete, clothes etc brush off immediately to avoid discolouration
- Observe label restrictions for interval before cutting after treatment

Environmental safety
- Harmful to fish or other aquatic life. Do not contaminate surface waters or ditches with chemical or used container [3, 4]

Hazard classification and safety precautions
Operator protection U20a [1, 2]; U20b [5, 8, 9]; U20c [3, 4, 6, 7]
Environmental protection E13c [3, 4]; E15a [1, 2, 5, 7-9]
Storage and disposal D01 [5, 6, 9]; D09a, D11a [1-9]

227 fipronil

A phenylpyrazole insecticide for horticulture

Products

Vi-Nil	Certis	0.1% w/w	FG	11077

Uses
- Vine weevil in *container-grown ornamentals*

Efficacy guidance
- Product may be used at any stage of plant propagation to potting on
- Treated compost will give control of vine weevil larvae into second yr after treatment
- Incorporate into fresh compost each time the subject is re-potted
- Control not guaranteed if untreated liners or plugs are potted into treated compost

Restrictions
- Maximum number of treatments 1 per batch of compost
- Check tolerance on sample plants before large scale treatment
- Do not use in compost for aquatic plants or marginals

Crop-specific information
- Latest use: before planting in treated compost
- Product may be used in all compost types but dose should be reduced in proportion if more than 20% inert material included

SEE SECTION 3 FOR PRODUCTS ALSO REGISTERED

Environmental safety
- Harmful to fish or other aquatic life. Do not contaminate surface waters or ditches with chemical or used container

Hazard classification and safety precautions
    Operator protection A, D; U13, U20b
    Environmental protection E13c
    Consumer protection C01
    Storage and disposal D01, D02, D11a

228 flonicamid

A nicotinamide aphicide

Products

Teppeki	Belchim	50% w/w	WG	12402

Uses
- Aphids in *apples, pears, potatoes, winter wheat*

Efficacy guidance
- Apply when warning systems forecast significant aphid infestations
- Persistence of action is 21 d
- Do not apply more than two consecutive treatments of flonicamid. If further treatment is needed use an insecticide with a different mode of action
- Must not be applied to winter wheat before 50% ear emerged stage (BBCH 53)

Restrictions
- Maximum number of treatments 2 per crop for potatoes, winter wheat; 3 per yr for apples, pears
- Consult processor before use on apples or pears for processing

Crop-specific information
- HI: potatoes 14 d; apples, pears 21 d; winter wheat 28 d

Environmental safety
- Dangerous for the environment

Hazard classification and safety precautions
    Hazard H11
    Risk phrases R52, R53a
    Operator protection A, H; U05a
    Environmental protection E15a
    Storage and disposal D01, D02, D07, D12b

229 florasulam

A triazolopyrimidine herbicide for cereals

See also 2,4-D + florasulam

Products

Boxer	Dow	50 g/l	SC	09819

Uses
- Annual dicotyledons in *spring barley, spring oats, spring wheat, winter barley, winter oats, winter wheat*
- Chickweed in *spring barley, spring oats, spring wheat, winter barley, winter oats, winter wheat*
- Cleavers in *spring barley, spring oats, spring wheat, winter barley, winter oats, winter wheat*
- Mayweeds in *spring barley, spring oats, spring wheat, winter barley, winter oats, winter wheat*
- Volunteer oilseed rape in *spring barley, spring oats, spring wheat, winter barley, winter oats, winter wheat*

FOR FULL CONDITIONS OF USE ALWAYS READ THE PRODUCT LABEL

SECTION 2

Approval information
- Florasulam included in Annex I under EC Directive 91/414
- Accepted by BBPA for use on malting barley

Efficacy guidance
- Best results obtained from treatment of small actively growing weeds in good conditions
- Apply in autumn or spring once crop has 3 leaves
- Product is mainly absorbed by leaves of weeds and is effective on all soil types
- Florasulam is a member of the ALS-inhibitor group of herbicides

Restrictions
- Maximum total dose on any crop equivalent to one full dose treatment
- Do not roll or harrow within 7 d before or after application
- Do not spray when crops under stress from cold, drought, pest damage, nutrient deficiency or any other cause
- Do not spray in tank mixture with a product containing any other ALS-inhibitor (eg sulfonylurea herbicides) except those containing iodosulfuron-methyl-sodium + mesosulfuron-methyl, metsulfuron-methyl, thifensulfuron-methyl or tribenuron-methyl. Product may be sprayed in sequence with any of the above ALS-inhibitors or flupyrsulfuron-methyl but not with any other ALS-inhibitor
- Only one other product with an ALS-inhibitor mode of action may be applied to a treated cereal crop

Crop-specific information
- Latest use: before flag leaf sheath extending stage (GS 41) for all crops

Following crops guidance
- Only cereals, oilseed rape, field beans, grass or vegetable brassicas as transplants may be sown as a following crop in the same calendar yr as treatment. Oilseed rape may show some temporary reduction of vigour after a dry summer, but yields are not affected
- In addition to the above, linseed, peas, sugar beet, potatoes, maize, clover (for use in grass/clover mixtures) or carrots may be sown in the calendar yr following treatment
- In the event of failure of a treated crop in spring only spring wheat, spring barley, spring oats, maize or ryegrass may be sown

Environmental safety
- Dangerous for the environment
- Very toxic to aquatic organisms
- See label for detailed instructions on tank cleaning

Hazard classification and safety precautions
  Hazard H11
  Risk phrases R50, R53a
  Operator protection U05a
  Environmental protection E15a, E34, E38
  Storage and disposal D01, D02, D05, D09a, D10b, D12a

---

230   florasulam + fluroxypyr

A post-emergence herbicide mixture for cereals

Products

1	GF 184	Dow	2.5:100 g/l	SE	10878
2	Hiker	Dow	1.0:100 g/l	SE	11451
3	Starane Gold	Dow	1.0:100 g/l	SE	10879
4	Starane Vantage	Dow	1.0:100 g/l	SE	10922
5	Starane XL	Dow	2.5:100 g/l	SE	10921

Uses
- Annual dicotyledons in *spring barley*, *spring oats*, *spring wheat*, *winter barley*, *winter oats*, *winter wheat* [1, 5]
- Chickweed in *spring barley*, *spring oats*, *spring wheat*, *winter barley*, *winter oats*, *winter wheat* [1-5]

SEE SECTION 3 FOR PRODUCTS ALSO REGISTERED

- Cleavers in *spring barley, spring oats, spring wheat, winter barley, winter oats, winter wheat* [1-5]
- Mayweeds in *spring barley, spring oats, spring wheat, winter barley, winter oats, winter wheat* [1, 5]

Approval information
- Florasulam and fluroxypyr included in Annex I under EC Directive 91/414
- Accepted by BBPA for use on malting barley

Efficacy guidance
- Best results obtained when weeds are small and growing actively
- Products are mainly absorbed through weed foliage. Cleavers emerging after application will not be controlled
- Florasulam is a member of the ALS-inhibitor group of herbicides

Restrictions
- Maximum total dose equivalent to one full dose treatment
- Product must not be applied before 1 Mar in yr of harvest
- Do not roll or harrow 7 d before or after application
- Do not spray when crops are under stress from cold, drought, pest damage or nutrient deficiency
- Do not apply through CDA applicators
- These products may be applied in sequence with products containing iodosulfuron-methyl-sodium + mesosulfuron-methyl, or products containing only one of flupyrsulfuron-methyl, metsulfuron-methyl, thifensulfuron-methyl or tribenuron-methyl
- Apart from the above, these products must not be applied in tank mix or sequence with any other ALS-inhibitor (eg sulfonylurea herbicides)
- Only one other product with an ALS-inhibitor mode of action may be applied to a treated cereal crop

Crop-specific information
- Latest use: before flag leaf sheath extended for barley and wheat; before second node detectable for oats

Following crops guidance
- Cereals, oilseed rape, field beans or grass may follow treated crops in the same yr. Oilseed rape may suffer temporary vigour reduction after a dry summer
- In addition to the above, linseed, peas, sugar beet, potatoes, maize or clover may be sown in the calendar yr following treatment
- In the event of failure of a treated crop in the spring, only spring cereals, maize or ryegrass may be planted

Environmental safety
- Dangerous for the environment
- Toxic to aquatic organisms
- Take extreme care to avoid drift onto non-target crops or plants

Hazard classification and safety precautions
    Hazard H04, H11
    Risk phrases R36, R38, R51, R53a, R67
    Operator protection A, C; U05a, U11, U19a
    Environmental protection E15a, E34, E38
    Storage and disposal D01, D02, D09a, D10c, D12a

231   fluazifop-P-butyl

A phenoxypropionic acid grass herbicide for broadleaved crops

Products

Fusilade Max	Syngenta	125 g/l	EC	11519

Uses
- Annual grasses in *blackcurrants, broad beans* (off-label), *carrots, chinese cabbage* (off-label), *collards* (off-label), *combining peas, farm forestry, field margins, flax, flax for industrial use, fodder beet, garlic* (off-label), *gooseberries, haricot beans* (off-label), *hops, kale* (stockfeed only), *linseed, linseed for industrial use, navy beans* (off-label), *onions, ornamental plant*

SECTION 2

production (off-label), parsnips (off-label), protected ornamentals (off-label), raspberries, red beet (off-label), spring field beans, spring oilseed rape, spring oilseed rape for industrial use, strawberries, sugar beet, swedes (stockfeed only), turnips (stockfeed only), vining peas, winter field beans, winter oilseed rape, winter oilseed rape for industrial use

- Annual meadow grass in *lucerne (off-label)*
- Barren brome in *field margins*
- Blackgrass in *hops, lucerne (off-label), spring field beans, winter field beans*
- Green cover in *land not being used for crop production*
- Perennial grasses in *blackcurrants, broad beans (off-label), carrots, chinese cabbage (off-label), collards (off-label), combining peas, farm forestry, flax, flax for industrial use, fodder beet, garlic (off-label), gooseberries, haricot beans (off-label), hops, kale (stockfeed only), linseed, linseed for industrial use, navy beans (off-label), onions, ornamental plant production (off-label), parsnips (off-label), protected ornamentals (off-label), raspberries, red beet (off-label), spring field beans, spring oilseed rape, spring oilseed rape for industrial use, strawberries, sugar beet, swedes (stockfeed only), turnips (stockfeed only), vining peas, winter field beans, winter oilseed rape, winter oilseed rape for industrial use*
- Ryegrass in *lucerne (off-label)*
- Volunteer cereals in *blackcurrants, carrots, combining peas, farm forestry, field margins, flax, flax for industrial use, fodder beet, gooseberries, hops, kale (stockfeed only), linseed, linseed for industrial use, lucerne (off-label), onions, raspberries, spring field beans, spring oilseed rape, spring oilseed rape for industrial use, strawberries, sugar beet, swedes (stockfeed only), turnips (stockfeed only), vining peas, winter field beans, winter oilseed rape, winter oilseed rape for industrial use*
- Wild oats in *blackcurrants, carrots, combining peas, farm forestry, field margins, flax, flax for industrial use, fodder beet, gooseberries, hops, kale (stockfeed only), linseed, linseed for industrial use, lucerne (off-label), onions, raspberries, spring field beans, spring oilseed rape, spring oilseed rape for industrial use, strawberries, sugar beet, swedes (stockfeed only), turnips (stockfeed only), vining peas, winter field beans, winter oilseed rape, winter oilseed rape for industrial use*

Specific Off-Label Approvals (SOLAs)
- *broad beans, chinese cabbage, collards, garlic, haricot beans, navy beans, ornamental plant production, parsnips, protected ornamentals, red beet (OLA 032138) Dec 2008* [1]
- *lucerne (OLA 041321) Dec 2008* [1]

Approval information
- Accepted by BBPA for use on hops

Efficacy guidance
- Best results achieved by application when weed growth active under warm conditions with adequate soil moisture. Agral or other specified adjuvant must always be added to spray. See label
- Spray weeds from 2-expanded leaf stage to fully tillered, couch from 4 leaves when majority of shoots have emerged, with a second application if necessary
- Control may be reduced under dry conditions. Do not cultivate for 2 wk after spraying couch
- Annual meadow grass is not controlled
- May also be used to remove grass cover crops

Restrictions
- Maximum number of treatments normally 2 per crop for winter oilseed rape (including crops for industrial use); 2 per yr for farm forestry; 1 per crop or yr for spring oilseed rape (including crops for industrial use) and other crops
- Do not sow cereals for at least 8 wk after application of high rate or 2 wk after low rate
- Do not apply through CDA sprayer, with hand-held equipment or from air
- Avoid treatment before spring growth has hardened or when buds opening
- Do not treat bush and cane fruit or hops between flowering and harvest
- Oilseed rape, linseed and flax for industrial use must not be harvested for human or animal consumption nor grazed
- Do not use for forestry establishment on land not previously under arable cultivation or improved grassland
- Treated vegetation in field margins, land temporarily removed from production etc, must not be grazed or harvested for human or animal consumption and unprotected persons must be kept out of treated areas for at least 24 h

SEE SECTION 3 FOR PRODUCTS ALSO REGISTERED

Crop-specific information
- Latest use: before 5 leaf stage for spring oilseed rape; before flowering for blackcurrants, gooseberries, hops, raspberries, strawberries; before flower buds visible for field beans, peas, linseed, flax, oilseed rape; 2 wk before sowing cereals or grass for field margins, land temporarily removed from production, oilseed rape for industrial use, grass crops for farm forestry
- HI beet crops, brassicas, carrots, parsnips, 8 wk; garlic, onions 4 wk
- Apply to sugar and fodder beet from 1-true leaf to 50% ground cover
- Apply to winter oilseed rape from 1-true leaf to established plant stage
- Apply to spring oilseed rape from 1-true leaf but before 5-true leaves
- Apply in fruit crops after harvest. See label for timing details on other crops
- Before using on onions or peas use crystal violet test to check that leaf wax is sufficient

Environmental safety
- Dangerous for the environment
- Very toxic to aquatic organisms

Hazard classification and safety precautions
Hazard H03, H11
Risk phrases R38, R50, R53a, R61, R63
Operator protection A, C, H, M; U05a, U08, U20b
Environmental protection E15a, E38
Storage and disposal D01, D02, D09a, D10c, D12a

232   fluazinam

A pyridinamine fungicide for use in potatoes

Products
1	Greencrop Solanum	Greencrop	500 g/l	SC	11052
2	Landgold Fluazinam	Landgold	500 g/l	SC	08060
3	Landgold Fluazinam	Teliton	500 g/l	SC	12119
4	Shirlan	Syngenta	500 g/l	SC	10573

Uses
- Blight in *potatoes* [1-4]
- Root rot in *protected blackberries* (off-label), *protected raspberries* (off-label), *protected rubus hybrids* (off-label) [4]

Specific Off-Label Approvals (SOLAs)
- *protected blackberries, protected raspberries, protected rubus hybrids* (OLA 032168) Dec 2008 [4]

Efficacy guidance
- Commence treatment at the first blight risk warning (before blight enters the crop). Products are rainfast within 1 h
- In the absence of a warning, treatment should start before foliage of adjacent plants meets in the rows
- Spray at 7-14 d intervals (5-10 d intervals [4]) depending on severity of risk (see label)
- Ensure complete coverage of the foliage and stems, increasing volume as haulm growth progresses, in dense crops and if blight risk increases

Restrictions
- Maximum number of treatments 10 per crop on potatoes; 2 per yr on protected cane fruit
- Do not use with hand-held sprayers

Crop-specific information
- HI 7 d (0 d [4])
- Latest Use: before end Mar in yr of harvest for protected blackberries, protected raspberries, protected Rubus hybrids

Environmental safety
- Dangerous for the environment
- Very toxic to aquatic organisms
- LERAP Category B

FOR FULL CONDITIONS OF USE ALWAYS READ THE PRODUCT LABEL

Hazard classification and safety precautions
    Hazard H04, H11
    Risk phrases R36, R43, R50, R53a
    Operator protection A, C, H; U02a, U04a, U05a, U08 [1-4]; U14, U15, U20a [2-4]; U20b [1]
    Environmental protection E15a, E16a [1-4]; E16b [2-4]; E34, E38 [4]
    Storage and disposal D01, D02, D09a [1-4]; D05, D10b [1-3]; D10c, D12a [4]
    Medical advice M03 [1-4]; M05a [4]

233   fluazinam + metalaxyl-M

A mixture of contact and systemic fungicides for potatoes

Products

1	Emrald Flumet	Me2	400:200 g/l	EC	12000
2	Epok	Belchim	400:200 g/l	EC	11997

Uses
- Blight in **potatoes**

Efficacy guidance
- Commence treatment at the first blight risk warning (before blight enters the crop). Products are rainfast within 1 h
- In the absence of a warning, treatment should start before foliage of adjacent plants meets in the rows
- Spray at 7-14 d intervals depending on severity of risk (see label)
- Ensure complete coverage of the foliage and stems, increasing volume as haulm growth progresses, in dense crops and if blight risk increases
- Metalaxyl-M works best on young actively growing foliage and efficacy declines with the onset of senescence. Therefore it is recommended that the product is only used for the first part of the blight control program, which should be completed with a reliable protectant fungicide

Restrictions
- Maximum number of treatments 5 per crop

Crop-specific information
- HI 7d for potatoes

Environmental safety
- Dangerous for the environment
- Very toxic to aquatic organisms
- LERAP Category B

Hazard classification and safety precautions
    Hazard H03, H11
    Risk phrases R20, R36, R38, R40, R43, R50, R53a
    Operator protection A, C, H; U05a, U07, U11, U14, U15, U19a, U20a
    Environmental protection E15a, E16a, E16b, E34, E38
    Storage and disposal D01, D02, D09a, D10b, D12a
    Medical advice M05a

234   fludioxonil

A cyanopyrrole fungicide seed treatment for wheat and barley

See also cymoxanil + fludioxonil + metalaxyl-M

Products

Beret Gold	Syngenta	25 g/l	FS	11635

Uses
- Bunt in **winter wheat** (seed treatment)
- Covered smut in **spring barley** (seed treatment), **winter barley** (seed treatment)
- Fusarium foot rot and seedling blight in **spring barley** (seed treatment), **spring oats** (seed treatment), **spring wheat** (seed treatment), **winter barley** (seed treatment), **winter oats** (seed treatment), **winter wheat** (seed treatment)

SEE SECTION 3 FOR PRODUCTS ALSO REGISTERED

- Leaf stripe in **spring barley** *(seed treatment - reduction)*, **winter barley** *(seed treatment - reduction)*
- Pyrenophora leaf spot in **spring oats** *(seed treatment)*, **winter oats** *(seed treatment)*
- Snow mould in **spring barley** *(seed treatment)*, **winter barley** *(seed treatment)*, **winter wheat** *(seed treatment)*

Approval information
- Accepted by BBPA for use on malting barley

Efficacy guidance
- Apply direct to seed using conventional seed treatment equipment. Continuous flow treaters should be calibrated using product before use
- Effective against benzimidazole-resistant strains of *Fusarium nivale*

Restrictions
- Maximum number of treatments 1 per seed batch
- Do not apply to cracked, split or sprouted seed
- Sow treated seed within 6 mth

Crop-specific information
- Latest use: before drilling
- Product may reduce flow rate of seed through drill. Recalibrate with treated seed before drilling

Environmental safety
- Harmful to aquatic organisms
- Do not use treated seed as food or feed
- Treated seed harmful to game and wildlife

Hazard classification and safety precautions
Risk phrases R52, R53a
Operator protection A, H; U05a, U20b
Environmental protection E03, E15a
Storage and disposal D01, D02, D09a, D11a
Treated seed S02, S05, S07

235   flufenacet

A broad spectrum oxyacetamide herbicide available only in mixtures

See also diflufenican + flufenacet

236   flufenacet + metribuzin

A herbicide mixture for potatoes

Products

Artist	Bayer CropScience	24:17.5% w/w	WP	11239

Uses
- Annual dicotyledons in **early potatoes**, **maincrop potatoes**
- Annual meadow grass in **early potatoes**, **maincrop potatoes**

Efficacy guidance
- Product acts through root uptake and needs sufficient soil moisture at and shortly after application
- Effectiveness is reduced under dry soil conditions
- Residual activity is reduced on mineral soils with a high organic matter content and on peaty or organic soils
- Ensure application is made evenly to both sides of potato ridges
- Perennial weeds are not controlled

Restrictions
- Maximum total dose equivalent to one full dose treatment
- Potatoes must be sprayed before emergence of crop and weeds
- See label for list of tolerant varieties. Do not treat Maris Piper grown on Sands or Very Light soils

- Do not use on Sands
- On stony or gravelly soils there is risk of crop damage especially if heavy rain falls soon after application

Crop-specific information
- Latest use: before potato crop emergence
- Consult processor before use on crops for processing

Following crops guidance
- Before drilling or planting any succeeding crop soil must be mouldboard ploughed to at least 15 cm as soon as possible after lifting and no later than end Dec
- In W Cornwall on soils with more than 5% organic matter treated early potatoes may be followed by summer planted brassica crops 14 wk after treatment and after mouldboard ploughing. Elsewhere cereals or winter beans may be grown in the same year if at least 16 wk have elapsed since treatment
- In the yr following treatment any crop may be grown except lettuce or radish, or vegetable brassica crops on silt soils in Lincs

Environmental safety
- Dangerous for the environment
- Very toxic to aquatic organisms
- Take care to avoid spray drift onto neighbouring crops, especially lettuce or brassicas

Hazard classification and safety precautions
> **Hazard** H03, H11
> **Risk phrases** R22a, R43, R48, R50, R53a
> **Operator protection** A, D, H; U05a, U08, U13, U14, U19a, U20b
> **Environmental protection** E15a, E38
> **Storage and disposal** D01, D02, D09a, D11a, D12a
> **Medical advice** M03

237 flufenacet + pendimethalin

A broad spectrum residual and contact herbicide mixture for winter cereals

Products

1	Crystal	BASF	60:300 g/l	EC	10657
2	Emrald Berg	Me2	60:300 g/l	EC	12086
3	Greencrop Duet 603	Greencrop	60:300 g/l	EC	12171
4	Ice	BASF	60:300 g/l	EC	10681
5	Landgold Bedrock	Teliton	60:300 g/l	EC	12145
6	Standon FFA60 Plus	Standon	60:300 g/l	EC	11649
7	Trooper	BASF	60:300 g/l	EC	10682

Uses
- Annual dicotyledons in *winter barley, winter wheat*
- Annual grasses in *winter barley, winter wheat*
- Annual meadow grass in *winter barley, winter wheat*
- Blackgrass in *winter barley, winter wheat*
- Chickweed in *winter barley, winter wheat*
- Corn marigold in *winter barley, winter wheat*
- Speedwells in *winter barley, winter wheat*

Approval information
- Pendimethalin included in Annex I under EC Directive 91/414
- Accepted by BBPA for use on malting barley

Efficacy guidance
- Best results achieved when applied from pre-emergence of weeds to 2-leaf stage but post emergence treatment is not recommended on clay soils
- Product requires some soil moisture to be activated ideally from rain within 7 d of application. Prolonged dry conditions may reduce residual control
- Product is slow acting and final level of weed control may take some time to appear
- For effective weed control seed bed preparations should ensure even incorporation of any trash, straw and ash to 15 cm

SEE SECTION 3 FOR PRODUCTS ALSO REGISTERED

- Efficacy may be reduced on soils with more than 6% organic matter
- Always follow WRAG guidelines for preventing and managing herbicide resistant weeds. Section 5 for more information

Restrictions
- Maximum total dose equivalent to one full dose treatment
- For pre-emergence treatments seed should be covered with min 32 mm settled soil. Shallow drilled crops should be treated post-emergence only
- Do not treat undersown crops
- Avoid spraying during periods of prolonged or severe frosts
- Do not use on stony or gravelly soils or those with more than 10% organic matter
- Pre-emergence treatment may only be used on crops drilled before 30 Nov. All crops must be treated before 31 Dec in yr of planting.
- Concentrated or diluted product may stain clothing or skin

Crop-specific information
- Latest use: before third tiller stage (GS 23) and before 31 Dec in yr of planting
- Very wet weather before and after treatment may result in loss of crop vigour and reduced yield, particularly where soils become waterlogged

Following crops guidance
- Any crop may follow a failed or normally harvested treated crop provided ploughing to at least 15 cm is carried out beforehand

Environmental safety
- Dangerous for the environment
- Very toxic to aquatic organisms
- Risk to certain non-target insects or other arthropods
- LERAP Category B
- Some products supplied in small volume returnable packs. Follow instructions for use

Hazard classification and safety precautions
>   **Hazard** H03, H11
>   **Risk phrases** R22a, R22b, R38, R40, R50, R53a
>   **Operator protection** A, C, H; U02a, U05a, U20c
>   **Environmental protection** E15a, E16a, E16b, E34 [1-7]; E22b, E38 [1-5, 7]; E22c [6]
>   **Storage and disposal** D01, D02, D09a, D10a [1-7]; D12a [1-5, 7]
>   **Medical advice** M03, M05b

238   fluoxastrobin

A protectant stobilurin fungicide available in mixtures

239   fluoxastrobin + prothioconazole

A strobilurin and triazole fungicide mixture for cereals

Products

1	Fandango	Bayer CropScience	100:100 g/l	EC	12276
2	Maestro	Bayer CropScience	100:100 g/l	EC	12307
3	Redigo Twin	Bayer CropScience	37.5:37.5 g/l	FS	12314

Uses
- Brown rust in *spring barley*, *winter barley*, *winter rye*, *winter wheat* [1, 2]
- Bunt in *winter wheat* (seed treatment) [3]
- Eyespot in *spring barley* (reduction), *winter barley* (reduction), *winter rye* (reduction), *winter wheat* (reduction) [1, 2]
- Fusarium foot rot and seedling blight in *winter wheat* (seed treatment) [3]
- Late ear diseases in *spring barley*, *winter barley*, *winter wheat* [1, 2]
- Loose smut in *winter wheat* (seed treatment) [3]
- Net blotch in *spring barley*, *winter barley* [1, 2]
- Powdery mildew in *spring barley*, *winter barley*, *winter rye*, *winter wheat* [1, 2]
- Rhynchosporium in *spring barley*, *winter barley*, *winter rye* [1, 2]

FOR FULL CONDITIONS OF USE ALWAYS READ THE PRODUCT LABEL

- Septoria diseases in **winter wheat** [1, 2]
- Tan spot in **winter wheat** [1, 2]
- Yellow rust in **winter wheat** [1, 2]

Approval information
- Accepted by BBPA for use on malting barley

Efficacy guidance
- Seed treatments must be applied by manufacturer's recommended treatment application equipment [3]
- Treated seed should preferably be drilled in the same season [3]
- Follow-up treatments will be needed later in the season to give protection against air-borne and splash-borne diseases [3]
- Best results on foliar diseases obtained from treatment at early stages of disease development. Further treatment may be needed if disease attack is prolonged [1, 2]
- Foliar applications to established infections of any disease are likely to be less effective [1, 2]
- Best control of cereal ear diseases obtained by treatment during ear emergence [1, 2]
- Fluoxastrobin is a member of the QoI cross resistance group. Foliar product should be used preventatively and not relied on for its curative potential [1, 2]
- Use product as part of an Integrated Crop Management strategy incorporating other methods of control, including where appropriate other fungicides with a different mode of action. Do not apply more than two foliar applications of QoI containing products to any cereal crop
- There is a significant risk of widespread resistance occurring in *Septoria tritici* populations in UK. Failure to follow resistance management action may result in reduced levels of disease control
- Strains of wheat and barley powdery mildew resistant to QoIs are common in the UK. Control of wheat mildew can only be relied on from the triazole component
- Where specific control of wheat mildew is required this should be achieved through a programme of measures including products recommended for the control of mildew that contain a fungicide from a different cross-resistance group and applied at a dose that will give robust control

Restrictions
- Maximum number of seed treatments one per batch [3]
- Maximum total dose of foliar sprays equivalent to two full dose treatments [1, 2]
- Seed treatment must be fully re-dispersed and homogeneous before use [3]
- Do not use on seed with more than 16% moisture content, or on sprouted, cracked or skinned seed [3]
- All seed batches should be tested to ensure they are suitable for treatment [3]

Crop-specific information
- Latest use: pre-drilling for seed treatment [3]; before grain milky ripe for wheat and rye; beginning of flowering for barley [1, 2]
- Some transient leaf chlorosis may occur after treatment of wheat or barley but this has not been found to affect yield [1, 2]

Environmental safety
- Dangerous for the environment
- Toxic to aquatic organisms
- Risk to non-target insects or other arthropods. Avoid spraying within 6 m of the field boundary to reduce the effects on non-target insects or other arthropods [1, 2]
- LERAP Category B [1, 2]

Hazard classification and safety precautions
Hazard H04 [3]; H11 [1-3]
Risk phrases R43 [3]; R51, R53a [1-3]
Operator protection A [1-3]; H [3]; U05a [1-3]; U09b, U20a [1, 2]; U14 [3]
Environmental protection E03 [3]; E15a, E34, E38 [1-3]; E16a, E22c [1, 2]
Storage and disposal D01, D05, D09a, D12a [1-3]; D02, D11a [3]; D10b [1, 2]
Treated seed S01, S02, S03, S04a, S04b, S05, S06a, S07, S08 [3]
Medical advice M03 [1, 2]

SECTION 2

SEE SECTION 3 FOR PRODUCTS ALSO REGISTERED

240   fluoxastrobin + prothioconazole + trifloxystrobin

A triazole and strobilurin fungicide mixture for cereals

Products

Jaunt	Bayer CropScience	75:150:75 g/l	EC	12350

Uses

* Brown rust in **spring barley, winter barley, winter wheat**
* Eyespot in **spring barley** *(reduction)*, **winter barley** *(reduction)*, **winter wheat** *(reduction)*
* Late ear diseases in **spring barley, winter barley, winter wheat**
* Net blotch in **spring barley, winter barley**
* Powdery mildew in **spring barley, winter barley, winter wheat**
* Rhynchosporium in **spring barley, winter barley**
* Septoria diseases in **winter wheat**
* Tan spot in **winter wheat**
* Yellow rust in **winter wheat**

Approval information

* Trifloxystrobin included in Annex I under EC Directive 91/414
* Accepted by BBPA for use on malting barley

Efficacy guidance

* Best results obtained from treatment at early stages of disease development. Further treatment may be needed if disease attack is prolonged
* Applications to established infections of any disease are likely to be less effective
* Best control of cereal ear diseases obtained by treatment during ear emergence
* Fluoxastrobin and trifloxystrobin are members of the QoI cross resistance group. Product should be used preventatively and not relied on for its curative potential
* Use product as part of an Integrated Crop Management strategy incorporating other methods of control, including where appropriate other fungicides with a different mode of action. Do not apply more than two foliar applications of QoI containing products to any cereal crop
* There is a significant risk of widespread resistance occurring in *Septoria tritici* populations in UK. Failure to follow resistance management action may result in reduced levels of disease control
* Strains of wheat and barley powdery mildew resistant to QoIs are common in the UK. Control of wheat mildew can only be relied on from the triazole component
* Where specific control of wheat mildew is required this should be achieved through a programme of measures including products recommended for the control of mildew that contain a fungicide from a different cross-resistance group and applied at a dose that will give robust control

Restrictions

* Maximum total dose equivalent to two full dose treatments

Crop-specific information

* Latest use: before grain milky ripe for winter wheat; up to beginning of anthesis (GS 61) for barley

Environmental safety

* Dangerous for the environment
* Very toxic to aquatic organisms
* Risk to non-target insects or other arthropods. Avoid spraying within 6 m of the field boundary to reduce the effects on non-target insects or other arthropods
* LERAP Category B

Hazard classification and safety precautions

Hazard H04, H11
Risk phrases R37, R50, R53a
Operator protection A, C, H; U05a, U09b, U19a, U20b
Environmental protection E15a, E16a, E22c, E34, E38
Storage and disposal D01, D02, D05, D09a, D10b, D12a
Medical advice M03

FOR FULL CONDITIONS OF USE ALWAYS READ THE PRODUCT LABEL

241   flupyrsulfuron-methyl

A sulfonylurea herbicide for winter wheat and winter barley

See also carfentrazone-ethyl + flupyrsulfuron-methyl

Products

1 Bullion	DuPont	50% w/w	WG	12050
2 Lexus 50 DF	DuPont	50% w/w	WG	09026

Uses
- Annual dicotyledons in **winter wheat** [1, 2]
- Blackgrass in **winter barley** [1]; **winter wheat** [1, 2]

Approval information
- Flupyrsulfuron-methyl included in Annex I under EC Directive 91/414
- Approval expiry 2 Nov 2006 [2]

Efficacy guidance
- Best results on winter wheat achieved from applications made in good growing conditions
- Good spray cover of weeds must be obtained
- Best control of blackgrass in wheat obtained from application from the 1-leaf stage
- Winter barley must be treated pre-emergence and used in mixture. See label [1]
- Growth of weeds is inhibited within hours of treatment but visible symptoms may not be apparent for up to 4 wk
- Product has moderate residual life in soil. Under normal moisture conditions susceptible weeds germinating soon after treatment will be controlled
- Product may be used on all soil types but residual activity and weed control is reduced on highly alkaline soils
- Flupyrsulfuron-methyl is a member of the ALS-inhibitor group of herbicides and products should be used in a planned Resistance Management strategy. See Section 5 for more information

Restrictions
- Maximum number of treatments 1 per crop
- Do not use on wheat undersown with grasses or legumes, or any other broad-leaved crop
- Do not apply within 7 d of rolling
- Do not treat any crop suffering from drought, waterlogging, pest or disease attack, nutrient deficiency, or any other stress factors
- Specific restrictions apply to use in sequence or tank mixture with other sulfonylurea or ALS-inhibiting herbicides. See label for details

Crop-specific information
- Latest use: before 1st node detectable (GS 31) for wheat; pre-emergence for barley [1]
- Chlorosis, stunting and loss of vigour may occur in certain conditions. Crops normally recover but yield reductions are possible [1]

Following crops guidance
- Only cereals, oilseed rape, field beans, clover or grass may be sown in the yr of harvest of a treated crop
- In the event of crop failure only winter or spring wheat may be sown within 3 mth of treatment. Land should be ploughed and cultivated to 15 cm minimum before resowing. After 3 mth crops may be sown as shown above

Environmental safety
- Dangerous for the environment
- Very toxic to aquatic organisms
- Extremely dangerous to fish or other aquatic life. Do not contaminate surface waters or ditches with chemical or used container [2]
- Take extreme care to avoid drift onto broad-leaved plants outside the target area or onto ponds, waterways or ditches, or onto land intended for cropping
- Spraying equipment should not be drained or flushed onto land planted, or to be planted, with trees or crops other than cereals and should be thoroughly cleansed after use - see label for instructions

SEE SECTION 3 FOR PRODUCTS ALSO REGISTERED

SECTION 2

Hazard classification and safety precautions
 Hazard H11
 Risk phrases R50, R53a
 Operator protection U08 [1]; U09a [2]; U19a, U20b [1, 2]
 Environmental protection E13a [2]; E15a [1]; E38 [1, 2]
 Storage and disposal D09a, D11a, D12a

242   flupyrsulfuron-methyl + picolinafen

A mixture of sulfonylurea and pyridinecarboxamide herbicides for cereals

Products
 DP 945   DuPont   16.7:50% w/w   WG   11447

Uses
 • Annual dicotyledons in *winter wheat*
 • Blackgrass in *winter wheat*

Approval information
 • Flupyrsulfuron-methyl included in Annex I under EC Directive 91/414

Efficacy guidance
 • Best results obtained when applied to small actively growing weeds
 • Good spray cover of weeds must be obtained
 • Increased degradation of active ingredient in high soil temperatures reduces residual activity
 • Product has moderate residual life in soil. Under normal moisture conditions susceptible weeds germinating soon after treatment will be controlled
 • Product may be used on all soil types but residual activity and weed control is reduced on highly alkaline soils
 • Blackgrass should be treated from 2 leaves up to first node detectable
 • Flupyrsulfuron-methyl is a member of the ALS-inhibitor group of herbicides and products should be used in a planned Resistance Management strategy. See Section 5 for more information

Restrictions
 • Maximum number of treatments 1 per crop
 • Do not use on wheat undersown with grasses or legumes, or any other broad-leaved crop
 • Do not apply within 7 d of rolling
 • Do not treat any crop suffering from drought, waterlogging, pest or disease attack, nutrient deficiency, or any other stress factors
 • Specific restrictions apply to use in sequence or tank mixture with other sulfonylurea or ALS-inhibiting herbicides. See label for details

Crop-specific information
 • Latest use: before pseudo-stem erect stage (GS 30) of wheat
 • Slight chlorosis and stunting may occur in certain conditions. Recovery is rapid and yield not affected

Following crops guidance
 • Only cereals, oilseed rape, field beans, clover or grass may be sown in the yr of harvest of a treated crop
 • In the event of crop failure only winter or spring wheat may be sown within 3 mth of treatment. Land should be ploughed and cultivated to 15 cm minimum before resowing. After 3 mth crops may be sown as indicated above

Environmental safety
 • Dangerous for the environment
 • Very toxic to aquatic organisms
 • Extremely dangerous to fish or other aquatic life. Do not contaminate surface waters or ditches with chemical or used container
 • LERAP Category B
 • Take extreme care to avoid drift onto broad-leaved plants outside the target area or onto ponds, waterways or ditches, or onto land intended for cropping

FOR FULL CONDITIONS OF USE ALWAYS READ THE PRODUCT LABEL

- Spraying equipment should not be drained or flushed onto land planted, or to be planted, with trees or crops other than cereals and should be thoroughly cleansed after use - see label for instructions

Hazard classification and safety precautions
  Hazard H11
  Risk phrases R50, R53a
  Operator protection A; U08, U19a, U20b
  Environmental protection E13a, E16a, E38
  Storage and disposal D01, D09a, D11a, D12a

SECTION 2

243   flupyrsulfuron-methyl + thifensulfuron-methyl

A sulfonylurea herbicide mixture for winter wheat and winter oats

Products
  Lexus Millenium   DuPont   10:40% w/w   WG   09206

Uses
- Annual dicotyledons in **winter oats, winter wheat**
- Blackgrass in **winter oats, winter wheat**

Approval information
- Flupyrsulfuron-methyl and thifensulfuron-methyl included in Annex I under EC Directive 91/414

Efficacy guidance
- Best results obtained when applied to small actively growing weeds
- Good spray cover of weeds must be obtained
- Increased degradation of active ingredient in high soil temperatures reduces residual activity
- Product has moderate residual life in soil. Under normal moisture conditions susceptible weeds germinating soon after treatment will be controlled
- Product may be used on all soil types but residual activity and weed control is reduced on highly alkaline soils
- Blackgrass should be treated from 1 leaf up to first node stage
- Flupyrsulfuron-methyl and thifensulfuron-methyl are members of the ALS-inhibitor group of herbicides and products should be used in a planned Resistance Management strategy. See Section 5 for more information

Restrictions
- Maximum number of treatments 1 per crop
- Do not use on wheat undersown with grasses or legumes, or any other broad-leaved crop
- Do not apply within 7 d of rolling
- Do not treat any crop suffering from drought, waterlogging, pest or disease attack, nutrient deficiency, or any other stress factors
- Contact contract agents before use on crops grown for seed
- Specific restrictions apply to use in sequence or tank mixture with other sulfonylurea or ALS-inhibiting herbicides. See label for details

Crop-specific information
- Latest use: before 1st node detectable (GS 31) on wheat; before 31 Dec in yr of sowing on oats
- Slight chlorosis, speckling and stunting may occur in certain conditions. Recovery is rapid and yield not affected

Following crops guidance
- Only cereals, oilseed rape, field beans or grass may be sown in the yr of harvest of a treated crop. Any crop may be sown the following spring
- In the event of crop failure only winter wheat may be sown before normal harvest date. Land should be ploughed and cultivated to 15 cm minimum before resowing

Environmental safety
- Dangerous for the environment
- Very toxic to aquatic organisms
- LERAP Category B

SEE SECTION 3 FOR PRODUCTS ALSO REGISTERED

Hazard classification and safety precautions
    Hazard H04, H11
    Risk phrases R43, R50, R53a
    Operator protection A, H; U08, U14, U19a, U20b
    Environmental protection E15a, E16a, E38
    Storage and disposal D01, D02, D09a, D11a, D12a

244   flupyrsulfuron-methyl + tribenuron-methyl

A foliar applied sulfonylurea herbicide mixture

Products

Excalibur	DuPont	30:30 % w/w	WG	12047

Uses
- Annual dicotyledons in *winter oats, winter wheat*
- Blackgrass in *winter oats, winter wheat*

Approval information
- Flupyrsulfuron and tribenuron-methyl included in Annex I under EC Directive 91/414

Efficacy guidance
- Best results obtained when applied to small actively growing weeds
- Good spray cover of weeds must be obtained
- Increased degradation of active ingredient in high soil temperatures reduces residual activity
- Product has moderate residual life in soil. Under normal moisture conditions susceptible weeds germinating soon after treatment will be controlled
- Product may be used on all soil types but residual activity and weed control is reduced on highly alkaline soils
- Blackgrass should be treated from 2 leaf up to before first node detectable
- Flupyrsulfuron-methyl and tribenuron-methyl are members of the ALS-inhibitor group of herbicides and products should be used in a planned Resistance Management strategy. See Section 5 for more information

Restrictions
- Maximum number of treatments 1 per crop on oats and wheat
- Do not use on wheat undersown with grasses or legumes, or any other broad-leaved crop
- Do not apply within 7 d of rolling
- Do not treat any crop suffering from drought, waterlogging, pest or disease attack, nutrient deficiency, or any other stress factors
- Specific restrictions apply to use in sequence or tank mixture with other sulfonylurea or ALS-inhibiting herbicides. See label for details

Crop-specific information
- Latest Use: 31 Dec for winter oats; before first node detectable stage (GS 31) for winter wheat
- Slight chlorosis and stunting may occur in certain conditions. Recovery is rapid and yield not affected

Following crops guidance
- Only cereals, oilseed rape or field beans may be sown in the yr of harvest of a treated crop
- In the event of crop failure only winter or spring wheat may be sown within 3 mth of treatment. Land should be ploughed and cultivated to 15 cm minimum before resowing. After 3 mth crops may be sown as indicated above

Environmental safety
- Dangerous for the environment
- Extremely dangerous to fish or other aquatic life. Do not contaminate surface waters or ditches with chemical or used container
- Take extreme care to avoid drift onto broad-leaved plants outside the target area or onto ponds, waterways or ditches, or onto land intended for cropping
- Spraying equipment should not be drained or flushed onto land planted, or to be planted, with trees or crops other than cereals and should be thoroughly cleansed after use - see label for instructions

FOR FULL CONDITIONS OF USE ALWAYS READ THE PRODUCT LABEL

Hazard classification and safety precautions
> **Hazard** H04, H11
> **Risk phrases** R43, R50, R53a
> **Operator protection** A; U05a, U08, U20b
> **Environmental protection** E13a, E38
> **Storage and disposal** D01, D02, D09a, D11a, D12a

245   fluquinconazole

A protectant, eradicant and systematic triazole fungicide for winter cereals

Products

1	Flamenco	BASF	100 g/l	SC	11699
2	Galmano	Bayer CropScience	167 g/l	FS	11650
3	Jockey F	BASF	167 g/l	FS	11690
4	Sahara	Bayer CropScience	100 g/l	SC	11905

Uses
- Brown rust in **winter wheat** [1, 4]
- Bunt in **winter wheat** *(seed treatment)* [2, 3]
- Covered smut in **winter barley** *(seed treatment)* [3]
- Loose smut in **winter barley** *(seed treatment)* [3]
- Powdery mildew in **winter wheat** [1, 4]
- Septoria diseases in **winter wheat** [1, 4]
- Septoria leaf spot in **winter wheat** *(seed treatment)* [2, 3]
- Take-all in **winter barley** *(seed treatment - reduction)* [3]; **winter wheat** *(seed treatment - reduction)* [2, 3]
- Yellow rust in **winter barley** *(seed treatment)* [3]; **winter wheat** [1, 4]; **winter wheat** *(seed treatment)* [2, 3]

Efficacy guidance
- Best results obtained from spray treatments when disease first becomes active in crop but before infection spreads to younger leaves [1, 4]
- Adequate disease protection throughout the season will usually require a programme of at least two fungicide treatments [1, 4]
- May also be applied as a protectant at end of ear emergence (but no later) if crop still disease free [1, 4]
- With seed treatments ensure good even coverage of seed to obtain reliable disease control [2, 3]
- Seed treatment provides early control of *Septoria* leaf spot and yellow rust but may require foliar treatment for later infection [2, 3]

Restrictions
- Maximum total spray dose equivalent to two full doses [1, 4]
- Maximum number of seed treatments 1 per batch [2, 3]
- Do not apply spray treatments after the start of anthesis (GS 59) [1, 4]
- Do not mix with any other products [2, 3]
- Do not treat cracked, split or sprouted seed [2, 3]
- Do not use treated seed as food or feed [2, 3]

Crop-specific information
- Latest use: before beginning of anthesis (GS 59) for spray treatment; before drilling for seed treatment
- Ensure good spray coverage and increase volume in dense crops [1, 4]
- Delayed emergence may result when treated seed is drilled into heavy or poorly drained soils which then become wet or waterlogged [2, 3]

Environmental safety
- Dangerous for the environment
- Toxic to aquatic organisms
- Dangerous to fish or other aquatic life. Do not contaminate surface waters or ditches with chemical or used container
- Special PPE requirements and precautions apply where product supplied in returnable packs. Check label

SEE SECTION 3 FOR PRODUCTS ALSO REGISTERED

Hazard classification and safety precautions

Hazard H02, H11

Risk phrases R22a, R48, R51, R53a [1-4]; R36, R43 [1, 4]

Operator protection A, H [1-4]; C [1, 4]; D [2, 3]; M [1]; U05a [1-4]; U07 [2]; U07 [3] (1000 l containers); U11, U14, U15 [1, 4]; U20b [2, 3]

Environmental protection E13b [2, 4]; E15a [1, 3]; E34, E38 [1-4]; E36 [2, 3]

Storage and disposal D01, D02, D12a [1-4]; D05 [1, 2, 4]; D08, D09a [2, 3]; D10c [1, 4]; D11a [3] (25-200 l containers); D14 [2]; D14 [3] (1000 l containers)

Treated seed S01, S02, S03, S04b, S05, S06a, S07, S08 [2, 3]

Medical advice M03 [1-4]; M04 [1, 4]

246   fluquinconazole + prochloraz

A broad-spectrum triazole mixture for use as a spray and seed treatment in winter cereals

Products

1	Epona	BASF	167:31 g/l	FS	12444
2	Foil	BASF	54:174 g/l	SE	11700
3	Galmano Plus	Bayer CropScience	167:31 g/l	FS	11645
4	Jockey	BASF	167:31 g/l	FS	11689

Uses

- Brown rust in *winter barley*, *winter wheat* [2]; *winter wheat* (seed treatment - moderate control) [1]
- Bunt in *winter wheat* (seed treatment) [1, 3, 4]
- Covered smut in *winter barley* (seed treatment) [4]
- Fusarium foot rot and seedling blight in *winter barley* (seed treatment) [4]
- Fusarium root rot in *winter wheat* (seed treatment) [1, 3, 4]
- Loose smut in *winter barley* (seed treatment) [4]
- Powdery mildew in *winter barley*, *winter wheat* (moderate control) [2]
- Rhynchosporium in *winter barley* [2]
- Septoria leaf spot in *winter wheat* [2]; *winter wheat* (seed treatment) [3, 4]; *winter wheat* (seed treatment - moderate control) [1]
- Take-all in *winter barley* (seed treatment - reduction) [4]; *winter wheat* (seed treatment - reduction) [1, 3, 4]
- Yellow rust in *winter barley* (seed treatment) [4]; *winter wheat* [2]; *winter wheat* (seed treatment) [3, 4]; *winter wheat* (seed treatment - moderate control) [1]

Approval information

- Accepted by BBPA for use on malting barley (before ear emergence only)

Efficacy guidance

- Best results obtained from spay treatment when disease first becomes active in crop but before infection spreads to younger leaves [2]
- When treating seed ensure good even coverage of seed to obtain reliable disease control [1, 3, 4]
- Seed treatment provides early control of *Septoria* leaf spot and rust but may require foliar treatment for later infection [1, 3, 4]

Restrictions

- Maximum number of treatments 1 per seed batch (seed dressing)
- Maximum total dose equivalent to two full doses (spray)
- Do not treat cracked, split or sprouted seed [1, 3, 4]
- Do not use treated seed as food or feed [1, 3, 4]

Crop-specific information

- Latest use: before 1st awns visible (GS 47) for winter barley; before beginning of anthesis (GS 59) for winter wheat; before drilling (seed treatment)
- Ensure good spray coverage and increase volume in dense crops [2]
- Delayed emergence may result when treated seed is drilled into heavy or poorly drained soils which then become wet or waterlogged [1, 3, 4]

Environmental safety

- Dangerous for the environment
- Toxic to aquatic organisms

FOR FULL CONDITIONS OF USE ALWAYS READ THE PRODUCT LABEL

- Dangerous to fish or other aquatic life. Do not contaminate surface waters or ditches with chemical or used container
- Product supplied in returnable packs for which special PPE requirements and precautions apply. Check label

Hazard classification and safety precautions
Hazard H02 [1, 3, 4]; H03, H04 [2]; H11 [1-4]
Risk phrases R22a, R48, R51, R53a [1-4]; R36 [2, 3]; R43, R66 [2]
Operator protection A, H [1-4]; C [2]; D [3, 4]; G [1]; U05a [1-4]; U07 [3]; U07 [4] (1000 l containers only); U20b [1, 3, 4]
Environmental protection E13b, E36 [3]; E15a [1, 2, 4]; E34, E38 [1-4]; E36 [4] (1000 l containers only)
Storage and disposal D01, D02, D09a, D12a [1-4]; D05 [2, 3]; D08 [1, 3, 4]; D10c [2]; D11a [1]; D11a [4] (25-200 l containers only); D14 [3]; D14 [4] (1000 l containers only)
Treated seed S01, S03, S04b, S06a [3, 4]; S02, S05, S07, S08 [1, 3, 4]
Medical advice M03 [2, 3]; M04 [1, 4]; M05b [2]

SECTION 2

247 fluroxypyr

A post-emergence aryloxyalkanoic acid herbicide

See also 2,4-D + dicamba + fluroxypyr
clopyralid + fluroxypyr + MCPA
clopyralid + fluroxypyr + triclopyr
florasulam + fluroxypyr

Products

1	Greencrop Reaper	Greencrop	200 g/l	EC	12261
2	Starane 2	Dow	200 g/l	EC	12018
3	Tomahawk	Makhteshim	200 g/l	EC	09249

Uses
- Annual dicotyledons in **bulb onions** *(off-label)*, **leeks** *(off-label)*, **maize** [2]; **durum wheat, established grassland, seedling leys, spring barley, spring oats, spring rye, spring wheat, triticale, winter barley, winter oats, winter rye, winter wheat** [1-3]; **forage maize** [1, 3]
- Black bindweed in **durum wheat, established grassland, seedling leys, spring barley, spring oats, spring rye, spring wheat, triticale, winter barley, winter oats, winter rye, winter wheat** [1-3]; **forage maize** [1, 3]; **maize** [2]
- Chickweed in **durum wheat, established grassland, seedling leys, spring barley, spring oats, spring rye, spring wheat, triticale, winter barley, winter oats, winter rye, winter wheat** [1-3]; **forage maize** [1, 3]; **maize** [2]
- Cleavers in **apple orchards** *(off-label)*, **maize, pear orchards** *(off-label)* [2]; **durum wheat, established grassland, seedling leys, spring barley, spring oats, spring rye, spring wheat, triticale, winter barley, winter oats, winter rye, winter wheat** [1-3]; **forage maize** [1, 3]
- Docks in **apple orchards** *(off-label)*, **maize, pear orchards** *(off-label)* [2]; **durum wheat, established grassland, seedling leys, spring barley, spring oats, spring rye, spring wheat, triticale, winter barley, winter oats, winter rye, winter wheat** [1-3]; **forage maize** [1, 3]
- Forget-me-not in **durum wheat, established grassland, seedling leys, spring barley, spring oats, spring rye, spring wheat, triticale, winter barley, winter oats, winter rye, winter wheat** [1-3]; **forage maize** [1, 3]; **maize** [2]
- Hemp-nettle in **durum wheat, established grassland, seedling leys, spring barley, spring oats, spring rye, spring wheat, triticale, winter barley, winter oats, winter rye, winter wheat** [1-3]; **forage maize** [1, 3]; **maize** [2]
- Stinging nettle in **apple orchards** *(off-label)*, **pear orchards** *(off-label)* [2]
- Volunteer potatoes in **bulb onions** *(off-label)*, **leeks** *(off-label)*, **sweetcorn** *(off-label)* [2]; **winter barley, winter wheat** [1-3]

Specific Off-Label Approvals (SOLAs)
- **apple orchards, pear orchards** *(OLA 040988) May 2007* [2]
- **bulb onions** *(OLA 040986) May 2007* [2]
- **leeks** *(OLA 040987) Dec 2008* [2]
- **sweetcorn** *(OLA 051696) Nov 2010* [2]

SEE SECTION 3 FOR PRODUCTS ALSO REGISTERED

Approval information
- Fluroxypyr included in Annex I under EC Directive 91/414
- Accepted by BBPA for use on malting barley

Efficacy guidance
- Best results achieved under good growing conditions in a strongly competing crop
- A number of tank mixtures with HBN and other herbicides are recommended for use in autumn and spring to extend range of species controlled. See label for details
- Spray is rainfast in 1 h

Restrictions
- Maximum number of treatments 1 per crop or yr or maximum total dose equivalent to one full dose treatment
- Do not apply in any tank-mix on triticale or forage maize
- Do not use on crops undersown with clovers or other legumes
- Do not treat crops suffering stress caused by any factor
- Do not roll or harrow for 7 d before or after treatment
- Do not spray if frost imminent
- Straw from treated crops must not be returned directly to the soil but must be removed and used only for livestock bedding

Crop-specific information
- Latest use: before flag leaf sheath opening (GS 47) for winter wheat and barley; before flag leaf sheath extending (GS 41) for spring wheat and barley; before second node detectable (GS 32) for oats, rye, triticale and durum wheat; before 7 leaves unfolded and before buttress roots appear for maize; before flower buds exposed for poppies
- HI apples, pears 4 wk; onions 11 wk
- Apply to new leys from 3 expanded leaf stage
- Timing varies in tank mixtures. See label for details
- Crops undersown with grass may be sprayed provided grasses are tillering

Following crops guidance
- Clovers, peas, beans and other legumes must not be sown for 12 mth following treatment at the highest dose

Environmental safety
- Dangerous for the environment
- Toxic to aquatic organisms
- Keep livestock out of treated areas for at least 3 d following treatment and until poisonous weeds, such as ragwort, have died down and become unpalatable
- Wash spray equipment thoroughly with water and detergent immediately after use. Traces of product can damage susceptible plants sprayed later

Hazard classification and safety precautions
Hazard H03, H08, H11
Risk phrases R22b, R51, R53a, R67 [1-3]; R37 [2, 3]
Operator protection U08, U19a, U20b
Environmental protection E07a [1-3] (3 d); E15a [1-3]; E34 [3]; E38 [2, 3]
Storage and disposal D05 [1, 3]; D09a, D10b [1-3]; D12a [2, 3]
Medical advice M05b

248   fluroxypyr + mecoprop-P

A post-emergence herbicide for broadleaved weeds in amenity turf

Products
Bastion T	Rigby Taylor	72:300 g/l	ME	06011

Uses
- Annual dicotyledons in *amenity turf, managed amenity turf*
- Perennial dicotyledons in *amenity turf, managed amenity turf*
- Slender speedwell in *amenity turf, managed amenity turf*

Approval information
- Fluroxypyr and mecoprop-P included in Annex I under EC Directive 91/414

FOR FULL CONDITIONS OF USE ALWAYS READ THE PRODUCT LABEL

Efficacy guidance
- Best results achieved when soil moist and weeds in active growth, normally Apr-Sep

Restrictions
- Maximum number of treatments 2 per yr. The total amount of mecoprop-P applied in a single yr must not exceed the maximum total dose approved for any single product for the crop/situation
- Do not treat turf under stress from any cause, if night temperatures are low, if ground frost imminent or during prolonged cold weather
- Do not apply in drought period unless irrigation applied
- Do not spray if turf wet

Crop-specific information
- Avoid mowing for 3 d before or after spraying (5 d before for young turf)
- Young turf must only be treated in spring provided that at least 2 mth have elapsed between sowing and application

Environmental safety
- Dangerous for the environment
- Toxic to aquatic organisms
- Wash spray equipment thoroughly with water and detergent immediately after use. Traces of product can damage susceptible plants sprayed later
- Avoid drift onto all broad-leaved plants outside the target area

Hazard classification and safety precautions
Hazard H03, H11
Risk phrases R21, R22a, R22b, R41, R51, R53a
Operator protection A, C; U05a, U08, U11, U19a, U20b
Environmental protection E15a, E34, E38
Storage and disposal D01, D02, D05, D09a, D10b, D12a
Medical advice M03, M05a

---

249 fluroxypyr + thifensulfuron-methyl + tribenuron-methyl

A foliar applied herbicide mixture for cereals with some residual activity

Products

GEX 353	DuPont	200 g/l + 50:25% w/w	KK	11474

Uses
- Annual dicotyledons in *spring barley*, *spring wheat*, *winter barley*, *winter wheat*

Approval information
- Fluroxypyr and thifensulfuron-methyl included in Annex I under EC Directive 91/414
- Accepted by BBPA for use on malting barley

Efficacy guidance
- Best results obtained when weeds small and actively growing. This is particularly important for cleavers
- Ensure good spray cover
- Susceptible species cease growth immediately after application but may take 2 wk to show symptoms
- Weed control may be reduced in very dry conditions
- Rain within 4 hr may reduce effectiveness
- Thifensulfuron-methyl and tribenuron-methyl are members of the ALS-inhibitor group of herbicides and products should be used in a planned Resistance Management strategy. See Section 5 for more information

Restrictions
- Maximum number of treatments 1 per crop for all cereals
- Do not use on any crop suffering stress from drought, waterlogging, frost, deficiency, pest or disease attack or any other cause
- Do not use on crops undersown with grasses, clover or legumes, or any other broad leaved crop
- Specific restrictions apply to use in sequence or tank mixture with other sulfonylurea or ALS-inhibiting herbicides. See label for details
- Do not apply within 7 d of rolling

SEE SECTION 3 FOR PRODUCTS ALSO REGISTERED

Crop-specific information
- Latest use: before flag leaf extending (GS 39) for all crops

Following crops guidance
- Only cereals, field beans or oilseed rape may be sown in the same calendar yr as harvest of a treated crop
- In the event of failure of a treated crop only a cereal crop may be sown within 3 mth

Environmental safety
- Dangerous for the environment
- Very toxic to aquatic organisms
- Extremely dangerous to fish or other aquatic life. Do not contaminate surface waters or ditches with chemical or used container
- Keep livestock out of treated areas for at least 3 days and until foliage of any poisonous weeds such as ragwort has died and become unpalatable
- Take extreme care to avoid damage by drift onto broad-leaved plants outside the target area, or onto ponds, waterways or ditches
- Spraying equipment should not be drained or flushed onto land planted, or to be planted, with trees or crops other than cereals and should be thoroughly cleansed after use - see label for instructions

Hazard classification and safety precautions
Hazard H03, H08, H11
Risk phrases R22b, R37, R43, R50, R53a, R67
Operator protection U08, U19a, U20b
Environmental protection E07c (3 d); E13a, E16a, E16b, E38
Storage and disposal D09a, D10b, D12a
Medical advice M05b

250   fluroxypyr + triclopyr

A foliar acting herbicide for docks in grassland

Products

Doxstar	Dow	100:100 g/l	EC	11063

Uses
- Chickweed in *newly sown grass*
- Docks in *established grassland, newly sown grass*

Approval information
- Fluroxypyr included in Annex I under EC Directive 91/414

Efficacy guidance
- Seedling docks in new leys are controlled up to 50 mm diameter. In established grass apply in spring or autumn or, at lower dose, in spring and autumn on docks up to 200 mm. A second application in the subsequent yr may be needed
- Allow 2-3 wk after cutting or grazing to allow sufficient regrowth of docks to occur before spraying
- Control may be reduced if rain falls within 2 h of application
- To allow maximum translocation to the roots of docks do not cut grass for 28 d after spraying

Restrictions
- Maximum total dose equivalent to one full dose treatment
- Do not roll or harrow for 10 d before or 7 d after spraying
- Do not spray in drought, very hot or very cold weather

Crop-specific information
- Latest use: 7 d before grazing or harvest of grass
- Grass less than one yr old may be treated at half dose from the third leaf visible stage
- Clover will be killed or severely checked by treatment

Following crops guidance
- Do not sow kale, turnips, swedes or grass mixtures containing clover by direct drilling or minimum cultivation techniques within 6 wk of application

FOR FULL CONDITIONS OF USE ALWAYS READ THE PRODUCT LABEL

Environmental safety
- Dangerous for the environment
- Toxic to aquatic organisms
- Keep livestock out of treated areas for at least 7 d following treatment and until poisonous weeds, such as ragwort, have died down and become unpalatable
- Do not allow drift to come into contact with crops, amenity plantings, gardens, ponds, lakes or watercourses
- Wash spray equipment thoroughly with water and detergent immediately after use. Traces of product can damage susceptible plants sprayed later

Hazard classification and safety precautions
Hazard H03, H08, H11
Risk phrases R22a, R22b, R37, R38, R43, R51, R53a, R67
Operator protection A, C; U02a, U05a, U08, U14, U19a, U20b
Environmental protection E07a (7 d); E15a, E34, E38
Consumer protection C01
Storage and disposal D01, D02, D05, D09a, D10b, D12a
Medical advice M05b

251 flurtamone

A carotenoid synthesis inhibitor available only in mixtures

See also diflufenican + flurtamone
diflufenican + flurtamone + isoproturon

252 flusilazole

A systemic, protective and curative conazole fungicide for cereals and oilseed rape

See also carbendazim + flusilazole
famoxadone + flusilazole
fenpropimorph + flusilazole

Products

1	Capitan 40	DuPont	400 g/l	EC	10914
2	Genie 25	DuPont	250 g/l	EW	10285
3	Lyric	DuPont	250 g/l	EW	08252
4	Sanction 25	DuPont	250 g/l	EW	10284

Uses
- Brown rust in *spring barley, winter barley, winter wheat* [1-4]
- Eyespot in *spring barley, winter barley, winter wheat* [1-4]
- Light leaf spot in *spring oilseed rape, winter oilseed rape* [1-4]
- Net blotch in *spring barley, winter barley* [1-4]
- Powdery mildew in *spring barley, winter barley, winter wheat* [1-4]; *sugar beet* [2-4]
- Rhynchosporium in *spring barley, winter barley* [1-4]
- Rust in *sugar beet* [2-4]
- Septoria in *spring barley, winter barley, winter wheat* [1-4]
- Yellow rust in *spring barley, winter barley, winter wheat* [1-4]

Approval information
- Accepted by BBPA for use on malting barley

Efficacy guidance
- On cereals use as a routine preventative spray or when disease first develops
- Best control of eyespot achieved by spraying between leaf-sheath erect and second node detectable stages (GS 30-32)
- Product active against both MBC-sensitive and MBC-resistant strains of eyespot
- Rain occurring within 2 h after application may reduce effectiveness
- See label for recommended tank-mixes to give broader spectrum control

SECTION 2

Restrictions

- Maximum number of treatments (including other products containing flusilazole) on cereals and oilseed rape depends on dose and timing - see labels for details. High rate must not be used more than once in any crop
- Maximum total dose on sugar beet equivalent to one full dose
- Do not apply to crops under stress or during frosty weather

Crop-specific information

- Latest use: high dose before 3rd node detectable (GS 33) plus reduced dose before early milk stage (GS 72) on winter wheat and barley; before first flowers open (GS 4,0) on oilseed rape
- HI 7 wk for sugar beet
- Treat oilseed rape in autumn and/or spring at stem extension stage
- On sugar beet apply at an early stage of disease development, usually in early Aug

Environmental safety

- Dangerous for the environment
- Toxic to aquatic organisms
- Harmful to fish or other aquatic life. Do not contaminate surface waters or ditches with chemical or used container

Hazard classification and safety precautions

Hazard H02, H11
Risk phrases R22a, R40, R51, R53a, R61 [1-4]; R38 [2-4]
Operator protection A, C; U05a, U11, U19a [1-4]; U20a [2-4]; U20b [1]
Environmental protection E13c, E34, E38
Storage and disposal D01, D02, D05, D09a, D10b, D12a
Medical advice M04

---

253   flutolanil

An oxathiin fungicide for treatment of potato seed tubers

Products

Rhino	Certis	460 g/l	FS	11802

Uses

- Black scurf in **potatoes** *(tuber treatment)*
- Rhizoctonia in **potatoes** *(off-label - chitted seed treatment)*
- Stem canker in **potatoes** *(tuber treatment)*

Specific Off-Label Approvals (SOLAs)

- **potatoes** *(chitted seed treatment) (OLA 041135) Dec 2008* [1]

Efficacy guidance

- Apply to clean tubers before chitting, prior to planting, or at planting
- Apply through canopied, hydraulic or spinning disc equipment (with or without electrostatics) mounted on a rolling conveyor or table
- May be diluted with water up to 2.0 l per tonne to improve tuber coverage. Disease in areas not covered by spray will not be controlled

Restrictions

- Maximum number of treatments 1 per batch of seed tubers

Crop-specific information

- Latest use: at planting
- HI 12 wk for potatoes
- Seed tubers should be of good quality and free from bacterial rots, physical damage or virus infection, and should not be sprouted to such an extent that mechanical damage to the shoots will occur during treatment or planting

Environmental safety

- Harmful to aquatic organisms

Hazard classification and safety precautions

Hazard H04
Risk phrases R43, R52

FOR FULL CONDITIONS OF USE ALWAYS READ THE PRODUCT LABEL

Operator protection A, H; U04a, U05a, U19a, U20a
Environmental protection E15a, E34
Storage and disposal D01, D02, D05, D09a, D10a, D12a
Treated seed S01, S03, S04a, S05
Medical advice M03

254   flutriafol

A broad-spectrum conazole fungicide for cereals

Products

1 Consul	Headland	125 g/l	SC	11523
2 Pointer	Headland	125 g/l	SC	11522

Uses
- Brown rust in *spring barley, winter barley, winter wheat*
- Powdery mildew in *spring barley, winter barley, winter wheat*
- Rhynchosporium in *spring barley, winter barley*
- Septoria leaf spot in *winter wheat*
- Yellow rust in *spring barley, winter barley, winter wheat*

Approval information
- Accepted by BBPA for use on malting barley

Efficacy guidance
- Best results obtained from treatment in early stages of disease development. See label for detailed guidance on spray timing for specific diseases and the need for repeat treamtments
- Tank mix options available on the label to broaden activity spectrum
- Good spray coverage essential for optimum performance

Restrictions
- Maximum number of treatments 2 per crop (including other products containing flutriafol)

Crop-specific information
- Latest use: before early grain milky ripe stage (GS 73)
- Flag leaf tip scorch on wheat caused by stress may be increased by fungicide treatment

Environmental safety
- Harmful to aquatic organisms
- Harmful to fish or other aquatic life. Do not contaminate surface waters or ditches with chemical or used container

Hazard classification and safety precautions
Hazard H03
Risk phrases R43 [1]; R48, R52 [1, 2]
Operator protection A, H; U05a, U09a, U19a, U20b
Environmental protection E13c, E38
Storage and disposal D01, D02, D09a, D10c

255   fomesafen

A contact and residual diphenyl ether herbicide for use in leguminous crops

Products

Flex	Syngenta	250 g/l	SL	08885

Uses
- Annual dicotyledons in *dwarf beans*
- Black nightshade in *soya beans* *(off-label)*
- Volunteer oilseed rape in *dwarf beans, soya beans* *(off-label)*

Specific Off-Label Approvals (SOLAs)
- *soya beans* *(OLA 041849) Dec 2007* [1]

SEE SECTION 3 FOR PRODUCTS ALSO REGISTERED

Approval information

- Products containing this active ingredient have been granted derogations for specified 'Essential Uses' for use until 31 December 2007. Sale and supply must cease by 30 June 2007 but growers have no guarantee that the products will continue to be available until then.
  For more information see 'The Review Programme' under 'Pesticide Legislation' in Section 5

Efficacy guidance

- Fomesafen combines contact activity and residual soil uptake
- Best results obtained when seedling weeds are actively growing in warm moist conditions with adequate soil moisture
- Weeds should be treated at seedling stage. Larger weeds, and those hardened off by adverse conditions, may be less well controlled
- Residual activity gives control of later germinating weed seedlings
- Product may be used on all soil types but residual control of weeds germinating after treatment may be reduced on soils with high organic matter
- Weed spectrum may be widened by use in a programme with other herbicides. See label for details

Restrictions

- Maximum total dose equivalent to one full dose treatment
- Apply to healthy crops only. Crops growing under stress from drought, waterlogging etc should not be treated
- This product must not be used on any crops other than those listed, including any extrapolations that would normally be permissible under the Long Term Arrangements for Extension of Use (see Section 5)

Crop-specific information

- HI green beans 5 wk; soya beans 8 wk
- Some crop damage, such as leaf crinkling, may occur. Effect is transient and should not affect yield

Environmental safety

- Dangerous for the environment
- Very toxic to aquatic organisms

Hazard classification and safety precautions

Hazard H04, H11
Risk phrases R36, R43, R50, R53a
Operator protection A, C, H; U04a, U05a, U11, U14, U19a, U20b
Environmental protection E15a, E38
Storage and disposal D01, D02, D09a, D10c, D12a
Medical advice M03

256   fomesafen + terbutryn

A residual herbicide for use in leguminous crops

Products

Reflex T	Syngenta	80:400 g/l	SC	08884

Uses

- Annual dicotyledons in **broad beans** *(spring sown)*, **combining peas** *(spring sown)*, **spring field beans**, **vining peas** *(spring sown)*

Approval information

- Products containing fomesafen and terbutryn have been granted derogations for specified 'Essential Uses' for use until 31 December 2007. Sale and supply must cease by 30 June 2007 but growers have no guarantee that the products will continue to be available until then.
  For more information see 'The Review Programme' under 'Pesticide Legislation' in Section 5

Efficacy guidance

- Best results achieved when applied to moist, fine, firm tilth
- Rain soon after application is essential for optimum weed control
- Results may be unsatisfactory in dry, or excessively wet, soil conditions
- Weed control may be reduced on soils with more than 10% organic matter

FOR FULL CONDITIONS OF USE ALWAYS READ THE PRODUCT LABEL

Restrictions
- Maximum number of treatments 1 per crop
- Use only once every 5 yr
- All varieties of spring sown peas may be treated but forage varieties may be damaged from which recovery may not be complete. All varieties of spring sown field and broad beans may be treated
- Do not use on Sands
- This product must not be used on any crops other than those listed, including any extrapolations that would normally be permissible under the Long Term Arrangements for Extension of Use (see Section 5)

Crop-specific information
- Latest use: pre-emergence of crop
- Apply after planting, but before the crop emerges
- Emerged crop leaves may be severely scorched or killed

Following crops guidance
- Plough or cultivate to at least 150 mm before drilling or planting another crop. Only cereals should be planted in the calendar year of use with a minimum interval of 4 mth after treatment

Environmental safety
- Dangerous for the environment
- Very toxic to aquatic organisms

Hazard classification and safety precautions
  Hazard H03, H11
  Risk phrases R22a, R36, R43, R50, R53a
  Operator protection A, C, H; U04a, U05a, U08, U19a, U20b
  Environmental protection E15a, E38
  Storage and disposal D01, D02, D10c, D12a

SECTION 2

257 formaldehyde (commodity substance)

An agricultural/horticultural and animal husbandry fungicide

Products

1 formaldehyde	various	38-40%	SL
2 paraformaldehyde	various	-	ZZ

Uses
- Fungus diseases in **flower bulbs** *(dip)*, **glasshouses** *(spray, dip or fumigant)*, **mushroom houses** *(spray or fumigant)* [1]; **livestock houses** [2]
- Soil-borne diseases in **soil and compost** *(drench)* [1]

Approval information
- Approval for the use of formaldehyde as a commodity substance was granted on 1 March 1991 by Ministers under regulation 5 of the Control of Pesticides Regulations 1986

Efficacy guidance
- Use as a dip to sterilize flower bulbs [1]
- Use as drench to sterilize soil and compost, indoors and outdoors [1]
- Use as spray or fumigant in mushroom houses [1]
- Use as spray, dip or fumigant for glasshouse hygiene [1]
- Use as fumigant to sterilize animal houses [2]

Restrictions
- Formaldehyde is subject to the Poisons Rules 1982 and the Poisons Act 1972. See notes in Section 5
- Operators must observe Occupational Exposure Standard as set out in HSE Guidance Note EH40/90 and ACOP 30 *Control of Substances Hazardous to Health in Fumigation*
- Operators must be supplied with a Section 6 (HSW) Safety Data Sheet before commencing work
- Observe maximum permitted concentrations. See PR 12, 1990, pp 9-10

Hazard classification and safety precautions
  Operator protection A, D, P [1]

SEE SECTION 3 FOR PRODUCTS ALSO REGISTERED

258   fosetyl-aluminium

A systemic phosphonic acid fungicide for various horticultural crops

Products

1 Aliette 80 WG	Certis	80% w/w	WG	11213
2 Standon Fosetyl-AL 80 WG	Standon	80% w/w	WG	10667

Uses
- Collar rot in *apples* [1, 2]
- Crown rot in *apples* [1, 2]; *protected strawberries* (off-label), *strawberries* (off-label) [1]
- Damping off in *broccoli* (off-label), *brussels sprouts* (off-label), *cabbages* (off-label), *calabrese* (off-label), *cauliflowers* (off-label), *chinese cabbage* (off-label), *collards* (off-label), *kale* (off-label), *protected broccoli* (off-label), *protected brussels sprouts* (off-label), *protected cabbages* (off-label), *protected calabrese* (off-label), *protected cauliflowers* (off-label), *protected chinese cabbage* (off-label), *protected collards* (off-label), *protected kale* (off-label) [1]
- Downy mildew in *broad beans*, *broccoli* (off-label - seedlings), *brussels sprouts* (off-label - seedlings), *cabbages* (off-label - seedlings), *calabrese* (off-label - seedlings), *cauliflowers* (off-label - seedlings), *chinese cabbage* (off-label - seedlings), *chives* (off-label), *herbs (see appendix 6)* (off-label), *hops*, *kale* (off-label - seedlings), *leaf spinach* (off-label), *protected chives* (off-label), *protected herbs (see appendix 6)* (off-label), *protected lettuce*, *spinach beet* (off-label), *watercress* (off-label - during propagation) [1, 2]; *broccoli* (off-label), *brussels sprouts* (off-label), *cabbages* (off-label), *calabrese* (off-label), *cauliflowers* (off-label), *chinese cabbage* (off-label), *collards* (off-label), *combining peas* (off-label - seed treatment), *grapevines* (off-label), *kale* (off-label), *protected brassica seedlings* (off-label), *protected broccoli* (off-label), *protected brussels sprouts* (off-label), *protected cabbages* (off-label), *protected calabrese* (off-label), *protected cauliflowers* (off-label), *protected chinese cabbage* (off-label), *protected collards* (off-label), *protected kale* (off-label), *salad onions* (off-label), *vining peas* (off-label - seed treatment) [1]; *combining peas* (off-label), *lettuce* (off-label), *parsley* (off-label), *protected broccoli* (off-label - seedlings), *protected brussels sprouts* (off-label - seedlings), *protected cabbages* (off-label - seedlings), *protected calabrese* (off-label - seedlings), *protected cauliflowers* (off-label - seedlings), *protected chinese cabbage* (off-label - seedlings), *protected kale* (off-label - seedlings), *protected parsley* (off-label), *vining peas* (off-label) [2]
- Phytophthora in *chicory* (off-label - for forcing) [1]; *chicory* (off-label - in forcing sheds) [2]; *watercress* (off-label - during propagation) [1, 2]
- Phytophthora root rot in *capillary benches* [1]; *protected pot plants* [1, 2]
- Phytophthora stem rot in *capillary benches* [1]; *protected pot plants* [1, 2]
- Phytophthora wilt in *hardy ornamental nursery stock* [1, 2]
- Pythium in *watercress* (off-label - during propagation) [1, 2]
- Red core in *protected strawberries* (off-label), *strawberries* (off-label) [1]; *strawberries* [1, 2]
- Root rot in *capillary benches* [1]; *protected pot plants* [1, 2]

Specific Off-Label Approvals (SOLAs)
- *broccoli, brussels sprouts, cabbages, calabrese, cauliflowers, chinese cabbage, collards, kale, protected broccoli, protected brussels sprouts, protected cabbages, protected calabrese, protected cauliflowers, protected chinese cabbage, protected collards, protected kale* (OLA 040149) Jul 2008 [1]
- *broccoli, brussels sprouts, cabbages, calabrese, cauliflowers, chinese cabbage, kale* (seedlings) (OLA 030863) Dec 2008 [1]
- *broccoli, brussels sprouts, cabbages, calabrese, cauliflowers, chinese cabbage, kale, protected broccoli, protected brussels sprouts, protected cabbages, protected calabrese, protected cauliflowers, protected chinese cabbage, protected kale* (seedlings) (OLA 030366) Dec 2008 [2]
- *chicory* (for forcing) (OLA 030866) Dec 2008 [1]
- *chicory* (in forcing sheds) (OLA 030366) Dec 2008 [2]
- *chives, herbs (see appendix 6), protected chives, protected herbs (see appendix 6)* (OLA 030868) Dec 2008 [1]
- *chives, herbs (see appendix 6), leaf spinach, lettuce, parsley, protected chives, protected herbs (see appendix 6), protected parsley, spinach beet* (OLA 030366) Dec 2008 [2]
- *combining peas, vining peas* (seed treatment) (OLA 030861) Dec 2008 [1]
- *combining peas, vining peas* (OLA 030367) Dec 2008 [2]

FOR FULL CONDITIONS OF USE ALWAYS READ THE PRODUCT LABEL

- **grapevines** *(OLA 030862) Dec 2008* [1]
- **leaf spinach**, **spinach beet** *(OLA 030864) Dec 2008* [1]
- **protected brassica seedlings** *(OLA 030863) Dec 2008* [1]
- **protected strawberries**, **strawberries** *(OLA 030579) Dec 2008* [1]
- **salad onions** *(OLA 050548) Dec 2008* [1]
- **watercress** *(during propagation) (OLA 030867) Dec 2008* [1]
- **watercress** *(during propagation) (OLA 030366) Dec 2008* [2]

Approval information
- Accepted by BBPA for use on hops

Restrictions
- Maximum number of treatments 1 per seed batch for peas; 1 per batch of compost for lettuce; 1 per yr for strawberries (root dip or foliar spray); 1 per yr for apples (bark paste), 2 per yr for apples (foliar spray); 2 per crop for broad beans, peas; 2 per yr for hops (basal spray); 6 per yr for hops (foliar spray); 1 per crop for brassicas and salad brassicas; 1 per crop during propagation, 2 per crop after planting out for lettuce; 7 per yr for grapevines
- Check tolerance of ornamental species before large-scale treatment

Crop-specific information
- Latest use: pre-planting for dipping autumn planted strawberry runners; up to 31 Dec for spraying autumn planted strawberry runners; pre-sowing for protected lettuce, peas; before transplanting or pre-emergence for brassicas; 24 h after seeding watercress
- HI apples 5 mth (bark paste) or 4 wk (spray); grapevines 35 d; chicory 21 d; broad beans 17 d; hops, strawberries, herbs and salad crops 14 d; leaf spinach, spinach beet 7 d
- Spray young orchards for crown rot protection after blossom when first leaves fully open and repeat after 4-6 wk. Apply as paste to bark of apples to control collar rot
- Apply to broad beans when infection appears (usually at flowering) and 14 d later. Consult before treating crops to be processed
- Use on strawberries only between harvest and 31 Dec
- Spray autumn-planted strawberry runners 2-3 wk after planting or use dip treatment at planting. Spray established crops in late summer/early autumn after picking and repeat annually
- Apply to hops as early season basal spray or as foliar spray every 10-14 d from when training is completed
- Use by wet incorporation in blocking compost for protected lettuce only from Sep to Apr and follow all directions carefully to avoid severe crop injury. Crop maturity may be delayed by a few days
- Apply as drench to rooted cuttings of hardy nursery stock after first potting and repeat mthly. Up to 6 applications may be needed
- See label for details of application to capillary benches and trays of young plants.

Environmental safety
- Harmful to aquatic organisms

Hazard classification and safety precautions
  Risk phrases R52 [2]
  Operator protection A, C, H; U20c
  Environmental protection E15a
  Storage and disposal D09a, D11a

259   fosthiazate

An organophosphorus contact nematicide for potatoes

Products

Nemathorin 10G	Syngenta	10% w/w	FG	11003

Uses
- Potato cyst nematode in **potatoes**
- Wireworm in **potatoes** *(reduction)*

Approval information
- Fosthiazate included in Annex I under EC Directive 91/414

SEE SECTION 3 FOR PRODUCTS ALSO REGISTERED

SECTION 2

Efficacy guidance
- Apply and incorporate granules in one operation to a uniform depth of 10-15 cm. Deeper incorporation will reduce control
- Application best achieved using equipment such as Horstine Farmery Microband Applicator, Matco or Stocks Micrometer applicators together with a rear mounted powered rotary cultivator
- Granules must not become wet or damp before use

Restrictions
- Contains an anticholinesterase organophosphorus compound. Do not use if under medical advice not to work with such compounds
- Maximum number of treatments 1 per crop
- Product must only be applied using tractor-mounted/drawn direct placement machinery. Do not use air assisted broadcast machinery
- Do not allow granules to stand overnight in the application hopper
- Do not apply more than once every four years on the same area of land
- Consult before using on crops intended for processing

Crop-specific information
- Latest use: at planting
- HI 17 wk for potatoes

Environmental safety
- Dangerous for the environment
- Toxic to aquatic organisms
- Dangerous to game, wild birds and animals
- Dangerous to livestock. Keep all livestock out of treated areas for at least 13 wk
- Risk of explosion if dust in the air
- Incorporation to 10-15 cm and ridging up of treated soil must be carried out immediately after application. Powered rotary cultivators are preferred implements for incorporation but discs, power, spring tine or Dutch harrows may be used provided two passes are made at right angles
- Failure completely to bury granules immediately after application is hazardous to wildlife

Hazard classification and safety precautions
Hazard H03, H11
Risk phrases R22a, R43, R51, R53a
Operator protection A, E, G, H, K, M; U02a, U04a, U05a, U07, U09a, U13, U14, U20a
Environmental protection E06b (13 wk); E10a, E15a, E34, E38
Consumer protection C02 (17 wk)
Storage and disposal D01, D02, D05, D09a, D11a, D12a, D14
Medical advice M01, M03, M05a

260 fuberidazole

A benzimidazole (MBC) fungicide available only in mixtures

See also bitertanol + fuberidazole
bitertanol + fuberidazole + imidacloprid

261 fuberidazole + imidacloprid + triadimenol

A broad spectrum systemic fungicide and insecticide seed treatment for winter cereals
Products

Baytan Secur	Bayer CropScience	15:117:125 g/l	FS	11253

Uses
- Blue mould in **winter wheat** *(seed treatment)*
- Brown foot rot in **winter barley** *(seed treatment)*
- Brown rust in **winter barley** *(seed treatment)*, **winter wheat** *(seed treatment)*
- Bunt in **winter wheat** *(seed treatment)*
- Covered smut in **winter barley** *(seed treatment)*

FOR FULL CONDITIONS OF USE ALWAYS READ THE PRODUCT LABEL

- Fusarium foot rot and seedling blight in **winter barley** *(seed treatment - reduction)*, **winter oats** *(seed treatment - reduction)*, **winter wheat** *(seed treatment - reduction)*
- Lodging control in **winter wheat** *(seed treatment - reduction)*
- Loose smut in **winter barley** *(seed treatment)*, **winter oats** *(seed treatment)*, **winter wheat** *(seed treatment)*
- Net blotch in **winter barley** *(seed treatment - seed-borne only)*
- Powdery mildew in **winter barley** *(seed treatment)*, **winter oats** *(seed treatment)*, **winter wheat** *(seed treatment)*
- Pyrenophora leaf spot in **winter oats** *(seed treatment)*
- Septoria leaf spot in **winter wheat** *(seed treatment)*
- Septoria seedling blight in **winter wheat** *(seed treatment - reduction)*
- Snow rot in **winter barley** *(seed treatment - reduction)*
- Virus vectors in **winter barley** *(seed treatment)*, **winter oats** *(seed treatment)*, **winter wheat** *(seed treatment)*
- Wireworm in **winter barley** *(reduction of damage)*, **winter oats** *(reduction of damage)*, **winter wheat** *(reduction of damage)*
- Yellow rust in **winter barley** *(seed treatment)*, **winter wheat** *(seed treatment)*

Approval information
- Accepted by BBPA for use on malting barley

Efficacy guidance
- Apply through recommended seed treatment machinery
- Evenness of seed cover improved by simultaneous application of equal volumes of product and water
- Calibrate drill for treated seed and drill at 4 cm into firm, well prepared seedbed
- Use minimum 125 kg treated seed per ha and increase seed rate as drilling season progresses. Lower drilling rates and/or early drilling affect duration of BYDV protection needed and may require follow-up aphicide treatment
- When aphid activity unusually late, or is heavy and prolonged in areas of high risk, and mild weather predominates, follow-up treatment may be required
- In addition to seed-borne diseases early attacks of various foliar, air-borne diseases are controlled or suppressed. See label for details

Restrictions
- Maximum number of treatments 1 per batch of seed
- Do not use on naked oats
- Do not drill treated winter wheat seed after end of Nov
- Do not handle seed unnecessarily
- Do not use treated seed as food or feed
- Treated seed should not be broadcast, but drilled to a depth of 4 cm in a well prepared seedbed
- Germination tests should be done on all batches of seed to be treated to ensure seed viability and suitability for treatment
- Do not use on seed with moisture content above 16%, on sprouted, cracked or skinned seed or on seed already treated with another seed treatment

Crop-specific information
- Latest use: before drilling
- Store treated seed in cool, dry, well-ventilated store and drill as soon as possible, preferably in season of purchase
- Treatment may accentuate effects of adverse seedbed conditions on crop emergence

Environmental safety
- Dangerous to fish or other aquatic life. Do not contaminate surface waters or ditches with chemical or used container
- Dangerous to birds, game and other wildlife. Treated seed should not be left on the soil surface. Bury spillages
- Seed should be drilled to a depth of 4 cm into a well-prepared seed bed
- If seed is present on the soil surface, or if spills have occurred, the field should be harrowed and rolled if conditions are appropriate to ensure good incorporation

Hazard classification and safety precautions
    Risk phrases R53a
    Operator protection A, H; U04a, U05a, U07, U13, U14, U20b

SEE SECTION 3 FOR PRODUCTS ALSO REGISTERED

Environmental protection E03, E13b, E36, E38
Storage and disposal D01, D02, D05, D09a, D12a
Treated seed S01, S02, S03, S04c, S05, S06a, S07, S08

262 fuberidazole + triadimenol

A broad spectrum systemic fungicide seed treatment for cereals

Products

Baytan Flowable	Makhteshim	22.5:187.5 g/l	FS	11714

Uses

- Blue mould in *spring wheat* (seed treatment), *triticale* (seed treatment), *winter wheat* (seed treatment)
- Brown foot rot in *spring barley* (seed treatment), *spring oats* (seed treatment), *spring rye* (seed treatment), *spring wheat* (seed treatment), *triticale* (seed treatment), *winter barley* (seed treatment), *winter oats* (seed treatment), *winter rye* (seed treatment), *winter wheat* (seed treatment)
- Brown rust in *spring barley* (seed treatment), *spring rye* (seed treatment), *spring wheat* (seed treatment), *winter barley* (seed treatment), *winter rye* (seed treatment), *winter wheat* (seed treatment)
- Bunt in *spring wheat* (seed treatment), *winter wheat* (seed treatment)
- Covered smut in *spring barley* (seed treatment), *winter barley* (seed treatment)
- Crown rust in *spring oats* (seed treatment), *winter oats* (seed treatment)
- Leaf stripe in *spring barley* (seed treatment)
- Loose smut in *spring barley* (seed treatment), *spring oats* (seed treatment), *spring wheat* (seed treatment), *winter barley* (seed treatment), *winter oats* (seed treatment), *winter wheat* (seed treatment)
- Mildew in *spring barley* (seed treatment), *spring wheat* (seed treatment), *winter barley* (seed treatment), *winter wheat* (seed treatment)
- Pyrenophora leaf spot in *spring oats* (seed treatment), *winter oats* (seed treatment)
- Septoria in *spring wheat* (seed treatment), *winter wheat* (seed treatment)
- Yellow rust in *spring barley* (seed treatment), *spring wheat* (seed treatment), *winter barley* (seed treatment), *winter wheat* (seed treatment)

Approval information

- Accepted by BBPA for use on malting barley

Efficacy guidance

- Apply through recommended seed treatment machinery
- Calibrate drill for treated seed and drill at 2.5-4 cm into firm, well prepared seedbed
- In addition to seed-borne diseases early attacks of various foliar, air-borne diseases are controlled or suppressed. See label for details

Restrictions

- Maximum number of treatments 1 per batch of seed
- Do not use on naked oats
- Do not drill treated winter wheat or rye seed after end of Nov. Seed rate should be increased as drilling season progresses
- Germination tests should be done on all batches of seed to be treated to ensure seed viability and suitability for treatment
- Do not use on seed with moisture content above 16%, on sprouted, cracked or skinned seed or on seed already treated with another seed treatment
- Do not use treated seed as food or feed

Crop-specific information

- Latest use: before drilling
- Treated spring wheat may be drilled in autumn up to end of Nov or from Feb onwards
- Store treated seed in cool, dry, well-ventilated store and drill as soon as possible, preferably in season of purchase
- Treatment may accentuate effects of adverse seedbed conditions on crop emergence

---

FOR FULL CONDITIONS OF USE ALWAYS READ THE PRODUCT LABEL

Environmental safety
- Dangerous for the environment
- Harmful to aquatic organisms

Hazard classification and safety precautions
    Hazard H11
    Risk phrases R52, R53a
    Operator protection A; U20b
    Environmental protection E03, E15a, E34
    Storage and disposal D05, D09a, D11a
    Treated seed S01, S02, S03, S04a, S05, S06a, S07

SECTION 2

263   gibberellins

A plant growth regulator for use in top fruit

Products
Novagib	Fine	10 g/l	SL	08954

Uses
- Growth regulation in **pears** *(off-label)*
- Reducing fruit russeting in **apples**

Specific Off-Label Approvals (SOLAs)
- **pears** *(OLA 012756) Dec 2008* [1]

Efficacy guidance
- Apply to apples at completion of petal fall and repeat 3 or 4 times at 7-10 d intervals. Number of sprays and spray interval depend on dose (see labels)

Restrictions
- Maximum total dose on apples equivalent to four full dose treatments
- Good results achieved on Cox's Orange Pippin, Discovery, Golden Delicious and Karmijn apples. For other cultivars test on a small number of trees
- Return bloom may be reduced in yr following treatment
- Consult before treating apple crops grown for processing

Crop-specific information
- HI zero for apples

Hazard classification and safety precautions
    Operator protection U20b
    Environmental protection E15a
    Storage and disposal D01, D03, D05, D07, D09a, D10b

264   glufosinate-ammonium

A non-selective, non-residual phosphinic acid contact herbicide

Products
1	Challenge 60	Fargro	60 g/l	SL	08236
2	Finale	Bayer Environ.	120 g/l	SL	10092
3	Harvest	Bayer CropScience	150 g/l	SL	07321
4	Kaspar	Certis	150 g/l	SL	11214

Uses
- Annual and perennial weeds in **apple orchards, bilberries, blueberries, cane fruit, cherries, cranberries, cultivated land/soil, currants, damsons, fallows, forest, grapevines, non-crop farm areas, pear orchards, plums, potatoes** *(not for seed)***, strawberries, stubbles, sugar beet, tree nuts, vegetables** [3]
- Annual dicotyledons in **apple orchards, cherries, currants, damsons, fallows, forest, grapevines, headlands** *(uncropped)***, non-crop farm areas, pear orchards, plums, strawberries, sugar beet, tree nuts, vegetables, ware potatoes** [4]; **bush fruit, fruit trees, ornamental trees, shrubs** [1]; **cane fruit, cultivated land/soil** [1, 4]; **nursery stock, woody ornamentals** [2]

SEE SECTION 3 FOR PRODUCTS ALSO REGISTERED

- Annual grasses in *apple orchards, cherries, currants, damsons, fallows, forest, grapevines, headlands (uncropped), non-crop farm areas, pear orchards, plums, strawberries, sugar beet, tree nuts, vegetables, ware potatoes* [4]; *bush fruit, fruit trees, ornamental trees, shrubs* [1]; *cane fruit, cultivated land/soil* [1, 4]; *nursery stock, woody ornamentals* [2]
- Green cover in *land not being used for crop production* [3, 4]
- Harvest management/desiccation in *combining peas, ware potatoes* [4]; *combining peas (not for seed), potatoes* [3]; *linseed, spring field beans, spring oilseed rape, winter field beans, winter oilseed rape* [3, 4]
- Line marking preparation in *managed amenity turf* [2]
- Perennial dicotyledons in *apple orchards, cherries, currants, damsons, fallows, forest, grapevines, headlands (uncropped), land not being used for crop production, non-crop farm areas, pear orchards, plums, strawberries, tree nuts* [4]; *bush fruit, cultivated land/soil, fruit trees, ornamental trees, shrubs* [1]; *cane fruit* [1, 4]; *nursery stock, woody ornamentals* [2]
- Perennial grasses in *apple orchards, cherries, currants, damsons, fallows, forest, grapevines, headlands (uncropped), land not being used for crop production, non-crop farm areas, pear orchards, plums, strawberries, tree nuts* [4]; *bush fruit, cultivated land/soil, fruit trees, ornamental trees, shrubs* [1]; *cane fruit* [1, 4]; *nursery stock, woody ornamentals* [2]
- Sward destruction in *grassland* [3, 4]

Approval information
- Accepted by BBPA for use on malting barley and hops

Efficacy guidance
- Activity quickest under warm, moist conditions. Light rainfall 3-4 h after application will not affect activity. Do not spray wet foliage or if rain likely within 6 h
- For weed control uses treat when weeds growing actively. Deep rooted weeds may require second treatment [1]
- On uncropped headlands apply in May/Jun to prevent weeds invading field
- Glufosinate kills all green tissue but does not harm mature bark

Restrictions
- Maximum number of treatments 4 per crop for potatoes (including 2 desiccant uses); 2 per crop (including 1 desiccant use) for oilseed rape, dried peas, field beans, linseed, wheat and barley; 3 (2 [1]) per yr for fruit and forestry; 2 per yr for strawberries, non-crop land and land temporarily removed from production, round fruit trees, canes or bushes; 1 per crop for sugar beet, vegetables and other crops; 1 per yr for grassland destruction, on cultivated land prior to planting edible crops
- Pre-harvest desiccation sprays should not be used on seed crops of wheat, barley, peas or potatoes but may be used on seed crops of oilseed rape, field beans and linseed [3, 4]
- Do not desiccate potatoes in exceptionally wet weather or in saturated soil. See label for details
- Do not spray potatoes after emergence if grown from small or diseased seed or under very dry conditions
- Application for weed control and line-marking uses in sports turf must be between 1 Mar and 30 Sep [2]
- Do not allow spray to contact dormant or green buds, suckers, damaged or green bark and foliage of wanted plants [1]
- Do not use straw from treated crops as animal feed or bedding [3]
- Do not spray hedge bottoms [3]

Crop-specific information
- Latest use: 30 Sep for use on non-crop land, top fruit, soft fruit, cane fruit, bush fruit, forestry; pre-drilling, pre-planting or pre-emergence in sugar beet, vegetables and other crops; before winter dormancy for grassland destruction.
- HI potatoes, oilseed rape, combining peas, field beans, linseed 7 d; wheat, barley 14 d
- Ploughing or other cultivations can follow 4 h after spraying
- Crops can normally be sown/planted immediately after spraying or sprayed post-drilling. On sand, very light or immature peat soils allow at least 3 d before sowing/planting or expected emergence
- For weed control in potatoes apply pre-emergence or up to 10% emergence on earlies and seed crops, up to 40% on maincrop, on plants up to 15 cm high
- In sugar beet and vegetables apply just before crop emergence, using stale seedbed technique
- In top and soft fruit, grapevines and forestry apply between 1 Mar and 30 Sep as directed sprays
- For grass destruction apply before winter dormancy occurs. Heavily grazed fields should show active regrowth. Plough from the day after spraying

- Apply pre-harvest desiccation treatments 10-21 d before harvest (14-21 d for oilseed rape). See label for timing details on individual crops
- For potato haulm desiccation apply to listed varieties (not seed crops) at onset of senescence, 14-21 d before harvest

Environmental safety

- Harmful to fish or other aquatic life. Do not contaminate surface waters or ditches with chemical or used container
- Keep livestock out of treated areas until foliage of any poisonous weeds such as ragwort has died and become unpalatable
- Treated pea haulm may be fed to livestock from 7 d after spraying, treated grain from 14 d [3]
- Product supplied in small volume returnable container - see label for filling and mixing instructions [3]

Hazard classification and safety precautions

Hazard H03 [3, 4]; H04 [2]

Risk phrases R21, R22a, R41 [3, 4]; R36 [2]

Operator protection A [1, 3, 4]; B [2]; C, H, M; U02a [1]; U04a, U11, U13, U14, U15, U19a [3, 4]; U05a [2-4]; U08 [1-4]; U20a [1, 3, 4]; U20b [2]

Environmental protection E07a, E13c [1-4]; E22b [4]; E22c [3]; E34 [1, 3, 4]

Consumer protection C01 [1]

Storage and disposal D01, D02, D09a, D10b [1-4]; D05 [1, 2]; D12a [3, 4]

Medical advice M03 [3, 4]

265   glyphosate

A translocated non-residual phosphonic acid herbicide

See also diflufenican + glyphosate

Products

1	Amega Pro TMF	Nufarm UK	480 g/l	SL	09630
2	Asteroid	Headland	360 g/l	SL	11118
3	Azural	Cardel	360 g/l	SL	11668
4	Barclay Barbarian	Barclay	360 g/l	SL	11362
5	Barclay Gallup 360	Barclay	360 g/l	SL	11369
6	Barclay Gallup Amenity	Barclay	360 g/l	SL	06753
7	Barclay Gallup Biograde Amenity	Barclay	360 g/l	SL	11363
8	Barclay Gallup Hi-Aktiv	Barclay	490 g/l	SL	11371
9	Buggy SG	Sipcam	36% w/w	WG	08573
10	CDA Vanquish Biactive	Bayer Environ.	120 g/l	RH	10950
11	Clinic	Nufarm UK	360 g/l	SL	09378
12	Dow Agrosciences Glyphosate 360	Dow	360 g/l	SL	11552
13	Envision	Headland	450 g/l	SL	10569
14	Flame	Dow	360 g/l	SL	11583
15	Glyfos	Headland	360 g/l	SL	10995
16	Glyfos Gold	Nomix Enviro	360 g/l	SL	10570
17	Glyfos Proactive	Nomix Enviro	360 g/l	SL	11976
18	Glyphogan	Makhteshim	360 g/l	SL	05784
19	Glyphosate 360	SumiAgro Amenity	360 g/l	SL	11060
20	Glyphosate 360	Monsanto	360 g/l	SL	11726
21	Greenaway Gly-490	Greenaway	490 g/l	SL	11064
22	Greencrop Gypsy	Greencrop	360 g/l	SL	09432
23	Habitat	SumiAgro Amenity	360 g/l	SL	10498
24	Hilite	Nomix Enviro	144 g/l	RH	06261
25	Kernel	Headland	480 g/l	SL	10993
26	KN 540	Monsanto	540 g/l	SL	12009
27	Landgold Glyphosate 360	Landgold	360 g/l	SL	05929
28	Landgold Glyphosate 360	Teliton	360 g/l	SL	12120
29	Manifest	Headland	360 g/l	SL	11041
30	Nufosate	Nufarm UK	360 g/l	SL	10096
31	Reliance	Cardel	360 g/l	SL	11667
32	Romany	Greencrop	360 g/l	SL	11000
33	Roundup Amenity	Monsanto	360 g/l	SL	10319
34	Roundup Biactive	Monsanto	360 g/l	SL	10320

SEE SECTION 3 FOR PRODUCTS ALSO REGISTERED

Products – continued

35	Roundup Gold	Monsanto	450 g/l	SL	10975
36	Roundup Max	Monsanto	68% w/w	SG	10231
37	Roundup Pro Biactive	Monsanto	360 g/l	SL	10330
38	Roundup Pro-Green	Monsanto	450 g/l	SL	11907
39	Roundup ProVide	Monsanto	68% w/w	SG	11721
40	Roundup Rail	Monsanto	360 g/l	SL	11874
41	Roundup Ultra ST	Monsanto	450 g/l	SL	10199
42	Samurai	Monsanto	360 g/l	SL	10334
43	Scorpion	Cardel	360 g/l	SL	11656
44	Spasor	Bayer Environ.	360 g/l	SL	09945
45	Spasor Biactive	Bayer Environ.	360 g/l	SL	09940
46	Stacato	Sipcam	360 g/l	SL	05892
47	Sting ECO	Monsanto	120 g/l	SL	10337
48	Stirrup	Nomix Enviro	144 g/l	RH	06132
49	Tangent	Headland Amenity	450 g/l	SL	11872
50	Touchdown	Syngenta	330 g/l	SL	10538
51	Touchdown Quattro	Syngenta	360 g/l	SL	10608
52	Trustee Elite	Barclay	450 g/l	SL	11556
53	Tumbleweed Pro-Active	Scotts	450 g/l	SL	11963
54	Typhoon 360	Makhteshim	360 g/l	SL	11175

Uses

- Annual and perennial weeds in *all top fruit, hedges, managed amenity turf* (pre-establishment), *ornamental plant production* [24, 48]; *amenity areas* [6-8, 21, 37, 45]; *amenity areas* (wiper application), *broad-leaved trees, conifers, farm buildings/yards, land clearance* [37]; *amenity grass, amenity grass* (wiper application), *amenity trees and shrubs* (wiper application), *woody ornamentals* [37, 44]; *amenity trees and shrubs* [24, 37, 44, 48]; *amenity vegetation* [10, 24, 37, 48]; *apple orchards, cherries, damsons, pear orchards, plums* [2-5, 8, 9, 11-15, 18, 20-22, 25, 26, 29, 31, 32, 34, 35, 41-43, 52, 54]; *asparagus* (off-label), *blackcurrants* (off-label), *blueberries* (off-label), *chestnuts* (off-label), *cob nuts* (off-label), *grapevines* (off-label), *hazel nuts* (off-label), *rhubarb* (off-label), *walnuts* (off-label) [2, 4, 34]; *bilberries* (off-label), *cranberries* (off-label), *gooseberries* (off-label), *vaccinium spp.* (off-label) [2, 34]; *combining peas, durum wheat, spring barley, spring field beans, spring oilseed rape, spring wheat, winter barley, winter field beans, winter oilseed rape, winter wheat* [2, 4, 51]; *cultivated land/soil* [1-3, 11, 12, 14, 20, 22, 32, 42]; *farm forestry* [39]; *fencelines, walls* [37, 45]; *field crops* (wiper application) [3, 11, 12, 14, 20, 22, 32, 42]; *forest* [4-9, 11-13, 15-19, 21-25, 29, 30, 32, 33, 37-39, 48, 49, 52-54]; *grassland* [1, 3, 9, 11, 12, 14, 18, 20, 31, 32, 34, 35, 42, 43]; *hard surfaces* [16, 21, 37-40, 45, 49, 51, 53]; *industrial sites* [6-8, 16, 17, 19, 23, 30, 33, 37]; *land immediately adjacent to aquatic areas* [21, 38, 39, 53]; *land not being used for crop production* [2, 9, 26, 34, 35, 51]; *land not intended to bear vegetation* [3-10, 12-15, 17, 19-21, 23-25, 29-34, 36, 37, 41-43, 48, 50, 52, 54]; *land prior to cultivation* [49]; *linseed* [2, 51]; *mustard* [51]; *natural surfaces not intended to bear vegetation, permeable surfaces overlying soil* [16, 21, 38, 39, 49, 51, 53]; *non-crop areas* [35, 37, 45]; *non-crop farm areas* [3-5, 11, 12, 20, 22, 32, 42, 46]; *paths and drives* [16, 23, 30, 33, 37, 45]; *railway tracks* [40]; *redcurrants* (off-label), *whitecurrants* (off-label) [34]; *road verges* [16, 17, 23, 30, 33, 37, 45]; *spring oats, winter oats* [2, 4]; *stubbles* [1-4, 9, 11, 12, 14, 18, 20-22, 26-28, 31, 32, 34-36, 42, 43, 46, 50, 51]
- Annual dicotyledons in *all edible crops, all non-edible crops, hard surfaces, land not intended to bear vegetation, permeable surfaces overlying soil* [47]; *combining peas, durum wheat, leeks, linseed, mustard, onions, spring field beans, spring oilseed rape, sugar beet, swedes, turnips, vining peas, winter field beans, winter oilseed rape* [36]; *cultivated land/soil* [4, 5, 8, 21]; *spring barley, spring oats, spring wheat, winter barley, winter oats, winter wheat* [13, 15, 25, 29, 36, 52, 54]; *stubbles* [5, 8, 41, 47]
- Annual grasses in *all edible crops, all non-edible crops, hard surfaces, land not intended to bear vegetation, permeable surfaces overlying soil* [47]; *combining peas, durum wheat, leeks, linseed, mustard, onions, spring barley, spring field beans, spring oats, spring oilseed rape, spring wheat, sugar beet, swedes, turnips, vining peas, winter barley, winter field beans, winter oats, winter oilseed rape, winter wheat* [36]; *cultivated land/soil* [4, 5, 8, 13, 15, 21, 25, 29, 41, 52, 54]; *stubbles* [5, 8, 41, 47]
- Annual weeds in *durum wheat, leeks, linseed, mustard, onions, spring barley, spring field beans, spring oats, spring oilseed rape, spring wheat, sugar beet, swedes, turnips, winter barley, winter field beans, winter oats, winter oilseed rape, winter wheat* [26, 34, 35]; *grassland* [26, 31, 34, 35, 43, 46]; *peas* [34, 35]; *vining peas* [26]

FOR FULL CONDITIONS OF USE ALWAYS READ THE PRODUCT LABEL

- Aquatic weeds in **open waters** [21, 38, 39, 53]
- Black bent in **combining peas, spring field beans, spring oilseed rape, stubbles, winter field beans, winter oilseed rape** [4, 5, 8, 13, 15, 21, 25, 29, 41, 52, 54]; **durum wheat, spring barley, spring oats, spring wheat, winter barley, winter oats, winter wheat** [4, 5, 8, 21, 41]; **linseed** [8, 13, 15, 25, 29, 41, 52, 54]
- Bolters in **sugar beet** *(wiper application)* [3-5, 8, 11, 12, 14, 20-22, 26, 31, 32, 34, 35, 41-43]
- Bracken in **amenity trees and shrubs, fencelines, land clearance, non-crop areas, paths and drives** [37]; **farm forestry** [39]; **forest** [4-9, 11-13, 15-19, 21-23, 25, 29, 30, 32, 33, 37-39, 49, 52-54]; **grassland** [50]; **tolerant conifers** [9, 17, 37]
- Brambles in **tolerant conifers** [9, 17, 37]
- Chemical thinning in **forest** [6-8, 11, 13, 15, 16, 18, 19, 21, 23, 25, 29, 30, 32, 33, 37, 38, 49, 52-54]
- Couch in **combining peas, spring field beans, winter field beans** [1-5, 8, 9, 11-15, 18, 20-22, 25-29, 31, 32, 34, 35, 41-43, 51, 52, 54]; **durum wheat** [1, 2, 4, 5, 8, 21, 26-28, 31, 34, 35, 41, 43, 51]; **farm forestry** [39]; **forest** [6-8, 11, 18, 21, 22, 32, 37-39, 53]; **grassland** [1, 4, 9]; **land not intended to bear vegetation, oilseed rape for industrial use** [50]; **linseed** [1, 2, 8, 9, 13, 15, 18, 25, 26, 29, 31, 34, 35, 41, 43, 51, 52, 54]; **mustard** [9, 18, 26, 31, 34, 35, 43, 51]; **spring barley, spring wheat, winter barley, winter wheat** [1-5, 8, 9, 11, 12, 14, 18, 20-22, 26-28, 31, 32, 34, 35, 41-43, 46, 51]; **spring oats, winter oats** [1-5, 8, 9, 11, 12, 14, 18, 20-22, 26-28, 31, 32, 34, 35, 41-43, 46]; **spring oilseed rape, stubbles, winter oilseed rape** [1-5, 8, 9, 11-15, 18, 20-22, 25-29, 31, 32, 34, 35, 41-43, 46, 50-52, 54]
- Creeping bent in **combining peas, spring field beans, spring oilseed rape, stubbles, winter field beans, winter oilseed rape** [4, 5, 8, 13, 15, 21, 25, 29, 41, 52, 54]; **cultivated land/soil** [1]; **durum wheat, spring barley, spring oats, spring wheat, winter barley, winter oats, winter wheat** [4, 5, 8, 21, 41]; **linseed** [8, 13, 15, 25, 29, 41, 52, 54]
- Green cover in **land not being used for crop production** [2, 5, 8, 13-15, 21, 25, 29, 36, 41, 50, 52, 54]
- Harvest management/desiccation in **combining peas, spring field beans, winter field beans** [1, 3, 9, 11, 12, 14, 18, 20, 22, 26-28, 31, 32, 34, 35, 42, 43, 50]; **durum wheat** [1, 3, 11, 12, 14, 20, 22, 26-28, 31, 32, 34, 35, 42, 43]; **linseed** [1, 3, 9, 11, 12, 18, 20, 21, 26, 31, 32, 34, 35, 41-43, 50]; **mustard** [1, 3, 9, 11, 12, 14, 18, 20, 26, 31, 32, 34, 35, 42, 43, 50, 51]; **oilseed rape for industrial use** [50]; **spring barley, winter barley** [1, 9, 13, 15, 18, 22, 25-29, 31, 34, 35, 43, 46, 50, 52, 54]; **spring oats, winter oats** [1, 3, 9, 11-15, 18, 20, 22, 25-29, 31, 32, 34, 35, 42, 43, 46, 52, 54]; **spring oilseed rape, winter oilseed rape** [1-4, 9, 11-15, 18, 20-22, 25-29, 31, 32, 34, 35, 41-43, 46, 50-52, 54]; **spring wheat, winter wheat** [1, 3, 9, 11-15, 18, 20, 22, 25-29, 31, 32, 34, 35, 42, 43, 46, 50, 52, 54]
- Heather in **farm forestry** [39]; **forest** [4-9, 12, 13, 15-17, 19, 21, 23, 25, 29, 30, 32, 33, 38, 39, 49, 52-54]
- Japanese knotweed in **land not intended to bear vegetation** [50]
- Perennial dicotyledons in **combining peas, spring field beans, spring oilseed rape, winter field beans, winter oilseed rape** [5, 8, 13, 15, 25, 29, 41, 52, 54]; **durum wheat, spring barley, spring oats, spring wheat, winter barley, winter oats, winter wheat** [5, 8, 41]; **forest** [16, 19, 23, 30, 33, 49]; **forest** *(wiper application)* [37]; **grassland** *(wiper application)* [26, 31, 34, 35, 43]; **linseed** [8, 13, 15, 25, 29, 41, 52, 54]
- Perennial grasses in **aquatic situations** [2, 4-9, 11-19, 21, 23, 25, 29, 30, 32, 33, 37-39, 44, 45, 49, 52-54]; **enclosed waters, land immediately adjacent to aquatic areas** [16, 49]; **tolerant conifers** [9, 17, 37]
- Pre-harvest desiccation in **linseed** [8]; **poppies** *(off-label - for morphine production)* [1, 34]; **spring oilseed rape, winter oilseed rape** [5, 8]
- Reeds in **aquatic situations** [2, 4-9, 11-19, 21, 23, 25, 29, 30, 32-34, 37-39, 44, 45, 49, 52-54]; **enclosed waters, land immediately adjacent to aquatic areas** [16, 21, 38, 39, 49, 53]
- Rhododendrons in **farm forestry** [39]; **forest** [6-9, 12, 13, 15-17, 19, 21, 23, 25, 29, 30, 32, 33, 37-39, 49, 52-54]
- Rushes in **aquatic situations** [2, 4-9, 11-19, 21, 23, 25, 29, 30, 32-34, 37-39, 44, 45, 49, 52-54]; **enclosed waters, land immediately adjacent to aquatic areas** [16, 21, 38, 39, 49, 53]; **forest** [18]; **grassland** [50]
- Sedges in **aquatic situations** [2, 4-9, 11-19, 21, 23, 25, 29, 30, 32-34, 37-39, 44, 45, 49, 52-54]; **enclosed waters, land immediately adjacent to aquatic areas** [16, 21, 38, 39, 49, 53]
- Sucker control in **apple orchards, cherries, damsons, pear orchards, plums** [20, 31, 34, 35, 42, 43]

SEE SECTION 3 FOR PRODUCTS ALSO REGISTERED

- Sward destruction in **amenity grass** [37]; **grassland** [1-5, 8, 9, 11-15, 18, 20-22, 25-29, 31, 32, 34, 35, 41-43, 46, 50-52, 54]; **rotational grassland** [47]
- Total vegetation control in **amenity areas** [37, 44]; **amenity vegetation, land not intended to bear vegetation** [10, 17, 37]; **fencelines, road verges** [17, 37, 44]; **hard surfaces** [17, 37]; **industrial sites** [17, 37, 44, 45]
- Volunteer cereals in **all edible crops, all non-edible crops** [47]; **combining peas** [36]; **cultivated land/soil** [4, 5, 8, 13, 15, 21, 25, 29, 41, 52, 54]; **durum wheat, leeks, linseed, mustard, onions, spring barley, spring field beans, spring oats, spring oilseed rape, spring wheat, sugar beet, swedes, turnips, winter barley, winter field beans, winter oats, winter oilseed rape, winter wheat** [26, 34-36]; **peas** [34, 35]; **stubbles** [1-5, 8, 9, 11-15, 18, 20-22, 25-29, 31, 32, 34, 35, 41-43, 46, 47, 50-52, 54]; **vining peas** [26, 36]
- Volunteer oilseed rape in **stubbles** [50]
- Volunteer potatoes in **stubbles** [1-5, 8, 9, 11, 12, 14, 18, 20-22, 26-28, 31, 32, 34, 35, 41-43, 46, 50, 51]
- Waterlilies in **aquatic situations** [2, 4-9, 11-19, 21, 23, 25, 29, 30, 32, 33, 37-39, 44, 45, 49, 52-54]; **enclosed waters** [21, 38, 39, 53]
- Weed beet in **sugar beet** *(wiper application)* [3, 11, 14, 20, 22, 32, 42]
- Wild oats in **durum wheat** [3, 12, 14, 20, 32, 42]; **spring barley, spring oats, spring wheat, winter barley, winter oats, winter wheat** [3, 11, 12, 14, 20, 32, 42]
- Woody weeds in **farm forestry** [39]; **forest** [4-9, 11-13, 15-19, 21-23, 25, 29, 30, 32, 33, 37-39, 49, 52-54]; **tolerant conifers** [9, 17, 37]

Specific Off-Label Approvals (SOLAs)

- **asparagus** *(OLA 051162) Jul 2007* [2]
- **asparagus, blackcurrants, blueberries, chestnuts, cob nuts, grapevines, hazel nuts, rhubarb, walnuts** *(OLA 032129) Dec 2008* [4]
- **asparagus** *(OLA 042122) Dec 2008* [34]
- **bilberries, blackcurrants, blueberries, cranberries, gooseberries, vaccinium spp.** *(OLA 051160) Jul 2007* [2]
- **bilberries, blackcurrants, blueberries, cranberries, gooseberries, redcurrants, vaccinium spp., whitecurrants** *(OLA 042119) Dec 2008* [34]
- **chestnuts, cob nuts, hazel nuts, walnuts** *(OLA 051161) Jul 2007* [2]
- **chestnuts, cob nuts, hazel nuts, walnuts** *(OLA 042121) Dec 2008* [34]
- **grapevines** *(OLA 051163) Jul 2007* [2]
- **grapevines** *(OLA 042120) Dec 2008* [34]
- **poppies** *(for morphine production) (OLA 050871) Dec 2008* [1]
- **poppies** *(for morphine production) (OLA 031409) Dec 2008* [34]
- **rhubarb** *(OLA 051164) Jul 2007* [2]
- **rhubarb** *(OLA 042123) Dec 2008* [34]

Approval information

- Accepted by BBPA for use on malting barley and hops
- Approval expiry 31 Dec 2006 [14]
- Approval expiry 31 Dec 2006 [30]

Efficacy guidance

- For best results apply to actively growing weeds with enough leaf to absorb chemical
- A rainfree period of at least 6 h (preferably 24 h) should follow spraying
- Adjuvants are obligatory for some products and recommended for some uses with others. See labels
- Mixtures with other pesticides or fertilizers may lead to reduced control.
- Products are formulated as isopropylamine, ammonium, potassium, or trimesium salts of glyphosate and may vary in the details of efficacy claims. See individual product labels
- Some products in ready-to-use formulations for use through hand-held applicators. See label for instructions [10, 24, 48]
- When applying products through rotary atomisers the spray droplet spectra must have a minimum Volume Median Diameter (VMD) of 200 microns
- With wiper application weeds should be at least 10 cm taller than crop
- Annual weed grasses should have at least 5 cm of leaf and annual broad-leaved weeds at least 2 expanded true leaves

FOR FULL CONDITIONS OF USE ALWAYS READ THE PRODUCT LABEL

- Perennial grass weeds should have 4-5 new leaves and be at least 10 cm long when treated. Perennial broad-leaved weeds should be treated at or near flowering but before onset of senescence
- Volunteer potatoes and polygonums are not controlled by harvest-aid rates
- Bracken must be treated at full frond expansion
- Fruit tree suckers best treated in late spring
- Chemical thinning treatment can be applied as stump spray or stem injection
- In order to allow translocation, do not cultivate before spraying and do not apply other pesticides, lime, fertilizer or farmyard manure within 5 d of treatment
- Recommended intervals after treatment and before cultivation vary. See labels

Restrictions
- Maximum number of treatments normally 1 per crop or season on field and edible crops and unrestricted for non-crop uses. Check labels for details
- Do not treat cereals grown for seed or undersown crops
- Consult grain merchant before treating crops grown on contract or intended for malting
- Do not use treated straw as a mulch or growing medium for horticultural crops
- Do not use on grassland if the crop is to be used as animal feed or bedding
- For use in nursery stock, shrubberies, orchards, grapevines and tree nuts care must be taken to avoid contact with the trees. Do not use in orchards established less than 2 yr and keep off low-lying branches
- Certain conifers may be sprayed overall in dormant season. See label for details
- Use a tree guard when spraying in established forestry plantations
- Do not spray root suckers in orchards in late summer or autumn
- Do not use under glass or polythene as damage to crops may result
- Do not use wiper techniques in soft fruit crops
- Do not use trimesium salt products in aquatic situations or forestry
- Do not mix, store or apply in galvanised or unlined mild steel containers or spray tanks
- Do not leave diluted chemical in spray tanks for long periods and make sure that tanks are well vented

Crop-specific information
- Harvest intervals: 4 wk for blackcurrants, blueberries; 14 d for linseed, oilseed rape; 8 d for mustard; 5-7 d for all other edible crops. Check label for exact details
- Latest use: 2-14 d before cultivating, drilling or planting a crop in treated land; after harvest (post-leaf fall) but before bud formation in the following season for nuts and most fruit and vegetable crops; before fruit set for grapevines. See labels for details

Following crops guidance
- Decaying remains of plants killed by spraying must be dispersed before direct drilling

Environmental safety
- Products differ in their hazard and environmental safety classification. See labels
- Harmful to fish or other aquatic life. Do not contaminate surface waters or ditches with chemical or used container
- Do not dump surplus herbicide in water or ditch bottoms or empty into drains
- Maximum permitted concentration in treated water 0.2 ppm
- The Environment Agency or Local River Purification Authority must be consulted before use in or near water
- LERAP Category B [26]
- Take extreme care to avoid drift and possible damage to neighbouring crops or plants
- Treated poisonous plants must be removed before grazing or conserving
- Do not use in covered areas such as greenhouses or under polythene
- For field edge treatment direct spray away from hedge bottoms
- Exclude livestock from treated areas. Livestock may not graze or be fed the treated forage nor may it be used for hay, silage or bedding

Hazard classification and safety precautions
Hazard H03 [12, 14, 50]; H04 [1, 3-6, 9, 11, 16, 18-20, 22, 26-28, 30-33, 36, 39, 40, 42-44, 46-48, 54]; H11 [3-6, 11, 12, 14, 16, 18, 20, 22, 25-33, 36, 39, 40, 42, 43, 47, 48, 54]
Risk phrases R20 [12, 14]; R22a [50]; R36 [1, 3, 6, 11, 12, 14, 16, 18-20, 30-33, 40, 42-44, 46, 48]; R38 [1, 6, 11, 19, 22, 26-28, 32, 44, 46]; R41 [4, 5, 9, 22, 27, 28, 32, 36, 39, 54]; R50 [26, 47]; R51

SEE SECTION 3 FOR PRODUCTS ALSO REGISTERED

[3-6, 11, 12, 14, 16, 18, 20, 22, 25, 27-33, 36, 39, 40, 42, 43, 48, 54]; R52 [1, 7, 23, 38, 50]; R53a [1, 3-6, 11, 12, 14, 16, 18, 20, 22, 25-33, 36, 38-40, 42, 43, 48, 50, 54]

Operator protection A [1-46, 48-54]; C [1-9, 11-25, 27-40, 42-44, 46, 48-50, 52-54]; D [2, 13, 15-17, 25, 29, 38, 49, 52, 53]; F [3, 11, 14, 18-20, 30, 31, 33, 42, 43]; H [2, 3, 7-11, 13-18, 20-23, 25, 26, 29, 31, 33-39, 41-43, 45, 49, 50, 52-54]; M [2-5, 7-26, 29-39, 41-45, 48-54]; N [3-5, 11, 12, 14, 18-20, 22, 24, 30-33, 42-44, 48, 51, 54]; U02a [1, 4-9, 11, 12, 14, 15, 18-25, 27-30, 32, 36, 41, 44, 46, 48, 52, 53]; U05a [1, 3-6, 9, 11, 12, 14, 16, 18-20, 22, 24, 26-28, 30-33, 36, 39, 40, 42-44, 46, 48, 50, 51, 54]; U07 [40, 50]; U08 [1, 3-9, 11, 12, 14-16, 18-25, 27-33, 36, 40-44, 46, 48, 52, 53]; U09a [24, 39, 48, 54]; U11 [3, 11, 12, 14, 16, 18, 20, 22, 27, 28, 30-33, 36, 39, 40, 42, 43, 47, 48, 54]; U15 [3, 11, 16, 18, 20, 30, 31, 33, 40, 42, 43]; U19a [1, 3-12, 14-16, 18-25, 27-33, 36, 39-44, 46, 48, 52-54]; U20a [6, 7, 23, 24, 26, 34, 35, 45, 50, 51]; U20b [1-5, 8-22, 24, 25, 27-33, 35-44, 46-49, 52-54]

Environmental protection E06a [51] (unstipulated period); E13c [1, 8, 9, 11, 15, 18, 19, 21, 24, 35, 44, 46, 48]; E15a [2-7, 10-14, 16-18, 20, 22, 23, 25-43, 45, 47-53]; E16a, E16b [26]; E19a [2, 4, 5, 7-9, 11-15, 21, 23, 25, 29, 30, 32, 44, 49, 52]; E19b [54]; E23 [53]; E34 [1, 10, 18, 22, 40, 50]; E36 [40, 50]; E38 [3, 11, 12, 14, 16, 18, 20, 26, 30, 31, 33, 36, 39, 40, 42, 43, 47, 50]

Storage and disposal D01 [1-12, 14-20, 22-40, 42-46, 48-51, 53, 54]; D02 [1, 3-12, 14, 16-20, 22-24, 26-28, 30-40, 42-46, 48, 50, 51, 53, 54]; D05 [1, 3-8, 10, 16-24, 26-28, 30-33, 36-43, 47, 48, 50, 51, 53]; D07 [36, 39, 50]; D09a [1-54]; D10a [10]; D10b [1, 2, 4-9, 11-14, 18, 19, 21-23, 25, 27, 28, 32, 45, 46, 49, 52, 54]; D10c [3, 15-17, 20, 26, 29-31, 33-35, 37, 38, 41-43, 47, 51, 53]; D11a [24, 36, 39, 48]; D12a [3, 11, 12, 14, 16, 18, 20, 26, 30, 31, 33, 36, 39, 40, 42, 43, 47, 50, 54]; D12b [25, 29]; D14 [40, 50]

Medical advice M03 [50]; M05a [54]

266　guazatine

A guanidine fungicide seed dressing for cereals

Products

1	Panoctine	Makhteshim	300 g/l	LS	11404
2	Ravine	Makhteshim	300 g/l	LS	12211

Uses

- Brown foot rot in *spring barley* (seed treatment - reduction), *spring oats* (seed treatment - reduction), *winter barley* (seed treatment - reduction), *winter oats* (seed treatment - reduction) [1]
- Foot rot in *spring barley* (seed treatment), *winter barley* (seed treatment) [1]
- Fusarium foot rot and seedling blight in *spring barley* (seed treatment - reduction), *spring oats* (seed treatment - reduction), *winter barley* (seed treatment - reduction), *winter oats* (seed treatment - reduction) [1, 2]; *spring wheat* (seed treatment - reduction), *winter wheat* (seed treatment - reduction) [2]
- Leaf stripe in *spring barley* (seed treatment), *winter barley* (seed treatment) [1]
- Net blotch in *spring barley* (seed treatment), *winter barley* (seed treatment) [1]
- Pyrenophora leaf spot in *spring oats* (seed treatment), *winter oats* (seed treatment) [1]
- Septoria seedling blight in *spring wheat* (seed treatment), *winter wheat* (seed treatment) [2]

Approval information

- Accepted by BBPA for use on malting barley

Efficacy guidance

- Apply with conventional seed treatment machinery

Restrictions

- Maximum number of treatments 1 per batch
- Do not treat grain with moisture content above 16% and do not allow moisture content of treated seed to exceed 16%
- Do not apply to cracked, split or sprouted seed
- Do not use treated seed as food or feed

FOR FULL CONDITIONS OF USE ALWAYS READ THE PRODUCT LABEL

Crop-specific information
- Latest use: pre-drilling
- After treating, bag seed immediately and keep in dry, draught-free store
- Treatment may lower germination capacity, particularly if seed grown, harvested or stored under adverse conditions

Environmental safety
- Dangerous for the environment
- Very toxic to aquatic organisms
- Harmful to fish or other aquatic life. Do not contaminate surface waters or ditches with chemical or used container

Hazard classification and safety precautions
Hazard H03, H11
Risk phrases R20, R22a, R37, R41, R43, R50, R53a
Operator protection A, C, D, H; U05a, U20b [1, 2]; U09a, U14, U15 [1]
Environmental protection E13c [1]; E15a [2]; E34, E38 [1, 2]
Storage and disposal D01, D02, D05, D09a, D10a [1, 2]; D14 [1] (returnable container)
Treated seed S01, S02, S04b, S05, S06a, S07
Medical advice M03 [1]; M04 [1, 2]

267 guazatine + imazalil

A fungicide seed treatment for barley and oats

Products
1	Panoctine Plus	Makhteshim	300:25 g/l	LS	11757
2	Ravine Plus	Makhteshim	300:25 g/l	LS	11832

Uses
- Brown foot rot in **spring barley** *(seed treatment)*, **winter barley** *(seed treatment)*
- Covered smut in **spring barley** *(seed treatment)*, **winter barley** *(seed treatment)*
- Leaf stripe in **spring barley** *(seed treatment)*, **winter barley** *(seed treatment)*
- Net blotch in **spring barley** *(seed treatment - moderate control)*, **winter barley** *(seed treatment - moderate control)*
- Pyrenophora leaf spot in **spring oats** *(seed treatment)*, **winter oats** *(seed treatment)*
- Seedling blight and foot rot in **spring barley** *(seed treatment - moderate control)*, **winter barley** *(seed treatment - moderate control)*
- Snow mould in **spring barley** *(seed treatment)*, **winter barley** *(seed treatment)*

Approval information
- Imazalil included in Annex I under EC Directive 91/414
- Accepted by BBPA for use on malting barley

Efficacy guidance
- Apply with conventional seed treatment machinery

Restrictions
- Maximum number of treatments 1 per batch
- Do not mix with other formulations
- Do not treat grain with moisture content above 16% and do not allow moisture content of treated seed to exceed 16%
- Do not apply to cracked, split or sprouted seed
- Do not store treated seed for more than 6 mth
- Treatment may lower germination capacity, particularly if seed grown, harvested or stored under adverse conditions

Crop-specific information
- Latest use: pre-drilling

Environmental safety
- Dangerous for the environment
- Very toxic to aquatic organisms
- Do not use treated seed as food or feed

SEE SECTION 3 FOR PRODUCTS ALSO REGISTERED

Hazard classification and safety precautions
   Hazard H02, H11
   Risk phrases R22a, R23, R37, R41, R43, R50, R53a
   Operator protection A, C, D, H; U05a, U10, U11, U14, U19a, U20b
   Environmental protection E15a, E34, E38
   Storage and disposal D01, D02, D09a, D10b
   Treated seed S02, S04b, S05, S07
   Medical advice M04

268   hymexazol

A systemic isoxazole fungicide for pelleting sugar beet seed

Products

Tachigaren 70 WP	SumiAgro	70% w/w	WP	10496

Uses
- Black leg in **sugar beet** *(seed treatment)*

Efficacy guidance
- Incorporate into pelleted seed using suitable seed pelleting machinery

Restrictions
- Maximum number of treatments 1 per batch of seed
- Do not use treated seed as food or feed

Crop-specific information
- Latest use: before planting sugar beet seed

Environmental safety
- Harmful to aquatic organisms
- Harmful to fish or other aquatic life. Do not contaminate surface waters or ditches with chemical or used container
- Treated seed harmful to game and wildlife

Hazard classification and safety precautions
   Hazard H04, H07
   Risk phrases R41, R52, R53a
   Operator protection A, C, F; U05a, U11, U20b
   Environmental protection E13c
   Storage and disposal D01, D02, D09a, D11a
   Treated seed S01, S02, S05, S06a

269   imazalil

A systemic and protectant conazole fungicide

Products

1	Fungazil 100 SL	BASF	100 g/l	LS	11762
2	Magnate 100 SL	Makhteshim	100 g/l	SL	11705

Uses
- Dry rot in **seed potatoes**, **ware potatoes**
- Gangrene in **seed potatoes**, **ware potatoes**
- Silver scurf in **seed potatoes**, **ware potatoes**
- Skin spot in **seed potatoes**, **ware potatoes**

Approval information
- Imazalil included in Annex I under EC Directive 91/414
- Approval expiry 26 Oct 2006 [2]

Efficacy guidance
- For best control of skin and wound diseases of ware potatoes treat as soon as possible after harvest, preferably within 7-10 d, before any wounds have healed

FOR FULL CONDITIONS OF USE ALWAYS READ THE PRODUCT LABEL

Restrictions
- Maximum number of treatments 2 per batch of seed tubers [1, 2]; 1 per batch of ware tubers [1]
- Consult processor before treating potatoes for processing

Crop-specific information
- Latest use: during storage and before chitting for seed potatoes
- HI edible crops 1 d
- Apply to clean soil-free potatoes post-harvest before putting into store, or at first grading. A further treatment may be applied in early spring before planting
- Apply through canopied hydraulic or spinning disc equipment preferably diluted with up to two litres water per tonne of potatoes to obtain maximum skin cover and penetration
- Use on ware potatoes subject to discharges of imazalil from potato washing plants being within emission limits set by the UK monitoring authority

Environmental safety
- Dangerous for the environment
- Very toxic to aquatic organisms
- Harmful to fish or other aquatic life. Do not contaminate surface waters or ditches with chemical or used container
- Do not empty into drains [1]
- Personal protective equipment requirements may vary for each pack size. Check label

Hazard classification and safety precautions
Hazard H04 [1, 2]; H11 [2]
Risk phrases R36, R50 [2]; R41, R52 [1]; R53a [1, 2]
Operator protection A, C [1, 2]; D [2]; H [1]; U04a, U05a, U19a, U20a [1, 2]; U11, U14 [1]
Environmental protection E13c [2]; E15a, E19b [1]; E34 [1, 2]
Storage and disposal D01, D02, D05, D09a [1, 2]; D10a [2]; D10c, D12a [1]
Treated seed S01, S03, S04a, S05, S06a [1, 2]; S02 [1]
Medical advice M03 [2]

270 imazalil + pencycuron

A fungicide mixture for treatment of seed potatoes

Products

Monceren IM	Bayer CropScience	0.6:12.5% w/w	DS	11426

Uses
- Black scurf in **potatoes** *(tuber treatment)*
- Silver scurf in **potatoes** *(tuber treatment - reduction)*
- Stem canker in **potatoes** *(tuber treatment - reduction)*

Approval information
- Imazalil included in Annex I under EC Directive 91/414

Efficacy guidance
- Apply to clean seed tubers during planting (see label for suitable method) or sprinkle over tubers in chitting trays before loading into planter. It is essential to obtain an even distribution over tubers
- If seed tubers become damp from light rain distribution of product should not be affected. Tubers in the hopper should be covered if a shower interrupts planting

Restrictions
- Maximum number of treatments 1 per batch
- Operators must wear suitable respiratory equipment and gloves when handling product and when riding on planter. Wear gloves when handling treated tubers
- Do not use on tubers which have previously been treated with a dry powder seed treatment or hot water
- Do not use treated tubers for human or animal consumption

Crop-specific information
- Latest use: immediately before planting
- May be used on seed tubers previously treated with a liquid fungicide but not before 8 wk have elapsed if this contained imazalil

Environmental safety
- Harmful to aquatic organisms
- Harmful to fish or other aquatic life. Do not contaminate surface waters or ditches with chemical or used container
- Treated seed harmful to game and wildlife

Hazard classification and safety precautions
Risk phrases R52, R53a
Operator protection A, D; U20b
Environmental protection E03, E13c
Storage and disposal D09a, D11a
Treated seed S01, S02, S03, S04a, S05, S06a

271   imazalil + thiabendazole

A fungicide mixture for treatment of seed potatoes

Products

Extratect Flowable	Banks Cargill	100:300 g/l	FS	08704

Uses
- Dry rot in **seed potatoes** *(tuber treatment - reduction)*
- Gangrene in **seed potatoes** *(tuber treatment - reduction)*
- Silver scurf in **potatoes** *(tuber treatment - reduction)*, **seed potatoes** *(tuber treatment - reduction)*
- Skin spot in **potatoes** *(tuber treatment - reduction)*, **seed potatoes** *(tuber treatment - reduction)*
- Stem canker in **potatoes** *(tuber treatment - reduction)*

Approval information
- Imazalil and thiabendazole included in Annex I under EC Directive 91/414
- Approved for use through ULV equipment [1]

Efficacy guidance
- Apply immediately after lifting with suitable spinning disc or low volume hydraulic applicator for disease reduction during storage
- Apply within 2 mth of planting for maximum disease reduction in the progeny crop

Restrictions
- Maximum number of treatments 1 per batch
- Do not mix with any other product
- May not be used on seed tubers previously treated with a product containing imazalil
- Do not use treated seed as food or feed

Crop-specific information
- Latest use: pre-planting
- Delayed emergence noted under certain conditions but with no final effect on yield

Environmental safety
- Harmful to fish or other aquatic life. Do not contaminate surface waters or ditches with chemical or used container

Hazard classification and safety precautions
Hazard H04
Risk phrases R37, R41
Operator protection A, C, D, H; U05a, U09a, U11, U19a, U20c
Environmental protection E13c
Storage and disposal D01, D02, D09a, D10a
Treated seed S01, S02, S05, S06a

272   imazalil + triticonazole

A conazole fungicide mixture for seed treatment of barley

Products

Robust	BASF	12.5:12.5 g/l	FS	11807

FOR FULL CONDITIONS OF USE ALWAYS READ THE PRODUCT LABEL

Uses

- Covered smut in **spring barley** *(seed treatment)*, **winter barley** *(seed treatment)*
- Leaf stripe in **spring barley** *(seed treatment)*, **winter barley** *(seed treatment)*
- Loose smut in **spring barley** *(seed treatment)*, **winter barley** *(seed treatment)*
- Net blotch in **spring barley** *(seed treatment - moderate control)*, **winter barley** *(seed treatment - moderate control)*
- Seedling blight and foot rot in **spring barley** *(seed treatment - moderate control)*, **winter barley** *(seed treatment - moderate control)*

Approval information

- Imazalil included in Annex I under EC Directive 91/414
- Accepted by BBPA for use on malting barley

Efficacy guidance

- Apply undiluted through conventional seed treatment machine
- Treated seed may be stored for up to 18 mth but germination should be checked before use and sowing rate adjusted if necessary
- Keep treated seed in a dry, draught free store
- Calibrate seed drill before sowing

Restrictions

- Maximum number of treatments 1 per batch of seed
- Do not treat seed with moisture content above 16%
- Do not apply to cracked, split or sprouted seed
- Do not handle treated seed unnecessarily
- Do not use treated seed as food or feed

Crop-specific information

- Latest use: pre-drilling

Environmental safety

- Toxic to aquatic organisms
- Product available in large (1000 l) returnable containers. They should be re-circulated for 30 min before use and operated as directed on the label

Hazard classification and safety precautions

Hazard H04
Risk phrases R43, R51, R53a
Operator protection A, C, H; U05a, U07, U09a, U14, U20b
Environmental protection E15a, E34, E36
Storage and disposal D01, D02, D09a, D14
Treated seed S02, S04b, S05, S07
Medical advice M03

273   imazaquin

An imidazolinone herbicide and plant growth regulator available only in mixtures

See also chlormequat + imazaquin

274   imidacloprid

A neonicotinoid insecticide for seed, soil, peat or foliar treatment

See also beta-cyfluthrin + imidacloprid
bitertanol + fuberidazole + imidacloprid
fuberidazole + imidacloprid + triadimenol

Products

1	Admire	Bayer CropScience	70% w/w	WG	11234
2	Gaucho	Bayer CropScience	70% w/w	WS	11281
3	Intercept 5GR	Scotts	5% w/w	GR	08126
4	Intercept 70WG	Scotts	70% w/w	WG	08585

SEE SECTION 3 FOR PRODUCTS ALSO REGISTERED

Uses

- Aphids in **bedding plants**, **hardy ornamental nursery stock**, **herbaceous perennials**, **pot plants** [3, 4]; **lettuce** *(off-label)* [2]
- Beet virus yellows vectors in **sugar beet** *(seed treatment)* [2]
- Cabbage aphid in **collards** *(off-label - seed treatment)*, **kale** *(off-label - seed treatment)* [2]
- Damson-hop aphid in **hops** [1]
- Flea beetle in **sugar beet** *(seed treatment)* [2]
- Glasshouse whitefly in **bedding plants**, **hardy ornamental nursery stock**, **herbaceous perennials**, **pot plants** [3, 4]
- Mangold fly in **sugar beet** *(seed treatment)* [2]
- Millipedes in **sugar beet** *(seed treatment)* [2]
- Peach-potato aphid in **broccoli** *(off-label - seed treatment)*, **brussels sprouts** *(off-label - seed treatment)*, **cabbages** *(off-label - seed treatment)*, **calabrese** *(off-label - seed treatment)*, **cauliflowers** *(off-label - seed treatment)*, **collards** *(off-label - seed treatment)*, **kale** *(off-label - seed treatment)* [2]
- Pygmy beetle in **sugar beet** *(seed treatment)* [2]
- Sciarid flies in **bedding plants**, **hardy ornamental nursery stock**, **herbaceous perennials**, **pot plants** [3, 4]
- Springtails in **sugar beet** *(seed treatment)* [2]
- Symphylids in **sugar beet** *(seed treatment)* [2]
- Tobacco whitefly in **bedding plants**, **hardy ornamental nursery stock**, **herbaceous perennials**, **pot plants** [3, 4]
- Vine weevil in **bedding plants**, **hardy ornamental nursery stock**, **herbaceous perennials**, **pot plants** [3, 4]

Specific Off-Label Approvals (SOLAs)

- **broccoli**, **brussels sprouts**, **cabbages**, **calabrese**, **cauliflowers**, **collards**, **kale** *(seed treatment)* *(OLA 023927) Dec 2008* [2]
- **lettuce** *(OLA 031886) Jan 2006* [2]

Approval information

- Accepted by BBPA for use on hops

Efficacy guidance

- Treated seed should be drilled within the season of purchase [2]
- Base of hop plants should be free of weeds and debris at application [1]
- Hop bines emerging away from the main stock or adjacent to poles may require a separate application [1]
- Uptake and movement within hops requires soil moisture and good growing conditions [1]
- Control may be impaired in plantations greater than 3640 plants/ha [1]
- To minimise likelihood of resistance do not treat all the hop crop in any one yr [1] and adopt a planned programme to alternate with pesticides of different types or use other measures when using in compost [3, 4]
- When applied as drench or incorporated as granules in moist compost compound is readily absorbed and translocated to aerial parts of plant [3, 4]

Restrictions

- Maximum number of treatments 1 per batch of seed [2] or growing medium [3, 4]; 1 per yr [1]
- Product must not be used in compost that has already been treated with an imidacloprid-containing product [4]
- Product formulated for use only as a compost incorporation treatment into peat-based growing media using suitable automated equipment [3]
- For use only on container grown ornamentals [3, 4]
- The safety of seeds sown into treated compost should not be assumed. Test treat before full-scale use [3]
- Product must not be used on crops for human or animal consumption and treated compost must not be re-used for this purpose [3, 4]
- Do not use treated seed as food or feed [2]

Crop-specific information

- Latest use: before bines reach 2 m or before end 1st wk Jun for hops; before drilling for sugar beet; before sowing or planting for brassicas, bedding plants, hardy ornamental nursery stock, pot plants

---

FOR FULL CONDITIONS OF USE ALWAYS READ THE PRODUCT LABEL

- Apply to sugar beet seed as part of the normal commercial pelleting process using special treatment machinery [2]
- Apply to hops as a directed stem base spray before most bines reach a height of 2 m. If necessary treat both sides of the crown at half the normal concentration [1]
- The rate used per hop plant must be adjusted in accordance with the plant population if greater than 3640 per ha. See label for details [1]

Environmental safety
- Harmful to aquatic organisms [1]
- High risk to bees. Do not apply to crops in flower or to those in which bees are actively foraging. Do not apply when flowering weeds are present [1, 4]
- Avoid spillage or other environmental contamination when incorporating into compost [3]
- Treated seed harmful to game and wildlife [2]

Hazard classification and safety precautions
Hazard H03 [1, 4]; H04 [2]
Risk phrases R22a [1, 4]; R43 [2]; R52 [1]
Operator protection A [1-3]; C, D, H [2]; U05a [1, 4]; U20a [2]; U20b [1, 3, 4]
Environmental protection E12a, E34 [1, 4]; E15a [1-4]
Storage and disposal D01, D02 [1, 4]; D07 [3]; D09a, D11a [1-4]
Treated seed S02, S04b, S05, S06a, S07 [2]

275 imidacloprid + tebuconazole + triazoxide

A broad spectrum fungicide and insecticide seed treatment for winter barley

Products
Raxil Secur	Bayer CropScience	233:20:20 g/l	LS	11298

Uses
- Leaf stripe in **winter barley** *(seed treatment)*
- Loose smut in **winter barley** *(seed treatment)*
- Net blotch in **winter barley** *(seed treatment)*
- Virus vectors in **winter barley** *(seed treatment)*
- Wireworm in **winter barley** *(reduction of damage)*

Approval information
- Accepted by BBPA for use on malting barley

Efficacy guidance
- Best applied through recommended seed treatment machines
- Evenness of seed cover improved by simultaneous application of equal volumes of product and water or dilution of product with an equal volume of water
- Drill treated seed in the same season. Use minimum 125 kg treated seed per ha. Lower drilling rates and/or early drilling affect duration of BYDV protection needed and may require follow-up aphicide treatment
- In high risk areas where aphid activity is heavy and prolonged a follow-up aphicide treatment may be required
- Only the seed-borne phase of net blotch is controlled by seed treatment
- Protection against foliar air-borne and splash-borne diseases later in the season will require appropriate fungicide follow-up sprays

Restrictions
- Maximum number of treatments 1 per batch of seed
- Do not use on seed with more than 16% moisture content, or on sprouted, cracked or skinned seed
- Do not use treated seed as food or feed

Crop-specific information
- Latest use: before drilling
- Slightly delayed and reduced emergence may occur but this is normally outgrown
- Any delay in field emergence, for whatever reason, may be accentuated by treatment

SEE SECTION 3 FOR PRODUCTS ALSO REGISTERED

Environmental safety
- Harmful to aquatic organisms
- Harmful to fish or other aquatic life. Do not contaminate surface waters or ditches with chemical or used container
- Dangerous to birds, game and other wildlife. Treated seed should not be left on the soil surface. Bury spillages
- Treated seed should not be broadcast, but drilled to a depth of 4 cm in a well prepared seedbed
- If seed is left on the soil surface the field should be harrowed and rolled to ensure good incorporation

Hazard classification and safety precautions
Hazard H03
Risk phrases R22a, R43, R52, R53a
Operator protection A, H; U04a, U05a, U07, U13, U20b
Environmental protection E03, E13c, E34, E36
Storage and disposal D01, D02, D05, D09a, D14
Treated seed S01, S02, S03, S04c, S05, S06a, S07, S08
Medical advice M03

276   indol-3-ylacetic acid

A plant growth regulator for promoting rooting of cuttings

Products

1	Rhizopon A Powder	Rhizopon	1% w/w	DP	09087
2	Rhizopon A Tablets	Rhizopon	50 mg a.i.	WT	09088

Uses
- Rooting of cuttings in ***ornamental specimens***

Efficacy guidance
- Apply by dipping end of prepared cuttings into powder or dissolved tablet solution
- Shake off excess powder and make planting holes to prevent powder stripping off [1]
- Consult manufacturer for details of application by spray or total immersion

Restrictions
- Maximum number of treatments 1 per cutting
- Store product in a cool, dark and dry place
- Use solutions once only. Discard after use [2]
- Use plastic, not metallic, container for solutions [2]

Crop-specific information
- Latest use: before cutting insertion

Hazard classification and safety precautions
Operator protection U19a [1]; U20a [1, 2]
Environmental protection E15a
Storage and disposal D09a, D11a

277   4-indol-3-yl-butyric acid

A plant growth regulator promoting the rooting of cuttings

Products

1	Chryzoplus Grey 0.8%	Rhizopon	0.8% w/w	DP	09094
2	Chryzopon Rose 0.1%	Rhizopon	0.1% w/w	DP	09092
3	Chryzosan White 0.6%	Rhizopon	0.6% w/w	DP	09093
4	Chryzotek Beige	Rhizopon	0.4% w/w	DP	09081
5	Chryzotop Green	Rhizopon	0.25% w/w	DP	09085
6	Rhizopon AA Powder (0.5%)	Rhizopon	0.5% w/w	DP	09082
7	Rhizopon AA Powder (1%)	Rhizopon	1% w/w	DP	09084
8	Rhizopon AA Powder (2%)	Rhizopon	2% w/w	DP	09083
9	Rhizopon AA Tablets	Rhizopon	50 mg a.i.	WT	09086
10	Seradix 1	Certis	0.1% w/w	DP	11330
11	Seradix 2	Certis	0.3% w/w	DP	11331
12	Seradix 3	Certis	0.8% w/w	DP	11332

FOR FULL CONDITIONS OF USE ALWAYS READ THE PRODUCT LABEL

Uses
- Rooting of cuttings in **ornamental specimens**

Efficacy guidance
- Dip base of cuttings into powder immediately before planting
- Powders or solutions of different concentration are required for different types of cutting. Lowest concentration for softwood, intermediate for semi-ripe, highest for hardwood
- See label for details of concentration and timing recommended for different species
- Use of planting holes recommended for powder formulations to ensure product is not removed on insertion of cutting. Cuttings should be watered in if necessary

Restrictions
- Maximum number of treatments 1 per situation
- Use of too strong a powder or solution may cause injury to cuttings
- No unused moistened powder should be returned to container

Crop-specific information
- Latest use: before cutting insertion for ornamental specimens

Hazard classification and safety precautions
Operator protection A, C [10-12]; U14, U15 [10-12]; U19a, U20a [1-12]
Environmental protection E15a
Storage and disposal D01 [10-12]; D09a, D11a [1-12]

278 4-indol-3-yl-butyric acid + 2-(1-naphthyl)acetic acid with dichlorophen

A plant growth regulator for promoting rooting of cuttings

Products

Synergol	Certis	5.0:5.0 g/l	SL	07386

Uses
- Rooting of cuttings in **ornamental specimens**

Efficacy guidance
- Dip base of cuttings into diluted concentrate immediately before planting
- Suitable for hardwood and softwood cuttings
- See label for details of concentration and timing for different species

Hazard classification and safety precautions
Hazard H03
Risk phrases R22a
Operator protection A, C; U05a, U13, U20b
Environmental protection E15a, E34
Storage and disposal D01, D02, D09a, D11a, D12a
Medical advice M03

279 iodosulfuron-methyl-sodium

A post-emergence sulfonylurea herbicide for winter sown cereals

See also amidosulfuron + iodosulfuron-methyl-sodium

Products

Hussar	Bayer CropScience	5% w/w	WG	11205

Uses
- Chickweed in **spring barley, triticale, winter rye, winter wheat**
- Cleavers in **triticale, winter rye, winter wheat**
- Fat hen in **spring barley**
- Field pansy in **spring barley**
- Field speedwell in **spring barley, triticale, winter rye, winter wheat**
- Hemp-nettle in **spring barley**
- Italian ryegrass in **triticale** *(from seed)*, **winter rye** *(from seed)*, **winter wheat** *(from seed)*

SEE SECTION 3 FOR PRODUCTS ALSO REGISTERED

SECTION 2

- Mayweeds in **spring barley**, **triticale**, **winter rye**, **winter wheat**
- Perennial ryegrass in **triticale** *(from seed)*, **winter rye** *(from seed)*, **winter wheat** *(from seed)*
- Red dead-nettle in **spring barley**, **triticale**, **winter rye**, **winter wheat**
- Speedwells in **triticale**, **winter rye**, **winter wheat**
- Volunteer oilseed rape in **spring barley**

Approval information
- Accepted by BBPA for use on malting barley
- Approval expiry 30 Jun 2006 [1]

Efficacy guidance
- Best results obtained from treatment in warm weather when soil is moist and the weeds are growing actively
- Weeds must be present at application to be controlled
- Weed control is slow especially under cool dry conditions
- Dry conditions resulting in moisture stress may reduce effectiveness
- Occasionally weeds may only be stunted but they will normally have little or no competitive effect on the crop
- Iodosulfuron is a member of the ALS-inhibitor group of herbicides and products should be used in a planned Resistance Management strategy. See Section 5 for more information

Restrictions
- Maximum number of treatments 1 per crop
- Must only be applied between 1 Feb in yr of harvest and specified latest time of application
- Do not apply to undersown crops, or crops to be undersown
- Do not roll or harrow within 1 wk of spraying
- Do not spray crops under stress from any cause or if the soil is compacted
- Treat broadcast crops after the plants have a well-established root system
- Do not spray if rain imminent or frost expected
- Do not apply in mixture or in sequence with any other ALS inhibitor

Crop-specific information
- Latest use: before third node detectable (GS 33)

Following crops guidance
- No restrictions apply to the sowing of cereal crops or sugar beet in the spring of the yr following treatment

Environmental safety
- Dangerous for the environment
- Very toxic to aquatic organisms
- Dangerous to fish or other aquatic life. Do not contaminate surface waters or ditches with chemical or used container
- LERAP Category B
- Take extreme care to avoid damage by drift onto broad-leaved plants outside the target area or onto ponds, waterways and ditches
- Observe carefully label instructions for sprayer cleaning

Hazard classification and safety precautions
Hazard H04, H11
Risk phrases R41, R50, R53a
Operator protection A, C, H; U05a, U08, U11, U14, U15, U20b
Environmental protection E13b, E16a, E16b, E38
Storage and disposal D01, D02, D10a, D12a

280   iodosulfuron-methyl-sodium + mesosulfuron-methyl

A sulfonyl urea herbicide mixture for winter wheat

Products

1 Atlantis WG	Bayer CropScience	0.6:3.0% w/w	WG	11679
2 Atlantis WG	Bayer CropScience	0.6:3.0% w/w	WG	12478
3 Greencrop Doonbeg WG	Greencrop	0.6:3.0% w/w	WG	12199
4 Landgold Minoan	Teliton	0.6:3.0% w/w	WG	12223
5 Pacifica	Bayer CropScience	1.0:3.0% w/w	WG	12049

FOR FULL CONDITIONS OF USE ALWAYS READ THE PRODUCT LABEL

SECTION 2

Products – continued

| 6 | Qdos Rockall 2 | Me2 | 0.6:3.0% w/w | WG | 12488 |
| 7 | Standon Mimas WG | Standon | 0.6:3.0% w/w | WG | 12075 |

Uses

- Annual meadow grass in *winter wheat*
- Blackgrass in *winter wheat*
- Chickweed in *winter wheat*
- Italian ryegrass in *winter wheat*
- Mayweeds in *winter wheat*
- Perennial ryegrass in *winter wheat*
- Rough meadow grass in *winter wheat*
- Wild oats in *winter wheat*

Approval information

- Approval expiry 31 Aug 2006 [3, 4]

Efficacy guidance

- Optimum grass weed control obtained when all grass weeds are emerged at spraying. Activity is primarily via foliar uptake and good spray coverage of the target weeds is essential
- Translocation occurs readily within the target weeds and growth is inhibited within hours of treatment but symptoms may not be apparent for up to 4 wk, depending on weed species, timing of treatment and weather conditions
- Residual activity is important for best results and is optimised by treatment on fine moist seedbeds. Avoid application under very dry conditions
- Residual efficacy may be reduced by high soil temperatures and cloddy seedbeds
- Iodosulfuron-methyl and mesosulfuron-methyl are both members of the ALS-inhibitor group of herbicides and products should be used in a planned Resistance Management strategy. See Section 5 for more information

Restrictions

- Maximum number of treatments 1 per crop with a maximum total dose equivalent to one full dose treatment
- Do not use on crops undersown with grasses, clover or other legumes or any other broad-leaved crop
- Do not use as a stand-alone treatment for blackgrass, ryegrass or chickweed control
- Do not use as the sole means of weed control in successive crops
- Do not use in mixture or in sequence with any other ALS-inhibitor herbicide except those specified on the label
- Do not apply earlier than 1 Feb in the yr of harvest [5]
- Do not apply to crops under stress from any cause
- Do not apply when rain is imminent or during periods of frosty weather
- Specified adjuvant must be used. See label

Crop-specific information

- Latest use: flag leaf just visible (GS 39)
- Winter wheat may be treated from the two-leaf stage of the crop
- Safety to crops grown for seed not established

Following crops guidance

- In the event of crop failure sow only winter or spring wheat in the same cropping season
- Only winter wheat or winter barley (or winter oilseed rape [2]) may be sown in the year of harvest of a treated crop
- Spring wheat, spring barley or sugar beet may be drilled in the following spring

Environmental safety

- Dangerous for the environment
- Very toxic to aquatic organisms
- Dangerous to fish or other aquatic life. Do not contaminate surface waters or ditches with chemical or used container
- LERAP Category B
- Take extreme care to avoid drift onto plants outside the target area or on to ponds, waterways or ditches

SEE SECTION 3 FOR PRODUCTS ALSO REGISTERED

Hazard classification and safety precautions
  Hazard H04, H11
  Risk phrases R40 [5]; R41, R53a [1-7]; R50 [1-3, 5, 6]; R51 [4, 7]
  Operator protection A, C, H; U05a, U11, U20b
  Environmental protection E13b [1, 3, 6]; E15a [2, 4, 5, 7]; E16a [1-7]; E16b [3, 4, 7]; E38 [1-3, 5, 6]
  Storage and disposal D01, D02, D09a, D10a [1-7]; D05, D12a [1-3, 5, 6]

281   ioxynil

A contact acting HBN herbicide for use in turf and onions

See also bromoxynil + diflufenican + ioxynil
      bromoxynil + ethofumesate + ioxynil
      bromoxynil + ioxynil
      bromoxynil + ioxynil + mecoprop-P

Products

Totril	Bayer CropScience	225 g/l	EC	10026

Uses
  • Annual dicotyledons in **bulb onions**, **chives** (off-label), **garlic**, **leeks**, **salad onions**, **shallots**

Specific Off-Label Approvals (SOLAs)
  • **chives** (OLA 040509) Dec 2008 [1]

Approval information
  • Ioxynil was reviewed in 1995 and approvals for home garden use, and most hand held applications revoked

Efficacy guidance
  • Best results on seedling to 4-leaf stage weeds in active growth during mild weather

Restrictions
  • Maximum number of treatments up to 4 per crop at split doses on onions; 1 per crop on carrots, garlic, leeks, parsnips, shallots, chives
  • Do not apply by hand-held equipment or at concentrations higher than those recommended

Crop-specific information
  • Latest use: pre-emergence for carrots, parsnips
  • HI onions, chives, shallots, garlic, leeks 14 d
  • Apply to sown onion crops as soon as possible after plants have 3 true leaves or to transplanted crops when established

Environmental safety
  • Dangerous for the environment
  • Very toxic to aquatic organisms
  • Harmful to bees. Do not apply to crops in flower or to those in which bees are actively foraging. Do not apply when flowering weeds are present
  • Keep livestock out of treated areas for at least 6 wk after treatment and until foliage of any poisonous weeds such as ragwort has died and become unpalatable

Hazard classification and safety precautions
  Hazard H03, H08, H11
  Risk phrases R22a, R22b, R36, R43, R50, R53a, R63, R66, R67
  Operator protection A, C; U05a, U08, U19a, U20b, U23a
  Environmental protection E06a (6 wk); E12f, E34, E38
  Storage and disposal D01, D02, D05, D09a, D10b, D12a
  Medical advice M03, M05b

282   iprodione

A protectant dicarboximide fungicide with some eradicant activity

See also carbendazim + iprodione

Products

1	Amenitywise Iprodione Green	Standon	250 g/l	SC	12237
2	Chipco Green	Bayer Environ.	250 g/l	SC	11211
3	Rovral Flo	BASF	255 g/l	SC	11702
4	Rovral Liquid FS	BASF	500 g/l	FS	11703
5	Rovral WP	BASF	50% w/w	WP	11694
6	Standon Iprodione 50 WP	Standon	50% w/w	WP	12322

Uses

- Alternaria in **brassica seed crops**, **mustard** *(seed treatment)*, **stored cabbages**, **stubble turnips** *(seed treatment)*, **swedes** *(seed treatment)*, **turnips** *(seed treatment)* [5, 6]; **broccoli, brussels sprouts**, **cabbage seed crops, calabrese, cauliflower seed crops, cauliflowers, mustard seed crops, spring oilseed rape, stubble turnips, winter oilseed rape, winter wheat** [3]; **chinese cabbage** *(seed treatment)*, **edible brassicas** *(seed treatment)*, **flower seeds** *(seed treatment)*, **kohlrabi** *(seed treatment)* [5]; **flax** *(seed treatment)* [4, 5]; **fodder rape** *(seed treatment)*, **leaf brassicas** *(seed treatment)*, **ornamental specimens** *(seed treatment)* [6]; **poppies** *(off-label - for morphine production - seed treatment)* [4]; **spring oilseed rape** *(seed treatment)*, **winter oilseed rape** *(seed treatment)* [4-6]
- Black scurf in **seed potatoes** *(seed treatment)* [4, 6]
- Black scurf and stem canker in **potatoes** *(seed treatment)* [5]
- Botrytis in **cabbage seed crops, cauliflower seed crops, cress** *(off-label)*, **grapevines** *(off-label)*, **lamb's lettuce** *(off-label)*, **lettuce** *(off-label)*, **mustard seed crops, radicchio** *(off-label)*, **scarole** *(off-label)*, **spring oilseed rape, winter oilseed rape, winter wheat** [3]; **chicory** *(off-label)*, **chinese cabbage** *(off-label)*, **choi sum** *(off-label)*, **combining peas** *(off-label)*, **dwarf beans** *(off-label)*, **endives** *(off-label)*, **fennel** *(off-label)*, **french beans** *(off-label)*, **mange-tout peas** *(off-label)*, **pak choi** *(off-label)*, **pears** *(off-label)*, **protected courgettes** *(off-label)*, **protected endives** *(off-label)*, **protected radicchio** *(off-label)*, **protected rhubarb** *(off-label)*, **red beet** *(off-label)*, **runner beans** *(off-label)*, **vining peas** *(off-label)* [5]; **lettuce** *(outdoor crops)*, **outdoor tomatoes, pot plants, protected lettuce, protected tomatoes, raspberries, stored cabbages** [5, 6]; **strawberries** [3, 5, 6]
- Botrytis bunch rot in **grapevines** *(off-label)* [3]
- Brown patch in **amenity grass** [2]; **managed amenity turf** [1, 2]
- Canker in **horseradish** *(off-label)* [5]
- Chocolate spot in **spring field beans, winter field beans** [3]
- Collar rot in **onions, salad onions** [3]
- Dollar spot in **amenity grass** [2]; **managed amenity turf** [1, 2]
- Fusarium patch in **amenity grass** [2]; **managed amenity turf** [1, 2]
- Glume blotch in **winter wheat** [3]
- Grey mould in **protected cucumbers** *(off-label)* [5]
- Leaf rot in **onions, salad onions** [3]
- Melting out in **amenity grass** [2]; **managed amenity turf** [1, 2]
- Net blotch in **spring barley, winter barley** [3]
- Phoma in **red beet** *(off-label)* [5]
- Red thread in **amenity grass** [2]; **managed amenity turf** [1, 2]
- Sclerotinia in **chicory** *(off-label)*, **combining peas** *(off-label)*, **dwarf beans** *(off-label)*, **french beans** *(off-label)*, **mange-tout peas** *(off-label)*, **runner beans** *(off-label)*, **vining peas** *(off-label)* [5]; **cress** *(off-label)*, **lamb's lettuce** *(off-label)*, **lettuce** *(off-label)*, **radicchio** *(off-label)*, **scarole** *(off-label)* [3]
- Sclerotinia stem rot in **spring oilseed rape, winter oilseed rape** [3]
- Snow mould in **amenity grass** [2]; **managed amenity turf** [1, 2]
- Stemphylium in **asparagus** *(off-label)* [5]

Specific Off-Label Approvals (SOLAs)

- **asparagus, combining peas, vining peas** *(OLA 001902) Dec 2008* [5]
- **chicory** *(OLA 001906) Dec 2008* [5]

SEE SECTION 3 FOR PRODUCTS ALSO REGISTERED

SECTION 2

- ***chinese cabbage, choi sum, pak choi*** *(OLA 051065) Dec 2008* [5]
- ***cress, lamb's lettuce, lettuce, radicchio, scarole*** *(OLA 040513) Dec 2008* [3]
- ***dwarf beans, french beans, mange-tout peas, runner beans*** *(OLA 001904) Dec 2008* [5]
- ***endives, fennel, protected courgettes, protected endives, protected radicchio*** *(OLA 001908) Dec 2008* [5]
- ***grapevines*** *(OLA 040514) Dec 2008* [3]
- ***horseradish*** *(OLA 040515) Dec 2008* [5]
- ***pears*** *(OLA 012525) Dec 2008* [5]
- ***poppies*** *(for morphine production - seed treatment) (OLA 040682) Dec 2008* [4]
- ***protected cucumbers*** *(OLA 003070) Dec 2008* [5]
- ***protected rhubarb*** *(OLA 050089) Dec 2012* [5]
- ***red beet*** *(OLA 001910) Dec 2008* [5]

Approval information
- Iprodione included in Annex I under EC Directive 91/414
- Accepted by BBPA for use on malting barley

Efficacy guidance
- Many diseases require a programme of 2 or more sprays at intervals of 2-4 wk. Recommendations vary with disease and crop - see label for details
- Use as a drench to control cabbage storage diseases. Spray ornamental pot plants and cucumbers to run-off
- Apply turf and amenity grass treatments after mowing to dry grass, free of dew. Do not mow again for at least 24 h
- Seed treatments on oilseed rape and linseed should be applied to dry seed prior to sowing
- Best results on seed potatoes achieved using hydraulic sprayer with solid or hollow cone jets with air or liquid pressure atomisation
- Apply to seed potatoes after harvest or after grading out and before traying out for chitting

Restrictions
- Maximum number of treatments 1 per batch for seed treatments; 1 per crop on cereals, vining peas, cabbage (as drench), chicory, borage; 2 per crop on field beans and stubble turnips; 3 per crop on brassicas (including seed crops), oilseed rape, protected winter lettuce (Oct-Feb); 4 per crop on strawberries, grapevines, salad onions, cucumbers; 5 per crop on raspberries; 6 per crop on bulb onions, tomatoes, curcurbits, peppers, aubergines, turf, nuts; 7 per crop on lettuce (Mar-Sep)
- A minimum of 3 wk must elapse between treatments on leaf brassicas
- See label for pot plants showing good tolerance. Check other species before applying on a large scale
- Do not treat oilseed rape seed that is cracked or broken or of low viability
- Do not excessively wet seed potato skins
- Do not treat oats
- Do not use treated seed as food or feed [4]

Crop-specific information
- Latest use: pre-planting for seed treatments; at planting for potatoes; before grain watery ripe (GS 69) for wheat and barley; 5 wk before forcing for chicory; 6 wk before release of pears for human consumption; 16 wk before processing or sale for red beet
- HI strawberries, protected tomatoes 1 d; outdoor tomatoes, cucurbits, aubergines 2 d; lettuce, raspberries, nuts, cress, dwarf beans, endives, fennel, French beans, runner beans, lamb's lettuce (Mar-Sep), lettuce, mange-tout peas, onions, radicchio, scarole 7 d; grapevines, protected rhubarb 14 d; brassicas, brassica seed crops, oilseed rape, mustard, field beans, peas, stubble turnips 21 d; protected lettuce (Oct-Feb) 28 d; asparagus 5 mth
- Turf and amenity grass may become temporarily yellowed if frost or hot weather follows treatment
- Personal protective equipment requirements may vary for each pack size. Check label

Environmental safety
- Dangerous for the environment
- Very toxic to aquatic organisms
- Treated brassica seed crops not to be used for human or animal consumption
- See label for guidance on disposal of spent drench liquor
- Treatment harmless to *Encarsia* or *Phytoseiulus* being used for integrated pest control

FOR FULL CONDITIONS OF USE ALWAYS READ THE PRODUCT LABEL

Hazard classification and safety precautions
>**Hazard** H03, H11
>**Risk phrases** R36, R38 [1, 2]; R40, R53a [1-6]; R50 [1-4]; R51 [5, 6]
>**Operator protection** A [1-4]; C [1, 2]; H [4]; U04a [1, 2]; U05a [1-4]; U08 [1-3]; U19a [1, 5, 6]; U20b [3]; U20c [1, 2, 4-6]
>**Environmental protection** E15a [1-6]; E34 [1, 2, 4-6]; E38 [2-5]
>**Storage and disposal** D01, D02 [1-4]; D05 [1, 2, 4]; D09a [1-6]; D10b [1-3]; D11a [4-6]; D12a [2-5]
>**Treated seed** S01, S05, S06a, S07 [4-6]; S02 [5, 6]; S03, S04a [4]
>**Medical advice** M03 [1, 2]

283   iprodione + thiophanate-methyl

A protectant and systemic fungicide for oilseed rape

Products

1	Compass	BASF	167:167 g/l	SC	11740
2	Snooker	BASF	150:200 g/l	SC	11744

Uses
- Alternaria in *carrots (off-label)*, *horseradish (off-label)*, *parsnips (off-label)*, *spring oilseed rape*, *winter oilseed rape* [1, 2]
- Botrytis in *limnanthes alba (meadowfoam) (off-label)* [1]
- Chocolate spot in *spring field beans*, *winter field beans* [1, 2]
- Crown rot in *carrots (off-label)*, *horseradish (off-label)*, *parsnips (off-label)* [1, 2]
- Grey mould in *spring oilseed rape*, *winter oilseed rape* [1, 2]
- Light leaf spot in *spring oilseed rape*, *winter oilseed rape* [1, 2]
- Sclerotinia stem rot in *spring oilseed rape*, *winter oilseed rape* [1, 2]
- Stem canker in *spring oilseed rape*, *winter oilseed rape* [1]

Specific Off-Label Approvals (SOLAs)
- *carrots, horseradish, parsnips (OLA 001936) Dec 2008* [1]
- *carrots, horseradish, parsnips (OLA 041186) Dec 2008* [2]
- *limnanthes alba (meadowfoam) (OLA 040507) Dec 2008* [1]

Approval information
- Iprodione and thiophanate-methyl included in Annex I under EC Directive 91/414

Efficacy guidance
- Timing of sprays on oilseed rape varies with disease, see label for details
- For season-long control product should be applied as part of a disease control programme

Restrictions
- Maximum number of treatments (refers to total sprays containing benomyl, carbendazim and thiophanate methyl) 2 per crop for oilseed rape, field beans, meadowfoam; 3 per crop for carrots, horseradish, parsnips

Crop-specific information
- Latest use: before end of flowering for oilseed rape, field beans
- HI meadowfoam, field beans, oilseed rape 3 wk; carrots, horseradish, parsnips 28 d
- Treatment may extend duration of green leaf in winter oilseed rape

Environmental safety
- Dangerous for the environment
- Very toxic to aquatic organisms

Hazard classification and safety precautions
>**Hazard** H03, H11
>**Risk phrases** R36, R38 [2]; R40, R43, R50, R53a, R68 [1, 2]
>**Operator protection** A, C, H, M; U05a, U20c [1, 2]; U09a [2]
>**Environmental protection** E15a, E38
>**Storage and disposal** D01, D02, D09a, D10b, D12a [1, 2]; D05 [2]; D08 [1]
>**Medical advice** M05a [1]

284   isoproturon

A residual urea herbicide for use in cereals

See also *bifenox + isoproturon*
chlorotoluron + isoproturon
diflufenican + flurtamone + isoproturon
diflufenican + isoproturon

Products

1	Aligran	Nufarm UK	83% w/w	WG	11761
2	Alpha IPU 500	Makhteshim	500 g/l	SC	10733
3	Alpha Isoproturon 500	Makhteshim	500 g/l	SC	05882
4	Arelon 500	Nufarm UK	500 g/l	SC	11639
5	Bison 83 WG	Nufarm UK	83% w/w	WG	11580
6	Emrald Wotsit	Me2	500 g/l	SC	12060
7	Fieldgard	Nufarm UK	500 g/l	SC	11770
8	IPU Minrinse	Makhteshim	80% w/w	WG	12457
9	Isotron 500	DAPT	500 g/l	SC	10668
10	Luxan Isoproturon 500 Flowable	Luxan	500 g/l	SC	12426
11	Primer	AgriGuard	500 g/l	SC	11782
12	Protugan	Makhteshim	500 g/l	SC	10043
13	Protugan 80 WDG	Makhteshim	80% w/w	WG	10430

Uses

- Annual dicotyledons in **spring barley** *(off-label)* [3]; **spring wheat** [4, 6, 7]; **spring wheat** *(autumn sown)* [8, 13]; **triticale**, **winter rye** [2, 4, 6-8, 10, 12, 13]; **winter barley**, **winter wheat** [1-13]
- Annual grasses in **spring wheat** [4, 6, 7]; **spring wheat** *(autumn sown)* [8, 13]; **triticale**, **winter rye** [2, 4, 6-8, 10, 12, 13]; **winter barley**, **winter wheat** [1-13]
- Annual meadow grass in **spring barley** *(off-label)* [1-5, 7, 12, 13]
- Blackgrass in **spring wheat** [4, 6, 7]; **spring wheat** *(autumn sown)* [8, 13]; **triticale**, **winter rye** [2, 4, 6-8, 10, 12, 13]; **winter barley**, **winter wheat** [2-4, 6-13]
- Rough meadow grass in **spring wheat** [4, 6, 7]; **spring wheat** *(autumn sown)* [8, 13]; **triticale**, **winter rye** [2, 4, 6-8, 10, 12, 13]; **winter barley**, **winter wheat** [2-4, 6-13]
- Wild oats in **spring wheat** [4, 6, 7]; **spring wheat** *(autumn sown)* [8, 13]; **triticale**, **winter rye** [2, 4, 6-8, 10, 12, 13]; **winter barley**, **winter wheat** [2-4, 6-13]

Specific Off-Label Approvals (SOLAs)

- *spring barley (OLA 050045) Dec 2008* [1]
- *spring barley (OLA 012166) Dec 2008* [2]
- *spring barley (OLA 011481) Dec 2008* [3]
- *spring barley (OLA 031860) Dec 2008* [4]
- *spring barley (OLA 032358) Dec 2008* [5]
- *spring barley (OLA 042320) Dec 2008* [7]
- *spring barley (OLA 011488) Dec 2008* [12]
- *spring barley (OLA 023410) Dec 2008* [13]

Approval information

- Isoproturon included in Annex I under EC Directive 91/414
- All approvals for aerial use of isoproturon were revoked in 1995

Efficacy guidance

- Best control normally achieved by early post-emergence treatment
- See label for details of rates, timings and tank mixes for different weed problems
- Apply to moist soils. Effectiveness may be reduced in seasons of above average rainfall or by prolonged dry weather
- Residual activity reduced on soils with more than 10% organic matter. Only use on such soils in spring
- Always follow WRAG guidelines for preventing and managing herbicide resistant weeds. See Section 5 for more information

Restrictions

- Maximum number of treatments 1 per crop

FOR FULL CONDITIONS OF USE ALWAYS READ THE PRODUCT LABEL

- Maximum total dose 2.5 kg a.i./ha for any crop. The IPU stewardship programme guidelines recommend that this maximum dose should not be used after the end of Oct, and that where possible it should be reduced to 1.5 kg a.i./ha by using mixtures or sequences with other herbicides
- Do not apply pre-emergence to wheat or barley. On triticale and winter rye use pre-emergence only and on named varieties
- Do not use on durum wheats, oats, undersown cereals or crops to be undersown
- Do not apply to very wet or waterlogged soils, or when heavy rainfall is forecast
- Do not use on very cloddy soils or if frost is imminent or after onset of frosty weather
- Do not roll for 1 wk before or after treatment or harrow for 1 wk before or any time after treatment

Crop-specific information
- Latest use: not later than second node detectable stage (GS 32)
- Recommended timing of treatment varies depending on crop to be treated, method of sowing, season of application, weeds to be controlled and product being used. See label for details
- Crop damage may occur on free draining, stony or gravelly soils if heavy rain falls soon after spraying. Early sown crops may be damaged if spraying precedes or coincides with a period of rapid growth

Environmental safety
- Dangerous for the environment
- Very toxic to aquatic organisms
- Dangerous to fish or other aquatic life. Do not contaminate surface waters or ditches with chemical or used container
- Do not apply to dry, cracked or waterlogged soils where heavy rain may lead to contamination of drains by isoproturon
- Do not empty into drains

Hazard classification and safety precautions
Hazard H03, H11 [1-8, 10-13]

Risk phrases R22a, R51 [1, 5]; R40 [1-5, 8, 10-13]; R50 [2-4, 6-8, 10-13]; R53a [1-8, 10-13]; R68 [6, 7]

Operator protection A, C, H [1-13]; D [1, 5, 8, 13]; U05a [2, 3, 7-10, 12, 13]; U08 [1-10, 12, 13]; U14, U15 [8, 13]; U19a [2, 3, 8-10, 12, 13]; U20a [3, 11]; U20b [1, 2, 4-8, 10, 12, 13]

Environmental protection E13b [9]; E13c [3]; E15a [1, 2, 4-8, 10-13]; E19b [1, 5]; E34 [2, 3, 10, 12]

Storage and disposal D01, D02 [1-3, 5, 8-10, 12, 13]; D05, D09a [1-13]; D10b [2-4, 6, 7, 9-12]; D11a [1, 5, 8, 13]; D12a [4, 6]; D12b [2, 3, 7, 10, 12]

Treated seed S06a [9]

285   isoproturon + pendimethalin

A contact and residual herbicide for use in winter cereals

Products

1	Encore	BASF	125:250 g/l	SC	10375
2	Trump	BASF	236:236 g/l	SC	10376

Uses
- Annual dicotyledons in **spring wheat** *(autumn sown)* [2]; **triticale, winter barley, winter rye, winter wheat** [1, 2]
- Annual grasses in **spring wheat** *(autumn sown)* [2]; **triticale, winter barley, winter rye, winter wheat** [1, 2]
- Blackgrass in **spring wheat** *(autumn sown)* [2]; **triticale, winter barley, winter rye, winter wheat** [1, 2]
- Wild oats in **spring wheat** *(autumn sown)* [2]; **triticale, winter barley, winter rye, winter wheat** [1, 2]

Approval information
- Isoproturon and pendimethalin included in Annex I under EC Directive 91/414

SEE SECTION 3 FOR PRODUCTS ALSO REGISTERED

Efficacy guidance
- May be applied in autumn or spring (but only post-emergence on wheat or barley)
- Annual grasses controlled from pre-emergence to 4-leaf stage, extended to 3-tiller stage with blackgrass by tank mixing with additional isoproturon. Best results on wild oats post-emergence in autumn
- Annual dicotyledons controlled pre-emergence and up to 8-leaf stage [1], up to 12-leaf stage [2]. See label for details
- Apply to moist soils. Best results achieved by application to fine, firm seedbed. Trash, ash or straw should have been incorporated evenly
- Contact activity is reduced by rain within 6 h of application
- Activity may be reduced on soil with more than 6% organic matter or ash. Do not use on soils with more than 10% organic matter
- Always follow WRAG guidelines for preventing and managing herbicide resistant weeds. See Section 5 for more information

Restrictions
- Maximum number of treatments 1 per crop. Maximum total dose of isoproturon 2.5 kg a.i./ha per crop
- Do not apply pre-emergence to wheat or barley. On winter rye and triticale use pre-emergence only on named cultivars
- Do not use on durum wheat or spring cereals
- Do not use pre-emergence on winter rye or triticale drilled after 30 Nov [2]
- Do not use on crops suffering stress from disease, drought, waterlogging, poor seedbed conditions or other causes or apply post-emergence when frost imminent
- Do not apply to very wet or waterlogged soils, or when heavy rain is forecast
- Do not undersow treated crops
- Do not roll or harrow after application

Crop-specific information
- Latest use: pre-emergence for winter rye and triticale; before leaf sheath erect (GS 30) [2] or before 1st node detectable (GS 31) [1] for winter wheat and barley
- Apply before first node detectable (GS 31) [1], to before leaf sheath erect (GS 30) [2]
- May be used on autumn sown spring wheat [2]

Following crops guidance
- In the event of autumn crop failure spring wheat, spring barley, maize, potatoes, beans or peas may be grown following ploughing to at least 150 mm

Environmental safety
- Dangerous for the environment
- Very toxic to aquatic organisms
- Do not apply to dry, cracked or waterlogged soils where heavy rain may lead to contamination of drains by isoproturon

Hazard classification and safety precautions
Hazard H03, H11
Risk phrases R40, R50, R53a [1, 2]; R68 [2]
Operator protection A, C, H; U05a, U08, U13, U19a, U20b
Environmental protection E15a, E38 [1, 2]; E34 [1]
Storage and disposal D01, D02, D05, D09a, D10b, D12a
Medical advice M05a

286 isoproturon + trifluralin

A residual early post-emergence herbicide for cereals

Products

1 Autumn Kite	Nufarm UK	300:200 g/l	EC	11549
2 Snitch	Nufarm UK	300:200 g/l	EC	12466

FOR FULL CONDITIONS OF USE ALWAYS READ THE PRODUCT LABEL

Uses
- Annual dicotyledons in **winter barley**, **winter wheat**
- Annual grasses in **winter barley**, **winter wheat**
- Blackgrass in **winter barley**, **winter wheat**

Efficacy guidance
- Provides contact and residual control of weeds germinating in the surface layers of the soil
- Best results achieved in warm moist weather when weeds are growing actively and the leaves are dry and rain not expected for 2 h
- Apply to dicotyledons pre-emergence or up to 2-leaf stage, to blackgrass up to 2-3 tillers,
- Effectiveness may be reduced by prolonged dry or sunny weather after application
- If used in conjunction with minimum cultivation ensure that all trash and burnt straw is removed, buried or dispersed before spraying
- After mild wet winters, or where early sown winter barley has been treated, residual life may not be sufficient to control later germinating weeds and post-emergence treatment may be needed in spring
- Always follow WRAG guidelines for preventing and managing herbicide resistant weeds. See Section 5 for more information

Restrictions
- Maximum number of treatments 1 per crop. Maximum total dose of isoproturon 2.5 kg a.i./ha per crop
- Do not use on durum wheat or crops to be undersown
- Do not use on Sands and do not incorporate in soil. On very stony, gravelly or other free draining soils crops may be damaged if heavy rain falls soon after treatment
- Do not spray when heavy rain is forecast or to crops stressed by frost, waterlogging, deficiency or pest attack
- Do not roll after treatment until following spring
- Do not use on soils with more than 10% organic matter
- Do not harrow treated crops

Crop-specific information
- Latest use: before 6 leaf stage (GS 16) for wheat and barley
- Where climatic conditions in autumn have favoured rapid, lush growth treatment may cause scorch or stunting. Such effect may be increased by certain tank mixtures. Recovery is normally complete

Following crops guidance
- Any crop may follow normal harvest of a treated crop
- In case of crop failure only carrots, peas or sunflowers may be sown within 5 mth and after ploughing to at least 15 cm. Winter cereals, spring cereals, maize, grass or spring oilseed rape may follow after an interval of 5 mth.
- Do not sow sugar beet in spring following autumn application

Environmental safety
- Dangerous for the environment
- Very toxic to aquatic organisms
- Dangerous to fish or other aquatic life. Do not contaminate surface waters or ditches with chemical or used container
- Keep in original container, tightly closed, in a safe place, under lock and key
- Do not spray where soils are cracked, to avoid run-off through drains

Hazard classification and safety precautions
Hazard H03, H11
Risk phrases R22b, R36, R40, R43, R50, R53a, R66, R67
Operator protection A, C; U05a, U08, U11, U13, U19a, U20b
Environmental protection E13b [2]; E15a [1]; E34 [1, 2]
Storage and disposal D01, D02, D05, D08, D09a, D10b, D12a
Medical advice M05b

SECTION 2

SEE SECTION 3 FOR PRODUCTS ALSO REGISTERED

287   isoxaben

A soil-acting amide herbicide for use in grass and fruit

Products

Flexidor 125	Landseer	125 g/l	SC	10946

Uses
- Annual dicotyledons in **amenity vegetation**, **apple orchards**, **blackberries**, **blackcurrants**, **canary grass** *(off-label - grown for game cover)*, **carrots** *(off-label)*, **cherries**, **clovers** *(off-label - grown for game cover)*, **container-grown woody ornamentals**, **courgettes** *(off-label)*, **forage maize** *(off-label - grown for game cover)*, **forestry transplants**, **gooseberries**, **grapevines**, **hardy ornamental nursery stock**, **hops**, **marrows** *(off-label)*, **millet** *(off-label - grown for game cover)*, **pear orchards**, **plums**, **protected carrots** *(off-label - temporary protection)*, **protected courgettes** *(off-label)*, **protected marrows** *(off-label)*, **protected ornamentals** *(off-label)*, **protected parsnips** *(off-label - temporary protection)*, **protected pumpkins** *(off-label)*, **protected squashes** *(off-label)*, **pumpkins** *(off-label)*, **quinoa** *(off-label - grown for game cover)*, **raspberries**, **squashes** *(off-label)*, **strawberries**, **sunflowers** *(off-label - grown for game cover)*
- Cleavers in **asparagus** *(off-label)*
- Fat hen in **asparagus** *(off-label)*
- Groundsel in **asparagus** *(off-label)*
- Knotgrass in **asparagus** *(off-label)*

Specific Off-Label Approvals (SOLAs)
- **asparagus** *(OLA 050893) Dec 2008* [1]
- **canary grass**, **clovers**, **forage maize**, **millet**, **quinoa**, **sunflowers** *(grown for game cover) (OLA 050896) Dec 2008* [1]
- **carrots** *(OLA 050895) Dec 2008* [1]
- **courgettes**, **marrows**, **protected courgettes**, **protected marrows**, **protected pumpkins**, **protected squashes**, **pumpkins**, **squashes** *(OLA 050894) Dec 2008* [1]
- **protected carrots**, **protected parsnips** *(temporary protection) (OLA 050892) Dec 2008* [1]
- **protected ornamentals** *(OLA 050891) Dec 2008* [1]

Approval information
- Accepted by BBPA for use on hops

Efficacy guidance
- When used alone apply pre-weed emergence
- Effectiveness is reduced in dry conditions. Weed seeds germinating at depth are not controlled
- Activity reduced on soils with more than 10% organic matter. Do not use on peaty soils
- Various tank mixtures are recommended for early post-weed emergence treatment (especially for grass weeds). See label for details

Restrictions
- Maximum number of treatments 1 per crop for all edible crops; 2 per yr on amenity vegetation and non-edible crops

Crop-specific information
- Latest use: 3 true leaves for cucurbits; before 1 Apr in yr of harvest for other edible crops

Following crops guidance
- See label for details of crops which may be sown in the event of failure of a treated crop

Environmental safety
- Keep all livestock out of treated areas for at least 50 d

Hazard classification and safety precautions
    Operator protection A, C; U05a, U20b
    Environmental protection E06a (50 d); E15a
    Storage and disposal D01, D05, D09a, D11a

FOR FULL CONDITIONS OF USE ALWAYS READ THE PRODUCT LABEL

288   isoxaben + terbuthylazine

A contact and residual herbicide for use in peas and spring field beans

Products

Skirmish	Syngenta	75:420 g/l	SC	08444

Uses
- Annual dicotyledons in **combining peas**, **spring field beans**, **vining peas**

Efficacy guidance
- May be applied pre- or post-emergence of crop but before second node stage (GS 102)
- Product will give residual control of germinating weeds on mineral soils for up to 8 wk
- Product is slow acting and control may not be evident for 7-10 d or more after spraying
- Best results achieved when soil surface damp and with a fine, firm tilth. Do not use on very cloddy or stony soil
- Rain after spraying will normally improve weed control but excessive rainfall, very dry conditions or unusually low soil temperatures may lead to unsatisfactory control

Restrictions
- Maximum total dose equivalent to one full dose treatment on peas and spring field beans
- Do not use on forage peas
- Do not use on soils lighter than Coarse Sandy Loam, on very stony soils or soils with more than 10% organic matter
- Pea seed should be covered by at least 25 mm of soil
- Heavy rain after application may cause some crop damage, especially on light soils. Do not use on soils where surface water is likely to accumulate
- For post-emergence treatment a crystal violet test for cuticle wax is advised

Crop-specific information
- Latest use: before second node stage (GS 102) for peas; pre-emergence for spring field beans
- Product may be used on all varieties of spring sown vining and combining peas and spring field beans but pea varieties Vedette and Printana may be damaged from which recovery may not be complete

Environmental safety
- Dangerous for the environment
- Very toxic to aquatic organisms

Hazard classification and safety precautions
   Hazard H03, H11
   Risk phrases R22a, R50, R53a
   Operator protection A; U05a, U20a
   Environmental protection E15a, E38
   Storage and disposal D01, D02, D05, D07, D09a, D10c, D12a

289   isoxaben + trifluralin

A residual herbicide mixture for amenity and horticultural use

Products

1	Axit GR	Fargro	0.5:2.0% w/w	GR	08892
2	Premiere	Dow	0.5:2.0% w/w	GR	10866

Uses
- Annual dicotyledons in **amenity trees and shrubs**, **amenity vegetation** [2]; **container-grown ornamentals** (outdoor stock only), **hardy ornamental nursery stock** (outdoor stock only) [1]
- Annual meadow grass in **amenity trees and shrubs**, **amenity vegetation** [2]; **container-grown ornamentals** (outdoor stock only), **hardy ornamental nursery stock** (outdoor stock only) [1]
- Fat hen in **amenity trees and shrubs**, **amenity vegetation** [2]; **container-grown ornamentals** (outdoor stock only), **hardy ornamental nursery stock** (outdoor stock only) [1]
- Groundsel in **amenity trees and shrubs**, **amenity vegetation** [2]
- Hairy bittercress in **amenity trees and shrubs**, **amenity vegetation** [2]; **container-grown ornamentals** (outdoor stock only), **hardy ornamental nursery stock** (outdoor stock only) [1]

SECTION 2

SEE SECTION 3 FOR PRODUCTS ALSO REGISTERED

- Shepherd's purse in **hardy ornamental nursery stock** *(outdoor stock only)* [1]
- Sowthistle in **amenity trees and shrubs**, **amenity vegetation** [2]; **container-grown ornamentals** *(outdoor stock only)*, **hardy ornamental nursery stock** *(outdoor stock only)* [1]

Efficacy guidance

- Best results obtained when granules applied to firm, moist soil, free from clods. 20-30 mm irrigation or rainfall required within 3 d after treatment
- Weed control reduced under dry soil conditions
- Hoeing after application will reduce weed control
- Treatment should be made before mulching and before weed emergence
- Uniform cover with granules optimises results. Improved uniformity of cover may be achieved by spreading the granules twice at half dose at right angles
- Control may be reduced on soils with high organic matter or where organic manure has been applied
- Perennial weeds regenerating from underground rootstocks or stolons are not controlled

Restrictions

- Maximum number of treatments 1 per yr
- Products contain up to 1% crystalline silica. The MEL for respirable silica dust is 0.4 mg/cu m
- Do not use on soils with more than 10% organic matter (except in container grown hardy nursery stock)
- Do not mix in the soil or compost
- Do not use on ornamentals grown under protection
- Do not apply to plants with wet foliage or where granules can lodge in the foliage
- Do not apply to heathers or *Potentilla spp.*

Crop-specific information

- Ensure soil has settled and no cracks are present before application to transplanted trees and shrubs
- May be used on Light, Medium and Heavy soils, but on Light soils early growth may be reduced under adverse conditions
- For newly planted containerised stock first irrigate and allow compost to settle and foliage and stems to dry before application [1]

Following crops guidance

- Land must be mouldboard ploughed to at least 20 cm before drilling or planting succeeding plants except well rooted forestry trees and ornamentals
- Do not use in the 12 mth prior to lifting field grown nursery stock if edible crops are to be grown as a following crop
- After an autumn application cereals or grass crops should not be sown until the following autumn
- Crops most sensitive to treatment residues are oilseed rape, stubble turnips, other brassicae, sugar beet, fodder beet, herbage seed, grass leys and amenity grass

Environmental safety

- Harmful to fish or other aquatic life. Do not contaminate surface waters or ditches with chemical or used container
- Do not allow direct applications of granules from ground based/vehicle mounted applicators to fall within 6 m, or from hand-held applicators to within 2 m, of surface waters or ditches. Direct applications away from water

Hazard classification and safety precautions

Operator protection A; U20b
Environmental protection E13c, E16f, E34
Storage and disposal D01, D09a, D11a

290   kresoxim-methyl

A protectant strobilurin fungicide for apples

*See also epoxiconazole + fenpropimorph + kresoxim-methyl
epoxiconazole + kresoxim-methyl
epoxiconazole + kresoxim-methyl + pyraclostrobin
fenpropimorph + kresoxim-methyl*

Products

Stroby WG   BASF   50% w/w   WG   08653

Uses
- Black spot in *roses*
- Powdery mildew in *apples* *(reduction)*, *blackcurrants*, *protected strawberries*, *roses*, *strawberries*
- Scab in *apples*

Approval information
- Kresoxim-methyl included in Annex I under EC Directive 91/414
- Approved for use in ULV systems

Efficacy guidance
- Activity is protectant. Best results achieved from treatments prior to disease development. See label for timing details on each crop. Treatments should be repeated at 10-14 d intervals but note limitations below
- To minimise the likelihood of development of resistance to strobilurin fungicides these products should be used in a planned Resistance Management strategy. See Section 5 for more information
- Product may be applied in ultra low volumes (ULV) but disease control may be reduced

Restrictions
- Maximum number of treatments 4 per yr on apples; 3 per yr on other crops. See notes in Efficacy about limitations on consecutive treatments
- Consult before using on crops intended for processing

Crop-specific information
- HI 14 d for blackcurrants, protected strawberries, strawberries; 35 d for apples
- On apples do not spray product more than twice consecutively and separate each block of two consecutive treatments with at least two applications from a different cross-resistance group. For all other crops do not apply consecutively and use a maximum of once in every three fungicide sprays
- Product should not be used as final spray of the season on apples

Environmental safety
- Dangerous for the environment
- Very toxic to aquatic organisms
- Broadcast air-assisted LERAP (5 m)
- Harmless to ladybirds and predatory mites
- Harmless to honey bees and may be applied during flowering. Nevertheless local beekeepers should be notified when treatment of orchards in flower is to occur

Hazard classification and safety precautions

Hazard H03, H11
Risk phrases R40, R50, R53a
Operator protection U05a, U20b
Environmental protection E15a, E38; E17b (5 m)
Storage and disposal D01, D02, D08, D09a, D10c, D12a
Medical advice M05a

291   lambda-cyhalothrin

A quick-acting contact and ingested pyrethroid insecticide

Products

1 Hallmark with Zeon Technology	Syngenta	100 g/l	CS	10480
2 Landgold Lambda-Z	Teliton	100 g/l	CS	12383
3 Me2 Lambda 2	Me2	100 g/l	CS	12437
4 Stealth	Syngenta	2.5% w/w	WG	11514

Uses

- Aphids in **chestnuts** *(off-label)*, **cob nuts** *(off-label)*, **hazel nuts** *(off-label)*, **short rotation coppice willow** *(off-label)*, **walnuts** *(off-label)* [1]; **durum wheat**, **potatoes**, **spring barley**, **spring oats**, **spring wheat**, **winter barley**, **winter oats**, **winter wheat** [1-3]
- Barley yellow dwarf virus vectors in **durum wheat**, **winter barley**, **winter oats**, **winter wheat** [1-4]; **spring barley**, **spring wheat** [1-3]
- Beet leaf miner in **sugar beet** [1-3]
- Beet virus yellows vectors in **spring oilseed rape** [1-3]; **winter oilseed rape** [1-4]
- Beetles in **short rotation coppice willow** *(off-label)* [1]
- Cabbage seed weevil in **spring oilseed rape**, **winter oilseed rape** [1-3]
- Cabbage stem flea beetle in **spring oilseed rape** [1-3]; **winter oilseed rape** [1-4]
- Cabbage stem weevil in **radishes** *(off-label)* [1]
- Carrot fly in **carrots** *(off-label)*, **celeriac** *(off-label)*, **celery** *(off-label)*, **fennel** *(off-label)*, **parsnips** *(off-label)* [1]
- Caterpillars in **broccoli**, **brussels sprouts**, **cabbages**, **calabrese**, **cauliflowers** [1-3]; **chestnuts** *(off-label)*, **cob nuts** *(off-label)*, **hazel nuts** *(off-label)*, **walnuts** *(off-label)* [1]
- Cutworms in **carrots**, **lettuce**, **sugar beet** [1-3]; **celeriac** *(off-label)*, **chicory** *(off-label - for forcing)*, **fennel** *(off-label)*, **red beet** *(off-label)* [1]
- Flea beetle in **poppies** *(off-label - for morphine production)* [1]; **spring oilseed rape**, **sugar beet** [1-3]; **winter oilseed rape** [1-4]
- Frit fly in **sweetcorn** *(off-label)* [1]
- Pea and bean weevil in **peas**, **spring field beans**, **winter field beans** [1-3]
- Pea aphid in **peas** [1-3]
- Pea moth in **peas** [1-3]
- Pear sucker in **pears** [1-3]
- Pod midge in **spring oilseed rape**, **winter oilseed rape** [1-3]
- Pollen beetle in **poppies** *(off-label - for morphine production)* [1]; **spring oilseed rape**, **winter oilseed rape** [1-3]
- Sawflies in **short rotation coppice willow** *(off-label)* [1]
- Silver Y moth in **celery** *(off-label)*, **dwarf beans** *(off-label)*, **navy beans** *(off-label)*, **red beet** *(off-label)*, **runner beans** *(off-label)* [1]
- Thrips in **bulb onions** *(off-label)*, **leeks** *(off-label)*, **salad onions** *(off-label)* [1]
- Whitefly in **broccoli**, **brussels sprouts**, **cabbages**, **calabrese**, **cauliflowers** [1-3]
- Yellow cereal fly in **winter wheat** [1-3]

Specific Off-Label Approvals (SOLAs)

- **bulb onions**, **leeks**, **salad onions** *(OLA 051610) Jan 2007* [1]
- **carrots**, **parsnips** *(OLA 051597) Jan 2007* [1]
- **celeriac**, **radishes** *(OLA 051596) Dec 2008* [1]
- **celery** *(OLA 051599) Jan 2007* [1]
- **chestnuts**, **cob nuts**, **hazel nuts**, **walnuts** *(OLA 051602) Jan 2007* [1]
- **chicory** *(for forcing) (OLA 051605) Jan 2007* [1]
- **dwarf beans**, **navy beans**, **runner beans** *(OLA 051601) Jan 2007* [1]
- **fennel** *(OLA 051604) Jan 2007* [1]
- **poppies** *(for morphine production) (OLA 051608) Jan 2007* [1]
- **red beet** *(OLA 051598) Jan 2007* [1]
- **short rotation coppice willow** *(OLA 051609) Jan 2007* [1]
- **sweetcorn** *(OLA 051603) Jan 2007* [1]

FOR FULL CONDITIONS OF USE ALWAYS READ THE PRODUCT LABEL

Approval information
- Lambda-cyhalothrin included in Annex I under EC Directive 91/414
- Accepted by BBPA for use on malting barley

Efficacy guidance
- Best results normally obtained from treatment when pest attack first seen. See label for detailed recommendations on each crop
- Timing for control of barley yellow dwarf virus vectors depends on specialist assessment of the level of risk in the area
- Repeat applications recommended in some crops where prolonged attack occurs, up to maximum total dose. See label for details
- Where strains of aphids resistant to lambda-cyhalothrin occur control is unlikely to be satisfactory
- Addition of wetter recommended for control of certain pests in brassicas and oilseed rape
- Use of sufficient water volume to ensure thorough crop penetration recommended for optimum results
- Use of drop-legged sprayer gives improved results in crops such as Brussels sprouts

Restrictions
- Maximum number of applications or maximum total dose per crop varies - see labels
- Do not apply to a cereal crop if any product containing a pyrethroid insecticide or dimethoate has been applied to the crop after the start of ear emergence (GS 51)
- Do not spray cereals in the spring/summer (ie after 1 Apr) within 6 m of edge of crop

Crop-specific information
- Latest use on cereals and oilseed rape depends on product used. See labels
- HI 3 d for lettuce, radishes, red beet; 7 d for celery, chicory, dwarf beans, fennel, navy beans, runner beans, sweetcorn; 14 d for carrots, celeriac, nuts, parsnips; 6 wk for spring oilseed rape; 8 wk for sugar beet
- If using mixtures, add lambda-cyhalothrin product to the tank last

Environmental safety
- Dangerous for the environment
- Very toxic to aquatic organisms
- High risk to non-target insects or other arthropods. Do not spray within 6 m of the field boundary [4]
- LERAP Category A
- Broadcast air-assisted LERAP (38 m in pears) [1-3]

Hazard classification and safety precautions
    Hazard H03, H11
    Risk phrases R20, R22a, R43, R50, R53a [1-4]; R36, R38 [4]
    Operator protection A, C, H, J, K, M [1-4]; L [1-3]; U02a, U05a, U08, U14, U20b [1-4]; U11 [4]
    Environmental protection E15a, E16c, E16d, E34, E38 [1-4]; E17b [1-3] (pears 38 m); E22a [4]
    Storage and disposal D01, D02, D09a, D12a [1-4]; D05, D10c [4]; D10b [1-3]
    Medical advice M03

292   lambda-cyhalothrin + pirimicarb

An insecticide mixture combining translaminar, contact, fumigant and stomach activity

Products

Dovetail	Syngenta	5:100 g/l	EC	10479

Uses
- Aphids in *broccoli, brussels sprouts, cabbages, calabrese, carrots, cauliflowers, durum wheat, lettuce, peas, potatoes, spring barley, spring oats, spring rye, spring wheat, sugar beet, triticale, winter barley, winter oats, winter rye, winter wheat*
- Beet virus yellows vectors in *spring oilseed rape, winter oilseed rape*
- Cabbage seed weevil in *spring oilseed rape, winter oilseed rape*
- Cabbage stem flea beetle in *spring oilseed rape, winter oilseed rape*
- Caterpillars in *broccoli, brussels sprouts, cabbages, calabrese, cauliflowers*
- Cutworms in *carrots, lettuce, potatoes, sugar beet*
- Flea beetle in *spring oilseed rape, sugar beet, winter oilseed rape*
- Leaf miner in *sugar beet*

SEE SECTION 3 FOR PRODUCTS ALSO REGISTERED

- Mealy aphid in *broccoli, brussels sprouts, cabbages, calabrese, cauliflowers, spring oilseed rape, winter oilseed rape*
- Pea and bean weevil in *peas, spring field beans, winter field beans*
- Pea midge in *peas*
- Pea moth in *peas*
- Pod midge in *spring oilseed rape, winter oilseed rape*
- Pollen beetle in *broccoli, brussels sprouts, cabbages, calabrese, cauliflowers, spring oilseed rape, winter oilseed rape*
- Whitefly in *broccoli, brussels sprouts, cabbages, calabrese, cauliflowers*

Approval information
- Lambda-cyhalothrin included in Annex I under EC Directive 91/414
- Accepted by BBPA for use on malting barley

Efficacy guidance
- Best results obtained from treatment when pest attack first seen or after warning issued
- Repeat applications recommended in some crops where prolonged attacks occur, up to maximum total dose
- Addition of Agral recommended for certain uses in brassicas and oilseed rape. See label
- Use of drop-leg sprayers improves efficacy in crops such as Brussels sprouts
- Control unlikely to be satisfactory if aphids resistant to lambda-cyhalothrin or pirimicarb present

Restrictions
- Maximum total dose equivalent to two full dose treatments on carrots, lettuce, peas, field beans; three full dose treatments on brassicas, oilseed rape, cereals; four full dose treatments on sugar beet; eight full dose treatments on potatoes
- Must not be applied to cereals if any product containing a pyrethroid insecticide or dimethoate has been sprayed after the start of ear emergence (GS 51)

Crop-specific information
- Latest use: before late milky ripe (GS 77) for cereals; before end of flowering for oilseed rape
- HI sugar beet 8 wk; carrots 14 d; brassicas, lettuce, potatoes, peas, field beans 3 d

Environmental safety
- Dangerous for the environment
- Very toxic to aquatic organisms
- Do not spray cereals after 1 Apr within 6 m of the edge of the crop
- Harmful to livestock. Keep all livestock out of treated areas for at least 7 d following treatment
- Keep in original container, tightly closed, in a safe place, under lock and key
- LERAP Category A

Hazard classification and safety precautions
Hazard H03, H08, H11
Risk phrases R20, R22a, R22b, R37, R38, R41, R43, R50, R53a
Operator protection A, C, E, H, J, K, M; U02a, U05a, U08, U11, U19a, U20b
Environmental protection E06c (7 d); E15a, E16c, E16d, E34, E38
Storage and disposal D01, D02, D09b, D10c, D12a
Medical advice M02, M03, M05b

293　Ienacil

A residual, soil-acting uracil herbicide for beet crops

See also chloridazon + lenacil

Products

1	Agriguard Lenacil	AgriGuard	80% w/w	WP	10488
2	Fernpath Lenzo Flo	AgriGuard	440 g/l	SC	11919
3	Lenacil FL	United Phosphorus	440 g/l	SC	12183
4	Lenazar Flo	Hermoo	440 g/l	SC	12286
5	Lenazar Flowable	Hermoo	440 g/l	SC	11068
6	Venzar Flowable	DuPont	440 g/l	SC	06907

FOR FULL CONDITIONS OF USE ALWAYS READ THE PRODUCT LABEL

Uses
- Annual dicotyledons in **farm woodland** (off-label) [6]; **fodder beet**, **mangels**, **sugar beet** [1-6]; **red beet** [1, 2, 4, 6]
- Annual meadow grass in **farm woodland** (off-label) [6]; **fodder beet**, **mangels**, **sugar beet** [1-6]; **red beet** [1, 2, 4, 6]

Specific Off-Label Approvals (SOLAs)
- **farm woodland** (OLA 971282) Dec 2008 [6]

Efficacy guidance
- On beet crops may be used pre- or post-emergence, alone or in mixture to broaden weed spectrum
- Apply overall or as band spray to beet crops pre-drilling incorporated, pre- or post-emergence
- All labels have limitations on soil types that may be treated. Residual activity reduced on soils with high OM content
- Best results achieved on a firm moist tilth. Continuing presence of moisture from rain or irrigation gives improved residual control of later germinating weeds. Effectiveness may be reduced by dry conditions

Restrictions
- Maximum number of treatments in beet crops 1 pre-emergence + 3 post-emergence per crop; 2 per yr on established woody ornamentals and roses
- See label for soil type restrictions
- Do not apply other residual herbicides within 3 mth before or after spraying
- Do not treat crops under stress from drought, low temperatures, nutrient deficiency, pest or disease attack, or waterlogging

Crop-specific information
- Latest use: pre-emergence for red beet, fodder beet, spinach, spinach beet, mangels; before leaves meet over rows when used on these crops post-emergence
- Heavy rain after application may cause damage especially if followed by very hot weather
- Reduction in beet stand may occur where crop emergence or vigour is impaired by soil capping or pest attack

Following crops guidance
- Succeeding crops should not be planted or sown for at least 4 mth (6 mth on organic soils) after treatment following ploughing to at least 150 mm

Environmental safety
- Dangerous for the environment
- Very toxic to aquatic organisms
- Harmful to fish or other aquatic life. Do not contaminate surface waters or ditches with chemical or used container

Hazard classification and safety precautions
Hazard H04 [1-3, 5]; H11 [1-6]
Risk phrases R36, R38 [1-3, 5]; R37 [1, 2]; R50, R53a [2-6]
Operator protection A [1-6]; C [1-5]; U05a, U08, U19a, U20a [1-6]; U14, U15 [1-3]
Environmental protection E13c, E38 [2, 4-6]; E15a, E34 [1, 3]
Storage and disposal D01, D02, D09a [1-6]; D05 [2-6]; D10a, D12a [2, 4-6]; D10b [1, 3]; D12b [3]

294   lenacil + phenmedipham

A contact and residual herbicide for use in beet crops

Products
Lens	Interfarm	66.6:95 g/l	SC	12182

Uses
- Annual dicotyledons in **fodder beet**, **mangels**, **sugar beet**

Efficacy guidance
- Apply at any stage of crop when weeds at cotyledon stage
- Germinating weeds controlled by root uptake for several weeks
- Best results achieved under warm, moist conditions on a fine, firm seedbed
- Treatment may be repeated up to a maximum of 3 applications on later weed flushes

SEE SECTION 3 FOR PRODUCTS ALSO REGISTERED

- Residual activity may be reduced on highly organic soils or under very dry conditions
- Rain falling 1 h after application does not reduce activity

Restrictions
- Maximum number of treatments 3 per crop
- Do not apply when temperature above or likely to exceed 21°C on day of spraying or under conditions of high light intensity
- Do not spray any crop under stress from drought, waterlogging, cold, wind damage or any other cause

Crop-specific information
- Latest use: before crop leaves meet across rows
- Heavy rain after application may reduce stand of crop particularly in very hot weather. In severe cases yield may be reduced

Following crops guidance
- Do not sow or plant any crop within 4 mth of treatment. In case of crop failure only sow or plant beet crops, strawberries or other tolerant horticultural crop within 4 mth

Environmental safety
- Dangerous for the environment
- Toxic to aquatic organisms

Hazard classification and safety precautions
  Hazard H03, H11
  Risk phrases R21, R22a, R22b, R36, R37, R40, R41, R43, R51, R53a
  Operator protection A, C, H; U02a, U05a, U08, U11, U14, U19a, U20b
  Environmental protection E15a, E34
  Storage and disposal D01, D02, D09a, D10b
  Medical advice M04, M05b

295   lenacil + triflusulfuron-methyl

A foliar and residual herbicide mixture for sugar beet

Products

Safari Lite WSB	DuPont	71.4:5.4% w/w	WG	12169

Uses
- Annual dicotyledons in *sugar beet*
- Volunteer oilseed rape in *sugar beet*

Efficacy guidance
- Best results obtained when weeds are small and growing actively
- Product recommended for use in a programme of treatments in tank mixture with a suitable herbicide partner to broaden the weed spectrum
- Ensure good spray cover of weeds. Apply when first weeds have emerged provided crop has reached cotyledon stage
- Susceptible plants cease to grow almost immediately after treatment and symptoms can be seen 5-10 d later
- Weed control may be reduced in very dry soil conditions
- Triflusulfuron-methyl is a member of the ALS-inhibitor group of herbicides

Restrictions
- Maximum number of treatments on sugar beet 3 per crop and do not apply more than 4 applications of any product containing triflusulfuron-methyl
- Do not apply to crops suffering from stress caused by drought, water-logging, low temperatures, pest or disease attack, nutrient deficiency or any other factors affecting crop growth
- Do not use on Sands, stony or gravelly soils or on soils with more than 10% organic matter
- Do not apply when temperature above or likely to exceed 21°C on day of spraying or under conditions of high light intensity

Crop-specific information
- Latest Use: before crop leaves meet between rows for sugar beet

FOR FULL CONDITIONS OF USE ALWAYS READ THE PRODUCT LABEL

Following crops guidance
- Only cereals may be sown in the same calendar yr as a treated sugar beet crop. Any crop may be sown in the following spring
- In the event of crop failure sow only sugar beet within 4 mth of treatment

Environmental safety
- Dangerous for the environment
- Very toxic to aquatic organisms
- LERAP Category B
- Take extreme care to avoid drift onto broad-leaved plants outside the target area or onto surface waters or ditches, or land intended for cropping
- Spraying equipment should not be drained or flushed onto land planted, or to be planted, with trees or crops other than cereals and should be thoroughly cleansed after use - see label for instructions

Hazard classification and safety precautions
Hazard H04, H11
Risk phrases R36, R37, R38, R43, R50, R53a
Operator protection A, C; U05a, U08, U11, U19a, U20b, U22a
Environmental protection E15a, E16a, E38
Storage and disposal D01, D02, D09a, D12a

296   lindane

An organochlorine insecticide all approvals for which were revoked in 2001 following non-inclusion of the substance in Annex I of EC Directive 91/414

297   linuron

A contact and residual urea herbicide for various field crops

See also 2,4-DB + linuron + MCPA

Products

1 Afalon	Makhteshim	450 g/l	SC	11665
2 Alpha Linuron 50 SC	Makhteshim	500 g/l	SC	06967
3 Lincon 50	DAPT	500 g/l	SC	10639
4 Linuron 500	Nufarm UK	480 g/l	SC	11590
5 Luas	AgriGuard	500 g/l	SC	12357
6 UPL Linuron 45% Flowable	United Phosphorus	450 g/l	SC	07435

Uses

- Annual dicotyledons in **bulb onions** *(off-label)*, **celeriac** *(off-label)*, **chives** *(off-label)*, **dwarf beans** *(off-label)*, **garlic** *(off-label)*, **herbs (see appendix 6)** *(off-label)*, **leeks** *(off-label)*, **runner beans** *(off-label)* [2]; **carrots, potatoes** [1-6]; **celery** [1]; **parsnips** [1-3, 5, 6]; **spring barley, spring wheat** [3-6]; **spring oats** [2, 3, 5]
- Annual meadow grass in **carrots, potatoes** [1-6]; **celery** [1]; **chives** *(off-label)*, **herbs (see appendix 6)** *(off-label)* [2]; **parsnips** [1-3, 5, 6]; **spring barley, spring wheat** [2-6]; **spring oats** [3, 5]
- Black bindweed in **carrots, potatoes** [1-6]; **celery** [1]; **chives** *(off-label)*, **herbs (see appendix 6)** *(off-label)* [2]; **parsnips** [1-3, 5, 6]; **spring barley, spring wheat** [2-6]; **spring oats** [2, 3, 5]
- Chickweed in **carrots, potatoes** [1-6]; **celery** [1]; **chives** *(off-label)*, **herbs (see appendix 6)** *(off-label)* [2]; **parsnips** [1-3, 5, 6]; **spring barley, spring wheat** [2-6]; **spring oats** [2, 3, 5]
- Corn marigold in **carrots, potatoes, spring barley, spring wheat** [2-5]; **chives** *(off-label)*, **herbs (see appendix 6)** *(off-label)* [2]; **parsnips, spring oats** [2, 3, 5]
- Fat hen in **carrots, potatoes** [1-6]; **celery** [1]; **chives** *(off-label)*, **herbs (see appendix 6)** *(off-label)* [2]; **parsnips** [1-3, 5, 6]; **spring barley, spring wheat** [2-6]; **spring oats** [2, 3, 5]
- General weed control in **bulb onions** *(off-label)*, **celeriac** *(off-label)*, **garlic** *(off-label)*, **leeks** *(off-label)* [1]; **herbs for medicinal uses (see appendix 6)** *(off-label)* [2]
- Redshank in **carrots, potatoes** [1-6]; **celery** [1]; **chives** *(off-label)*, **herbs (see appendix 6)** *(off-label)* [2]; **parsnips** [1-3, 5, 6]; **spring barley, spring wheat** [2-6]; **spring oats** [2, 3, 5]

SEE SECTION 3 FOR PRODUCTS ALSO REGISTERED

Specific Off-Label Approvals (SOLAs)
- *bulb onions, celeriac, garlic, leeks* *(OLA 041755) Dec 2008* [1]
- *bulb onions, celeriac, garlic, leeks* *(OLA 021748) Dec 2008* [2]
- *chives, herbs (see appendix 6)* *(OLA 050595) Dec 2008* [2]
- *dwarf beans, runner beans* *(OLA 051698) Dec 2008* [2]
- *herbs for medicinal uses (see appendix 6)* *(OLA 050154) Dec 2008* [2]

Approval information
- Linuron included in Annex I under EC Directive 91/414
- All approvals for use of linuron on parsley were revoked in June 2005
- Accepted by BBPA for use on malting barley

Efficacy guidance
- Many weeds controlled pre-emergence or post-emergence to 2-3 leaf stage, some (annual meadow grass, mayweed) only susceptible pre-emergence. See label for details
- Best results achieved by application to firm, moist soil of fine tilth
- Little residual effect on soil with more than 10% organic matter

Restrictions
- Maximum number of treatments 1 per crop on most crops or 1 pre-em + 1 post-em for carrots, parsley, parsnips
- Maximum total dose equivalent to one full dose treatment
- Do not use on undersown cereals or crops grown on Sands or Very Light soils or soils heavier than Sandy Clay Loam or with more than 10% organic matter
- Do not apply to emerged crops of carrots, parsnips or parsley under stress
- Do not apply by hand-held sprayers

Crop-specific information
- Latest use: pre-emergence for most crops; pre- or up to 40% emergence for potatoes (products differ, see label for details).
- HI onions, garlic 8 wk; celeriac 12 wk; leeks 16 wk
- Drill spring cereals at least 3 cm deep and apply pre-emergence of crop or weeds
- Apply to potatoes well earthed up to a rounded ridge pre-crop emergence and do not cultivate after spraying
- Apply to carrots at any time after drilling on organic soils, and within 4 d of drilling on other soils. Apply post-emergence as soon as weeds appear but after first rough leaf stage.
- Recommendations for parsnips, parsley and celery vary. See label for details

Following crops guidance
- Potatoes, carrots and parsnips may be planted at any time after application. Lettuce should not be grown within 12 mth of treatment. Transplanted brassicas may be grown from 3 mth after treatment

Environmental safety
- Dangerous for the environment [4]
- Very toxic to aquatic organisms
- Keep livestock out of treated areas for at least 5 mth [4]
- LERAP Category B

Hazard classification and safety precautions
    Hazard H03, H11
    Risk phrases R22a [1, 2, 6]; R40 [1-3, 5, 6]; R48, R50, R53a [1-6]; R68 [4]
    Operator protection A, H [1-6]; C [2-6]; U05a [1-6]; U08, U20b [1, 2, 4, 6]; U09a, U20a [3, 5]; U19a [2-6]; U23a [1]
    Environmental protection E07b [4] (5 mth); E15a, E16a [1-6]; E34 [3, 5]; E38 [4]
    Storage and disposal D01, D05, D09a [1-6]; D02 [1-3, 5, 6]; D08 [1]; D10b, D12a [1, 4]; D10c [3, 5]; D11a [2, 6]; D12b [1, 2, 6]

298   linuron + trifluralin

A residual pre-emergence herbicide for use in winter cereals

Products

1	Arizona	Makhteshim	120:240 g/l	EC	11722
2	Blois	Makhteshim	128:256 g/l	EC	12376

FOR FULL CONDITIONS OF USE ALWAYS READ THE PRODUCT LABEL

Products – continued

3 Uranus	Makhteshim	120:240 g/l	EC	11637
4 Uranus	Makhteshim	120:240 g/l	EC	12506

Uses

- Annual dicotyledons in **broad beans**, **combining peas**, **spring field beans**, **vining peas**, **winter field beans** [2]; **triticale**, **winter barley**, **winter wheat** [1, 3, 4]
- Annual grasses in **broad beans**, **combining peas**, **spring field beans**, **vining peas**, **winter field beans** [2]; **triticale**, **winter barley**, **winter wheat** [1, 3, 4]
- Annual meadow grass in **winter oats** *(off-label)* [3]
- Blackgrass in **spring oats** *(off-label)* [3]

Specific Off-Label Approvals (SOLAs)

- **spring oats**, **winter oats** *(OLA 051924) Dec 2008* [3]

Approval information

- Linuron included in Annex I under EC Directive 91/414
- Accepted by BBPA for use on malting barley

Efficacy guidance

- Effective against weeds germinating near soil surface. Best results achieved by application to fine, firm, moist seedbed free of clods, crop residues or established weeds
- Effectiveness reduced by long dry periods after application or on waterlogged soil
- Autumn application residual effects normally last until spring but further herbicide treatment may be needed on thin or backward crops
- Results in loose seedbeds on lighter soils improved by rolling after drilling

Restrictions

- Maximum number of treatments 1 per crop
- Do not treat durum wheat or undersown crops
- Do not use on soils classed as Sands. Do not harrow after treatment
- Do not use on peaty soils or where organic matter exceeds 10%

Crop-specific information

- Latest use: pre-emergence of crop
- Apply without incorporation as soon as possible after drilling and before crop emergence, within 3 d on early drilled crops
- Crop seed must be well covered to a minimum depth of 30 mm

Following crops guidance

- In the event of failure of a treated crop, only peas, carrots or sunflowers may be sown within 5 mth of application. Grass and cereals should not be planted until the following autumn, and sugar beet not for one yr
- Before drilling or planting a following crop mouldboard plough to at least 150 mm

Environmental safety

- Dangerous for the environment
- Very toxic to aquatic organisms
- LERAP Category B
- Do not store near heat or flame

Hazard classification and safety precautions

Hazard H03, H08, H11
Risk phrases R20, R21 [4]; R22a, R36, R40, R43, R48, R50, R53a [1-4]; R22b [3]
Operator protection A, C, H; U05a, U08, U11, U13, U19a, U20b [1-4]; U14, U15 [4]
Environmental protection E15a, E16a, E16b
Storage and disposal D01, D02, D05, D09a, D10b, D12b
Medical advice M05b

299 magnesium phosphide

A phosphine generating compound used to control insect pests in stored commodities

Products

Degesch Plates	Rentokil	56% w/w	GE	07603

SEE SECTION 3 FOR PRODUCTS ALSO REGISTERED

SECTION 2

Uses
- Insect pests in **stored grain**

Approval information
- Accepted by BBPA for use in stores for malting barley

Efficacy guidance
- Product acts as fumigant by releasing poisonous hydrogen phosphide gas on contact with moisture in the air
- Place plates on the floor or wall of the building or on the surface of the commodity. Exposure time varies depending on temperature and pest. See label

Restrictions
- Magnesium phosphide is subject to the Poisons Rules 1982 and the Poisons Act 1972. See notes in Section 5
- Only to be used by professional operators trained in the use of magnesium phosphide and familiar with the precautionary measures to be observed. See label for full precautions

Environmental safety
- Highly flammable
- Prevent access to buildings under fumigation by livestock, pets and other non-target mammals and birds
- Dangerous to fish or other aquatic life. Do not contaminate surface waters or ditches with chemical or used container
- Keep in original container, tightly closed, in a safe place, under lock and key
- Do not allow plates or their spent residues to come into contact with food other than raw cereal grains
- Remove used plates after treatment. Do not bulk spent plates and residues: spontaneous ignition could result
- Keep livestock out of treated areas

Hazard classification and safety precautions
    Hazard H01, H07
    Risk phrases R21, R26, R28
    Operator protection A, D, H; U01, U05b, U07, U13, U19a, U20a
    Environmental protection E02a (4 h min); E02b, E13b, E34
    Storage and disposal D01, D02, D05, D07, D09b, D11b
    Medical advice M04

300   malathion

A broad-spectrum contact organophosphorus insecticide and acaricide

See also bifenthrin + malathion

Products

Fyfanon 440	Headland	440 g/l	EW	11134

Uses
- Aphids in **carnations, chrysanthemums, dahlias, gladioli, outdoor cyclamen, protected carnations, protected chrysanthemums, protected cyclamen, protected dahlias, protected gladioli, protected roses, protected tulips, roses, tulips**

Efficacy guidance
- Spray when pest first seen and repeat as necessary, usually at 7-14 d intervals
- Number and timing of sprays vary with crop and pest. See label for details
- Repeat spray routinely for scale insect and whitefly control in glasshouses
- Where aphids resistant to malathion occur control is unlikely to be satisfactory

Restrictions
- Malathion is an anticholinesterase organophosphorus compound. Do not use if under medical advice not to work with such compounds
- Do not use on antirrhinums, crassula, ferns, fuchsias, gerberas, petunias, pileas, sweet peas or zinnias as damage may occur
- Do not apply by tractor mounted/drawn or broadcast air assisted sprayers

FOR FULL CONDITIONS OF USE ALWAYS READ THE PRODUCT LABEL

Environmental safety
- Dangerous for the environment
- Very toxic to aquatic organisms
- Dangerous to bees. Do not apply to crops in flower or to those in which bees are actively foraging. Do not apply when flowering weeds are present
- Risk to certain non-target insects or other arthropods. For advice on risk management and use in Integrated Pest Management (IPM) see directions for use

Hazard classification and safety precautions
> **Hazard** H11
> **Risk phrases** R50, R53a
> **Operator protection** A, H, M, P; U08, U19a, U20b
> **Environmental protection** E12d, E15a, E22b, E34, E38
> **Storage and disposal** D05, D09a, D10a
> **Medical advice** M01

301   maleic hydrazide

A pyridazinone plant growth regulator suppressing sprout and bud growth

Products

1	Fazor	Dow	60% w/w	SG	05558
2	Mazide 25	Vitax	250 g/l	SL	02067
3	Regulox K	Bayer Environ.	250 g/l	SL	09937
4	Royal MH 180	Nufarm UK	180 g/l	SL	10982
5	Royal MH 180	Certis	180 g/l	SL	11062
6	Source II	Chiltern	60% w/w	WB	08314

Uses
- Growth retardation in *amenity grass*, *hedges*, *roadside grass* [2]
- Growth suppression in *amenity grass* [3-5]; *areas around farm buildings*, *golf courses*, *road verges* [4]; *carrots* (off-label), *parsnips* (off-label) [1]; *grass near water* [3, 4]; *industrial sites*, *roadside grass*, *waste ground* [3]; *managed amenity turf* [5]
- Sprout suppression in *bulb onions* [1, 2, 5, 6]; *ware potatoes* [1, 6]
- Sucker inhibition in *amenity trees and shrubs* [2, 3]
- Volunteer suppression in *ware potatoes* [1, 6]

Specific Off-Label Approvals (SOLAs)
- *carrots*, *parsnips* (OLA 012159) Dec 2008 [1]

Approval information
- Maleic hydrazide included in Annex I under EC Directive 91/414
- Approved for use on grass near water [4] . See notes in Section 5 on use of herbicides in or near water

Efficacy guidance
- Apply to grass at any time of yr when growth active, best when growth starting in Apr-May and repeated when growth recommences
- Uniform coverage and dry weather necessary for effective results
- Accurate timing essential for good results on potatoes but rain or irrigation within 24 h may reduce effectiveness on onions and potatoes
- Mow 2-3 d before and 5-10 d after spraying for best results. Need for mowing reduced for up to 6 wk
- When used for suppression of volunteer potatoes treatment may also give some suppression of sprouting in store but separate treatment will be necessary if sprouting occurs
- To control suckers wet trunks thoroughly, especially pruned and basal bud areas [2]

Restrictions
- Maximum number of treatments 2 per yr on amenity grass, land not intended for cropping and land adjacent to aquatic areas; 1 per crop on onions, potatoes and on or around tree trunks
- Do not apply in drought or when crops are suffering from pest, disease or herbicide damage. Do not treat fine turf or grass seeded less than 8 mth previously
- Do not treat potatoes within 3 wk of applying a haulm desiccant or if temperatures above 26°C

SEE SECTION 3 FOR PRODUCTS ALSO REGISTERED

- Consult processor before use on potato crops for processing
- Do not apply by knapsack sprayer [1]

Crop-specific information
- Latest use: 3 wk before haulm destruction for potatoes; before 50% necking for onions
- HI onions 4-7 d; potatoes, carrots, parsnips 3 wk
- Apply to onions at 10% necking and not later than 50% necking stage when the tops are still green
- Only treat onions in good condition and properly cured, and do not treat more than 2 wk before maturing. Treated onions may be stored until Mar but must then be removed to avoid browning
- Apply to second early or maincrop potatoes at least 3 wk before haulm destruction
- Only treat potatoes of good keeping quality; not on seed, first earlies or crops grown under polythene
- Spray hawthorn hedges in full leaf, privet 7 d after cutting, in Apr-May [2]
- May be applied to grass along water courses but not to water surface [3, 4]

Environmental safety
- Only apply to grass not to be used for grazing
- Do not use treated water for irrigation purposes within 3 wk of treatment or until concentration in water falls below 0.02 ppm [3-5]
- Maximum permitted concentration in water 2 ppm
- Do not dump surplus product in water or ditch bottoms
- Avoid drift onto nearby vegetables, flowers or other garden plants

Hazard classification and safety precautions
> **Hazard** H04 [6]
> **Risk phrases** R41 [6]
> **Operator protection** A [1, 6]; C [6]; H [1]; U02a, U05a, U13, U14, U15, U19a, U20a, U22a [6]; U08 [1]; U20b [1, 4]; U20c [2, 3, 5]
> **Environmental protection** E15a [1-6]; E19a [3-5]; E21 [3-5] (3 wk)
> **Storage and disposal** D01, D02 [6]; D05 [3, 6]; D07, D10c [3]; D09a [1-6]; D10a [2, 4, 5]; D11a [1]

302   mancozeb

A protective dithiocarbamate fungicide for potatoes and other crops

See also   benalaxyl + mancozeb
chlorothalonil + mancozeb
cymoxanil + mancozeb
dimethomorph + mancozeb
fenamidone + mancozeb

Products

1	Dithane 945	Interfarm	80% w/w	WP	09897
2	Dithane 945	Dow	80% w/w	WP	10985
3	Dithane NT	Dow	75% w/w	WG	10986
4	Dithane NT Dry Flowable	Interfarm	75% w/w	WG	09898
5	Dithane Superflo	Dow	455 g/l	SC	10799
6	Karamate Dry Flo Newtec	Landseer	75% w/w	WG	09759
7	Manzate 75 WG	Headland	75% w/w	WG	12070
8	Micene 80	Sipcam	80% w/w	WP	09112
9	Micene DF	Sipcam	77% w/w	WG	09957
10	Penncozeb WDG	Nufarm UK	77% w/w	WG	12156
11	Quell Flo	Interfarm	455 g/l	SC	09894

Uses
- Black spot in *roses* [6]
- Blight in *potatoes* [1-5, 7-11]
- Brown rust in *spring barley*, *winter barley* [8-10]; *spring wheat*, *winter wheat* [1-5, 9-11]
- Celery leaf spot in *celeriac* (off-label), *celery* (off-label) [1, 2]
- Currant leaf spot in *blackcurrants*, *gooseberries* [6]
- Downy mildew in *bulb onions* (off-label), *garlic* (off-label) [2, 9, 10]; *chicory* (off-label), *courgettes* (off-label), *marrows* (off-label), *pumpkins* (off-label), *red beet* (off-label), *squashes* (off-label), *witloof* (off-label) [1, 2]; *lettuce, protected lettuce* [6]; *poppies* (off-label - for morphine production) [4, 10]; *winter oilseed rape* [1-5, 11]

FOR FULL CONDITIONS OF USE ALWAYS READ THE PRODUCT LABEL

- Fire in **tulips** [6]
- Fungus diseases in **grapevines** *(off-label)* [6, 10]
- Net blotch in **spring barley**, **winter barley** [8-10]
- Ray blight in **chrysanthemums** [6]
- Rhynchosporium in **spring barley** [9, 10]; **winter barley** [1-3, 9, 10]
- Root malformation disorder in **red beet** *(off-label)* [1, 2]
- Rust in **asparagus** *(off-label)*, **chicory** *(off-label)*, **witloof** *(off-label)* [1, 2]; **carnations**, **geraniums**, **roses** [6]
- Scab in **apples** [6, 8-10]; **pears** [6]
- Septoria diseases in **spring wheat**, **winter wheat** [9, 10]
- Septoria leaf spot in **spring wheat**, **winter wheat** [1-5, 8, 11]
- Sooty moulds in **spring barley**, **winter barley** [9, 10]; **spring wheat**, **winter wheat** [1-5, 9-11]
- Stemphylium in **asparagus** *(off-label)* [1, 2]
- White mould in **daffodils** *(off-label - for galanthamine production)* [2, 10]
- White tip in **leeks** *(off-label)* [10]
- Yellow rust in **spring barley**, **spring wheat**, **winter barley** [9, 10]; **winter wheat** [1-5, 9-11]

Specific Off-Label Approvals (SOLAs)
- **asparagus** *(OLA 051676) Dec 2008* [1]
- **asparagus** *(OLA 051728) Dec 2008* [2]
- **bulb onions**, **garlic** *(OLA 021430) Jun 2006* [2]
- **bulb onions**, **garlic** *(OLA 020158) Jun 2006* [9]
- **bulb onions**, **garlic** *(OLA 051477) Jun 2006* [10]
- **celeriac** *(OLA 051678) Dec 2008* [1]
- **celeriac** *(OLA 051731) Dec 2008* [2]
- **celery**, **chicory**, **witloof** *(OLA 051677) Dec 2008* [1]
- **celery** *(OLA 051729) Dec 2008* [2]
- **chicory**, **witloof** *(OLA 051730) Dec 2008* [2]
- **courgettes**, **marrows**, **pumpkins**, **squashes** *(OLA 051674) Dec 2008* [1]
- **courgettes**, **marrows**, **pumpkins**, **squashes** *(OLA 051732) Dec 2008* [2]
- **daffodils** *(for galanthamine production) (OLA 041519) Dec 2008* [2]
- **daffodils** *(for galanthamine production) (OLA 051476) Dec 2008* [10]
- **grapevines** *(OLA 021426) Dec 2008* [6]
- **grapevines** *(OLA 051478) Dec 2008* [10]
- **leeks** *(OLA 051479) Dec 2008* [10]
- **poppies** *(for morphine production) (OLA 040162) Dec 2008* [4]
- **poppies** *(for morphine production) (OLA 051475) Dec 2008* [10]
- **red beet** *(OLA 051679) Dec 2008* [1]
- **red beet** *(OLA 051727) Dec 2008* [2]

Approval information
- Mancozeb included in Annex I under EC Directive 91/414
- Approved for aerial application on potatoes [1-5, 8-11]. See notes in Section 5
- Accepted by BBPA for use on malting barley
- Approval expiry 31 Dec 2006 [5]

Efficacy guidance
- Mancozeb is a protectant fungicide and will give moderate control, suppression or reduction of the cereal diseases listed if treated before they are established but in many cases mixture with carbendazim is essential to achieve satisfactory results. See labels for details
- May be recommended for suppression or control of mildew in cereals depending on product and tank mix. See label for details

Restrictions
- Maximum number of treatments varies with crop and product used - check labels for details
- Check labels for minimum interval that must elapse between treatments
- On protected lettuce only 2 post-planting applications of mancozeb or of any combination of products containing EBDC fungicide (mancozeb, maneb, thiram, zineb) either as a spray or a dust are permitted within 2 wk of planting out and none thereafter.
- Avoid treating wet cereal crops or those suffering from drought or other stress
- Keep dry formulations away from fire and sparks
- Use dry formulations immediately. Do not store

SEE SECTION 3 FOR PRODUCTS ALSO REGISTERED

Crop-specific information
- Latest use: before early milk stage (GS 73) for cereals; before 6 true leaf stage and before 31 Dec for winter oilseed rape.
- HI potatoes 7 d; outdoor lettuce 14 d; protected lettuce 21 d; apples, blackcurrants, bulb onions, garlic, gooseberries, pears 4 wk; grapevines 9 wk
- Apply to potatoes before haulm meets across rows (usually mid-Jun) or at earlier blight warning, and repeat every 7-14 d depending on conditions and product used (see label)
- May be used on potatoes up to desiccation of haulm
- On oilseed rape apply as soon as disease develops between cotyledon and 5-leaf stage (GS 1,0-1,5)
- Apply to cereals from 4-leaf stage to before early milk stage (GS 71). Recommendations vary, see labels for details
- Treat winter oilseed rape before 6 true leaf stage (GS 1,6) and before 31 Dec

Environmental safety
- Dangerous for the environment
- Very toxic to aquatic organisms
- Harmful to fish or other aquatic life. Do not contaminate surface waters or ditches with chemical or used container
- Do not empty into drains

Hazard classification and safety precautions
Hazard H03 [10]; H04 [1-9, 11]; H11 [1-7, 10, 11]

Risk phrases R36 [7-9]; R37 [1-6, 8-11]; R38 [8, 9]; R42 [10]; R43 [1-6, 10, 11]; R50, R53a [1-7, 10, 11]

Operator protection A [1-11]; C [7-9]; D [2-4, 6, 8, 9]; U05a [1-11]; U08 [5, 7-9, 11]; U11 [7]; U13 [8, 9]; U14 [1-6, 11]; U19a [7-10]; U20a [1, 8]; U20b [2-7, 9, 11]

Environmental protection E13c [8, 9]; E15a [1-7, 10, 11]; E19b [10]; E34 [5, 7, 11]; E38 [1-7, 11]

Storage and disposal D01, D02, D09a [1-11]; D05 [3-9, 11]; D07, D10a [7]; D11a [1-6, 8-11]; D12a [1-7, 11]

303   mancozeb + metalaxyl-M

A systemic and protectant fungicide mixture

Products
Fubol Gold WG	Syngenta	64:4% w/w	WG	10184

Uses
- Blight in **potatoes**
- Downy mildew in **rhubarb** *(off-label)*, **salad onions** *(off-label)*
- Fungus diseases in **chives** *(off-label)*, **herbs (see appendix 6)** *(off-label)*, **lettuce** *(off-label)*, **parsley** *(off-label)*, **protected chives** *(off-label)*, **protected herbs (see appendix 6)** *(off-label)*, **protected lettuce** *(off-label)*, **protected parsley** *(off-label)*
- Phytophthora fruit rot in **apples** *(off-label - applied to orchard floor)*
- White blister in **cabbages** *(off-label)*, **protected radishes** *(off-label)*, **radishes** *(off-label)*

Specific Off-Label Approvals (SOLAs)
- **apples** *(applied to orchard floor) (OLA 011610) Dec 2008* [1]
- **cabbages, protected radishes, radishes** *(OLA 011610) Dec 2008* [1]
- **chives, herbs (see appendix 6), lettuce, parsley, protected chives, protected herbs (see appendix 6), protected lettuce, protected parsley** *(OLA 032142) Dec 2008* [1]
- **rhubarb** *(OLA 022044) Aug 2006* [1]
- **salad onions** *(OLA 032324) Dec 2008* [1]

Approval information
- Mancozeb and metalaxyl-M included in Annex I under EC Directive 91/414

Efficacy guidance
- Commence potato blight programme before risk of infection occurs as crops begin to meet along the rows and repeat every 10-14 d according to blight risk. Do not exceed a 14 d interval between sprays
- If infection risk conditions occur earlier than the above growth stage commence spraying potatoes immediately

FOR FULL CONDITIONS OF USE ALWAYS READ THE PRODUCT LABEL

- Complete the potato blight programme using a protectant fungicide starting no later than 10 d after the last phenylamide spray. At least 2 such sprays should be applied
- Monitor sprayed potato crops and stop using any phenylamide product if active blight is identified. Switch to a tin-based product programme within 7 d and continue up to haulm destruction or harvest
- To minimise the likelihood of development of resistance these products should be used in a planned Resistance Management strategy. See Section 5 for more information

Restrictions
- Maximum number of treatments or maximum total dose varies with crop. See labels

Crop-specific information
- Latest use: before end of active potato haulm growth or before end Aug, whichever is earlier; before transplanting protected herbs; before dormancy in yr of harvest for rhubarb
- HI 14 d for outdoor herbs; 21 d for protected herbs; 28 d for apples
- After treating early potatoes destroy and remove any remaining haulm after harvest to minimise blight pressure on neighbouring maincrop potatoes

Environmental safety
- Dangerous for the environment
- Very toxic to aquatic organisms
- Do not harvest crops for human consumption for at least 7 d after final application

Hazard classification and safety precautions
Hazard H04, H11
Risk phrases R37, R43, R50, R53a
Operator protection A; U05a, U08, U20b
Environmental protection E15a, E34, E38
Consumer protection C02 (7 d)
Storage and disposal D01, D02, D09a, D11a, D12a

304   mancozeb + propamocarb hydrochloride

A systemic and contact protectant fungicide for potato blight control

Products
Tattoo	Bayer CropScience	301.6:248 g/l	SC	07293

Uses
- Blight in **potatoes**

Approval information
- Mancozeb included in Annex I under EC Directive 91/414

Efficacy guidance
- Commence treatment as soon as there is risk of blight infection
- In the absence of a warning treatment should start before the crop meets within the row
- Repeat treatments every 10-14 d depending on degree of infection risk
- Irrigated crops should be regarded as at high risk and treated every 10 d
- Do not spray when rainfall is imminent and apply only to dry foliage
- To complete the spray programme after the end of Aug, make subsequent treatments up to haulm destruction with a protectant fungicide, preferably fentin based
- To minimise the likelihood of development of resistance these products should be used in a planned Resistance Management strategy. See Section 5 for more information

Restrictions
- Maximum number of treatments 5 per crop
- Do not use once blight has become readily visible

Crop-specific information
- HI potatoes 14 d

Environmental safety
- Dangerous for the environment
- Very toxic to aquatic organisms
- Harmful to fish or other aquatic life. Do not contaminate surface waters or ditches with chemical or used container

SEE SECTION 3 FOR PRODUCTS ALSO REGISTERED

Hazard classification and safety precautions
    Hazard H04, H11
    Risk phrases R43, R50, R53a
    Operator protection A, H; U05a, U08, U14, U19a, U20a
    Environmental protection E13c, E38
    Consumer protection C02 (14 d)
    Storage and disposal D01, D02, D05, D09a, D10b, D12a

305   mancozeb + zoxamide

A protectant fungicide mixture for potatoes

Products

1	Electis 75 WG	Dow	66.7:8.3% w/w	WG	11013
2	Roxam 75 WG	Dow	66.7:8.3% w/w	WG	11017
3	Unikat 75 WG	Landseer	66.7:8.3% w/w	WG	10567

Uses
- Blight in *potatoes* [1, 2]
- Downy mildew in *grapevines* [3]

Approval information
- Mancozeb included in Annex I under EC Directive 91/414

Efficacy guidance
- Apply as protectant spray on potatoes immediately risk of blight in district or as crops begin to meet along the rows and repeat every 7-14 d according to blight risk [1, 2]
- Do not use if potato blight present in crop. Products are not curative [1, 2]
- Spray irrigated potato crops as soon as possible after irrigation once the crop leaves are dry [1, 2]
- In the absence of an official warning commence treatment of grapevines at the 5-6 leaf stage and repeat at a minimum interval of 10 d [3]
- Increase spray volume with growth of grapevines but avoid excessive application [3]

Restrictions
- Maximum number of treatments 6 per yr for grapevines; 10 per crop for potatoes
- Do not use on grapevines for plant propagation [3]
- Consult processors before use on crops for juice production [3]

Crop-specific information
- HI 7 d for potatoes [1, 2]; 56 d for grapevines [3]

Environmental safety
- Dangerous for the environment
- Very toxic to aquatic organisms
- LERAP Category B
- Broadcast air-assisted LERAP (20 m) [3]
- Keep away from fire and sparks

Hazard classification and safety precautions
    Hazard H04, H11
    Risk phrases R37, R43, R50, R53a
    Operator protection A, H; U05a, U14, U20b
    Environmental protection E15a, E16b, E38 [1-3]; E16a [1, 2]; E17b [3] (20 m)
    Storage and disposal D01, D02, D05, D09a, D11a, D12a

FOR FULL CONDITIONS OF USE ALWAYS READ THE PRODUCT LABEL

306　MCPA

A translocated phenoxyacetic herbicide for cereals and grassland

See also 2,4-D + dichlorprop-P + MCPA + mecoprop-P
2,4-D + MCPA
2,4-DB + linuron + MCPA
2,4-DB + MCPA
bentazone + MCPA + MCPB
clopyralid + 2,4-D + MCPA
clopyralid + diflufenican + MCPA
clopyralid + fluroxypyr + MCPA
dicamba + dichlorprop-P + ferrous sulphate + MCPA
dicamba + dichlorprop-P + MCPA
dicamba + MCPA + mecoprop-P
dichlorprop-P + MCPA
dichlorprop-P + MCPA + mecoprop-P

Products

1	Agricorn 500 II	FCC	500 g/l	SL	09155
2	Agritox 50	Nufarm UK	500 g/l	SL	07400
3	Agritox Dry	Nufarm UK	80% w/w	SG	10554
4	Agroxone	Headland	500 g/l	SL	09947
5	Campbell's MCPA 50	United Phosphorus	500 g/l	SL	00381
6	Headland Spear	Headland	500 g/l	SL	07115
7	HY-MCPA	Agrichem	500 g/l	SL	06293
8	Tasker 75	Headland	750 g/l	SL	10544

Uses

- Annual dicotyledons in **amenity grass, managed amenity turf** [1-4, 6]; **established grassland, spring barley, spring oats, spring wheat, winter barley, winter oats, winter wheat** [1-8]; **grass seed crops** [1-4, 8]; **land not intended to bear vegetation** [2, 4]; **linseed** [1-4, 7, 8]; **newly sown grass** [4]; **rotational grassland** [8]; **spring rye** [1-3]; **undersown barley** (red clover or grass), **undersown wheat** (red clover or grass) [4, 8]; **undersown spring cereals, undersown winter cereals** [1, 7]; **winter rye** [1-3, 5, 6]
- Buttercups in **amenity grass, managed amenity turf** [3, 6, 7]; **established grassland** [5-7]
- Charlock in **established grassland** [1-4, 8]; **grass seed crops** [8]; **linseed, rotational grassland** [7, 8]; **newly sown grass** [4]; **spring barley, spring oats, spring wheat, winter barley, winter oats, winter wheat** [1-8]; **spring rye** [1-3]; **undersown barley** (red clover or grass), **undersown wheat** (red clover or grass) [4, 8]; **winter rye** [1-3, 5, 6]
- Dandelions in **amenity grass, managed amenity turf** [3]; **established grassland** [5, 7]
- Docks in **established grassland** [5-7]
- Fat hen in **amenity grass, hard surfaces, managed amenity turf, natural surfaces not intended to bear vegetation, permeable surfaces overlying soil** [7]; **established grassland** [1-4, 8]; **grass seed crops** [8]; **linseed, rotational grassland** [7, 8]; **newly sown grass** [4]; **spring barley, spring oats, spring wheat, winter barley, winter oats, winter wheat** [1-8]; **spring rye** [1-3]; **undersown barley** (red clover or grass), **undersown wheat** (red clover or grass) [4, 8]; **winter rye** [1-3, 5, 6]
- Hemp-nettle in **established grassland** [1-4, 8]; **grass seed crops, rotational grassland** [8]; **linseed** [7, 8]; **newly sown grass** [4]; **spring barley, spring oats, spring wheat, winter barley, winter oats, winter wheat** [1-8]; **spring rye** [1-3]; **undersown barley** (red clover or grass), **undersown wheat** (red clover or grass) [4, 8]; **winter rye** [1-3, 5, 6]
- Perennial dicotyledons in **amenity grass, managed amenity turf** [1-4, 6]; **established grassland, spring barley, spring oats, spring wheat, winter barley, winter oats, winter wheat** [1-8]; **grass seed crops, linseed, rotational grassland** [8]; **land not intended to bear vegetation** [2, 4]; **spring rye** [1-3]; **undersown barley** (red clover or grass), **undersown wheat** (red clover or grass) [4, 8]; **winter rye** [1-3, 5, 6]
- Plantains in **amenity grass, established grassland, grass seed crops, managed amenity turf** [7]
- Ragwort in **established grassland** [6]
- Rushes in **established grassland** [6]

- Stinging nettle in **hard surfaces, natural surfaces not intended to bear vegetation, permeable surfaces overlying soil** [7]; **industrial sites, land not intended to bear vegetation** [6]
- Thistles in **established grassland** [5-7]; **hard surfaces, natural surfaces not intended to bear vegetation, permeable surfaces overlying soil** [7]; **industrial sites, land not intended to bear vegetation** [6]
- Wild radish in **established grassland** [1-4, 8]; **grass seed crops, rotational grassland** [8]; **linseed** [7, 8]; **newly sown grass** [4]; **spring barley, spring oats, spring wheat, winter barley, winter oats, winter wheat** [1-8]; **spring rye** [1-3]; **undersown barley** (red clover or grass), **undersown wheat** (red clover or grass) [4, 8]; **winter rye** [1-3, 5, 6]

Approval information
- MCPA included in Annex I under EC Directive 91/414
- Accepted by BBPA for use on malting barley

Efficacy guidance
- Best results achieved by application to weeds in seedling to young plant stage under good growing conditions when crop growing actively
- Spray perennial weeds in grassland before flowering. Most susceptible growth stage varies between species. See label for details
- Do not spray during cold weather, drought, if rain or frost expected or if crop wet

Restrictions
- Maximum number of treatments normally 1 per crop or yr except grass (2 per yr) for some products. See label
- Do not treat grass within 3 mth of germination and preferably not in the first yr of a direct sown ley or after reseeding
- Do not use on cereals before undersowing
- Do not roll, harrow or graze for a few days before or after spraying; see label
- Do not use on grassland where clovers are an important part of the sward
- Do not use on any crop suffering from stress or herbicide damage
- Avoid spray drift onto nearby susceptible crops

Crop-specific information
- Latest use: before 1st node detectable (GS 31) for cereals; 4-6 wk before heading for grass seed crops; before crop 15-25 cm high for linseed
- Apply to winter cereals in spring from fully tillered, leaf sheath erect stage to before first node detectable (GS 31)
- Apply to spring barley and wheat from 5-leaves unfolded (GS 15), to oats from 1-leaf unfolded (GS 11) to before first node detectable (GS 31)
- Apply to cereals undersown with grass after grass has 2-3 leaves unfolded
- Recommendations for crops undersown with legumes vary. Red clover may withstand low doses after 2-trifoliate leaf stage, especially if shielded by taller weeds but white clover is more sensitive. See label for details
- Apply to grass seed crops from 2-3 leaf stage to 5 wk before head emergence
- Temporary wilting may occur on linseed but without long term effects

Following crops guidance
- Do not direct drill brassicas or legumes within 6 wk of spraying grassland

Environmental safety
- Harmful to aquatic organisms
- MCPA is active at low concentrations. Take extreme care to avoid drift onto neighbouring crops, especially beet crops, brassicas, most market garden crops including lettuce and tomatoes under glass, pears and vines
- Harmful to fish or other aquatic life. Do not contaminate surface waters or ditches with chemical or used container [1]
- Keep livestock out of treated areas until foliage of poisonous weeds such as ragwort has died and become unpalatable

Hazard classification and safety precautions
Hazard H03
Risk phrases R20, R21 [4, 5, 7]; R22a [4-8]; R41 [1-3, 5-8]; R52 [1, 6]
Operator protection A, C [1-7]; U05a, U08, U11 [1-8]; U14 [7]; U15 [6, 7]; U20a [1, 5]; U20b [2-4, 6-8]
Environmental protection E07a, E15a, E34 [1-8]; E38 [6]

FOR FULL CONDITIONS OF USE ALWAYS READ THE PRODUCT LABEL

Storage and disposal D01, D02, D05, D09a [1-8]; D10a [1-5, 7, 8]; D10b [6]
Medical advice M03 [1-8]; M05a [5]

307   MCPA + MCPB

A translocated herbicide for undersown cereals, grassland and various legumes

Products

1 Bellmac Plus	United Phosphorus	38:262 g/l	SL	07521
2 Impetus	Headland	25:275 g/l	SL	11021
3 Trifolex-Tra	BASF	34:216 g/l	SL	10396
4 Tropotox Plus	Nufarm UK	37.5:262.5 g/l	SL	11142

Uses
* Annual dicotyledons in *all cereals, clover seed crops, direct-sown seedling clovers, dredge corn containing peas, established leys* [4]; *combining peas, red clover, spring barley, spring oats, spring wheat, vining peas, white clover, winter barley, winter oats, winter wheat* [2]; *established grassland, peas* [3]; *permanent pasture* [2, 4]; *rotational grassland* [1-3]; *sainfoin* [1, 4]; *undersown spring cereals, undersown winter cereals* [1-4]
* Perennial dicotyledons in *all cereals, clover seed crops, direct-sown seedling clovers, dredge corn containing peas, established leys* [4]; *combining peas, red clover, spring barley, spring oats, spring wheat, vining peas, white clover, winter barley, winter oats, winter wheat* [2]; *established grassland, peas* [3]; *permanent pasture* [2, 4]; *rotational grassland* [1-3]; *sainfoin* [1, 4]; *undersown spring cereals, undersown winter cereals* [1-4]

Approval information
* MCPA and MCPB included in Annex I under EC Directive 91/414
* Accepted by BBPA for use on malting barley

Efficacy guidance
* Best results achieved by application to weeds in seedling to young plant stage under good growing conditions when crop growing actively. Spray perennials when adequate leaf surface before flowering. Retreatment often needed in following year
* Spray leys and sainfoin before crop provides cover for weeds
* Rain, cold or drought may reduce effectiveness

Restrictions
* Maximum number of treatments 1 per crop or yr
* Do not spray clovers for seed
* Do not spray peas [1]
* Do not roll or harrow for a few days before or after spraying

Crop-specific information
* Latest use: before first node detectable (GS 31) for cereals; before flower buds visible for peas
* Apply to cereals from 2-expanded leaf stage to before jointing (GS 12-30) and, where undersown, after 1-trifoliate leaf stage of clover
* Apply to direct-sown seedling clover after 1-trifoliate leaf stage
* Apply to mature white clover for fodder at any stage. Do not spray red clover after flower stalk has begun to form
* Apply to sainfoin after first trifoliate leaf stage

Environmental safety
* Harmful to aquatic organisms
* Harmful to fish or other aquatic life. Do not contaminate surface waters or ditches with chemical or used container
* Keep livestock out of treated areas until foliage of any poisonous weeds such as ragwort has died and become unpalatable
* Take extreme care to avoid drift onto neighbouring crops, especially beet crops, brassicas, most market garden crops including lettuce and tomatoes under glass, pears and vines

Hazard classification and safety precautions
Hazard H03
Risk phrases R20, R21 [2]; R22a [1-4]; R38, R53a [1]; R41, R52 [1, 2]
Operator protection A, C [2]; U05a [2-4]; U08, U20b [1-4]; U11, U14, U15 [1, 2]; U19a [3]
Environmental protection E07a, E34 [1-4]; E13c [1]; E15a [2-4]

SEE SECTION 3 FOR PRODUCTS ALSO REGISTERED

Storage and disposal D01 [2-4]; D02 [2, 3]; D05, D09a, D10b [1-4]
Medical advice M03 [2]; M05a [1, 2, 4]

308   MCPA + mecoprop-P

A translocated selective herbicide for amenity grass

Products

1	Cleanrun Pro	Scotts	0.49:0.29% w/w	GR	12083
2	Greenmaster Extra	Scotts	0.49:0.29 % w/w	GR	11563

Uses
- Annual dicotyledons in *managed amenity turf*
- Perennial dicotyledons in *managed amenity turf*

Approval information
- MCPA and mecoprop-P included in Annex I under EC Directive 91/414

Efficacy guidance
- Apply from Apr to Sep, when weeds growing actively and have large leaf area available for chemical absorption

Restrictions
- Maximum total dose 105 g/sq m per yr. The total amount of mecoprop-P applied in a single yr must not exceed the maximum total dose approved for any single product for use on turf
- Avoid contact with cultivated plants
- Do not use first 4 mowings as compost or mulch unless composted for 6 mth
- Do not treat newly sown or turfed areas for at least 6 mth
- Do not reseed bare patches for 8 wk after treatment
- Do not apply when heavy rain expected or during prolonged drought. Irrigate after 1-2 d unless rain has fallen
- Do not mow within 2-3 d of treatment
- Treat areas planted with bulbs only after the foliage has died down
- Avoid walking on treated areas until it has rained or irrigation has been applied

Crop-specific information
- Granules contain NPK fertilizer to encourage grass growth

Environmental safety
- Take extreme care to avoid drift onto neighbouring crops, especially beet crops, brassicas, most market garden crops including lettuce and tomatoes under glass, pears and vines
- Harmful to fish or other aquatic life. Do not contaminate surface waters or ditches with chemical or used container
- Keep livestock out of treated areas for at least 2 wk and until foliage of any poisonous weeds such as ragwort has died and become unpalatable

Hazard classification and safety precautions
Operator protection A, C, H, M; U20b
Environmental protection E07a, E13c, E19b
Storage and disposal D01, D09a, D12a
Medical advice M05a

309   MCPA + mecoprop-P + ferrous sulphate

A herbicide/fertilizer combination for moss and weed control in turf

Products

Renovator Pro	Scotts	0.49:0.29:16.3% w/w	GR	12204

Uses
- Annual dicotyledons in *managed amenity turf*
- Moss in *managed amenity turf*
- Perennial dicotyledons in *managed amenity turf*

Approval information
- MCPA and mecoprop-P included in Annex I under EC Directive 91/414

FOR FULL CONDITIONS OF USE ALWAYS READ THE PRODUCT LABEL

Efficacy guidance
- Apply from Apr to Sep when weeds are growing
- For best results apply when light showers or heavy dews are expected
- Apply with a suitable calibrated fertilizer distributor
- Retreatment may be necessary after 6 wk if weeds or moss persist
- For best control of moss scarify vigorously after 2 wk to remove dead moss
- Where regrowth of moss or weeds occurs a repeat treatment may be made after 6 wk
- Avoid treatment of wet grass or during drought. If no rain falls within 48 h water in thoroughly

Restrictions
- Maximum number of treatments 3 per yr
- Do not treat new turf until established for 6 mth
- The first 4 mowings after treatment should not be used to mulch cultivated plants unless composted at least 6 mth
- Avoid walking on treated areas until it has rained or they have been watered
- Do not re-seed or turf within 8 wk of last treatment
- Do not cut grass for at least 3 d before and at least 4 d after treatment
- Do not apply during freezing conditions or when rain imminent

Environmental safety
- Keep livestock out of treated areas for at least 2 wk following treatment and until poisonous weeds, such as ragwort, have died down and become unpalatable
- Harmful to fish or other aquatic life. Do not contaminate surface waters or ditches with chemical or used container
- Do not empty into drains

Hazard classification and safety precautions
Operator protection U20b
Environmental protection E07a, E13c, E19b
Storage and disposal D01, D09a, D12a
Medical advice M05a

310   MCPB

A translocated phenoxybutyric herbicide

See also bentazone + MCPA + MCPB
bentazone + MCPB
MCPA + MCPB

Products
1	Bellmac Straight	United Phosphorus	400 g/l	SL	07522
2	Butoxone	Headland	400 g/l	SL	10501
3	Tropotox	Nufarm UK	400 g/l	SL	11141

Uses
- Annual dicotyledons in *blackberries, combining peas, gooseberries, loganberries, permanent pasture, raspberries, spring barley, spring oats, spring wheat, vining peas, white clover* (seed crops), *winter barley, winter oats, winter wheat* [2]; *blackcurrants* [1-3]; *clover seed crops, peas* [1, 3]; *rotational grassland, undersown spring cereals, undersown winter cereals* [1, 2]
- Perennial dicotyledons in *blackberries, combining peas, gooseberries, loganberries, permanent pasture, raspberries, spring barley, spring oats, spring wheat, vining peas, white clover* (seed crops), *winter barley, winter oats, winter wheat* [2]; *blackcurrants* [1-3]; *clover seed crops, peas* [1, 3]; *rotational grassland, undersown spring cereals, undersown winter cereals* [1, 2]

Approval information
- MCPB included in Annex I under EC Directive 91/414
- Accepted by BBPA for use on malting barley

Efficacy guidance
- Best results achieved by spraying young seedling weeds in good growing conditions

SEE SECTION 3 FOR PRODUCTS ALSO REGISTERED

- Best results on perennials by spraying before flowering
- Effectiveness may be reduced by rain within 12 h, by very cold or dry conditions

Restrictions

- Maximum number of treatments 1 per crop or yr. One label allows two treatments on blackcurrants [3]
- Do not roll or harrow for 7-10 d before or after treatment (check label)

Crop-specific information

- Latest use: first node detectable stage (GS 31) for cereals; before flower buds appear in terminal leaf (GS 201) for peas; before flower buds form for clover; before weeds damaged by frost for cane and bush fruit
- Apply to undersown cereals from 2-leaves unfolded to first node detectable (GS 12-31), and after first trifoliate leaf stage of clover
- Red clover seedlings may be temporarily damaged but later growth is normal
- Apply to white clover seed crops in Mar to early Apr, not after mid-May, and allow 3 wk before cutting and closing up for seed
- Apply to peas from 3-6 leaf stage but before flower bud detectable (GS 103-201). Consult PGRO (see Appendix 2) or label for information on susceptibility of cultivars. Do not treat peas grown for seed [2]
- Do not use on leguminous crops not mentioned on the label
- Apply to cane and bush fruit after harvest and after shoot growth ceased but before weeds are damaged by frost, usually in late Aug or Sep; direct spray onto weeds as far as possible

Environmental safety

- Harmful to aquatic organisms
- Harmful to fish or other aquatic life. Do not contaminate surface waters or ditches with chemical or used container
- Keep livestock out of treated areas until foliage of any poisonous weeds such as ragwort has died and become unpalatable
- Take extreme care to avoid drift onto neighbouring sensitive crops

Hazard classification and safety precautions

Hazard H03

Risk phrases R22a [1-3]; R38, R41 [1, 2]; R52, R53a [1]

Operator protection A, C; U05a, U08, U20b [1-3]; U11, U14, U15 [1, 2]; U19a [3]

Environmental protection E07a, E34 [1-3]; E13c [1, 2]; E15a [3]

Storage and disposal D01, D02, D09a, D10b [1-3]; D05 [1, 2]

Medical advice M03, M05a

311 mecoprop-P

A translocated phenoxypropionic herbicide for cereals and grassland

See also 2,4-D + dichlorprop-P + MCPA + mecoprop-P
2,4-D + mecoprop-P
bromoxynil + ioxynil + mecoprop-P
carfentrazone-ethyl + mecoprop-P
dicamba + MCPA + mecoprop-P
dicamba + mecoprop-P
dichlorprop-P + MCPA + mecoprop-P
fluroxypyr + mecoprop-P
MCPA + mecoprop-P
MCPA + mecoprop-P + ferrous sulphate

Products

1	Clenecorn Super	FCC	600 g/l	SL	09818
2	Clovotox	Bayer Environ.	142.5 g/l	SL	09928
3	Compitox Plus	Nufarm UK	600 g/l	SL	10077
4	Duplosan KV	Nufarm UK	600 g/l	SL	12073
5	Isomec	Nufarm UK	600 g/l	SL	11156
6	Landgold Mecoprop-P 600	Landgold	600 g/l	SL	09014
7	Landgold Mecoprop-P 600	Teliton	600 g/l	SL	12122
8	Optica	Headland	600 g/l	SL	09963

FOR FULL CONDITIONS OF USE ALWAYS READ THE PRODUCT LABEL

Uses

- Annual dicotyledons in *established grassland* [6-8]; *grass seed crops, young leys* [1, 3-5, 8]; *grassland* [1, 3-5]; *managed amenity turf* [2, 3, 5, 8]; *rotational grassland* [6, 7]; *spring barley, spring oats, spring wheat, winter barley, winter oats, winter wheat* [1, 3-8]; *undersown spring cereals, undersown winter cereals* [1]
- Chickweed in *established grassland* [6-8]; *grass seed crops, young leys* [1, 3-5, 8]; *grassland* [1, 3-5]; *managed amenity turf* [2, 3, 5, 8]; *rotational grassland* [6, 7]; *spring barley, spring oats, spring wheat, winter barley, winter oats, winter wheat* [1, 3-8]; *undersown spring cereals, undersown winter cereals* [1]
- Cleavers in *established grassland* [6-8]; *grass seed crops, young leys* [1, 3-5, 8]; *grassland* [1, 3-5]; *managed amenity turf* [3, 5, 8]; *rotational grassland* [6, 7]; *spring barley, spring oats, spring wheat, winter barley, winter oats, winter wheat* [1, 3-8]; *undersown spring cereals, undersown winter cereals* [1]
- Clovers in *managed amenity turf* [2]
- Perennial dicotyledons in *established grassland, rotational grassland* [6, 7]; *grass seed crops, grassland, young leys* [3-5]; *managed amenity turf* [2, 3, 5]; *spring barley, spring oats, spring wheat, winter barley, winter oats, winter wheat* [3-7]

Approval information

- Mecoprop-P included in Annex I under EC Directive 91/414
- Accepted by BBPA for use on malting barley

Efficacy guidance

- Best results achieved by application to seedling weeds which have not been frost hardened, when soil warm and moist and expected to remain so for several days

Restrictions

- Maximum number of treatments normally 1 per crop for spring cereals and newly sown grass; 2 per crop for winter cereals and established grass. Check labels for details
- The total amount of mecoprop-P applied in a single yr must not exceed the maximum total dose approved for any single product for the crop/situation
- Do not spray cereals undersown with clovers or legumes or to be undersown with legumes or grasses
- Do not spray grass seed crops within 5 wk of seed head emergence
- Do not spray crops suffering from herbicide damage or physical stress
- Do not spray during cold weather, periods of drought, if rain or frost expected or if crop wet
- Do not roll or harrow for 7 d before or after treatment

Crop-specific information

- Latest use: generally before 1st node detectable (GS 31) for spring cereals and before 3rd node detectable (GS 33) for winter cereals, but individual labels vary; 5 wk before emergence of seed head for grass seed crops
- Spray winter cereals from 1 leaf stage in autumn up to and including first node detectable in spring (GS 10-31) or up to second node detectable (GS 32) if necessary. Apply to spring cereals from first fully expanded leaf stage (GS 11) but before first node detectable (GS 31)
- Spray cereals undersown with grass after grass starts to tiller
- Spray newly sown grass leys when grasses have at least 3 fully expanded leaves and have begun to tiller. Any clovers will be damaged

Environmental safety

- Harmful to aquatic organisms
- Harmful to fish or other aquatic life. Do not contaminate surface waters or ditches with chemical or used container
- Keep livestock out of treated areas for at least 2 wk and until foliage of any poisonous weed, such as ragwort, has died and become unpalatable
- Take extreme care to avoid drift onto neighbouring crops, especially beet crops, brassicas, most market garden crops including lettuce and tomatoes under glass, pears and vines

Hazard classification and safety precautions

Hazard H03 [1, 3-8]; H04 [2]

Risk phrases R21 [6, 7]; R22a [1, 3-8]; R36 [2, 6, 7]; R38 [4, 6-8]; R41 [1, 3-5, 8]; R43 [1, 3, 5]; R52 [8]

Operator protection A, C [1-8]; B [1]; H [2, 4, 8]; M [1, 4, 8]; U05a, U08, U20b [1-8]; U11 [1, 3-5]; U14 [1, 3, 5]; U15 [8]

SEE SECTION 3 FOR PRODUCTS ALSO REGISTERED

Environmental protection E07a [1-8]; E13c [1, 2, 6-8]; E15a [3-5]; E34 [1, 3-8]; E38 [8]
Storage and disposal D01, D02, D05 [1-8]; D08 [1, 5]; D09a [1-7]; D10a [3, 6, 7]; D10b [1, 5, 8]; D10c [2, 4]
Medical advice M03 [1, 3-8]; M05a [4]

312   mepanipyrim

An anilinopyrimidine fungicide for use in horticulture

Products

Frupica SC	Certis	450 g/l	SC	12067

Uses
• Botrytis in **courgettes** *(off-label)*, **protected strawberries**, **strawberries**

Specific Off-Label Approvals (SOLAs)
• **courgettes** *(OLA 051383) Jul 2007* [1]

Approval information
• Mepanipyrim included in Annex I under Directive 91/414

Efficacy guidance
• Product is protectant and should be applied as a preventative spray when conditions favourable for Botrytis development occur
• To maintain Botrytis control use as part of a programme with other fungicides that control the disease
• To minimise the possibility of development of resistance adopt resistance management procedures by using products from different chemical groups as part of a mixed spray programme

Restrictions
• Maximum number of treatments 2 per crop (including other anilinopyrimidine products)
• Consult processor before use on crops for processing
• Use spray mixture immediately after preparation

Crop-specific information
• HI 3 d

Environmental safety
• Dangerous for the environment
• Very toxic to aquatic organisms
• LERAP Category B

Hazard classification and safety precautions
Hazard H11
Risk phrases R50, R53a
Operator protection A, H; U20c
Environmental protection E16a, E16b, E34
Storage and disposal D01, D02, D05, D10c, D11a, D12b

313   mepiquat chloride

A quaternary ammonium plant growth regulator available only in mixtures

See also 2-chloroethylphosphonic acid + mepiquat chloride
chlormequat + 2-chloroethylphosphonic acid + mepiquat chloride
chlormequat + mepiquat chloride

314   mesosulfuron-methyl

A sulfonyl urea herbicide for cereals available in mixtures

See also iodosulfuron-methyl-sodium + mesosulfuron-methyl

FOR FULL CONDITIONS OF USE ALWAYS READ THE PRODUCT LABEL

315   mesotrione

A foliar applied triketone herbicide for maize

Products

Callisto	Syngenta	100 g/l	SC	12323

Uses
- Annual dicotyledons in *forage maize*
- Volunteer oilseed rape in *forage maize*

Efficacy guidance
- Best results obtained from treatment of young actively growing weed seedlings in the presence of adequate soil moisture
- Treatment in poor growing conditions or in dry soil may give less reliable control
- Activity is mostly by foliar uptake with some soil uptake
- To minimise the possible development of resistance where continuous maize is grown the product should not be used for more than two consecutive seasons

Restrictions
- Maximum number of treatments 1 per crop of forage maize
- Do not use on seed crops or on sweetcorn varieties
- Do not spray when crop foliage wet or when excessive rainfall is expected to follow application
- Do not treat crops suffering from stress from cold or drought conditions, or when wide temperature fluctuations are anticipated

Crop-specific information
- Latest use: 8 leaves unfolded stage (GS 18) for forage maize
- Treatment under adverse conditions may cause mild to moderate chlorosis. The effect is transient and does not affect yield

Following crops guidance
- Winter wheat, durum wheat, winter barley or ryegrass may follow a normally harvested treated crop of maize. Oilseed rape may be sown provided it is preceded by deep ploughing to more than 15 cm
- In the spring following application only forage maize, ryegrass, spring wheat or spring barley may be sown
- In the event of crop failure maize may be re-seeded immediately. Some slight crop effects may be seen soon after emergence but these are normally transient

Environmental safety
- Dangerous for the environment
- Very toxic to aquatic organisms
- LERAP Category B
- Take extreme care to avoid drift onto all plants outside the target area

Hazard classification and safety precautions

Hazard H04, H11

Risk phrases R36, R50, R53a

Operator protection A, C; U05a, U09a, U20b

Environmental protection E15b, E16a, E16b, E38

Storage and disposal D01, D02, D05, D09a, D10c, D11a, D12a

316   mesotrione + terbuthylazine

A foliar and soil acting herbicide mixture for maize

Products

Calaris	Syngenta	70:330 g/l	SC	12405

Uses
- Annual dicotyledons in *forage maize*
- Annual meadow grass in *forage maize*

SEE SECTION 3 FOR PRODUCTS ALSO REGISTERED

Efficacy guidance

- Best results obtained from treatment of young actively growing weed seedlings in the presence of adequate soil moisture
- Treatment in poor growing conditions or in dry soil may give less reliable control
- Residual weed control is reduced on soils with more than 10% organic matter
- To minimise the possible development of resistance where continuous maize is grown the product should not be used for more than two consecutive seasons

Restrictions

- Maximum number of treatments 1 per crop of forage maize
- Do not use on seed crops or on sweetcorn varieties
- Do not spray when crop foliage wet or when excessive rainfall is expected to follow application
- Do not treat crops suffering from stress from cold or drought conditions, or when wide temperature fluctuations are anticipated
- Do not apply on Sands or Very Light soils

Crop-specific information

- Latest use: 8 leaves unfolded stage (GS 18) for forage maize
- Treatment under adverse conditions may cause mild to moderate chlorosis. The effect is transient and does not affect yield

Following crops guidance

- Winter wheat, durum wheat, winter barley or ryegrass may follow a normally harvested treated crop of maize. Oilseed rape may be sown provided it is preceded by deep ploughing to more than 15 cm
- In the spring following application forage maize, ryegrass, spring wheat or spring barley may be sown
- Spinach, beet crops, peas, beans, lettuce and cabbages must not be sown in the yr following application
- In the event of crop failure maize may be re-seeded immediately. Some slight crop effects may be seen soon after emergence but these are normally transient

Environmental safety

- Dangerous for the environment
- Very toxic to aquatic organisms
- LERAP Category B
- Take extreme care to avoid drift onto all plants outside the target area

Hazard classification and safety precautions

Hazard H03, H11

Risk phrases R22a, R50, R53a

Operator protection A; U05a

Environmental protection E15a, E16a, E16b, E34, E38

Storage and disposal D01, D02, D05, D09a, D10c, D12a

317 metalaxyl-M

A phenylamide systemic fungicide

See also chlorothalonil + metalaxyl-M
cymoxanil + fludioxonil + metalaxyl-M
mancozeb + metalaxyl-M

Products

1	SL 567A	Syngenta	480 g/l	EC	10594
2	SL 567A	Syngenta	465.2 g/l	SL	12380
3	Subdue	Fargro	480 g/l	EC	12358
4	Subdue	Fargro	465 g/l	SL	12503

Uses

- Cavity spot in **carrots**, **parsnips** *(off-label)* [1, 2]
- Crown rot in **water lilies** *(off-label)* [1, 2]

- Downy mildew in **grapevines** *(off-label)*, **hops** *(off-label)*, **leaf spinach** *(off-label)*, **protected cucumbers** *(off-label)*, **protected spinach** *(off-label)*, **salad onions** *(off-label)*, **spinach beet** *(off-label)* [1, 2]
- Phytophthora in **asparagus** *(off-label)* [1, 2]
- Phytophthora root rot in **ornamental plant production**, **protected ornamentals** [3, 4]; **raspberries** *(off-label)* [1, 2]
- Pythium in **ornamental plant production**, **protected ornamentals** [3, 4]; **watercress** *(off-label)* [1, 2]
- Root malformation disorder in **red beet** *(off-label)* [1, 2]
- Storage rots in **apples** *(off-label)*, **cabbages** *(off-label)*, **pears** *(off-label)* [1, 2]
- White blister in **horseradish** *(off-label)* [1, 2]

Specific Off-Label Approvals (SOLAs)
- **apples**, **cabbages**, **pears** *(OLA 040723) Apr 2007* [1]
- **apples**, **cabbages**, **pears** *(OLA 051514) Apr 2007* [2]
- **asparagus** *(OLA 040611) Sep 2012* [1]
- **asparagus** *(OLA 051502) Apr 2008* [2]
- **grapevines** *(OLA 040949) Sep 2012* [1]
- **grapevines** *(OLA 051504) Apr 2008* [2]
- **hops** *(OLA 050144) Sep 2012* [1]
- **hops** *(OLA 051500) Apr 2008* [2]
- **horseradish** *(OLA 040722) Sep 2012* [1]
- **horseradish** *(OLA 051499) Apr 2008* [2]
- **leaf spinach, protected spinach, spinach beet** *(OLA 040948) Sep 2012* [1]
- **leaf spinach, protected spinach, spinach beet** *(OLA 051507) Apr 2008* [2]
- **parsnips** *(OLA 040613) Sep 2012* [1]
- **parsnips** *(OLA 051508) Apr 2008* [2]
- **protected cucumbers** *(OLA 041136) Sep 2012* [1]
- **protected cucumbers** *(OLA 051503) Apr 2008* [2]
- **raspberries, salad onions** *(OLA 040720) Apr 2007* [1]
- **raspberries, salad onions** *(OLA 051515) Apr 2007* [2]
- **red beet** *(OLA 040835) Sep 2012* [1]
- **red beet** *(OLA 051307) Apr 2008* [2]
- **water lilies** *(OLA 040721) Sep 2012* [1]
- **water lilies** *(OLA 051501) Apr 2008* [2]
- **watercress** *(OLA 040719) Apr 2007* [1]
- **watercress** *(OLA 051516) Apr 2007* [2]

Approval information
- Metalaxyl-M included in Annex I under EC Directive 91/414
- Accepted by BBPA for use on hops

Efficacy guidance
- Best results achieved when applied to damp soil or potting media
- Efficacy may be reduced in prolonged dry weather. Irrigation immediately before or after application may be necessary
- Results may not be satisfactory on soils with high organic matter content
- Control of cavity spot on carrots overwintered in the ground or lifted in winter may be lower than expected
- Use in an integrated pest management strategy and, where appropriate, alternate with products from different chemical groups
- Product should ideally be used preventatively and the number of phenylamide applications should be limited to 1-2 consecutive treatments depending on the crop

Restrictions
- Maximum number of treatments or maximum total dose varies with crop and nature of treatment. See label for details
- Do not use where carrots have been grown on the same site within the previous eight yrs
- Consult before use on crops intended for processing
- Do not re-use potting media from treated plants for subsequent crops [3]
- Disinfect pots thoroughly prior to re-use [3]

SECTION 2

SEE SECTION 3 FOR PRODUCTS ALSO REGISTERED

Crop-specific information
- Latest use: 4 wk before removal of apples or pears from store; 6 wk after drilling for carrots; 10 wk before sale or supply (after rinsing) of water lilies
- HI raspberries 3 mth; horseradish 8 wk; grapevines 30 d; red beet 28 d; hops, leaf spinach, protected spinach, salad onions, spinach beet 14 d; asparagus 7 d; protected cucumbers 2 d
- Because of the large number species and ornamental cultivars susceptibility should be checked before large scale treatment [3]
- Treatment of *Viburnum* and *Prunus* species not recommended [3]

Environmental safety
- Dangerous for the environment
- Harmful to aquatic organisms

Hazard classification and safety precautions
Hazard H03 [1-4]; H11 [3, 4]

Risk phrases R22a, R52, R53a [1-4]; R36, R43 [1, 3, 4]; R37 [2]

Operator protection A [1-4]; B [4]; C [1, 3, 4]; H [1, 3]; U02a, U04a, U05a, U10, U20b [1-4]; U19a [2]

Environmental protection E15a, E34, E38

Storage and disposal D01, D02, D10c, D12a [1-4]; D05, D07 [2-4]; D09a [1, 2]

Medical advice M03

318   metaldehyde

A molluscicide bait for controlling slugs and snails

Products

1	Allure	Chiltern	1.5% w/w	PT	11089
2	Appeal	Chiltern	1.5% w/w	PT	12022
3	Attract	Chiltern	1.5% w/w	PT	12023
4	Bristol Blues	Doff Portland	6% w/w	PT	11798
5	CDP Minis	De Sangosse	5% w/w	PT	12374
6	Chiltern Blues	Chiltern	6% w/w	PT	10071
7	Chiltern Hundreds	Chiltern	3% w/w	PT	10072
8	Condor	Doff Portland	6% w/w	PT	11791
9	Dixie 6	Greencrop	6% w/w	PT	11790
10	Doff Horticultural Slug Killer Blue Mini Pellets	Doff Portland	3% w/w	PT	11463
11	Entice	De Sangosse	4% w/w	RB	11096
12	Escar-Go 6	Chiltern	6% w/w	PT	06076
13	ESP	De Sangosse	6% w/w	PT	10938
14	Hardy	Chiltern	6% w/w	PT	06948
15	Lincoln VI	Doff Portland	6% w/w	PT	11793
16	Luxan Deal	Luxan	6% w/w	PT	11101
17	Luxan Metaldehyde 5	Luxan	5% w/w	PT	12401
18	Luxan Red 5	Luxan	5% w/w	PT	12390
19	Luxan Trigger 5	Luxan	5% w/w	PT	12389
20	Metarex Amba	De Sangosse	5% w/w	PT	11157
21	Metarex Green	De Sangosse	5% w/w	PT	10113
22	Metarex RG	De Sangosse	5% w/w	PT	10115
23	Molotov	Chiltern	3% w/w	PT	08295
24	Optimol	De Sangosse	4% w/w	PT	10939
25	Pathfinder Excel	Barclay	5% w/w	PT	12501
26	Regel	De Sangosse	5% w/w	PT	10114
27	Rowent	Chiltern	6% w/w	PT	12024
28	Steam	Chiltern	3% w/w	PT	12473
29	Super Six	Doff Portland	6% w/w	PT	11794

Uses
- Slugs in *all edible crops*, *all non-edible crops* [1-29]; *amenity grass*, *managed amenity turf* [17-19]; *cultivated land/soil* [1-3, 6, 7, 10, 12, 14, 23, 25, 27, 28]; *natural surfaces not intended to bear vegetation* [4, 5, 8, 9, 11, 15, 17-19, 24, 29]; *ornamental plant production* [20-22, 26]; *protected crops* [1-3, 6, 7, 12, 14, 20-23, 25-28]
- Snails in *all edible crops*, *all non-edible crops* [1-29]; *amenity grass*, *managed amenity turf* [17-19]; *cultivated land/soil* [1-3, 6, 7, 10, 12, 14, 23, 25, 27, 28]; *natural surfaces not intended*

to bear vegetation [4, 5, 8, 9, 11, 15, 17-19, 24, 29]; **ornamental plant production** [20-22, 26]; **protected crops** [1-3, 6, 7, 12, 14, 20-23, 25-28]

Approval information
- Approved for aerial application on all edible crops, all non-edible crops [9, 10, 16]; on natural surfaces not intended to be cultivated [9]; on bare soil [10]. See notes in Section 5
- Accepted by BBPA for use on malting barley and hops

Efficacy guidance
- Apply pellets by hand, fiddle drill, fertilizer distributor, by air (check label) or in admixture with seed. See labels for rates and timing.
- Best results achieved from an even spread of granules applied during mild, damp weather when slugs and snails most active. May be applied in standing crops
- Varieties of oilseed rape low in glucosinolates can be more acceptable to slugs than "single low" varieties and control may not be as good
- To prevent slug build up apply at end of season to brassicas and other leafy crops
- To reduce tuber damage in potatoes apply twice in Jul and Aug

Restrictions
- Do not apply when rain imminent or water glasshouse crops within 4 d of application
- Take care to avoid lodging of pellets in the foliage when making late applications to edible crops

Environmental safety
- Dangerous to game, wild birds and animals
- Some products contain proprietary cat and dog deterrent
- Keep poultry out of treated areas for at least 7 d

Hazard classification and safety precautions
Operator protection A, H [1-5, 7-9, 11, 13, 15-29]; J [4, 5, 8, 9, 11, 13, 15, 16, 24, 29]; U05a [1-3, 5-7, 11, 12, 14, 16-19, 23, 25, 27]; U15, U20a [20-22, 26]; U20b [6, 7, 12, 14, 23, 27]; U20c [1-5, 8-11, 13, 15-19, 24, 25, 28, 29]
Environmental protection E05b (7 d); E10a, E15a
Storage and disposal D01 [1-8, 11-29]; D02 [1-3, 5-7, 11, 12, 14, 16-23, 25-27]; D05 [5, 11, 16-19]; D07 [1-3, 6, 7, 10, 12, 14, 20-23, 26-28]; D09a [1-9, 11-29]; D11a [1-29]; D12a [20-22, 26]
Treated seed S04a [4, 8, 13, 15, 20-22, 24-26, 29]
Medical advice M04 [5, 11, 16-19]; M05a [20-22, 26]

319   metamitron

A contact and residual triazinone herbicide for use in beet crops

See also chloridazon + chlorpropham + metamitron
  chloridazon + metamitron
  chlorpropham + metamitron

Products

1	Agriguard Metamitron	AgriGuard	70% w/w	WG	09859
2	Alpha Metamitron	Makhteshim	70% w/w	WG	11081
3	Barclay Seismic	Barclay	70% w/w	ZZ	11377
4	Bettix 70 WG	United Phosphorus	70% w/w	WG	11154
5	Bettix Flo	United Phosphorus	700 g/l	SC	11959
6	Defiant SC	United Phosphorus	700 g/l	SC	12302
7	Defiant WG	United Phosphorus	70% w/w	WG	12300
8	Fernpath Haptol	AgriGuard	70% w/w	WG	11951
9	Fernpath Haptol Flo	AgriGuard	700 g/l	SC	11481
10	Goldbeet	Makhteshim	90% w/w	WG	11538
11	Goltix 90	Makhteshim	90% w/w	WG	11578
12	Goltix Flowable	Makhteshim	700 g/l	SC	11540
13	Goltix WG	Makhteshim	70% w/w	WG	11539
14	Landgold Metamitron	Landgold	70% w/w	WG	06287
15	Landgold Metamitron	Teliton	70% w/w	WG	12123
16	Marquise	Makhteshim	70% w/w	WG	08738
17	Mitron 70 WG	Hermoo	70% w/w	WG	11516
18	Mitron 90 WG	Hermoo	90% w/w	WG	10888
19	Mitron SC	Hermoo	700 g/l	SC	11643
20	MM 70	Nufarm UK	70% w/w	WG	11582
21	MM 70 Flo	Nufarm UK	700 g/l	SC	11197

SEE SECTION 3 FOR PRODUCTS ALSO REGISTERED

Products – continued

22	Skater	Makhteshim	700 g/l	SC	11159
23	Target SC	Unicrop	700 g/l	SC	11070
24	Volcan	Sipcam	70% w/w	WG	09295

Uses

- Annual dicotyledons in **asparagus** *(off-label)*, **forest** *(off-label)* [16, 22]; **fodder beet** [1-19, 21-24]; **mangels** [1-16, 18, 19, 21-24]; **red beet** [1-16, 18, 19, 21, 23, 24]; **sugar beet** [1-24]
- Annual grasses in **asparagus** *(off-label)*, **forest** *(off-label)* [16, 22]; **fodder beet** [1-19, 21-24]; **mangels** [1-16, 18, 19, 21-24]; **red beet** [1-16, 18, 19, 21, 23, 24]; **sugar beet** [1-24]
- Annual meadow grass in **fodder beet** [1-19, 21-24]; **mangels** [1-16, 18, 19, 21-24]; **red beet** [1-16, 18, 19, 21, 23, 24]; **sugar beet** [1-24]
- Fat hen in **fodder beet** [1-19, 21-24]; **mangels** [1-16, 18, 19, 21-24]; **red beet** [1-16, 18, 19, 21, 23, 24]; **sugar beet** [1-24]
- General weed control in **asparagus** *(off-label)*, **forestry transplants** *(off-label)* [12, 13]
- Groundsel in **chives** *(off-label)*, **herbs (see appendix 6)** *(off-label)*, **parsley** *(off-label)* [12, 13, 16, 22]

Specific Off-Label Approvals (SOLAs)

- **asparagus** *(OLA 042188) Dec 2008* [12]
- **asparagus** *(OLA 041758) Dec 2008* [13]
- **asparagus, chives, forest, herbs (see appendix 6), parsley** *(OLA 030737) Dec 2008* [16]
- **asparagus, chives, forest, herbs (see appendix 6), parsley** *(OLA 050318) Dec 2008* [22]
- **chives, herbs (see appendix 6), parsley** *(OLA 042189) Dec 2008* [12]
- **chives, herbs (see appendix 6), parsley** *(OLA 041756) Dec 2008* [13]
- **forestry transplants** *(OLA 042190) Dec 2008* [12]
- **forestry transplants** *(OLA 041757) Dec 2008* [13]

Efficacy guidance

- May be used pre-emergence alone or post-emergence in tank mixture or with an authorised adjuvant oil
- Low dose programme (LDP). Apply a series of low-dose post-weed emergence sprays, including adjuvant oil, timing each treatment according to weed emergence and size. See label for details and for recommended tank mixes and sequential treatments. On mineral soils the LDP should be preceded by pre-drilling or pre-emergence treatment
- Traditional application. Apply either pre-drilling before final cultivation with incorporation to 8-10 cm, or pre-crop emergence at or soon after drilling into firm, moist seedbed to emerged weeds from cotyledon to first true leaf stage
- On emerged weeds at or beyond 2-leaf stage addition of adjuvant oil advised
- Up to 3 post-emergence sprays may be used on soils with over 10% organic matter
- For control of wild oats and certain other weeds, tank mixes with other herbicides or sequential treatments are recommended. See label for details

Restrictions

- Maximum total dose equivalent to three full dose treatments for most products. Check label
- Using traditional method post-crop emergence on mineral soils do not apply before first true leaves have reached 1 cm long

Crop-specific information

- Latest use: before crop foliage meets across rows for beet crops; 12 wk after transplanting asparagus
- HI herbs 6 wk
- Crop tolerance may be reduced by stress caused by growing conditions, effects of pests, disease or other pesticides, nutrient deficiency etc

Following crops guidance

- Only sugar beet, fodder beet or mangels may be drilled within 4 mth after treatment. Winter cereals may be sown in same season after ploughing, provided 16 wk passed since last treatment

Environmental safety

- Dangerous for the environment
- Very toxic to aquatic organisms
- Dangerous to fish or other aquatic life. Do not contaminate surface waters or ditches with chemical or used container
- Do not empty into drains

FOR FULL CONDITIONS OF USE ALWAYS READ THE PRODUCT LABEL

Hazard classification and safety precautions
Hazard H03 [1, 2, 4-7, 9-16, 18-21]; H04 [3, 8]; H11 [1, 2, 4-23]
Risk phrases R22a [1, 2, 4-7, 9-16, 18-21]; R41 [1, 3, 8, 9]; R43 [1-3, 5, 6, 8, 9, 16]; R50 [2, 4-7, 10-12, 14-19, 21]; R51 [1, 8, 13, 20, 22, 23]; R53a [1, 2, 4-8, 10-23]
Operator protection A [1-3, 5, 6, 8, 9, 12, 13, 16, 19, 21, 23]; C [1, 3, 8]; H [1-3, 8, 16]; U05a [1-4, 7-9, 16, 17, 21]; U08 [2, 9, 16, 17, 21]; U09a [5, 6]; U14 [2, 5, 6, 9, 16, 17, 21]; U19a [1, 2, 4, 7, 9, 13, 16, 17, 20, 21]; U20a [1, 2, 8-11, 13-18, 20, 21]; U20b [3, 4, 7, 12, 19]; U20c [5, 6, 22-24]
Environmental protection E13b [4-7]; E15a [1-3, 8-24]; E19b [10-12, 18, 19, 22, 23]; E34 [9, 12, 19, 21]; E38 [1, 13, 17, 20]
Storage and disposal D01 [1-12, 16-19, 21]; D02 [1-9, 12, 16, 17, 19, 21]; D05 [5, 6, 9, 21]; D09a [1-24]; D10c [5, 6, 22, 23]; D11a [1-4, 7-21, 24]; D12a [1, 4-7, 10-13, 18-20, 22, 23]; D12b [2, 16, 21]
Medical advice M03 [4-7, 9, 12, 19]; M04 [18]; M05a [4-7, 10-12, 14, 15, 19]

320   metam-sodium

A methyl isothiocyanate producing sterilant for glasshouse, nursery and outdoor soils

Products
1	Discovery	United Phosphorus	510 g/l	SL	10416
2	Sistan 51	Unicrop	510 g/l	SL	10046

Uses
* Nematodes in *glasshouse soils, nursery soils, outdoor soils, potting soils*
* Soil pests in *glasshouse soils, nursery soils, outdoor soils, potting soils*
* Soil-borne diseases in *glasshouse soils, nursery soils, outdoor soils, potting soils*
* Weed seeds in *glasshouse soils, nursery soils, outdoor soils, potting soils*

Efficacy guidance
* Metam-sodium is a partial soil sterilant and acts by breaking down in contact with soil to release methyl isothiocyanate (MIT)
* Apply to glasshouse soils as a drench, or inject undiluted to 20 cm at 30 cm intervals and seal immediately, or apply to surface and rotavate
* May also be used by mixing into potting soils
* Apply when soil temperatures exceed 7°C, preferably above 10°C, between 1 Apr and 31 Oct. Soil must be of fine tilth, free from debris and with 'potting moisture' content. If soil is too dry postpone treatment and water soil

Restrictions
* No plants must be present during treatment
* Crops must not be planted until a cress germination test has been completed satisfactorily
* Do not treat glasshouses within 2 m of growing crops. Fumes are damaging to all plants
* Avoid using in equipment incorporating natural rubber parts

Crop-specific information
* Latest use: pre-planting of crop

Following crops guidance
* Do not plant until soil is entirely free of fumes

Environmental safety
* Keep unprotected persons, livestock and pets out of treated areas for at least 24 h following treatment
* When diluted breakdown commences almost immediately. Only quantities for immediate use should be made up
* Divert or block drains which could carry solution under untreated glasshouses
* After treatment allow sufficient time (several weeks) for residues to dissipate and aerate soil by forking. Time varies with soil and season. Soils with high clay or organic matter content will retain gas longer than lighter soils

Hazard classification and safety precautions
Hazard H03, H04
Risk phrases R22a, R37, R38, R43
Operator protection A, C, H, M; U02a, U05a, U08, U19a, U20b [1, 2]; U04a [2]; U07 [1]
Environmental protection E02a [1, 2] (24 h); E15a, E34 [1, 2]; E36 [1]

Storage and disposal D01, D02 [1, 2]; D09a, D10a [2]; D14 [1]
Medical advice M03

321 metazachlor

A residual anilide herbicide for use in brassicas, nurseries and forestry

Products

1	Agriguard Metazachlor	AgriGuard	500 g/l	SC	10417
2	Alpha Metazachlor 50 SC	Makhteshim	500 g/l	SC	10669
3	Butisan S	BASF	500 g/l	SC	11733
4	Greencrop Monogram	Greencrop	500 g/l	SC	12048
5	Landgold Metazachlor 50	Landgold	500 g/l	SC	09726
6	Landgold Metazachlor 50	Teliton	500 g/l	SC	12124
7	Landgold Metazachlor 50 SC	Teliton	500 g/l	SC	12133
8	Me2 Booty	Me2	500 g/l	SC	10659
9	Me2 Booty 2	Me2	500 g/l	SC	12010
10	Rapsan 500 SC	Nufarm UK	500 g/l	SC	12365
11	Standon Metazachlor 50	Standon	500 g/l	SC	05581
12	Standon Metazachlor 500	Standon	500 g/l	SC	12012
13	Sultan 50 SC	Makhteshim	500 g/l	SC	10418

Uses

- Annual dicotyledons in **broccoli** [2-9, 11-13]; **brussels sprouts, cabbages, cauliflowers, spring oilseed rape, swedes, turnips, winter oilseed rape** [1-13]; **calabrese** [1-3, 5-13]; **farm forestry** [7]; **farm woodland** [1-3, 8-10, 12, 13]; **forest** [1-4, 7-10, 12, 13]; **nursery fruit trees and bushes, ornamental trees, shrubs** [1-4, 8-10, 12, 13]
- Annual grasses in **farm forestry** [7]; **farm woodland** [1-3, 8-10, 12, 13]; **forest** [1-4, 7-10, 12, 13]; **ornamental trees, shrubs** [1-4, 8-10, 12, 13]
- Annual meadow grass in **broccoli, brussels sprouts, cabbages, cauliflowers, spring oilseed rape, swedes, turnips, winter oilseed rape** [1-13]; **calabrese** [1-3, 5-13]; **nursery fruit trees and bushes, ornamental trees, shrubs** [1-4, 8-10, 12, 13]
- Blackgrass in **broccoli** [2-9, 11-13]; **brussels sprouts, cabbages, cauliflowers, spring oilseed rape, swedes, turnips, winter oilseed rape** [1-13]; **calabrese** [1-3, 5-13]; **nursery fruit trees and bushes, ornamental trees, shrubs** [1-4, 8-10, 12, 13]
- Groundsel in **chinese cabbage** *(off-label)*, **choi sum** *(off-label)*, **collards** *(off-label)*, **kale** *(off-label)*, **pak choi** *(off-label)* [3]
- Mayweeds in **chinese cabbage** *(off-label)*, **choi sum** *(off-label)*, **collards** *(off-label)*, **kale** *(off-label)*, **pak choi** *(off-label)* [3]
- Shepherd's purse in **chinese cabbage** *(off-label)*, **choi sum** *(off-label)*, **collards** *(off-label)*, **kale** *(off-label)*, **pak choi** *(off-label)* [3]

Specific Off-Label Approvals (SOLAs)

- ***chinese cabbage, choi sum, collards, kale, pak choi*** *(OLA 050344) Dec 2008* [3]

Efficacy guidance

- Activity is dependent on root uptake. For pre-emergence use apply to firm, moist, clod-free seedbed
- Some weeds (chickweed, mayweed, blackgrass etc) susceptible up to 2- or 4-leaf stage. Moderate control of cleavers achieved provided weeds not emerged and adequate soil moisture present
- Split pre- and post-emergence treatments recommended for certain weeds in winter oilseed rape on light and/or stony soils
- Effectiveness is reduced on soils with more than 10% organic matter
- Various tank-mixtures recommended to broaden spectrum. See label for details
- Always follow WRAG guidelines for preventing and managing herbicide resistant weeds. Section 5 for more information

Restrictions

- Maximum number of treatments 1 per crop for spring oilseed rape, swedes, turnips and brassicas; 2 per crop for winter oilseed rape and honesty (split dose treatment); 3 per yr for ornamentals, nursery stock, nursery fruit trees, forestry and farm forestry
- Do not use on sand, very light or poorly drained soils
- Do not treat protected crops or spray overall on ornamentals with soft foliage

FOR FULL CONDITIONS OF USE ALWAYS READ THE PRODUCT LABEL

- Do not spray crops suffering from wilting, pest or disease
- Do not spray broadcast crops or if a period of heavy rain forecast
- When used on nursery fruit trees any fruit harvested within 1 yr of treatment must be destroyed

Crop-specific information
- Latest use: pre-emergence for swedes and turnips; before 10 leaf stage for spring oilseed rape; before end of Jan for winter oilseed rape; before 6 pairs of true leaves visible for honesty
- HI brassicas 6 wk
- On winter oilseed rape may be applied pre-emergence from drilling until seed chits, post-emergence after fully expanded cotyledon stage (GS 1,0) or by split dose technique depending on soil and weeds. See label for details
- On spring oilseed rape, swedes, turnips metazachlor recommended as a pre-emergence sequential treatment following trifluralin
- On spring oilseed rape may also be used pre-weed-emergence from cotyledon to 10-leaf stage of crop (GS 1,0-1,10)
- With pre-emergence treatment ensure seed covered by 15 mm of well consolidated soil. Harrow across slits of direct-drilled crops
- Ensure brassica transplants have roots well covered and are well established. Direct drilled brassicas should not be treated before 3 leaf stage
- In ornamentals and hardy nursery stock apply after plants established and hardened off as a directed spray or, on some subjects, as an overall spray. See label for list of tolerant subjects. Do not treat plants in containers

Following crops guidance
- Any crop can follow normally harvested treated winter oilseed rape. See label for details of crops which may be planted after spring treatment and in event of crop failure

Environmental safety
- Dangerous for the environment
- Very toxic to aquatic organisms
- Keep livestock out of treated areas until foliage of any poisonous weeds such as ragwort has died and become unpalatable
- Keep livestock out of treated areas of swede and turnip for at least 5 wk following treatment

Hazard classification and safety precautions
Hazard H03, H11
Risk phrases R22a, R43, R50 [1-13]; R38 [2-9, 11, 13]; R53a [3, 5-9, 11]
Operator protection A, C, H, M; U05a, U08, U19a, U20b [1-13]; U14 [2, 3, 13]; U15 [2, 13]
Environmental protection E06a [1-13] (5 wk for swedes, turnips); E07a, E15a, E34 [1-13]; E38 [3]
Storage and disposal D01, D02, D09a [1-13]; D05 [1, 2, 4-13]; D10b [1, 4-9, 11, 12]; D10c [2, 3, 10, 13]; D12a [3]; D12b [2, 13]
Medical advice M03 [1-13]; M05a [3]

322   metazachlor + quinmerac

A residual herbicide mixture for oilseed rape

Products

1	Katamaran	BASF	375:125 g/l	SC	11732
2	Novall	BASF	400:100 g/l	SC	12031
3	Standon Metazachlor-Q	Standon	375:125 g/l	SC	09676

Uses
- Annual dicotyledons in *winter oilseed rape*
- Annual meadow grass in *winter oilseed rape*
- Blackgrass in *winter oilseed rape*
- Cleavers in *winter oilseed rape*
- Poppies in *winter oilseed rape*

Efficacy guidance
- Activity is dependent on root uptake. Pre-emergence treatments should be applied to firm moist seedbeds. Applications to dry soil do not become effective until after rain has fallen
- Maximum activity achieved from treatment before weed emergence for some species

SEE SECTION 3 FOR PRODUCTS ALSO REGISTERED

SECTION 2

- Weed control may be reduced if excessive rain falls shortly after application especially on light soils
- May be used on all soil types except Sands, Very Light Soils, and soils containing more than 10% organic matter. Crop vigour and/or plant stand may be reduced on brashy and stony soils

Restrictions
- Maximum total dose equivalent to one full dose treatment
- Damage may occur in waterlogged conditions. Do not use on poorly drained soils
- Do not treat stressed crops. In frosty conditions transient scorch may occur

Crop-specific information
- Latest use: end Jan in yr of harvest
- To ensure crop safety it is essential that crop seed is well covered with soil to 15 mm. Loose or puffy seedbeds must be consolidated before treatment. Do not use on broadcast crops
- Crop vigour and possibly plant stand may be reduced if excessive rain falls shortly after treatment especially on light soils

Following crops guidance
- In the event of crop failure after use, wheat or barley may be sown in the autumn after ploughing to 15 cm. Spring cereals or brassicas may be planted after ploughing in the spring

Environmental safety
- Dangerous for the environment
- Very toxic to aquatic organisms
- Keep livestock out of treated areas until foliage of any poisonous weeds such as ragwort has died and become unpalatable
- To reduce risk of movement to water do not apply to dry soil or if heavy rain is forecast. On clay soils create a fine consolidated seedbed
- LERAP Category B

Hazard classification and safety precautions
Hazard H04, H11
Risk phrases R43, R50, R53a
Operator protection A [1-3]; C, H [2]; U04a [2]; U05a, U08, U14, U19a, U20b [1-3]
Environmental protection E07a, E15a, E16a [1-3]; E16b [3]; E38 [1, 2]
Storage and disposal D01, D02, D09a [1-3]; D05, D10b [3]; D08, D10c, D12a [1, 2]
Medical advice M05a [1, 2]

323   metconazole

A conazole fungicide for cereals and oilseed rape

Products
1	Caramba	BASF	60 g/l	SL	10213
2	Sunorg Pro	BASF	90 g/l	SL	11112

Uses
- Alternaria in **spring oilseed rape, winter oilseed rape** [1, 2]
- Brown rust in **spring barley, winter barley, winter wheat** [1]
- Fusarium ear blight in **winter wheat** *(reduction)* [1]
- Net blotch in **spring barley** *(reduction)*, **winter barley** *(reduction)* [1]
- Phoma in **spring oilseed rape** *(foliar disease - reduction)*, **winter oilseed rape** *(foliar disease - reduction)* [1]; **spring oilseed rape** *(reduction)*, **winter oilseed rape** *(reduction)* [2]
- Powdery mildew in **spring barley** *(moderate control)*, **winter barley**, **winter wheat** *(moderate control)* [1]
- Rhynchosporium in **spring barley, winter barley** [1]
- Sclerotinia stem rot in **spring oilseed rape** *(reduction)*, **winter oilseed rape** *(reduction)* [1, 2]
- Septoria leaf spot in **winter wheat** [1]
- Yellow rust in **winter wheat** [1]

Approval information
- Accepted by BBPA for use on malting barley

FOR FULL CONDITIONS OF USE ALWAYS READ THE PRODUCT LABEL

Efficacy guidance

- Best results from application to healthy, vigorous crops when disease starts to develop
- Treatment for leaf blotch should be made after GS 33 and when weather favouring development of the disease has occurred
- Treat mildew infections before 3% infection on any green leaf. A specific mildewicide will improve control of established infections
- Treat yellow rust before 1% infection on any leaf or as preventive treatment after GS 39
- For brown rust spray susceptible varieties before any of top 3 leaves has more than 2% infection
- Sclerotinia in oilseed rape should be treated at petal fall. A tank mixture with carbendazim may be needed for fully effective control

Restrictions

- Maximum total dose equivalent to two full dose treatments
- Do not apply to oilseed rape crops that are damaged or stressed from previous treatments, adverse weather, nutrient deficiency or pest attack. Spring application may lead to reduction of crop height
- The addition of adjuvants is neither advised nor necessary and can lead to enhanced growth regulatory effects on stressed crops of oilseed rape
- Ensure sprayer is free from residues of previous treatments that may harm the crop, especially oilseed rape. Use of a detergent cleaner is advised before and after use
- Do not apply with pyrethroid insecticides on oilseed rape at flowering [2]

Crop-specific information

- Latest use: up to and including milky ripe stage (GS 71) for cereals; 10% pods at final size for oilseed rape

Following crops guidance

- Only cereals, oilseed rape, sugar beet, linseed, maize, clover, beans, peas, carrots, potatoes or onions may be sown as following crops after treatment

Environmental safety

- Dangerous for the environment
- Very toxic to aquatic organisms
- Risk to certain non-target insects or other arthropods
- LERAP Category B but avoid treatment close to field boundary, even if permitted by LERAP assessment, to reduce effects on non-target insects or other arthropods

Hazard classification and safety precautions

Hazard H03, H08 [1]; H04 [2]; H11 [1, 2]
Risk phrases R22b, R38, R41, R43, R50 [1]; R36, R51 [2]; R53a [1, 2]
Operator protection A, C [1, 2]; H [1]; U02a, U05a, U20c [1, 2]; U07, U11, U14 [1]
Environmental protection E15a, E34, E38 [1, 2]; E16a, E16b [1]
Storage and disposal D01, D02, D05, D09a, D10a, D12a
Medical advice M05b [1]

324   methiocarb

A stomach acting carbamate molluscicide and insecticide

Products

1 Decoy Wetex	Bayer CropScience	2% w/w	PT	11266
2 Exit Wetex	Bayer CropScience	3% w/w	PT	11276
3 Huron	Bayer CropScience	3% w/w	PT	11288
4 Karan	Bayer CropScience	3% w/w	PT	11289
5 Lupus	Bayer CropScience	3% w/w	PT	11291
6 New Draza	Bayer CropScience	3% w/w	PT	11294
7 Rivet	Bayer CropScience	3% w/w	PT	11300

Uses

- Cutworms in **sugar beet** (reduction)
- Leatherjackets in **all cereals** (reduction), **potatoes** (reduction), **ryegrass** (seed admixture), **sugar beet** (reduction)
- Millipedes in **sugar beet** (reduction)

- Slugs in **all cereals**, **all non-edible crops** *(outdoor only)*, **brussels sprouts**, **cabbages**, **cauliflowers**, **leaf spinach**, **lettuce**, **maize**, **potatoes**, **ryegrass** *(seed admixture)*, **spring oilseed rape**, **strawberries**, **sugar beet**, **sunflowers**, **winter oilseed rape**
- Strawberry seed beetle in **strawberries**

Approval information
- Accepted by BBPA for use on malting barley

Efficacy guidance
- Use as a surface, overall application when pests active (normally mild, damp weather), pre-drilling or post-emergence. May also be used on cereals or ryegrass in admixture with seed at time of drilling
- Also reduces populations of cutworms and millipedes
- Best on potatoes in late Jul to Aug
- Apply to strawberries before strawing down to prevent seed beetles contaminating crop
- See label for details of suitable application equipment

Restrictions
- This product contains an anticholinesterase carbamate compound. Do not use if under medical advice not to work with such compounds
- Maximum number of treatments 3 per crop for potatoes; 2 per crop for cereals, Brussels sprouts, cauliflowers, maize, oilseed rape, sunflowers; 1 per crop for cabbages, lettuce, leaf spinach, sugar beet; 1 per yr for strawberries
- Do not treat any protected crops
- Do not allow pellets to lodge in edible crops

Crop-specific information
- Latest use: before first node detectable for cereals, maize; before three visibly extended internodes for oilseed rape, sunflowers
- HI 7 d for spinach, strawberrries; 14 d for Brussels sprouts, cabbages, cauliflowers, lettuce; 18 d for potatoes; 6 mth for sugar beet

Environmental safety
- Harmful to aquatic organisms
- Dangerous to fish or other aquatic life. Do not contaminate surface waters or ditches with chemical or used container
- Dangerous to game, wild birds and animals
- Risk to certain non-target insects or other arthropods
- Avoid surface broadcasting application within 6 m of field boundary to reduce effects on non-target species
- Admixed seed should be drilled and not broadcast, and may not be applied from the air

Hazard classification and safety precautions
Hazard H03
Risk phrases R22a, R52, R53a
Operator protection A, H, J; U05a, U20b
Environmental protection E05b [1-7] (7 d); E10a, E13b, E34 [1-7]; E22b [2-7]; E22c [1]
Storage and disposal D01, D02, D09a, D11a
Medical advice M02, M03

325 methoxyfenozide

A moulting accelerating diacylhydrazine insecticide

Products

Runner	Bayer CropScience	240 g/l	SC	11470

Uses
- Codling moth in **apples**, **pears**
- Tortrix moths in **apples**, **pears**
- Winter moth in **apples**, **pears**

Efficacy guidance
- To achieve best results uniform coverage of the foliage and full spray penetration of the leaf canopy is important, particularly when spraying post-blossom

- For maximum effectiveness on winter moth and tortrix spray pre-blossom when first signs of active larvae are seen, followed by a further spray in June if larvae of the summer generation are present
- For codling moth spray post-blossom to coincide with early to peak egg deposition. Follow-up treatments will normally be needed
- Methoxyfenozide is a moulting accelerating compound (MAC) and may be used in an anti-resistance strategy with other top fruit insecticides (including chitin biosynthesis inhibitors and juvenile hormones) which have a different mode of action
- To reduce further the likelihood of resistance development use at full recommended dose in sufficient water volume to achieve required spray penetration

Restrictions
- Maximum number of treatments 3 per yr but no more than two should be sprayed consecutively

Crop-specific information
- HI 14 d for apples, pears

Environmental safety
- LERAP Category B
- Broadcast air-assisted LERAP (5 m)

Hazard classification and safety precautions
  Operator protection U20b
  Environmental protection E16b; E17b (5 m)
  Storage and disposal D05, D09a, D11a

326   1-methylcyclopropene

An inhibitor of ethylene production for use in stored apples

Products

SmartFresh	Landseer	3.3% w/w	SP	11799

Uses
- Ethylene inhibition in **apples** *(post-harvest use)*
- Scald in **apples** *(post-harvest use)*

Approval information
- Approval expiry 2 Sep 2006 [1]

Efficacy guidance
- Best results obtained from treatment of fruit in good condition and of proper quality for long-term storage
- Effects may be reduced in fruit that is in poor condition or ripe prior to storage or harvested late
- Product acts by releasing vapour into store when mixed with water
- Apply as soon as possible after harvest
- Treatment controls superficial scald and maintains fruit firmness and acid content for 3-6 mth in normal air, and 6-9 mth in controlled atmosphere storage
- Ethylene production recommences after removal from storage

Restrictions
- Maximum number of treatments 1 per batch of apples
- Must only be used by suitably trained and competent persons in fumigation operations
- Consult processors before treatment of fruit destined for processing or cider making
- Do not apply in mixture with other products
- Ventilate all areas thoroughly with all refrigeration fans operating at maximum power for at least 15 min before re-entry

Crop-specific information
- Latest use: 7d after harvest of apples
- Product tested on Granny Smith, Gala, Jonagold, Bramley and Cox. Consult distributor or supplier before treating other varieties

Environmental safety
- Unprotected persons must be kept out of treated stores during the 24 h treatment period
- Prior to application ensure that the store can be properly and promptly sealed

SEE SECTION 3 FOR PRODUCTS ALSO REGISTERED

Hazard classification and safety precautions
    Operator protection U05a
    Environmental protection E02a (24 h); E15a, E34
    Consumer protection C12
    Storage and disposal D01, D02, D07, D14

327   metoxuron

A contact and residual urea herbicide for carrots

Products
    Dosaflo   Syngenta   500 g/l   SC   09351

Uses
- Annual dicotyledons in **carrots**
- Mayweeds in **carrots**

Approval information
- Products containing this active ingredient have been granted derogations for specified 'Essential Uses' for use until 31 December 2007. Sale and supply must cease by 30 June 2007 but growers have no guarantee that the products will continue to be available until then.
  For more information see 'The Review Programme' under 'Pesticide Legislation' in Section 5

Efficacy guidance
- Best results achieved by application to weeds, especially mayweeds, in seedling to young plant stage
- Product will suppress growth of annual meadow-grass and wild oats between emergence and the 3 leaf stage of growth

Restrictions
- Maximum number of treatments not specified
- Carrots should be drilled to an even depth of not less than 1.5 cms otherwise some plants may be treated at cotyledon stage and their growth checked
- Do not spray when soil very dry or wet, when heavy rain is imminent, or when shade temperature exceeds 25°C
- Do not spray carrots on soils with more than 80% sand or less than 1% organic matter
- Do not apply during prolonged frosty weather, when temperatures are below freezing, or when frost is imminent
- Do not spray crops checked by pests, wind, frost or waterlogging until recovered
- This product must not be used on any crops other than those listed, including any extrapolations that would normally be permissible under the Long Term Arrangements for Extension of Use (see Section 5)

Crop-specific information
- Latest use: not specified for carrots

Following crops guidance
- No crop should be sown within 6 wk of treatment

Environmental safety
- Dangerous for the environment
- Very toxic to aquatic organisms

Hazard classification and safety precautions
    Hazard H11
    Risk phrases R50, R53a
    Operator protection U05a, U08, U19a, U20b
    Environmental protection E15a, E38
    Storage and disposal D01, D02, D05, D09a, D10c, D12a

FOR FULL CONDITIONS OF USE ALWAYS READ THE PRODUCT LABEL

328 metrafenone

A benzophenone protectant and curative fungicide for cereals

Products

1	Attenzo	BASF	300 g/l	SC	11917
2	Flexity	BASF	300 g/l	SC	11775

Uses
- Eyespot in **spring wheat** *(reduction)*, **winter wheat** *(reduction)*
- Powdery mildew in **spring barley**, **spring wheat**, **winter barley**, **winter wheat**

Approval information
- Accepted by BBPA for use on malting barley

Efficacy guidance
- Best results obtained from treatment at the start of foliar disease attack
- Activity against mildew in wheat is mainly protectant with moderate curative control in the latent phase; activity is entirely protectant in barley
- Useful reduction of eyespot in wheat is obtained if treatment applied at GS 30-32
- Should be used as part of a resistance management strategy that includes mixtures or sequences effective against mildew and non-chemical methods

Restrictions
- Maximum number of treatments 2 per crop
- Do not use sequential treatments containing metrafenone

Crop-specific information
- Latest use: beginning of flowering (GS 31) for wheat, barley

Following crops guidance
- Cereals, oilseed rape, sugar beet, linseed, maize, clover, field beans, peas, turnips, carrots, cauliflowers, onions, lettuce or potatoes may follow a treated cereal crop

Environmental safety
- Dangerous for the environment
- Toxic to aquatic organisms

Hazard classification and safety precautions
    Hazard H04, H11
    Risk phrases R43, R51, R53a
    Operator protection A, H; U05a, U20b
    Environmental protection E15a, E34
    Storage and disposal D01, D02, D05, D09a, D10c
    Medical advice M03

329 metribuzin

A contact and residual triazinone herbicide for use in potatoes

See also flufenacet + metribuzin

Products

1	Ag-Chem Metribuzin	Ag-Chem	70% w/w	WG	12007
2	Agriguard Metribuzin	AgriGuard	70% w/w	WG	09853
3	Citation WG	United Phosphorus	70% w/w	WG	11929
4	Landgold Metribuzin WG	Teliton	70% w/w	WG	12468
5	Lanxess Metribuzin 70 WG	Lanxess	70% w/w	WG	12487
6	Lexone 2	DuPont	70% w/w	WG	04991
7	Python	Makhteshim	70% w/w	WG	11166
8	Sencorex WG	Bayer CropScience	70% w/w	WG	11304
9	Shotput	Makhteshim	70% w/w	WG	11960
10	Tuberon	Unicrop	480 g/l	SC	11061

Uses
- Annual and perennial weeds in **asparagus** *(off-label)* [8]
- Annual dicotyledons in **asparagus** *(off-label)*, **carrots** *(off-label)*, **parsnips** *(off-label)* [7]; **early potatoes**, **maincrop potatoes** [1-10]

SEE SECTION 3 FOR PRODUCTS ALSO REGISTERED

- Annual grasses in **asparagus** *(off-label)*, **carrots** *(off-label)*, **parsnips** *(off-label)* [7]; **early potatoes, maincrop potatoes** [1-10]
- Perennial dicotyledons in **asparagus** *(off-label)*, **carrots** *(off-label)*, **parsnips** *(off-label)* [7]
- Volunteer oilseed rape in **early potatoes, maincrop potatoes** [1-9]

Specific Off-Label Approvals (SOLAs)
- **asparagus, carrots, parsnips** *(OLA 030368) Dec 2008* [7]
- **asparagus** *(OLA 050102) Dec 2008* [8]

Efficacy guidance
- Best results achieved on weeds at cotyledon to 1-leaf stage
- Water dispersible granule formulations may be applied pre- or post-emergence of crop [6-8]
- Suspension concentrate product may be applied pre-emergence only [10]
- Apply to moist soil with well-rounded ridges and few clods
- Activity reduced by dry conditions and on soils with high organic matter content
- On fen and moss soils pre-planting incorporation to 10-15 cm gives increased activity. Incorporate thoroughly and evenly
- With named maincrop and second early varieties on soils with more than 10% organic matter shallow pre- or post-planting incorporation may be used. See label for details
- Effective control using a programme of reduced doses is made possible by using a spray of smaller droplets, thus improving retention. See label for details [6-8]

Restrictions
- Maximum number of treatments 3 per crop on potatoes subject to maximum permitted dose for water dispersible granule formulations [6-8]; 2 per crop on carrots; 1 per crop on asparagus. See labels for details
- Maximum number of treatments 2 per crop on potatoes for suspension concentrate product [10]
- Apply pre-emergence only on named first earlies, pre- or post-emergence on named second earlies. On named maincrop varieties apply pre-emergence (except for certain varieties on Sands or Very Light soils) or post-emergence before longest shoots reach 15 cm. See labels for details [6-8]
- Do not treat any variety on Light soils [10]
- Do not cultivate after treatment
- Some varieties may be sensitive to post-emergence treatment if crop under stress

Crop-specific information
- Latest use: pre-emergence for early potatoes; before most advanced shoots reach 15 cm for maincrop potatoes [6-8]; whilst crop still dormant in yr of harvest for asparagus
- Latest use for suspension concentrate product: pre-emergence for all crops [10]
- HI carrots, parsnips 4 wk
- On stony or gravelly soils there is risk of crop damage, especially if heavy rain falls soon after application
- When days are hot and sunny delay spraying until evening

Following crops guidance
- Ryegrass, cereals or winter beans may be sown in same season provided at least 16 wk elapsed after treatment and ground ploughed to 15 cm and thoroughly cultivated as soon as possible after harvest and no later than end Dec
- In W Cornwall on soil with more than 5% organic matter early potatoes treated as recommended may be followed by summer planted brassica crops provided the soil has been ploughed, spring rainfall has been normal and at least 14 wk have elapsed since treatment
- Do not grow any vegetable brassicas on silt soils in Lincs, and lettuces or radishes anywhere in UK on land treated the previous yr. Other crops may be sown normally in spring of next yr

Environmental safety
- Dangerous for the environment
- Very toxic to aquatic organisms
- Dangerous to fish or other aquatic life. Do not contaminate surface waters or ditches with chemical or used container
- Do not empty into drains
- LERAP Category B

Hazard classification and safety precautions
Hazard H03, H11
Risk phrases R22a [1-10]; R43 [7, 9]; R50, R53a [1, 4-10]

FOR FULL CONDITIONS OF USE ALWAYS READ THE PRODUCT LABEL

Operator protection A [7, 9, 10]; H [7, 9]; U04a [10]; U05a [7, 9, 10]; U08, U13, U19a [1-10]; U14 [7, 9]; U20a [1, 4, 6, 8]; U20b [2, 3, 5, 7, 9, 10]
Environmental protection E13b, E38 [1, 4, 8]; E15a [2, 3, 5-7, 9, 10]; E16a, E16b [1-10]; E19b [7, 9]; E34 [10]
Storage and disposal D01, D02 [5, 7, 9, 10]; D05 [2, 3, 10]; D07 [2, 3, 5]; D09a [1-10]; D10c [10]; D11a [1-9]; D12a [1, 4, 7-9]
Medical advice M03 [10]; M05a [7, 9]

330   metsulfuron-methyl

A contact and residual sulfonylurea herbicide

See also carfentrazone-ethyl + metsulfuron-methyl

Products

1	Agriguard Metsulfuron	AgriGuard	20% w/w	WG	09569
2	Ally SX	DuPont	20% w/w	SG	12059
3	Jubilee SX	DuPont	20% w/w	SG	12203
4	Landgold Metsulfuron	Landgold	20% w/w	WG	06280

Uses

- Annual dicotyledons in *linseed* [2, 3]; *spring barley, spring oats, spring wheat, winter barley, winter oats, winter wheat* [1-4]; *triticale* [1-3]
- Chickweed in *linseed* [2, 3]; *spring barley, spring oats, spring wheat, winter barley, winter oats, winter wheat* [1-4]; *triticale* [1-3]
- Green cover in *land not being used for crop production* [2, 3]
- Mayweeds in *linseed* [2, 3]; *spring barley, spring oats, spring wheat, winter barley, winter oats, winter wheat* [1-4]; *triticale* [1-3]

Approval information

- Metsulfuron-methyl included in Annex I under EC Directive 91/414
- Accepted by BBPA for use on malting barley

Efficacy guidance

- Best results achieved on small, actively growing weeds up to 6-true leaf stage. Good spray cover is important
- Commonly used in tank-mixture on wheat and barley with other cereal herbicides to improve control of resistant dicotyledons (cleavers, fumitory, ivy-leaved speedwell), larger weeds and grasses. See label for recommended mixtures
- Metsulfuron-methyl is a member of the ALS-inhibitor group of herbicides and products should be used in a planned Resistance Management strategy. See Section 5 for more information

Restrictions

- Maximum number of treatments 1 per crop
- Product must only be used after 1 Feb
- Do not use on cereal crops undersown with grass or legumes
- Consult contract agents before use on a cereal crop for seed
- Do not use on any crop suffering stress from drought, waterlogging, frost, deficiency, pest or disease attack or apply within 7 d of rolling
- Specific restrictions apply to use in sequence or tank mixture with other sulfonylurea or ALS-inhibiting herbicides. See label for details
- Spraying equipment should not be drained or flushed onto land planted, or to be planted, with trees or crops other than cereals and should be thoroughly cleansed after use - see label for instructions

Crop-specific information

- Latest use: before flag leaf sheath extending stage for cereals (GS 41)
- Apply to wheat and oats from 2-leaf (GS 12), to barley and triticale from 3-leaf stage (GS 13) until flag-leaf fully emerged (GS 39). Do not spray Igri barley before leaf sheath erect stage (GS 30)
- Recommendations for oats, triticale and linseed apply to product alone

Following crops guidance

- Only cereals, oilseed rape, field beans or grass may be sown in same calendar year after treating cereals with the product alone. Other restrictions apply to tank mixtures. See label for details

SEE SECTION 3 FOR PRODUCTS ALSO REGISTERED

SECTION 2

- Only cereals should be planted within 16 mth of applying to a linseed crop or set-aside
- In the event of failure of a treated crop sow only wheat within 3 mth after treatment

Environmental safety

- Dangerous for the environment
- Very toxic to aquatic organisms
- Extremely dangerous to fish or other aquatic life. Do not contaminate surface waters or ditches with chemical or used container
- Extremely dangerous to aquatic higher plants. Do not contaminate surface waters or ditches with chemical or used container [1]
- Take extreme care to avoid damage by drift onto broad-leaved plants outside the target area, onto surface waters or ditches or onto land intended for cropping
- LERAP Category B
- A range of broad leaved species will be fully or partially controlled when used in land temporarily removed from production, hence product may be suitable where wild flower borders or other forms of conservation headland are being developed
- Before use on land temporarily removed from production as part of grant-aided scheme, ensure compliance with the management rules
- Green cover on land temporarily removed from production must not be grazed by livestock or harvested for human or animal consumption or used for animal bedding

Hazard classification and safety precautions

Hazard H11
Risk phrases R50 [2-4]; R51 [1]; R53a [1-4]
Operator protection U08, U19a [1, 4]; U20a [4]; U20b [1]
Environmental protection E13a, E34 [1]; E15a, E16b [1, 4]; E15b [2, 3]; E16a [1-4]; E38 [1-3]
Storage and disposal D09a, D11a [1, 4]; D12a [1-3]

---

331   metsulfuron-methyl + thifensulfuron-methyl

A contact residual and translocated sulfonylurea herbicide mixture for use in cereals

Products

1	Concert SX	DuPont	4:40% w/w	SG	12288
2	Finish SX	DuPont	6.6:33.3% w/w	SG	12259
3	Harmony M SX	DuPont	4:40% w/w	SG	12258
4	Presite SX	DuPont	6.7:33.3% w/w	WG	12291

Uses

- Annual dicotyledons in *spring barley, spring wheat, winter wheat* [1-4]; *winter barley, winter oats* [2, 4]
- Chickweed in *spring barley, spring wheat, winter barley, winter oats, winter wheat* [2, 4]
- Cleavers in *spring barley, spring wheat, winter wheat* [1, 3]
- Field pansy in *spring barley, spring wheat, winter wheat* [1-4]; *winter barley, winter oats* [2, 4]
- Knotgrass in *spring barley, spring wheat, winter barley, winter oats, winter wheat* [2, 4]
- Mayweeds in *spring barley, spring wheat, winter barley, winter oats, winter wheat* [2, 4]
- Polygonums in *spring barley, spring wheat, winter wheat* [1, 3]
- Speedwells in *spring barley, spring wheat, winter wheat* [1-4]; *winter barley, winter oats* [2, 4]

Approval information

- Metsulfuron-methyl and thifensulfuron-methyl included in Annex I under EC Directive 91/414
- Accepted by BBPA for use on malting barley

Efficacy guidance

- Best results by application to small, actively growing weeds up to 6-true leaf stage
- Ensure good spray cover
- Susceptible weeds stop growing almost immediately but symptoms may not be visible for about 2 wk
- Effectiveness may be reduced by heavy rain or if soil conditions very dry
- Metsulfuron-methyl and thifensulfuron-methyl are members of the ALS-inhibitor group of herbicides and products should be used in a planned Resistance Management strategy. See Section 5 for more information

---

FOR FULL CONDITIONS OF USE ALWAYS READ THE PRODUCT LABEL

Restrictions
- Maximum number of treatments 1 per crop
- Products may only be used after 1 Feb
- Do not use on any crop suffering stress from drought, waterlogging, frost, deficiency, pest or disease attack or any other cause
- Do not use on crops undersown with grasses, clover or legumes, or any other broad leaved crop
- Specific restrictions apply to use in sequence or tank mixture with other sulfonylurea or ALS-inhibiting herbicides. See label for details
- Do not apply within 7 d of rolling
- Consult contract agents before use on a cereal crop grown for seed

Crop-specific information
- Latest use: before flag leaf sheath extending (GS 39) for barley and wheat; before second node detectable (GS 32) for oats

Following crops guidance
- Only cereals, oilseed rape, field beans or grass may be sown in same calendar year after treatment
- Additional constraints apply after use of certain tank mixtures. See label
- In the event of crop failure sow only winter wheat within 3 mth after treatment and after ploughing and cultivating to a depth of at least 15 cm

Environmental safety
- Dangerous for the environment
- Very toxic to aquatic organisms
- LERAP Category B
- Take extreme care to avoid damage by drift onto broad-leaved plants outside the target area, or onto ponds, waterways or ditches
- Spraying equipment should not be drained or flushed onto land planted, or to be planted, with trees or crops other than cereals and should be thoroughly cleansed after use - see label for instructions

Hazard classification and safety precautions
Hazard H11
Risk phrases R50, R53a
Operator protection U05a [2, 4]; U08, U19a, U20b [1-4]
Environmental protection E15a [2, 4]; E15b [1, 3]; E16a, E38 [1-4]
Storage and disposal D01, D02 [2, 4]; D09a, D11a, D12a [1-4]

332　metsulfuron-methyl + tribenuron-methyl

A sulfonylurea herbicide mixture for winter wheat

Products

1 Biplay SX	DuPont	11.1:22.2 % w/w	SG	12246
2 Traton SX	DuPont	11.1:22.2% w/w	SG	12270

Uses
- Annual dicotyledons in *spring barley, spring wheat, triticale, winter barley, winter oats, winter wheat*
- Chickweed in *spring barley, spring wheat, triticale, winter barley, winter oats, winter wheat*
- Field pansy in *spring barley, spring wheat, triticale, winter barley, winter oats, winter wheat*
- Hemp-nettle in *spring barley, spring wheat, triticale, winter barley, winter oats, winter wheat*
- Mayweeds in *spring barley, spring wheat, triticale, winter barley, winter oats, winter wheat*
- Red dead-nettle in *spring barley, spring wheat, triticale, winter barley, winter oats, winter wheat*

Approval information
- Metsulfuron-methyl included in Annex I under EC Directive 91/414
- Accepted by BBPA for use on malting barley

Efficacy guidance
- Best results obtained when applied to small actively growing weeds

- Product acts by foliar and root uptake. Good spray cover essential but performance may be reduced when soil conditions are very dry and residual effects may be reduced by heavy rain
- Weed growth inhibited within hours of treatment and many show marked colour changes as they die back
- Metsulfuron-methyl and tribenuron-methyl are members of the ALS-inhibitor group of herbicides and products should be used in a planned Resistance Management strategy. See Section 5 for more information

Restrictions
- Maximum number of treatments 1 per crop
- Product must only be used after 1 Feb and after crop has three leaves
- Do not apply to a crop suffering from drought, water-logging, low temperatures, pest or disease attack, nutrient deficiency, soil compaction or any other stress
- Do not use on crops undersown with grasses, clover or other legumes
- Do not apply within 7 d of rolling
- Specific restrictions apply to use in sequence or tank mixture with other sulfonylurea or ALS-inhibiting herbicides. See label for details

Crop-specific information
- Latest use: before flag leaf sheath extending

Following crops guidance
- Only cereals, field beans or oilseed rape may be sown in the same calendar yr as harvest of a treated crop
- In the event of failure of a treated crop only winter wheat may sown within 3 mth of treatment, and only after ploughing and cultivation to 15 cm min

Environmental safety
- Dangerous for the environment
- Very toxic to aquatic organisms
- Some non-target crops are highly sensitive. Take extreme care to avoid drift outside the target area, or onto ponds, waterways or ditches
- Spraying equipment should be thoroughly cleaned in accordance with manufacturer's instructions

Hazard classification and safety precautions
Hazard H04, H11
Risk phrases R43, R50, R53a
Operator protection A, H; U08, U19a, U20b
Environmental protection E15a, E38
Storage and disposal D01, D02, D09a, D11a, D12a

333   myclobutanil

A systemic, protectant and curative conazole fungicide

Products

1 Aristocrat	AgriGuard	200 g/l	EW	12343
2 Masalon	Rigby Taylor	45 g/l	EW	12385
3 Systhane 20EW	Landseer	200 g/l	EW	09396

Uses
- American gooseberry mildew in **blackcurrants, gooseberries** [1, 3]
- Black spot in **ornamental plant production, roses** [1, 3]
- Blossom wilt in **cherries** *(off-label)*, **mirabelles** *(off-label)* [3]
- Fusarium patch in **managed amenity turf** [2]
- Plum rust in **plums** *(off-label)* [3]
- Powdery mildew in **apples, ornamental plant production, pears, roses, strawberries** [1, 3]; **grapevines** *(off-label)*, **hops** *(off-label)*, **protected blackberries** *(off-label)*, **protected raspberries** *(off-label)*, **protected rubus hybrids** *(off-label)* [3]
- Rust in **ornamental plant production, roses** [1, 3]
- Scab in **apples, pears** [1, 3]

FOR FULL CONDITIONS OF USE ALWAYS READ THE PRODUCT LABEL

Specific Off-Label Approvals (SOLAs)
- *cherries, mirabelles* (OLA 991535) Dec 2008 [3]
- *hops* (OLA 021412) Dec 2008 [3]
- *plums* (OLA 012459) Dec 2008 [3]
- *protected blackberries, protected raspberries* (OLA 051189) Dec 2008 [3]
- *protected rubus hybrids* (OLA 023195) Dec 2008 [3]

Approval information
- Accepted by BBPA for use on hops

Efficacy guidance
- Best results achieved when used as part of routine preventive spray programme from bud burst to end of flowering in apples and pears and from just before the signs of mildew infection in blackcurrants and gooseberries [3]
- In strawberries commence spraying at, or just prior to, first flower. Post-harvest sprays may be required on mildew-susceptible varieties where mildew is present and likely to be damaging [3]
- Spray at 7-14 d intervals depending on disease pressure and dose applied [3]
- For improved scab control on apples in post-blossom period tank-mix with mancozeb or captan [3]
- Apply alone from mid-Jun for control of secondary mildew on apples and pears
- On roses spray at first signs of disease and repeat every 2 wk. In high risk areas spray when leaves emerge in spring, repeat 1 wk later and then continue normal programme [3]
- Treatment of managed amenity turf may be carried out at any time of year before, or at, the first sign of disease [2]
- Myclobutanil products should be used in conjunction with other fungicides with a different mode of action to reduce the possibility of resistance developing

Restrictions
- Maximum total dose equivalent to ten full dose treatments in apples and pears; six full dose treatments in blackcurrants, gooseberries, strawberries; 5 full dose treatments in plums; 3.5 full dose treatments in cherries, mirabelles
- Maximum number of treatments on managed amenity turf: two per yr [2]
- Do not mow turf within 24 hr after treatment [2]

Crop-specific information
- HI 28 d (grapevines); 21 d (cherries, mirabelles); 14 d (apples, pears, blackcurrants, gooseberries, hops); 3 d (plums, protected blackberries, protected raspberries, protected Rubus hybrids, strawberries)
- Product may be applied to newly sown turf after the two-leaf stage but in view of the large number of turf grass cultivars a safety test on a small area is recommended before large scale treatment [2]

Environmental safety
- Dangerous for the environment
- Toxic (harmful [2]) to aquatic organisms
- LERAP Category B [2]

Hazard classification and safety precautions
Hazard H03, H11 [1, 3]
Risk phrases R22b, R51, R63 [1, 3]; R52 [2]; R53a [1-3]
Operator protection A, H [1-3]; J [1, 3]; U08, U20a [1-3]; U15 [1, 3]
Environmental protection E15a [1-3]; E16a [2]
Storage and disposal D01, D02 [1, 3]; D05, D10c, D12a [1-3]
Medical advice M05b [1, 3]

334   2-(1-naphthyl)acetic acid

A plant growth regulator to promote rooting of cuttings

See also 4-indol-3-yl-butyric acid + 2-(1-naphthyl)acetic acid with dichlorophen

Products

1	Rhizopon B Powder (0.1%)	Rhizopon	0.1% w/w	DP	09089
2	Rhizopon B Powder (0.2%)	Rhizopon	0.2% w/w	DP	09090
3	Rhizopon B Tablets	Rhizopon	25 mg a.i.	WT	09091

SEE SECTION 3 FOR PRODUCTS ALSO REGISTERED

SECTION 2

Uses
- Rooting of cuttings in *ornamental specimens*

Efficacy guidance
- Dip moistened base of cuttings into powder immediately before planting [1, 2]
- See label for details of concentrations recommended for promotion of rooting in cuttings of different species [3]
- Dip prepared cuttings in solution for 4-24 h depending on species [3]

Restrictions
- Maximum number of treatments 1 per cutting

Crop-specific information
- Latest use: before cutting insertion for ornamental specimens

Hazard classification and safety precautions
   Operator protection U08, U19a, U20a
   Environmental protection E15a, E34
   Storage and disposal D09a, D11a

---

335   (2-naphthyloxy)acetic acid

A plant growth regulator for setting tomato fruit

Products

Betapal Concentrate	Vitax	16 g/l	SL	00234

Uses
- Increasing fruit set in *outdoor tomatoes, protected tomatoes*

Efficacy guidance
- Apply with any fine sprayer or syringe when first half dozen flowers are open

Restrictions
- Do not spray growing head of tomato plants

Crop-specific information
- Spray actual trusses only once as they develop when about half the flowers are open
- Take care not to spray the growing head of the plant

Hazard classification and safety precautions
   Operator protection U20c
   Environmental protection E15a
   Storage and disposal D09a, D10a

---

336   napropamide

A soil applied amide herbicide for oilseed rape, fruit and woody ornamentals

Products

1	AC 650	United Phosphorus	450 g/l	SC	11102
2	Devrinol	United Phosphorus	450 g/l	SC	09374

Uses
- Annual dicotyledons in *blackcurrants, broccoli, brussels sprouts, cabbages, calabrese, cauliflowers, container-grown woody ornamentals, forest nurseries, gooseberries, kale, raspberries, strawberries, woody ornamentals* [2]; *winter oilseed rape* [1]
- Annual grasses in *blackcurrants, broccoli, brussels sprouts, cabbages, calabrese, cauliflowers, container-grown woody ornamentals, forest nurseries, gooseberries, kale, raspberries, strawberries, woody ornamentals* [2]; *winter oilseed rape* [1]
- Cleavers in *blackcurrants, broccoli, brussels sprouts, cabbages, calabrese, cauliflowers, container-grown woody ornamentals, forest nurseries, gooseberries, kale, raspberries, strawberries, woody ornamentals* [2]
- Groundsel in *blackcurrants, broccoli, brussels sprouts, cabbages, calabrese, cauliflowers, container-grown woody ornamentals, forest nurseries, gooseberries, kale, raspberries, strawberries, woody ornamentals* [2]

FOR FULL CONDITIONS OF USE ALWAYS READ THE PRODUCT LABEL

Efficacy guidance
- Best results obtained from treatment pre-emergence of weeds but product may be used in conjunction with contact herbicide such as paraquat for control of emerged weeds. Otherwise remove existing weeds before application
- Weed control may be reduced where spray is mixed too deeply in the soil
- Manure, crop debris or other organic matter may reduce weed control
- Napropamide broken down by sunlight, so application during such conditions not recommended. Most crops recommended for treatment between Nov and end-Feb
- Apply to winter oilseed rape pre-emergence or as pre-drilling treatment in tank-mixture with trifluralin and incorporate within 30 min [1]
- Post-emergence use of specific grass weedkilller recommended where volunteer cereals are a serious problem
- Increase water volume to ensure adequate dampening of compost when treating containerised nursery stock with dense leaf canopy [2]

Restrictions
- Maximum number of treatments 1 per crop or yr
- Do not use on Sands
- Do not use on soils with more than 10% organic matter
- Strawberries, blackcurrants, gooseberries and raspberries must be treated between 1 Nov and end Feb
- Applications to strawberries must only be made where the mature foliage has been previously removed
- Applications to ornamentals and forest nurseries must be made after 1 Nov and before end Apr
- Do not treat ornamentals in containers of less than 1 litre
- Some phytotoxicity seen on yellow and golden varieties of conifers and container grown alpines. On any ornamental variety treat only a small number of plants in first season
- Consult processors before use on any crop for processing

Crop-specific information
- Latest use: before transplanting for brassicas; pre-emergence for winter oilseed rape [1]; before end Feb for strawberries, bush and cane fruit; before end of Apr for field and container grown ornamental trees and shrubs
- Where minimal cultivation used to establish oilseed rape, tank-mixture may be applied directly to stubble and mixed into top 25 mm as part of surface cultivations [1]
- Apply up to 14 d prior to drilling winter oilseed rape [1]
- Apply to strawberries established for at least one season or to maiden crops as long as planted carefully and no roots exposed, between 1 Nov and end Feb. Do not treat runners of poor vigour or with shallow roots, or runner beds
- Newly planted ornamentals should have no roots exposed. Do not treat stock of poor vigour or with shallow roots. Treatments made in Mar and Apr must be followed by 25 mm irrigation within 24 h
- Bush and cane fruit must be established for at least 10 mth before spraying and treated between 1 Nov and end Feb

Following crops guidance
- After use in fruit or ornamentals no crop can be drilled within 7 mth of treatment. Leaf, flowerhead, root and fodder brassica crops may be drilled after 7 mth; potatoes, maize, peas or dwarf beans after 9 mth; autumn sown wheat or grass after 18 mth; any crop after 2 yr
- After use in oilseed rape only oilseed rape, swedes, fodder turnips, brassicas or potatoes should be sown within 12 mth of application
- Soil should be mould-board ploughed to a depth of at least 200 mm before drilling or planting any following crop

Environmental safety
- Dangerous for the environment
- Toxic to aquatic organisms

Hazard classification and safety precautions
Hazard H04, H11
Risk phrases R36, R38, R51, R53a
Operator protection A [1, 2]; C [2]; U05a, U09a, U11, U20b

SEE SECTION 3 FOR PRODUCTS ALSO REGISTERED

Environmental protection E34, E38
Storage and disposal D01, D02, D05, D09a, D10c, D12a, D12b

337   natural plant extracts

A contact insecticide that works by physical action

Products

Majestik	Certis	SL	-

Uses
- Aphids in *fruit crops, ornamental plant production, protected fruit, protected ornamentals, protected vegetables, vegetables*
- Leafhoppers in *fruit crops, ornamental plant production, protected fruit, protected ornamentals, protected vegetables, vegetables*
- Mites in *fruit crops, ornamental plant production, protected fruit, protected ornamentals, protected vegetables, vegetables*
- Thrips in *fruit crops, ornamental plant production, protected fruit, protected ornamentals, protected vegetables, vegetables*
- Whitefly in *fruit crops, ornamental plant production, protected fruit, protected ornamentals, protected vegetables, vegetables*

Approval information
- Product not controlled by Control of Pesticides Regulations/Plant Protection Products Regulations because it acts by physical means only

Efficacy guidance
- Product acts by coating target pests and inhibiting respiration or movement by physical action
- Effective on all growth stages of mites, aphids, thrips, whitefly and leafhoppers. Treat as soon as insects are seen and repeat as necessary
- Ensure thorough spray coverage of plant, paying special attention to growing points and the underside of leaves
- Consult manufacturer for guidance on efficacy before treating outdoor crops for the first time
- Product may be used in conjunction with biological control agents. Spray 24 h before they are introduced

Restrictions
- No limit on the number of treatments and no minimum interval between applications
- Before large scale use on a new crop, treat a few plants to check for crop safety
- Do not treat ornamental crops when in flower
- Avoid treating fruit or vegetables close to harvest as product may leave a sticky deposit

Crop-specific information
- HI zero. Product is not taken into plant

Environmental safety
- If bees are being used for pollination ensure hives are closed before treatment and do not re-open before spray has completely dried on the plants

Hazard classification and safety precautions
Operator protection U05a, U09a, U19a, U20c

338   nicosulfuron

A sulfonylurea herbicide for maize

Products

1 Amaize	AgriGuard	40 g/l	SC	12298
2 Samson	Syngenta	40 g/l	SC	10433
3 Samson	Syngenta	40 g/l	SC	12141

Uses
- Annual dicotyledons in *maize* [1-3]
- Annual meadow grass in *maize* [1-3]
- Chickweed in *maize* [1-3]

FOR FULL CONDITIONS OF USE ALWAYS READ THE PRODUCT LABEL

- Couch in *maize* [1-3]
- Grass weeds in *sweetcorn* *(off-label)* [2]
- Mayweeds in *maize* [1-3]
- Ryegrass in *maize* [1-3]
- Shepherd's purse in *maize* [1-3]
- Volunteer oilseed rape in *maize* [1-3]

Specific Off-Label Approvals (SOLAs)
- *sweetcorn* *(OLA 040840) Dec 2008* [2]

Efficacy guidance
- Product should be applied post-emergence between 2 and 8 crop leaf stage to emerged weeds from the 2-leaf stage
- Product acts mainly by foliar activity. Ensure good spray cover of the weeds
- Always follow WRAG guidelines for preventing and managing herbicide resistant weeds, especially when using in continuously grown maize. Section 5 for more information

Restrictions
- Maximum total dose equivalent to one full dose treatment
- Do not use if an organophosphorus soil insecticide has been used or if an organophosphorus foliar insecticide will be used on the same crop
- Do not mix with foliar or liquid fertilisers or specified herbicides. See label for details
- Do not apply in mixture, or sequence, with any other sulfonyl-urea containing product
- Do not treat crops under stress
- Do not apply if rainfall is forecast to occur within 6 h of application

Crop-specific information
- Latest Use: up to and including 8 true leaves for maize
- Some transient yellowing may be seen from 1-2 wk after treatment
- Only healthy maize crops growing in good field conditions should be treated

Following crops guidance
- Winter wheat (not undersown) may be sown 4 mth after treatment; maize may be sown in the spring following treatment; all other crops may be sown from the next autumn

Environmental safety
- Dangerous for the environment [3]
- Very toxic to aquatic organisms [3]
- May cause long-term adverse effects in the aquatic environment [3]
- Dangerous to fish or other aquatic life. Do not contaminate surface waters or ditches with chemical or used container [1]
- LERAP Category B

Hazard classification and safety precautions
 Hazard H04 [1-3]; H11 [3]
 Risk phrases R38 [1-3]; R50, R53a [3]
 Operator protection A, H; U02a, U05a, U14, U20b [1-3]; U10 [2, 3]; U15 [1]
 Environmental protection E13b [1]; E15a [2, 3]; E16a, E16b [1-3]
 Storage and disposal D01, D02 [1-3]; D10c [2, 3]
 Medical advice M05a [2, 3]

339 nicotine

A general purpose, non-persistent, contact, alkaloid insecticide

Products

1	Nicotine 40% Shreds	Dow	40% w/w	FU	05725
2	No-Fid	Certis	75 g/l	LI	11183
3	Stalwart	United Phosphorus	75 g/l	SL	11877
4	XL-All Insecticide	Vitax	70 g/l	SL	02369
5	XL-All Nicotine 95%	Vitax	950 g/l	SL	07402

Uses
- Aphids in *all soft fruit, protected asparagus, protected celery, protected marrows, vegetables* [4]; *all top fruit, flower bulbs, leaf brassicas, navy beans, nursery stock, ornamental specimens, parsley, peas, protected carnations, protected chrysanthemums,*

SEE SECTION 3 FOR PRODUCTS ALSO REGISTERED

protected crops, protected flowers, protected melons, protected roses [4, 5]; *apples, artichokes, asparagus, broccoli, brussels sprouts, cabbages, calabrese, cauliflowers, celeriac, chicory, chinese cabbage, chives, combining peas, courgettes, cress, cucumbers, dwarf beans, edible podded peas, endives, fennel, frise, garlic, kale, lamb's lettuce, marrows, ornamental plant production, outdoor tomatoes, peppers, potatoes, pumpkins, radicchio, radishes, shallots, spring field beans, squashes, sweetcorn, vining peas, winter field beans* [2, 3]; *aubergines, protected herbs (see appendix 6), protected potatoes, protected salad onions* [1]; *broad beans, carrots, celery, leaf spinach, lettuce, parsnips, red beet, runner beans, swedes, turnips* [2-5]; *bush fruit, protected figs, protected grapevines, protected peaches* [5]; *leeks, onions* [2-4]; *protected courgettes, protected peppers* [1, 4]; *protected cucumbers, protected lettuce* [1, 4, 5]; *protected ornamentals, protected tomatoes* [1-5]; *strawberries* [2, 3, 5]

- Capsids in *all soft fruit, vegetables* [4]; *all top fruit, flower bulbs, protected crops* [4, 5]; *apples* [2, 3]; *bush fruit* [5]
- Caterpillars in *artichokes, asparagus, broad beans, broccoli, brussels sprouts, cabbages, calabrese, carrots, cauliflowers, celeriac, celery, chicory, chinese cabbage, chives, combining peas, courgettes, cress, cucumbers, dwarf beans, edible podded peas, endives, fennel, frise, garlic, kale, lamb's lettuce, leaf spinach, leeks, lettuce, marrows, onions, ornamental plant production, outdoor tomatoes, parsnips, peppers, potatoes, protected ornamentals, protected tomatoes, pumpkins, radicchio, radishes, red beet, runner beans, shallots, spring field beans, squashes, swedes, sweetcorn, turnips, vining peas, winter field beans* [2, 3]
- Glasshouse whitefly in *aubergines, protected courgettes, protected cucumbers, protected herbs (see appendix 6), protected lettuce, protected ornamentals, protected peppers, protected potatoes, protected salad onions, protected tomatoes* [1]
- Green leafhopper in *aubergines, protected courgettes, protected cucumbers, protected herbs (see appendix 6), protected lettuce, protected ornamentals, protected peppers, protected potatoes, protected salad onions, protected tomatoes* [1]
- Insect pests in *french beans (off-label)* [4]
- Leaf miner in *artichokes, asparagus, broad beans, broccoli, brussels sprouts, cabbages, calabrese, carrots, cauliflowers, celeriac, celery, chicory, chinese cabbage, chives, combining peas, courgettes, cress, cucumbers, dwarf beans, edible podded peas, endives, fennel, frise, garlic, kale, lamb's lettuce, leaf spinach, leeks, lettuce, marrows, onions, ornamental plant production, outdoor tomatoes, parsnips, peppers, potatoes, protected ornamentals, protected tomatoes, pumpkins, radicchio, radishes, red beet, runner beans, shallots, spring field beans, squashes, swedes, sweetcorn, turnips, vining peas, winter field beans* [2, 3]; *flower bulbs, leaf brassicas, protected crops* [4, 5]
- Leafhoppers in *all soft fruit, vegetables* [4]; *bush fruit* [5]; *flower bulbs, protected crops* [4, 5]
- Potato virus vectors in *chitting potatoes* [1]
- Sawflies in *all soft fruit, vegetables* [4]; *all top fruit, flower bulbs, protected crops* [4, 5]; *bush fruit* [5]; *gooseberries* [2, 3]
- Thrips in *aubergines, protected courgettes, protected cucumbers, protected herbs (see appendix 6), protected lettuce, protected ornamentals, protected peppers, protected potatoes, protected salad onions, protected tomatoes* [1]; *flower bulbs, protected crops* [4, 5]; *vegetables* [4]

Specific Off-Label Approvals (SOLAs)
- *french beans (OLA 920080) Dec 2008* [4]

Efficacy guidance
- Apply as foliar spray, taking care to cover undersides of leaves and repeat as necessary or dip young plants, cuttings or strawberry runners before planting out
- Best results achieved by spraying at air temperatures above 16°C but do not treat in bright sunlight or windy weather [1]
- Fumigate glasshouse crops at temperatures of at least 16°C. See label for details of recommended fumigation procedure [1]
- Potatoes may be fumigated in chitting houses to control virus-spreading aphids [1]
- May be used in integrated control systems to give partial control of organophosphorus-, organochlorine- and pyrethroid-resistant whitefly

FOR FULL CONDITIONS OF USE ALWAYS READ THE PRODUCT LABEL

Restrictions

- Nicotine is subject to the Poisons Rules 1982 and the Poisons Act 1972. See notes in Section 5 [1, 5]
- Maximum number of treatments not specified
- On plants of unknown sensitivity test first on a small scale
- Keep unprotected persons out of treated glasshouses for at least 12 h
- Wear overall, hood, rubber gloves and respirator if entering glasshouse within 12 h of fumigating [1]

Crop-specific information

- HI 2 d for most crops - check labels

Environmental safety

- Dangerous for the environment [5]
- Toxic to aquatic organisms [5]
- Dangerous to fish or other aquatic life. Do not contaminate surface waters or ditches with chemical or used container
- Dangerous to livestock. Keep all livestock out of treated areas/away from treated water for at least 12 h
- Dangerous to game, wild birds and animals
- Harmful to bees. Do not apply to crops in flower or to those in which bees are actively foraging. Do not apply when flowering weeds are present
- Keep in original container, tightly closed, in a safe place, under lock and key [1, 5]

Hazard classification and safety precautions

Hazard H01 [2, 3, 5]; H02 [4]; H03, H07 [1]; H11 [5]

Risk phrases R20, R21 [1]; R22a [1, 2, 4]; R24 [4]; R25, R27 [2, 3, 5]; R36 [2, 3]; R37 [2]; R38 [3]; R51 [5]; R52 [2-4]; R53a [2-5]

Operator protection A [1-5]; C [2, 3, 5]; D, J [1]; H [1, 5]; K, M [5]; U02a, U04a, U05a, U10, U13 [2-5]; U05b [1]; U08 [4, 5]; U09a [2, 3]; U19a, U20a [1-5]

Environmental protection E02a, E06b [1-5] (12 h); E10a, E12f [1-5]; E13b [1-3]; E15a [4, 5]; E34 [1, 4, 5]

Consumer protection C02 [1] (24 h); C02 [4, 5] (2 d)

Storage and disposal D01, D02 [1, 3-5]; D05, D10b [2, 3]; D09a [2-4]; D09b [1, 5]; D10a [4, 5]; D11a [1]

Medical advice M03 [1]; M04 [2-5]

340   oxadiazon

A residual and contact oxadiazolone herbicide for fruit and ornamentals

Products

1	Ronstar 2G	Certis	2% w/w	GR	10011
2	Ronstar Liquid	Certis	250 g/l	EC	11215

Uses

- Annual dicotyledons in **apple orchards, blackcurrants, gooseberries, grapevines, hops, pear orchards, raspberries, woody ornamentals** [2]; **container-grown ornamentals** [1]
- Annual grasses in **apple orchards, blackcurrants, gooseberries, grapevines, hops, pear orchards, raspberries, woody ornamentals** [2]; **container-grown ornamentals** [1]
- Bindweeds in **apple orchards, blackcurrants, gooseberries, grapevines, hops, pear orchards, raspberries, woody ornamentals** [2]
- Cleavers in **apple orchards, blackcurrants, gooseberries, grapevines, hops, pear orchards, raspberries, woody ornamentals** [2]
- Knotgrass in **apple orchards, blackcurrants, gooseberries, grapevines, hops, pear orchards, raspberries, woody ornamentals** [2]

Approval information

- Accepted by BBPA for use on hops

SEE SECTION 3 FOR PRODUCTS ALSO REGISTERED

SECTION 2

Efficacy guidance
- Rain or overhead watering is needed soon after application for effective results
- Pre-emergence activity reduced on soils with more than 10% organic matter and, in these conditions, post-emergence treatment is more effective [2]
- Best results on bindweed when first shoots are 10-15 cm long

Restrictions
- See label for list of ornamental species which may be treated with granules. Treat small numbers of other species to check safety. Do not treat hydrangea, spiraea or genista
- Do not cultivate after treatment [2]
- Do not treat container stock under glass or use on plants rooted in media with high sand or non-organic content. Do not apply to plants with wet foliage
- Do not use more than 8 l/ha in any 12 mth period [2]

Crop-specific information
- Latest use: Jul for apples, grapevines, hops, pears; Jun for raspberries, woody ornamentals
- Apply spray to apples and pears from Jan to Jul, avoiding young growth
- Treat bush fruit from Jan to bud-break, avoiding bushes, grapevines in Feb/Mar before start of new growth or in Jun/Jul avoiding foliage
- Treat hops cropped for at least 2 yr in Feb or in Jun/Jul after deleafing
- Treat woody ornamentals from Jan to Jun, avoiding young growth. Do not spray container stock overall

Environmental safety
- Dangerous for the environment
- Very toxic to aquatic organisms

Hazard classification and safety precautions
    Hazard H03, H08, H11 [2]
    Risk phrases R22b, R36, R38, R41, R50, R67 [2]; R52 [1]; R53a [1, 2]
    Operator protection A, C; U05a, U14, U15, U20b [1, 2]; U08 [2]; U09a, U19a [1]
    Environmental protection E15a [1, 2]; E38 [2]
    Storage and disposal D01, D02, D09a, D12a [1, 2]; D10a [2]; D11a [1]
    Medical advice M05b [2]

341 oxamyl

A soil-applied, systemic oxime carbamate nematicide and insecticide

Products

Vydate 10G	DuPont	10% w/w	GR	02322

Uses
- American serpentine leaf miner in **aubergines** *(off-label)*, **ornamental plant production** *(off-label)*, **protected broad beans** *(off-label)*, **protected ornamentals** *(off-label)*, **protected peppers** *(off-label)*, **protected soya beans** *(off-label)*, **protected tomatoes** *(off-label)*
- Aphids in **potatoes**, **sugar beet**
- Docking disorder vectors in **sugar beet**
- Free-living nematodes in **potatoes**
- Leaf miner in **aubergines** *(off-label)*, **ornamental plant production** *(off-label)*, **protected broad beans** *(off-label)*, **protected ornamentals** *(off-label)*, **protected peppers** *(off-label)*, **protected soya beans** *(off-label)*, **protected tomatoes** *(off-label)*
- Mangold fly in **sugar beet**
- Millipedes in **sugar beet**
- Potato cyst nematode in **potatoes**
- Pygmy beetle in **sugar beet**
- South American leaf miner in **aubergines** *(off-label)*, **ornamental plant production** *(off-label)*, **protected broad beans** *(off-label)*, **protected ornamentals** *(off-label)*, **protected peppers** *(off-label)*, **protected soya beans** *(off-label)*, **protected tomatoes** *(off-label)*
- Spraing vectors in **potatoes**
- Stem nematodes in **carrots** *(off-label)*, **garlic** *(off-label)*, **parsnips** *(off-label)*, **protected garlic** *(off-label)*, **protected onions** *(off-label)*

FOR FULL CONDITIONS OF USE ALWAYS READ THE PRODUCT LABEL

Specific Off-Label Approvals (SOLAs)
- *aubergines, ornamental plant production, protected broad beans, protected ornamentals, protected peppers, protected soya beans, protected tomatoes* (OLA 930020) Dec 2008 [1]
- *carrots, parsnips* (OLA 040617) Dec 2008 [1]
- *garlic* (OLA 920163) Dec 2008 [1]
- *protected garlic, protected onions* (OLA 940925) Dec 2008 [1]

Approval information
- Oxamyl included in Annex under EC Directive 91/414

Efficacy guidance
- Apply granules with suitable applicator before drilling or planting. See label for details of recommended machines
- In potatoes incorporate thoroughly to 10 cm and plant within 3-4 d
- In sugar beet apply in seed furrow at drilling

Restrictions
- Oxamyl is subject to the Poisons Rules 1982 and the Poisons Act 1972. See notes in Section 5
- This product contains an anticholinesterase carbamate compound. Do not use if under medical advice not to work with such compounds
- Maximum number of treatments 1 per crop or yr
- Keep in original container, tightly closed, in a safe place, under lock and key
- Wear protective gloves if handling treated compost or soil within 2 wk after treatment
- Allow at least 12 h, followed by at least 1 h ventilation, before entry of unprotected persons into treated glasshouses

Crop-specific information
- Latest use: at drilling/planting for vegetables; before drilling/planting for potatoes and peas.
- HI tomatoes, ornamentals 2 wk

Environmental safety
- Dangerous for the environment
- Toxic to aquatic organisms
- Dangerous to fish or other aquatic life. Do not contaminate surface waters or ditches with chemical or used container
- Dangerous to game, wild birds and animals. Bury spillages

Hazard classification and safety precautions
Hazard H02, H11
Risk phrases R23, R25, R51, R53a
Operator protection A, B, H, K, M [1]; C [1] (or D); U02a, U04a, U05a, U09a, U13, U19a, U20a
Environmental protection E10a, E13b, E34, E38
Consumer protection C02 (3 wk)
Storage and disposal D01, D02, D09a, D11a, D12a
Medical advice M02, M04

342 oxycarboxin

A protectant, eradicant and systemic carboxamide fungicide for use on ornamentals

Products
Plantvax 75	Fargro	75% w/w	WP	01601

Uses
- Rust in *ornamental specimens, protected ornamentals*

Approval information
- Products containing this active ingredient have been granted derogations for specified 'Essential Uses' for use until 31 December 2007. Sale and supply must cease by 30 June 2007 but growers have no guarantee that the products will continue to be available until then.
  For more information see 'The Review Programme' under 'Pesticide Legislation' in Section 5

Efficacy guidance
- Apply as a routine to established carnation beds from Sep to Feb at 7-10 d intervals. On rust-susceptible species or if rust already established add authorised wetting agent and repeat every 10-14 d until infection pressure ceases

SEE SECTION 3 FOR PRODUCTS ALSO REGISTERED

SECTION 2

- On stock plants or cuttings of carnations or geraniums apply after striking and repeat 3 times at 10-14 d intervals. Remove and burn any infected cuttings
- Can be used as a drench on carnations grown in peat bags

Restrictions
- This product must not be used on any crops other than those listed, including any extrapolations that would normally be permissible under the Long Term Arrangements for Extension of Use (see Section 1)

Crop-specific information
- Spray product alone and preferably not within 2 d of other sprays
- Can cause marginal leaf scorch when taken up by roots. Avoid spraying to run-off
- Product leaves visible deposit on plants so avoid treating when blooms are open

Environmental safety
- Harmful to aquatic organisms
- May cause long-term adverse effects in the aquatic environment
- Dangerous to fish or other aquatic life. Do not contaminate surface waters or ditches with chemical or used container

Hazard classification and safety precautions
Hazard H03
Risk phrases R22a, R38, R52, R53a
Operator protection A, C; U02a, U05a, U09a, U19a, U20b
Environmental protection E13b, E38
Storage and disposal D01, D02, D09a, D11a
Medical advice M03

343   paclobutrazol

A conazole plant growth regulator for ornamentals and fruit

Products
1 Bonzi	Syngenta Bioline	4 g/l	SC	10517
2 Cultar	Syngenta	250 g/l	SC	10523

Uses
- Control of shoot growth in **apples**, **pears** [2]
- Growth regulation in **cherries** *(off-label)*, **plums** *(off-label)* [2]
- Improving colour in **poinsettias** [1]
- Increasing flowering in **azaleas**, **bedding plants**, **begonias**, **kalanchoes**, **lilies**, **roses**, **tulips** [1]
- Increasing fruit set in **apples**, **pears** [2]
- Stem shortening in **azaleas**, **bedding plants**, **begonias**, **kalanchoes**, **lilies**, **poinsettias**, **roses**, **tulips** [1]

Specific Off-Label Approvals (SOLAs)
- **cherries**, **plums** *(OLA 031235) Dec 2008* [2]

Efficacy guidance
- Chemical is active via both foliage and root uptake. For best results apply in dull weather when relative humidity not high
- Apply as spray to produce compact pot plants and to improve bract colour of poinsettias [1]
- Apply as compost drench to reduce flower stem length of potted tulips and Mid-Century hybrid lilies [1]
- Timing is critical and varies with species. See label for details [1]

Restrictions
- Maximum number of treatments 1 per specimen for some species [1]
- Maximum total dose equivalent to 3 full dose treatments for pears and 4 full dose treatments for apples [2]
- Some varietal restrictions apply in top fruit (see label for details) [2]
- Do not use on trees of low vigour or under stress [2]
- Do not use on trees from green cluster to 2 wk after full petal fall [2]
- Do not use in underplanted orchards or those recently interplanted [2]

FOR FULL CONDITIONS OF USE ALWAYS READ THE PRODUCT LABEL

Crop-specific information
- HI apples, pears 14 d [2]
- Apply to apple and pear trees under good growing conditions as pre-blossom spray (apples only) and post-blossom at 7-14 d intervals [2]
- Timing and dose of orchard treatments vary with species and cultivar. See label [2]

Following crops guidance
- Chemical has residual soil activity which can affect growth of following crops. Withhold treatment from orchards due for grubbing to allow the following intervals between the last treatment and planting the next crop: apples, pears 1 yr; beans, peas, onions 2 yr; stone fruit 3 yr; cereals, grass, oilseed rape, carrots 4 yr; market brassicas 6 yr; potatoes and other crops 7 yr [2]

Environmental safety
- Harmful to aquatic organisms
- Harmful to fish or other aquatic life. Do not contaminate surface waters or ditches with chemical or used container
- Do not use on food crops [1]
- Permissible in organic systems
- Keep livestock out of treated areas for at least 2 yr following treatment

Hazard classification and safety precautions
Hazard H04 [2]
Risk phrases R36, R52, R53a [2]
Operator protection A, H, M [2]; U05a [2]; U08 [1]; U20a [1, 2]
Environmental protection E06a [2] (2 yr); E13c [1]; E15a [1, 2]; E34, E38 [2]
Consumer protection C01 [1]; C02 [2] (2 wk)
Storage and disposal D01, D02, D10c, D12a [2]; D05, D10b [1]; D09a [1, 2]
Medical advice M03 [2]

344   paraffin oil (commodity substance)

An agent for the control of birds by egg treatment

Products

paraffin oil	various	OL

Uses
- Birds in **miscellaneous pest control situations** *(egg treatment)*

Approval information
- Approval for the use of paraffin oil as a commodity substance was granted on 25 April 1995 by Ministers under regulation 5 of the Control of Pesticides Regulations 1986
- Only to be used where a licence has been approved in accordance with Section 16(1) of the Wildlife and Countryside Act (1981)

Efficacy guidance
- Egg treatment should be undertaken as soon as clutch is complete
- Eggs should be treated by complete immersion in liquid paraffin

Restrictions
- Use to control eggs of birds covered by licences issued by the Agriculture and Environment Departments under Section 16(1) of the Wildlife and Countryside Act (1981)
- Treat eggs once only

Hazard classification and safety precautions
Operator protection A, C

SECTION 2

345   paraquat

A non-selective, non-residual contact bipyridilium herbicide

See also diquat + paraquat
diuron + paraquat

Products

1	Agriguard Paraquat	AgriGuard	200 g/l	SL	09951
2	Dextrone X	Nomix Enviro	200 g/l	SL	00687
3	Fernpath Graminite	AgriGuard	200 g/l	SL	11646
4	Gramoxone 100	Syngenta	200 g/l	SL	10526

Uses

- Annual dicotyledons in **all top fruit**, **aubergines** *(off-label - inter-row directed treatment)*, **forest nursery beds**, **grapevines**, **grassland** *(sward destruction/direct drilling)*, **hardy ornamentals**, **lettuce** *(off-label - inter-row directed treatment)*, **lucerne** *(off-label)*, **protected celery** *(off-label - inter-row directed treatment)*, **protected courgettes** *(off-label - inter-row directed treatment)*, **protected cucumbers** *(off-label - inter-row directed treatment)*, **protected gherkins** *(off-label - inter-row directed treatment)*, **protected lettuce** *(off-label - inter-row directed treatment)*, **protected marrows** *(off-label - inter-row directed treatment)*, **protected melons** *(off-label - inter-row directed treatment)*, **protected onions** *(off-label - inter-row directed treatment)*, **protected ornamentals** *(off-label - inter-row directed treatment)*, **protected peppers** *(off-label - inter-row directed treatment)*, **protected pumpkins** *(off-label - inter-row directed treatment)*, **protected squashes** *(off-label - inter-row directed treatment)*, **protected tomatoes** *(off-label - inter-row directed treatment)*, **rhubarb** *(off-label)*, **row crops** *(stale seedbed/inter-row)* [4]; **blackcurrants**, **cultivated land/soil** *(minimum cultivation)*, **gooseberries**, **hops**, **potatoes**, **raspberries**, **stubbles**, **sugar beet** [1, 3, 4]; **flower bulbs**, **forest**, **forestry transplants**, **non-crop areas** [1-4]; **forest nursery beds** *(stale seedbed)* [1-3]; **orchards**, **redcurrants**, **row crops**, **whitecurrants** [1, 3]; **woody ornamentals** [2]
- Annual grasses in **all top fruit**, **aubergines** *(off-label - inter-row directed treatment)*, **grapevines**, **hardy ornamentals**, **lettuce** *(off-label - inter-row directed treatment)*, **lucerne** *(off-label)*, **protected celery** *(off-label - inter-row directed treatment)*, **protected courgettes** *(off-label - inter-row directed treatment)*, **protected cucumbers** *(off-label - inter-row directed treatment)*, **protected gherkins** *(off-label - inter-row directed treatment)*, **protected lettuce** *(off-label - inter-row directed treatment)*, **protected marrows** *(off-label - inter-row directed treatment)*, **protected melons** *(off-label - inter-row directed treatment)*, **protected onions** *(off-label - inter-row directed treatment)*, **protected ornamentals** *(off-label - inter-row directed treatment)*, **protected peppers** *(off-label - inter-row directed treatment)*, **protected pumpkins** *(off-label - inter-row directed treatment)*, **protected squashes** *(off-label - inter-row directed treatment)*, **protected tomatoes** *(off-label - inter-row directed treatment)*, **rhubarb** *(off-label)*, **row crops** *(stale seedbed/inter-row)* [4]; **blackcurrants**, **cultivated land/soil** *(minimum cultivation)*, **gooseberries**, **grassland** *(sward destruction/direct drilling)*, **hops**, **potatoes**, **raspberries**, **stubbles**, **sugar beet** [1, 3, 4]; **flower bulbs**, **forest**, **forest nursery beds** *(stale seedbed)*, **forestry transplants**, **non-crop areas** [1-4]; **orchards**, **redcurrants**, **row crops**, **whitecurrants** [1, 3]; **woody ornamentals** [2]
- Barren brome in **field crops** *(stubble treatment)* [1, 3]; **stubbles** [4]
- Chemical stripping in **hops** [1, 3, 4]
- Creeping bent in **all top fruit**, **hardy ornamentals** [4]; **blackcurrants**, **cultivated land/soil** *(minimum cultivation)*, **gooseberries**, **grassland** *(sward destruction/direct drilling)*, **hops**, **raspberries**, **stubbles** [1, 3, 4]; **flower bulbs**, **forest**, **forestry transplants**, **non-crop areas** [1-4]; **orchards**, **redcurrants**, **whitecurrants** [1, 3]; **woody ornamentals** [2]
- Firebreak desiccation in **forest** [1-4]
- Green cover in **field margins** [4]; **land not being used for crop production** [1, 3, 4]
- Perennial ryegrass in **all top fruit** [4]; **blackcurrants**, **gooseberries**, **grassland** *(sward destruction/direct drilling)*, **hops**, **raspberries** [1, 3, 4]; **non-crop areas** [1-4]; **orchards**, **redcurrants**, **whitecurrants** [1, 3]
- Rough meadow grass in **all top fruit** [4]; **blackcurrants**, **gooseberries**, **grassland** *(sward destruction/direct drilling)*, **hops**, **raspberries** [1, 3, 4]; **non-crop areas** [1-4]; **orchards**, **redcurrants**, **whitecurrants** [1, 3]
- Runner desiccation in **strawberries** [1, 3, 4]

FOR FULL CONDITIONS OF USE ALWAYS READ THE PRODUCT LABEL

- Volunteer cereals in **cultivated land/soil** *(minimum cultivation)*, **potatoes**, **stubbles**, **sugar beet** [1, 3, 4]; **row crops** [1, 3]; **row crops** *(stale seedbed/inter-row)* [4]
- Wild oats in **stubbles** [1, 3, 4]

Specific Off-Label Approvals (SOLAs)
- **aubergines, lettuce, protected celery, protected courgettes, protected cucumbers, protected gherkins, protected lettuce, protected marrows, protected melons, protected onions, protected ornamentals, protected peppers, protected pumpkins, protected squashes, protected tomatoes** *(inter-row directed treatment) (OLA 020225) Dec 2008* [4]
- **lucerne, rhubarb** *(OLA 020225) Dec 2008* [4]

Approval information
- Paraquat included in Annex I under EC Directive 91/414
- Accepted by BBPA for use on hops

Efficacy guidance
- Apply to green weeds preferably less than 15 cm high. Spray is rainfast after 10 min
- Addition of wetter recommended with lower application rates
- Spray in autumn to suppress couch when shoots have 2 leaves, repeat as necessary and plough after last treatment
- For direct-drilling land should be free of perennial weeds and protection against slugs should be provided. Allow 7-10 d before drilling into sprayed grass
- Best time for use in top fruit is Nov-Apr
- For forestry fire-break use apply in Jul-Aug and fire 7-10 d later
- For interrow use in row crops apply with guarded, no-drift sprayer
- Use as a carefully directed spray in blackcurrants, gooseberries, grapevines and other fruit crops, in raspberries only when dormant
- For stawberry runner control use guarded sprayer, not when flowers or fruit present
- Apply to bulbs pre-emergence (at least 3 d pre-emergence on sandy soils) or at end of season, provided no attached foliage and bulbs well covered (not on sandy soils)
- In forestry seedbed apply up to 3 d before seedling emergence

Restrictions
- Paraquat is subject to the Poisons Rules 1982 and the Poisons Act 1972. See notes in Section 5
- Maximum number of treatments not stated for many situations but normally 1 per crop or yr; 3 per yr on leafy herbs
- Products in this profile are for professional use only
- Do not put in a food or drinks container
- Do not use on straw or other artificial growing media
- Do not use on potatoes under hot, dry conditions or on hops under drought conditions
- If using around glasshouses ensure vents and doors closed
- Only use clean water for mixing up spray
- Allow at least 4 h before cultivating, leave overnight if possible
- Observe label restrictions for maximum permitted concentration when spraying at low volume

Crop-specific information
- Latest use: before shoots 15 cm high and before 10% of shoots emerged for early potatoes, before 40% of shoots emerged for maincrop potatoes [1, 3, 4]; before 31 Mar for lucerne; see labels for other crops and situations
- HI mint, leafy herbs 8 wk
- Apply to lucerne in late Feb/early Mar when crop dormant

Following crops guidance
- Chemical is inactivated on contact with soil. Crops can be sown or planted soon after spraying on most soils but 3 d should be allowed on sandy or immature peat soils

Environmental safety
- Dangerous for the environment
- Very toxic to aquatic organisms
- Paraquat may be harmful to hares; stubbles must be sprayed early in the day [1, 3, 4]
- Keep in original container, tightly closed, in a safe place, under lock and key
- Harmful to livestock
- Keep all livestock out of treated areas for at least 24 h
- Do not contaminate surface waters or ditches with chemical or used container

SEE SECTION 3 FOR PRODUCTS ALSO REGISTERED

Hazard classification and safety precautions
    Hazard H02, H11
    Risk phrases R20 [1, 3]; R21, R36, R37, R38 [1-4]; R22a [4]; R25 [1-3]; R48, R50, R53a [2, 4]
    Operator protection A, C; U02a [1-3]; U04a, U20a [1, 3]; U04b, U20b [2, 4]; U04c [4]; U05a, U09a, U19a [1-4]; U08 [3]
    Environmental protection E06c [1-4] (24 h); E11, E15a, E34 [1-4]; E38 [2, 4]
    Storage and disposal D01, D02, D09b [1-4]; D05, D10a [1, 3]; D10c [4]; D11a [2]; D12a [2, 4]
    Medical advice M04 [1-4]; M05c [3, 4]

346   penconazole

A protectant conazole fungicide with antisporulant activity

Products
    Topas   Syngenta   100 g/l   EC   09717

Uses
  - Powdery mildew in **apples**, **blackcurrants**, **hops**, **ornamental trees**
  - Rust in **roses**
  - Scab in **ornamental trees**

Approval information
  - Accepted by BBPA for use on hops

Efficacy guidance
  - Use as a protectant fungicide by treating at the earliest signs of disease
  - Treat crops every 10-14 d (every 7-10 d in warm, humid weather) at first sign of infection or as a protective spray ensuring complete coverage. See label for details of timing
  - Increase dose and volume with growth of hops but do not exceed 2000 l/ha. Little or no activity will be seen on established powdery mildew
  - Antisporulant activity reduces development of secondary mildew in apples

Restrictions
  - Maximum number of treatments 10 per yr for apples, 6 per yr for hops, 4 per yr for blackcurrants
  - Check for varietal susceptibility in roses. Some defoliation may occur after repeat applications on Dearest

Crop-specific information
  - HI apples, hops 14 d; blackcurrants 4 wk

Environmental safety
  - Dangerous for the environment
  - Toxic to aquatic organisms

Hazard classification and safety precautions
    Hazard H04, H11
    Risk phrases R36, R51, R53a
    Operator protection A, C; U02a, U05a, U08, U20b
    Environmental protection E15a, E38
    Storage and disposal D01, D02, D05, D09a, D10c, D12a
    Medical advice M05b

347   pencycuron

A non-systemic urea fungicide for use on seed potatoes

See also imazalil + pencycuron

Products

1 Agriguard Pencycuron	AgriGuard	12.5% w/w	DS	10445
2 Me2 Penny	Me2	12.5% w/w	DS	10369
3 Monceren DS	Bayer CropScience	12.5% w/w	DS	11292
4 Monceren Flowable	Bayer CropScience	250 g/l	FS	11425
5 Standon Pencycuron DP	Standon	12.5% w/w	DS	08774

FOR FULL CONDITIONS OF USE ALWAYS READ THE PRODUCT LABEL

SECTION 2

pendimethalin

Uses
- Black scurf in **potatoes** *(tuber treatment)*
- Stem canker in **potatoes** *(tuber treatment)*

Efficacy guidance
- Provides control of tuber-borne disease and gives some reduction of stem canker
- Seed tubers should be of good quality and vigour, and free from soil deposits when treated
- Dust formulations [1-3, 5] should be applied immediately before, or during, the planting process by treating seed tubers in chitting trays, in bulk bins immediately before planting or in hopper at planting. Liquid formulations [4] may be applied by misting equipment at any time, into or out of store but treatment over a roller table at the end of grading out, or at planting, is usually most convenient
- Whichever method is chosen, complete and even distribution on the tubers is essential for optimum efficacy
- Apply in accordance with detailed guidelines in manufacturer's literature
- If rain interrupts planting cover tubers in hopper

Restrictions
- Maximum number of treatments 1 per batch of seed potatoes
- Treated tubers must be used only as seed and not for human or animal consumption
- Do not use on tubers previously treated with a dry powder seed treatment or hot water
- Use of suitable dust mask is mandatory when applying dust, filling the hopper or riding on planter [1-3, 5]
- Treated tubers in bulk bins must be allowed to dry before being stored, especially when intended for cold storage [4]
- Some internal staining of boxes used to store treated tubers may occur. Ware tubers should not be stored or supplied to packers in such stained boxes [4]

Crop-specific information
- Latest use: at planting

Environmental safety
- Do not use treated seed as food or feed
- Treated seed harmful to game and wildlife

Hazard classification and safety precautions
Hazard H04 [4]
Risk phrases R43 [4]
Operator protection A [1-5]; F [1-3, 5]; U14, U20a [4]; U20b [1-3, 5]
Environmental protection E03, E15a
Storage and disposal D05 [4]; D07 [1, 2, 5]; D09a, D11a [1-5]
Treated seed S01, S02, S03, S04a, S05, S06a

348 pendimethalin

A residual dinitroaniline herbicide for cereals and other crops

See also bentazone + pendimethalin
cyanazine + pendimethalin
flufenacet + pendimethalin
isoproturon + pendimethalin

Products

1 Agriguard Pendimethalin	AgriGuard	400 g/l	SC	10864
2 Alpha Pendimethalin 330 EC	Makhteshim	330 g/l	EC	12214
3 Bema	Makhteshim	40% w/w	WG	11756
4 Blazer	Headland	330 g/l	EC	11217
5 Claymore	BASF	400 g/l	SC	11779
6 Greencrop Estuary	Greencrop	400 g/l	SC	12095
7 Landgold Pendimethalin 400	Landgold	400 g/l	SC	10288
8 Landgold Pendimethalin 400	Teliton	400 g/l	SC	12125
9 Me2 Pendimethalin	Me2	400 g/l	SC	12420
10 Stomp 400 SC	BASF	400 g/l	SC	11777

SEE SECTION 3 FOR PRODUCTS ALSO REGISTERED

Uses

- Annual dicotyledons in **apple orchards**, **blackberries**, **blackcurrants**, **broccoli**, **brussels sprouts**, **cabbages**, **calabrese**, **carrots**, **cauliflowers**, **cherries**, **forage maize**, **gooseberries**, **hops**, **leeks**, **loganberries**, **parsley**, **parsnips**, **pear orchards**, **plums**, **raspberries**, **strawberries**, **sunflowers** [1-6, 9, 10]; **asparagus** (off-label), **bulb onions** (off-label - pre + post emergence treatment), **fennel** (off-label), **garlic** (off-label - pre + post emergence treatment), **leeks** (off-label), **shallots** (off-label - pre + post emergence treatment), **sweetcorn** (off-label - under covers) [10]; **bulb onions**, **chives** (off-label), **lettuce** (off-label) [5, 10]; **combining peas**, **durum wheat**, **potatoes**, **spring barley**, **triticale**, **winter barley**, **winter rye**, **winter wheat** [1-10]; **cress** (off-label), **radicchio** (off-label), **scarole** (off-label), **sweetcorn** (off-label - under covers and mulches) [5]; **evening primrose** (off-label), **farm forestry** (off-label) [3, 10]; **herbs (see appendix 6)** (off-label), **runner beans** (off-label) [3, 5, 10]; **maize** (off-label), **onion sets** (off-label), **sweetcorn** (off-label) [3]; **onions** [1-4, 6, 9]; **rubus hybrids** [1, 3, 5, 6, 9, 10]; **tayberries** [2, 4]; **winter field beans** (off-label) [3, 5]
- Annual grasses in **combining peas**, **durum wheat**, **potatoes**, **spring barley**, **triticale**, **winter barley**, **winter rye**, **winter wheat** [7, 8]; **evening primrose** (off-label), **farm forestry** (off-label), **fennel** (off-label), **lettuce** (off-label), **runner beans** (off-label), **sweetcorn** (off-label - under covers) [10]
- Annual meadow grass in **apple orchards**, **blackberries**, **blackcurrants**, **broccoli**, **brussels sprouts**, **cabbages**, **calabrese**, **carrots**, **cauliflowers**, **cherries**, **combining peas**, **durum wheat**, **forage maize**, **gooseberries**, **hops**, **leeks**, **loganberries**, **parsley**, **parsnips**, **pear orchards**, **plums**, **potatoes**, **raspberries**, **spring barley**, **strawberries**, **sunflowers**, **triticale**, **winter barley**, **winter rye**, **winter wheat** [1-6, 9, 10]; **bulb onions** [5, 10]; **bulb onions** (off-label - pre + post emergence treatment), **garlic** (off-label - pre + post emergence treatment), **leeks** (off-label), **shallots** (off-label - pre + post emergence treatment) [10]; **onions** [1-4, 6, 9]; **rubus hybrids** [1, 3, 5, 6, 9, 10]; **tayberries** [2, 4]
- Blackgrass in **apple orchards**, **blackberries**, **blackcurrants**, **broccoli**, **brussels sprouts**, **cabbages**, **calabrese**, **carrots**, **cauliflowers**, **cherries**, **combining peas**, **durum wheat**, **gooseberries**, **hops**, **leeks**, **loganberries**, **parsley**, **parsnips**, **pear orchards**, **plums**, **raspberries**, **spring barley**, **strawberries**, **sunflowers**, **triticale**, **winter barley**, **winter rye**, **winter wheat** [1-6, 9, 10]; **bulb onions** [5, 10]; **bulb onions** (off-label - pre + post emergence treatment), **garlic** (off-label - pre + post emergence treatment), **leeks** (off-label), **shallots** (off-label - pre + post emergence treatment) [10]; **onions** [1-4, 6, 9]; **rubus hybrids** [1, 3, 5, 6, 9, 10]; **tayberries** [2, 4]
- Cleavers in **winter field beans** (off-label) [10]
- Field pansy in **daffodils** (off-label - for galanthamine production) [10]
- Knotgrass in **daffodils** (off-label - for galanthamine production), **edible podded peas** (off-label), **vining peas** (off-label) [10]
- Rough meadow grass in **apple orchards**, **blackberries**, **blackcurrants**, **broccoli**, **brussels sprouts**, **cabbages**, **calabrese**, **carrots**, **cauliflowers**, **cherries**, **combining peas**, **durum wheat**, **forage maize**, **gooseberries**, **hops**, **leeks**, **loganberries**, **parsley**, **parsnips**, **pear orchards**, **plums**, **potatoes**, **raspberries**, **spring barley**, **strawberries**, **sunflowers**, **triticale**, **winter barley**, **winter rye**, **winter wheat** [1-6, 9, 10]; **bulb onions** [5, 10]; **bulb onions** (off-label - pre + post emergence treatment), **garlic** (off-label - pre + post emergence treatment), **leeks** (off-label), **shallots** (off-label - pre + post emergence treatment) [10]; **onions** [1-4, 6, 9]; **rubus hybrids** [1, 3, 5, 6, 9, 10]; **tayberries** [2, 4]
- Speedwells in **daffodils** (off-label - for galanthamine production) [10]
- Volunteer oilseed rape in **apple orchards**, **blackberries**, **blackcurrants**, **broccoli**, **brussels sprouts**, **cabbages**, **calabrese**, **carrots**, **cauliflowers**, **cherries**, **combining peas**, **durum wheat**, **forage maize**, **gooseberries**, **hops**, **leeks**, **loganberries**, **parsley**, **parsnips**, **pear orchards**, **plums**, **potatoes**, **raspberries**, **spring barley**, **strawberries**, **sunflowers**, **triticale**, **winter barley**, **winter rye**, **winter wheat** [1-6, 9, 10]; **bulb onions** [5, 10]; **bulb onions** (off-label - pre + post emergence treatment), **daffodils** (off-label - for galanthamine production), **garlic** (off-label - pre + post emergence treatment), **leeks** (off-label), **shallots** (off-label - pre + post emergence treatment) [10]; **onions** [1-4, 6, 9]; **rubus hybrids** [1, 3, 5, 6, 9, 10]; **tayberries** [2, 4]
- Wild oats in **durum wheat**, **triticale**, **winter barley**, **winter rye**, **winter wheat** [1-6, 9, 10]

Specific Off-Label Approvals (SOLAs)

- **asparagus** (OLA 050934) Dec 2008 [10]
- **bulb onions**, **garlic**, **shallots** (pre + post emergence treatment) (OLA 042311) Dec 2008 [10]

FOR FULL CONDITIONS OF USE ALWAYS READ THE PRODUCT LABEL

- **chives, cress, herbs (see appendix 6), lettuce, radicchio, runner beans, scarole, winter field beans** *(OLA 052022) Dec 2008* [5]
- **chives, herbs (see appendix 6), lettuce** *(OLA 051927) Sep 2008* [10]
- **daffodils** *(for galanthamine production) (OLA 041515) Dec 2008* [10]
- **edible podded peas, vining peas** *(OLA 050520) Dec 2008* [10]
- **evening primrose** *(OLA 041057) Dec 2008* [3]
- **evening primrose** *(OLA 042301) Dec 2008* [10]
- **farm forestry** *(OLA 041059) Dec 2008* [3]
- **farm forestry** *(OLA 042309) Dec 2008* [10]
- **fennel** *(OLA 042076) Dec 2008* [10]
- **herbs (see appendix 6), onion sets, runner beans, sweetcorn, winter field beans** *(OLA 041058) Dec 2008* [3]
- **leeks** *(OLA 042311) Dec 2008* [10]
- **maize** *(OLA 041060) Dec 2008* [3]
- **runner beans** *(OLA 042310) Dec 2008* [10]
- **sweetcorn** *(under covers and mulches) (OLA 052022) Dec 2008* [5]
- **sweetcorn** *(under covers) (OLA 042305) Dec 2008* [10]
- **winter field beans** *(OLA 042312) Dec 2008* [10]

Approval information
- Pendimethalin included in Annex I under EC Directive 91/414
- Accepted by BBPA for use on malting barley and hops

Efficacy guidance
- Apply as soon as possible after drilling. Weeds are controlled as they germinate and emerged weeds will not be controlled by use of the product alone
- For effective blackgrass control apply not more than 2 d after final cultivation and before weed seeds germinate
- Tank mixes with approved formulations of isoproturon or chlorotoluron recommended for improved pre- and post-emergence control of blackgrass in cereals
- Best results by application to fine firm, moist, clod-free seedbeds when rain follows treatment. Effectiveness reduced by prolonged dry weather after treatment
- Effectiveness reduced on soils with more than 6% organic matter. Do not use where organic matter exceeds 10%
- Any trash, ash or straw should be incorporated evenly during seedbed preparation
- Do not disturb soil after treatment
- On peas drilled after end Mar (mid-Apr in Scotland) tank-mix with cyanazine
- Apply to potatoes as soon as possible after planting and ridging in tank-mix with metribuzin but note that this will restrict the varieties that may be treated

Restrictions
- Maximum number of treatments 1 per crop or yr
- Maximum total dose equivalent to one full dose treatment on most crops; 2 full dose treatments on leeks
- May be applied pre-emergence of cereal crops sown before 30 Nov provided seed covered by at least 32 mm soil, or post-emergence to early tillering stage (GS 23)
- Do not undersow treated crops
- Do not use on crops suffering stress due to disease, drought, waterlogging, poor seedbed conditions or chemical treatment or on soils where water may accumulate

Crop-specific information
- Latest use: pre-emergence for spring barley, carrots, lettuce, fodder maize, parsnips, parsley, sage, peas, runner beans, potatoes, onions and leeks; before transplanting for brassicas; before leaf sheaths erect for winter cereals; before bud burst for blackcurrants, gooseberries, cane fruit, hops; before flower trusses emerge for strawberries; 14 d after transplanting for leaf herbs
- Do not use on spring barley after end Mar (mid-Apr in Scotland on some labels) because dry conditions likely. Do not apply to dry seedbeds in spring unless rain imminent
- Apply to combining peas as soon as possible after sowing. Do not spray if plumule less than 13 mm below soil surface
- Apply to potatoes up to 7 d before first shoot emerges
- Apply to drilled crops as soon as possible after drilling but before crop and weed emergence
- Apply in top fruit, bush fruit and hops from autumn to early spring when crop dormant

SEE SECTION 3 FOR PRODUCTS ALSO REGISTERED

- In cane fruit apply to weed free soil from autumn to early spring, immediately after planting new crops and after cutting out canes in established crops
- Apply in strawberries from autumn to early spring (not before Oct on newly planted bed). Do not apply pre-planting or during flower initiation period (post-harvest to mid-Sep)
- Apply pre-emergence in drilled onions or leeks, not on Sands, Very Light, organic or peaty soils or when heavy rain forecast
- Apply to brassicas after final plant-bed cultivation but before transplanting. Avoid unnecessary soil disturbance after application and take care not to introduce treated soil into the root zone when transplanting. Follow transplanting with specified post-planting treatments - see label
- Do not use on protected crops or in greenhouses

Following crops guidance
- Before ryegrass is drilled after a very dry season plough or cultivate to at least 15 cm. If treated spring crops are to be followed by crops other than cereals, plough or cultivate to at least 15 cm
- In the event of crop failure land must be ploughed or thoroughly cultivated to at least 15 cm. See label for minimum intervals that should elapse between treatment and sowing a range of replacement crops

Environmental safety
- Dangerous for the environment
- Very toxic to aquatic organisms
- Dangerous to fish or other aquatic life. Do not contaminate surface waters or ditches with chemical or used container
- Do not empty into drains [3]
- LERAP Category B [2, 4]
- Flammable [2, 4]

Hazard classification and safety precautions
Hazard H03, H08 [2, 4]; H04 [3, 5, 9, 10]; H11 [2-10]

Risk phrases R22b, R51 [2, 4]; R43 [3]; R50 [3, 5-10]; R53a [2-10]

Operator protection A [3, 6]; H [3]; U05a [2, 4, 5, 9, 10]; U08, U19a [1-10]; U13 [1, 3, 5-8, 10]; U14, U15 [9]; U20b [1, 3, 5-10]

Environmental protection E13b [1]; E15a [2-10]; E16a, E34 [2, 4]; E19b [3]; E38 [2, 4, 5, 9, 10]

Storage and disposal D01, D12a [2-5, 9, 10]; D02 [2, 4, 5, 9, 10]; D05, D10b [1, 2, 4-10]; D09a [1-10]; D11a [3]

Medical advice M03, M05b [2, 4]; M05a [3]

349   pendimethalin + picolinafen

A post-emergence broad-spectrum herbicide mixture for winter cereals

Products

1	Flight	BASF	330:7.5 g/l	SC	12534
2	Pico Stomp	BASF	320:16 g/l	SC	11987
3	PicoMax	BASF	320:16 g/l	SC	10739
4	Picona	BASF	320:16 g/l	SC	10715
5	PicoPro	BASF	320:16 g/l	SC	10740

Uses
- Annual dicotyledons in *winter barley, winter wheat* [1-5]
- Annual meadow grass in *winter barley, winter wheat* [1-5]
- Chickweed in *winter barley, winter wheat* [1-5]
- Cleavers in *winter barley, winter wheat* [2-5]
- Rough meadow grass in *winter barley, winter wheat* [1-5]
- Speedwells in *winter barley, winter wheat* [1-5]

Approval information
- Pendimethalin and picolinafen included in Annex I under EC Directive 91/414

Efficacy guidance
- Best results obtained on crops growing in a fine, firm tilth and when rain falls within 7 d of application
- Loose or cloddy seed beds must be consolidated prior to application otherwise reduced weed control may occur

FOR FULL CONDITIONS OF USE ALWAYS READ THE PRODUCT LABEL

- Residual weed control may be reduced on soils with more than 6% organic matter or under prolonged dry conditions
- Always follow WRAG guidelines for preventing and managing herbicide resistant weeds. Section 5 for more information

Restrictions
- Maximum number of treatments 1 per crop
- Do not treat undersown cereals or those to be undersown
- Do not roll emerged crops before treatment nor autumn treated crops until the following spring
- Do not use on stony or gravelly soils, on soils that are waterlogged or prone to waterlogging, or on soils with more than 10% organic matter
- Do not treat crops under stress from any cause
- Do not apply pre-emergence to crops drilled after 30 Nov [1]
- Consult processor before treating crops for processing [1]

Crop-specific information
- Latest use: before pseudo-stem erect stage (GS 30)
- Transient bleaching may occur after treatment but it does not lead to yield loss
- May be used post-emergence from first leaf unfolded [2-5]; pre- and post-emergence [1]

Following crops guidance
- In the event of failure of a treated crop plough to at least 15 cm to ensure any residues are evenly dispersed
- In the event of failure of a treated crop allow at least 8 wk from the time of treatment before re-drilling either spring wheat or spring barley. After a normally harvested crop there are no restrictions on following crops apart from rye-grass, as indicated below [2-5]
- In a very dry season, plough or cultivate to at least 15 cm before drilling rye-grass
- A minimum of 5 mth must follow autumn applications before sowing spring barley, spring wheat, maize, peas, potatoes, spring field beans, dwarf beans, turnips, Brussels sprouts, cabbage, calabrese, linseed, carrots, parsnips, parsley, cauliflower [1]
- After spring or summer applications a minimum of 2 mth must follow before sowing spring field beans, dwarf beans, broad beans, peas, turnips, Brussels sprouts, cabbage, calabrese, cauliflower, carrots, parsnips, parsley, linseed and any crop may be sown after 5 mth except red beet, sugar beet and spinach, for which an interval of 12 mth must be allowed [1]

Environmental safety
- Dangerous for the environment
- Very toxic to aquatic organisms
- Product binds strongly to soil minimising likelihood of movement into groundwater
- LERAP Category B [1]

Hazard classification and safety precautions
Hazard H04 [2]; H11 [1-5]
Risk phrases R43 [2]; R50, R53a [1-5]
Operator protection A, H; U02a, U05a, U20c
Environmental protection E15a, E34, E38 [1-5]; E16a [1]
Storage and disposal D01, D02, D05, D09a, D10c, D12a
Medical advice M03

350　pentanochlor

A contact anilide herbicide

See also chlorpropham + pentanochlor

Products
1 Croptex Bronze	Certis	400 g/l	EC	11329
2 Solan 40	Nufarm UK	400 g/l	EC	11897

Uses
- Annual dicotyledons in **anemones, carnations, celeriac, foxgloves, freesias, larkspur, roses, sweet williams, wallflowers** [2]; **carrots, celery, chrysanthemums, nursery stock, parsley, parsnips, sweet peas** [1, 2]

- Annual meadow grass in *anemones, carnations, celeriac, foxgloves, freesias, larkspur, roses, sweet williams, wallflowers* [2]; *carrots, celery, chrysanthemums, nursery stock, parsley, parsnips, sweet peas* [1, 2]

Approval information
- Products containing this active ingredient have been granted derogations for specified 'Essential Uses' for use until 31 December 2007. Sale and supply must cease by 30 June 2007 but growers have no guarantee that the products will continue to be available until then.
  For more information see 'The Review Programme' under 'Pesticide Legislation' in Section 5

Efficacy guidance
- Pentanochlor acts mainly through contact activity with some residual effect in moist soils. Best results achieved on young weed seedlings under warm, moist conditions
- Weeds most susceptible in cotyledon to 2-3 leaf stage although some important species controlled at later stages
- Effectiveness reduced by very cold weather or drought

Restrictions
- Maximum number of treatments 2 per crop for edible crops and sweet peas; 1 per yr for other horticultural ornamental crops
- Not recommended for flower crops on light sandy soils
- Do not treat crops suffering from stress, drought, mineral deficiency, pest attack or damage from previous herbicide applications
- Do not harvest crops for human or animal consumption for at least 28 d after last application
- These products must not be used on any crops other than those listed, including any extrapolations that would normally be permissible under the Long Term Arrangements for Extension of Use (see Section 5), except on herbs and other ornamentals

Crop-specific information
- HI 28 d for all edible crops
- Apply pre-emergence in anemones, freesias, foxgloves, larkspur

Following crops guidance
- In the event of failure of a treated crop only recommended crops should be re-planted
- Any crop may be planted 4-5 wk after application following mould-board ploughing or repeated cultivation

Environmental safety
- Dangerous to fish or other aquatic life. Do not contaminate surface waters or ditches with chemical or used container

Hazard classification and safety precautions
  Hazard H03 [2]; H04 [1]
  Risk phrases R21, R22a, R22b, R37, R40 [2]; R36 [1, 2]; R38 [1]
  Operator protection A, C; U05a, U08, U11, U19a [1, 2]; U14, U15, U20b [1]; U20a [2]
  Environmental protection E13b [1]; E15a [2]
  Consumer protection C02 [1] (28 d)
  Storage and disposal D01, D02, D09a, D10b
  Medical advice M05b [2]

351   permethrin

A broad spectrum, contact and ingested pyrethroid insecticide

Products

Fortefog 'P' Fumer	Agropharm	13.5% w/w	FU	H7187

Uses
- Flies in *agricultural premises*
- Grain storage pests in *grain stores*

Approval information
- All approvals for plant protection uses of permethrin were revoked in Jun 2001. Off-label uses on young plants in forestry remained valid until 31 Dec 2003. Several grain store and non-agricultural uses remain approved

FOR FULL CONDITIONS OF USE ALWAYS READ THE PRODUCT LABEL

Efficacy guidance
- For best results in farm buildings fumigate in the late afternoon and leave closed overnight
- Make areas to be fumigated as smoke tight as possible
- Place smoke generators as low as possible
- Space treatment controls insects present at the time by direct contact action

Restrictions
- For use only by professional operators
- Maximum number of treatments up to 3 fumigations at 5-7 d intervals

Environmental safety
- Extremely dangerous to fish or other aquatic life. Do not contaminate surface waters or ditches with chemical or used container
- Domestic animals, birds and fish should be removed from the vicinity of buildings to be treated. Foodstuffs should be protected from smoke
- Before treating any structure used by bats consult English Nature, Scottish Natural Heritage or the Countryside Council for Wales. All bats are protected under the Wildlife and Countryside Act 1981

Hazard classification and safety precautions
Operator protection A, E, H; U02b, U14, U18, U19a, U20b
Environmental protection E02c, E13a
Consumer protection C04, C09, C11
Storage and disposal D01, D09a, D11a

SECTION 2

352   peroxyacetic acid (commodity substance)

An inorganic fungicide for use on potato tubers

Products
peroxyacetic acid   various   5% w/w   SL

Uses
- Fungus diseases in **potatoes** *(tuber treatment)*

Approval information
- Approval for the use of peroxyacetic acid as a commodity substance was granted on 8 January 2003 by Ministers under regulation 5 of the Control of Pesticides Regulations 1986

Efficacy guidance
- Seed potato tubers must be immersed for one minute in a 50:1 dilution of proprietary 5% w/w peroxyacetic acid formulation

Environmental safety
- Used dip must be disposed of in accordance with Part Five of the Code of Practice for the Safe Use of Pesticides on Farms and Holdings

Hazard classification and safety precautions
Operator protection A, C, D, H, M

353   petroleum oil

An insecticidal and acaricidal hydrocarbon oil

Products
Certis Spraying Oil   Certis   710 g/l   EC   -

Uses
- Mealybugs in **bush fruit, cane fruit, fruit trees, grapevines, hops, nursery stock, protected cucumbers, protected pot plants, protected tomatoes**
- Red spider mites in **bush fruit, cane fruit, fruit trees, grapevines, hops, nursery stock, protected cucumbers, protected pot plants, protected tomatoes**
- Scale insects in **bush fruit, cane fruit, fruit trees, grapevines, hops, nursery stock, protected cucumbers, protected pot plants, protected tomatoes**

Approval information
- Product not controlled by Control of Pesticides Regulations because it acts by physical means only

SEE SECTION 3 FOR PRODUCTS ALSO REGISTERED

Efficacy guidance

- Spray at 1% (0.5% on tender foliage) to wet plants thoroughly, particularly the underside of leaves, and repeat as necessary
- Petroleum oil acts by blocking pest breathing pores and making leaf surfaces inhospitable to pests seeking to attack or attach to the leaf surface
- On outdoor crops apply when dormant or on plants with a known tolerance
- On plants of unknown sensitivity test first on a small scale
- Mixtures with certain pesticides may damage crop plants. If mixing, spray a few plants to test for tolerance before treating larger areas

Restrictions

- Do not mix with sulphur or iprodione products or use such mixtures within 28 d of treatment
- Do not treat protected crops in bright sunshine unless the glass is well shaded
- Test species tolerance before large scale treatment

Crop-specific information

- HI zero
- Treat grapevines before flowering

Hazard classification and safety precautions

Hazard H03
Risk phrases R22a, R36, R37, R38
Operator protection A, C; U05a, U08, U11, U19a
Environmental protection E34
Storage and disposal D01, D02, D09a, D10a, D12a
Medical advice M03

354 phenmedipham

A contact carbamate herbicide for beet crops and strawberries

*See also desmedipham + ethofumesate + phenmedipham
desmedipham + phenmedipham
ethofumesate + phenmedipham
lenacil + phenmedipham*

Products

1	Agriguard Phenmedipham	AgriGuard	118 g/l	EC	12148
2	Alpha Phenmedipham 320 SC	Makhteshim	320 g/l	SC	12320
3	Beetup	United Phosphorus	114 g/l	EC	07520
4	Betanal Flow	Bayer CropScience	160 g/l	SE	10997
5	Cleancrop Phenmedipham	United Phosphorus	114 g/l	EC	11586
6	Crotale	Sipcam	471 g/l	SC	11593
7	Dancer	Sipcam	160 g/l	SC	10048
8	Herbasan Flow	Nufarm UK	160 g/l	SC	11448
9	Mandolin Flow	Nufarm UK	160 g/l	SE	11162

Uses

- Annual dicotyledons in **fodder beet**, **mangels**, **sugar beet** [1-9]; **leaf spinach** *(off-label)*, **spinach beet** *(off-label)* [4]; **red beet** [2-9]; **strawberries** [3, 5]

Specific Off-Label Approvals (SOLAs)

- **leaf spinach**, **spinach beet** *(OLA 050633) Dec 2008* [4]

Approval information

- Approval expiry 31 Aug 2006 [1, 7]

Efficacy guidance

- Best results achieved by application to young seedling weeds, preferably cotyledon stage, under good growing conditions when low doses are effective
- If using a low-dose programme 2-3 repeat applications at 7-10 d intervals are recommended on mineral soils, 3-5 applications may be needed on organic soils
- Addition of adjuvant oil may improve effectiveness on some weeds

FOR FULL CONDITIONS OF USE ALWAYS READ THE PRODUCT LABEL

- Various tank-mixtures with other beet herbicides recommended. See label for details
- Use of certain pre-emergence herbicides is recommended in combination with post-emergence treatment. See label for details

Restrictions
- Maximum number of treatments and maximum total dose varies with crop and product used. See label for details
- At high temperatures (above 21°C) reduce rate and spray after 5 pm
- Do not apply immediately after frost or if frost expected
- Do not spray wet foliage or if rain imminent
- Do not spray crops stressed by wind damage, nutrient deficiency, pest or disease attack etc. Do not roll or harrow for 7 d before or after treatment
- Do not use on strawberries under cloches or polythene tunnels
- Do not use after 31 Jul in yr of harvest [4, 6-9]
- Consult processor before use on crops for processing

Crop-specific information
- Latest use: before crop leaves meet between rows for beet crops; before flowering for strawberries
- Apply to beet crops at any stage as low dose/low volume spray or from fully developed cotyledon stage with full rate. Apply to red beet after fully developed cotyledon stage
- Apply to strawberries at any time when weeds in susceptible stage, except in period from start of flowering to picking

Following crops guidance
- Beet crops may follow at any time after a treated crop. 3 mth must elapse from treatment before any other crop is sown and must be preceded by mould-board ploughing to 15 cm

Environmental safety
- Dangerous for the environment
- Very toxic to aquatic organisms
- Harmful (dangerous [7]) to fish or other aquatic life. Do not contaminate surface waters or ditches with chemical or used container
- Do not empty into drains [2]

Hazard classification and safety precautions

Hazard H03 [1, 3, 5]; H04 [2, 6]; H11 [1-5, 8, 9]

Risk phrases R21, R22a, R36, R37, R40 [1, 3, 5]; R38 [1, 2]; R43 [2, 6]; R50 [2, 4, 8, 9]; R51 [3, 5]; R52 [1]; R53a [1-5, 8, 9]

Operator protection A [1-4, 6, 9]; C [1, 3]; H [2, 6]; U05a [1-3, 5, 6]; U08 [1-3, 5]; U09a [7]; U13 [1, 3, 5]; U14, U20a [2]; U19a [1-3, 5, 7]; U20b [1, 3-9]

Environmental protection E13b [7]; E13c [4, 6]; E15a [1-3, 5, 8, 9]; E19b [2]; E34 [1-3, 5]; E38 [1, 2, 4, 8, 9]

Storage and disposal D01 [1-3, 5, 6]; D02 [1, 3, 5, 6]; D05 [1-5, 7, 9]; D09a [1-9]; D10b [1-6, 8, 9]; D10c [7]; D12a [1, 2, 4, 8, 9]

Medical advice M03 [1, 3, 5]; M05a [2]

355 d-phenothrin

A non-systemic pyrethroid insecticide available only in mixtures

356 d-phenothrin + tetramethrin

A pyrethroid insecticide mixture for control of flying insects

Products

Killgerm ULV 500	Killgerm	36.8:18.4 g/l	UL	H4647

Uses
- Flies in *agricultural premises*
- Grain storage mite in *grain stores*
- Mosquitoes in *agricultural premises*
- Wasps in *agricultural premises*

SEE SECTION 3 FOR PRODUCTS ALSO REGISTERED

SECTION 2

Approval information
- Product approved for ULV application. See label for details [1]

Efficacy guidance
- Close doors and windows and spray in all directions for 3-5 sec. Keep room closed for at least 10 min

Restrictions
- For use only by professional operators
- Do not use space sprays containing pyrethrins or pyrethroid more than once per week in intensive or controlled environment animal houses in order to avoid development of resistance. If necessary, use a different control method or product

Crop-specific information
- May be used in the presence of poultry and livestock

Environmental safety
- Dangerous for the environment
- Toxic to aquatic organisms
- Do not apply directly to livestock/poultry
- Remove exposed milk and collect eggs before application. Protect milk machinery and containers from contamination

Hazard classification and safety precautions
Hazard H03, H11
Risk phrases R22b, R51, R53a
Operator protection A, C, D, E, H; U02b, U09b, U14, U19a, U20a, U20b
Environmental protection E05a, E15a, E38
Consumer protection C06, C07, C08, C09, C11, C12
Storage and disposal D01, D05, D06a, D09a, D10a, D12a
Medical advice M05b

357   picloram

A persistent, translocated pyridine carboxylic acid herbicide for non-crop areas

See also clopyralid + picloram

Products

Tordon 22K	Nomix Enviro	240 g/l	SL	05790

Uses
- Annual dicotyledons in **land not intended for cropping**
- Bracken in **land not intended for cropping**
- Japanese knotweed in **land not intended for cropping**
- Perennial dicotyledons in **land not intended for cropping**
- Woody weeds in **land not intended for cropping**

Efficacy guidance
- May be applied at any time of year. Best results achieved by application as foliage spray in late winter to early spring
- For bracken control apply 2-4 wk before frond emergence
- Clovers are highly sensitive and eliminated at very low doses
- Persists in soil for up to 2 yr

Restrictions
- Maximum number of treatments 1 per yr on land not intended for cropping
- Do not apply around desirable trees or shrubs where roots may absorb chemical
- Do not apply on slopes where chemical may be leached onto areas of desirable plants

Environmental safety
- Harmful to fish or other aquatic life. Do not contaminate surface waters or ditches with chemical or used container
- Keep livestock out of treated areas for at least 2 wk and until foliage of any poisonous weeds such as ragwort has died and become unpalatable

FOR FULL CONDITIONS OF USE ALWAYS READ THE PRODUCT LABEL

Hazard classification and safety precautions
Hazard H04
Risk phrases R43
Operator protection A, H; U05a, U08, U14, U20b
Environmental protection E07a, E13c
Storage and disposal D01, D02, D09a, D10b

358   picolinafen

An aryloxypicolinamide herbicide for cereals available only in mixtures

See also flupyrsulfuron-methyl + picolinafen
pendimethalin + picolinafen

359   picoxystrobin

A broad spectrum strobilurin fungicide for cereals

See also chlorothalonil + picoxystrobin
cyprodinil + picoxystrobin

Products
Acanto	Syngenta	250 g/l	SC	10978

Uses
- Brown rust in **spring barley**, **spring wheat**, **winter barley**, **winter wheat**
- Crown rust in **spring oats**, **winter oats**
- Eyespot in **spring wheat** (reduction), **winter wheat** (reduction)
- Late ear diseases in **spring wheat**, **winter wheat**
- Net blotch in **spring barley**, **winter barley**
- Powdery mildew in **spring barley**, **spring oats**, **winter barley**, **winter oats**
- Rhynchosporium in **spring barley**, **winter barley**
- Septoria diseases in **spring wheat**, **winter wheat**
- Yellow rust in **spring wheat**, **winter wheat**

Approval information
- Accepted by BBPA for use on malting barley

Efficacy guidance
- Best results achieved from protectant treatments made before disease establishes in the crop
- Use of tank mixtures is recommended where disease has become established at the time of treatment
- Persistence of yellow rust control is less than for other foliar diseases but can be improved by use of an appropriate tank mixture
- Eyespot control is not reliable and an appropriate tank mix should be used where crops are at risk
- Picoxystrobin is a member of the QoI cross resistance group. Product should be used preventatively and not relied on for its curative potential
- Use product as part of an Integrated Crop Management strategy incorporating other methods of control, including where appropriate other fungicides with a different mode of action. Do not apply more than two foliar applications of QoI containing products to any cereal crop
- There is a significant risk of widespread resistance occurring in *Septoria tritici* populations in UK. Failure to follow resistance management action may result in reduced levels of disease control
- On cereal crops product must always be used in mixture with another product, recommended for control of the same target disease, that contains a fungicide from a different cross resistance group and is applied at a dose that will give robust control
- Strains of barley powdery mildew resistant to QoIs are common in the UK

Restrictions
- Maximum number of treatments 2 per crop per yr

Crop-specific information
- Latest use: grain watery ripe (GS 71)

Environmental safety
- Dangerous for the environment
- Very toxic to aquatic organisms

Hazard classification and safety precautions
Hazard H11
Risk phrases R50, R53a
Operator protection U05a, U09a, U14, U15, U20a
Environmental protection E15a, E38
Storage and disposal D01, D02, D05, D09a, D10c, D12a

360　pirimicarb

A carbamate insecticide for aphid control

See also lambda-cyhalothrin + pirimicarb

Products

1 Agriguard Pirimicarb	AgriGuard	50% w/w	WG	09620
2 Aphox	Syngenta	50% w/w	WG	10515
3 Greencrop Glenroe	Greencrop	50% w/w	WG	09903
4 Landgold Pirimicarb 50	Landgold	50% w/w	WG	09018
5 Landgold Pirimicarb 50	Teliton	50% w/w	WG	12126
6 Phantom	Syngenta	50% w/w	WG	11954
7 Standon Pirimicarb 50	Standon	50% w/w	WG	08878

Uses
- Aphids in **apples**, **broad beans**, **broccoli**, **brussels sprouts**, **cabbages**, **calabrese**, **cauliflowers**, **durum wheat**, **pears**, **peas**, **potatoes**, **spring barley**, **spring field beans**, **spring oats**, **spring rye**, **spring wheat**, **strawberries**, **sugar beet**, **swedes**, **triticale**, **turnips**, **winter barley**, **winter field beans**, **winter oats**, **winter rye**, **winter wheat** [1-7]; **blackcurrants**, **carrots**, **chinese cabbage**, **collards**, **dwarf beans**, **gooseberries**, **maize**, **parsnips**, **raspberries**, **redcurrants**, **runner beans**, **spring oilseed rape**, **sweetcorn**, **winter oilseed rape** [1-3, 6, 7]; **carnations**, **chrysanthemums**, **cinerarias**, **cyclamen**, **protected spinach** *(off-label)*, **roses**, **salad brassicas**, **sweetcorn** *(off-label)* [2]; **celeriac** *(off-label)*, **celery** *(off-label)*, **chicory** *(off-label - for forcing)*, **courgettes** *(off-label)*, **endives** *(off-label)*, **fennel** *(off-label)*, **gherkins** *(off-label)*, **honesty** *(off-label)*, **horseradish** *(off-label)*, **leaf spinach** *(off-label)*, **marrows** *(off-label)*, **parsley** *(off-label)*, **parsley root** *(off-label)*, **protected courgettes** *(off-label)*, **protected endives** *(off-label)*, **protected gherkins** *(off-label)*, **protected radishes** *(off-label)*, **protected spinach beet** *(off-label)*, **radishes** *(off-label)*, **red beet** *(off-label)*, **salad brassicas** *(off-label - for baby leaf production)*, **spinach beet** *(off-label)* [2, 6]; **cherries** [1-3, 6]; **chives** *(off-label)*, **herbs (see appendix 6)** *(off-label)*, **poppies** *(off-label - for morphine production)*, **protected horseradish** *(off-label)*, **protected leaf spinach** *(off-label)*, **protected parsley** *(off-label)*, **rhubarb** *(off-label)*, **salsify** *(off-label)* [6]; **cucumbers**, **forest nurseries**, **lettuce** *(outdoor crops)*, **outdoor tomatoes**, **peppers**, **protected carnations**, **protected chrysanthemums**, **protected cinerarias**, **protected cyclamen**, **protected lettuce**, **protected roses**, **protected tomatoes** [2, 3, 6]; **grassland** [1]; **kale** [1, 2, 4-7]; **ornamental specimens** [3, 6]
- Blackfly in **cherries** [7]
- Carrot willow aphid in **celery** [1]
- Leaf curling plum aphid in **plums** *(off-label)* [2, 6]
- Mealy plum aphid in **plums** *(off-label)* [2, 6]
- Thrips in **endives** *(off-label)*, **parsley** *(off-label)*, **protected endives** *(off-label)* [2, 6]; **protected parsley** *(off-label)* [6]

Specific Off-Label Approvals (SOLAs)
- **celeriac, fennel, red beet** *(OLA 011292) Dec 2008* [2]
- **celeriac, fennel, red beet, salsify** *(OLA 051747) Dec 2008* [6]
- **celery** *(OLA 040680) Dec 2008* [2]
- **celery, rhubarb** *(OLA 051744) Dec 2008* [6]
- **chicory** *(for forcing) (OLA 011283) Dec 2008* [2]
- **chicory** *(for forcing) (OLA 051749) Dec 2008* [6]

FOR FULL CONDITIONS OF USE ALWAYS READ THE PRODUCT LABEL

- *chives, courgettes, gherkins, herbs (see appendix 6), horseradish, leaf spinach, marrows, parsley root, protected horseradish, protected leaf spinach, protected radishes, protected spinach beet, radishes, spinach beet* (OLA 051746) Dec 2008 [6]
- *courgettes, gherkins, horseradish, leaf spinach, marrows, parsley root, protected radishes, protected spinach, protected spinach beet, radishes, spinach beet, sweetcorn* (OLA 011298) Dec 2008 [2]
- *endives, parsley, protected endives* (OLA 011286) Dec 2008 [2]
- *endives, parsley, protected endives, protected parsley* (OLA 051741) Dec 2008 [6]
- *honesty* (OLA 011289) Dec 2008 [2]
- *honesty* (OLA 051742) Dec 2008 [6]
- *plums* (OLA 012687) Dec 2008 [2]
- *plums* (OLA 051738) Dec 2008 [6]
- *poppies* (for morphine production) (OLA 051748) Dec 2008 [6]
- *protected courgettes, protected gherkins* (OLA 011301) Dec 2008 [2]
- *protected courgettes, protected gherkins* (OLA 051745) Dec 2008 [6]
- *salad brassicas* (for baby leaf production) (OLA 011295) Dec 2008 [2]
- *salad brassicas* (for baby leaf production) (OLA 051743) Dec 2008 [6]

Approval information
- Pirimicarb has been included in a comprehensive review of anticholinesterase compounds and considered by the ACP. As a result approvals for some SOLAs, and for uses on grassland and celery were revoked and approvals expire in Dec 2003
- Approved for aerial application on cereals [1-3, 6]. See notes in Section 5
- Accepted by BBPA for use on malting barley

Efficacy guidance
- Chemical has contact, fumigant and translaminar activity
- Best results achieved under warm, calm conditions when plants not wilting and spray does not dry too rapidly. Little vapour activity at temperatures below 15°C
- Apply as soon as aphids seen or warning issued and repeat as necessary
- Addition of non-ionic wetter recommended for use on brassicas
- On cucumbers and tomatoes a root drench is preferable to spraying when using predators in an integrated control programme
- Where aphids resistant to pirimicarb occur control is unlikely to be satisfactory

Restrictions
- This product contains an anticholinesterase carbamate compound. Do not use if under medical advice not to work with such compounds
- Maximum number of treatments not specified in some cases but normally 2-6 depending on crop
- When treating ornamentals check safety by treating a small number of plants first
- Spray equipment must only be used where the operator's normal working position is within a closed cab on a tractor or self-propelled sprayer when making air-assisted applications to apples or other top fruit

Crop-specific information
- Latest use in accordance with harvest intervals below
- HI oilseed rape, cereals, maize, sweetcorn, lettuce under glass 14 d (sweetcorn off-label 3 d); grassland, chicory, plums 7 d; cucumbers, tomatoes and peppers under glass 2 d; protected courgettes and gherkins 24 h; other edible crops 3 d; flowers and ornamentals zero

Environmental safety
- Dangerous for the environment
- Very toxic to aquatic organisms
- Dangerous to fish or other aquatic life. Do not contaminate surface waters or ditches with chemical or used container
- Chemical has little effect on bees, ladybirds and other insects and is suitable for use in integrated control programmes on apples and pears
- Keep all livestock out of treated areas for at least 7 d. Bury or remove spillages

Hazard classification and safety precautions
Hazard H02, H11 [2-7]; H03 [1]
Risk phrases R20 [2, 6]; R22a [1]; R25, R36, R50, R53a [2-7]
Operator protection A, C, D, H, J, M; U05a, U08, U19a [1-7]; U20a [3, 6]; U20b [1, 2, 4, 5, 7]
Environmental protection E06c [1-7] (7 d); E13b [1]; E15a [2-7]; E34 [1-7]; E38 [2, 6]

SEE SECTION 3 FOR PRODUCTS ALSO REGISTERED

Storage and disposal D01, D02, D09a, D11a [1-7]; D05, D07 [3]; D12a [2, 6]
Medical advice M02 [1-7]; M03 [1]; M04 [2-7]

361 pirimiphos-methyl

A contact, fumigant and translaminar organophosphorus insecticide

Products

1	Actellic D	Syngenta	250 g/l	EC	10509
2	Actellic Smoke Generator No. 20	Syngenta	22.5% w/w	FU	10540
3	Fumite Pirimiphos Methyl Smoke	Certis	22.5% w/w	FU	00941

Uses

* Ants in *protected cucumbers, protected ornamentals, protected peppers, protected tomatoes* [3]
* Aphids in *protected cucumbers, protected ornamentals, protected peppers, protected tomatoes* [3]
* Capsids in *protected cucumbers, protected ornamentals, protected peppers, protected tomatoes* [3]
* Earwigs in *protected cucumbers, protected ornamentals, protected peppers, protected tomatoes* [3]
* Flour beetle in *stored grain* [1]
* Flour moth in *stored grain* [1]
* Grain beetle in *stored grain* [1]
* Grain storage mite in *stored grain* [1]
* Grain storage pests in *grain stores* [1, 2]
* Grain weevil in *stored grain* [1]
* Leaf miner in *protected cucumbers, protected ornamentals, protected peppers, protected tomatoes* [3]
* Red spider mites in *protected cucumbers, protected ornamentals, protected peppers, protected tomatoes* [3]
* Sawflies in *protected cucumbers, protected ornamentals, protected peppers, protected tomatoes* [3]
* Thrips in *protected cucumbers, protected ornamentals, protected peppers, protected tomatoes* [3]
* Warehouse moth in *stored grain* [1]
* Whitefly in *protected cucumbers, protected ornamentals, protected peppers, protected tomatoes* [3]

Approval information

* Accepted by BBPA for use in stores for malting barley

Efficacy guidance

* Chemical acts rapidly and has short persistence in plants but persists for long periods on inert surfaces
* Best results for protection of stored grain achieved by cleaning store thoroughly before use and employing a combination of pre-harvest and grain/seed treatments
* Best results for admixture treatment obtained when grain stored at 15% moisture or less. Dry and cool moist grain coming into store but then treat as soon as possible, ideally as it is loaded
* Surface admixture can be highly effective on localised surface infestations but should not be relied on for long-term control unless application can be made to the full depth of the infestation
* For control of whitefly and other glasshouse pests apply as smoke or by thermal fogging. See label for details of techniques and suitable machines [3]
* Where insect pests resistant to pirimiphos-methyl occur control is unlikely to be satisfactory

Restrictions

* This product contains an organophosphorus anticholinesterase compound. Do not use if under medical advice not to work with such compounds
* Maximum number of treatments 2 per grain store; 1 per batch of stored grain
* Do not mix with grain store disinfectants because of risk of chemical interaction [1]
* Do not apply surface admixture or complete admixture using hand held equipment [1]

FOR FULL CONDITIONS OF USE ALWAYS READ THE PRODUCT LABEL

- Do not apply treated seed from the air [1]
- Do not fumigate glasshouses in bright sunshine or when foliage is wet or roots dry
- Do not fumigate young seedlings or plants being hardened off
- Do not fog open flowers of ornamentals without first consulting firm
- Do not fog mushrooms when wet as slight spotting may occur

Crop-specific information
- Latest use: before storing grain
- HI for edible crops zero [3]
- Disinfect empty grain stores by spraying surfaces and/or fumigation and treat grain by full or surface admixture. Treat well before harvest in late spring or early summer and repeat 6 wk later or just before harvest if heavily infested. See label for details of treatment and suitable application machinery
- Treatment volumes on structural surfaces should be adjusted according to surface porosity. See label for guidelines
- Treat inaccessible areas with smoke generating product used in conjunction with spray treatment of remainder of store

Environmental safety
- Dangerous for the environment
- Toxic to aquatic organisms
- Ventilate fumigated or fogged spaces thoroughly before re-entry
- Unprotected persons must be kept out of fumigated areas within 3 h of ignition and for 4 h after treatment
- Do not harvest for human or animal consumption for at least 5 d after last application
- Extremely dangerous to bees. Do not apply to crops in flower or to those in which bees are actively foraging. Do not apply when flowering weeds are present [3]
- Wildlife must be excluded from buildings during treatment [1]
- Grain treated by admixture as specified may be consumed by humans and livestock

Hazard classification and safety precautions
Hazard H03, H11 [1-3]; H07 [2, 3]; H08 [1]
Risk phrases R20, R51, R53a [1-3]; R22b, R38 [1]; R36, R37 [1, 3]
Operator protection A, C [1]; D, H [1, 2]; U05a [1-3]; U10, U20b [2]; U14, U19a [2, 3]; U20a [3]
Environmental protection E02a [2, 3] (4 h); E12c [3]; E15a [1-3]; E34 [2, 3]; E38 [1]
Consumer protection C02 [3] (5 d); C11 [2]; C12 [2, 3]
Storage and disposal D01, D02, D09a [1-3]; D10c, D12a [1]; D11a [2, 3]
Treated seed S05 [1]
Medical advice M01 [1-3]; M03 [2]; M05b [1]

362   potassium bicarbonate (commodity substance)

An inorganic fungicide for use in horticulture

Products
potassium bicarbonate	various	>99% w/w	SL

Uses
- Fungus diseases in *outdoor crops*, *protected crops*

Approval information
- Approval for the use of potassium bicarbonate as a commodity substance was granted on 26 July 2005 by Ministers under regulation 5 of the Control of Pesticides Regulations 1986

Efficacy guidance
- Adjuvants authorised by PSD may be used in conjunction with potassium bicarbonate
- Crop phytotoxicity can occur

Restrictions
- Food grade potassium bicarbonate must only be used

Hazard classification and safety precautions
Operator protection A, C, G, H

SECTION 2

363   prochloraz

A broad-spectrum protectant and eradicant conazole fungicide

See also fluquinconazole + prochloraz

Products

1	Alpha Prochloraz 40 EC	Makhteshim	400 g/l	EC	11002
2	Barclay Eyetak 40	Barclay	400 g/l	EC	11352
3	Mirage 40 EC	Makhteshim	400 g/l	EC	06770
4	Panache 40	DAPT	400 g/l	EC	10632
5	Poraz	BASF	450 g/l	EC	11701
6	Prelude 20LF	Agrichem	200 g/l	LS	04371
7	Prospero	Makhteshim	400 g/l	EC	11931
8	Scotts Octave	Scotts	46% w/w	WP	09275

Uses

- Alternaria in *spring oilseed rape* [1, 3]; *winter oilseed rape* [1, 3, 5, 7]
- Eyespot in *spring wheat* [1-4]; *winter barley, winter wheat* [1-5, 7]; *winter rye* [1, 3, 5, 7]
- Fungus diseases in *container-grown ornamentals, hardy ornamentals, woody ornamentals* [8]
- Glume blotch in *spring wheat* [1, 3]; *winter wheat* [1, 3, 5, 7]
- Grey mould in *spring oilseed rape* [1, 3]; *winter oilseed rape* [1, 3, 5, 7]
- Light leaf spot in *spring oilseed rape* [1, 3, 4]; *winter oilseed rape* [1, 3-5, 7]
- Net blotch in *spring barley, winter barley* [1, 3, 5, 7]
- Phoma in *spring oilseed rape* [1, 3, 4]; *winter oilseed rape* [1, 3-5, 7]
- Powdery mildew in *spring barley, spring wheat, winter barley, winter wheat* [5, 7]; *winter rye* [1, 3, 5, 7]
- Rhynchosporium in *spring barley, winter barley, winter rye* [1, 3, 5, 7]
- Ring spot in *chives* (off-label), *herbs (see appendix 6)* (off-label), *lettuce* (off-label), *parsley* (off-label), *protected chives* (off-label), *protected herbs (see appendix 6)* (off-label), *protected lettuce* (off-label), *protected parsley* (off-label) [3]
- Sclerotinia stem rot in *spring oilseed rape* [1, 3]; *winter oilseed rape* [1, 3, 5, 7]
- Seed-borne diseases in *flax* (seed treatment), *linseed* (seed treatment) [6]
- Septoria leaf spot in *spring wheat* [1, 3]; *winter rye* [5, 7]; *winter wheat* [1, 3, 5, 7]
- White leaf spot in *spring oilseed rape* [1, 3]; *winter oilseed rape* [1, 3, 5, 7]

Specific Off-Label Approvals (SOLAs)

- *chives, herbs (see appendix 6), lettuce, parsley, protected chives, protected herbs (see appendix 6), protected lettuce, protected parsley* (OLA 032623) Dec 2008 [3]

Approval information

- Accepted by BBPA for use on malting barley

Efficacy guidance

- Spray cereals at first signs of disease. Protection of winter crops through season usually requires at least 2 treatments. See label for details of rates and timing. Treatment active against strains of eyespot resistant to benzimidazole fungicides
- Tank mixes with other fungicides recommended to improve control of rusts in wheat and barley. See label for details
- A period of at least 3 h without rain should follow spraying
- Can be used through most seed treatment machines if good even seed coverage is obtained. Check drill calibration before drilling treated seed. Use inert seed flow agent supplied by manufacturer [6]
- Use methylated spirits, rather than water, to clean residual material from seed treatment machinery, then carry out final rinse with water and detergent [6]
- Apply as drench against soil diseases, as a spray against aerial diseases, as a dip for cuttings, or as a drench at propagation. Spray applications may be repeated at 10-14 d intervals. Under mist propagation use 7 d intervals [8]

Restrictions

- Maximum number of treatments 1 per batch of seed for flax, linseed [6]; varies with dose and crop for foliar spray uses - see labels for details

---

FOR FULL CONDITIONS OF USE ALWAYS READ THE PRODUCT LABEL

- Do not treat linseed varieties Linda, Bolas, Karen, Laura, Mikael, Norlin, Moonraker, Abbey, Agriace [6]
- Do not use treated seed as food or feed [6]

Crop-specific information
- Latest use: milky ripe stage (GS 77) for cereals; before drilling flax or linseed [6]
- HI 6 wk for oilseed rape and cereals; 21 d for herbs
- Application with other fungicides may cause cereal crop scorch [1-5, 7]

Environmental safety
- Dangerous for the environment
- Very toxic to aquatic organisms
- Dangerous to fish or other aquatic life. Do not contaminate surface waters or ditches with chemical or used container
- Treated seed harmful to game and wildlife. Product contains red dye for easy identification of treated seed [6]

Hazard classification and safety precautions
Hazard H03 [5, 6]; H04 [1-4, 7, 8]; H08 [2, 4]; H11 [1-3, 5-8]
Risk phrases R22a [5, 6]; R36 [1-8]; R38 [1-5, 7, 8]; R43 [2, 4]; R50 [2, 5, 8]; R51 [1, 3, 6, 7]; R53a [1-3, 5-8]
Operator protection A [1-8]; C [1-6, 8]; U05a, U20b [1-8]; U08 [1-7]; U09a [8]; U11 [1, 3, 6, 7]; U14 [1, 3, 5, 7]; U15 [1, 3, 7]; U19a [2, 8]
Environmental protection E13b [4]; E15a [1-3, 5, 7, 8]; E34 [5]; E38 [5, 6]
Storage and disposal D01, D09a [1-8]; D02, D10b [1-5, 7, 8]; D05 [1-4, 6, 7]; D10a [6]; D12a [5, 6, 8]
Treated seed S01, S02, S03, S04b, S05, S06a, S07 [6]
Medical advice M03, M05a [5]; M05b [1, 3, 7]

364 prochloraz + propiconazole

A broad spectrum fungicide mixture for wheat, barley and oilseed rape

Products
1	Bumper P	Makhteshim	400:90 g/l	EC	08548
2	Greencrop Twinstar	Greencrop	400:90 g/l	EC	09516

Uses
- Eyespot in **spring barley, spring wheat** [1]; **winter barley, winter wheat** [1, 2]
- Light leaf spot in **spring oilseed rape, winter oilseed rape** [1]
- Phoma in **spring oilseed rape, winter oilseed rape** [1]
- Rhynchosporium in **spring barley** [1]; **winter barley** [1, 2]
- Septoria leaf spot in **spring wheat** [1]; **winter wheat** [1, 2]

Approval information
- Propiconazole included in Annex I under EC Directive 91/414
- Accepted by BBPA for use on malting barley

Efficacy guidance
- Best results obtained from treatment when disease is active but not well established
- Treat wheat normally from flag leaf ligule just visible stage (GS 39) but earlier if there is a high risk of Septoria
- Treat barley from the first node detectable stage (GS 31)
- If disease pressure persists a second application may be necessary

Restrictions
- Maximum number of treatments 2 per crop

Crop-specific information
- Latest use: before grain watery ripe (GS 71) for cereals; before most seeds green for winter oilseed rape

Environmental safety
- Dangerous for the environment
- Very toxic to aquatic organisms

SEE SECTION 3 FOR PRODUCTS ALSO REGISTERED

SECTION 2

Hazard classification and safety precautions
> **Hazard** H04, H11
> **Risk phrases** R36, R53a [1, 2]; R50 [2]; R51 [1]
> **Operator protection** A, C; U05a, U08, U20b [1, 2]; U11, U15 [1]
> **Environmental protection** E15a
> **Storage and disposal** D01, D02, D05, D09a, D10b
> **Medical advice** M05a [1]

365   prochloraz + tebuconazole

A broad spectrum systemic fungicide mixture for cereals, oilseed rape and amenity use

Products

1 Agate	Makhteshim	267:133 g/l	EC	12016
2 Lunar	Sherriff Amenity	267:132 g/l	EC	12366

Uses
- Brown rust in *spring barley, spring rye, spring wheat, winter barley, winter rye, winter wheat* [1]
- Eyespot in *spring barley, spring rye, spring wheat, winter barley, winter rye, winter wheat* [1]
- Fusarium patch in *managed amenity turf* [2]
- Glume blotch in *spring wheat, winter wheat* [1]
- Late ear diseases in *spring wheat, winter wheat* [1]
- Net blotch in *spring barley, winter barley* [1]
- Powdery mildew in *spring barley, spring rye, spring wheat, winter barley, winter rye, winter wheat* [1]
- Rhynchosporium in *spring barley, spring rye, winter barley, winter rye* [1]
- Sclerotinia stem rot in *spring oilseed rape, winter oilseed rape* [1]
- Septoria leaf spot in *spring wheat, winter wheat* [1]
- Yellow rust in *spring barley, spring rye, spring wheat, winter barley, winter rye, winter wheat* [1]

Approval information
- Accepted by BBPA for use on malting barley

Efficacy guidance
- Best results achieved from applications at an early stage of disease development
- Adequate protection of winter cereals will usually require a programme of at least two treatments [1]
- Two treatments separated by a 2 wk gap may be needed for difficult Fusarium patch attacks [2]
- Optimum application timing is normally when disease first seen but varies with main target disease - see labels
- To minimise possibility of development of resistance repeated applications should not be made against the same pathogen on cereals. See notes on Resistance Management in Section 5

Restrictions
- Maximum total dose on cereals equivalent to two full dose treatments
- Amenity product for use by professionals only [2]
- Maximum number of treatments on managed amenity turf 2 per yr
- Do not apply during drought or to frozen turf [2]
- Turf product for professional use only

Crop-specific information
- Latest use: before grain milky ripe (GS 73) for cereals; most seeds green for oilseed rape
- HI 6 wk
- Occasionally transient leaf speckling may occur after treating wheat. Yield responses should not be affected

Environmental safety
- Dangerous for the environment
- Very toxic to aquatic organisms
- Dangerous to fish or other aquatic life. Do not contaminate surface waters or ditches with chemical or used container

FOR FULL CONDITIONS OF USE ALWAYS READ THE PRODUCT LABEL

Hazard classification and safety precautions
Hazard H03, H11
Risk phrases R22a, R36, R43, R50, R53a [1, 2]; R63 [2]
Operator protection A, C, H [1, 2]; M [2]; U05a, U14 [1, 2]; U09b, U20a [1]; U11 [2]
Environmental protection E13b [1]; E15a [2]; E34, E38 [1, 2]
Storage and disposal D01, D02, D05, D09a, D12a [1, 2]; D10b [2]; D10c [1]
Medical advice M03 [1, 2]; M05a [2]

366   prochloraz + thiram

A seed treatment fungicide mixture for oilseed rape

Products
| Agrichem Hy-Pro Duet | Agrichem | 150:333 g/l | FS | 12469 |

Uses
• Phoma in **winter oilseed rape** *(seed treatment)*

Efficacy guidance
• Can be applied through most types of seed treatment machinery
• Good coverage of the seed surface essential for best results
• Product is active against seed-borne disease. It will not control leaf spotting from leaf spores on trash on the seed bed

Restrictions
• Maximum number of treatments 1 per crop
• Seed must be of satisfactory quality and moisture content
• Sow seed as soon as possible after treatment
• Do not store treated seed from one season to the next
• Do not co-apply product with water or mix with any other seed treatment product

Crop-specific information
• Latest use: pre-sowing
• Product will also control soil-borne fungi that cause damping-off of rape seedlings

Following crops guidance
• Any crop may be grown following a treated winter oilseed rape crop

Hazard classification and safety precautions
Hazard H03, H11
Risk phrases R20, R22a, R36, R38, R43, R50, R53a
Operator protection A, C, H; U02a, U08, U14, U15, U20a
Environmental protection E15a, E36
Storage and disposal D01, D02, D05, D08, D09a, D12b, D13
Treated seed S01, S02, S03, S04d, S05, S08
Medical advice M04

367   prochloraz + triticonazole

A broad spectrum fungicide seed treatment mixture for winter cereals

Products
| Kinto | BASF | 60:20 g/l | FS | 12038 |

Uses
• Bunt in **winter wheat** *(seed treatment)*
• Covered smut in **winter barley** *(seed treatment)*
• Fusarium foot rot and seedling blight in **winter barley** *(seed treatment)*, **winter wheat** *(seed treatment)*
• Leaf stripe in **winter barley** *(seed treatment)*
• Loose smut in **winter barley** *(seed treatment)*, **winter wheat** *(seed treatment)*
• Septoria seedling blight in **winter wheat** *(seed treatment)*

Approval information
• Accepted by BBPA for use on malting barley

SEE SECTION 3 FOR PRODUCTS ALSO REGISTERED

Efficacy guidance
- Apply undiluted through conventional seed treatment machine
- Calibrate seed drill with treated seed before sowing
- Bag treated seed immediately and keep in a dry, draught free store
- Ensure good even coverage of seed to achieve best results
- Treatment may cause some slight delay in crop emergence

Restrictions
- Maximum number of treatments 1 per batch of seed
- Do not treat seed with moisture content above 16% and do not allow moisture content of treated seed to exceed 16%
- Do not treat cracked, split or sprouted seed
- Do not handle treated seed unnecessarily
- Do not use treated seed as food or feed

Crop-specific information
- Latest use: before drilling wheat or barley

Environmental safety
- Dangerous for the environment
- Toxic to aquatic organisms

Hazard classification and safety precautions
Hazard H11
Risk phrases R51, R53a
Operator protection A, H; U05a, U07, U20b
Environmental protection E15a, E34, E38
Storage and disposal D01, D02, D08, D09a, D11a, D12a
Treated seed S01, S02, S03, S04a, S04b, S05, S07

368   prohexadione-calcium

A cyclohexanecarboxylate growth regulator for use in apples

Products

Regalis	BASF	10% w/w	WG	12414

Uses
- Control of shoot growth in *apples*

Efficacy guidance
- For a standard orchard treat at the start of active growth and again after 3-5 wk depending on growth conditions
- Alternatively follow the first application with four reduced rate applications at two wk intervals

Restrictions
- Maximum total dose equivalent to two full dose treatments
- Do not apply in conjunction with calcium based foliar fertilisers

Crop-specific information
- HI 55 d for apples

Environmental safety
- Harmful to aquatic organisms
- Avoid spray drift onto neighbouring crops and other non-target plants
- Product does not harm natural insect predators when used as directed

Hazard classification and safety precautions
Risk phrases R52, R53a
Operator protection U05a
Environmental protection E38
Storage and disposal D01, D02, D09a, D10c, D12a

FOR FULL CONDITIONS OF USE ALWAYS READ THE PRODUCT LABEL

369   prometryn

A contact and residual triazine herbicide

Products

1	Alpha Prometryne 50 WP	Makhteshim	50% w/w	WP	04871
2	Gesagard	Syngenta	50% w/w	WP	08410
3	Pennine 50	DAPT	50% w/w	WP	10638

Uses

- Annual dicotyledons in **bulb onions** (off-label), **chives** (off-label), **leeks** (off-label), **salad onions** (off-label), **transplanted leeks** (off-label) [2]; **carrots, celery, parsley** [1-3]; **herbs (see appendix 6)** (off-label), **protected chives** (off-label), **protected herbs (see appendix 6)** (off-label) [1, 2]; **protected parsley** (off-label) [1]; **transplanted leeks** [1, 3]
- Annual grasses in **bulb onions** (off-label), **leeks** (off-label), **salad onions** (off-label), **transplanted leeks** (off-label) [2]; **carrots, celery, parsley** [1-3]; **chives** (off-label), **herbs (see appendix 6)** (off-label), **protected chives** (off-label), **protected herbs (see appendix 6)** (off-label) [1, 2]; **protected parsley** (off-label) [1]; **transplanted leeks** [1, 3]

Specific Off-Label Approvals (SOLAs)

- **bulb onions, salad onions** (OLA 011658) Dec 2008 [2]
- **chives, herbs (see appendix 6), protected chives, protected herbs (see appendix 6), protected parsley** (OLA 031794) Dec 2007 [1]
- **chives, herbs (see appendix 6), protected chives, protected herbs (see appendix 6)** (OLA 023394) Dec 2007 [2]
- **leeks, transplanted leeks** (OLA 011659) Dec 2008 [2]

Approval information

- Products containing this active ingredient have been granted derogations for specified 'Essential Uses' for use until 31 December 2007. Sale and supply must cease by 30 June 2007 but growers have no guarantee that the products will continue to be available until then.
  For more information see 'The Review Programme' under 'Pesticide Legislation' in Section 5

Efficacy guidance

- Best results achieved by application to young seedling weeds up to 5 cm high (cotyledon stage for knotgrass, mayweed and corn marigold) on fine, moist seedbed when rain falls afterwards. Do not use on very cloddy soils
- On organic soils only contact action effective and repeat application may be needed (certain crops only)
- Excessive rain after treatment may check crop

Restrictions

- Maximum number of treatments 1 per crop for transplanted leeks; 1 per crop for transplanted celery
- Maximum total dose equivalent to one full dose treatment on all crops
- This product must not be used on any crops other than those listed, including any extrapolations that would normally be permissible under the Long Term Arrangements for Extension of Use (see Section 5)

Crop-specific information

- HI onions 10 wk; carrots, parsley, celery, leeks, herbs 6 wk
- Apply to carrots, celery, parsley post-emergence after 2-rough leaf stage or after transplants established
- Apply to transplanted leeks or celery after transplants established. Do not use on drilled leeks

Following crops guidance

- In the event of crop failure within 8 wk of treatment only peas, potatoes, carrots, parsley or celery may be planted. After 8 wk the land should be cultivated or ploughed to 150 mm min before planting the next crop

Environmental safety

- Dangerous for the environment
- Toxic to aquatic organisms
- Harmful to fish or other aquatic life. Do not contaminate surface waters or ditches with chemical or used container

SEE SECTION 3 FOR PRODUCTS ALSO REGISTERED

Hazard classification and safety precautions
   Hazard H11 [1, 2]
   Risk phrases R50 [1]; R51 [2]; R53a [1, 2]
   Operator protection U05a [1, 2]; U14, U15 [1]; U20b [1-3]
   Environmental protection E13c [3]; E15a [1, 2]; E38 [2]
   Consumer protection C02 [1, 3] (6 wk)
   Storage and disposal D01, D02 [1, 2]; D09a, D11a [1-3]; D12a [2]; D12b [1]

370   propachlor

A pre-emergence chloroacetanilide herbicide for various horticultural crops

See also chloridazon + propachlor
   chlorthal-dimethyl + propachlor

Products

1	Alpha Propachlor 50 SC	Makhteshim	500 g/l	SC	04873
2	Brasson	Monsanto	480 g/l	SC	10560
3	Ramrod Flowable	Monsanto	480 g/l	SC	10314

Uses
   • Annual dicotyledons in ***blackcurrants*** *(off-label)*, ***blueberries*** *(off-label)*, ***celery*** *(off-label)*, ***chinese cabbage*** *(off-label)*, ***chives*** *(off-label)*, ***gooseberries*** *(off-label)*, ***herbs (see appendix 6)*** *(off-label)*, ***kohlrabi*** *(off-label)*, ***lettuce*** *(off-label)*, ***parsley*** *(off-label)*, ***radicchio*** *(off-label)*, ***redcurrants*** *(off-label)*, ***rhubarb*** *(off-label)*, ***salad brassicas*** *(off-label - for baby leaf production)*, ***strawberries*** *(off-label)*, ***whitecurrants*** *(off-label)* [1, 3]; ***broccoli, brussels sprouts, cabbages, calabrese, cauliflowers, fodder rape, kale, leeks, onions, ornamental plant production, spring oilseed rape, swedes, turnips*** [1-3]; ***brown mustard, sage, white mustard*** [2, 3]; ***cress*** *(off-label)*, ***frise*** *(off-label - under crop covers)*, ***lamb's lettuce*** *(off-label - under crop covers)*, ***lettuce*** *(off-label - under crop covers)*, ***radicchio*** *(off-label - under crop covers)*, ***ribes hybrids*** *(off-label)*, ***scarole*** *(off-label - under crop covers)* [3]; ***mustard, rubus hybrids*** *(off-label)*, ***salad onions, strawberries, winter oilseed rape*** [1]
   • Annual grasses in ***blackcurrants*** *(off-label)*, ***blueberries*** *(off-label)*, ***chinese cabbage*** *(off-label)*, ***chives*** *(off-label)*, ***gooseberries*** *(off-label)*, ***herbs (see appendix 6)*** *(off-label)*, ***kohlrabi*** *(off-label)*, ***lettuce*** *(off-label)*, ***parsley*** *(off-label)*, ***radicchio*** *(off-label)*, ***redcurrants*** *(off-label)*, ***strawberries*** *(off-label)*, ***whitecurrants*** *(off-label)* [1, 3]; ***broccoli, brussels sprouts, cabbages, calabrese, cauliflowers, fodder rape, kale, leeks, onions, ornamental plant production, spring oilseed rape, swedes, turnips*** [1-3]; ***brown mustard, sage, white mustard*** [2, 3]; ***celery*** *(off-label)*, ***mustard, rhubarb*** *(off-label)*, ***rubus hybrids*** *(off-label)*, ***salad brassicas*** *(off-label - for baby leaf production)*, ***salad onions, strawberries, winter oilseed rape*** [1]; ***cress*** *(off-label)*, ***frise*** *(off-label - under crop covers)*, ***lamb's lettuce*** *(off-label - under crop covers)*, ***lettuce*** *(off-label - under crop covers)*, ***radicchio*** *(off-label - under crop covers)*, ***ribes hybrids*** *(off-label)*, ***scarole*** *(off-label - under crop covers)* [3]

Specific Off-Label Approvals (SOLAs)
   • *blackcurrants, blueberries, chinese cabbage, chives, gooseberries, herbs (see appendix 6), kohlrabi, lettuce, parsley, radicchio, redcurrants, rubus hybrids, strawberries, whitecurrants (OLA 002424) Dec 2008* [1]
   • *blackcurrants, blueberries, chinese cabbage, chives, cress, gooseberries, herbs (see appendix 6), kohlrabi, lettuce, parsley, radicchio, redcurrants, ribes hybrids, strawberries, whitecurrants (OLA 021159) Dec 2008* [3]
   • *celery, rhubarb (OLA 051699) Dec 2008* [1]
   • *celery, rhubarb (OLA 051070) Dec 2008* [3]
   • *frise, lamb's lettuce, lettuce, radicchio, scarole (under crop covers) (OLA 021159) Dec 2008* [3]
   • *salad brassicas (for baby leaf production) (OLA 051700) Dec 2008* [1]
   • *salad brassicas (for baby leaf production) (OLA 030436) Dec 2008* [3]

Efficacy guidance
   • Controls germinating (not emerged) weeds for 6-8 wk
   • Best results achieved by application to fine, firm, moist seedbed free of established weeds in spring, summer and early autumn

---

FOR FULL CONDITIONS OF USE ALWAYS READ THE PRODUCT LABEL

- Use higher rate on soils with more than 10% organic matter
- Recommended as tank-mix with glyphosate or chlorthal-dimethyl on onions, leeks, swedes, turnips. See label for details

Restrictions
- Maximum number of treatments 1 or 2 per crop, varies with product and crop - see label for details
- Maximum total dose equivalent to two full dose treatments on brassicas and leeks; one full dose treatment on most other crops [1]
- Do not use after products that de-wax the leaves of plants, especially brassica crops
- Do not use on crops under glass or polythene
- Do not use under extremely wet, dry or other adverse growth conditions

Crop-specific information
- Latest use: ranges from before crop or weed emergence to 3-4 leaf stage of brassica crops and pre-blossom for strawberries. Consult label for details
- HI salad brassicas 28 d; other salad vegetables, herbs 6 wk
- Apply in brassicas from drilling to time seed chits or after 3-4 true leaf stage but before weed emergence, in swedes and turnips pre-emergence only
- Apply in transplanted brassicas within 48 h of planting in warm weather. Plants must be hardened off and special care needed with block sown or modular propagated plants
- Apply in onions and leeks pre-emergence or from post-crook to young plant stage
- Apply to newly planted strawberries soon after transplanting, to weed-free soil in established crops in early spring before new weeds emerge
- Granules may be applied to onion, leek and brassica nurseries and to most flower crops after bedding out and hardening off
- Do not use pre-emergence of drilled wallflower seed

Following crops guidance
- In the event of crop failure only replant recommended crops in treated soil

Environmental safety
- Dangerous for the environment
- Very toxic to aquatic organisms

Hazard classification and safety precautions
Hazard H03, H11
Risk phrases R21, R36 [1]; R22a, R38, R43, R50, R53a [1-3]
Operator protection A, C; U02a, U04a, U05a, U08, U13, U19a, U20b [1-3]; U10, U11 [1]; U14 [2, 3]
Environmental protection E15a, E34
Storage and disposal D01, D02, D05, D09a [1-3]; D10b, D12b [1]; D10c, D12a [2, 3]
Medical advice M03 [1-3]; M04 [1]

---

371 propamocarb hydrochloride

A translocated protectant carbamate fungicide

*See also chlorothalonil + propamocarb hydrochloride
fenamidone + propamocarb hydrochloride
mancozeb + propamocarb hydrochloride*

Products

1	Filex	Scotts	722 g/l	SL	07631
2	Flash	AgriGuard	722 g/l	SL	11989
3	Proplant	Fargro	722 g/l	SL	08572

Uses
- Blight in *outdoor tomatoes, protected tomatoes* [2]
- Botrytis in *chives* (off-label), *herbs (see appendix 6)* (off-label), *parsley* (off-label), *protected herbs (see appendix 6)* (off-label) [1, 3]; *cress* (off-label) [1]; *lettuce, protected chives* (off-label), *protected lettuce, protected parsley* (off-label), *protected radishes* (off-label) [3]
- Damping off in *inert substrate cucumbers* (off-label), *inert substrate peppers* (off-label), *inert substrate tomatoes* (off-label), *nft peppers* (off-label), *nft tomatoes* (off-label) [1, 3]; *protected peppers* (off-label), *protected tomatoes* (off-label) [1]

SEE SECTION 3 FOR PRODUCTS ALSO REGISTERED

- Downy mildew in **broccoli, brussels sprouts, cabbages, calabrese, cauliflowers, chinese cabbage** [1-3]; **chives** *(off-label)*, **herbs (see appendix 6)** *(off-label)*, **parsley** *(off-label)*, **protected herbs (see appendix 6)** *(off-label)*, **protected radishes** *(off-label)*, **radishes** *(off-label)* [1, 3]; **cress** *(off-label)*, **watercress** *(off-label - under protection)* [1]; **lettuce, protected chives** *(off-label)*, **protected lettuce, protected parsley** *(off-label)*, **watercress** *(off-label - during propagation)* [3]
- Phytophthora in **aubergines, bedding plants, broccoli, brussels sprouts, cabbages, calabrese, cauliflowers, chinese cabbage, cucumbers, leeks, nursery stock, onions, peppers, pot plants** [1-3]; **container-grown ornamentals, ornamental specimens, watercress** *(off-label - under protection)* [1]; **flower bulbs, outdoor tomatoes** [2, 3]; **ornamental plant production, watercress** *(off-label - during propagation)* [3]; **protected tomatoes, rockwool aubergines, rockwool cucumbers, rockwool peppers, rockwool tomatoes** [2]; **tulips** [1, 2]
- Pythium in **aubergines, bedding plants, broccoli, brussels sprouts, cabbages, calabrese, cauliflowers, chinese cabbage, cucumbers, flower bulbs, leeks, nursery stock, onions, outdoor tomatoes, peppers, pot plants** [1-3]; **ornamental plant production, watercress** *(off-label - during propagation)* [3]; **ornamental specimens, watercress** *(off-label - under protection)* [1]; **protected tomatoes, rockwool aubergines, rockwool cucumbers, rockwool peppers, rockwool tomatoes, tulips** [1, 2]
- Root malformation disorder in **red beet** *(off-label)* [1, 3]
- Root rot in **inert substrate cucumbers** *(off-label)*, **inert substrate peppers** *(off-label)*, **inert substrate tomatoes** *(off-label)*, **nft peppers** *(off-label)*, **nft tomatoes** *(off-label)* [3]
- White blister in **horseradish** *(off-label)*, **protected horseradish** *(off-label)* [1, 3]

Specific Off-Label Approvals (SOLAs)

- **chives, cress, herbs (see appendix 6), parsley, protected herbs (see appendix 6)** *(OLA 040626)* Dec 2008 [1]
- **chives, herbs (see appendix 6), horseradish, parsley, protected chives, protected herbs (see appendix 6), protected horseradish, protected parsley, protected radishes, radishes** *(OLA 040625)* Dec 2008 [3]
- **horseradish, protected horseradish** *(OLA 992036)* Dec 2008 [1]
- **inert substrate cucumbers, inert substrate peppers, inert substrate tomatoes, nft peppers, nft tomatoes, protected peppers, protected tomatoes** *(OLA 992032)* Dec 2008 [1]
- **inert substrate cucumbers, inert substrate peppers, inert substrate tomatoes, nft peppers, nft tomatoes** *(OLA 002527)* Dec 2008 [3]
- **protected radishes, radishes** *(OLA 992030)* Dec 2008 [1]
- **red beet** *(OLA 031247)* Dec 2008 [1]
- **red beet** *(OLA 031453)* Jul 2008 [3]
- **watercress** *(under protection)* *(OLA 010439)* Dec 2008 [1]
- **watercress** *(during propagation)* *(OLA 040625)* Dec 2008 [3]

Efficacy guidance

- Chemical is absorbed through roots and translocated throughout plant
- Incorporate in compost before use or drench moist compost or soil before sowing, pricking out, striking cuttings or potting up
- Drench treatment can be repeated at 3-6 wk intervals
- Concentrated solution is corrosive to all metals other than stainless steel
- May also be applied in trickle irrigation systems
- To prevent root rot in tulip bulbs apply as dip for 20 min but full protection only achieved by soil drench treatment as well

Restrictions

- Maximum number of treatments 4 per crop for cucumbers, tomatoes, peppers, aubergines; 3 per crop for outdoor and protected lettuce; 1 per crop for listed brassicas; 1 compost incorporation and/or 1 drench treatment for leeks, onions; 1 dip and 1 soil drench for flower bulbs
- Do not apply to lettuce crops grown in non-soil systems
- When applied over established seedlings rinse off foliage with water (except lettuce) and do not apply under hot, dry conditions
- On plants of unknown tolerance test first on a small scale
- Do not apply in a recirculating irrigation/drip system
- Store away from seeds and fertilizers

Crop-specific information

- Latest use: before transplanting for brassicas, herbs; before crop emergence for watercress

FOR FULL CONDITIONS OF USE ALWAYS READ THE PRODUCT LABEL

- HI 2 d for inert substrate and NFT vegetable crops; 14 d for cucumbers, tomatoes, peppers, aubergines, horseradish, radishes, herbs and protected herbs; 4 wk for calabrese, cauliflower, sprouting broccoli, Brussels sprouts, Chinese cabbage. watercress; 19 wk for leeks, onions

Hazard classification and safety precautions
Operator protection A, C, H, K, M [1-3]; B [1, 2]; U02a, U19a [3]; U05a [2, 3]; U08, U20b [1-3]
Environmental protection E15a [1-3]; E34 [3]
Storage and disposal D01, D02, D05 [2, 3]; D07 [2]; D09a [1-3]; D10b [3]; D11a [1, 2]

372 propaquizafop

A phenoxy alkanoic acid foliar acting herbicide for grass weeds in a range of crops

Products

1	Bulldog	Makhteshim	100 g/l	EC	11723
2	Cleancrop GYR	Makhteshim	100 g/l	EC	10646
3	Emrald Eyetort	Me2	100 g/l	EC	12034
4	Falcon	Makhteshim	100 g/l	EC	10585
5	Greencrop Satchmo	Greencrop	100 g/l	EC	10748
6	Landgold PQF 100	Landgold	100 g/l	EC	10763
7	Landgold PQF 100	Teliton	100 g/l	EC	12127
8	Raptor	Makhteshim	100 g/l	EC	11092
9	Shogun	Makhteshim	100 g/l	EC	10584
10	Standon Propaquizafop	Standon	100 g/l	EC	11536

Uses

- Annual grasses in *bulb onions, carrots, combining peas, early potatoes, farm forestry, fodder beet, linseed, maincrop potatoes, parsnips, spring field beans, spring oilseed rape, sugar beet, swedes, turnips, winter field beans, winter oilseed rape* [1-10]; *cut logs, forest, forest nurseries* [1, 2, 4, 8, 9]; *poppies* (off-label - for morphine production) [8]
- Annual meadow grass in *chives* (off-label), *parsley* (off-label) [2, 4, 8, 9]; *herbs (see appendix 6)* (off-label), *leaf spinach* (off-label), *leeks* (off-label), *red beet* (off-label) [1, 2, 4, 8, 9]
- Couch in *chives* (off-label), *parsley* (off-label) [2, 4, 8, 9]; *herbs (see appendix 6)* (off-label), *leaf spinach* (off-label), *leeks* (off-label), *red beet* (off-label) [1, 2, 4, 8, 9]
- Perennial grasses in *bulb onions, carrots, combining peas, early potatoes, farm forestry, fodder beet, linseed, maincrop potatoes, parsnips, spring field beans, spring oilseed rape, sugar beet, swedes, turnips, winter field beans, winter oilseed rape* [1-10]; *celery* (off-label) [2, 4, 8, 9]; *cut logs, forest, forest nurseries* [1, 2, 4, 8, 9]
- Volunteer cereals in *celery* (off-label), *chives* (off-label), *parsley* (off-label) [2, 4, 8, 9]; *herbs (see appendix 6)* (off-label), *leaf spinach* (off-label), *leeks* (off-label), *red beet* (off-label) [1, 2, 4, 8, 9]; *poppies* (off-label - for morphine production) [8]
- Wild oats in *celery* (off-label) [2, 4, 8, 9]

Specific Off-Label Approvals (SOLAs)

- *celery, chives, herbs (see appendix 6), leaf spinach, leeks, parsley, red beet* (OLA 021634) Dec 2008 [2]
- *celery* (OLA 021502) Dec 2008 [4]
- *celery, chives, herbs (see appendix 6), leaf spinach, leeks, parsley, red beet* (OLA 021632) Dec 2008 [8]
- *celery, chives, herbs (see appendix 6), leaf spinach, leeks, parsley, red beet* (OLA 021629) Dec 2008 [9]
- *chives, herbs (see appendix 6), leaf spinach, leeks, parsley, red beet* (OLA 020148) Dec 2008 [4]
- *herbs (see appendix 6), leaf spinach, leeks, red beet* (OLA 032022) Dec 2008 [1]
- *poppies* (for morphine production) (OLA 030770) Dec 2008 [8]

Efficacy guidance

- Apply to emerged weeds when they are growing actively with adequate soil moisture
- Activity is slower under cool conditions
- Broad-leaved weeds and any weeds germinating after treatment are not controlled
- Annual meadow grass up to 3 leaves checked at low doses and severely checked at highest dose
- Spray barley cover crops when risk of wind blow has passed and before there is serious competition with the crop

SEE SECTION 3 FOR PRODUCTS ALSO REGISTERED

- Various tank mixtures and sequences recommended for broader spectrum weed control in oilseed rape, peas and sugar beet. See label for details
- Severe couch infestations may require a second application at reduced dose when regrowth has 3-4 leaves unfolded
- Products contain surfactants. Tank mixing with adjuvants not required or recommended
- Always follow WRAG guidelines for preventing and managing herbicide resistant weeds. Section 5 for more information

Restrictions
- Maximum total dose varies according to crop treated. See label for details
- See label for list of tolerant tree species
- Do not treat seed potatoes

Crop-specific information
- Latest use: before crop flower buds visible for winter oilseed rape, linseed, field beans; before 8 fully expanded leaf stage for spring oilseed rape, mustard; before weeds are covered by the crop for potatoes, sugar beet, fodder beet; when flower buds visible for peas
- HI herbs, chives, leaf spinach, parsley 3 wk; celery, leeks, red beet, onions, carrots, early potatoes, parsnips 4 wk; combining peas, early potatoes 7 wk; sugar beet, fodder beet, maincrop potatoes, swedes, turnips 8 wk; field beans 14 wk
- Application in high temperatures and/or low soil moisture content may cause chlorotic spotting especially on combining peas and field beans
- Overlaps at the highest dose can cause damage from early applications to carrots and parsnips

Following crops guidance
- In the event of a failed treated crop an interval of 2 wk must elapse between the last application and redrilling with winter wheat or winter barley. 4 wk must elapse before sowing oilseed rape, peas or field beans, and 16 wk before sowing ryegrass or oats

Environmental safety
- Dangerous for the environment
- Toxic to aquatic organisms
- Risk to certain non-target insects or other arthropods. See directions for use

Hazard classification and safety precautions
Hazard H04, H11
Risk phrases R36, R38, R51, R53a
Operator protection A, C [3-7, 10]; U02a, U05a, U08, U19a, U20b [1-10]; U11 [3, 4, 6, 7, 10]; U13 [1, 2, 4, 8, 9]; U14, U15 [1-4, 6-10]; U23a [5]
Environmental protection E15a, E22b [1-10]; E34 [1, 2, 4, 8, 9]
Storage and disposal D01, D02, D05, D09a, D10b [1-10]; D12b [1, 2, 4, 8, 9]; D14 [1, 4, 8, 9] (returnable container)
Medical advice M05b

373   propiconazole

A systemic, curative and protectant conazole fungicide

See also cyproconazole + propiconazole
prochloraz + propiconazole

Products

1 Bumper 250 EC	Makhteshim	250 g/l	EC	09039
2 Tonik	AgriGuard	250 g/l	EC	12247

Uses
- Brown rust in **spring barley, spring wheat, winter barley, winter rye, winter wheat** [1, 2]
- Crown rust in **grass for ensiling, grass seed crops, spring oats, winter oats** [1, 2]
- Drechslera leaf spot in **grass for ensiling, grass seed crops** [1, 2]
- Leaf blotch in **garlic** (off-label), **onions** (off-label), **shallots** (off-label) [1]
- Light leaf spot in **spring oilseed rape** (reduction), **winter oilseed rape** (reduction) [1, 2]
- Mildew in **grass for ensiling, grass seed crops** [1, 2]
- Powdery mildew in **spring barley, spring oats, spring wheat, winter barley, winter oats, winter rye, winter wheat** [1, 2]

FOR FULL CONDITIONS OF USE ALWAYS READ THE PRODUCT LABEL

- Ramularia leaf spots in **sugar beet** *(reduction)* [1, 2]
- Rhynchosporium in **grass for ensiling**, **grass seed crops**, **spring barley**, **winter barley**, **winter rye** [1, 2]
- Rust in **garlic** *(off-label)* [1]; **sugar beet** [1, 2]
- Septoria in **spring wheat**, **winter rye**, **winter wheat** [1, 2]
- Sooty moulds in **spring wheat**, **winter wheat** [1, 2]
- White blister in **honesty** *(off-label)* [1]
- White rust in **chrysanthemums** *(off-label)*, **protected chrysanthemums** *(off-label)* [1]
- Yellow rust in **spring barley**, **spring wheat**, **winter barley**, **winter wheat** [1, 2]

Specific Off-Label Approvals (SOLAs)
- **chrysanthemums, garlic, honesty, onions, protected chrysanthemums, shallots** *(OLA 012142) Dec 2008* [1]

Approval information
- Propiconazole included in Annex I under EC Directive 91/414
- Accepted by BBPA for use on malting barley

Efficacy guidance
- Best results achieved by applying at early stage of disease. Recommended spray programmes vary with crop, disease, season, soil type and product. See label for details

Restrictions
- Maximum number of treatments 4 per crop for wheat (up to 3 in yr of harvest); 3 per crop for barley, oats, rye, triticale (up to 2 in yr of harvest); 2 per crop or yr for oilseed rape, sugar beet, grass seed crops, honesty; 1 per yr on grass for ensiling
- Propiconazole may produce an allergic reaction
- On oilseed rape do not apply during flowering. Apply first treatment before flowering, the second afterwards
- Grass seed crops must be treated in yr of harvest
- A minimum interval of 14 d must elapse between treatments on leeks and 21 d on sugar beet
- Avoid spraying crops under stress, eg during cold weather or periods of frost

Crop-specific information
- Latest use: before 4 pairs of true leaves for honesty
- HI cereals, grass seed crops 35 d; oilseed rape, sugar beet, garlic, onions, shallots, grass for ensiling 28 d

Environmental safety
- Dangerous for the environment
- Toxic to aquatic organisms

Hazard classification and safety precautions
  Hazard H11
  Risk phrases R51, R53a
  Operator protection A, C; U02a, U05a, U08, U19a, U20b
  Environmental protection E15a, E34
  Storage and disposal D01, D02, D05, D09a, D10b
  Treated seed S06a
  Medical advice M05a

374 propoxycarbazone-sodium

A sulfonylaminocarbonyltriazolinone residual grass weed herbicide for winter wheat

Products

1	Attribut	Bayer CropScience	70% w/w	SG	11221
2	Ethos	Bayer CropScience	70% w/w	SG	11223

Uses
- Blackgrass in **winter wheat**
- Couch in **winter wheat**

Approval information
- Approval expiry 31 Aug 2006 [1, 2]

SEE SECTION 3 FOR PRODUCTS ALSO REGISTERED

Efficacy guidance

- Activity by root and foliar absorption but depends on presence of sufficient soil moisture to ensure root uptake by weeds. Control is enhanced by use of an adjuvant to encourage foliar uptake
- Best results obtained from treatments applied when weed grasses are growing actively. Symptoms may not become apparent for 3-4 wk after application
- Ensure good even spray coverage
- To achieve best control of couch and reduction of infestation in subsequent crop treat between 2 true leaves and first node stage of the weed. A split dose programme may improve control
- Blackgrass should be treated after using specialist blackgrass herbicides with a different mode of action
- Proxycarbazone-sodium is a member of the ALS-inhibitor group of herbicides and products should be used in a planned Resistance Management strategy. See Section 5 for more information

Restrictions

- Maximum total dose equivalent to one full dose treatment
- Do not use in a programme with other aceto-lactase synthesis (ALS) inhibitors
- Do not apply when temperature near or below freezing
- Avoid treatment under dry soil conditions

Crop-specific information

- Third node detectable (GS 33)

Following crops guidance

- Only winter wheat, field beans or winter barley may be sown in the autumn following spring treatment
- Any crop may be grown in the spring on land treated during the previous calendar yr

Environmental safety

- Dangerous to fish or other aquatic life. Do not contaminate surface waters or ditches with chemical or used container
- LERAP Category B
- Take extreme care to avoid drift onto adjacent plants or land as this could result in severe damage

Hazard classification and safety precautions

Operator protection A; U05a, U20b
Environmental protection E13b, E16a, E16b
Storage and disposal D01, D02, D09a, D11a

375 propyzamide

A residual amide herbicide for use in a wide range of crops

Products

1	Agriguard Propyzamide 50 WP	AgriGuard	50% w/w	WP	10187
2	Barclay Piza 400FL	Barclay	400 g/l	SC	11495
3	Bulwark Flo	Interfarm	400 g/l	SC	10161
4	Flomide	Interfarm	400 g/l	SC	11171
5	Greencrop Saffron FL	Greencrop	400 g/l	SC	10244
6	Kerb 50 W	Dow	50% w/w	WP	10771
7	Kerb Flo	Dow	400 g/l	SC	10768
8	Kerb Granules	SumiAgro	4% w/w	GR	08917
9	Kerb Pro Flo	Dow	400 g/l	SC	10760
10	Kerb Pro Granules	SumiAgro Amenity	4% w/w	GR	09600
11	Menace 80 EDF	Dow	80% w/w	WG	10766
12	Mithras 80 EDF	Dow	80% w/w	WG	10767
13	Precis	Dow	400 g/l	SC	10762
14	Propose 50 W	Interfarm	50% w/w	WP	11170
15	Quaver Flo	Dow	400 g/l	SC	10769
16	Standon Propyzamide 400 SC	Standon	400 g/l	SC	10255
17	Standon Propyzamide 50	Standon	50% w/w	WP	10814

Uses

- Annual dicotyledons in *apple orchards, blackberries, blackcurrants, clover seed crops, gooseberries, loganberries, lucerne, plums, redcurrants, strawberries, sugar beet seed crops, winter field beans, winter oilseed rape* [1, 2, 4-7, 11-17]; *brassica seed crops, lettuce*

FOR FULL CONDITIONS OF USE ALWAYS READ THE PRODUCT LABEL

(outdoor crops), **rhubarb** *(outdoor)* [11, 12]; **camomile** *(off-label)*, **fenugreek** *(off-label)*, **radicchio** *(off-label)*, **sage** *(off-label)*, **tarragon** *(off-label)* [6]; **fodder rape seed crops, kale seed crops, lettuce, rhubarb, turnip seed crops** [1, 2, 4-7, 13-17]; **forest, trees and shrubs** [1, 3, 6, 8-10, 14]; **ornamental plant production, raspberries** [17]; **pear orchards** [1, 2, 4-7, 11, 13-17]; **pears** [12]; **raspberries** *(England only)* [1, 2, 4-7, 11-16]; **roses** [1, 6, 14, 17]; **woody ornamentals** [1, 3, 6, 8-12, 14]

- Annual grasses in **apple orchards, blackberries, blackcurrants, clover seed crops, gooseberries, loganberries, lucerne, plums, redcurrants, strawberries, sugar beet seed crops, winter field beans, winter oilseed rape** [1, 2, 4-7, 11-17]; **brassica seed crops, lettuce** *(outdoor crops)*, **rhubarb** *(outdoor)* [11, 12]; **camomile** *(off-label)*, **cherries** *(off-label)*, **chicory** *(off-label)*, **courgettes** *(off-label)*, **evening primrose** *(off-label)*, **fenugreek** *(off-label)*, **grapevines** *(off-label)*, **honesty** *(off-label)*, **hops** *(off-label)*, **marrows** *(off-label)*, **mirabelles** *(off-label)*, **protected chives** *(off-label)*, **protected courgettes** *(off-label)*, **protected herbs (see appendix 6)** *(off-label)*, **protected lamb's lettuce** *(off-label)*, **protected lettuce** *(off-label)*, **protected marrows** *(off-label)*, **protected parsley** *(off-label)*, **protected pumpkins** *(off-label)*, **protected radicchio** *(off-label)*, **protected scarole** *(off-label)*, **protected squashes** *(off-label)*, **pumpkins** *(off-label)*, **radicchio** *(off-label)*, **sage** *(off-label)*, **squashes** *(off-label)*, **tarragon** *(off-label)* [6]; **fodder rape seed crops, kale seed crops, lettuce, rhubarb, turnip seed crops** [1, 2, 4-7, 13-17]; **forest, trees and shrubs** [1, 3, 6, 8-10, 14]; **ornamental plant production, raspberries** [17]; **pear orchards** [1, 2, 4-7, 11, 13-17]; **pears** [12]; **raspberries** *(England only)* [1, 2, 4-7, 11-16]; **roses** [1, 6, 14, 17]; **woody ornamentals** [1, 3, 6, 8-12, 14]
- Horsetails in **forest** [3, 9, 11, 12]; **woody ornamentals** [3, 9]
- Perennial grasses in **apple orchards, blackberries, blackcurrants, gooseberries, loganberries, plums, redcurrants** [1, 2, 4-7, 11-17]; **camomile** *(off-label)*, **cherries** *(off-label)*, **chicory** *(off-label)*, **courgettes** *(off-label)*, **evening primrose** *(off-label)*, **fenugreek** *(off-label)*, **grapevines** *(off-label)*, **honesty** *(off-label)*, **hops** *(off-label)*, **marrows** *(off-label)*, **mirabelles** *(off-label)*, **protected chives** *(off-label)*, **protected courgettes** *(off-label)*, **protected herbs (see appendix 6)** *(off-label)*, **protected lamb's lettuce** *(off-label)*, **protected lettuce** *(off-label)*, **protected marrows** *(off-label)*, **protected parsley** *(off-label)*, **protected pumpkins** *(off-label)*, **protected radicchio** *(off-label)*, **protected scarole** *(off-label)*, **protected squashes** *(off-label)*, **pumpkins** *(off-label)*, **radicchio** *(off-label)*, **sage** *(off-label)*, **squashes** *(off-label)*, **tarragon** *(off-label)* [6]; **clover seed crops, fodder rape seed crops, kale seed crops, lettuce, lucerne, rhubarb, strawberries, sugar beet seed crops, turnip seed crops, winter field beans, winter oilseed rape** [1, 2, 4-7, 13-17]; **forest, woody ornamentals** [1, 3, 6, 8-12, 14]; **ornamental plant production, raspberries** [17]; **pear orchards** [1, 2, 4-7, 11, 13-17]; **pears** [12]; **raspberries** *(England only)* [1, 2, 4-7, 11-16]; **rhubarb** *(outdoor)* [11, 12]; **roses** [1, 6, 14, 17]; **trees and shrubs** [1, 3, 6, 8-10, 14]
- Sedges in **forest** [3, 9, 11, 12]; **woody ornamentals** [3, 9]
- Volunteer cereals in **sugar beet seed crops, winter field beans, winter oilseed rape** [11, 12]
- Wild oats in **sugar beet seed crops, winter field beans, winter oilseed rape** [11, 12]

Specific Off-Label Approvals (SOLAs)

- **camomile, cherries, chicory, courgettes, evening primrose, fenugreek, grapevines, honesty, hops, marrows, mirabelles, protected chives, protected courgettes, protected herbs (see appendix 6), protected lamb's lettuce, protected lettuce, protected marrows, protected parsley, protected pumpkins, protected radicchio, protected scarole, protected squashes, pumpkins, radicchio, sage, squashes, tarragon** *(OLA 021416) Dec 2008* [6]

Approval information

- Propyzamide included in Annex I under EC Directive 91/414
- Some products may be applied through CDA equipment. See labels for details
- Accepted by BBPA for use on hops

Efficacy guidance

- Active via root uptake. Weeds controlled from germination to young seedling stage, some species (including many grasses) also when established
- Best results achieved by winter application to fine, firm, moist soil. Rain is required after application if soil dry
- Uptake is slow and and may take up to 12 wk
- Excessive organic debris or ploughed-up turf may reduce efficacy
- For heavy couch infestations a repeat application may be needed in following winter
- Always follow WRAG guidelines for preventing and managing herbicide resistant weeds. Section 5 for more information

SEE SECTION 3 FOR PRODUCTS ALSO REGISTERED

Restrictions
- Maximum number of treatments 1 per crop or yr
- Maximum total dose equivalent to one full dose treatment for all crops
- Do not treat protected crops
- Apply to listed edible crops only between 1 Oct and the date specified as the latest time of application (except lettuce)
- Do not apply in windy weather and avoid drift onto non-target crops
- Do not use on soils with more than 10% organic matter except in forestry

Crop-specific information
- Latest use: labels vary but normally before 31 Dec for rhubarb, lucerne, strawberries and winter field beans; before 31 Jan for other crops; pre-emergence for chicory
- HI normally 6 wk for edible crops and 3 mth for protected herbs but labels vary in detail
- Apply as soon as possible after 3-true leaf stage of oilseed rape (GS 1,3) and seed brassicas, after 4-leaf stage of sugar beet for seed, within 7 d after sowing but before emergence for field beans, after perennial crops established for at least 1 season, strawberries after 1 yr
- Only apply to strawberries on heavy soils. Do not use on matted row crops
- Only apply to field beans on medium and heavy soils
- Only apply to established lucerne not less than 7 d after last cut
- In lettuce lightly incorporate in top 25 mm pre-drilling or irrigate on dry soil
- See label for lists of ornamental and forest species which may be treated

Following crops guidance
- Following an application between 1 Apr and 31 Jul at any dose the following minimum intervals must be observed before sowing the next crop: lettuce 0 wk; broad beans, chicory, clover, field beans, lucerne, radishes, peas 5 wk; brassicas, celery, leeks, oilseed rape, onions, parsley, parsnips 10 wk
- Following an application between 1 Aug and 31 Mar at any dose the following minimum intervals must be observed before sowing the next crop: lettuce 0 wk; broad beans, chicory, clover, field beans, lucerne, radishes, peas 10 wk; brassicas, celery, leeks, oilseed rape, onions, parsley, parsnips 25 wk or after 15 Jun, whichever occurs sooner
- Cereals or grasses or other crops not listed may be sown 30 wk after treatment up to 840 g ai/ha between 1 Aug and 31 Mar or 40 wk after treatment at higher doses at any time and after mouldboard ploughing to at least 15 cm
- A period at least 9 mth must elapse between applications of propyzamide to the same land

Environmental safety
- Dangerous for the environment
- Very toxic to aquatic organisms

Hazard classification and safety precautions
Hazard H03, H11 [1-8, 11, 13, 14, 16, 17]
Risk phrases R40, R53a [1-8, 11, 13, 14, 16, 17]; R50 [2-5, 7, 8, 11, 13, 16, 17]; R51 [1, 6, 14]
Operator protection A, H [3, 14]; U20a [8, 10]; U20c [1-7, 9, 11-17]
Environmental protection E15a [1-17]; E34 [1, 10]; E38 [2-4, 6-8, 11, 13, 14]
Consumer protection C02 [1, 2, 4-7, 10-17] (6 wk)
Storage and disposal D01 [5, 10, 16, 17]; D05 [1-7, 9, 11-17]; D09a [1-17]; D10a [8]; D10b [5, 16]; D11a [1-4, 6, 7, 9-15, 17]; D12a [2-4, 6-8, 11, 13, 14]

376   prothioconazole

A systemic, protectant and curative triazole fungicide

See also   clothianidin + prothioconazole
fluoxastrobin + prothioconazole
fluoxastrobin + prothioconazole + trifloxystrobin

Products

1	Proline	Bayer CropScience	250 g/l	EC	12084
2	Redigo	Bayer CropScience	100 g/l	FS	12085
3	Standon Win-Win	Standon	250 g/l	EC	12386

FOR FULL CONDITIONS OF USE ALWAYS READ THE PRODUCT LABEL

Uses

- Brown rust in *spring barley*, *winter barley*, *winter rye*, *winter wheat* [1, 3]
- Bunt in *spring wheat* *(seed treatment)*, *winter wheat* *(seed treatment)* [2]
- Covered smut in *winter barley* *(seed treatment)* [2]
- Eyespot in *spring barley*, *winter barley*, *winter rye*, *winter wheat* [1, 3]
- Fusarium foot rot and seedling blight in *spring wheat* *(seed treatment)*, *winter barley* *(seed treatment)*, *winter wheat* *(seed treatment)* [2]
- Late ear diseases in *spring barley*, *winter barley*, *winter wheat* [1, 3]
- Leaf stripe in *winter barley* *(seed treatment)* [2]
- Loose smut in *spring wheat* *(seed treatment)*, *winter barley* *(seed treatment)*, *winter wheat* *(seed treatment)* [2]
- Net blotch in *spring barley*, *winter barley* [1, 3]
- Powdery mildew in *spring barley*, *winter barley*, *winter rye*, *winter wheat* [1, 3]
- Rhynchosporium in *spring barley*, *winter barley*, *winter rye* [1, 3]
- Sclerotinia stem rot in *winter oilseed rape* [1, 3]
- Septoria diseases in *winter wheat* [1, 3]
- Tan spot in *winter wheat* [1, 3]
- Yellow rust in *winter barley*, *winter wheat* [1, 3]

Approval information

- Accepted by BBPA for use on malting barley

Efficacy guidance

- Seed treatments must be applied by manufacturer's recommended treatment application equipment [2]
- Treated seed should preferably be drilled in the same season [2]
- Follow-up treatments will be needed later in the season to give protection against air-borne and splash-borne diseases [2]
- Best results on foliar diseases obtained from treatment at early stages of disease development. Further treatment may be needed if disease attack is prolonged [1, 3]
- Foliar applications to established infections of any disease are likely to be less effective [1, 3]
- Best control of cereal ear diseases obtained by treatment during ear emergence [1, 3]
- Treat oilseed rape at early to full flower [1, 3]

Restrictions

- Maximum number of seed treatments one per batch [2]
- Maximum total dose equivalent to one full dose treatment on oilseed rape, two full dose treatments on barley, three full dose treatments on wheat [1, 3]
- Seed treatment must be fully re-dispersed and homogeneous before use [2]
- Do not use on seed with more than 16% moisture content, or on sprouted, cracked or skinned seed [2]
- All seed batches should be tested to ensure they are suitable for treatment [2]
- Treated winter barley seed must be used within the season of treeatment; treated winter wheat seed should preferably be drilled in the same season [2]
- Do not make repeated treatments of the product alone to the same crop against pathogens such as powdery mildew. Use tank mixtures or alternate with fungicides having a different mode of action (12085]
- Do not treat oilseed rape crops to be used for seed production [1, 3]

Crop-specific information

- Latest use: pre-drilling for seed treatments; before grain milky ripe for foliar sprays on winter rye, winter wheat; beginning of flowering for barley [1, 3]
- HI 56 d [1] or 65 d [3] for winter oilseed rape

Environmental safety

- Dangerous for the environment [1, 3]
- Toxic to aquatic organisms [1, 3]
- Harmful to aquatic organisms [2]
- LERAP Category B [1, 3]

Hazard classification and safety precautions

Hazard H04 [1-3]; H11 [1, 3]
Risk phrases R36, R51 [1, 3]; R43, R52 [2]; R53a [1-3]
Operator protection A, H [1-3]; C [1, 3]; U05a, U09b, U20b [1, 3]; U07, U14, U20a [2]

SEE SECTION 3 FOR PRODUCTS ALSO REGISTERED

Environmental protection E15a, E34 [1-3]; E16a, E38 [1, 3]; E36 [2]
Storage and disposal D01, D02, D05, D09a, D12a [1-3]; D10b [1, 3]; D14 [2]
Treated seed S01, S02, S03, S04a, S04b, S05, S06b, S07, S08 [2]
Medical advice M03 [1, 3]

377   prothioconazole + spiroxamine

A broad spectrum fungicide mixture for cereals

Products

Helix	Bayer CropScience	160:300 g/l	EC	12264

Uses

- Brown rust in **spring barley**, **winter barley**, **winter rye**, **winter wheat**
- Eyespot in **spring barley**, **winter barley**, **winter rye**, **winter wheat**
- Late ear diseases in **winter rye**, **winter wheat**
- Net blotch in **spring barley**, **winter barley**
- Powdery mildew in **spring barley**, **winter barley**, **winter rye**, **winter wheat**
- Rhynchosporium in **spring barley**, **winter barley**, **winter rye**
- Septoria diseases in **winter wheat**
- Tan spot in **winter wheat**
- Yellow rust in **spring barley**, **winter barley**, **winter wheat**

Approval information

- Spiroxamine included in Annex I under EC Directive 91/414
- Accepted by BBPA for use on malting barley

Efficacy guidance

- Best results obtained from treatment at early stages of disease development. Further treatment may be needed if disease attack is prolonged
- Applications to established infections of any disease are likely to be less effective
- Best control of cereal ear diseases obtained by treatment during ear emergence

Restrictions

- Maximum total dose equivalent to two full dose treatments on barley and three full dose treatments on wheat and rye

Crop-specific information

- Latest use: before grain watery ripe for rye, winter wheat; up to beginning of anthesis for barley

Environmental safety

- Dangerous for the environment
- Very toxic to aquatic organisms
- Risk to non-target insects or other arthropods. Avoid spraying within 6 m of the field boundary to reduce the effects on non-target insects or other arthropods

Hazard classification and safety precautions

Hazard H03, H11
Risk phrases R20, R22a, R36, R38, R50, R53a
Operator protection A, C, H; U05a, U09b, U11, U19a, U20b
Environmental protection E15a, E22c, E34, E38
Storage and disposal D01, D02, D05, D09a, D10b, D12a
Medical advice M03

378   prothioconazole + tebuconazole

A triazole fungicide mixture for cereals

Products

Prosaro	Bayer CropScience	125:125 g/l	EC	12263

Uses

- Brown rust in **spring barley**, **winter barley**, **winter rye**, **winter wheat**
- Eyespot in **spring barley** (reduction), **winter barley** (reduction), **winter rye** (reduction), **winter wheat** (reduction)

FOR FULL CONDITIONS OF USE ALWAYS READ THE PRODUCT LABEL

- Late ear diseases in **spring barley**, **winter barley**, **winter wheat**
- Net blotch in **spring barley**, **winter barley**
- Powdery mildew in **spring barley**, **winter barley**, **winter rye**, **winter wheat**
- Rhynchosporium in **spring barley**, **winter barley**, **winter rye**
- Septoria diseases in **winter wheat**
- Tan spot in **winter wheat**
- Yellow rust in **spring barley**, **winter barley**, **winter wheat**

Approval information
- Accepted by BBPA for use on malting barley

Efficacy guidance
- Best results on foliar diseases obtained from treatment at early stages of disease development. Further treatment may be needed if disease attack is prolonged
- Applications to established infections of any disease are likely to be less effective
- Best control of cereal ear diseases obtained by treatment during ear emergence

Restrictions
- Maximum total dose equivalent to two full dose treatments on barley and three full dose treatments on wheat and rye

Crop-specific information
- Latest use: before grain milky ripe for rye, wheat; beginning of flowering for barley

Environmental safety
- Dangerous for the environment
- Toxic to aquatic organisms
- Risk to non-target insects or other arthropods. Avoid spraying within 6 m of the field boundary to reduce the effects on non-target insects or other arthropods
- LERAP Category B

Hazard classification and safety precautions
  Hazard H04, H11
  Risk phrases R38, R43, R51, R53a, R63
  Operator protection A, H; U05a, U09b, U19a, U20b
  Environmental protection E15a, E16a, E22c, E34, E38
  Storage and disposal D01, D02, D05, D09a, D10b, D12a
  Medical advice M03

379   prothioconazole + tebuconazole + triazoxide

A broad spectrum fungicide seed treatment for cereals

Products

Raxil Pro	Bayer CropScience	25:15:10 g/l	FS	12432

Uses
- Covered smut in **spring barley** (seed treatment), **winter barley** (seed treatment)
- Fusarium foot rot and seedling blight in **spring barley** (seed treatment), **winter barley** (seed treatment)
- Leaf stripe in **spring barley** (seed treatment), **winter barley** (seed treatment)
- Loose smut in **spring barley** (seed treatment), **winter barley** (seed treatment)
- Net blotch in **spring barley** (seed treatment), **winter barley** (seed treatment)

Approval information
- Accepted by BBPA for use on malting barley

Efficacy guidance
- Treatments must be applied by manufacturer's recommended treatment application equipment
- Treated seed should preferably be drilled in the same season
- Follow-up treatments will be needed later in the season to give protection against air-borne and splash-borne diseases

Restrictions
- Maximum number of seed treatments one per batch

Crop-specific information
- Latest use for seed treatment: pre-drilling

SEE SECTION 3 FOR PRODUCTS ALSO REGISTERED

Environmental safety
- Harmful to aquatic organisms

Hazard classification and safety precautions
    Hazard H04
    Risk phrases R37, R43, R52, R53a
    Operator protection A, H; U07, U14, U19a, U20a
    Environmental protection E15a, E34, E36, E38
    Storage and disposal D05, D09a, D14
    Treated seed S01, S02, S03, S04b, S05, S06b, S07, S08

380   prothioconazole + trifloxystrobin

A triazole and strobilurin fungicide mixture for cereals

Products

Mobius	Bayer CropScience	175:150 g/l	SC	12324

Uses
- Brown rust in **spring barley, winter barley, winter wheat**
- Eyespot in **spring barley** *(reduction)*, **winter barley** *(reduction)*, **winter wheat**
- Late ear diseases in **spring barley, winter barley, winter wheat**
- Net blotch in **spring barley, winter barley**
- Powdery mildew in **spring barley, winter barley, winter wheat**
- Rhynchosporium in **spring barley, winter barley**
- Septoria diseases in **winter wheat**
- Yellow rust in **winter barley, winter wheat**

Approval information
- Trifloxystrobin included in Annex I under EC Directive 91/414
- Accepted by BBPA for use on malting barley

Efficacy guidance
- Best results obtained from treatment at early stages of disease development. Further treatment may be needed if disease attack is prolonged
- Applications to established infections of any disease are likely to be less effective
- Best control of cereal ear diseases obtained by treatment during ear emergence
- Trifloxystrobin is a member of the QoI cross resistance group. Product should be used preventatively and not relied on for its curative potential
- Use product as part of an Integrated Crop Management strategy incorporating other methods of control, including where appropriate other fungicides with a different mode of action. Do not apply more than two foliar applications of QoI containing products to any cereal crop
- There is a significant risk of widespread resistance occurring in *Septoria tritici* populations in UK. Failure to follow resistance management action may result in reduced levels of disease control
- Strains of wheat and barley powdery mildew resistant to QoIs are common in the UK. Control of wheat mildew can only be relied on from the triazole component
- Where specific control of wheat mildew is required this should be achieved through a programme of measures including products recommended for the control of mildew that contain a fungicide from a different cross-resistance group and applied at a dose that will give robust control

Restrictions
- Maximum total dose equivalent to two full dose treatments

Crop-specific information
- Latest use: before grain milky ripe for wheat; beginning of flowering for barley

Environmental safety
- Dangerous for the environment
- Very toxic to aquatic organisms
- LERAP Category B

Hazard classification and safety precautions
    Hazard H04, H11
    Risk phrases R37, R50, R53a
    Operator protection A, C, H; U05a, U09b, U19a, U20b

FOR FULL CONDITIONS OF USE ALWAYS READ THE PRODUCT LABEL

Environmental protection E15a, E16a, E34, E38
Storage and disposal D01, D02, D05, D09a, D10c, D12a
Medical advice M03

381 pymetrozine

A novel azomethine insecticide

Products

1	Chess WG	Syngenta Bioline	50% w/w	WG	10651
2	Plenum WG	Syngenta	50% w/w	WG	10652

Uses
- Aphids in **blackberries** *(off-label)*, **blackcurrants** *(off-label)*, **celeriac** *(off-label)*, **celery** *(off-label)*, **chives** *(off-label)*, **gooseberries** *(off-label)*, **herbs (see appendix 6)** *(off-label)*, **leaf spinach** *(off-label)*, **lettuce** *(off-label)*, **loganberries** *(off-label)*, **parsley** *(off-label)*, **potatoes, raspberries** *(off-label)*, **redcurrants** *(off-label)*, **ribes hybrids** *(off-label)*, **rubus hybrids** *(off-label)*, **salad brassicas** *(off-label - for baby leaf production)*, **strawberries** *(off-label)*, **sweetcorn** *(off-label)* [2]; **ornamental plant production, protected blackberries** *(off-label)*, **protected blackcurrants** *(off-label)*, **protected celery** *(off-label)*, **protected chinese cabbage** *(off-label)*, **protected chives** *(off-label)*, **protected choi sum** *(off-label)*, **protected cucumbers, protected gooseberries** *(off-label)*, **protected herbs (see appendix 6)** *(off-label)*, **protected lettuce** *(off-label)*, **protected loganberries** *(off-label)*, **protected ornamentals, protected pak choi** *(off-label)*, **protected parsley** *(off-label)*, **protected raspberries** *(off-label)*, **protected redcurrants** *(off-label)*, **protected ribes hybrids** *(off-label)*, **protected rubus hybrids** *(off-label)*, **protected salad brassicas** *(off-label - for baby leaf production)*, **protected strawberries** *(off-label)* [1]
- Damson-hop aphid in **hops** *(off-label)* [2]
- Glasshouse whitefly in **protected tomatoes** *(off-label)* [1]
- Peach-potato aphid in **broccoli** *(off-label)*, **brussels sprouts** *(off-label)*, **cabbages** *(off-label)*, **calabrese** *(off-label)*, **cauliflowers** *(off-label)*, **collards** *(off-label)*, **kale** *(off-label)* [2]; **protected aubergines** *(off-label)*, **protected peppers** *(off-label)* [1]
- Tobacco whitefly in **protected tomatoes** *(off-label)* [1]

Specific Off-Label Approvals (SOLAs)
- **blackberries, blackcurrants, gooseberries, loganberries, raspberries, redcurrants, ribes hybrids, rubus hybrids, strawberries** *(OLA 031073) Oct 2011* [2]
- **broccoli, brussels sprouts, cabbages, calabrese, cauliflowers** *(OLA 031478) Oct 2011* [2]
- **celeriac, celery** *(OLA 051062) Oct 2011* [2]
- **chives, herbs (see appendix 6), lettuce, parsley** *(OLA 030844) Oct 2011* [2]
- **collards, kale** *(OLA 050385) Oct 2011* [2]
- **hops** *(OLA 031423) Oct 2011* [2]
- **leaf spinach** *(OLA 050384) Oct 2011* [2]
- **protected aubergines, protected peppers, protected tomatoes** *(OLA 030845) Oct 2011* [1]
- **protected blackberries, protected blackcurrants, protected gooseberries, protected logan-berries, protected raspberries, protected redcurrants, protected ribes hybrids, protected rubus hybrids, protected strawberries** *(OLA 031072) Oct 2011* [1]
- **protected celery** *(OLA 051063) Oct 2011* [1]
- **protected chinese cabbage, protected choi sum, protected pak choi** *(OLA 051384) Oct 2011* [1]
- **protected chives, protected herbs (see appendix 6), protected lettuce, protected parsley** *(OLA 030843) Oct 2011* [1]
- **protected salad brassicas** *(for baby leaf production) (OLA 030843) Oct 2011* [1]
- **salad brassicas** *(for baby leaf production) (OLA 030844) Oct 2011* [2]
- **sweetcorn** *(OLA 041318) Oct 2011* [2]

Approval information
- Pymetrozine included in Annex I under EC Directive 91/414
- Accepted by BBPA for use on hops

Efficacy guidance
- Pymetrozine moves systemically in the plant and acts by preventing feeding leading to death by starvation in 1-4 d. There is no immediate knockdown

SEE SECTION 3 FOR PRODUCTS ALSO REGISTERED

- Aphids controlled include those resistant to organophosphorus and carbamate insecticides
- To prevent development of resistance do not use continuously or as the sole method of control
- Best results achieved by starting spraying as soon as aphids seen in crop and repeating as necessary.
- To limit spread of persistent viruses such as potato leaf roll virus, apply from 90% crop emergence

Restrictions

- Maximum total dose equivalent to two full dose treatments on potatoes; three full dose treatments on edible crops; four full dose treatments on ornamentals
- Consult processors before use on potatoes for processing
- Check tolerance of ornamental species before large scale use. See label for list of species known to have been treated without damage. Visible spray deposits may be seen on leaves of some species [1]

Crop-specific information

- HI aubergines, peppers, cucumbers 3 d; potatoes, collards, kale, lettuce, outdoor herbs 7 d, brassicas, sweetcorn, protected herbs 14 d; protected cane fruit, bush fruit, soft fruit 12 wk

Environmental safety

- High risk to bees (outdoor use only). Do not apply to crops in flower or to those in which bees are actively foraging. Do not apply when flowering weeds are present
- Avoid spraying within 6 m of field boundaries to reduce effects on non-target insects or arthropods

Hazard classification and safety precautions

Hazard H03 [2]
Risk phrases R40 [2]
Operator protection A; U05a, U20c
Environmental protection E12a (outdoor use only); E15a
Storage and disposal D01, D02 [1]; D09a, D11a [1, 2]

382　pyraclostrobin

A protectant and curative strobilurin fungicide for cereals

See also boscalid + pyraclostrobin
epoxiconazole + fenpropimorph + pyraclostrobin
epoxiconazole + kresoxim-methyl + pyraclostrobin
epoxiconazole + pyraclostrobin

Products

1	BAS 500 06F	BASF	200 g/l	EC	12338
2	Comet	BASF	250 g/l	EC	10875
3	Insignia	Vitax	20% w/w	WG	11865
4	Platoon	BASF	200 g/l	EC	12325
5	Tucana	BASF	250 g/l	EC	10899
6	Vivid	BASF	250 g/l	EC	10898

Uses

- Brown rust in **spring barley**, **spring wheat**, **winter barley**, **winter wheat** [1, 2, 4-6]
- Crown rust in **spring oats**, **winter oats** [1, 2, 4-6]
- Dollar spot in **managed amenity turf** *(reduction)* [3]
- Fusarium patch in **managed amenity turf** [3]
- Net blotch in **spring barley**, **winter barley** [1, 2, 4-6]
- Red thread in **managed amenity turf** [3]
- Rhynchosporium in **spring barley** *(moderate)*, **winter barley** *(moderate)* [1, 2, 4-6]
- Septoria diseases in **spring wheat**, **winter wheat** [1, 2, 4-6]
- Yellow rust in **spring barley**, **spring wheat**, **winter barley**, **winter wheat** [1, 2, 4-6]

Approval information

- Pyraclostrobin included in Annex I under EC Directive 91/414
- Accepted by BBPA for use on malting barley (before ear emergence only)

Efficacy guidance

- For best results apply at the start of disease attack. This is especially for the control of Fusarium Patch [3] as severe damage to turf can occur once it is established

FOR FULL CONDITIONS OF USE ALWAYS READ THE PRODUCT LABEL

- Regular turf aeration, appropriate scarification and judicious use of nitrogenous fertiliser will assist the control of Fusarium Patch [3]
- Best results on Septoria glume blotch achieved when used as a protective treatment and against Septoria leaf blotch when treated in the latent phase [1, 2, 4-6]
- Yield response may be obtained in the absence of visual disease symptoms [1, 2, 4-6]
- Pyraclostrobin is a member of the QoI cross resistance group. Product should be used preventatively and not relied on for its curative potential
- Use product as part of an Integrated Crop Management strategy incorporating other methods of control, including where appropriate other fungicides with a different mode of action. Do not apply more than two foliar applications of QoI containing products to any cereal crop or to grass
- There is a significant risk of widespread resistance occurring in *Septoria tritici* populations in UK. Failure to follow resistance management action may result in reduced levels of disease control [1, 2, 4-6]
- On cereal crops product must always be used in mixture with another product, recommended for control of the same target disease, that contains a fungicide from a different cross resistance group and is applied at a dose that will give robust control [1, 2, 4-6]

Restrictions
- Maximum total dose on cereals or turf equivalent to two full dose treatments
- Do not apply during drought conditions or to frozen turf [3]

Crop-specific information
- Latest use: before grain watery ripe (GS 71) for wheat; up to and including emergence of ear just complete (GS 59) for barley and oats
- Avoid applying to turf immediately after cutting or 48 hr before mowing [3]

Environmental safety
- Dangerous for the environment
- Very toxic to aquatic organisms
- LERAP Category B

Hazard classification and safety precautions
Hazard H03 [1-6]; H11 [2, 3, 5, 6]
Risk phrases R20, R50 [1-6]; R22a, R38, R53a [1, 2, 4-6]
Operator protection A [1-6]; D [3]; U05a, U14, U20b [1, 2, 4-6]
Environmental protection E15a, E16a, E16b, E38 [1-6]; E34 [1, 2, 4-6]
Storage and disposal D01, D02 [1, 2, 4-6]; D05 [2, 5, 6]; D08 [1, 3, 4]; D09a, D10c, D12a [1-6]
Medical advice M03 [1, 2, 4-6]; M05a [1, 4]

383 pyrethrins

A non-persistent, contact acting insecticide extracted from Pyrethrum

Products

1 Dairy Fly Spray	B H & B	0.75 g/l	AL	H5579
2 Killgerm ULV 400	Killgerm	30 g/l	UL	H4838
3 Turbair Super Flydown	SumiAgro	2.5 g/l	RH	H7225

Uses
- Flies in *dairies*, *farm buildings* [1, 2]; *livestock houses*, *poultry houses* [1-3]

Approval information
- Products formulated for ULV application [1, 2]. May be applied through fogging machine or sprayer [1]. See label for details

Efficacy guidance
- For fly control close doors and windows and spray or apply fog as appropriate
- For best results outdoors spray during early morning or late afternoon and evening when conditions are still
- Ensure spray is directed to all areas where flies are seen to be numerous
- To avoid possibility of development of resistance do not spray more frequently than once per week

Restrictions
- For use only by professional operators

SEE SECTION 3 FOR PRODUCTS ALSO REGISTERED

- Do not allow spray to contact open food products or food preparing equipment or utensils
- Remove exposed milk and collect eggs before application
- Do not treat plants
- Do not use space sprays containing pyrethrins or pyrethroid more than once per week in intensive or controlled environment animal houses in order to avoid development of resistance. If necessary, use a different control method or product [2]

Crop-specific information
- Product formulated for use through Turbair Flydowner machines [3]

Environmental safety
- Dangerous for the environment
- Toxic to aquatic organisms
- Harmful to fish or other aquatic life. Do not contaminate surface waters or ditches with chemical or used container
- Do not apply directly to livestock
- Exclude all persons and animals during treatment [1]

Hazard classification and safety precautions
   Hazard H03 [1-3]; H04 [1]; H11 [2]
   Risk phrases R22a, R36, R38 [1]; R22b, R53a [2, 3]; R51 [2]; R66 [3]
   Operator protection A [1-3]; B [1]; C [2, 3]; D, H [2]; E [1, 2]; P [3]; U02b, U14 [2, 3]; U05a [1, 3]; U09a [1]; U09b, U20a [2]; U11, U15 [3]; U19a, U20b [1-3]
   Environmental protection E02a [3] (until surfaces dry); E05a [1, 2]; E13c [1]; E15a, E38 [2]; E34 [3]
   Consumer protection C04, C10 [1]; C06, C11 [1-3]; C07, C08, C09 [1, 2]; C12 [2]
   Storage and disposal D01, D09a [1-3]; D02 [1, 3]; D05, D10a, D12a [2]; D06a [2, 3]; D11a [1]; D12b [3]
   Medical advice M05b [2, 3]

384   pyridate

A contact pyridazine herbicide for cereals, maize and brassicas

Products

Lentagran WP	Syngenta	45% w/w	WB	08478

Uses
- Annual dicotyledons in **spring oilseed rape** *(off-label)*, **winter oilseed rape** *(off-label)*
- Black nightshade in **brussels sprouts**, **cabbages**, **maize**, **onions**
- Cleavers in **brussels sprouts**, **cabbages**, **maize**, **onions**, **spring oilseed rape** *(off-label)*, **winter oilseed rape** *(off-label)*
- Fat hen in **brussels sprouts**, **cabbages**, **maize**, **onions**

Specific Off-Label Approvals (SOLAs)
- **spring oilseed rape**, **winter oilseed rape** *(OLA 051558) Jan 2007* [1]

Approval information
- Pyridate included in Annex I under EC Directive 91/414

Efficacy guidance
- Best results achieved by application to actively growing weeds at 6-8 leaf stage when temperatures are above 8°C before crop foliage forms canopy

Restrictions
- Maximum number of treatments 1 per crop
- Do not apply in mixture with or within 14 d of any other product which may result in dewaxing of crop foliage
- Apply to maize after first-leaf stage. Do not use on cv. Meritos, Sunrise or Tainon 236
- Do not use on crops suffering stress from frost, drought, disease or pest attack

Crop-specific information
- Latest use: before 7 leaf stage for maize; before 5 true leaves for onions; before flower buds visible for oilseed rape

FOR FULL CONDITIONS OF USE ALWAYS READ THE PRODUCT LABEL

- HI 6 wk for Brussels sprouts, cabbages
- Apply to cabbages and Brussels sprouts after 4 fully expanded leaf stage. Allow 2 wk after transplanting before treating

Environmental safety
- Dangerous for the environment
- Toxic to aquatic organisms

Hazard classification and safety precautions
  Hazard H04, H11
  Risk phrases R43, R51, R53a
  Operator protection A, C; U05a, U14, U22a
  Environmental protection E15a, E38
  Storage and disposal D01, D02, D09a, D10c, D12a

385 pyrimethanil

An anilinopyrimidine fungicide for apples and strawberries

See also chlorothalonil + pyrimethanil

Products

Scala	BASF	400 g/l	SC	11695

Uses
- Botrytis in **aubergines** *(off-label)*, **bilberries** *(off-label)*, **blackberries** *(off-label)*, **blackcurrants** *(off-label)*, **cranberries** *(off-label)*, **gooseberries** *(off-label)*, **grapevines** *(off-label)*, **outdoor tomatoes** *(off-label)*, **protected blackberries** *(off-label)*, **protected boysenberries** *(off-label)*, **protected lettuce** *(off-label)*, **protected loganberries** *(off-label)*, **protected marionberries** *(off-label)*, **protected raspberries** *(off-label)*, **protected sunberries** *(off-label)*, **protected tayberries** *(off-label)*, **protected tomatoes** *(off-label)*, **raspberries** *(off-label)*, **redcurrants** *(off-label)*, **strawberries**, **whitecurrants** *(off-label)*
- Scab in **apples**

Specific Off-Label Approvals (SOLAs)
- *aubergines, outdoor tomatoes, protected tomatoes (OLA 040516) Dec 2008* [1]
- *bilberries, blackcurrants, cranberries, gooseberries, redcurrants, whitecurrants (OLA 040519) Dec 2008* [1]
- *blackberries, raspberries (OLA 051737) Dec 2008* [1]
- *grapevines (OLA 040517) Dec 2008* [1]
- *protected blackberries, protected boysenberries, protected loganberries, protected marionberries, protected raspberries, protected sunberries, protected tayberries (OLA 023411) Dec 2008* [1]
- *protected lettuce (OLA 040518) Dec 2008* [1]

Efficacy guidance
- On apples a programme of sprays will give early season control of scab. Season long control can be achieved by continuing programme with other approved fungicides
- In strawberries product should be used as part of a programme of disease control treatments which should alternate with other materials to prevent or limit development of less sensitive strains of grey mould

Restrictions
- Maximum number of treatments 5 per yr for apples; 4 per yr for protected crops; 3 per yr for grapevines; 2 per yr for cane fruit, bush fruit, strawberries, tomatoes
- Product does not taint apples. Processors should be consulted before use on strawberries

Crop-specific information
- Latest use: before end of flowering for apples
- HI protected cane fruit, strawberries 1 d; aubergines, tomatoes 3 d; protected lettuce 14 d; bush fruit, grapevines 21 d
- All varieties of apples and strawberries may be treated
- Treat apples from bud burst at 10-14 d intervals

SEE SECTION 3 FOR PRODUCTS ALSO REGISTERED

- In strawberries start treatments at white bud to give maximum protection of flowers against grey mould and treat every 7-10 d. Product should not be used more than once in a 3 or 4 spray programme

Environmental safety
- Harmful to aquatic organisms
- Product has negligible effect on hoverflies and lacewings. Limited evidence indicates some margin of safety to *Typhlodromus pyri*
- Broadcast air-assisted LERAP (20 m)

Hazard classification and safety precautions
  Risk phrases R52, R53a
  Operator protection U08, U20b
  Environmental protection E15a; E17b (20 m)
  Storage and disposal D01, D02, D05, D08, D09a, D10b, D12a

386 quinmerac

A residual herbicide available only in mixtures

See also chloridazon + quinmerac
metazachlor + quinmerac

387 quinoxyfen

A systemic protectant fungicide for cereals

See also fenpropimorph + quinoxyfen

Products

1	Apres	Dow	500 g/l	SC	08881
2	Erysto	Dow	250 g/l	SC	08697
3	Fortress	Dow	500 g/l	SC	08279

Uses
- Powdery mildew in **durum wheat, spring barley, spring oats, spring rye, spring wheat, sugar beet, triticale, winter barley, winter oats, winter rye, winter wheat** [1-3]; **protected strawberries** (off-label), **strawberries** (off-label) [3]

Specific Off-Label Approvals (SOLAs)
- **protected strawberries, strawberries** (OLA 041923) Feb 2007 [3]

Approval information
- Quinoxyfen included in Annex I under Directive 91/414
- Accepted by BBPA for use on malting barley (before ear emergence only)
- Approval expiry 20 Dec 2006 [1, 2]

Efficacy guidance
- For best results treat at early stage of disease development before infection spreads to new crop growth. Further treatment may be necessary if disease pressure remains high
- Product not curative and will not control latent or established disease infections
- For broad spectrum control in cereals use in tank mixtures - see label
- Product rainfast after 1 h
- Systemic activity may be reduced in severe drought

Restrictions
- Maximum total dose equivalent to two full dose treatments in cereals; see labels for split dose recommendation in sugar beet
- On cereals apply only in the spring from mid-tillering stage (GS 25)

Crop-specific information
- Latest use: first awns visible stage (GS 49) for cereals
- HI sugar beet 28 d

Environmental safety
- Dangerous for the environment
- Very toxic to aquatic organisms

Hazard classification and safety precautions
    Hazard H04, H11
    Risk phrases R43, R50, R53a
    Operator protection A, C, H; U05a, U14
    Environmental protection E15a, E16a, E16b, E34, E38
    Storage and disposal D01, D02, D12a

388   quizalofop-P-ethyl

An aryl phenoxypropionic acid post-emergence herbicide for grass weed control

Products

CoPilot	Interfarm	100 g/l	EC	08042

Uses
- Annual grasses in *combining peas, fodder beet, linseed, mangels, red beet, spring field beans, spring oilseed rape, sugar beet, vining peas, winter field beans, winter oilseed rape*
- Couch in *combining peas, fodder beet, linseed, mangels, red beet, spring field beans, spring oilseed rape, sugar beet, vining peas, winter field beans, winter oilseed rape*
- Perennial grasses in *combining peas, fodder beet, linseed, mangels, red beet, spring field beans, spring oilseed rape, sugar beet, vining peas, winter field beans, winter oilseed rape*
- Volunteer cereals in *combining peas, fodder beet, linseed, mangels, red beet, spring field beans, spring oilseed rape, sugar beet, vining peas, winter field beans, winter oilseed rape*

Efficacy guidance
- Best results achieved by application to emerged weeds growing actively in warm conditions with adequate soil moisture. Use split treatment to extend period of control in oilseed rape
- Weed control may be reduced under conditions such as drought that limit uptake and translocation
- Annual meadow-grass is not controlled
- Various spray programmes and tank-mixtures recommended to control mixed dicotyledon/grass weed populations. See label for details
- For effective couch control do not hoe beet crops within 21 d after spraying
- At least 2 h without rain should follow application otherwise results may be reduced

Restrictions
- Maximum number of treatments 1 on peas and beans, 2 as split dose on oilseed rape, 2 for couch control in all other recommended crops. See labels for maximum total dose of individual products
- Do not spray crops under stress from any cause or in frosty weather
- May cause taint in peas. Consult processor before use on peas or beans for processing
- An interval of at least 3 d must elapse between treatment and use of another herbicide on beet crops, 14 d on oilseed rape, 21 d on linseed and other recommended crops

Crop-specific information
- HI 16 wk for beet crops; 11 wk for oilseed rape, linseed; 8 wk for field beans; 5 wk for peas
- In some situations treatment can cause yellow patches on foliage of peas, especially vining varieties. Symptoms usually rapidly and completely outgrown

Following crops guidance
- In the event of crop failure broad-leaved crops may be resown at any time, cereals, onions, leeks or maize may be sown after 2-6 wk depending on dose used. See label for details. Onions, leeks and maize are not recommended to follow a failed treated crop

Environmental safety
- Dangerous for the environment
- Toxic to aquatic organisms
- Dangerous to fish or other aquatic life. Do not contaminate surface waters or ditches with chemical or used container

SEE SECTION 3 FOR PRODUCTS ALSO REGISTERED

Hazard classification and safety precautions
    Hazard H03, H08, H11
    Risk phrases R22b, R37, R41, R51, R53a, R66, R67
    Operator protection A, C; U05a, U11, U20b
    Environmental protection E13b, E38
    Storage and disposal D01, D02, D05, D09a, D10b, D12a
    Medical advice M05b

389   quizalofop-P-tefuryl

An aryl phenoxypropionate herbicide for grass weed control

Products

Panarex	Certis	40 g/l	EC	12532

Uses
- Blackgrass in *cereal cover crops, combining peas, fodder beet, potatoes, spring field beans, spring linseed, spring oilseed rape, sugar beet, winter field beans, winter linseed, winter oilseed rape*
- Couch in *cereal cover crops, combining peas, fodder beet, potatoes, spring field beans, spring linseed, spring oilseed rape, sugar beet, winter field beans, winter linseed, winter oilseed rape*
- Italian ryegrass in *cereal cover crops, combining peas, fodder beet, potatoes, spring field beans, spring linseed, spring oilseed rape, sugar beet, winter field beans, winter linseed, winter oilseed rape*
- Perennial ryegrass in *cereal cover crops, combining peas, fodder beet, potatoes, spring field beans, spring linseed, spring oilseed rape, sugar beet, winter field beans, winter linseed, winter oilseed rape*
- Volunteer cereals in *cereal cover crops, combining peas, fodder beet, potatoes, spring field beans, spring linseed, spring oilseed rape, sugar beet, winter field beans, winter linseed, winter oilseed rape*
- Wild oats in *cereal cover crops, combining peas, fodder beet, potatoes, spring field beans, spring linseed, spring oilseed rape, sugar beet, winter field beans, winter linseed, winter oilseed rape*

Efficacy guidance
- Best results on annual grass weeds when growing actively and treated from 2 leaves to the start of tillering
- Treat cover crops when they have served their purpose and the threat of wind blow has passed
- Best results on couch achieved when the weed is growing actively and commencing new rhizome growth
- Grass weeds germinating after treatment will not be controlled
- Treatment quickly stops growth and visible colour changes to the leaf tips appear after about 7 d. Complete kill takes 3-4 wk under good growing conditions

Restrictions
- Maximum number of treatments 1 per crop
- Consult processors before use on peas or potatoes for processing
- Do not treat crops and weeds growing under stress from any cause

Crop-specific information
- HI 60 d for all crops
- Treat oilseed rape from the fully expanded cotyledon stage
- Treat linseed, peas and field beans from 2-3 unfolded leaves
- Treat sugar beet from 2 unfolded leaves

Following crops guidance
- In the event of failure of a treated crop any broad-leaved crop may be planted at any time. Cereals may be drilled from 4 wk after treatment

Environmental safety
- Dangerous for the environment
- Very toxic to aquatic organisms

FOR FULL CONDITIONS OF USE ALWAYS READ THE PRODUCT LABEL

- Risk to certain non-target insects or other arthropods. For advice on risk management and use in Integrated Pest Management (IPM) see directions for use
- Avoid spraying within 6 m of the field boundary to reduce effects on non-target insects and other arthropods

Hazard classification and safety precautions
  Hazard H04, H11
  Risk phrases R41, R43, R50, R53a
  Operator protection A, C, H; U02a, U04a, U05a, U10, U11, U14, U15, U19a, U20b
  Environmental protection E15a, E22b, E38
  Storage and disposal D01, D02, D09a, D10b, D12a

390 rimsulfuron

A selective systemic sulfonylurea herbicide

Products

1 Me2 Aducksbackside	Me2	25% w/w	SG	10065
2 Rigid	AgriGuard	25% w/w	SG	12319
3 Standon Rimsulfuron	Standon	25% w/w	SG	09955
4 Tarot	Makhteshim	25% w/w	SG	11896
5 Titus	Makhteshim	25% w/w	SG	11895

Uses
- Annual dicotyledons in *forage maize, potatoes*
- Volunteer oilseed rape in *forage maize, potatoes*

Efficacy guidance
- Product should be used with a suitable adjuvant or a suitable herbicide tank-mix partner. See label for details
- Product acts by foliar action. Best results obtained from good spray cover of small actively growing weeds. Effectiveness is reduced in very dry conditions
- Split application provides control over a longer period and improves control of fat hen and polygonums
- Weed spectrum can be broadened by tank mixture with other herbicides. See label for details
- Susceptible weeds cease growth immediately and symptoms can be seen 10 d later
- Rimsulfuron is a member of the ALS-inhibitor group of herbicides

Restrictions
- Maximum number of treatments 2 per crop
- Do not treat maize previously treated with organophosphorus insecticides
- Do not apply to potatoes grown for certified seed
- Consult processor before use on crops grown for processing
- Avoid high light intensity (full sunlight) and high temperatures on the day of spraying
- Do not treat during periods of substantial diurnal temperature fluctuation or when frost anticipated
- Do not apply to any crop stressed by drought, water-logging, low temperatures, pest or disease attack, nutrient or lime deficiency

Crop-specific information
- Latest use: before most advanced potato plants are 25 cm high; before 4-collar stage of fodder maize
- All varieties of ware potatoes may be treated, but variety restrictions of any tank-mix partner must be observed
- Only certain named varieties of forage maize may be treated. See label

Following crops guidance
- Only winter wheat should follow a treated crop in the same calendar yr
- Only barley, wheat or maize should be sown in the spring of the yr following treatment
- In the second autumn after treatment any crop except brassicas or oilseed rape may be drilled

Environmental safety
- Dangerous for the environment
- Toxic to aquatic organisms
- Extremely dangerous to fish or other aquatic life. Do not contaminate surface waters or ditches with chemical or used container

SEE SECTION 3 FOR PRODUCTS ALSO REGISTERED

- Herbicide is very active. Take particular care to avoid drift onto plants outside the target area
- Spraying equipment should not be drained or flushed onto land planted, or to be planted, with trees or crops other than potatoes or forage maize and should be thoroughly cleansed after use - see label for instructions

Hazard classification and safety precautions
Hazard H11 [1, 3]
Risk phrases R51, R53a [1, 3]
Operator protection U08, U19a [1-5]; U20a [2, 4, 5]; U20b [1, 3]
Environmental protection E13a [2, 4, 5]; E15a [1, 3]; E34 [1-5]
Storage and disposal D01, D09a, D11a [1-5]; D05 [2]

391   rotenone

A natural, contact insecticide of low persistence

Products

FS Liquid Derris	Ford Smith	50 g/l	EC	01213

Uses
- Aphids in *all soft fruit, all top fruit, miscellaneous flowers, ornamental specimens, protected crops, vegetables*
- Raspberry beetle in *cane fruit*
- Sawflies in *gooseberries*
- Slug sawfly in *pears, roses*

Efficacy guidance
- Apply as high volume spray when pest first seen and repeat as necessary
- Spray raspberries at first pink fruit, loganberries when most of blossom over, blackberries as first blossoms open. Repeat 2 wk later if necessary
- Spray to obtain thorough coverage, especially on undersurfaces of leaves

Crop-specific information
- HI 1 d

Environmental safety
- Flammable
- Dangerous to fish or other aquatic life. Do not contaminate surface waters or ditches with chemical or used container

Hazard classification and safety precautions
Hazard H08
Operator protection U20a
Environmental protection E13b, E34
Consumer protection C02 (1 d)
Storage and disposal D09a, D10a

392   silthiofam

A benzamide fungicide seed dressing for cereals

Products

Latitude	Monsanto	125 g/l	FS	10695

Uses
- Take-all in *spring wheat* (seed treatment), *winter barley* (seed treatment), *winter wheat* (seed treatment)

Approval information
- Accepted by BBPA for use on malting barley

Efficacy guidance
- Apply using approved seed treatment equipment which has been accurately calibrated
- Apply simultaneously (i.e. not in mixture) with a standard seed treatment
- Drill treated seed in the season of purchase. The viability of treated seed and fungicide activity may be reduced by physical storage

FOR FULL CONDITIONS OF USE ALWAYS READ THE PRODUCT LABEL

- Drill treated seed at 2.5-4 cm into a well prepared firm seedbed
- As precaution against possible development of disease resistance do not treat more than three consecutive susceptible cereal crops in any one rotation

Restrictions
- Maximum number of treatments one per batch of seed
- Do not use on seed with more than 16% moisture content, or on sprouted, cracked or skinned seed
- Test germination of all seed batches before treatment

Crop-specific information
- Latest use: immediately prior to drilling

Hazard classification and safety precautions
  Operator protection A, H; U20b
  Environmental protection E03, E15a, E34
  Storage and disposal D05, D09a, D10a
  Treated seed S01, S02, S04b, S05, S06a, S07

393   simazine

A soil-acting triazine herbicide with restricted permitted uses

Products

1	Alpha Simazine 50 SC	Makhteshim	500 g/l	SC	04801
2	Gesatop	Syngenta	500 g/l	SC	08412
3	Sipcam Simazine Flowable	Sipcam	500 g/l	SC	07622

Uses
- Annual dicotyledons in *asparagus, rhubarb, roses* (field grown nursery stock), *trees and shrubs* (field grown nursery stock) [3]; *asparagus* (off-label), *rhubarb* (off-label) [1, 2]; *broad beans, hops, spring field beans, strawberries, winter field beans* [1-3]; *ornamental plant production, roses* [2]; *ornamental plant production* (for resale only) [1, 3]
- Annual grasses in *asparagus, rhubarb, roses* (field grown nursery stock), *trees and shrubs* (field grown nursery stock) [3]; *asparagus* (off-label), *rhubarb* (off-label) [1, 2]; *broad beans, hops, spring field beans, strawberries, winter field beans* [1-3]; *ornamental plant production, roses* [2]; *ornamental plant production* (for resale only) [1, 3]

Specific Off-Label Approvals (SOLAs)
- *asparagus, rhubarb* (OLA 021255) Dec 2007 [1]
- *asparagus, rhubarb* (OLA 011660) Dec 2008 [2]

Approval information
- Approvals for sale, supply and use of simazine on non-crop land were revoked in 1992 and 1993
- Products containing this active ingredient have been granted derogations for specified 'Essential Uses' for use until 31 December 2007. Sale and supply must cease by 30 June 2007 but growers have no guarantee that the products will continue to be available until then.
- Approval for all remaining uses expired on 10 September 2005 following non-inclusion of simazine in Annex I of EC Directive 91/414

Efficacy guidance
- Active via root uptake. Best results achieved by application to fine, firm, moist soil, free of established weeds, when rain falls after treatment
- Do not use on highly organic soils as efficacy will be severely impaired
- Following repeated use of simazine or other triazine herbicides resistant strains of groundsel and some other annual weeds may develop

Restrictions
- Use must be restricted to one product containing simazine or atrazine applied either as a single application at the maximum approved dose or (subject to any existing maximum permitted number of treatments) to several applications at lower doses up to the maximum approved dose for a single application
- Do not spray beans on sandy or gravelly soils or cultivate after treatment
- Allow at least 7 mth before drilling or planting other crops, longer if weather dry

SEE SECTION 3 FOR PRODUCTS ALSO REGISTERED

- Do not sow oats in autumn following spring application in maize
- Do not apply through hand-held equipment

Crop-specific information
- Latest use: before emergence of spears for asparagus; 7 d after drilling for sweetcorn; before end of Feb (autumn planted), 10 d after drilling for spring planted field and broad beans; before end of Mar for top, bush and cane fruit; after harvest, before end of Nov for strawberries; after harvest, before 1 May for hops.
- HI mint and sage 4 mth; rhubarb 3 mth; herbs 14 wk; asparagus 5 wk [1, 2]
- Apply in top, bush and cane fruit and woody ornamentals established at least 12 mth in Feb-Mar. Roses may be sprayed immediately after planting. See label for lists of resistant and susceptible species
- Do not treat broad bean varieties Beryl, Feligreen or Rowena
- Apply to strawberries in Jul-Dec, not in spring. Do not treat spring-planted crops established less than 6 mth, or winter-planted crops less than 9 mth. Do not spray varieties Huxley Giant, Madame Montal or Regina
- Apply to hops overall in Feb-Apr before weeds emerge. Newly planted sets must be covered with minimum of 50 mm soil
- Only use on roses intended for resale, whether bare-rooted or container-grown stock
- Apply to forest nursery seedbed in second yr or to transplant lines after plants 5 cm tall. Do not treat Norway spruce (Christmas trees)
- On Sands, stony or gravelly soils there is risk of crop damage, especially with heavy rain

Environmental safety
- Dangerous for the environment
- Very toxic to aquatic organisms
- LERAP Category B
- To reduce soil run-off on gradients especially in orchard and forest plantations, users are advised to plant grass strips or leave 6 m wide strips between treated areas and surface waters

Hazard classification and safety precautions
Hazard H03, H11
Risk phrases R40, R50, R53a [1-3]; R43 [2, 3]
Operator protection A, C, H [1-3]; B [3]; D, M [1, 3]; U05a, U14, U20c [1-3]; U15 [1]; U23a [2, 3]
Environmental protection E15a, E16a [1-3]; E38 [2, 3]
Storage and disposal D01, D02, D10b, D12b [1]; D05, D10c, D12a [2, 3]; D09a [1-3]
Medical advice M05a [2, 3]

394   sodium chlorate

A non-selective inorganic herbicide for total vegetation control

Products

Doff Sodium Chlorate Weedkiller	Doff Portland	53% w/w	SP	06049

Uses
- Total vegetation control in **non-crop areas**, **paths and drives**

Efficacy guidance
- Active through foliar and root uptake
- Apply as overall spray at any time during growing season. Best results obtained from application to moist soil in spring or early summer

Restrictions
- Maximum number of treatments 2 per yr for non-crop areas, paths and drives
- Treated ground must not be replanted for at least 6 mth after treatment
- Do not apply before heavy rain

Environmental safety
- Dangerous for the environment
- Contact with combustible material may cause fire
- Harmful to aquatic organisms
- Keep livestock out of treated areas for at least two weeks following treatment and until poisonous weeds, such as ragwort, have died down and become unpalatable

FOR FULL CONDITIONS OF USE ALWAYS READ THE PRODUCT LABEL

- Clothing, paper, plant debris etc become highly inflammable when dry if contaminated with sodium chlorate
- Fire risk has been reduced by inclusion of fire depressant but product should not be used in areas of exceptionally high fire risk eg oil installations, timber yards
- Wash clothing thoroughly after use
- If clothes become contaminated do not stand near an open fire
- Must not be sold to persons under 18

Hazard classification and safety precautions
> **Hazard** H03, H09
> **Risk phrases** R08, R22a
> **Operator protection** A, H, M; U05a, U20b
> **Environmental protection** E13c, E15a, E34
> **Storage and disposal** D01, D02, D07, D09a, D11a
> **Medical advice** M03, M05a

395   sodium chloride (commodity substance)

An inorganic salt for use as a sugar beet herbicide

Products
sodium chloride	various	-	SL

Uses
- Polygonums in **sugar beet**
- Volunteer potatoes in **sugar beet**

Approval information
- Approval for the use of sodium chloride as a commodity substance was granted on 11 September 1996 by Ministers under regulation 5 of the Control of Pesticides Regulations 1986

Efficacy guidance
- Best results obtained from good spray cover treatment in hot, humid weather
- Use as a follow-up treatment where earlier sprays have failed to control large volunteer potatoes or polygonums
- Acts by contact scorch and has no direct effect on daughter potato tubers
- Saturated spray solution should contain 0.1% w/w non-ionic wetter and be applied at 1000 l/ha
- Spray between three true leaf stage of crops and end Jul
- Crop scorch may occur after treatment

396   sodium hypochlorite (commodity substance)

An inorganic horticultural bactericide for use in mushrooms

Products
sodium hypochlorite	various	100% w/v	ZZ

Uses
- Bacterial blotch in **mushrooms**

Approval information
- Approval for the use of sodium hypochlorite as a commodity substance was granted on 5 December 1996 by Ministers under regulation 5 of the Control of Pesticides Regulations 1986

Restrictions
- Maximum concentration 315 mg/litre of water
- Mixing and loading must only take place in a ventilated area
- Must only be used by suitably trained and competent operators

Crop-specific information
- HI 1 d

Environmental safety
- Harmful to fish or other aquatic life. Do not contaminate surface waters or ditches with chemical or used container

SEE SECTION 3 FOR PRODUCTS ALSO REGISTERED

Hazard classification and safety precautions
    Operator protection A, C, H
    Environmental protection E13c

397   sodium monochloroacetate

A contact herbicide for various horticultural crops

Products

1 Atlas Somon	Nufarm UK	96% w/w	SP	07727
2 Croptex Steel	Certis	95% w/w	SP	02418

Uses

- Annual dicotyledons in **broccoli** *(off-label)*, **calabrese** *(off-label)*, **cauliflowers** *(off-label)* [2]; **brussels sprouts, bulb onions, cabbages, kale, leeks** [1, 2]
- Basal defoliation in **hops** *(off-label)* [2]
- Sucker control in **blackberries** *(off-label)*, **loganberries** *(off-label)*, **raspberries** *(off-label)*, **rubus hybrids** *(off-label)* [2]
- Sucker inhibition in **blackberries** *(off-label)*, **loganberries** *(off-label)*, **raspberries** *(off-label)*, **rubus hybrids** *(off-label)* [1]

Specific Off-Label Approvals (SOLAs)

- **blackberries, loganberries, raspberries, rubus hybrids** *(OLA 041460) Dec 2007* [1]
- **blackberries, loganberries, raspberries, rubus hybrids** *(OLA 930115) Dec 2008* [2]
- **broccoli, calabrese, cauliflowers** *(OLA 992602) Dec 2008* [2]
- **hops** *(OLA 930357) Dec 2008* [2]

Approval information

- Products containing this active ingredient have been granted derogations for specified 'Essential Uses' for use until 31 December 2007. Sale and supply must cease by 30 June 2007 but growers have no guarantee that the products will continue to be available until then.
    For more information see 'The Review Programme' under 'Pesticide Legislation' in Section 5
- Accepted by BBPA for use on hops

Efficacy guidance

- Sodium monochloroacetate acts by contact action. Best results achieved by application to emerged weed seedlings up to young plant stage in warm humid weather when weeds are growing actively
- Effectiveness reduced by rain within 12 h

Restrictions

- Maximum number of treatments 1 per crop; 2 per yr for sucker control in cane fruit
- Do not spray cabbage that has begun to heart
- Do not treat transplanted onions or onion sets
- Safety on brassicas, onions and leeks depends on presence of adequate leaf wax, check by crystal violet wax test. Do not add wetters, pesticides or nutrients to spray
- These products must not be used on any crops other than those listed, including any extrapolations that would normally be permissible under the Long Term Arrangements for Extension of Use (see Section 5)
- Do not spray if frost likely or if temperature likely to exceed 27°C
- Do not apply any other pesticide within 7 d after use

Crop-specific information

- Latest use: before 4 true leaf stage for onions and leeks.
- HI 21 d for cabbage, kale, Brussels sprouts; 28 d for cane fruit; 14 wk for hops [2]
- Apply to brassicas from 2-4 leaf stage or after recovery from transplanting
- Apply to onions and leeks after crook stage but before 4-leaf stage
- Apply to cane fruit crops established for at least 1 yr as a directed spray
- Apply for sucker control in raspberries when canes 10-20 cm high with addition of Wayfarer adjuvant [2]

Environmental safety

- Corrosive. Causes burns
- Dangerous for the environment [2]

FOR FULL CONDITIONS OF USE ALWAYS READ THE PRODUCT LABEL

- Very toxic to aquatic organisms. May cause long-term adverse effects in the aquatic environment [2]
- Do not apply directly to livestock/poultry [1]
- Keep livestock, especially poultry, out of treated areas for at least 2 wk
- Harmful to bees. Do not apply to crops in flower or to those in which bees are actively foraging. Do not apply when flowering weeds are present

Hazard classification and safety precautions
Hazard H02, H11 [1, 2]; H05 [2]
Risk phrases R20, R21, R22a [2]; R25, R34, R41, R50, R53a [1, 2]; R38 [1]
Operator protection A, C, D, E; U05a, U10, U11, U19a, U20b
Environmental protection E05b, E06a [1, 2] (2 wk); E12f, E34 [1, 2]; E15a, E38 [1]
Storage and disposal D01, D02, D09a [1, 2]; D10a, D12a [1]; D11a [2]
Medical advice M04

398   spinosad

A selective insecticide derived from naturally occurring soil fungi (naturalyte)

Products
1 Conserve	Fargro	120 g/l	SC	12058
2 Tracer	Landseer	480 g/l	SC	12438

Uses
- Cabbage moth in **brussels sprouts, cabbages, cauliflowers** [2]
- Cabbage white butterfly in **brussels sprouts, cabbages, cauliflowers** [2]
- Codling moth in **apples, pears** [2]
- Diamond-back moth in **brussels sprouts, cabbages, cauliflowers** [2]
- Small white butterfly in **brussels sprouts, cabbages, cauliflowers** [2]
- Thrips in **bulb onions, leeks, protected strawberries** (off-label) [2]
- Tortrix moths in **apples, pears** [2]
- Western flower thrips in **protected cucumbers, protected ornamentals** [1]

Specific Off-Label Approvals (SOLAs)
- **protected strawberries** (OLA 051173) Sep 2007 [2]

Approval information
- Approval expiry 18 Dec 2006 [1]

Efficacy guidance
- Product enters insects by contact from a treated surface or ingestion of treated plant material therefore good spray coverage is essential
- Some plants, for example Fuchsia flowers, can provide effective refuges from spray deposits and whitefly control may be reduced [1]
- Apply to protected crops when whitefly nymphs or adults are first seen [1]
- Monitor whitefly development carefully to see whether further applications are necessary. A 2 or 3 spray programme at 5-7 d intervals may be needed when conditions favour rapid pest development [1]
- Treat top fruit and field crops when pests are first seen or at very first signs of crop damage [2]
- Ensure a rain-free period of 12 hr after treatment before applying irrigation [2]
- To reduce possibility of development of resistance, adopt resistance management measures. See label and Section 5

Restrictions
- Maximum number of treatments 3 per crop for protected cucumbers or 6 per structure per yr for protected ornamentals; 4 per crop for brassicas, onions, leeks; 1 pre-blossom and 3 post blossom for apples, pears
- Apply no more than 3 consecutive sprays. Rotate with another insecticide with a different mode of action or use no further treatment after applying the maximum number of treatments
- Establish whether any incoming plants have been treated and apply no more than 3 consecutive sprays. Maximum number of treatments 6 per structure per yr [1]
- Avoid application in bright sunlight or into open flowers [1]
- Contact processor before use on a crop for processing

SEE SECTION 3 FOR PRODUCTS ALSO REGISTERED

Crop-specific information
- HI 3 d for protected cucumbers, brassicas; 7 d for apples, pears, onions, leeks
- Test for tolerance on a small number of ornamentals or cucumbers before large scale treatment
- Some spotting of african violet flowers may occur

Environmental safety
- Dangerous for the environment
- Very toxic (toxic [1]) to aquatic organisms
- Whenever possible use an Integrated Pest Management system. Spinosad presents low risk to beneficial arthropods
- Product has low impact on many insect and mite predators but is harmful to adults of most parasitic wasps. Most beneficials may be introduced to treated plants when spray deposits are dry but an interval of 2 wk should elapse before introduction of parasitic wasps. See label for details
- Treatment may cause temporary reduction in abundance of insect and mite predators if present at application
- Exposure to direct spray is harmful to bees but dry spray deposits are harmless. Treatment of field crops should not be made in the heat of the day when bees are actively foraging
- LERAP Category and 40 m broadcast air-assisted LERAP [2]

Hazard classification and safety precautions
Hazard H11
Risk phrases R50 [2]; R51 [1]; R53a [1, 2]
Operator protection A, H; U05a [1]; U08, U20b [2]
Environmental protection E15a, E34, E38 [1, 2]; E16a [2]; E17b [2] (40 m)
Storage and disposal D01, D05 [1]; D02, D10b, D12a [1, 2]

399 spiromesifen

A ketoenol insecticide for protected tomatoes

Products

Oberon	Bayer CropScience	240 g/l	SC	11819

Uses
- Red spider mites in **protected aubergines** (off-label - on inert media or by NFT), **protected cucumbers** (off-label - on inert media or by NFT), **protected gherkins** (off-label - on inert media or by NFT), **protected strawberries** (off-label - on inert media or by NFT)
- Spider mites in **protected ornamentals** (off-label - container grown)
- Whitefly in **inert substrate tomatoes**, **nft tomatoes**, **protected aubergines** (off-label - on inert media or by NFT), **protected cucumbers** (off-label - on inert media or by NFT), **protected gherkins** (off-label - on inert media or by NFT), **protected ornamentals** (off-label - container grown), **protected strawberries** (off-label - on inert media or by NFT)

Specific Off-Label Approvals (SOLAs)
- **protected aubergines** (on inert media or by NFT) (OLA 050959) Dec 2007 [1]
- **protected cucumbers**, **protected gherkins** (on inert media or by NFT) (OLA 050958) Dec 2007 [1]
- **protected ornamentals** (container grown) (OLA 041718) Dec 2006 [1]
- **protected strawberries** (on inert media or by NFT) (OLA 050957) Dec 2007 [1]

Efficacy guidance
- Apply to run off and ensure that all sides of the crop are covered
- Apply when infestation first observed and repeat 7-10 d later if necessary
- Always follow guidelines for preventing and managing insect resistance. Section 5 for more information

Restrictions
- Maximum number of treatments 2 per cropping cycle
- Do not use on cherry tomatoes or on any outdoor crops
- Product should be used in rotation with compounds with different modes of action

Crop-specific information
- HI 3 d for tomatoes

FOR FULL CONDITIONS OF USE ALWAYS READ THE PRODUCT LABEL

Environmental safety
- Dangerous for the environment
- Very toxic to aquatic organisms

Hazard classification and safety precautions
Hazard H04, H11
Risk phrases R43, R50, R53a
Operator protection A, H; U05a, U14, U20b
Environmental protection E15a, E34, E38
Storage and disposal D01, D02, D05, D09a, D11a, D12a

400   spiroxamine

A spiroketal amine fungicide for cereals

See also prothioconazole + spiroxamine

Products

Torch Extra	Bayer CropScience	800 g/l	EC	12140

Uses
- Brown rust in **spring barley**, **spring rye**, **spring wheat**, **winter barley**, **winter rye**, **winter wheat**
- Powdery mildew in **spring barley**, **spring rye**, **spring wheat**, **winter barley**, **winter rye**, **winter wheat**
- Rhynchosporium in **spring barley** *(reduction)*, **spring rye** *(reduction)*, **winter barley** *(reduction)*, **winter rye** *(reduction)*
- Yellow rust in **spring barley**, **spring rye**, **spring wheat**, **winter barley**, **winter rye**, **winter wheat**

Approval information
- Spiroxamine included in Annex I under EC Directive 91/414
- Accepted by BBPA for use on malting barley

Efficacy guidance
- For best results treat at an early stage of disease development before infection spreads to new growth
- To reduce the risk of development of resistance avoid repeat treatments on diseases such as powdery mildew. If necessary tank mix or alternate with other non-morpholine fungicides

Restrictions
- Maximum total dose equivalent to two full dose treatments

Crop-specific information
- Latest use: before caryopsis watery ripe (GS 71) for wheat, rye; ear emergence complete (GS 59) for barley

Environmental safety
- Dangerous for the environment
- Very toxic to aquatic organisms
- LERAP Category B

Hazard classification and safety precautions
Hazard H03, H11
Risk phrases R20, R21, R22a, R41, R43, R50, R53a
Operator protection A, C, H; U05a, U11, U13, U14, U15, U20a
Environmental protection E15a, E16a, E34, E38
Storage and disposal D01, D02, D09a, D10b, D12a
Medical advice M03

401   spiroxamine + tebuconazole

A broad spectrum systemic fungicide mixture for cereals

Products

Sage	Bayer CropScience	250:133 g/l	EW	11303

SEE SECTION 3 FOR PRODUCTS ALSO REGISTERED

Uses
- Brown rust in **spring barley, spring rye, spring wheat, winter barley, winter rye, winter wheat**
- Fusarium ear blight in **spring wheat, winter wheat**
- Glume blotch in **spring wheat, winter wheat**
- Net blotch in **spring barley, winter barley**
- Powdery mildew in **spring barley, spring rye, spring wheat, winter barley, winter rye, winter wheat**
- Rhynchosporium in **spring barley, winter barley**
- Septoria leaf spot in **spring wheat, winter wheat**
- Sooty moulds in **spring wheat, winter wheat**
- Yellow rust in **spring barley, spring rye, spring wheat, winter barley, winter rye, winter wheat**

Approval information
- Spiroxamine included in Annex I under EC Directive 91/414
- Accepted by BBPA for use on malting barley

Efficacy guidance
- Best results achieved from treatment at early stage of disease development before infection spreads to new growth
- To protect flag leaf and ear from Septoria apply from flag leaf emergence to ear fully emerged (GS 37-59). Earlier treatment may be necessary where there is high disease risk
- Control of rusts, powdery mildew, leaf blotch and net blotch may require second treatment 2-3 wk later

Restrictions
- Maximum total dose equivalent to two full dose treatments
- Do not use on durum wheat

Crop-specific information
- Latest use: before caryopsis watery ripe (GS 71) for rye, wheat; ear emergence complete (GS 59) for barley
- Some transient leaf speckling may occur on wheat but this has not been shown to reduce yield reponse or disease control

Environmental safety
- Dangerous for the environment
- Very toxic to aquatic organisms
- Dangerous to fish or other aquatic life. Do not contaminate surface waters or ditches with chemical or used container
- LERAP Category B

Hazard classification and safety precautions
  Hazard H03, H11
  Risk phrases R20, R22a, R38, R41, R43, R50, R53a
  Operator protection A, C, H; U05a, U11, U13, U14, U15, U20b
  Environmental protection E13b, E16a, E16b, E34, E38
  Storage and disposal D01, D02, D05, D09a, D10b, D12a
  Medical advice M03

402   strychnine hydrochloride (commodity substance)

A vertebrate control agent for destruction of moles underground

Products

strychnine	various	-	ZZ

Uses
- Moles in **grassland** *(areas of restricted public access)*, **miscellaneous pest control situations** *(areas of restricted public access)*

Approval information
- Approval for the use of strychnine hydrochloride as a commodity substance was granted on 19 June 1997 by Ministers under regulation 5 of the Control of Pesticides Regulations 1986

FOR FULL CONDITIONS OF USE ALWAYS READ THE PRODUCT LABEL

- Strychnine was not supported in the EC Review Programme (see Section 5) and will not be authorised for mole control from September 2006. It has not been permitted as a plant protection product (e.g. to prevent crop damage) since 1 January 2005. Permits for use as a biocidal product under EC Directive 98/8 may continue to be issued by Defra and the agricultural departments for Scotland and Wales until 1 August 2006, but all product must be used up by 31 August 2006.

Efficacy guidance
- Only for use as poison bait against moles on commercial agricultural/horticultural land where public access restricted, on grassland associated with aircraft landing strips, horse paddocks, race and golf courses and other areas specifically approved by Defra in England and the appropriate authorities in Wales and Scotland

Restrictions
- Strychnine is subject to the Poisons Rules 1982 and the Poisons Act 1972. See notes in Section 5
- Must only be supplied to holders of a permit issued by Defra in England or the appropriate authorities in Wales and Scotland. Permits may only be issued to persons who satisfy the appropriate authority that they are trained and competent in its use
- Only to be supplied in original sealed pack in units up to 2 g. Quantities of more than 8 g may be supplied only to providers of a commercial service
- Other restrictions apply, see the PSD website at: www.pesticides.gov.uk/approvals.asp?id=310

Environmental safety
- Store in original container under lock and key and only on premises under control of a permit holder or of a named individual who satisfies the designated requirements of competence
- A written COSHH assessment must be made before use
- Must be prepared for application with great care so that there is no contamination of the ground surface
- Any prepared bait remaining at the end of the day must be buried
- Operators must be provided with a Chemicals (Hazard Information and Supply) (CHIP) Safety Data Sheet, obtainable from the supplier, before commencing work

Hazard classification and safety precautions
Operator protection A

403   sulfosulfuron

A sulfonylurea herbicide for grass and broad-leaved weed control in winter wheat

Products

1 Ag-Chem Prefect	Ag-Chem	80% w/w	WG	11985
2 Monitor ·	Monsanto	80% w/w	WG	10495
3 Safeguard	AgriGuard	80% w/w	WG	12020

Uses
- Brome grasses in **winter wheat**
- Chickweed in **winter wheat**
- Cleavers in **winter wheat**
- Loose silky bent in **winter wheat**
- Mayweeds in **winter wheat**
- Onion couch in **winter wheat**

Approval information
- Sulfosulfuron included in Annex I under EC Directive 91/414
- Approval expiry 31 Jul 2006 [2]

Efficacy guidance
- For best results treat in early spring when annual weeds are small and growing actively. Avoid treatment when weeds are dormant for any reason
- Best control of onion couch is achieved when the weed has more than two leaves. Effects on bulbils or on growth in the following yr have not been examined
- An extended period of dry weather before or after treatment may result in reduced control
- The addition of a recommended surfactant is essential for full activity
- Specific follow-up treatments may be needed for complete control of some weeds

SEE SECTION 3 FOR PRODUCTS ALSO REGISTERED

SECTION 2

- Use only where competitively damaging weed populations have emerged otherwise yield may be reduced
- Sulfosulfuron is a member of the ALS-inhibitor group of herbicides and products should be used in a planned Resistance Management strategy. See Section 5 for more information

Restrictions
- Maximum total dose equivalent to one treatment at full dose
- Do not treat crops under stress
- Apply only after 1 Feb and from 3 expanded leaf stage
- Do not treat durum wheat or any undersown wheat crop
- Do not use in mixture or in sequence with any other sulfonyl urea herbicide on the same crop

Crop-specific information
- Latest use: flag leaf ligule just visible (GS 39)

Following crops guidance
- In the autumn following treatment winter wheat, winter rye, winter oats, triticale, winter oilseed rape, winter peas or winter field beans may be sown on any soil, and winter barley on soils with less than 60% sand
- In the next spring following application crops of wheat, barley, oats, maize, peas, beans, linseed, oilseed rape, potatoes or grass may be sown
- In the second autumn following application winter linseed may be sown, and winter barley on soils with more than 60% sand
- Sugar beet or any other crop not mentioned above must not be drilled until the second spring following application
- Where winter oilseed rape is to be sown in the autumn following treatment soil cultivation to a minimum of 10 cm is recommended

Environmental safety
- Dangerous for the environment
- Very toxic to aquatic organisms
- Extremely dangerous to fish or other aquatic life. Do not contaminate surface waters or ditches with chemical or used container
- LERAP Category B
- Take extreme care to avoid drift onto broad leaved plants or other crops, or onto ponds, waterways or ditches.
- Follow detailed label instructions for cleaning the sprayer to avoid damage to sensitive crops during subsequent use

Hazard classification and safety precautions
> **Hazard** H11
> **Risk phrases** R50 [1-3]; R53a [3]
> **Operator protection** U20b
> **Environmental protection** E13a [3]; E15a, E38 [1, 2]; E16a, E16b, E34 [1-3]
> **Storage and disposal** D02, D10b [3]; D09a [1-3]; D11a, D12a [1, 2]

404   sulphur

A broad-spectrum inorganic protectant fungicide, foliar feed and acaricide

Products

1	Cosavet DF	Headland	80% w/w	WG	11477
2	Headland Sulphur	Headland	800 g/l	SC	03714
3	Kumulus DF	BASF	80% w/w	SG	04707
4	Luxan Micro-Sulphur	Luxan	80% w/w	WG	06565
5	Solfa WG	Nufarm UK	80% w/w	WG	11602
6	Sulphur Flowable	United Phosphorus	800 g/l	SC	07526
7	Thiovit Jet	Syngenta	80% w/w	WG	10928
8	Venus	Headland	80% w/w	WG	11856

Uses
- American gooseberry mildew in *gooseberries* [5]
- Foliar feed in *grassland* [4, 5, 7]; *spring barley, spring wheat, winter barley, winter wheat* [1, 3, 4, 7, 8]; *spring oilseed rape* [1, 4, 5, 7, 8]; *sugar beet, swedes* [4]; *winter oilseed rape* [1, 3-5, 7, 8]

FOR FULL CONDITIONS OF USE ALWAYS READ THE PRODUCT LABEL

- Gall mite in **blackcurrants** [1, 3, 5, 8]
- Powdery mildew in **apples**, **strawberries** [1-3, 5, 6, 8]; **aubergines** *(off-label)*, **parsnips** *(off-label)*, **protected chives** *(off-label)*, **protected cucumbers** *(off-label)*, **protected herbs (see appendix 6)** *(off-label)*, **protected parsley** *(off-label)*, **protected peppers** *(off-label)*, **protected tomatoes** *(off-label)* [7]; **gooseberries** [1, 3, 8]; **grapevines** [1, 3, 5, 8]; **hops** [1-3, 5-8]; **pears** [2, 6]; **spring barley**, **spring wheat**, **winter barley**, **winter wheat** [2, 4-6]; **spring oats**, **winter oats** [2, 5]; **spring oilseed rape**, **spring rye**, **winter oilseed rape**, **winter rye** [2]; **sugar beet** [1-8]; **swedes** [2, 5-7]; **turnips** [5]
- Scab in **apples** [1-3, 6, 8]; **pears** [2, 6]

Specific Off-Label Approvals (SOLAs)
- **aubergines**, **protected chives**, **protected cucumbers**, **protected herbs (see appendix 6)**, **protected parsley**, **protected peppers**, **protected tomatoes** *(OLA 023652) Dec 2008* [7]
- **parsnips** *(OLA 023654) Dec 2008* [7]

Approval information
- Accepted by BBPA for use on malting barley and hops (before burr)

Efficacy guidance
- Apply when disease first appears and repeat 2-3 wk later. Details of application rates and timing vary with crop, disease and product. See label for information
- Sulphur acts as foliar feed as well as fungicide and with some crops product labels vary in whether treatment recommended for disease control or growth promotion
- In grassland best results obtained at least 2 wk before cutting for hay or silage, 3 wk before grazing
- Treatment unlikely to be effective if disease already established in crop

Restrictions
- Maximum number of treatments normally 2 per crop for grassland, sugar beet, parsnips, swedes, protected herbs; 3 per yr for blackcurrants, gooseberries; 4 per crop on apples, hops; variable on cereals. See labels
- Do not use on sulphur-shy apples (Beauty of Bath, Belle de Boskoop, Cox's Orange Pippin, Lanes Prince Albert, Lord Derby, Newton Wonder, Rival, Stirling Castle) or pears (Doyenne du Comice)
- Do not use on gooseberry cultivars Careless, Early Sulphur, Golden Drop, Leveller, Lord Derby, Roaring Lion, or Yellow Rough
- Do not use on apples or gooseberries when young, under stress or if frost imminent
- Do not use on fruit for processing, on grapevines during flowering or near harvest on grapes for wine-making
- Do not use on hops at or after burr stage
- Do not spray top or soft fruit with oil or within 30 d of an oil-containing spray

Crop-specific information
- Latest use: before burr stage for hops; before end Sep for parsnips, swedes; before 1st wk of August [2] or end Sep (other products) for sugar beet; fruit swell for gooseberries
- HI cutting grass for hay or silage 2 wk; grazing grassland 3 wk

Environmental safety
- Sulphur products are attractive to livestock and must be kept out of their reach
- Do not empty into drains

Hazard classification and safety precautions
Hazard H04 [5]
Risk phrases R37 [5]
Operator protection A, H [7]; U02a, U08 [1, 7, 8]; U05a [5, 7]; U14 [1, 6-8]; U19a [5]; U20a [2]; U20b [6]; U20c [1, 3-5, 7, 8]
Environmental protection E15a [1-8]; E19b [5]; E34 [2, 4, 6]
Storage and disposal D01 [4-7]; D02 [5-7]; D05 [6, 7]; D09a [1-8]; D10b [2, 6]; D11a [1, 3, 4, 7, 8]

405   sulphuric acid (commodity substance)

A strong acid used as an agricultural desiccant

Products

sulphuric acid	various	77% w/w	SL

SEE SECTION 3 FOR PRODUCTS ALSO REGISTERED

Uses
- Haulm destruction in **potatoes**
- Pre-harvest desiccation in **bulbs/corms**, **peas**

Approval information
- Approval for the use of sulphuric acid as a commodity substance was granted on 23 November 1995 by Ministers under regulation 5 of the Control of Pesticides Regulations 1986

Efficacy guidance
- Apply with suitable equipment between 1 Mar and 15 Nov

Restrictions
- Sulphuric acid is subject to the Poisons Rules 1982 and the Poisons Act 1972. See notes in Section 5
- Must only be used by suitably trained operators competent in use of equipment for applying sulphuric acid
- Not to be applied using hand-held or pedestrian controlled applicators
- A written COSHH assessment must be made before use. Operators should observe OES set out in HSE guidance note EH40/90 or subsequent issues
- Operators must have liquid suitable for eye irrigation immediately available at all times throughout spraying operation
- Maximum number of treatments 3 per crop for potatoes; 1 per crop for peas; 1 per yr for bulbs and corms
- Only 'sulphur burnt' sulphuric acid to be used

Crop-specific information
- Latest use: 15 Nov for bulbs, corms, peas, potatoes

Environmental safety
- Spray must not be deposited within 1 m of public footpaths
- Written notice of any intended spraying must be given to owners of neighbouring land and readable warning notices posted beforehand and left in place for 96 h afterwards
- Unprotected persons must be kept out of treated areas for at least 96 h after treatment
- Do not apply to crops in which bees are actively foraging. Do not apply when flowering weeds are present

Hazard classification and safety precautions
Hazard H05
Risk phrases R34
Operator protection B, C, D, H, K, M

406   tau-fluvalinate

A contact pyrethroid insecticide for cereals and oilseed rape

Products

1 Greencrop Malin	Greencrop	240 g/l	EW	11787
2 Klartan	Makhteshim	240 g/l	EW	11074
3 Mavrik	Makhteshim	240 g/l	EW	10612

Uses
- Aphids in **spring barley**, **spring oilseed rape**, **spring wheat**, **winter barley**, **winter oilseed rape**, **winter wheat** [1-3]
- Barley yellow dwarf virus vectors in **winter barley**, **winter wheat** [1-3]
- Cabbage stem flea beetle in **winter oilseed rape** [2, 3]
- Pollen beetle in **spring oilseed rape**, **winter oilseed rape** [1-3]

Approval information
- Accepted by BBPA for use on malting barley

Efficacy guidance
- For BYDV control on winter cereals follow local warnings or spray high risk crops in mid-Oct and make repeat application in late autumn/early winter if aphid activity persists
- For summer aphid control on cereals spray once when aphids present on two thirds of ears and increasing

FOR FULL CONDITIONS OF USE ALWAYS READ THE PRODUCT LABEL

- On oilseed rape treat peach potato aphids in autumn in response to local warning and repeat if necessary
- Best control of pollen beetle in oilseed rape obtained from treatment at green to yellow bud stage and repeat if necessary
- Good spray cover of target essential for best results

Restrictions
- Maximum total dose equivalent to two full dose treatments on oilseed rape. See label for dose rates on cereals
- A minimum of 14 d must elapse between applications to cereals

Crop-specific information
- Latest use: before caryopsis watery ripe (GS 71) for barley; before flowering for oilseed rape; before kernel medium milk (GS 75) for wheat

Environmental safety
- Dangerous for the environment
- Very toxic to aquatic organisms
- High risk to non-target insects or other arthropods. Do not spray within 6 m of the field boundary [1-3]
- LERAP Category A
- Avoid spraying oilseed rape within 6 m of field boundary to reduce effects on certain non-target species or other arthropods
- Must not be applied to cereals if any product containing a pyrethroid insecticide or dimethoate has been sprayed after the start of ear emergence (GS 51)

Hazard classification and safety precautions
Hazard H04 [1]; H11 [1-3]
Risk phrases R36, R38, R51 [1]; R50 [2, 3]; R53a [1-3]
Operator protection A, C, H; U05a, U10, U11, U19a, U20a
Environmental protection E15a, E16c, E16d, E22a
Storage and disposal D01, D02, D05, D10c [1-3]; D09a [2, 3]; D09b [1]

407 tebuconazole

A systemic conazole fungicide for cereals and other field crops

See also imidacloprid + tebuconazole + triazoxide
prochloraz + tebuconazole
prothioconazole + tebuconazole
prothioconazole + tebuconazole + triazoxide
spiroxamine + tebuconazole

Products

1	Barclay Busker	Barclay	250 g/l	EC	11345
2	Folicur	Bayer CropScience	250 g/l	EW	11278
3	Greencrop Tabloid	Greencrop	250 g/l	EW	11969
4	Icon	AgriGuard	250 g/l	EW	12202
5	Me2 Tebuconazole	Me2	250 g/l	EW	09751
6	Mystique	Nufarm UK	250 g/l	EC	12348
7	Orius	Makhteshim	250 g/l	EW	12105
8	Standon Tebuconazole	Standon	250 g/l	EC	09056

Uses
- Alternaria in **cabbages** [1-4, 6, 7]; **carrots**, **horseradish** [2-4, 7]; **spring oilseed rape**, **winter oilseed rape** [1-8]
- Botrytis in **daffodils** (off-label - for galanthamine production) [2, 7]; **linseed** (reduction) [1-5, 7, 8]
- Brown rust in **spring barley**, **spring wheat**, **winter barley**, **winter wheat** [1-8]; **spring rye**, **winter rye** [1-5, 7, 8]
- Cane blight in **blackberries** (off-label), **raspberries** (off-label), **rubus hybrids** (off-label) [2]
- Canker in **apples** (off-label), **chestnuts** (off-label), **cob nuts** (off-label), **crab apples** (off-label), **pears** (off-label), **quinces** (off-label), **walnuts** (off-label) [2]
- Chocolate spot in **spring field beans**, **winter field beans** [1-8]
- Crown rust in **spring oats** (reduction), **winter oats** (reduction) [2-4, 7]

SEE SECTION 3 FOR PRODUCTS ALSO REGISTERED

- Fungus diseases in **kohlrabi** *(off-label)* [2, 7]
- Fusarium ear blight in **spring wheat**, **winter wheat** [2-4, 7]
- Glume blotch in **spring wheat**, **winter wheat** [1-8]
- Light leaf spot in **cabbages** [2-5, 7, 8]; **spring oilseed rape**, **winter oilseed rape** [1-8]
- Lodging control in **spring oilseed rape**, **winter oilseed rape** [2-4, 7]
- Net blotch in **spring barley**, **winter barley** [1-8]
- Phoma in **spring oilseed rape**, **winter oilseed rape** [2-4, 7]
- Powdery mildew in **cabbages**, **spring barley**, **spring oats**, **spring wheat**, **swedes**, **turnips**, **winter barley**, **winter oats**, **winter wheat** [1-8]; **carrots**, **linseed** [2-4, 7]; **chives** *(off-label)*, **herbs (see appendix 6)** *(off-label)*, **parsley** *(off-label)*, **parsnips** [2, 7]; **spring rye**, **winter rye** [1-5, 7, 8]
- Rhynchosporium in **spring barley**, **winter barley** [1-8]; **spring rye**, **winter rye** [1-5, 7, 8]
- Ring spot in **broccoli** *(off-label)*, **calabrese** *(off-label)*, **cauliflowers** *(off-label)*, **chinese cabbage** *(off-label)*, **collards** *(off-label)*, **kale** *(off-label)*, **salad brassicas** *(off-label - for baby leaf production)* [2, 7]; **cabbages** [1-8]; **spring oilseed rape** *(reduction)*, **winter oilseed rape** *(reduction)* [2-4, 7]
- Rust in **broad beans** *(off-label)*, **chives** *(off-label)*, **dwarf beans** *(off-label)*, **french beans** *(off-label)*, **herbs (see appendix 6)** *(off-label)*, **parsley** *(off-label)*, **runner beans** *(off-label)* [2, 7]; **leeks** [2-5, 7]; **spring field beans**, **winter field beans** [1-8]
- Sclerotinia in **carrots** [2-4, 7]
- Sclerotinia stem rot in **spring oilseed rape**, **winter oilseed rape** [1-8]
- Septoria leaf spot in **spring wheat**, **winter wheat** [1-8]
- Sooty moulds in **spring wheat**, **winter wheat** [2-4, 7]
- Stem canker in **spring oilseed rape**, **winter oilseed rape** [2-4, 7]
- White rot in **bulb onions** *(off-label)*, **garlic** *(off-label)*, **salad onions** *(off-label)*, **shallots** *(off-label)* [2, 7]
- Yellow rust in **spring barley**, **spring wheat**, **winter barley**, **winter wheat** [1-8]; **spring rye**, **winter rye** [1-5, 7, 8]

Specific Off-Label Approvals (SOLAs)
- **apples**, **crab apples**, **pears**, **quinces** *(OLA 051734) Dec 2008* [2]
- **blackberries**, **raspberries**, **rubus hybrids** *(OLA 050897) May 2009* [2]
- **broad beans** *(OLA 031878) Dec 2008* [2]
- **broad beans** *(OLA 051393) Dec 2008* [7]
- **broccoli**, **calabrese**, **cauliflowers**, **chinese cabbage** *(OLA 031874) Dec 2008* [2]
- **broccoli**, **calabrese**, **cauliflowers**, **chinese cabbage** *(OLA 051397) Dec 2008* [7]
- **bulb onions** *(OLA 031877) Dec 2008* [2]
- **bulb onions**, **garlic**, **shallots** *(OLA 031879) Dec 2008* [2]
- **bulb onions**, **garlic**, **shallots** *(OLA 051394) Dec 2008* [7]
- **bulb onions** *(OLA 051403) Dec 2008* [7]
- **chestnuts**, **cob nuts**, **walnuts** *(OLA 051735) Dec 2008* [2]
- **chives**, **herbs (see appendix 6)**, **parsley** *(OLA 031873) Dec 2008* [2]
- **chives**, **herbs (see appendix 6)**, **parsley** *(OLA 051396) Dec 2008* [7]
- **collards**, **kale** *(OLA 031872) Dec 2008* [2]
- **collards**, **kale** *(OLA 051395) Dec 2008* [7]
- **daffodils** *(for galanthamine production) (OLA 041516) Dec 2008* [2]
- **daffodils** *(for galanthamine production) (OLA 051402) Dec 2008* [7]
- **dwarf beans**, **french beans**, **runner beans** *(OLA 031880) Dec 2008* [2]
- **dwarf beans**, **french beans**, **runner beans** *(OLA 051401) Dec 2008* [7]
- **kohlrabi** *(OLA 031881) Dec 2008* [2]
- **kohlrabi** *(OLA 051400) Dec 2008* [7]
- **salad brassicas** *(for baby leaf production) (OLA 051398) Dec 2008* [2, 7]
- **salad onions** *(OLA 042408) Dec 2008* [2]
- **salad onions** *(OLA 051399) Dec 2008* [7]

Approval information
- Accepted by BBPA for use on malting barley

Efficacy guidance
- For best results apply at an early stage of disease development before infection spreads to new crop growth
- To protect flag leaf and ear from Septoria diseases apply from flag leaf emergence to ear fully emerged (GS 37-59). Earlier application may be necessary where there is a high risk of infection

FOR FULL CONDITIONS OF USE ALWAYS READ THE PRODUCT LABEL

- Improved control of established mildew can be obtained by tank mixture with fenpropimorph
- For light leaf spot control in oilseed rape apply in autumn/winter with a follow-up spray in spring/summer if required
- For control of most other diseases spray at first signs of infection with a follow-up spray 2-4 wk later if necessary. See label for details
- For disease control in cabbages a 3-spray programme at 21-28 d intervals will give good control

Restrictions
- Maximum total dose equivalent to 1 full dose treatment on linseed, green beans, runner beans; 2 full dose treatments on wheat, barley, rye, field beans, broad beans, swedes, turnips, onions, garlic, shallots; 2.25 full dose treatments on cabbages, herbs; 2.5 full dose treatments on oilseed rape; 3 full dose treatments on leeks, horseradish, parsnips, carrots
- Do not treat durum wheat
- Apply only to listed oat varieties (see label)
- Do not apply before swedes and turnips have a root diameter of 2.5 cm, or before heart formation in cabbages

Crop-specific information
- Latest use: before grain milky-ripe for cereals (GS 71); when most seed green-brown mottled for oilseed rape (GS 6,3); before brown capsule for linseed
- HI field beans, swedes, turnips 35 d; market brassicas, kale, carrots, parsnips, garlic, salad onions, salad brassicas, shallots 21 d; leeks, herbs 14 d; broad beans, green beans, runner beans 7 d
- Some transient leaf speckling on wheat or leaf reddening/scorch on oats may occur but this has not been shown to reduce yield response to disease control

Environmental safety
- Dangerous for the environment
- Toxic to aquatic organisms

Hazard classification and safety precautions
Hazard H03, H11
Risk phrases R20 [2, 7]; R22a [1-5, 7, 8]; R38 [1, 3-6, 8]; R41, R51, R53a [1-8]; R63 [6]
Operator protection A, C, H; U05a, U11, U20a [1-8]; U19a [1, 3-5, 8]
Environmental protection E15a [1-8]; E34 [2-4, 6, 7]; E38 [2, 6, 7]
Storage and disposal D01, D02, D05, D09a, D10b [1-8]; D12a [2, 6, 7]
Medical advice M03

408 tebuconazole + triadimenol

A broad spectrum systemic fungicide for cereals

Products
1	Silvacur	Bayer CropScience	250:125 g/l	EC	11309
2	Veto F	Bayer CropScience	225:75 g/l	EC	11317

Uses
- Brown rust in **spring barley, spring rye, spring wheat, winter barley, winter rye, winter wheat**
- Crown rust in **spring oats, winter oats**
- Fusarium ear blight in **spring wheat, winter wheat**
- Glume blotch in **spring wheat, winter wheat**
- Net blotch in **spring barley, winter barley**
- Powdery mildew in **spring barley, spring oats, spring rye, spring wheat, winter barley, winter oats, winter rye, winter wheat**
- Rhynchosporium in **spring barley, winter barley**
- Septoria leaf spot in **spring wheat, winter wheat**
- Sooty moulds in **spring wheat, winter wheat**
- Yellow rust in **spring barley, spring rye, spring wheat, winter barley, winter rye, winter wheat**

Approval information
- Accepted by BBPA for use on malting barley

SEE SECTION 3 FOR PRODUCTS ALSO REGISTERED

Efficacy guidance
- For best results apply at an early stage of disease development before infection spreads to new crop growth
- To protect flag leaf and ear from Septoria diseases apply from flag leaf emergence to ear fully emerged (GS 37-59). Earlier application may be necessary where there is a high risk of infection
- For control of rust, powdery mildew, leaf and net blotch apply at first signs of disease with a second application 2-3 wk later if necessary

Restrictions
- Maximum total dose equivalent to two full dose treatments
- Do not use on durum wheat
- Use only on listed varieties of oats. See label [1]

Crop-specific information
- Latest use: before grain milky-ripe (GS 71)
- Some transient leaf speckling may occur on wheat or leaf reddening/scorch on oats but this has not been shown to reduce yield response or disease control

Environmental safety
- Harmful to aquatic organisms
- Harmful to fish or other aquatic life. Do not contaminate surface waters or ditches with chemical or used container

Hazard classification and safety precautions
Hazard H03 [1]
Risk phrases R20, R36 [1]; R52, R53a [1, 2]
Operator protection A, C, H; U05a, U09b, U20a
Environmental protection E13c [1, 2]; E34 [1]
Storage and disposal D01, D02, D05, D09a [1, 2]; D10b [1]; D10c [2]
Medical advice M03 [2]

409   tebuconazole + triazoxide

A triazole and benzotriazine fungicide seed treatment for use in barley

Products

Raxil S	Bayer CropScience	20:20 g/l	LS	11296

Uses
- Leaf stripe in **spring barley** (seed treatment), **winter barley** (seed treatment)
- Loose smut in **spring barley** (seed treatment), **winter barley** (seed treatment)
- Net blotch in **spring barley** (seed treatment), **winter barley** (seed treatment)

Efficacy guidance
- Best applied through recommended seed treatment machines
- Treated seed should be drilled to a depth of 4 cm into a well-prepared seed bed
- Evenness of seed cover improved by simultaneous application of equal volumes of product and water or dilution of product with an equal volume of water
- Drill treated seed in the same season
- Only the seed-borne phase of net blotch is controlled by seed treatment

Restrictions
- Maximum number of treatments 1 per batch of seed
- Do not use on seed with more than 16% moisture content, or on sprouted, cracked or skinned seed
- If product has been left to stand for more than 2 mth full agitation may be necessary to re-suspend it before use
- Diluted product must be used immediately

Crop-specific information
- Latest use: before drilling
- Slightly delayed and reduced emergence may occur but this is normally outgrown
- Any delay in field emergence, for whatever reason, may be accentuated by treatment

Environmental safety
- Harmful to aquatic organisms

FOR FULL CONDITIONS OF USE ALWAYS READ THE PRODUCT LABEL

- Harmful to fish or other aquatic life. Do not contaminate surface waters or ditches with chemical or used container
- Treated seed harmful to game and wildlife
- Product also supplied in returnable containers. See label for guidance on handling, storage, protective clothing and precautions
- If seed is present on the soil surface, or if spills have occurred, the field should be harrowed and rolled if conditions are appropriate to ensure good incorporation

Hazard classification and safety precautions
Risk phrases R52, R53a
Operator protection A, H; U07, U20b
Environmental protection E03, E13c, E34, E36
Storage and disposal D05, D09a, D14
Treated seed S01, S02, S03, S04b, S05, S06a, S07

410   tebuconazole + trifloxystrobin

A conazole and stobilurin fungicide mixture for wheat

Products

Coronet	Bayer CropScience	200:100 g/l	SC	12267

Uses
- Brown rust in **winter wheat**
- Late ear diseases in **winter wheat**
- Powdery mildew in **winter wheat**
- Septoria diseases in **winter wheat**
- Yellow rust in **winter wheat**

Approval information
- Trifloxystrobin included in Annex I under EC Directive 91/414

Efficacy guidance
- Best results obtained from treatment at early stages of disease development. Further treatment may be needed if disease attack is prolonged
- Applications to established infections of any disease are likely to be less effective
- Best control of cereal ear diseases obtained by treatment during ear emergence
- Trifloxystrobin is a member of the QoI cross resistance group. Product should be used preventatively and not relied on for its curative potential
- Use product as part of an Integrated Crop Management strategy incorporating other methods of control, including where appropriate other fungicides with a different mode of action. Do not apply more than two foliar applications of QoI containing products to any cereal crop
- There is a significant risk of widespread resistance occurring in *Septoria tritici* populations in UK. Failure to follow resistance management action may result in reduced levels of disease control
- Strains of wheat powdery mildew resistant to QoIs are common in the UK. Control of wheat mildew can only be relied on from the triazole component
- Where specific control of wheat mildew is required this should be achieved through a programme of measures including products recommended for the control of mildew that contain a fungicide from a different cross-resistance group and applied at a dose that will give robust control

Restrictions
- Maximum total dose equivalent to two full dose treatments

Crop-specific information
- Latest use: before milky ripe stage
- HI 35 d

Environmental safety
- Dangerous for the environment
- Very toxic to aquatic organisms

Hazard classification and safety precautions
Hazard H03, H11
Risk phrases R37, R50, R53a, R63
Operator protection A, H; U09b, U19a, U20b

SEE SECTION 3 FOR PRODUCTS ALSO REGISTERED

Environmental protection E15a, E34, E38
Storage and disposal D01, D02, D05, D09a, D10b, D12a
Medical advice M03

411 tebufenpyrad

A pyrazole mitochondrial electron transport inhibitor (METI) aphicide and acaricide

Products

Masai	BASF	20% w/w	WB	10223

Uses
- Big-bud mite in *almonds* (off-label), *chestnuts* (off-label), *cob nuts* (off-label), *hazel nuts* (off-label), *walnuts* (off-label)
- Damson-hop aphid in *hops*, *plums* (off-label)
- Fruit tree red spider mite in *blackberries* (off-label), *raspberries* (off-label)
- Gall mite in *bilberries* (off-label), *blackcurrants* (off-label), *blueberries* (off-label), *cranberries* (off-label), *gooseberries* (off-label), *redcurrants* (off-label), *vaccinium spp.* (off-label), *whitecurrants* (off-label)
- Red spider mites in *apples*, *pears*
- Two-spotted spider mite in *blackberries* (off-label), *hops*, *protected roses*, *raspberries* (off-label), *strawberries*

Specific Off-Label Approvals (SOLAs)
- *almonds*, *chestnuts*, *cob nuts*, *hazel nuts*, *walnuts* (OLA 050512) Dec 2008 [1]
- *bilberries*, *blackcurrants*, *blueberries*, *cranberries*, *gooseberries*, *redcurrants*, *vaccinium spp.*, *whitecurrants* (OLA 031741) Aug 2007 [1]
- *blackberries*, *raspberries* (OLA 041498) Aug 2006 [1]
- *plums* (OLA 031742) Dec 2008 [1]

Approval information
- Accepted by BBPA for use on hops

Efficacy guidance
- Acts on eggs (except winter eggs) and all motile stages of spider mites up to adults
- Treat spider mites from 80% egg hatch but before mites become established
- For effective control total spray cover of the crop is required
- Product can be used in a programme to give season-long control of damson/hop aphids coupled with mite control
- Where aphids resistant to tebufenpyrad occur in hops control is unlikely to be satisfactory and repeat treatments may result in lower levels of control. Where possible use different active ingredients in a programme

Restrictions
- Maximum total dose equivalent to one full dose treatment on apples, pears, strawberries; 3 full dose treatments on hops
- Other mitochondrial electron transport inhibitor (METI) acaricides should not be applied to the same crop in the same calendar yr either separately or in mixture
- Do not treat apples before 90% petal fall
- Small-scale testing of rose varieties to establish tolerance recommended before use
- Inner liner of container must not be removed

Crop-specific information
- Latest use: end of burr stage for hops
- HI strawberries 3 d; apples, bilberries, blackcurrants, blueberries, cranberries, gooseberries, pears, redcurrants, whitecurrants 7d; blackberries, plums, raspberries 21 d
- Product has no effect on fruit quality or finish

Environmental safety
- Dangerous for the environment
- Very toxic to aquatic organisms
- Dangerous to fish or other aquatic life. Do not contaminate surface waters or ditches with chemical or used container

FOR FULL CONDITIONS OF USE ALWAYS READ THE PRODUCT LABEL

- High risk to bees. Do not apply to crops in flower or to those in which bees are actively foraging. Do not apply when flowering weeds are present
- LERAP Category B
- Broadcast air-assisted LERAP (18 m)

Hazard classification and safety precautions

Hazard H03, H11

Risk phrases R20, R22a, R37, R50, R53a

Operator protection A; U02a, U05a, U09a, U13, U14, U20b

Environmental protection E12a, E13b, E15a, E16a, E16b, E38; E17b (18 m)

Storage and disposal D01, D02, D09a, D11a, D12a

Medical advice M05a

412   teflubenzuron

A benzoylurea insecticide for use on ornamentals

Products

Nemolt	Fargro	150 g/l	SC	10226

Uses

- Browntail moth in *ornamental plant production*
- Caterpillars in *ornamental plant production*
- Whitefly in *ornamental plant production*

Efficacy guidance

- Product acts as larval stomach poison interfering with moulting process leading to cessation of feeding and larval death
- Apply as soon as first stage larvae seen. This will often coincide the peak of moth flight

Restrictions

- Maximum number of treatments 3 per yr
- Test specific varieties before carrying out extensive treatments

Environmental safety

- Dangerous for the environment
- Very toxic to aquatic organisms
- Limited evidence suggests some margin of safety to *Encarsia formosa*. Effects on other parasites and predators not fully tested

Hazard classification and safety precautions

Hazard H11

Risk phrases R50, R53a

Operator protection A, C; U05a, U20c

Environmental protection E15a, E38

Storage and disposal D01, D02, D05, D09a, D11a, D12a

413   tefluthrin

A soil acting pyrethroid insecticide seed treatment

Products

1 Evict	Bayer CropScience	100 g/l	CF	11673
2 Evict	Syngenta	100 g/l	CF	11735
3 Force ST	Syngenta	200 g/l	CF	11752

Uses

- Bean seed fly in *bulb onions* (off-label - seed treatment), *chard* (off-label - seed treatment), *chives* (off-label - seed treatment), *green mustard* (off-label - seed treatment), *herbs (see appendix 6)* (off-label - seed treatment), *leaf spinach* (off-label - seed treatment), *leeks* (off-label - seed treatment), *mibuna* (off-label - seed treatment), *mizuna* (off-label - seed treatment), *pak choi* (off-label - seed treatment), *parsley* (off-label - seed treatment), *red mustard* (off-label - seed treatment), *salad onions* (off-label - seed treatment), *spinach beet* (off-label - seed treatment), *tatsoi* (off-label - seed treatment) [3]
- Carrot fly in *carrots* (off-label - seed treatment), *parsnips* (off-label - seed treatment) [3]

SEE SECTION 3 FOR PRODUCTS ALSO REGISTERED

- Millipedes in **fodder beet** *(seed treatment)*, **sugar beet** *(seed treatment)* [3]
- Onion fly in **bulb onions** *(off-label - seed treatment)*, **leeks** *(off-label - seed treatment)*, **salad onions** *(off-label - seed treatment)* [3]
- Pygmy beetle in **fodder beet** *(seed treatment)*, **sugar beet** *(seed treatment)* [3]
- Springtails in **fodder beet** *(seed treatment)*, **sugar beet** *(seed treatment)* [3]
- Symphylids in **fodder beet** *(seed treatment)*, **sugar beet** *(seed treatment)* [3]
- Wheat bulb fly in **spring barley** *(seed treatment)*, **spring oats** *(seed treatment)*, **spring wheat** *(seed treatment)*, **winter barley** *(seed treatment)*, **winter oats** *(seed treatment)*, **winter wheat** *(seed treatment)* [1, 2]
- Wireworm in **spring barley** *(seed treatment)*, **spring oats** *(seed treatment)*, **spring wheat** *(seed treatment)*, **winter barley** *(seed treatment)*, **winter oats** *(seed treatment)*, **winter wheat** *(seed treatment)* [1, 2]

Specific Off-Label Approvals (SOLAs)
- **bulb onions, leeks, salad onions** *(seed treatment)* *(OLA 050546) Dec 2008* [3]
- **carrots, parsnips** *(seed treatment)* *(OLA 050547) Dec 2008* [3]
- **chard, chives, green mustard, herbs (see appendix 6), leaf spinach, mibuna, mizuna, pak choi, parsley, red mustard, spinach beet, tatsoi** *(seed treatment)* *(OLA 050545) Dec 2008* [3]

Approval information
- Accepted by BBPA for use on malting barley

Efficacy guidance
- Apply during process of pelleting beet seed. Consult manufacturer for details of specialist equipment required [3]
- Apply to cereal seed through a suitable liquid seed treater calibrated to achieve even coverage [1, 2]
- Micro-capsule formulation allows slow release to provide a protection zone around treated seed during establishment
- Product is non-systemic and may not protect against wireworm attack at the soil surface [1, 2]
- Where egg counts indicate a high risk of wheat bulb fly attack, a follow-up spray treatment may be needed [1, 2]
- Where wireworm populations exceed 1.25 million per ha a suitable spray treatment must also be applied
- Factors that adversely affect crop establishment may reduce the level of pest control [1, 2]
- Must be co-applied with a suitable fungicide seed treatment to protect against seed and soil-borne disease [1, 2]

Restrictions
- Maximum number of treatments 1 per batch of seed
- Sow treated seed as soon as possible. Do not store treated seed from one drilling season to next
- Do not use on seed that is sprouted, cracked, damaged or over 16% moisture content [1, 2]
- If used in areas where soil erosion by wind or water likely, measures must be taken to prevent this happening
- Can cause a transient tingling or numbing sensation to exposed skin. Avoid skin contact with product, treated seed and dust throughout all operations in the seed treatment plant and at drilling

Crop-specific information
- Latest use: before drilling seed
- Treated seed must be drilled within the season of treatment

Environmental safety
- Dangerous for the environment
- Very toxic to aquatic organisms [3]
- Toxic to aquatic organisms [1, 2]
- Extremely dangerous to fish or other aquatic life. Do not contaminate surface waters or ditches with chemical or used container
- Keep treated seed secure from people, domestic stock/pets and wildlife at all times during storage and use
- Treated seed harmful to game and wild life. Bury spillages
- In the event of seed spillage clean up as much as possible into the related seed sack and bury the remainder completely
- Do not apply treated seed from the air
- Keep livestock out of areas drilled with treated seed for at least 80 d [3]

FOR FULL CONDITIONS OF USE ALWAYS READ THE PRODUCT LABEL

Hazard classification and safety precautions
  Hazard H03, H11
  Risk phrases R20, R43, R53a [1-3]; R50 [3]; R51 [1, 2]
  Operator protection A, D, H [1-3]; C [1, 2]; E [3]; U02a, U04a, U07, U08 [1, 3]; U05a, U20b [1-3]; U14 [1, 2]
  Environmental protection E06a [3] (80 d); E13a [2]; E15a, E38 [1-3]; E34 [1, 3]; E36 [1]
  Storage and disposal D01, D02, D12a [1-3]; D05 [1, 3]; D09a [3]; D11a [2, 3]; D14 [1]
  Treated seed S01, S03, S08 [1, 2]; S02, S04b, S05, S07 [1-3]; S06a [1]
  Medical advice M05b [3]

414 tepraloxydim

A systemic post-emergence herbicide for control of annual grass weeds

Products

1 Aramo	BASF	50 g/l	EC	10280
2 Landgold Tepraloxydim	Landgold	50 g/l	EC	11808
3 Landgold Tepraloxydim	Teliton	50 g/l	EC	12129
4 Standon Tepraloxydim	Standon	50 g/l	EC	11119

Uses
  • Annual grasses in **bulb onions, cabbages, carrots, cauliflowers, combining peas, fodder beet, land not being used for crop production, leeks, linseed, spring field beans, sugar beet, vining peas, winter field beans, winter oilseed rape** [1-3]; **flax, linseed for industrial use, winter oilseed rape for industrial use** [1]
  • Annual meadow grass in **bulb onions, cabbages, carrots, cauliflowers, combining peas, fodder beet, leeks, linseed, spring field beans, spring oilseed rape, sugar beet, vining peas, winter field beans, winter oilseed rape** [4]; **golf courses** (off-label), **poppies** (off-label - for morphine production), **salad onions** (off-label) [1]
  • Blackgrass in **bulb onions, cabbages, carrots, cauliflowers, combining peas, fodder beet, leeks, linseed, spring field beans, spring oilseed rape, sugar beet, vining peas, winter field beans, winter oilseed rape** [4]
  • Green cover in **land not being used for crop production** [4]
  • Perennial grasses in **bulb onions, cabbages, carrots, cauliflowers, combining peas, fodder beet, land not being used for crop production, leeks, linseed, spring field beans, sugar beet, vining peas, winter field beans, winter oilseed rape** [1-3]; **flax, linseed for industrial use, winter oilseed rape for industrial use** [1]
  • Volunteer barley in **bulb onions, cabbages, carrots, cauliflowers, combining peas, fodder beet, leeks, linseed, spring field beans, spring oilseed rape, sugar beet, vining peas, winter field beans, winter oilseed rape** [4]
  • Volunteer cereals in **bulb onions, cabbages, carrots, cauliflowers, combining peas, fodder beet, land not being used for crop production, leeks, linseed, spring field beans, sugar beet, vining peas, winter field beans, winter oilseed rape** [1-3]; **flax, linseed for industrial use, winter oilseed rape for industrial use** [1]
  • Volunteer wheat in **bulb onions, cabbages, carrots, cauliflowers, combining peas, fodder beet, leeks, linseed, spring field beans, spring oilseed rape, sugar beet, vining peas, winter field beans, winter oilseed rape** [4]

Specific Off-Label Approvals (SOLAs)
  • **golf courses** (OLA 040612) Nov 2007 [1]
  • **poppies** (for morphine production) (OLA 050850) Nov 2007 [1]
  • **salad onions** (OLA 032474) Nov 2007 [1]

Approval information
  • Tepraloxydim included in Annex I under EC Directive 91/414

Efficacy guidance
  • Best results obtained from applications when weeds small and have not begun to compete with crop
  • Only emerged weeds are controlled
  • Cool conditions slow down activity and very dry conditions reduce activity by interfering with uptake and translocation

SEE SECTION 3 FOR PRODUCTS ALSO REGISTERED

- Foliar death of susceptible weeds evident after 3-4 wks in warm conditions
- Reduced doses must not be used on resistant grass weed populations

Restrictions
- Maximum total dose equivalent to one full dose treatment on all crops
- Consult processors or contract agents before treatment of crops intended for processing or for seed
- Do not apply to crops damaged or stressed by factors such as previous herbicide treatments, pest or disease attack
- Do not spray if rain or frost expected, or if foliage is wet
- Do not treat oilseed rape with very low vigour and poor yield potential. Overlapping on oilseed rape may cause damage and reduce yields
- On peas a satisfactory crystal violet leaf wax test must be carried out before treatment. Winter varieties may be treated only in the spring
- For sugar beet, fodder beet, linseed and green cover on land temporarily removed from production, applications are prohibited between 1 Nov and 31 Mar
- For field beans, peas, leeks, bulb onions, carrots, cabbages and cauliflowers applications are prohibited between 1 Nov and 1 Mar

Crop-specific information
- Latest use : before end Nov or before crop has 9 true leaves, whichever is first for winter oilseed rape; before crop has 6 true leaves for spring oilseed rape [4]; before flower buds visible for flax [1], linseed; before head formation for cabbages, cauliflowers
- HI 3 wk for carrots; 4 wk for bulb onions, leeks; 5 wk for peas; 8 wk for fodder beet, field beans, sugar beet
- Treatment prohibited between 1 Nov and 31 Mar on golf courses (off-label use) [1]

Following crops guidance
- In the event of failure of a treated crop, wheat or barley may be drilled after 2 wk following normal seedbed cultivations, maize or Italian rye-grass may be planted after 8 wk following cultivation to 20 cm
- Graminaceous crops other than those mentioned above should not follow a treated crop in the rotation.
- Broad leaved crops may be planted at any time following a failed or normally harvested treated crop

Environmental safety
- Dangerous for the environment
- Toxic to aquatic organisms

Hazard classification and safety precautions
Hazard H03, H11
Risk phrases R22b, R38, R51, R53a, R63, R66
Operator protection A; U05a, U08, U20b [1-4]; U23a [1-3]
Environmental protection E15a [1-4]; E38 [1]
Storage and disposal D01, D02, D08, D09a [1-4]; D05, D10b [2-4]; D07 [2, 3]; D10c, D12a [1]
Medical advice M05b

415   terbuthylazine

A triazine herbicide available only in mixtures

See also isoxaben + terbuthylazine
mesotrione + terbuthylazine

416   terbuthylazine + terbutryn

A pre-emergence herbicide for peas, beans and lupins

Products
1 Batallion	Makhteshim	150:350 g/l	SC	08305
2 Opogard	Syngenta	150:350 g/l	SC	08427

FOR FULL CONDITIONS OF USE ALWAYS READ THE PRODUCT LABEL

Uses

- Annual dicotyledons in **broad beans**, **combining peas**, **spring field beans**, **vining peas**, **winter field beans** [1, 2]; **edible podded peas** [1]; **lupins** [2]
- Annual grasses in **broad beans**, **combining peas**, **spring field beans**, **vining peas**, **winter field beans** [1, 2]; **edible podded peas** [1]; **lupins** [2]

Approval information

- Products containing terbuthylazine and terbutryn have been granted derogations for specified 'Essential Uses' for use until 31 December 2007. Sale and supply must cease by 30 June 2007 but growers have no guarantee that the products will continue to be available until then.
  For more information see 'The Review Programme' under 'Pesticide Legislation' in Section 5

Efficacy guidance

- Active via root uptake and with foliar activity on cotyledon stage weeds
- Best results achieved by application to fine, firm, moist seedbed, preferably at weed emergence, when rain falls after spraying
- Effectiveness may be reduced by excessive rain, drought or cold
- Residual control lasts for up to 8 wk on mineral soils. Effectiveness reduced on highly organic soils and subsequent use of post-emergence treatment recommended
- Do not use on soils which are very cloddy or have more than 10% organic matter otherwise weed control will be reduced

Restrictions

- Maximum number of treatments 1 per crop
- Vedette and Printana peas may be damaged by treatment. Do not use on forage peas
- Do not treat peas, beans or lupins on soils lighter than loamy fine sand, or lupins on silty clay soil
- Do not cultivate after treatment

Crop-specific information

- Latest use: 3 d before crop emergence
- Apply as soon as possible after drilling
- Crop seed must be covered by at least 25 mm of settled soil
- Heavy rain after application may cause damage to peas on light soils

Following crops guidance

- Treated land should be mould-board ploughed to 150 mm before any succeeding crop is planted
- A period of 12 wk (14 wk in dry conditions) since application must elapse before planting any succeeding crop

Environmental safety

- Dangerous for the environment
- Very toxic to aquatic organisms

Hazard classification and safety precautions

Hazard H03 [1]; H04 [2]; H11 [1, 2]
Risk phrases R22a [1]; R43 [2]; R50, R53a [1, 2]
Operator protection U02a, U08, U14, U15, U19a [1]; U05a, U20a [1, 2]
Environmental protection E15a [1, 2]; E34 [1]; E38 [2]
Storage and disposal D01, D02, D09a [1, 2]; D10b, D12b [1]; D10c, D12a [2]
Medical advice M05a [1]

417 terbutryn

A residual triazine for control of aquatic algae

See also fomesafen + terbutryn
terbuthylazine + terbutryn

Products

Clarosan	Scotts	1% w/w	GR	H7692

Uses

- Algae in **areas of water**

Approval information

- Approved for aquatic weed control. See notes in Section 5 on use of herbicides in or near water

SECTION 2

SEE SECTION 3 FOR PRODUCTS ALSO REGISTERED

- Certain products containing terbutryn have been granted derogations for specified 'Essential Uses' for use until 31 December 2007. Sale and supply must cease by 30 June 2007 but growers have no guarantee that the products will continue to be available until then.
  For more information see 'The Review Programme' under 'Pesticide Legislation' in Section 5

Efficacy guidance
- For best results apply when algal growth active and before heavy infestations have built up. Control is more rapid in warm weather
- Apply granules evenly across either from the bank using a machine with a long throw or through suitable equipment mounted on a boat
- Granular formulation permits early application to avoid deoxygenation caused by dead or dying vegetation
- Use only in static or sluggishly moving water. The flow in moving water should be stopped for at least 7 d after treatment otherwise weed control may be reduced. Treat small ponds before start of new growth, usually before end Apr
- Algal growth ceases immediately but visible signs may not be obvious for 2-4 wk after treatment. Re-growth should not occur for at least 3-4 mth
- Effectiveness reduced in water with peaty bottom

Restrictions
- This product must not be used for any purpose other than that listed, including any extrapolations that would normally be permissible under the Long Term Arrangements for Extension of Use (see Section 5)
- Product must only be used in static or sluggishly moving water where the maximum water flow is less than 1 metre per 3 minutes
- Do not use in trout farms or areas where fish are intensively reared and dissolved oxygen levels are critical

Crop-specific information
- Latest use: Aug for areas of water

Environmental safety
- Dangerous for the environment
- Toxic to aquatic organisms
- Do not dump surplus herbicide in water or ditch bottoms
- Do not use treated water for irrigation purposes within 7 d of treatment
- If dense weed growth in watercourses to be controlled without de-oxygenation treat in sections of about 400 m at a time with intervals of at least 14 d
- If the water body to be treated is subject to the Water Resources Act and the Control of Pollution Act, users must consult the approriate water regulatory body before use

Hazard classification and safety precautions
    Hazard H11
    Risk phrases R51, R53a
    Operator protection U20b
    Environmental protection E19a; E21 (7 d)
    Storage and disposal D07, D09a, D12b

418   tetraconazole

A systemic, protectant and curative triazole fungicide for cereals

See also chlorothalonil + tetraconazole

Products

Juggler	Sipcam	100 g/l	EC	09391

Uses
- Brown rust in **spring wheat**, **winter wheat**
- Crown rust in **spring oats**, **winter oats**
- Powdery mildew in **spring wheat**, **winter wheat**
- Septoria leaf spot in **spring wheat**, **winter wheat**
- Sooty moulds in **spring wheat**, **winter wheat**
- Yellow rust in **spring wheat**, **winter wheat**

FOR FULL CONDITIONS OF USE ALWAYS READ THE PRODUCT LABEL

Efficacy guidance
- Apply at any time from the tillering stage up to the latest times indicated for each crop
- Best results achieved from applications before disease becomes established
- If disease pressure persists a second treatment may be required to prevent late season attacks
- Best control of ear diseases in wheat obtained from application at or after the ear completely emerged stage

Restrictions
- Maximum total dose equivalent to three full dose applications on wheat and two full dose applications on other cereals

Crop-specific information
- Latest use: before end of ear emergence (GS 59) for oats; before end of flowering (GS 69) for barley; before grain watery ripe (GS 71) for wheat

Environmental safety
- Harmful to fish or other aquatic life. Do not contaminate surface waters or ditches with chemical or used container
- LERAP Category B

Hazard classification and safety precautions
Hazard H03, H04
Risk phrases R22a, R36, R38
Operator protection A, C, H; U05a, U19a, U20a
Environmental protection E13c, E16a, E16b
Storage and disposal D01, D02, D05, D09a, D10b
Medical advice M03

419 tetramethrin

A contact acting pyrethroid insecticide

See also d-phenothrin + tetramethrin

Products

Killgerm Py-Kill W	Killgerm	10.2% w/v	EC	H4632

Uses
- Flies in **agricultural premises**, **livestock houses**

Efficacy guidance
- Dilute in accordance with directions and apply as space or surface spray

Restrictions
- For use only by professional operators
- Maximum number of treatments 1 per wk in agricultural premises and livestock houses
- Do not apply directly on food or livestock
- Remove exposed milk before application. Protect milk machinery and containers from contamination
- Do not use space sprays containing pyrethrins or pyrethroid more than once per wk in intensive or controlled environment animal houses in order to avoid development of resistance. If necessary, use a different control method or product

Environmental safety
- Dangerous for the environment
- Toxic to aquatic organisms

Hazard classification and safety precautions
Hazard H03, H08, H11
Risk phrases R22b, R41, R51, R53a
Operator protection A, H; U02b, U09a, U19a, U20b
Environmental protection E15a
Consumer protection C05, C06, C07, C08, C09, C10, C11
Storage and disposal D01, D09a, D12a
Medical advice M05b

SECTION 2

SEE SECTION 3 FOR PRODUCTS ALSO REGISTERED

420 thiabendazole

A systemic, curative and protectant benzimidazole (MBC) fungicide

See also imazalil + thiabendazole

Products

1 Hykeep	Agrichem	2% ww	DP	06744
2 Storite Clear Liquid	Syngenta	220 g/l	SL	09503
3 Storite Excel	Syngenta	500 g/l	SC	09542
4 Storite Flowable	Banks Cargill	450 g/l	FS	08703

Uses

- Dry rot in **seed potatoes** *(off-label)* [3]; **seed potatoes** *(tuber treatment - post-harvest)* [2, 4]; **ware potatoes** *(tuber treatment - post-harvest)* [1-4]
- Dutch elm disease in **elm trees** [2]
- Fusarium basal rot in **narcissi** *(bulb dip or spray)* [2]
- Gangrene in **seed potatoes** *(tuber treatment - post-harvest)* [2, 4]; **ware potatoes** *(tuber treatment - post-harvest)* [1-4]
- Silver scurf in **seed potatoes** *(tuber treatment - post-harvest)* [2, 4]; **ware potatoes** *(tuber treatment - post-harvest)* [1-4]
- Skin spot in **seed potatoes** *(tuber treatment - post-harvest)* [2, 4]; **ware potatoes** *(tuber treatment - post-harvest)* [1-4]

Specific Off-Label Approvals (SOLAs)

- **seed potatoes** *(OLA 051701)* Jan 2007 [3]

Approval information

- Thiabendazole included in Annex I under EC Directive 91/414

Restrictions

- Maximum number of treatments 1 per batch for seed potato treatments; 2 per yr for narcissi bulbs; 1 per yr on elm trees
- Treated seed potatoes must not be used for food or feed
- Do not mix with any other product
- Injection of trees for dutch elm disease control must only be carried out by trained arborists or others trained in injection techniques and identification of the disease [2]

Crop-specific information

- Latest use: before planting for seed potatoes; 21 d before removal from store for sale, processing or consumption for ware potatoes
- Apply to potatoes as soon as possible after harvest using suitable equipment and always within 2 wk of lifting provided the skins are set. See label for details
- Potatoes should only be treated by systems that provide an accurate dose to tubers not carrying excessive quantities of soil
- Use as a post-lifting or hot water treatment to reduce basal and neck rot in narcissus bulbs. Ensure bulbs are clean [2]
- Some reduction in flower number may occur in the first spring following the double treatment [2]

Environmental safety

- Dangerous for the environment
- Toxic to aquatic organisms
- Harmful to fish or other aquatic life. Do not contaminate surface waters or ditches with chemical or used container

Hazard classification and safety precautions

Hazard H04 [3]; H11 [2, 3]
Risk phrases R43 [3]; R51 [2, 3]; R52 [1]; R53a [1-3]
Operator protection A, C, D, H [1-3]; B, K [2]; M [1, 2]; U05a [2, 3]; U14 [3]; U19a, U20c [1-4]
Environmental protection E13c [1, 4]; E15a, E38 [2, 3]
Storage and disposal D01, D09a [1-4]; D02, D12a [2, 3]; D05 [2-4]; D07, D11a [1]; D10a [3, 4]; D10c [2]
Treated seed S03, S04a, S05 [2]

421   thiabendazole + thiram

A fungicide seed dressing mixture for field and vegetable crops

Products

| Hy-TL | Agrichem | 225:300 g/l | FS | 06246 |

Uses
- Ascochyta in **broad beans** *(seed treatment)*, **peas** *(seed treatment)*, **spring field beans** *(seed treatment)*, **winter field beans** *(seed treatment)*
- Damping off in **broad beans** *(seed treatment)*, **peas** *(seed treatment)*, **spring field beans** *(seed treatment)*, **winter field beans** *(seed treatment)*
- Fusarium in **bulb onions** *(off-label - seed treatment)*
- Neck rot in **bulb onions** *(off-label - seed treatment)*

Specific Off-Label Approvals (SOLAs)
- **bulb onions** *(seed treatment) (OLA 021299) Dec 2008* [1]

Approval information
- Thiabendazole included in Annex I under EC Directive 91/414

Efficacy guidance
- Dress seed as near to sowing as possible
- Dilution may be needed with particularly absorbent types of seed. If diluted material used, seed may require drying before storage

Restrictions
- Maximum number of treatments 1 per batch
- Seed to be treated should be of satisfactory quality and moisture content
- Do not use treated seed as food or feed

Crop-specific information
- Latest use: before drilling

Environmental safety
- Dangerous for the environment
- Treated seed harmful to game and wildlife

Hazard classification and safety precautions
Hazard H03, H11
Risk phrases R20, R22a, R36, R38, R43, R48, R50, R53a
Operator protection A, C, D, H, M; U05a, U08, U14, U15, U20a
Environmental protection E03, E34
Storage and disposal D01, D02, D05, D09a, D10c
Treated seed S01, S02, S03, S04a, S04b, S05, S06a
Medical advice M03, M04

422   thiacloprid

A chloronicotinyl insecticide for use in apples

Products

| Calypso | Bayer CropScience | 480 g/l | SC | 11257 |

Uses
- Big-bud mite in **almonds** *(off-label)*, **chestnuts** *(off-label)*, **cob nuts** *(off-label)*, **hazel nuts** *(off-label)*, **walnuts** *(off-label)*
- Common green capsid in **blackberries** *(off-label)*, **raspberries** *(off-label)*, **rubus hybrids** *(off-label)*
- Rosy apple aphid in **apples**
- Tarnished plant bug in **blackberries** *(off-label)*, **raspberries** *(off-label)*, **rubus hybrids** *(off-label)*

Specific Off-Label Approvals (SOLAs)
- **almonds, chestnuts, cob nuts, hazel nuts, walnuts** *(OLA 050516) Jun 2006* [1]
- **blackberries, raspberries, rubus hybrids** *(OLA 041494) Dec 2008* [1]

Approval information
- Thiacloprid included in Annex I under EC Directive 91/414

SEE SECTION 3 FOR PRODUCTS ALSO REGISTERED

SECTION 2

Efficacy guidance
- Best results obtained from a programme of sprays commencing pre-blossom at the first sign of aphids
- Minimise the possibility of development of resistance by alternating insecticides with different modes of action in the programme
- In dense canopies and on larger trees increase water volume to ensure full coverage

Restrictions
- Maximum number of treatments 3 per yr on plums; 2 per yr on all other crops

Crop-specific information
- HI apples 14 d; protected vegetables 3 d

Environmental safety
- Harmful to aquatic organisms
- Harmful to fish or other aquatic life. Do not contaminate surface waters or ditches with chemical or used container
- Risk to certain non-target insects or other arthropods. See directions for use
- Broadcast air-assisted LERAP (30 m)

Hazard classification and safety precautions
Hazard H03
Risk phrases R20, R22a, R40, R43, R52, R53a
Operator protection A, H; U05a, U14, U20b
Environmental protection E13c, E22b, E34; E17b (30 m)
Storage and disposal D01, D02, D09a, D10b
Medical advice M03, M05b

423   thifensulfuron-methyl

A translocated sulfonylurea herbicide

See also flupyrsulfuron-methyl + thifensulfuron-methyl
fluroxypyr + thifensulfuron-methyl + tribenuron-methyl
metsulfuron-methyl + thifensulfuron-methyl

Products

1	DUK 118	Headland	75% w/w	WG	11923
2	Pinnacle	Headland	50% w/w	SG	12285

Uses
- Annual dicotyledons in **spring barley**, **spring wheat**, **winter barley**, **winter wheat** [1]
- Charlock in **spring barley**, **spring wheat**, **winter barley**, **winter wheat** [1]
- Chickweed in **spring barley**, **spring wheat**, **winter barley**, **winter wheat** [1]
- Docks in **established grassland**, **rotational grassland** [2]
- Green cover in **land not being used for crop production** [2]
- Knotgrass in **spring barley**, **spring wheat**, **winter barley**, **winter wheat** [1]
- Mayweeds in **spring barley**, **spring wheat**, **winter barley**, **winter wheat** [1]

Approval information
- Thifensulfuron-methyl included in Annex I under EC Directive 91/414
- Accepted by BBPA for use on malting barley

Efficacy guidance
- Best results achieved from application to small emerged weeds when growing actively. Broad-leaved docks are susceptible during the rosette stage up to onset of stem extension
- Ensure good spray coverage and apply to dry foliage
- Susceptible weeds stop growing almost immediately but symptoms may not be visible for about 2 wk
- Product sold in twin pack with mecoprop-P to provide option for improved control of cleavers [1]
- Only broad-leaved docks (*Rumex obtusifolius*) are controlled; curled docks (*Rumex crispus*) are resistant [2]
- Docks with developing or mature seed heads should be topped and the regrowth treated later [2]
- Established docks with large tap roots may require follow-up treatment [2]

FOR FULL CONDITIONS OF USE ALWAYS READ THE PRODUCT LABEL

- High populations of docks in grassland will require further treatment in following yr [2]
- Thifensulfuron-methyl is a member of the ALS-inhibitor group of herbicides and products should be used in a planned Resistance Management strategy. See Section 5 for more information

Restrictions

- Maximum number of treatments 1 per crop for cereals or one per yr for grassland and cereals. Must only be applied from 1 Feb in yr of harvest
- Do not apply to cereal crops undersown with grasses, clover or other legumes, or any other broad-leaved crop [1]
- Do not treat new leys in year of sowing [2]
- Do not treat where nutrient imbalances, drought, waterlogging, low temperatures, lime deficiency, pest or disease attack have reduced crop or sward vigour
- Do not roll or harrow within 7 d of spraying
- Do not graze grass crops within 7 d of spraying
- Specific restrictions apply to use in sequence or tank mixture with other sulfonylurea or ALS-inhibiting herbicides. See label for details
- Only one application of a sulfonylurea product may be applied per calendar yr to grassland and green cover on land temporarily removed from production

Crop-specific information

- Latest use: before 1 Aug on grass; flag leaf fully emerged stage (GS 39) on cereals
- On grass apply 7-10 d before grazing and do not graze for 7 d afterwards [2]
- Product may cause a check to both sward and clover which is usually outgrown [2]

Following crops guidance

- Only grass or cereals may be sown within 4 wk of application to grassland or setaside, or in the event of failure of any treated crop
- No restrictions apply after normal harvest of a treated cereal crop

Environmental safety

- Dangerous for the environment
- Very toxic to aquatic organisms
- LERAP Category B
- Keep livestock out of treated areas for at least 7 d following treatment
- Take extreme care to avoid drift onto broad-leaved plants outside the target area or onto surface waters or ditches, or land intended for cropping
- Spraying equipment should not be drained or flushed onto land planted, or to be planted, with trees or crops other than cereals and should be thoroughly cleansed after use - see label for instructions

Hazard classification and safety precautions

Hazard H11
Risk phrases R50, R53a
Operator protection U08 [1]; U19a, U20b [1, 2]
Environmental protection E06a [2] (7 d); E15a, E38 [1, 2]; E16a, E16b [1]
Storage and disposal D09a, D11a, D12a

424 thifensulfuron-methyl + tribenuron-methyl

A mixture of two sulfonylurea herbicides for cereals

Products

Calibre SX	DuPont	33.3:16.7% w/w	SG	12241

Uses

- Annual dicotyledons in **spring barley**, **spring wheat**, **winter barley**, **winter wheat**
- Charlock in **spring barley**, **spring wheat**, **winter barley**, **winter wheat**
- Chickweed in **spring barley**, **spring wheat**, **winter barley**, **winter wheat**
- Mayweeds in **spring barley**, **spring wheat**, **winter barley**, **winter wheat**

Approval information

- Thifensulfuron-methyl and tribenuron-methyl included in Annex I under EC Directive 91/414
- Accepted by BBPA for use on malting barley

SEE SECTION 3 FOR PRODUCTS ALSO REGISTERED

SECTION 2

Efficacy guidance
- Apply after 1 Feb when weeds are small and actively growing
- Ensure good spray cover of the weeds
- Susceptible weeds cease growth almost immediately after application and symptome become evident 2 wk later
- Effectiveness reduced by rain within 4 h of treatment and in very dry conditions
- Various tank mixtures recommended to broaden weed control spectrum
- Thifensulfuron-methyl and tribenuron-methyl are members of the ALS-inhibitor group of herbicides and products should be used in a planned Resistance Management strategy. See Section 5 for more information

Restrictions
- Maximum number of treatments 1 per crop
- Do not apply to cereals undersown with grass, clover or other legumes
- Do not apply within 7 d of rolling
- Specific restrictions apply to use in sequence or tank mixture with other sulfonylurea or ALS-inhibiting herbicides. See label for details
- Do not apply to any crop suffering from stress

Crop-specific information
- Latest use: before flag leaf ligule first visible (GS 39) for all crops

Following crops guidance
- Only cereals, field beans or oilseed rape may be sown in the same calendar year as harvest of a treated crop
- In the event of failure of a treated crop sow only a cereal crop within 3 mth of product application

Environmental safety
- Dangerous for the environment
- Very toxic to aquatic organisms
- Spraying equipment should not be drained or flushed onto land planted, or to be planted, with trees or crops other than cereals and should be thoroughly cleansed after use - see label for instructions
- Take particular care to avoid damage by drift onto broad-leaved plants outside the target area or onto surface waters or ditches

Hazard classification and safety precautions
Hazard H04, H11
Risk phrases R43, R50, R53a
Operator protection A, H; U05a, U08, U20a
Environmental protection E15a, E38
Storage and disposal D01, D02, D09a, D11a, D12a

425 thiodicarb

A carbamate insecticide with molluscicide uses

Products

1	Judge	Sipcam	4% w/w	RB	11847
2	Me2 Exodus	Me2	4% w/w	RB	09786
3	Toro	Sipcam	4% w/w	RB	12138

Uses
- Slugs in *brussels sprouts* [1, 3]; *durum wheat, potatoes, spring barley, spring oats, spring oilseed rape, spring wheat, triticale, winter barley, winter oats, winter oilseed rape, winter wheat* [1-3]

Approval information
- Accepted by BBPA for use on malting barley

Efficacy guidance
- Apply bait as broadcast treatment or admixed with seed
- Additional broadcast treatments may be needed after drilling admixed seed
- Sow admixed seed as soon as possible after treatment

FOR FULL CONDITIONS OF USE ALWAYS READ THE PRODUCT LABEL

Restrictions

- Product contains an anticholinesterase carbamate compound. Do not use if under medical advice not to work with such compounds
- Maximum number of treatments 1 per crop (admixture), 3 per crop (broadcast)
- Do not treat grain with more than 16% moisture content or allow treated seed to rise above this level
- May only be applied to oilseed rape and potatoes as broadcast application
- Must not be admixed with cereal seed already admixed with any other product containing thiodicarb
- Do not use treated seed as food or feed

Crop-specific information

- Latest use: before drilling when admixed with barley, durum wheat, oats, triticale, wheat; before second node detectable (GS 32) for barley, durum wheat, oats, triticale, wheat; before stem extension (GS 2,0) for oilseed rape
- HI potatoes, Brussels sprouts 3 wk

Environmental safety

- Dangerous for the environment
- Toxic to aquatic organisms
- Harmful to game, wild birds and animals

Hazard classification and safety precautions

Hazard H03 [2]; H04 [1, 3]; H11 [1-3]
Risk phrases R20, R53a [1, 3]; R22a, R36, R43, R51 [1-3]
Operator protection A, D, H [1-3]; J [1, 3]; U05a, U13, U20b
Environmental protection E10b, E34 [1-3]; E15a [2]
Storage and disposal D01, D02, D09a, D11a
Treated seed S01, S02, S05, S07 [2]; S04a [1, 3]
Medical advice M02 [1-3]; M03 [2]; M05a [1, 3]

426   thiophanate-methyl

A carbendazim precursor fungicide with protectant and curative activity

See also iprodione + thiophanate-methyl

Products

1 Mildothane Turf Liquid	Bayer Environ.	500 g/l	SC	09935
2 Snare	Headland Amenity	500 g/l	SC	11873

Uses

- Dollar spot in **managed amenity turf**
- Fusarium patch in **managed amenity turf**
- Red thread in **managed amenity turf**
- Wormcast formation in **managed amenity turf**

Approval information

- Thiophanate-methyl included in Annex I under EC Directive 91/414
- Following implementation of Directive 98/82/EC, approval for use of thiophanate-methyl on several crops was revoked in 1999

Efficacy guidance

- Treat turf at the first sign of disease and repeat at monthly intervals as necessary
- For earthworm cast suppression apply to moist turf at first sign of casting activity and repeat as necessary
- Apply to turf during period of active growth and do not mow for 48 h
- Control of Red Thread and Dollar Spot will be helped by application of a nitrogenous fertilizer 7 d before fungicide treatment

Restrictions

- On mown turf apply after cutting and at least 48 hr before the next cut
- Do not use when earthworms are inactive during periods of drought or when ground is frozen

SEE SECTION 3 FOR PRODUCTS ALSO REGISTERED

SECTION 2

Environmental safety
- Dangerous for the environment
- Very toxic to aquatic organisms

Hazard classification and safety precautions
    Hazard H03, H11
    Risk phrases R43, R50, R53a, R68
    Operator protection A, C, H, M; U05a, U14, U20c
    Environmental protection E15a, E34, E38
    Storage and disposal D01, D02, D05, D09a, D10a, D12a
    Medical advice M03

427   thiram

A protectant dithiocarbamate fungicide

See also carboxin + thiram
             prochloraz + thiram
             thiabendazole + thiram

Products

1	Agrichem Flowable Thiram	Agrichem	600 g/l	FS	10784
2	Thiraflo	Crompton	480 g/l	FS	11526
3	Thyram Plus	Agrichem	600 g/l	FS	10785
4	Unicrop Thianosan DG	Unicrop	80% w/w	WG	05454

Uses
- Botrytis in **chrysanthemums**, **freesias**, **lettuce** *(outdoor crops)*, **ornamental specimens** *(except Hydrangea)*, **outdoor tomatoes**, **protected lettuce**, **protected tomatoes**, **raspberries**, **strawberries** [4]
- Botrytis fruit rot in **apples** [4]
- Cane spot in **raspberries** [4]
- Damping off in **broad beans** *(seed treatment)*, **dwarf beans** *(seed treatment)*, **grass seed** *(seed treatment)*, **maize** *(seed treatment)*, **runner beans** *(seed treatment)*, **spring field beans** *(seed treatment)*, **spring oilseed rape** *(seed treatment)*, **winter field beans** *(seed treatment)*, **winter oilseed rape** *(seed treatment)* [1-3]; **cabbages** *(seed treatment)*, **carrots** *(seed treatment)*, **cauliflowers** *(seed treatment)*, **leeks** *(seed treatment)*, **lettuce** *(seed treatment)*, **onions** *(seed treatment)*, **peas** *(seed treatment)*, **radishes** *(seed treatment)*, **salad onions** *(seed treatment)*, **turnips** *(seed treatment)* [1, 3]; **combining peas** *(seed treatment)*, **vining peas** *(seed treatment)* [2]; **poppies** *(off-label - for morphine production - seed treatment)* [1]
- Downy mildew in **protected lettuce** [4]
- Fire in **tulips** [4]
- Gloeosporium in **apples** [4]
- Pythium in **poppies** *(off-label - for morphine production - seed treatment)* [1]
- Rust in **blackcurrants**, **carnations**, **chrysanthemums** [4]
- Scab in **apples**, **pears** [4]
- Seed-borne diseases in **carrots** *(seed soak)*, **celery** *(seed soak)*, **fodder beet** *(seed soak)*, **mangels** *(seed soak)*, **parsley** *(seed soak)*, **red beet** *(seed soak)*, **sugar beet** *(seed soak)* [1]
- Spur blight in **raspberries** [4]

Specific Off-Label Approvals (SOLAs)
- **poppies** *(for morphine production - seed treatment) (OLA 040681) Dec 2008* [1]

Efficacy guidance
- Spray before onset of disease and repeat every 7-14 d. Spray interval varies with crop and disease. See label for details [4]
- Apply as seed treatment for protection against damping off. May be applied through most types of seed treatment machinery, from automated continuous flow machines to smaller batch treating apparatus [1, 3]
- Co-application of 175 ml water per 100 kg of seed likely to improve evenness of seed coverage [1-3]
- Do not spray when rain imminent [4]
- For use on tulips, chrysanthemums and carnations add non-ionic wetter [4]

FOR FULL CONDITIONS OF USE ALWAYS READ THE PRODUCT LABEL

Restrictions

- Maximum number of treatments 3 per crop for protected winter lettuce (thiram based products only [4]; 2 per crop if sequence of thiram and other EBDC fungicides used) [4]; 2 per crop for protected summer lettuce [4]; 1 per batch of seed for seed treatments
- Do not apply to hydrangeas [4]
- Notify processor before dusting or spraying crops for processing [4]
- Do not dip roots of forestry transplants [4]
- Do not treat seed of tomatoes, peppers or aubergines [1, 3]

Crop-specific information

- Latest use: pre-drilling for broad beans, peas, dwarf beans, grass seed, maize, poppies, runner beans, field beans, oilseed rape, vining peas; 21 d after planting out or 21 d before harvest, whichever is earlier, for protected winter lettuce; 14 d after planting out or 21 d before harvest, whichever is earlier, for protected summer lettuce; before drilling for soya beans [1]
- HI protected lettuce 21 d; outdoor lettuce 14 d; apples, pears, blackcurrants, raspberries, strawberries, tomatoes 7 d
- Seed to be treated should be of satisfactory quality and moisture content [1-3]
- Follow label instructions for treating small quantities of seed [1, 3]

Environmental safety

- Dangerous for the environment
- Very toxic to aquatic organisms
- Dangerous to fish or other aquatic life. Do not contaminate surface waters or ditches with chemical or used container
- Do not use treated seed as food or feed [2]
- Treated seed harmful to game and wildlife [2]
- A red dye is available from manufacturer to colour treated seed. See label for details [1, 3]
- Additional precautions apply if using small volume returnable container- see labels [1, 3]

Hazard classification and safety precautions

Hazard H03, H11 [1-3]; H04 [4]
Risk phrases R20, R43 [1, 3]; R22a, R50, R53a [1-3]; R36, R38 [1, 3, 4]; R37 [4]; R48 [2]
Operator protection A; U02a, U14, U15 [1, 3]; U05a, U20b [1-4]; U07 [1, 3] (SVR only); U08 [2-4]; U09a [1]; U19a [4]
Environmental protection E13b [4]; E15a, E34 [1, 3]; E36 [1, 3] (SVR only); E38 [2]
Storage and disposal D01, D02, D09a [1-4]; D05 [1-3]; D10a [2]; D10c [1, 3]; D11a [4]; D14 [1, 3] (SVR only)
Treated seed S01, S02, S03, S04b, S05, S06a, S07 [1-3]; S04a [1, 3]
Medical advice M03 [2, 3]; M04 [1, 3]

428 tolclofos-methyl

A protectant organophosphorus fungicide for soil-borne diseases

Products

1	Basilex	Scotts	50% w/w	WP	07494
2	Me2 Cindy	Me2	500 g/l	FS	10634
3	Rizolex	Certis	10% w/w	DS	09673
4	Rizolex Flowable	Certis	500 g/l	FS	11399

Uses

- Black scurf and stem canker in **potatoes** *(tuber treatment)* [2-4]
- Bottom rot in **protected lettuce** [1]
- Damping off in **seedlings of ornamentals** [1]
- Damping off and wirestem in **brussels sprouts**, **cabbages**, **calabrese**, **cauliflowers**, **chinese cabbage** [1]
- Foot rot in **ornamental specimens**, **seedlings of ornamentals** [1]
- Rhizoctonia in **potatoes** *(off-label - chitted seed treatment)* [4]; **protected celery** *(off-label)*, **protected radishes** *(off-label)*, **swedes** *(with fleece or mesh covers)*, **swedes** *(without covers)*, **turnips** *(with fleece or mesh covers)*, **turnips** *(without covers)* [1]
- Root rot in **ornamental specimens**, **seedlings of ornamentals** [1]

SEE SECTION 3 FOR PRODUCTS ALSO REGISTERED

Specific Off-Label Approvals (SOLAs)
- *potatoes* *(chitted seed treatment)* *(OLA 041323) Dec 2008* [4]
- *protected celery* *(OLA 992009) Dec 2008* [1]
- *protected radishes* *(OLA 002006) Dec 2008* [1]

Approval information
- Tolclofos-methyl has been included in a comprehensive review of anticholinesterase compounds and considered by the ACP. As a result approvals have been allowed to continue subject to the imposition of additional operator protection requirements.
- May be applied by misting equipment mounted over roller table. See label for details [3, 4]

Efficacy guidance
- Apply dust to seed potatoes during hopper loading [3]
- Apply flowable formulation to clean tubers with suitable misting equipment over a roller table. Spray as potatoes taken into store (first earlies) or as taken out of store (second earlies, maincrop, crops for seed) pre-chitting [2, 4]
- Do not mix flowable formulation with any other product [2, 4]
- To control Rhizoctonia in vegetables and ornamentals apply as drench before sowing, pricking out or planting [1]
- On established seedlings and pot plants apply as drench and rinse off foliage [1]

Restrictions
- Tolclofos-methyl is an atypical organophosphorus compound which has weak anticholinesterase activity. Do not use if under medical advice not to work with such compounds
- Maximum number of treatments 1 per batch for seed potatoes; 1 per crop for lettuce, brassicas, celery, radishes, swedes, turnips; 1 at each stage of growth (ie sowing, pricking out, potting) to a maximum of 3, for ornamentals
- Only to be used with automatic planters [3]
- Not recommended for use on seed potatoes where hot water treatment used or to be used [3]
- Do not apply as overhead drench to vegetables or ornamentals when hot and sunny [1]
- Do not use on heathers [1]
- Must not be used via hand-held equipment
- Application to protected crops must only be made where the operator is outside the structure at the time of treatment

Crop-specific information
- Latest use: at planting of seed potatoes [2-4]; before planting celery [1]; before transplanting for lettuce, brassicas, ornamental specimens [1]; before 2 true lvs for swedes, turnips [1]
- HI potatoes 12 wk [4]

Environmental safety
- Dangerous for the environment [1-4]
- Very toxic to aquatic organisms
- LERAP Category B [1]
- Treated tubers to be used as seed only, not for food or feed [2, 3]

Hazard classification and safety precautions
Hazard H04, H11
Risk phrases R36, R37, R38 [1-4]; R50 [3]; R51, R53a [1, 2, 4]
Operator protection A, D, H [1-4]; C [2, 4]; E [2-4]; J, M [3]; U05a, U11 [3]; U16 [2, 4]; U19a, U20b [1-4]
Environmental protection E15a [2-4]; E16a [1]
Storage and disposal D01, D02, D12a [3]; D08 [2-4]; D09a [1-4]; D10a [2, 4]; D11a [1, 3]
Medical advice M01 [1]; M04 [2-4]

429   tolylfluanid

A multi-site protectant fungicide

See also fenhexamid + tolylfluanid

Products

Elvaron Multi	Bayer CropScience	50.5% w/w	WG	11422

Uses

- Botrytis in **blackberries**, **blackcurrants**, **gooseberries**, **loganberries**, **raspberries**, **redcurrants**, **strawberries**, **whitecurrants**
- Red spider mites in **apples** *(reduction)*, **pears** *(reduction)*
- Rust mite in **apples** *(reduction)*, **pears** *(reduction)*
- Scab in **apples**, **pears**

Approval information
- Tolylfluanid included in Annex I under EC Directive 91/414

Efficacy guidance
- Best results in soft fruit achieved by good spray coverage of flowers and young fruitlets
- To achieve required coverage of flowers and young fruitlets spray using a hand lance, drop arm attachments or a special strawberry boom with leaf rolling bar
- Best results using high volume directed over the crop rows

Restrictions
- Maximum total dose equivalent to 2.5 full dose treatments on currants, gooseberries; 4 full dose treatments on blackberries, loganberries, raspberries, strawberries; 8 full dose treatments on apples, pears

Crop-specific information
- HI apples, pears 7 d; blackberries, loganberries, raspberries, strawberries 14 d; currants, gooseberries 21 d

Environmental safety
- Dangerous for the environment
- Very toxic to aquatic organisms
- Dangerous to fish or other aquatic life. Do not contaminate surface waters or ditches with chemical or used container
- Risk to non-target insects or other arthropods
- LERAP Category B
- Broadcast air-assisted LERAP
- Use best available application technique that minimises off-target drift to reduce effects on non-target insects or other arthropods

Hazard classification and safety precautions
Hazard H04, H11
Risk phrases R36, R43, R50
Operator protection A, H, M; U02a, U05a, U14, U20b
Environmental protection E13b, E16a, E16b, E22c, E38 [1]; E17b [1] (30 m for apples pears; 10 m for cane fruit, currants, gooseberries)
Storage and disposal D01, D02, D05, D09a, D11a, D12a

430 tralkoxydim

A foliar applied oxime herbicide for grass weed control in cereals.

Products

1	Grasp	Syngenta	250 g/l	SC	10441
2	Grasp 400 SC	Syngenta	400 g/l	SC	12280
3	Greencrop Gweedore	Greencrop	250 g/l	SC	09882
4	Landgold Tralkoxydim	Landgold	250 g/l	SC	08604
5	Landgold Tralkoxydim	Teliton	250 g/l	SC	12130
6	Standon Tralkoxydim	Standon	250 g/l	SC	09579

Uses

- Awned canary grass in **spring barley** *(qualified minor use)*, **spring wheat** *(qualified minor use)* [1, 2]
- Blackgrass in **durum wheat**, **spring barley**, **spring wheat**, **triticale**, **winter barley**, **winter rye**, **winter wheat** [1-6]; **spring barley** *(autumn sown)*, **spring wheat** *(autumn sown)* [1]
- Rough meadow grass in **durum wheat**, **spring barley** *(autumn sown)*, **spring wheat**, **spring wheat** *(autumn sown)*, **triticale**, **winter barley**, **winter rye**, **winter wheat** [1]
- Ryegrass in **durum wheat**, **spring barley**, **spring wheat**, **triticale**, **winter barley**, **winter rye**, **winter wheat** [1, 2]; **spring barley** *(autumn sown)*, **spring wheat** *(autumn sown)* [1]

- Wild oats in *durum wheat, spring barley, spring wheat, triticale, winter barley, winter rye, winter wheat* [1-6]; *spring barley* (autumn sown), *spring wheat* (autumn sown) [1]
- Yorkshire fog in *durum wheat* (qualified minor use), *triticale* (qualified minor use), *winter barley* (qualified minor use), *winter rye* (qualified minor use), *winter wheat* (qualified minor use) [1, 2]; *spring barley* (autumn sown - qualified minor use), *spring wheat* (autumn sown - qualified minor use) [1]

Approval information
- Accepted by BBPA for use on malting barley

Efficacy guidance
- Product leaf-absorbed and translocated rapidly to growing points. Best results achieved after completion of emergence of grass weeds when growing actively in competitive crops under warm humid conditions with adequate soil moisture
- Activity not dependent on soil type or condition. Weeds germinating after application will not be controlled
- Best control of wild oats obtained from 2 leaf to 1st node detectable stage of weeds, and of blackgrass up to 3 tillers
- Always follow WRAG guidelines for preventing and managing herbicide resistant weeds. Section 5 for more information

Restrictions
- Maximum number of treatments 1 or 2 per crop. See label
- Authorised adjuvant must always be added. See label
- Do not use on oats
- Do not spray undersown crops or crops to be undersown
- Do not spray when foliage wet or covered in ice or crop otherwise under stress
- Do not spray if a protracted period of cold weather forecast
- Do not spray crops under stress from chemical treatment, grazing, pest attack, mineral deficiency or low fertility. Treatment of stressed crops may be followed by transient foliar discolouration
- Do not roll or harrow within 1 wk of spraying
- Restrictions apply to mixtures or sequences with phenoxy hormone or sulfonylurea herbicides. See label

Crop-specific information
- Latest use: before booting (GS 41)
- Apply to winter cereals from 2 leaves unfolded. If necessary winter cereals may be sprayed twice: once in autumn and once in spring
- Apply to spring cereals from end of tillering

Environmental safety
- Harmful to aquatic organisms. May cause long-term adverse effects to the aquatic environment [1]
- Take care to avoid drift onto neighbouring crops, especially oats

Hazard classification and safety precautions
Hazard H03 [1]; H04 [3-6]
Risk phrases R36 [3-6]; R43, R52, R53a [1, 3-6]
Operator protection A, H [1-6]; C [1, 3-6]; U02a, U04a, U05a, U08, U20b [1-6]; U14, U19a [3-5]
Environmental protection E15a [1-6]; E34 [6]; E38 [1]
Storage and disposal D01, D02, D09a [1-6]; D05, D10b [3-6]; D10c [1, 2]; D12a [1]
Medical advice M03 [6]

431   tri-allate

A soil-acting thiocarbamate herbicide for grass weed control

Products

Avadex Excel 15G	Gowan	15% w/w	GR	10299

Uses
- Blackgrass in *combining peas, durum wheat, fodder beet, lucerne, mangels, red beet, red clover, sainfoin, spring barley, spring field beans, sugar beet, triticale, vetches, vining peas, white clover, winter barley, winter field beans, winter wheat*

FOR FULL CONDITIONS OF USE ALWAYS READ THE PRODUCT LABEL

- Meadow grasses in **combining peas, durum wheat, fodder beet, lucerne, mangels, red beet, red clover, sainfoin, spring barley, spring field beans, sugar beet, triticale, vetches, vining peas, white clover, winter barley, winter field beans, winter wheat**
- Wild oats in **combining peas, durum wheat, fodder beet, lucerne, mangels, red beet, red clover, sainfoin, spring barley, spring field beans, sugar beet, triticale, vetches, vining peas, white clover, winter barley, winter field beans, winter wheat**

Approval information
- Approved for aerial application on wheat, barley, rye, triticale, field beans, peas, fodder beet, sugar beet, red beet, lucerne, sainfoin, vetches, clover. See notes in Section 5
- Accepted by BBPA for use on malting barley
- Approval expiry 28 Feb 2006 [1]

Efficacy guidance
- Incorporate or apply to surface pre-emergence (post-emergence application possible in winter cereals up to 2-leaf stage of wild oats)
- Do not use on soils with more than 10% organic matter
- Wild oats controlled up to 2-leaf stage
- If applied to dry soil, rainfall needed for full effectiveness, especially with granules on surface. Do not use if top 5-8 cm bone dry
- Do not apply with spinning disc granule applicator; see label for suitable types
- Do not apply to cloddy seedbeds
- Use sequential treatments to improve control of barren brome and annual dicotyledons (see label for details)

Restrictions
- Maximum number of treatments 1 per crop
- Consolidate loose, puffy seedbeds before drilling to avoid chemical contact with seed
- Drill cereals well below treated layer of soil (see label for safe drilling depths)
- Do not use on direct-drilled crops or undersow grasses into treated crops
- Do not sow oats or grasses within 1 yr of treatment

Crop-specific information
- Latest use: pre-drilling for beet crops; before crop emergence for field beans, spring barley, peas, forage legumes; before first node detectable stage (GS 31) for winter wheat, winter barley, durum wheat, triticale, winter rye

Environmental safety
- Irritating to eyes and skin
- May cause sensitization by skin contact
- Harmful to fish or other aquatic life. Do not contaminate surface waters or ditches with chemical or used container

Hazard classification and safety precautions
Hazard H04
Risk phrases R36, R38, R43
Operator protection A, C, H; U02a, U05a, U20a
Environmental protection E13c
Storage and disposal D01, D02, D09a, D11a

432   triazamate

A carbamoyl triazole insecticide for apples and specified field crops

Products
Aztec	BASF	140 g/l	EW	10211

Uses
- Aphids in **brussels sprouts, sugar beet** (including Myzus persicae)
- Green aphid in **apples**
- Pea aphid in **combining peas, vining peas**
- Rosy apple aphid in **apples**

SECTION 2

Approval information
- Triazamate is not supported in the EU Review Programme. All approvals will be revoked with a final use-by date of 4 January 2007

Efficacy guidance
- Treat as soon as aphids seen in crop or (on sugar beet) immediately after official warnings are issued. Treat vining peas when 15% of plants infested and combining peas when 20% of plants infested
- Products are systemic and act through contact and stomach action. Ensure foliage well covered with spray by using higher volumes in dense crops
- Controls aphids resistant to other chemical groups
- To reduce risk of development of resistance consider use of products with alternative modes of action in intensive pest control programmes. If tank mixtures used, do not reduce dose of either component

Restrictions
- Maximum number of treatments 1 per crop on peas; 2 per crop on apples, 3 per crop on sugar beet; 4 per crop on Brussels sprouts
- Maximum total dose on brassicas equivalent to one full dose treatment on peas, 2 on apples, 3 on sugar beet, 4 on Brussels sprouts
- Product must be used with recommended adjuvant, except on apples. Consult manufacturer
- Consult processor before use on crops for processing

Crop-specific information
- HI apples, Brussels sprouts, sugar beet 28 d; peas 21 d
- Label warns that damage may result unless all recommendations carefully followed
- Treatment may cause increased russetting of apples

Environmental safety
- Dangerous for the environment
- Very toxic to aquatic organisms
- Product not harmful to bees when used as directed but spraying in late evening/early morning or in dull weather recommended to avoid unnecessary stress on foraging bees. Do not apply to apples during flowering
- Do not store above 40°C

Hazard classification and safety precautions
    Hazard H03, H11
    Risk phrases R20, R22a, R43, R50, R53a
    Operator protection A, H; U02a, U05a, U09a, U13, U14, U19a, U20b
    Environmental protection E15a, E34, E38
    Storage and disposal D01, D02, D05, D09a, D11a, D12a
    Medical advice M03, M05a, M06

433   triazoxide

A benzotriazine fungicide available only in mixtures

See also imidacloprid + tebuconazole + triazoxide
            prothioconazole + tebuconazole + triazoxide
            tebuconazole + triazoxide

434   tribenuron-methyl

A foliar acting sulfonylurea herbicide with some root activity for use in cereals

See also flupyrsulfuron-methyl + tribenuron-methyl
fluroxypyr + thifensulfuron-methyl + tribenuron-methyl
thifensulfuron-methyl + tribenuron-methyl

Products

1	Landgold Tribenuron 75	Landgold	75% w/w	WG	11203
2	Landgold Tribenuron 75	Teliton	75% w/w	WG	12131
3	Quantum SX	DuPont	50% w/w	SG	12239
4	Rockett	AgriGuard	75% w/w	WG	12344

Uses

- Annual dicotyledons in *durum wheat* [1, 2, 4]; *spring barley, spring oats, spring wheat, triticale, winter barley, winter oats, winter rye, winter wheat* [1-4]
- Charlock in *durum wheat* [1, 2, 4]; *spring barley, spring oats, spring wheat, triticale, winter barley, winter oats, winter rye, winter wheat* [1-4]
- Chickweed in *durum wheat* [1, 2, 4]; *spring barley, spring oats, spring wheat, triticale, winter barley, winter oats, winter rye, winter wheat* [1-4]
- Mayweeds in *durum wheat* [1, 2, 4]; *spring barley, spring oats, spring wheat, triticale, winter barley, winter oats, winter rye, winter wheat* [1-4]

Approval information

- Tribenuron-methyl included in Annex I under EC Directive 91/414
- Accepted by BBPA for use on malting barley

Efficacy guidance

- Best control achieved when weeds small and actively growing
- Good spray cover must be achieved since larger weeds often become less susceptible
- Susceptible weeds cease growth almost immediately after treatment and symptoms can be seen in about 2 wk
- Weed control may be reduced when conditions very dry
- Tribenuron-methyl is a member of the ALS-inhibitor group of herbicides and products should be used in a planned Resistance Management strategy. See Section 5 for more information

Restrictions

- Maximum number of treatments 1 per crop
- Specific restrictions apply to use in sequence or tank mixture with other sulfonylurea or ALS-inhibiting herbicides. See label for details
- Do not apply to crops undersown with grass, clover or other broad-leaved crops
- Do not apply to any crop suffering stress from any cause or not actively growing
- Do not apply within 7 d of rolling

Crop-specific information

- Latest use: up to and including flag leaf ligule/collar just visible (GS 39)
- Apply in autumn or in spring from 3 leaf stage of crop

Following crops guidance

- Only cereals, field beans or oilseed rape may be sown in the same calendar yr as harvest of a treated crop
- In the event of crop failure sow only a cereal within 3 mth of application. After 3 mth field beans or oilseed rape may also be sown

Environmental safety

- Dangerous for the environment
- Very toxic to aquatic organisms
- Take extreme care to avoid drift onto broad-leaved plants outside the target area or onto surface waters or ditches, or land intended for cropping
- Spraying equipment should not be drained or flushed onto land planted, or to be planted, with trees or crops other than cereals and should be thoroughly cleansed after use - see label for instructions

SEE SECTION 3 FOR PRODUCTS ALSO REGISTERED

Hazard classification and safety precautions
   Hazard H04, H11
   Risk phrases R43, R50, R53a
   Operator protection A [1-4]; H [3]; U05a, U08, U20b [1-4]; U19a [1, 2, 4]
   Environmental protection E15a [1-4]; E38 [3]
   Storage and disposal D01, D02, D09a, D11a [1-4]; D12a [3]

435   triclopyr

An aryloxyalkanoic acid herbicide for perennial and woody weed control

See also 2,4-D + dicamba + triclopyr
*   clopyralid + fluroxypyr + triclopyr*
*   clopyralid + triclopyr*
*   fluroxypyr + triclopyr*

Products

1 Garlon 2	Syngenta	240 g/l	EC	10672
2 Garlon 4	Dow	480 g/l	EC	05090
3 Nomix Garlon 4	Nomix Enviro	480 g/l	EC	12081
4 Timbrel	Dow	480 g/l	EC	05815

Uses
   - Brambles in **established grassland, land not intended for cropping, non-crop areas** (directed treatment) [1]; **forest, land not intended to bear vegetation** [2-4]
   - Broom in **established grassland, land not intended for cropping, non-crop areas** (directed treatment) [1]; **forest, land not intended to bear vegetation** [2-4]
   - Brush clearance in **industrial sites** [2-4]
   - Docks in **established grassland, land not intended for cropping, non-crop areas** [1]; **forest, land not intended to bear vegetation** [2-4]
   - Gorse in **established grassland, land not intended for cropping, non-crop areas** (directed treatment) [1]; **forest, land not intended to bear vegetation** [2-4]
   - Hard rush in **established grassland, land not intended for cropping, non-crop areas** [1]
   - Perennial dicotyledons in **established grassland, land not intended for cropping, non-crop areas** [1]; **forest, industrial sites, land not intended to bear vegetation** [2-4]
   - Perennial weeds in **asparagus** (off-label - directed treatment) [4]
   - Rhododendrons in **forest, land not intended to bear vegetation** [2-4]
   - Scrub clearance in **industrial sites** [2-4]; **land not intended for cropping, non-crop areas** (directed treatment) [1]
   - Stinging nettle in **established grassland, land not intended for cropping, non-crop areas** [1]; **forest, land not intended to bear vegetation** [2-4]
   - Woody weeds in **established grassland, land not intended for cropping, non-crop areas** (directed treatment) [1]; **forest, industrial sites, land not intended to bear vegetation** [2-4]

Specific Off-Label Approvals (SOLAs)
   - **asparagus** (directed treatment) (OLA 050530) Dec 2008 [4]

Efficacy guidance
   - Apply in grassland as spot treatment or overall foliage spray when weeds in active growth in spring or summer. Details of dose and timing vary with species. See label
   - Apply to woody weeds as summer foliage, winter shoot, basal bark, cut stump or tree injection treatment
   - Apply foliage spray in water when leaves fully expanded but not senescent
   - Do not spray in drought, in very hot or cold conditions
   - Control may be reduced if rain falls within 2 h of application
   - Control of rhododendron can be variable. If higher than 1.8 m cut stump treatment recommended. A follow-up shoot treatment may be required
   - See label for maximum concentrations when applying in oil, water or via watering can

Restrictions
   - Maximum number of treatments 1 per yr on non-crop land (as directed spray); 2 per yr on established grassland (including land not intended for cropping) and forestry
   - Do not apply to grass leys less than 1 yr old

FOR FULL CONDITIONS OF USE ALWAYS READ THE PRODUCT LABEL

- Do not drill kale, swedes, turnips, grass or mixtures containing clover within 6 wk of treatment. Allow at least 6 wk before planting trees
- Not to be used on food crops
- Do not apply through hand held rotary atomisers
- Not to be applied in or near water

Crop-specific information
- Latest use: 6 wk before replanting; 7 d before grazing
- HI grass 6 wk
- Uses on land not intended for cropping include grassland of no agricultural interest such as roadside verges, railway and motorway embankments
- Apply winter shoot, basal bark or cut stump sprays in paraffin or diesel oil. Dose and timing vary with species. See label for details
- Inject undiluted or 1:1 dilution into cuts spaced every 7.5 cm round trunk
- Clover will be killed or severely checked by application in grassland

Environmental safety
- Dangerous for the environment
- Very toxic to aquatic organisms
- LERAP Category B
- Do not allow spray to drift onto agricultural or horticultural crops, amenity plantings, gardens, ponds, lakes or water courses. Vapour drift may occur under hot conditions
- Keep livestock out of treated areas for at least 7 d and until foliage of any poisonous weeds such as buttercups or ragwort has died and become unpalatable

Hazard classification and safety precautions
Hazard H03, H11 [1-4]; H08 [1]
Risk phrases R22a, R50 [2-4]; R22b, R38, R43, R53a [1-4]; R51 [1]
Operator protection A, C, H, M; U02a, U05a, U08, U14, U20b [1-4]; U19a [1]
Environmental protection E07a (7 d); E15a, E16a, E16b, E23, E34, E38
Consumer protection C01
Storage and disposal D01, D02, D09a, D10b, D12a
Medical advice M05b

436   trifloxystrobin

A protectant strobilurin fungicide for cereals and managed amenity turf

See also cyproconazole + trifloxystrobin
fluoxastrobin + prothioconazole + trifloxystrobin
prothioconazole + trifloxystrobin
tebuconazole + trifloxystrobin

Products
1	Scorpio	Bayer Environ.	50% w/w	WG	12293
2	Swift SC	Bayer CropScience	500 g/l	SC	11227
3	Twist	Bayer CropScience	125 g/l	EC	11230

Uses
- Brown rust in **spring barley, winter barley, winter wheat** [2, 3]
- Fusarium patch in **amenity grass, managed amenity turf** [1]
- Net blotch in **spring barley, winter barley** [2, 3]
- Red thread in **amenity grass, managed amenity turf** [1]
- Rhynchosporium in **spring barley, winter barley** [2, 3]
- Septoria diseases in **winter wheat** [2, 3]

Approval information
- Trifloxystrobin included in Annex I under EC Directive 91/414
- Accepted by BBPA for use on malting barley
- Approval expiry 31 Mar 2006 [3]

Efficacy guidance
- Should be used protectively before disease is established in crop. Further treatment may be necessary if disease attack prolonged

SEE SECTION 3 FOR PRODUCTS ALSO REGISTERED

SECTION 2

- Treat grass after cutting and do not mow for at least 48 h afterwards to allow adequate systemic movement [1]
- Trifloxystrobin is a member of the QoI cross resistance group. Product should be used preventatively and not relied on for its curative potential
- Use product as part of an Integrated Crop Management strategy incorporating other methods of control, including where appropriate other fungicides with a different mode of action. Do not apply more than two foliar applications of QoI containing products to any cereal crop
- There is a significant risk of widespread resistance occurring in *Septoria tritici* populations in UK. Failure to follow resistance management action may result in reduced levels of disease control
- On cereal crops product must always be used in mixture with another product, recommended for control of the same target disease, that contains a fungicide from a different cross resistance group and is applied at a dose that will give robust control
- Strains of barley powdery mildew resistant to QoIs are common in the UK

Restrictions
- Maximum number of treatments 2 per crop per yr
- Do not apply to turf during dought conditions or to frozen turf [1]

Crop-specific information
- HI barley, wheat 35 d

Environmental safety
- Dangerous for the environment
- Very toxic to aquatic organisms
- Dangerous to fish or other aquatic life. Do not contaminate surface waters or ditches with chemical or used container
- Do not harvest for human or animal consumption for at least 35 days after last application
- LERAP category B [1]

Hazard classification and safety precautions
Hazard H04 [1, 3]; H11 [1-3]
Risk phrases R36 [3]; R43 [1, 3]; R50, R53a [1-3]
Operator protection A, H [1-3]; C [3]; U02a, U09a, U19a [2]; U05a [1, 2]; U11, U13, U15 [1]; U14 [1, 3]; U20b [1-3]
Environmental protection E13b [2, 3]; E15b, E16a, E16b, E34 [1]; E38 [1-3]
Consumer protection C02 [2, 3] (35 d)
Storage and disposal D01, D02, D09a, D12a [1-3]; D05, D07, D10b, D11a [3]; D10c [1, 2]
Medical advice M03 [1]

---

437 trifluralin

A soil-incorporated dinitroaniline herbicide for use in various crops

See also clodinafop-propargyl + trifluralin
diflufenican + trifluralin
isoproturon + trifluralin
isoxaben + trifluralin

Products

1	Alpha Trifluralin 48 EC	Makhteshim	480 g/l	EC	07406
2	Tandril 48	DAPT	480 g/l	EC	10643
3	Treflan	Dow	480 g/l	EC	05817
4	Triflurex 48 EC	Makhteshim	480 g/l	EC	07947
5	Trimaran	Nufarm UK	480 g/l	EC	11400

Uses

- Annual dicotyledons in **broad beans, broccoli, brussels sprouts, cabbages, carrots, cauliflowers, french beans, kale, lettuce, parsnips, raspberries, runner beans, strawberries, swedes, turnips, winter barley, winter wheat** [1-5]; **calabrese** [1, 3-5]; **celeriac** *(off-label)*, **evening primrose** *(off-label)*, **ornamental plant production** *(off-label)*, **parsley, sugar beet** [3]; **chives** *(off-label)*, **combining peas** *(off-label)*, **herbs (see appendix 6)** *(off-label)*, **kohlrabi** *(off-label)*, **linseed** *(off-label)*, **nursery fruit trees and bushes** *(off-label)*, **ornamental specimens** *(off-label)*, **parsley** *(off-label)*, **radish seed crops** *(off-label)*, **soya beans** *(off-label)*, **sunflowers** *(off-label)* [1, 3]; **linseed** [1]; **mustard** [1-3]; **navy beans** [3, 5]; **spring field beans, winter field**

SECTION 2

beans [1, 2]; *spring linseed* [3-5]; *spring oilseed rape, winter oilseed rape* [1-4]; *winter linseed* [3, 4]

- Annual grasses in *broad beans, broccoli, brussels sprouts, cabbages, carrots, cauliflowers, french beans, kale, lettuce, parsnips, raspberries, runner beans, strawberries, swedes, turnips, winter barley, winter wheat* [1-5]; *calabrese* [1, 3-5]; *celeriac* (off-label), *evening primrose* (off-label), *ornamental plant production* (off-label), *parsley, sugar beet* [3]; *chives* (off-label), *combining peas* (off-label), *herbs (see appendix 6)* (off-label), *kohlrabi* (off-label), *linseed* (off-label), *nursery fruit trees and bushes* (off-label), *ornamental specimens* (off-label), *parsley* (off-label), *radish seed crops* (off-label), *soya beans* (off-label), *sunflowers* (off-label) [1, 3]; *linseed* [1]; *mustard* [1-3]; *navy beans* [3, 5]; *spring field beans, winter field beans* [1, 2]; *spring linseed* [3-5]; *spring oilseed rape, winter oilseed rape* [1-4]; *winter linseed* [3, 4]
- Fat hen in *chives* (off-label), *combining peas* (off-label), *herbs (see appendix 6)* (off-label), *kohlrabi* (off-label), *linseed* (off-label), *nursery fruit trees and bushes* (off-label), *ornamental specimens* (off-label), *parsley* (off-label), *radish seed crops* (off-label), *soya beans* (off-label), *sunflowers* (off-label) [1]

Specific Off-Label Approvals (SOLAs)
- *celeriac, combining peas, linseed, ornamental specimens, sunflowers* (OLA 981560) Dec 2008 [3]
- *chives, combining peas, herbs (see appendix 6), kohlrabi, linseed, nursery fruit trees and bushes, ornamental specimens, parsley, radish seed crops, soya beans, sunflowers* (OLA 012763) Dec 2008 [1]
- *chives, herbs (see appendix 6), nursery fruit trees and bushes, ornamental plant production, parsley, radish seed crops, soya beans* (OLA 930074) Dec 2008 [3]
- *evening primrose* (OLA 921080) Dec 2008 [3]
- *kohlrabi* (OLA 982341) Dec 2008 [3]

Approval information
- Accepted by BBPA for use on malting barley

Efficacy guidance
- Acts on germinating weeds and requires soil incorporation to 5 cm (10 cm for crops to be grown on ridges) within 30 min of spraying. See label for details of suitable application equipment
- Apply and incorporate at any time during 2 wk before sowing or planting
- Best results achieved by application to fine, firm seedbed, free of clods, crop residues and established weeds
- In winter cereals normally applied as surface treatment without incorporation in tank-mixture with other herbicides to increase spectrum of control. See label for details
- Follow-up herbicide treatment recommended with some crops. See label for details

Restrictions
- Maximum number of treatments 1 per crop or per yr
- Do not apply to brassica plant raising beds
- Minimum interval between application and drilling or planting may be up to 12 mth. See label for details
- Do not use on sand, fen soil or soils with more than 10% organic matter

Crop-specific information
- Latest use: varies between products; check labels. Normally before 4 leaves unfolded (GS 13) for cereals; up to 6, 8 or 10 leaves for sugar beet; pre-sowing/planting for other crops
- Transplants should be hardened off prior to transplanting
- Apply in sugar beet after plants 10 cm high with 4-8 leaves and harrow into soil
- Apply in cereals after drilling up to and including 3 leaf stage (GS 13)

Environmental safety
- Dangerous for the environment
- Very toxic to aquatic organisms
- Harmful to fish or other aquatic life. Do not contaminate surface waters or ditches with chemical or used container
- Keep livestock out of treated areas for at least two weeks following treatment

SEE SECTION 3 FOR PRODUCTS ALSO REGISTERED

Hazard classification and safety precautions
   Hazard H03, H08 [1-5]; H04 [2]; H11 [1, 3-5]
   Risk phrases R20, R22a, R66 [5]; R22b [1-5]; R36 [1, 2, 4, 5]; R37, R43, R67 [3, 5]; R38 [2]; R50, R53a [1, 3-5]
   Operator protection A, C [1, 2, 4]; U05a [1, 2, 4, 5]; U08, U13, U20b [1-5]; U11 [1, 4, 5]; U14 [3, 5]; U19a [3]
   Environmental protection E07a [5]; E13c [2]; E15a, E34 [1, 4, 5]; E38 [3]
   Storage and disposal D01 [1, 2, 4, 5]; D02, D07 [1, 2, 4]; D05, D06a [2, 5]; D06b [3]; D09a, D10b [1-5]; D12a [1, 3, 4]
   Medical advice M05b

438   triflusulfuron-methyl

A sulfonyl urea herbicide for beet crops

See also lenacil + triflusulfuron-methyl

Products
   Debut   DuPont   50% w/w   WG   07804

Uses
   • Annual dicotyledons in **chicory** *(off-label)*, **fodder beet**, **sugar beet**
   • Charlock in **red beet** *(off-label)*
   • Cleavers in **red beet** *(off-label)*
   • Flixweed in **red beet** *(off-label)*
   • Fool's parsley in **red beet** *(off-label)*
   • Nipplewort in **red beet** *(off-label)*
   • Ox-eye daisy in **red beet** *(off-label)*

Specific Off-Label Approvals (SOLAs)
   • **chicory** *(OLA 050937) Dec 2008* [1]
   • **red beet** *(OLA 023629) Dec 2008* [1]

Efficacy guidance
   • Product should be used with a recommended adjuvant or a suitable herbicide tank-mix partner - see label for details
   • Product acts by foliar action. Best results obtained from good spray cover of small actively growing weeds
   • Susceptible weeds cease growth immediately and symptoms can be seen 5-10 d later
   • Best results achieved from a programme of up to 4 treatments starting when first weeds have emerged with subsequent applications every 5-14 d when new weed flushes at cotyledon stage
   • Weed spectrum can be broadened by tank mixture with other herbicides. See label for details
   • Product may be applied overall or via band sprayer
   • Triflusulfuron-methyl is a member of the ALS-inhibitor group of herbicides

Restrictions
   • Maximum number of treatments 4 per crop
   • Do not apply to any crop stressed by drought, water-logging, low temperatures, pest or disease attack, nutrient or lime deficiency

Crop-specific information
   • Latest use: before crop leaves meet between rows
   • HI 4 wk for red beet
   • All varieties of sugar beet and fodder beet may be treated from early cotyledon stage until the leaves begin to meet between the rows

Following crops guidance
   • Only winter cereals should follow a treated crop in the same calendar yr. Any crop may be sown in the next calendar yr
   • After failure of a treated crop, sow only spring barley, linseed or sugar beet within 4 mth of spraying unless prohibited by tank-mix partner

Environmental safety
   • Dangerous for the environment
   • Very toxic to aquatic organisms

FOR FULL CONDITIONS OF USE ALWAYS READ THE PRODUCT LABEL

- Extremely dangerous to fish or other aquatic life. Do not contaminate surface waters or ditches with chemical or used container
- LERAP Category B
- Take extreme care to avoid drift onto broad-leaved plants outside the target area or onto surface waters or ditches, or land intended for cropping
- Spraying equipment should not be drained or flushed onto land planted, or to be planted, with trees or crops other than sugar beet and should be thoroughly cleansed after use - see label for instructions

Hazard classification and safety precautions
Hazard H11
Risk phrases R50, R53a
Operator protection A; U05a, U08, U19a, U20a
Environmental protection E13a, E15a, E16a, E16b, E38
Storage and disposal D01, D02, D05, D09a, D11a, D12a

439   trinexapac-ethyl

A novel cyclohexanecarboxylate plant growth regulator for cereals, turf and amenity grassland

Products

1 Moddus	Syngenta	250 g/l	EC	08801
2 Pan Tepee	Pan Agriculture	250 g/l	EC	12279
3 Primo Maxx	Scotts	121 g/l	SL	11878
4 Shortcut	Scotts	25% w/w	WB	09254

Uses

- Growth retardation in *amenity grass, managed amenity turf* [3, 4]
- Lodging control in *durum wheat, ryegrass seed crops, spring barley, spring oats, spring rye, triticale, winter barley, winter oats, winter rye, winter wheat* [1, 2]; *spring wheat* (off-label - cv A C Barrie only) [1]

Specific Off-Label Approvals (SOLAs)

- *spring wheat* (cv A C Barrie only) (OLA 042300) Dec 2008 [1]

Approval information

- Accepted by BBPA for use on malting barley

Efficacy guidance

- Best results on cereals and ryegrass seed crops obtained from treatment from the leaf sheath erect stage [1, 2]
- Best results on turf achieved from application to actively growing weed free turf grass that is adequately fertilized and watered and is not under stress. Adequate soil moisture is essential [3, 4]
- Turf should be dry and weed free before application [3, 4]
- Environmental conditions, management and cultural practices that affect turf growth and vigour will influence effectiveness of treatment [3, 4]
- Repeat treatments up to the maximum approved dose may be made as soon as turf growth resumes [3, 4]

Restrictions

- Maximum total dose equivalent to one full dose on cereals, ryegrass seed crops
- Maximum number of treatments on turf equivalent to five full dose treatments
- Do not apply if rain or frost expected or if crop wet. Products are rainfast after 12 h
- Only use on crops at risk of lodging [1, 2]
- Do not apply within 12 h of mowing [3, 4]
- Not recommended for closely mown fine turf [4]
- Do not treat newly sown turf [3, 4]
- Not to be used on food crops [3, 4]
- Do not compost or mulch grass clippings [3, 4]

Crop-specific information

- Latest use: before 2nd node detectable (GS 32) for oats, ryegrass seed crops; before 3rd node detectable (GS 33) for durum wheat, spring barley, triticale, rye; before flag leaf sheath extending (GS 41) for winter barley, winter wheat

- On wheat apply as single treatment between leaf sheath erect stage (GS 30) and flag leaf fully emerged (GS 39) [1, 2]
- On barley, rye, triticale and durum wheat apply as single treatment between leaf sheath erect stage (GS 30) and second node detectable (GS 32), or on winter barley at higher dose between flag leaf just visible (GS 37) and flag leaf fully emerged (GS 39) [1, 2]
- On oats and ryegrass seed crops apply between leaf sheath erect stage (GS 30) and first node detectable stage (GS 32) [1, 2]
- Treatment may cause ears of cereals to remain erect through to harvest [1, 2]
- Turf under stress when treated may show signs of damage [3, 4]
- Any weed control in turf must be carried out before application of the growth regulator [3, 4]

Environmental safety
- Dangerous for the environment
- Toxic to aquatic organisms
- Harmful to fish or other aquatic life. Do not contaminate surface waters or ditches with chemical or used container
- Avoid drift outside target area

Hazard classification and safety precautions
    Hazard H04, H11 [1, 2]
    Risk phrases R43, R51 [1, 2]; R53a [1, 2, 4]
    Operator protection A [1-4]; C [1-3]; H, K [3]; U05a [1-4]; U15, U20c [1, 2]; U20b [3, 4]
    Environmental protection E13c [4]; E15a [1, 2]; E38 [1, 2, 4]
    Consumer protection C01 [3, 4]
    Storage and disposal D01, D02, D05 [1-4]; D09a, D12a [1, 2, 4]; D10a [4]; D10c [1-3]

440   triticonazole

A conazole fungicide available only in mixtures

See also imazalil + triticonazole
*   prochloraz + triticonazole*

441   urea (commodity substance)

A fungicide for use in forestry

Products

urea	various	37% w/v	SL

Uses
- Fungus diseases in *cut stumps*

Approval information
- Approval for the use of urea as a commodity substance was granted on 25 July 2003 by Ministers under regulation 5 of the Control of Pesticides Regulations 1986

Efficacy guidance
- For the treatment of cut stumps of trees, the following dyes may be used at 0.04% w/v concentration when mixed with solution of urea: Kenacid Turquoise V5898 (CI Acid Blue 42045), Denacid Turquoise AN 200 (CI Acid Blue 9 42090), Duasyn Acid Blue AE-20 (CI Acid Blue 9 42090)

Restrictions
- Maximum number of treatments: 1 per stump per yr
- Operators should wear rubber or other chemical-proof gloves, a nuisance dust mask and a cotton/terylene overall when handling dye powder

Hazard classification and safety precautions
    Operator protection A

FOR FULL CONDITIONS OF USE ALWAYS READ THE PRODUCT LABEL

442   Verticillium lecanii

A fungal parasite of aphids and whitefly

Products

1	Mycotal	Koppert	16.1% w/w	WP	04782
2	Vertalec	Koppert	2.5% w/w	WP	04781

Uses

- Aphids in **aubergines, protected beans, protected chrysanthemums, protected cucumbers, protected lettuce, protected ornamentals, protected peppers, protected roses, protected tomatoes** [2]
- Whitefly in **aubergines, protected beans, protected cucumbers, protected lettuce, protected ornamentals, protected peppers, protected tomatoes** [1]

Efficacy guidance

- *Verticillium lecanii* is a pathogenic fungus that infects the target pests and destroys them
- Apply spore powder as spray as part of biological control programme keeping the spray liquid well agitated
- Pre-soak the product for 2-4 h before application to rehydrate the spores and assist in dispersion
- Treat before infestations build to high levels and repeat as directed on the label
- Spray during late afternoon and early evening directing spray onto underside of leaves and to growing points
- Best results require minimum 80% relative humidity and 18°C within the crop canopy
- Product highly infective to many aphid species except the chrysanthemum aphid. Follow specific label directions for this pest [2]

Restrictions

- Never use in tank mixture
- Do not use a fungicide within 3 d of treatment. Pesticides containing captan, chlorothalonil, fenarimol, dichlofluanid, imazalil, maneb, prochloraz, quinomethionate, thiram or tolylfluanid may not be used on the same crop
- Keep in a refrigerated store at 2-6°C but do not freeze

Environmental safety

- Products have negligible effects on commercially available natural predators or parasites but consult manufacturer before using with a particular biological control agent for the first time

Hazard classification and safety precautions

Operator protection U19a, U20b
Environmental protection E15a
Storage and disposal D09a, D11a

443   vinclozolin

A protectant dicarboximide fungicide

Products

1	Ronilan FL	BASF	500 g/l	SC	02960
2	Standon Vinclozolin	Standon	500 g/l	SC	07836

Uses

- Alternaria in **spring oilseed rape, winter oilseed rape** [1, 2]
- Ascochyta in **combining peas** [1, 2]
- Blossom wilt in **apples** [1]
- Botrytis in **combining peas, dwarf beans, navy beans, runner beans, spring oilseed rape, vining peas, winter oilseed rape** [1, 2]; **daffodils** *(off-label - for galanthamine production)* [1]
- Chocolate spot in **broad beans, spring field beans, winter field beans** [1, 2]
- Mycosphaerella in **combining peas** [1, 2]
- Sclerotinia stem rot in **spring oilseed rape, winter oilseed rape** [1, 2]

Specific Off-Label Approvals (SOLAs)

- **daffodils** *(for galanthamine production) (OLA 041517) Dec 2008* [1]

SEE SECTION 3 FOR PRODUCTS ALSO REGISTERED

SECTION 2

Efficacy guidance
- Timing of sprays varies with crop and disease. See label for details
- May be used at reduced rate on peas and field beans in tank-mix with chlorothalonil
- Where dicarboximide resistant strains have developed product may not be effective

Restrictions
- Maximum number of treatments 2 per crop or per yr
- Do not treat mange-tout varieties of peas
- Do not spray if crop wet or if rain or frost expected
- Operator must use a vehicle fitted with a cab and forced air filtration unit with a pesticide filter complying with HSE Guidance Note PM 74 or to an equally effective standard

Crop-specific information
- Before end of petal fall for apples
- HI beans, peas 2 wk; daffodils 4 wk; oilseed rape 7 wk

Environmental safety
- Dangerous for the environment
- Toxic to aquatic organisms
- Product presents a minimal hazard to bees when used as directed but consider informing local bee-keepers if intending to spray crops in flower

Hazard classification and safety precautions
Hazard H02 [1, 2]; H11 [2]
Risk phrases R40, R43, R53a, R60, R61 [1, 2]; R51 [2]; R52 [1]
Operator protection A, C, H, K [1, 2]; M [1]; U05a, U19a [1, 2]; U08, U14, U20b [1]; U09a, U20a [2]
Environmental protection E15a [1, 2]; E34 [2]
Storage and disposal D01, D02, D09a, D10c [1, 2]; D05, D10b [2]
Medical advice M04

444   warfarin

A hydroxycoumarin rodenticide

Products

1	Grey Squirrel Liquid Concentrate	Killgerm	0.5% w/w	CB	06455
2	Sakarat Ready-to-Use (Cut Wheat Base)	Killgerm	0.025% w/w	RB	H6807
3	Sakarat Ready-to-Use (Whole Wheat)	Killgerm	0.025% w/w	RB	H6808
4	Sakarat X	Killgerm	0.05% w/w	RB	H6809
5	Sewarin Extra	Killgerm	0.05% w/w	RB	H6810
6	Sewarin P	Killgerm	0.025% w/w	RB	H6811
7	Sewercide Cut Wheat Rat Bait	Killgerm	0.05% w/w	RB	H6805
8	Sewercide Whole Wheat Rat Bait	Killgerm	0.05% w/w	RB	H6806
9	Warfarin 0.5% Concentrate	B H & B	0.5% w/w	CB	H6815
10	Warfarin Ready Mixed Bait	B H & B	0.025% w/w	RB	H6816

Uses
- Grey squirrels in **agricultural premises**, **forest**, **industrial sites** [1]
- Rats in **agricultural premises** [2-10]

Approval information
- Warfarin included in Annex I under EC Directive 91/414
- Product not approved for use in N Ireland [1]

Efficacy guidance
- For rodent control place ready-to-use or prepared baits at many points wherever rats active. Out of doors shelter bait from weather
- Inspect baits frequently and replace or top up as long as evidence of feeding. Do not underbait
- For grey squirrel control mix with whole wheat and leave to stand for 2-3 h before use
- Use bait in specially constructed hoppers and inspect every 2-3 d. Replace as necessary

FOR FULL CONDITIONS OF USE ALWAYS READ THE PRODUCT LABEL

Restrictions
- For use only by local authorities, professional operators providing a pest control service and persons occupying industrial, agricultural or horticultural premises
- For use only between 15 Mar and 15 Aug for tree protection [1]
- Must not be used outdoors at all in Scotland, nor in areas of England or Wales where pine martens occur naturally [1]

Environmental safety
- Prevent access to baits by children and animals, especially cats, dogs and pigs
- Rodent bodies must be searched for and burned or buried, not placed in refuse bins or rubbish tips. Remains of bait and containers must be removed after treatment and burned or buried
- Bait must not be used where food, feed or water could become contaminated
- The use of warfarin to control grey squirrels is illegal unless the provisions of the Grey Squirrels Order 1973 are observed. See label for list of counties in which bait may not be used [1]

Hazard classification and safety precautions
Operator protection A, C, D, E, H [2-8]; U13 [1-10]; U20a [1-6]; U20b [7-10]
Storage and disposal D09a [1-10]; D10a [9, 10]; D11a [1-8]
Vertebrate/rodent control products V01a, V03a, V04a [1, 4, 7-10]; V01b, V03b, V04b [2, 3, 5, 6]; V02 [1-10]; V05 [1]
Medical advice M03 [2, 3, 5, 6]

445   zeta-cypermethrin

A contact and stomach acting pyrethroid insecticide

Products

1 Fury 10 EW	Belchim	100 g/l	EW	11608
2 Fury 10 EW	Belchim	100 g/l	EW	12248
3 Minuet EW	Belchim	100 g/l	EW	11610
4 Minuet EW	Belchim	100 g/l	EW	12304

Uses
- Aphids in *spring barley, spring oats, spring wheat, winter barley, winter oats, winter wheat*
- Barley yellow dwarf virus vectors in *spring barley, spring wheat, winter barley, winter wheat*
- Cabbage seed weevil in *spring oilseed rape, winter oilseed rape*
- Cabbage stem flea beetle in *spring oilseed rape, winter oilseed rape*
- Cutworms in *potatoes, sugar beet*
- Flax flea beetle in *linseed*
- Flea beetle in *spring oilseed rape, winter oilseed rape*
- Large flax flea beetle in *linseed*
- Pea and bean weevil in *combining peas, spring field beans, vining peas, winter field beans*
- Pea aphid in *combining peas, vining peas*
- Pea moth in *combining peas, vining peas*
- Pod midge in *spring oilseed rape, winter oilseed rape*
- Pollen beetle in *spring oilseed rape, winter oilseed rape*
- Rape winter stem weevil in *spring oilseed rape, winter oilseed rape*

Approval information
- Following implementation of Directive 98/82/EC, approval for use of zeta-cypermethrin on numerous crops was revoked in 1999
- Accepted by BBPA for use on malting barley
- Approval expiry 31 Jul 2006 [1, 3]

Efficacy guidance
- On winter cereals spray when aphids first found in the autumn for BYDV control. A second spray may be required on late drilled crops or in mild conditions
- For summer aphids on cereals spray when treatment threshold reached
- For listed pests in other crops spray when feeding damage first seen or when treatment threshold reached. Under high infestation pressure a second treatment may be necessary
- Best results for pod midge and seed weevil control in oilseed rape obtained from treatment after pod set but before 80% petal fall

SEE SECTION 3 FOR PRODUCTS ALSO REGISTERED

- Pea moth treatments should be applied according to ADAS/PGRO warnings or when economic thresholds reached as indicated by pheromone traps
- Treatments for cutworms should be made at egg hatch and repeated no sooner than 10 d later

Restrictions
- Maximum number of treatments 2 per crop
- Consult processors before use on crops for processing

Crop-specific information
- Latest use: before end of flowering for oilseed rape; before flowering completed (GS 69) for cereals
- HI potatoes, field beans 14 d; sugar beet 60 d

Environmental safety
- Dangerous for the environment
- Very toxic to aquatic organisms
- High risk to non-target insects or other arthropods. Do not spray within 6 m of the field boundary
- Risk to certain non-target insects or other arthropods. See directions for use
- LERAP Category A

Hazard classification and safety precautions
> **Hazard** H03, H11
> **Risk phrases** R20, R22a, R43, R50, R53a
> **Operator protection** A, C, H; U05a, U08, U14, U15, U19a, U20b
> **Environmental protection** E15a, E16c, E16d, E22a, E34, E38
> **Storage and disposal** D01, D02, D09a, D10b, D12a
> **Medical advice** M05a

446   ziram

A dithiocarbamate bird and animal repellent

Products

AAprotect	Unicrop	32% w/w	PA	03784

Uses
- Birds in *all top fruit, field crops, forest, ornamental specimens*
- Deer in *all top fruit, field crops, forest, ornamental specimens*
- Hares in *all top fruit, field crops, forest, ornamental specimens*
- Rabbits in *all top fruit, field crops, forest, ornamental specimens*

Efficacy guidance
- Apply undiluted to main stems up to knee height to protect against browsing animals at any time of yr or spray 1:1 dilution on stems and branches in dormant season
- Use dilute spray on fully dormant fruit buds to protect against bullfinches
- Only apply to dry stems, branches or buds
- Use of diluted spray can give limited protection to field crops in areas of high risk during establishment period

Restrictions
- Maximum number of treatments 3 per season as spray for all top fruit, field crops, forest and ornamental specimens; 1 per season as paste treatment on trees and ornamental specimens
- Do not spray elongating shoots or buds about to open
- Do not apply concentrated spray to foliage, fruit buds or field crops

Crop-specific information
- HI edible crops 8 wk

Environmental safety
- Harmful to fish or other aquatic life. Do not contaminate surface waters or ditches with chemical or used container

Hazard classification and safety precautions
> **Hazard** H04
> **Risk phrases** R36, R37, R38
> **Operator protection** A, C; U05a, U08, U19a, U20b
> **Environmental protection** E13c

FOR FULL CONDITIONS OF USE ALWAYS READ THE PRODUCT LABEL

Consumer protection C02 (8 wk)
Storage and disposal D01, D02, D09a, D10c

447   zoxamide

A substituted benzamide available only in mixtures

See also mancozeb + zoxamide

SEE SECTION 3 FOR PRODUCTS ALSO REGISTERED

SECTION 3
PRODUCTS ALSO REGISTERED

Products also Registered

Products listed in the table below have not been notified for inclusion in Section 2 of this edition of the *Guide*. However they have extant approval until 31 December 2008 unless an earlier expiry date is shown. These products may legally be stored and used in accordance with their label until their approval expires, but they may not still be available for purchase.

Product	Approval holder	Reg. No.	Expiry Date
abamectin			
Dynamec	Syngenta	12539	
acetic acid			
Natural Weed Spray No. 1	Punya	12328	
acibenzolar-S-methyl			
Bion	Syngenta	09803	
aldicarb			
Agriguard Aldicarb	Tronsan	10481	31-12-07
Me2 New Aldee	Me2	10434	31-12-07
Standon Aldicarb 10G	Standon	11024	31-12-07
alpha-cypermethrin			
Alert	BASF	10272	
Alpha C 6 ED	Techneat	11838	
Alphaguard 100 EC	Nufarm UK	12003	31-08-06
Alphathrin	Nufarm UK	11163	31-08-06
Alphatop 100 EC	Interfarm	12102	31-08-06
Bestseller 100 EC	Chimac-Agriphar	10461	31-08-06
Cleancrop Acymet	United Agri	10497	31-08-06
Cleancrop Scope	United Agri	10273	
Fastac	BASF	10220	
Fastac Dry	BASF	10221	
I T Alpha-Cyper	I T Agro	10483	31-08-06
aluminium phosphide			
Degesch Fumigation Pellets	Rentokil	11436	
Detia Gas Ex-P	Igrox	09802	
amidosulfuron			
Barclay Cleave	Barclay	11340	
2-aminobutane			
Hortichem 2-Aminobutane	Certis	06147	31-12-07
amitraz			
Bye Bye 20 EC	Chimac-Agriphar	09346	
Cleancrop Bye Bye 20EC	United Agri	11530	
Mitac HF	Bayer CropScience	07358	31-12-07

Product	Approval holder	Reg. No.	Expiry Date
amitrole			
Aminotriazole Technical	Nufarm UK	10855	
Loft	Marks	06030	
Weedazol Pro	Nufarm UK	11995	01-01-07
Weedazol-TL	Marks	02349	
Weedazol-TL	Bayer CropScience	11430	01-01-07
asulam			
Agrotech Asulam	Agrotech-Trading	12243	
Cleancrop Asulam	United Agri	10465	
I T Asulam	I T Agro	10186	
Milentus Asulam	Milentus	12245	
atrazine			
Alpha Atrazine 50 WP	Makhteshim	04793	
Atrazol	Sipcam	07598	31-12-07
DG90	Sipcam	11200	31-12-07
Unicrop Atrazine 50	Unicrop	02645	
Unicrop Atrazine FL	Unicrop	08045	
azaconazole + imazalil			
Nectec Paste	Certis	08510	31-12-07
azoxystrobin			
5504	Syngenta	12351	01-07-08
Barclay ZX	Barclay	11336	
Clayton Stobik	Clayton	09440	01-07-08
Olympus	Syngenta	10541	01-07-08
Ortiva	Syngenta	10542	01-07-08
Priori	Syngenta	10543	01-07-08
azoxystrobin + chlorothalonil			
Amistar Opti	Syngenta	11863	30-11-06
Olympus	Syngenta	12517	
benfuracarb			
Oncol 10G	Mirfield	08249	31-03-06
bentazone			
Basagran	BASF	00188	
Milentus Bentazone	Milentus	12257	01-08-06
Standon Bentazone	Standon	09204	01-08-06
Troy 480	Agrichem	12341	31-07-11
Troy 480	PG - Crop Protection	12213	31-10-06
bentazone + dichlorprop-P			
Quitt SL	BASF	08108	
benzoic acid			
Menno Florades	Menno	12472	31-05-14

Product	Approval holder	Reg. No.	Expiry Date
beta-cyfluthrin + imidacloprid			
Chinook	Bayer CropScience	11421	
bifenox + isoproturon			
RP 4169	Makhteshim	11190	
bifenox + MCPA + mecoprop-P			
Sirocco	Makhteshim	11192	
bifenthrin + malathion			
Prostore 420EC	Nickerson	12210	
bitertanol + fuberidazole			
Sibutol New Formula	Bayer CropScience	11307	
UK 743	Bayer CropScience	11315	
boscalid			
Me2 Succotash	Me2	12147	24-11-06
bromoxynil + clopyralid			
Vindex	Dow	05470	31-07-06
bromoxynil + diflufenican + ioxynil			
Capture	Nufarm UK	12514	
bromoxynil + fluroxypyr			
Tomahawk Plus	Makhteshim	09836	
bromoxynil + fluroxypyr + ioxynil			
Advance	Dow	05173	
Treble	Barclay	10577	
bromoxynil + ioxynil			
Deloxil	Bayer CropScience	09987	
Percept	MTM Agrochem.	05481	
Stellox	Nufarm UK	11551	
bromoxynil + ioxynil + triasulfuron			
Teal	Nufarm UK	11940	
bromoxynil + terbuthylazine			
Alpha Bromotril PT	Makhteshim	09435	
Cleancrop Amaize	United Agri	11990	
Templar	Makhteshim	10254	
bromuconazole			
Granit	Bayer CropScience	09995	
bupirimate			
Nimrod	Syngenta	10530	
captan			
Alpha Captan 50 WP	Makhteshim	04797	

SECTION 3

Product	Approval holder	Reg. No.	Expiry Date
PP Captan 80 WG	Calliope SAS	12435	
PP Captan 83	Calliope SAS	12330	
PP Captan 83	Tomen	08768	

captan + penconazole

Topas C	Syngenta	08459	
Topas C 50 WP	Novartis	08459	

carbendazim

Barclay Shelter	Barclay	09000	
BASF Turf Systemic Fungicide	BASF	05774	
Bavistin	BASF	00217	
Bavistin DF	BASF	03848	
Bavistin FL	BASF	00218	
Clayton Am-Carb	Clayton	11906	
Clayton Chizm FL	Clayton	11050	
Cleancrop Curve	United Agri	11774	
Goldazim 500 SC	Chimac-Agriphar	10675	
Headland Addstem	Headland	06755	
Headland Addstem DF	Headland	08904	
I.T. Carbendazim	I T Agro	12155	
Mascot Systemic	Scotts	09132	
Mascot Systemic	Rigby Taylor	08776	30-11-06
Occidor 500 SC	Chimac-Agriphar	10937	
S.M.I Carbendazim	Sub-Micron	12040	
Tripart Defensor FL	Tripart	02752	
Turf Systemic Fungicide	SumiAgro	10821	
Turf Systemic Fungicide	PBI	09349	

carbendazim + chlorothalonil

Bravocarb	Syngenta	10520	
Bravocarb	Zeneca	09105	
Greenshield	Scotts	07988	

carbendazim + flusilazole

Landgold Flusilazole MBC	Landgold	08528	
Standon Flusilazole Plus	Standon	07403	

carbendazim + iprodione

Calidan	BASF	11687	

carbendazim + maneb

Tripart 147	Tripart	07978	

carbendazim + prochloraz

Novak	BASF	11693	
Sportak Alpha HF	BASF	11698	

carbendazim + tebuconazole

Tricur	Bayer CropScience	11314	

Product	Approval holder	Reg. No.	Expiry Date
carbendazim + vinclozolin			
Konker	BASF	03988	
carbosulfan			
Marshal 10G	Belchim	11682	
carfentrazone-ethyl			
Aurora 40 WG	Belchim	11614	
Platform	Belchim	11615	
Spotlight Plus	Belchim	12436	24-08-08
carfentrazone-ethyl + mecoprop-P			
Platform S	FMC	10726	
carfentrazone-ethyl + thifensulfuron-methyl			
Harmony Express	DuPont	09467	
chloridazon			
Cleancrop AKAZON	United Agri	11903	
Gladiator DF	Tripart	06342	
Lidazone 65 WG	Globachem	10847	
Luxan Chloridazon	Luxan	06304	
Parador	United Phosphorus	12310	30-04-06
Portman Weedmaster	Agform	06018	
Pyramin FL	BASF	11628	
Questar	BASF	11629	
Starter Flowable	Truchem	03421	
Tripart Gladiator 2	Tripart	06618	
Weedmaster SC	Agform	08793	
chloridazon + ethofumesate			
Magnum	BASF	08635	
chloridazon + metamitron			
Volcan Combi FL	Sipcam	11442	
chlormequat			
Agriguard 5C Chlormequat 460	AgriGuard	09851	
Agriguard Chlormequat 700	AgriGuard	09782	
Agriguard Chlormequat 760	AgriGuard	10290	
Alpha Chlormequat 460	Makhteshim	04804	
Alpha Pentagan	Makhteshim	04794	
Alpha Pentagan Extra	Makhteshim	04796	
Atlas 5C Quintacel	Nufarm UK	11130	
Barclay Holdup	Barclay	11365	
Barclay Holdup	Barclay	06799	
Barclay Holdup 600	Barclay	11373	
Barclay Holdup 600	Barclay	08794	
Barclay Holdup 640	Barclay	08795	
Barclay Holdup 640	Barclay	11374	

SECTION 3

Product	Approval holder	Reg. No.	Expiry Date
Barclay Liffey	Barclay	11366	
Barclay Liffey	Barclay	09856	
Barclay Lucan	Barclay	09855	
Barclay Lucan	Barclay	11367	
Barclay Take 5	Barclay	08524	
Barleyquat B	Mandops	06001	
BASF 3C Chlormequat 600	BASF	04077	
BASF 3C Chlormequat 750	BASF	06878	
Belcocel	Taminco	11881	
Chlormequat 46	Nufarm UK	11504	
Ciba Chlormequat 460	Ciba Specialty	09525	
Ciba Chlormequat 5C 460:320	Ciba Specialty	09527	
Ciba Chlormequat 730	Ciba Specialty	09526	
Clayton CCC 750	Clayton	07952	
Clayton Manquat	Clayton	09916	
Clayton Squat	Clayton	12271	
Clayton Standup	Clayton	11760	
Cleancrop Chlormequat 700	United Agri	10143	
Cropsafe 5C Chlormequat	Certis	07897	
Cropsafe 5C Chlormequat	Certis	11179	
Hyquat 70	Agrichem	03364	
Intracrop Balance	Intracrop	08037	
Intracrop MCCC	Intracrop	08506	
Larke	Nufarm UK	11453	
Midget	Nufarm UK	12283	
MSS Mirquat	Mirfield	08166	
New 5C Cycocel	Cyanamid	01483	
Portman Chlormequat 400	Agform	01523	
Portman Chlormequat 460	Agform	02549	
Portman Chlormequat 700	Agform	03465	
Portman Supaquat	Agform	03466	
Stabilan 460	Nufarm UK	09304	
Stabilan 750	Nufarm UK	09303	
Supaquat 720	Agform	09381	
Terbine	Nufarm UK	11407	
Tripart 5C	Tripart	04726	
Tripart Brevis	Tripart	03754	
Tripart Brevis 2	Tripart	06612	
Tripart Chlormequat 460	Tripart	03685	
Uplift	United Phosphorus	07527	

chlormequat + 2-chloroethylphosphonic acid

Barclay Banshee XL	Barclay	11339	
Terpal C	BASF	07062	

chlormequat + 2-chloroethylphosphonic acid + imazaquin

Satellite	BASF	10395	

Product	Approval holder	Reg. No.	Expiry Date
chlormequat + imazaquin			
Upright	BASF	10404	
2-chloroethylphosphonic acid			
Agrotech Ethephon	Agrotech-Trading	12170	
Barclay Coolmore	Barclay	11349	
Dithane NT Dry Flowable	United Agri	09868	
Milentus Ethephon	Milentus	12185	
2-chloroethylphosphonic acid + mepiquat chloride			
Clayton Mepiquat	Clayton	12360	
CleanCrop Fonic M	United Agri	09553	
chlorothalonil			
Agriguard Chlorothalonil	Tronsan	09390	30-04-06
Barclay Corrib 500	Barclay	11350	
Bombardier	Unicrop	02675	
Bravo 720	Syngenta	10519	
Clayton Turret	Clayton	09400	
Clean Crop Wanderer	United Agri	12326	
Cleancrop Chlorothalonil 720	United Agri	10102	
Cleancrop Rover	United Agri	10500	
Cleancrop Rover 2	United Agri	11965	
Cracker	Tronsan	12149	
Expert Turf	AgriGuard	12571	
ISK 375	Zeneca	09103	
IT Chlorothalonil	I T Agro	12215	
Jupital	Zeneca	09109	
Jupital	Syngenta	10528	
Jupital DG	Zeneca	09181	
Mace	Nufarm UK	11957	
Mainstay	Quadrangle	05625	
Me2 Hooray	Me2	12546	
Mycoguard	Nufarm UK	12008	
Nuturf Chlorothalonil	Nufarm UK	11974	
Repulse	Zeneca	06705	
Repulse	Certis	07641	
Standon Chlorothalonil 500	Standon	08597	
Tripart Faber	Tripart	04549	
Ultrafaber	Tripart	05627	
Visclor 500 SC	Sipcam	09404	
Visclor 75 DF	Sipcam	09361	
Winner	Headland Amenity	12573	
chlorothalonil + cymoxanil			
Gex 44	Griffin	10168	
chlorothalonil + cyproconazole			
Bravo Xtra	Syngenta	11824	

SECTION 3

Product	Approval holder	Reg. No.	Expiry Date
Cleancrop Cyprothal	United Agri	09580	
Octolan	Syngenta	11675	
SAN 703	Syngenta	11676	

chlorothalonil + flutriafol

Argon	Headland	12312	
Prospa	Headland	11548	

chlorothalonil + mancozeb

Sipcam Flo	Sipcam	07601	

chlorothalonil + metalaxyl

Folio	Syngenta	08547	

chlorothalonil + propamocarb hydrochloride

Pan Wizard	Pan Agriculture	11953	

chlorothalonil + tetraconazole

Eminent Star	Isagro	10447	

chlorothalonil + vinclozolin

Curalan CL	BASF	07174	

chlorotoluron

Clayton Chloron	Clayton	08148	
Clayton Chloron 500 FL	Clayton	10791	
Luxan Chlorotoluron 500 Flowable	Luxan	09165	
NWA CTU 500	Nufarm UK	11140	
Talisman	FCC	03109	
Tripart Culmus	Tripart	06619	

chlorotoluron + pendimethalin

Totem	BASF	10485	

chlorpropham

Atlas CIPC 40	Atlas	07710	
Atlas Herbon Pabrac	Atlas	03997	
Atlas Herbon Pabrac	Atlas	07714	
Croptex Pewter	Certis	02507	
Jupiter 40 EC	Whyte Agrochemicals	12527	
Luxan Gro-Stop Innovator	Luxan	12340	
MSS CIPC 30M	Nufarm UK	10064	
Sprout Nip	Aceto	11786	
Triherbicide CIPC	Cerexagri	06874	
Whyte CIPC 40EC	Whyte Agrochemicals	11952	31-03-07

chlorpropham + fenuron

Croptex Chrome	Certis	02415	31-07-06

chlorpyrifos

Agriguard Chlorpyrifos	Tronsan	10626	

Product	Approval holder	Reg. No.	Expiry Date
Barclay Clinch II	Barclay	11346	
Choir	Nufarm UK	09778	
Cleancrop Pychlorex	United Agri	11681	
Crossfire 480	Dow	12516	
Crossfire 480	Dow	08141	
Dispatch	Dow	08139	
Equity	Dow	12465	
Lorsban 480	Dow	08076	
Lorsban WG	Dow	10139	
suSCon Indigo	Fargro	09902	

chlorthal-dimethyl

Dacthal W75	AMVAC	10289	
Dacthal W75	United Agri	10617	
Dacthal W-75	AMVAC	10623	28-02-06

cinidon-ethyl

Lotus	BASF	09231	30-09-12

clodinafop-propargyl

Cleancrop Theme	United Agri	12342	31-10-06
Viscount	Syngenta	12309	

clodinafop-propargyl + diflufenican

Amazon	Bayer CropScience	10266	
Amazon	Syngenta	08128	

clodinafop-propargyl + trifluralin

Hawk	Syngenta	12507	
Reserve	Syngenta	12077	

clomazone

Centium 360 CS	FMC	10720	
Centium 360 CS	FMC	10720	

clopyralid

Agrotech-Clopyralid 200 sl	Agrotech-Trading	12445	
Barclay Karaoke	Barclay	11357	
Cliophar	Chimac-Agriphar	09430	
Glopyr 200 SL	Globachem	10979	
Loncid	I T Agro	10832	
Lontrel 200	Dow	11558	
Milentus Clopyralid	Milentus	12448	

clopyralid + 2,4-D + MCPA

Esteem	Vitax	12555	

clopyralid + propyzamide

Matrikerb	Dow	10806	

Product	Approval holder	Reg. No.	Expiry Date
clothianidin			
Bayer UK 978	Bayer CropScience	11201	
Poncho 250	Bayer CropScience	12287	13-02-09
copper hydroxide			
Spin Out	DuPont	12069	
copper oxychloride + metalaxyl			
Ridomil Plus	Syngenta	08353	
copper silicate			
Socusil Slug Spray	Doff Portland	11817	
cyanazine			
Cleancrop Vectro	United Agri	10475	31-12-07
I T Cyanazine	I T Agro	10179	31-12-07
cycloxydim			
Stratos	BASF	06891	
cymoxanil			
Curzate 60DF	DuPont	11515	
Scribe 60WG	Syngenta	11971	
cymoxanil + mancozeb			
Clayton Krypton	Clayton	09398	
Cleancrop Cyman	United Agri	12232	
Cleancrop Xanilite	United Agri	10050	
Curzate M68 WSB	DuPont	08073	
Rhythm	Interfarm	09636	
Rhythm CM	Interfarm	11825	
Standon Cymoxanil Extra	Standon	09442	
Systol M	Quadrangle	08085	
Zetanil	Sipcam	11993	
cypermethrin			
Afrisect 10	Chimac-Agriphar	12027	
Arrivo	FMC	10731	
Brazil	Nufarm UK	11135	
Cleancrop Pyrimet	United Agri	10619	
Cyperguard 100 EC	Chimac-Agriphar	12227	
Cyperguard 100 EC	Nufarm UK	12039	
Cyperguard 100 EC	Gharda	10775	
Cyperkill 10	Chimac-Agriphar	12028	
Cyperkill 25	Chimac-Agriphar	12029	
Cyperkill 5	Chimac-Agriphar	12026	
I T Cyper	I T Agro	10557	
Mcc 25 Ec	Chimac-Agriphar	12030	
Nufarm Cypermethrin	Nufarm UK	12205	

Product	Approval holder	Reg. No.	Expiry Date
cyproconazole			
Alto 240 EC	Bayer CropScience	11236	
cyproconazole + quinoxyfen			
DOE 1762	Dow	08962	
cyprodinil			
Barclay Amtrak	Barclay	11338	
Cleancrop Cyprodinil	United Agri	09668	
Standon Cyprodinil	Standon	09345	
2,4-D			
Barclay Haybob II	Barclay	08532	
Damine	Chimac-Agriphar	12567	19-09-08
Dupont 24-D	DuPont	11737	
Easel	Nufarm UK	11146	
GroWell 2,4-D Amine	GroWell	11395	
Headland Staff	Headland	07189	
Herboxone 60	Marks	09693	
HY-D	Agrichem	06278	
Luxan 2,4-D	Luxan	09379	
Maton	Headland	10366	
MSS 2,4-D Amine	Nufarm UK	11139	
Ragox	Nufarm UK	11145	
2,4-D + dicamba			
Lawn Builder Plus Weed Control	Scotts	08499	
2,4-D + dicamba + triclopyr			
Cleancrop Broadshot	United Agri	11664	
2,4-D + MCPA			
Agroxone Combi	Marks	11025	
2,4-D + picloram			
Atladox HI	Nomix Enviro	05559	
daminozide			
B-Nine	Crompton	11465	
Dazide	Fine	02691	
dazomet			
Basamid	Kanesho	12418	
Basamid	BASF	00192	30-11-06
Basamid	Certis	07204	
2,4-DB			
Butoxone DB	Marks	10482	
2,4-DB + MCPA			
Butoxone DB Extra	Marks	12152	

Products also Registered

Product	Approval holder	Reg. No.	Expiry Date
Butoxone DB Extra	Marks	10455	31-03-06
Headland Cedar	Headland	12180	
MSS 2,4-DB + MCPA	Mirfield	01392	

deltamethrin

Agrotech Deltamethrin	Agrotech-Trading	12165	
Deleet	Rentokil	09312	
Delta-M 2.5 EC	O Endres	11334	
Milentus Deltamethrin	Milentus	12219	

desmedipham + ethofumesate + phenmedipham

Agrotech-Desmedipham Plus EC	Agrotech-Trading	12381	

dicamba

Cadence	Syngenta	08796	
Cadence	Barclay	09578	
I T Dicamba	I T Agro	10976	28-02-06

dicamba + MCPA + mecoprop-P

ALS Premier Selective Plus	Amenity Land	08940	
Mircam Super	Nufarm UK	11836	
Premier Amenity Selective	Amenity Land	10098	
Tribute	Nomix Enviro	06921	
Trireme	Nufarm UK	11524	

dicamba + mecoprop-P

Headland Swift	Headland	11945	
Optica Forte	Marks	11913	

dichlobenil

Casoron G	Rigby Taylor	09326	31-12-06
Casoron G	Crompton	09022	
Casoron G	Miracle	07926	
Casoron G4	Miracle	07927	
Casoron G4	Crompton	09215	
Casoron G-SR	Miracle	07925	
Dichlobenil Granules	Certis	11871	
Dicore	Agrichem	12542	
Osorno	Globachem	12486	
Scotts Dichlo G Macro	Scotts	12011	
Scotts Dichlo G Micro	Scotts	11857	
Sierraron G	Scotts	09675	
Sierraron G	Scotts	09263	
Sierraron G4	Scotts	10491	
Standon Dichlobenil 6G	Standon	08874	
Viking Granules	Nomix Enviro	11859	

dichlorophen

50/50 Liquid Mosskiller	Vitax	07191	31-12-07

Product	Approval holder	Reg. No.	Expiry Date
n2n Mosskiller	Nomix Enviro	11823	30-04-06
Nomix-Chipman Mosskiller	Nomix-Chipman	06271	
Panacide M	CMS Chemicals	12413	31-12-07
Panacide M	Coalite	05611	
Panacide TS	Coalite	05612	

dichlorophen + ferrous sulphate

Aitken's Lawn Sand Plus	Aitken	04542	31-12-07

1,3-dichloropropene

Telone 2000	Dow	05748	

dichlorprop-P

Headland Link	Headland	11091	
Optica DP	Marks	07818	
Optica DP	Marks	11067	

dichlorprop-P + ferrous sulphate + MCPA

Vitagrow Granular Feed, Weed & Mosskiller	Sinclair	10971	

dichlorprop-P + ioxynil

Duet	DuPont	11553	
Mextrol DP	Nufarm UK	11529	

dichlorprop-P + MCPA

Optica Duo	Marks	10343	

dichlorprop-P + MCPA + mecoprop-P

Optica Trio	Marks	09747	

dicloran

Fumite Dicloran Smoke	Certis	09291	30-04-06

difenoconazole

Difcor 250 EC	Globachem	12297	30-09-06
Landgold Difenoconazole	Landgold	09964	

diflubenzuron

Dimilin 25-WP	Crompton	08902	

diflufenican

Alliance Pro	Nufarm UK	12434	
Alpha DFF 500 SC	Makhteshim	12187	
DFF 50 WG	Agform	12362	
Diflan 500 SC	Q-Chem	12250	31-01-07
Diflanil 500 SC	Q-Chem	12489	
Diflufenican GL 500	Globachem	12531	
Hurricane SC	Makhteshim	12424	

diflufenican + flufenacet

Equinox	Bayer CropScience	12463	

Product	Approval holder	Reg. No.	Expiry Date
Regatta	Bayer CropScience	12054	
Zephyr	Bayer CropScience	12053	

diflufenican + flupyrsulfuron-methyl
Absolute	DuPont	12558	

diflufenican + isoproturon
Clayton Fenican 550	Clayton	11160	
Javelin Gold	Bayer CropScience	12536	
Panther WDG	Bayer CropScience	10650	

dimethoate
Danadim	Cheminova	09583	28-02-06
Sector	Cheminova	10492	

dimethomorph + mancozeb
Saracen	BASF	12005	

dimoxystrobin
BAS 505 04f	BASF	11926	27-01-07

diphenylamine
No-Scald DPA	Cerexagri	08312	

diquat
Clayton Diquat	Clayton	10730	
Standon Diquat	Standon	10674	

diquat + paraquat
Clayton Paradigm	Clayton	12001	

dithianon
Barclay Cluster	Barclay	11347	
Dithanon WG	BASF	12538	

diuron
Chipko Diuron 80	Chipman	00497	30-04-06
Chipman Diuron 80	Nomix-Chipman	08054	30-04-06
Chipman Diuron Flowable	Nomix-Chipman	05701	
Diuron 50 FL	Staveley	02814	
Diuron 80% WP	Staveley	00730	
I.T. Diuron	I T Agro	12240	
Karmex	DuPont	12068	
Karmex Flowable	DuPont	12284	
n2n Diuron 80	Nomix Enviro	11821	30-04-06
n2n Diuron Flowable	Nomix Enviro	11820	
n2n Diuron Flowable	Nomix Enviro	11822	
Nomix Diuron 80	Nomix Enviro	12079	
Rescind	RP Amenity	08036	
Sanuron	Dow	09236	

Product	Approval holder	Reg. No.	Expiry Date
diuron + glyphosate			
NX 2075	Nomix Enviro	11104	
NX 3083	Nomix Enviro	08712	
NX 3083	Nomix Enviro	11100	
Total	Nomix Enviro	07932	
Total	Nomix Enviro	11099	
Touche	Nomix Enviro	11097	
Trymark	Nomix Enviro	11867	
dodeca-8,10-dienyl acetate			
Exosex CM	Exosect	12103	
dodine			
Barclay Dodex	Barclay	11351	
Dodifun 400 SC	Hermoo	11657	
Syllit 400 SC	Chimac-Agriphar	11079	
endosulfan			
Thiodan 20 EC	Bayer CropScience	07335	
epoxiconazole			
Agrotech-Epoxiconazole 125 SC	Agrotech-Trading	12382	
Clayton Oust	Clayton	11588	
Cleancrop EPX	United Agri	12135	
Cleancrop EPX	United Agri	10381	
Epic	BASF	08320	31-08-06
Milentus Epoxiconazole	Milentus	12363	
Opus	BASF	08319	31-07-06
epoxiconazole + fenpropimorph			
Barclay Riverdance	Barclay	11341	
Landgold Epoxiconazole FM	Landgold	08806	
epoxiconazole + fenpropimorph + kresoxim-methyl			
Allegro Plus	BASF	12218	
BAS 493F	BASF	11748	
BAS 493F	BASF	10438	31-07-06
Bullseye	AgriGuard	12278	
Cleancrop Chant	United Agri	11746	
Cleancrop Chant	United Agri	10466	31-07-06
Mantra	BASF	08886	
Mastiff	BASF	11747	
epoxiconazole + kresoxim-methyl			
Allegro	BASF	12220	
Barclay Avalon	Barclay	11337	
Clayton Gantry	Clayton	09482	
Cleancrop Kresoxazole	United Agri	09698	
Landgold Strobilurin KE	Landgold	09908	

SECTION 3

Product	Approval holder	Reg. No.	Expiry Date
Landmark	BASF	08889	
Serial Duo	AgriGuard	12252	

epoxiconazole + kresoxim-methyl + pyraclostrobin

Covershield	BASF	10900	

epoxiconazole + pyraclostrobin

Ibex	BASF	12168	
Ibex	BASF	10901	31-05-07
Opera	BASF	10876	

esfenvalerate

Clayton Estate	Clayton	11155	01-08-06

ethofumesate

Barclay Keeper 500 Flow	Barclay	11843	
Barclay Keeper EC	Barclay	11887	
Cleancrop EC 200	Barclay	11886	
Cleancrop SC 500	Barclay	11844	
Ethofumesate 2000	Barclay	12406	
I T Ethofumesate	I T Agro	11108	
Kubist 2	Nufarm UK	11126	
Landgold Ethofumesate 200	Landgold	08980	
Linesman 200 EC	Barclay	11885	
Linesman 500 SC	Barclay.	11845	
Oblix 500	Agrichem	12349	28-02-13
Salute	United Phosphorus	07660	

ethofumesate + metamitron + phenmedipham

MAUK 540	Makhteshim	11545	

ethofumesate + phenmedipham

Agriguard Duo	AgriGuard	12356	31-08-06
Duo 400 sc	Makhteshim	12561	

etridiazole

Aaterra WP	Zeneca	06625	
Terraguard	AgriGuard	12321	
Terrazole 35 WP	Crompton	09800	

fatty acids

Safers Insecticidal Soap	Woodstream	07197	

fenarimol

Rubigan	Margarita	11069	30-04-06

fenazaquin

Matador 200 SC	Margarita	11058	

fenbuconazole

Indar 5EW	Whelehan	09644	

Product	Approval holder	Reg. No.	Expiry Date
Kruga 5EC	Interfarm	09863	31-12-06
Reward 5EC	Interfarm	09862	31-12-06
Surpass 5EC	Interfarm	09861	31-12-06

fenbuconazole + propiconazole

Graphic	Interfarm	10987	

fenbutatin oxide

Torque	BASF	10399	

fenpropathrin

Meothrin	Cyanamid	07206	

fenpropidin

Cleancrop Fulmar	United Agri	12033	
Landgold Fenpropidin 750	Landgold	08973	
Mallard	Syngenta	08662	
Patrol	Syngenta	10531	

fenpropidin + prochloraz

SL 552A	Syngenta	08673	
Sponsor	BASF	11902	

fenpropidin + propiconazole

Prophet	Syngenta	08433	
Sheen	Syngenta	08442	
Zulu	Syngenta	08464	

fenpropidin + propiconazole + tebuconazole

Gladio	Syngenta	08413	

fenpropidin + tebuconazole

SL 556 500 EC	Syngenta	08449	

fenpropimorph

BAS 421F	BASF	06127	
Clayton Spigot	Clayton	11560	
Cleancrop Fenpro	United Agri	09885	
Cleancrop Fenpropimorph	United Agri	09445	
Keetak	BASF	06950	

fenpropimorph + flusilazole

BAS 48500F	BASF	06784	
Colstar	DuPont	06783	30-04-06
Pluton	DuPont	10957	30-04-06

fenpropimorph + kresoxim-methyl

Cleancrop Duster	United Agri	12306	
Landgold Strobilurin KF	Landgold	09196	

fenpropimorph + metrafenone

Flexity TP	BASF	11855	

SECTION 3

Products also Registered

Product	Approval holder	Reg. No.	Expiry Date
fenpropimorph + pyraclostrobin			
BAS 528 00f	BASF	11444	
Jemker	BASF	12225	
fenpyroximate			
NNI 850 5SC	Nihon Nohyaku	09887	
ferric phosphate			
Growing Success Advanced Slug Killer	Growing Success	12378	31-10-11
Sluggo	Omex	12529	31-10-11
ferrous sulphate			
Aitken's Lawn Sand	Aitken	05253	
Fisons Greenmaster Autumn	Fisons	03211	28-02-06
Fisons Greenmaster Mosskiller	Fisons	00881	28-02-06
Greenmaster Autumn	Scotts	12196	
Greenmaster Autumn	Scotts	07508	28-02-06
Greenmaster Mosskiller	Scotts	12197	
Greenmaster Mosskiller	Scotts	07509	28-02-06
Maxicrop Moss Killer & Conditioner	Maxicrop	04635	
Moss Control Plus Lawn Fertilizer	Miracle	07912	
No More Moss Lawn Feed	Wolf	10754	
Pentagon Prestige Lawn Sand	Sinclair	10456	
Vitagrow Lawn Sand	Vitagrow	05097	
fipronil			
Regent 1 GR	BASF	11743	
florasulam			
Primus 25 SC	Dow	10175	
florasulam + triclopyr			
Gf-660	Dow	12417	
fluazifop-P-butyl			
Clayton Crowe	Clayton	12572	
Clayton Maximus	Clayton	12543	
Fusilade 250 EW	Zeneca	06531	
Fusilade 250 EW	Syngenta	10525	
PP 007	Zeneca	06533	
PP 007	Syngenta	10533	
Wizzard	Zeneca	06521	
Wizzard	Syngenta	10539	
fluazinam			
Barclay Cobbler	Barclay	11348	
Legacy	ISK Biosciences	09966	
Shirlan Programme	Zeneca	08761	

Product	Approval holder	Reg. No.	Expiry Date
Shirlan Programme	Syngenta	10574	
Standon Fluazinam 500	Standon	08670	

fludioxonil

Beret Gold	BASF	12013	

flufenacet + pendimethalin

Cleancrop Hector	United Agri	12163	

fluoxastrobin

Bayer UK 831	Bayer CropScience	12091	16-08-08

fluoxastrobin + prothioconazole + tebuconazole

Scenic	Bayer CropScience	12289	

flupyrsulfuron-methyl

DPX-KE459 WSB	DuPont	08540	
KE 459 DF	DuPont	09835	

flupyrsulfuron-methyl + metsulfuron-methyl

Standon Flupyrsulfuron MM	Standon	09098	30-06-06

flupyrsulfuron-methyl + picolinafen

BUK 960	BASF	11152	01-03-07
Buk 960 01H	BASF	11662	03-03-07
DP 944	DuPont	11120	01-03-07

flupyrsulfuron-methyl + thifensulfuron-methyl

Lexus Millenium WSB	DuPont	09207	

fluquinconazole

Diablo	BASF	11688	
Jockey Flexi	BASF	11691	
Triplex	BASF	12443	

fluquinconazole + prochloraz

Baron	BASF	11686	
Jockey Plus	BASF	11692	

fluroxypyr

Agrotech Fluroxypyr	Agrotech-Trading	12294	30-11-10
Barclay Hurler	Barclay	12425	30-11-10
Crescent	AgriGuard	12559	30-11-10
Gala	Dow	12019	24-05-07
Hatchet	AgriGuard	12524	30-11-10
Milentus Fluroxypyr	Milentus	12266	30-11-10

fluroxypyr + metosulam

EF 1166	Bayer CropScience	11381	

fluroxypyr + triclopyr

Evade	Dow	08071	

SECTION 3

Product	Approval holder	Reg. No.	Expiry Date
flusilazole			
Sanction	DuPont	08237	
flutolanil			
NNF-136	Nihon Nohyaku	11585	
flutriafol			
Impact	Headland	11521	
Pennant	Headland	12281	
fosetyl-aluminium			
I T Fosetyl-AL	I T Agro	11717	
gibberellins			
Berelex	Nufarm UK	08903	
Gibb 3	Globachem	12290	
GIBB Plus	Globachem	11875	
Gistar	Hermoo	11397	
Regulex	Sumitomo	10147	
glufosinate-ammonium			
Challenge	Bayer CropScience	07306	
Nomix Touchweed	Nomix Enviro	09596	
glyphosate			
Acrion	Bayer Environ.	11718	
Apache	Syngenta	10514	
Barbarian	Barclay	07625	
Barclay Barbarian	Barclay	09865	31-03-06
Barclay Dart	Barclay	05129	
Barclay Gallup	Barclay	05161	
Barclay Gallup 360	Barclay	09127	28-02-06
Barclay Gallup Biograde 450	Dalgety	10151	31-03-06
Barclay Gallup Biograde Amenity	Barclay	10203	
Barclay Garryowen	Barclay	11364	
Barclay Garryowen	Barclay	09869	31-03-06
Biactive 270	Monsanto	10031	
Biactive 270	Monsanto	10301	
Bioglyce	Austrital	11459	
CDA Vanquish Biactive	Bayer Environ.	12586	30-06-12
Clarion	Syngenta	10521	
Clayton Glyphosate	Clayton	06608	
Clayton Rhizeup	Clayton	08920	
Clayton Swath	Clayton	06715	
CleanCrop Egret	United Agri	10689	
CleanCrop Egret	United Agri	10689	
Cleancrop Hoedown	United Agri	10852	
Cleancrop Hoedown	Loveland	11194	
Clinic	Nufarm UK	09378	

Product	Approval holder	Reg. No.	Expiry Date
Do-Away	Barclay	11380	
Do-Away	NCH	10112	31-03-06
Dow Agrosciences Glyphosate 360	Dow	11552	
Economix	Nomix Enviro	08008	
First Line	Linemark	11386	
Glydate	Nufarm UK	10999	
Glyfos 480	Headland	10996	
Glyfos Supreme	Headland	12371	05-04-08
Glymark	Nomix Enviro	11868	
Glyper	Interfarm	11095	
Glyphosate 360	Austrital	11555	
Glyphosate 360	Applyworld	09151	28-02-06
Glyphosate 360	Danagri	09233	28-02-06
Glyphosate 360	Monsanto	11726	
Glyphosate 360	Cardel	11579	
Greenaway Gly-490	Greenaway	11064	
Helosate	Helm	06499	
Hilite	Nomix Enviro	06261	
Initial Line	Linemark	11384	
Mogul	Monsanto	12045	02-06-07
MSS Glyfield	Nufarm UK	11117	
New Total Weedkiller	Headland	11207	
Nomix Conqueror	Nomix Enviro	12370	10-04-08
Nomix G	Nomix Enviro	08781	
Nomix Nova	Nomix Enviro	11829	
Nomix Revenge	Nomix Enviro	11827	
Nomix Supernova	Nomix Enviro	09473	
Nufosate	Nufarm UK	12299	
NX 2031	Nomix Enviro	11105	
NX 2033	Nomix Enviro	11106	
Pontil 360	Mastra	11433	
Portman Glyphosate 360	Agform	04699	
Preline	Linemark	10977	
Preline Plus	Linemark	11385	
Qdos Glyfo	Me2	12574	
Rival	Monsanto	10316	
Rival	Monsanto	09220	
Roundup	Monsanto	10317	
Roundup	Monsanto	01828	
Roundup Biactive Dry	Monsanto	10321	
Roundup Express	Monsanto	12526	30-06-12
Roundup Express	Monsanto	10318	
Roundup Four 80	Monsanto	10325	
Roundup Greenscape	Monsanto	10599	
Roundup GT	Monsanto	10327	
Roundup Pro	Monsanto	10329	
Roundup Rapide	Monsanto	10331	

SECTION 3

Products also Registered

Product	Approval holder	Reg. No.	Expiry Date
Smart 360	Mastra	09476	
Stampede	Syngenta	10536	
Standon Glyphosate 360	Standon	05582	
Stride	Syngenta	10537	
Touchdown LA	Scotts	09270	
Trustee Elite	Dalgety	11556	

guazatine

Ravine	Aventis	10095	31-05-06

guazatine + triticonazole

Premis	BASF	11742	31-08-06

hymexazol

Tachigaren 70 WP	Summit Agro	12568	

imazalil

IT Imazalil 100 SL	I T Agro	12308	01-02-08
Sphinx	BASF	11764	

imazamethabenz-methyl

Assert	BASF	10209	
Dagger	BASF	10218	

imidacloprid

Gaucho FS	Bayer CropScience	11282	
Levington Professional Plus Intercept	Scotts	08569	

indol-3-ylacetic acid

Rhizopon A Powder	Fargro	07131	
Rhizopon A Tablets	Fargro	07132	

4-indol-3-yl-butyric acid

Chryzoplus Grey	Fargro	07984	
Chryzopon Rose	Fargro	07982	
Chryzosan White	Fargro	07983	
Chryzotek Beige	Fargro	07125	
Chryzotop Green	Fargro	07129	
Rhizopon AA Powder (0.5%)	Fargro	07126	
Rhizopon AA Powder (1%)	Fargro	07127	
Rhizopon AA Powder (2%)	Fargro	07128	
Rhizopon AA Tablets	Fargro	07130	
Seradix 1	Certis	10422	
Seradix 2	Certis	10423	
Seradix 3	Certis	10424	

iodosulfuron-methyl-sodium

Hussar	Bayer CropScience	12364	31-12-13

iodosulfuron-methyl-sodium + mesosulfuron-methyl

Greencrop Biscay	Greencrop	12132	30-04-06

Product	Approval holder	Reg. No.	Expiry Date
Me2 Rockall	Me2	12229	29-10-07
Neper	AgriGuard	12256	31-08-06
ioxynil			
Actrilawn 10	RP Amenity	05247	31-08-06
iprodione			
Agrotech-Iprodione 50 WP	Agrotech-Trading	12400	
isoproturon			
Alpha Isoproturon 650	Makhteshim	07034	
Arelon 2	Nufarm UK	11670	
Arelon 700	Nufarm UK	11542	
Bison	Gharda	08699	
Bison 83 WG	Nufarm UK	10063	28-02-06
Bison 83 WG	Gharda	10062	
Clayton Impasse	Clayton	12520	01-01-08
Clayton Siptu 50 FL	Clayton	08751	
Cordelia 2	Griffin	10238	30-04-06
Fieldgard	Nufarm UK	10710	31-03-06
Fieldgard	Nufarm UK	11541	
Griffin IPU 700	Griffin	10409	31-03-06
Isoguard	Nufarm UK	10829	
Isoguard 83 WG	Gharda	10061	
Isoproturon 500	Griffin	10261	31-03-06
Tolkan Liquid	Bayer CropScience	11994	
isoproturon + pendimethalin			
Artillery	BASF	10372	
Artillery SC	BASF	10385	
Encore SC	BASF	10386	
Jolt	BASF	10373	
Jolt SC	BASF	10387	
Trump SC	BASF	10389	
isoproturon + trifluralin			
Autumn Kite	Griffin	10294	
isoxaben			
Agriguard Isoxaben	AgriGuard	11652	
Flexidor	Dow	05121	
Flexidor 125	Dow	05104	
Gallery 125	Rigby Taylor	06889	
Knot Out	Vitax	05163	
isoxaben + trifluralin			
Premiere Granules	Dow	07987	30-04-06
kresoxim-methyl			
Kresoxy 50WG	Tronsan	12275	31-01-09

SECTION 3

Product	Approval holder	Reg. No.	Expiry Date
lambda-cyhalothrin			
Agrotech Lambda Cyhalothrin 100 CS	Agrotech-Trading	12519	01-01-07
Clayton Lanark	Clayton	12369	01-01-07
Jackpot	AgriGuard	12347	01-01-07
Zeus	Interfarm	11852	01-01-07
lambda-cyhalothrin + pirimicarb			
Clayton Groove	Clayton	12346	
Dovetail	Syngenta	12550	
lenacil			
Agricola Lenacil FL	Agricola	09481	31-05-06
Clayton Lenacil 80 W	Clayton	09488	
Clayton Lenaflo	Clayton	11031	
Cleancrop Lenflow	United Agri	10059	
Venzar 80 WP	DuPont	09981	
Volcano	United Phosphorus	12301	
lenacil + phenmedipham			
Agricola Lens	Agricola	10257	30-04-06
DUK 880	DuPont	04121	
linuron			
Alpha Linuron 50 WP	Makhteshim	04870	
Linurex 50 SC	Makhteshim	07950	
maleic hydrazide			
Cleancrop Malahide	United Agri	11066	
Fazor	Crompton	05461	
Rouge	Nufarm UK	11090	
Royal MH 180	Crompton	07043	
mancozeb			
Agrizeb	Chimac-Agriphar	10980	
Barclay Manzeb 455	Barclay	11358	
Cleancrop Mancozeb	United Agri	11193	
Cleancrop Mandrake	Indofil	12500	
Dithane 945	Dow	12545	
Dithane Dry Flowable Newtec	Dow	12564	
Laminator DG	Interfarm	11073	
Laminator FL	Interfarm	11072	
Laminator WP	Interfarm	11071	
Manconex	Griffin	09555	
Manfill 75 WG	Indofil	12419	
Micene 80	Sipcam	08560	
Penncozeb WDG	Nufarm UK	11065	
Penncozeb WDG	Nufarm UK	11622	
Sabero Mancozeb 80% WP	Sabero	11879	

Product	Approval holder	Reg. No.	Expiry Date
Trimanzone	Intracrop	09584	
Unicrop Mancozeb	Unicrop	05467	

mancozeb + zoxamide

Electis 75 WG	Interfarm	10565	
RH 7281/Mancozeb 75 WG	Dow	10807	
Roxam 75 WG	Interfarm	10566	
Unikat 75 WG	Dow	11018	

maneb

Luxan Maneb 80	Luxan	06570	
Trimangol 80	Cerexagri	06871	
Trimangol WDG	Cerexagri	06992	
X-Spor SC	United Phosphorus	08077	

MCPA

Agrichem MCPA-50	Agrichem	04097	
Agricorn 500	FCC	00055	
Agroxone 40	Marks	10264	
Agroxone 50	Mirfield	08345	
Agroxone 75	Marks	09208	
Barclay Meadowman	Barclay	07639	
Barclay Meadowman II	Barclay	08525	
BH MCPA 75	RP Amenity	05395	
Circium II	Nufarm UK	11801	
Dupont MCPA	DuPont	11738	
Empal	Unicrop	00795	
FCC Agricorn 50M	FCC	09032	
Luxan MCPA 500	Luxan	07470	
Marks MCPA P30	Marks	01292	
MCPA 25%	Nufarm UK	07998	
MCPA 500	Nufarm UK	08655	
Nufarm MCPA 750	Nufarm UK	11768	
Nufarm MCPA Amine 50	Nufarm UK	12046	
Nufarm MCPA DMA 500	Nufarm UK	09685	
Quadrangle MCPA 50	Quadrangle	10493	
Tripart MCPA 50	Tripart	02206	

MCPA + MCPB

Butoxone Plus	Marks	11080	
Trifolex-Tra	Cyanamid	07147	

MCPA + mecoprop-P

Greenmaster Extra	Scotts	07594	31-03-06
No More Weeds Lawn Feed	Wolf	10747	
Optica Combi	Marks	10118	

mecoprop-P

Duplosan 500	Nufarm UK	12108	

Product	Approval holder	Reg. No.	Expiry Date
Duplosan 500	BASF	07889	
Duplosan KV 500	Nufarm UK	12107	
Duplosan KV 500	BASF	08027	
Duplosan New System CMPP	Nufarm UK	12106	
Duplosan New System CMPP	BASF	04481	
Dupont Mecoprop-p	DuPont	11739	
Headland Charge	Headland	10981	
Landgold Mecoprop-p	Teliton	12121	
Landgold Mecoprop-p	Landgold	06052	

mecoprop-P + metribuzin

Optica Plus	Marks	09325	

mecoprop-P + metsulfuron-methyl

Headland Neptune	Headland	10230	

mesosulfuron-methyl

AEF 6012-33H	Bayer CropScience	11218	14-07-08

metalaxyl + thiabendazole

Apron T 69 WS	Novartis	08387	

metalaxyl-M

Fongarid Gold	Syngenta	12547	10-04-08

metaldehyde

Antares	CDP-Clartex	11654	
Aristo	De Sangosse	11797	
Aristo M	De Sangosse	11599	
Aristo M	De Sangosse	12394	
Barclay Metaldehyde Dry	Barclay	12446	
Barclay Metaldehyde Wet	Barclay	12447	28-02-07
Brits	Doff Portland	11792	
Clartex	CDP-Clartex	10942	
Clean Crop Hyde	United Agri	12025	
Clean Crop Jekyll D	United Agri	12490	
Cleancrop Jekyll	United Agri	11724	
Dixie 6 M	De Sangosse	12398	
Dixie 6 M	De Sangosse	11597	
Helimax	De Sangosse	11720	
Luxan 9363	Luxan	07359	
Luxan 9363 Blue	Luxan	12391	
Luxan 9363 Red	Luxan	11480	
Luxan Metaldehyde	Luxan	06564	
Luxan Trigger	Luxan	10419	
Lynx	De Sangosse	11767	
Lynx M	De Sangosse	11598	
Lynx S	Lonza	12397	
Metasec M	De Sangosse	11600	

Product	Approval holder	Reg. No.	Expiry Date
Metasec M	De Sangosse	12396	
Mifaslug	FCC	10292	
Mifaslug 5	Luxan	12392	
Optimol XL	CDP-Clartex	10940	
Pastel M	De Sangosse	11736	
Pastel M	De Sangosse	12395	
Pesta	De Sangosse	11796	
Pesta M	De Sangosse	11596	
Pesta M	De Sangosse	12399	
Regel Star	De Sangosse	12441	
Slug Pellets	CDP-Clartex	10941	
Slug-Lentils	Luxan	12533	
Super-Flor 6% Metaldehyde Slug Killer Mini Pellets	CMI	11789	
Unicrop 6% Mini Slug Pellets	Unicrop	11795	

metamitron

Product	Approval holder	Reg. No.	Expiry Date
Barclay Seismic	Barclay	11378	
Bettix 70 WG	United Phosphorus	11019	
Clayton Mitrex	Clayton	11921	
Cobra	Interfarm	10578	
Homer	Makhteshim	11544	
Inter Metamitron WG	I T Agro	12388	
IT Metamitron WG	I T Agro	12467	
Lektan	Makhteshim	11543	
Maymat 70 WDG	Global Crop Care	12477	
MM70	Gharda	09490	
Standon Metamitron	Standon	07885	

metam-sodium

Product	Approval holder	Reg. No.	Expiry Date
Fumetham	Chemical Nutrition	10047	
Metam 510	Taminco	09796	
Metham Sodium 400	United Phosphorus	08051	
Sistan	Unicrop	01957	
Sistan 38	Unicrop	08646	
Vapam	Willmot Pertwee	09194	

metazachlor

Product	Approval holder	Reg. No.	Expiry Date
Agrotech Metazachlor 500 SC	Agrotech-Trading	12454	
Barclay Metaza	Barclay	11359	
Butisan S	BASF	00357	
Clayton Buzz	Q-Chem	12509	30-09-06
Clayton Metazachlor	Clayton	09688	
Clayton Metazachlor 50 SC	Clayton	11719	
Clean Crop MTZ 500	United Agri	09222	
Fuego	Makhteshim	11177	
Fuego 50	Makhteshim	11473	
Gharda Bonanza	Nufarm UK	11571	

SECTION 3

Product	Approval holder	Reg. No.	Expiry Date
Luxan Taurus	Luxan	12442	
Metarap 500 SC	Global Crop Care	12540	
Rapsan 500 SC	Q-Chem	12088	30-09-06
Rapsan 500 SC	Globachem	11468	

metazachlor + quinmerac

Clayton Mazarac	Clayton	11161	
Syndicate	Tronsan	12262	

metconazole

Clayton Tunik	Clayton	10545	
Pan Metconazole	Pan Agriculture	10473	

methabenzthiazuron

Clayton Benson	Clayton	08687	

methiocarb

Barclay Poacher	Barclay	09031	
Decoy	Bayer CropScience	11264	
Draza 2	Bayer CropScience	11269	
Draza Wetex	Bayer CropScience	11271	

metosulam

EF 1077	Bayer CropScience	11479	

metribuzin

Citation 70	United Phosphorus	09370	
Clean Crop Solmet B	United Agri	12273	
Cleancrop Metribuzin	United Agri	09621	
Cleancrop Solmet	United Agri	10865	
Greencrop Majestic	Greencrop	12422	
Inter-Metribuzin WG	Lanxess	12235	28-02-07
Inter-Metribuzin WG	I T Agro	09801	
Lexone 2	DuPont	12051	
Metriphar	Chimac-Agriphar	12354	
Milentus Metribuzin 70 WG	Milentus	12224	

metsulfuron-methyl

Ally WSB	DuPont	06588	30-06-06
Clayton Metsulfuron	Clayton	06734	30-06-06
Standon Metsulfuron	Standon	05670	30-06-06

metsulfuron-methyl + thifensulfuron-methyl

Concert	DuPont	10984	
DP 928	DuPont	09632	
DP 928 PX	DuPont	10991	
Finish PX	DuPont	10989	31-03-06
Harmony M	DuPont	03990	

Product	Approval holder	Reg. No.	Expiry Date
metsulfuron-methyl + tribenuron-methyl			
Biplay PX	DuPont	10990	28-02-06
DP 911 PX	DuPont	10992	31-03-06
DP 911 WSB	DuPont	09867	
myclobutanil			
Systhane 20 EW	Whelehan	09397	
Systhane 6 Flo	Whelehan	06551	
Systhane 6 Flo	Landseer	07334	
Systhane 6 W	Dow	10808	
2-(1-naphthyl)acetic acid			
Rhizopon B Powder (0.1%)	Fargro	07133	
Rhizopon B Powder (0.2%)	Fargro	07134	
Rhizopon B Tablets	Fargro	07135	
Tipoff	Hatrim	12292	
Tipoff	Unicrop	05878	
napropamide			
Banweed	United Phosphorus	09376	
Devrinol	United Phosphorus	09375	
nicotine			
No-FID	Certis	07959	
paclobutrazol			
Bonzi	Zeneca	06640	
paraquat			
Barclay Total	Barclay	11122	01-05-06
Barclay Total	Barclay	11361	01-05-06
Clayton Paragon	Clayton	11210	
CleanCrop Parachute	United Agri	11876	
Speedway Liquid	Miracle	07744	
penconazole			
Topenco 100 EC	Globachem	11972	
pencycuron			
Curon SC	Globachem	12449	
Luxan Pencycuron	Luxan	12144	
Luxan Pencycuron 250 SC	Luxan	12142	
Pan Pencycuron P	Pan Agriculture	10494	
Tubercare 12.5 DS	Agrichem	12194	
Tubercare 12.5 DS	PG - Crop Protection	11147	
pendimethalin			
Claymore	BASF	10214	31-12-06
Cleancrop Stomp	United Agri	11778	
PDM 330 EC	BASF	11084	

SECTION 3

Product	Approval holder	Reg. No.	Expiry Date
Sovereign	BASF	11781	
Standon Pendimethalin 400 SC	Standon	10729	
Stomp 400 SC	BASF	10229	31-12-06
pendimethalin + picolinafen			
Orient	BASF	12541	30-09-12
Stomp Pico	BASF	10738	
Peniophora gigantea			
PG Suspension	Forest Research	11772	
pentanochlor			
Atlas Solan 40	Atlas	07726	31-12-07
Croptex Bronze	Certis	04087	31-12-07
phenmedipham			
Beetup Flo	United Phosphorus	11916	
Beetup Flo 160	United Phosphorus	10382	
Betasana SC	United Phosphorus	12353	
Betosip	Sipcam	06787	31-08-06
Betosip 114	Sipcam	05910	31-08-06
Dancer Flow	Sipcam	11405	
Kemifam E	Kemira	06104	
Luxan Phenmedipham	Luxan	06933	
Mandolin	Griffin	09456	
Mandolin	Nufarm UK	11127	
Mandolin Flo	Nufarm UK	11125	
Protrum G	Nufarm UK	11133	
Tripart Beta	Tripart	03111	31-08-06
picloram			
Tordon 22K	Dow	05083	
picolinafen			
AC 900001	BASF	10714	30-09-12
pinoxaden			
Axial	Syngenta	12521	19-09-09
pirimicarb			
Agrotech-Pirimicarb 50 WG	Agrotech-Trading	12269	
Barclay Pirimisect	Barclay	11360	
Clayton Pirimicarb 50 SG	Clayton	09221	
Cleancrop Miricide	United Agri	11776	
Milentus Pirimicarb	Milentus	12268	
Pirimate	Agform	09568	
prochloraz			
Mirage 45 SC	Makhteshim	10894	
Octave	BASF	11741	

Product	Approval holder	Reg. No.	Expiry Date
Prelude 20 LF	BASF	11904	
Sportak 45 EW	BASF	11696	
Sportak 45 HF	BASF	11697	

prochloraz + propiconazole

Bumper Excell	Makhteshim	11015	

prochloraz + tebuconazole

Agate	Bayer CropScience	11235	

prochloraz + thiram

Hy-Pro Duet	Agrichem	12334	

prochloraz + triticonazole

Premis Pro	BASF	12056	

prometryn

Alpha Prometryne 80 WP	Makhteshim	04795	31-12-07

propachlor

Alpha Propachlor 65 WP	Makhteshim	04807	
Portman Propachlor 50 FL	Portman	06892	
Ramrod 20 Granular	Monsanto	10313	
Tripart Sentinel	Tripart	03250	
Tripart Sentinel 2	Tripart	05140	

propamocarb hydrochloride

Previcur N	Bayer CropScience	08575	
Propeller	Agrichem	12192	
Propeller	PG - Crop Protection	11811	

propaquizafop

Agil	Makhteshim	11048	
Barclay Rebel	Barclay	11648	
Longhorn	AgChem Access	12512	
Zealot	AgriGuard	12548	

propiconazole + tebuconazole

Cogito	Syngenta	11833	

propiconazole + trifloxystrobin

Rombus	Bayer CropScience	11301	

propineb

Antracol	Bayer CropScience	11237	

propyzamide

Bulwark Flo	Dow	10755	
Clayton Propel	Clayton	09783	
Clayton Propel 80 WG	Clayton	11216	
Compose	Spectrum	12505	31-03-09
Conform	Spectrum	12508	31-03-09

SECTION 3

519

Product	Approval holder	Reg. No.	Expiry Date
Flomide	Alfa	11103	
Kerb 50 W	Dow	12296	
Kerb 80 EDF	Dow	10761	
Kerb Granules	Dow	10803	
Propose	Interfarm	12504	31-03-09
Propose	Interfarm	11195	
Redeem Flo	Dow	10764	
Setanta 50 WP	AgriGuard	12475	31-03-09
Setanta 50 WP	AgriGuard	12474	31-07-06
Setanta Flo	AgriGuard	12450	

pyraclostrobin

BASF Insignia	BASF	11900	31-05-14

pyrethrins

Advanced Bug Killer Concentrate	Growing Success	12062	

pyrimethanil

Standon Pyrimethanil	Standon	10576	

quinoxyfen

Clean Crop QFN	United Agri	11966	20-12-06

quizalofop-P-ethyl

EXP31922D	Bayer CropScience	12166	
Kingpin	Interfarm	12188	
Sceptre	Bayer CropScience	08043	
Sceptre Ultra	Bayer CropScience	12249	
Sniper 100	AgriGuard	12479	

rimsulfuron

Agrotech-Rimsulfuron 25 WG	Agrotech-Trading	12427	
Landgold Rimsulfuron	Landgold	08959	

rotenone

Devcol Liquid Derris	Nehra	06063	

silthiofam

Me2 Silthiofam	Me2	12513	31-12-13

simazine

Alpha Simazine 50 WP	Makhteshim	04879	
Alpha Simazine 80 WP	Makhteshim	04800	
Simazine 90WG	Sipcam	10641	
Unicrop Simazine 50	Unicrop	02646	
Unicrop Simazine FL	Unicrop	08032	

sodium chlorate

Aztec Complete Weedkiller	Aztec Chemicals	12484	28-02-06
Blitzweed	Micromix	12393	

Product	Approval holder	Reg. No.	Expiry Date
Cooke's Professional Sodium Chlorate Weedkiller with Fire Depressant	Cooke's Chemicals	06796	
Deosan Chlorate Weedkiller (Fire Suppressed)	JohnsonDiversey	08521	
Gem Sodium Chlorate Weedkiller	Joseph Metcalf	04276	
Growing Success Sodium Chlorate Weedkiller	Growing Success	10787	
Sodium Chlorate	Marlow Chemical	06294	
Strathclyde Sodium Chlorate Weedclear	Dremm	12015	
t.w.k	Apex	12453	
TWK Total Weedkiller	Reabrook	06393	28-02-06
sodium monochloroacetate			
Atlas Somon	Atlas	03045	31-12-07
spinosad			
Conserve	Dow	11011	26-02-06
spiroxamine			
Torch	Bayer CropScience	11258	01-09-09
Zenon	Bayer CropScience	11232	01-09-09
spiroxamine + tebuconazole			
Array	Bayer CropScience	11238	
Beam	Bayer CropScience	11255	
Draco	Bayer CropScience	11267	
Folicur Star	Bayer CropScience	11531	
sulfosulfuron			
Monitor	Monsanto	12236	05-01-08
sulfuryl difluoride			
Profume	Dow	12035	06-07-07
sulphur			
Headland Venus	Headland	10611	
Kumulus S	BASF	01170	
Microthiol Special	Cerexagri	06268	
Solfa	Nufarm UK	11390	31-03-06
Thiomex	Omex	11176	
Tripart Imber	Tripart	04050	
tebuconazole			
Agrotech-Tebuconazole 250 EW	Agrotech-Trading	12523	
Aurigen	Syngenta	12043	
Bayer UK226	Bayer CropScience	12412	
Clayton Tebucon	Clayton	08707	
Orius 20 EW	Makhteshim	12311	

SECTION 3

Product	Approval holder	Reg. No.	Expiry Date
Raxil	Bayer CropScience	11295	
Tebucur	Globachem	11890	
tebuconazole + triadimenol			
Garnet	Bayer CropScience	11280	
tebufenpyrad			
Masai G	BASF	10224	
teflubenzuron			
Nemolt	Cyanamid	07012	
tefluthrin			
Force ST	Bayer CropScience	11671	
tepraloxydim			
Clayton Lombard	Clayton	12575	30-11-07
Clayton Rally	Clayton	12549	30-11-07
Me2 I-Tora	Me2	12209	21-05-06
tetraconazole			
Digit	Monsanto	10503	
Domark	Isagro	10452	
Emerald	Isagro	10449	
Eminent	Monsanto	10504	
MON 10 EC	Monsanto	10505	
Omen	Sipcam	09413	
Tetra 5634	Monsanto	10507	
thiabendazole + thiram			
Hy-Vic	Agrichem	06247	
sHYlin	Agrichem	10030	
thiacloprid			
Biscaya	Bayer CropScience	12471	04-07-08
thifensulfuron-methyl			
Harmony 75DF	DuPont	11884	
Harmony SX	DuPont	12181	31-10-07
Prospect	Nufarm UK	09389	
Prospect SX	DuPont	12212	31-10-07
thifensulfuron-methyl + tribenuron-methyl			
Calibre	DuPont	07795	
thiodicarb			
Genesis	Sipcam	11846	
Genesis ST	Sipcam	11853	
thiophanate-methyl			
Cercobin Liquid	Certis	11941	
Mildothane Liquid	Certis	11942	

Product	Approval holder	Reg. No.	Expiry Date
thiram			
Thiraflo	Crompton	11225	
tolclofos-methyl			
Rizolex 50 WP	Aventis	07272	
tralkoxydim			
Clayton Lansdown	Clayton	12327	
triadimenol			
Bayfidan	Bayer CropScience	11417	
Hi-Shot	BASF	06508	
Hi-Shot	BASF	10397	
Spinnaker	BASF	10398	
Spinnaker	Cyanamid	07023	
tri-allate			
Avadex BW Granular	Gowan	12104	
Avadex BW Granular	Monsanto	10298	
Avadex BW Granular	Monsanto	00174	
Avadex Excel 15G	Gowan	12109	
triasulfuron			
Lo-Gran	Syngenta	08421	
tribenuron-methyl			
Quantum	DuPont	06270	
Quantum 75 DF	DuPont	09340	
Quantum PX	DuPont	10843	
triclopyr			
Agriguard Triclopyr	AgriGuard	10679	
Altix 240 EC	Chimac-Agriphar	10648	
Cleancrop Triptic 48 EC	United Agri	11568	
Cleancrop Unival	United Agri	11388	
Garlon 2	Dow	05682	
n2n Garlon 4	Nomix Enviro	11806	
Tricle	Agrichem	12193	
Tricle	PG - Crop Protection	11862	
trifloxystrobin			
Aprix	Bayer CropScience	11220	
Flint	Bayer CropScience	11259	
Twist 500 SC	Bayer CropScience	11231	
Zest SC	Bayer CropScience	11233	
trifluralin			
Das-320	Dow	12375	
Digermin	Sipcam	07221	
FCC Trigard	FCC	09030	

SECTION 3

Product	Approval holder	Reg. No.	Expiry Date
Ipifluor	I Pi Ci	04692	
Portman Trifluralin	Agform	05751	
Triflur	Nufarm UK	10750	
Trilogy	United Phosphorus	08996	
Triplen	Sipcam	05897	
triflusulfuron-methyl			
Debut WSB	DuPont	07809	
Landgold TFS 50	Landgold	08941	
Standon Triflusulfuron	Standon	09487	
trinexapac-ethyl			
Cleancrop Alatrin	United Agri	11805	
vinclozolin			
Barclay Flotilla	Barclay	11354	
Landgold Vinclozolin DG	Landgold	06500	
Landgold Vinclozolin SC	Landgold	06459	
Ronilan DF	BASF	04456	
zeta-cypermethrin			
Fury 10 EW	FMC	10718	
Fury 10 EW	FMC	10718	31-07-06
Minuet EW	FMC	10716	31-07-06
zineb-ethylene thiuram disulphide adduct			
Polyram DF	BASF	08234	

SECTION 4
ADJUVANTS

Adjuvants

Adjuvants are not themselves classed as pesticides and there is considerable misunderstanding over the extent to which they are legally controlled under the Food and Environment Protection Act. An adjuvant is a substance other than water which enhances the effectiveness of a pesticide with which it is mixed. Consent C(i)5 under the Control of Pesticides Regulations allows that an adjuvant can be used with a pesticide only if that adjuvant is authorised and on a list published from time to time on the PSD website. An authorised adjuvant has an *adjuvant number* and may have specific requirements about the circumstances in which it may be used.

Adjuvant product labels must be consulted for full details of authorised use, but the table below provides a summary of the label information to indicate the area of use of the adjuvant. The codes for protective clothing and label precautions refer to the keys given in Appendix 4. The table includes all adjuvants notified by suppliers as available in 2006.

Product	Supplier	Adj. No.	Type
Abacus	De Sangosse	A0434	adjuvant/wetter
Contains	505.2 g/l esterified rapeseed oil, 300 g/l alkylphenyl hydroxypolyoxyethylene and 60 g/l natural fatty acids		
Use with	All approved pesticides on edible and non-edible crops when used at half their approved dose or less. Also with approved pesticides on specified crops, up to specified growth stages, at up to their full approved dose		
Protective clothing	A, C, H		
Precautions	R36, R58, U05a, U11, U14, U19a, U20b, E15a, E19b, E34, D01, D02, D05, D10a, D12a, H04		
Activator 90	De Sangosse	A0337	non-ionic surfactant/spreader/wetter
Contains	750 g/l alkylphenyl hydroxypolyoxyethylene and 150 g/l natural fatty acids		
Use with	Any spray for which additional wetter is approved and recommended		
Protective clothing	A, C, H		
Precautions	R36, R58, U02a, U04a, U05a, U10, U11, U20b, E15a, E19b, D01, D02, D05, D10a, H04		
Admix-P	De Sangosse	A0301	wetter
Contains	80% w/w polyalkylene oxide modified heptamethyltrisiloxane and a maximum of 20 % w/w allyloxypolyethylene glycol methyl ether		
Use with	A wide range of pesticides applied as corm, tuber, onion and other bulb treatments in seed production and in seed potato treatment		
Protective clothing	A, C, H		
Precautions	R20, R21, R22a, R36, R43, R48, R51, R58, U11, U15, U19a, E15a, E19b, D01, D02, D05, D10a, D12a, H03, H11		
Agral	Syngenta	A0421	non-ionic surfactant/spreader/wetter
Contains	948 g/l alky phenol ethylene oxide		
Use with	Any spray for which additional wetter is approved and recommended		
Protective clothing	A, C		
Precautions	R36, R38, U05a, U08, U20c, E15a, E34, D01, D02, D05, D09a, D10a, H04, H08		

SECTION 4

Product	Supplier	Adj. No.	Type
Agropen	Intracrop	A0227	vegetable oil
Contains	95% refined rapeseed oil		
Use with	Recommended rates of approved pesticides up to the growth stage indicated for specified crops, and with half or less than the recommended rate on these crops after the stated growth stages. Also for use with pesticides on grassland at half their recommended rate		
Protective clothing	A, C		
Precautions	U08, U20b, E13c, E37, D09a, D10b		
Amber	Interagro	A0367	vegetable oil
Contains	95% w/w methylated rapeseed oil		
Use with	Sugar beet herbicides, oilseed rape herbicides, cereal graminicides and a wide range of other pesticides that have a label recommendation for use with authorised adjuvant oils on specified crops. See label for details		
Precautions	U05a, U20b, E15a, D01, D02, D05, D09a, D10a		
Arma	Interagro	A0306	penetrant
Contains	500 g/l alkoxylated fatty amine + 500 g/l polyoxyethylene monolaurate		
Use with	Cereal growth regulators, cereal herbicides, cereal fungicides, oilseed rape fungicides and a wide range of other pesticides on specified crops		
Precautions	R51, R58, U05a, E15a, E34, E37, D01, D02, D05, D09a, D10a, H11		
BackRow	Interagro	A0472	mineral oil
Contains	60% w/w refined paraffinic petroleum oil		
Use with	Pre-emergence herbicides. Refer to label or contact supplier for further details		
Protective clothing	A		
Precautions	R22b, R38, U02a, U05a, U08, U20b, E15a, D01, D02, D05, D09a, D10a, M05b, H03		
Ballista	Interagro	A0524	spreader/wetter
Contains	10% w/w alkoxylated triglycerides		
Use with	All approved pesticides and plant growth regulators for use in winter and spring cereals applied at full rate up to and including GS 52, and at half the approved rate thereafter. May be applied with approved pesticides for other specified crops at full rate up to specified growth stages and at half rate thereafter		
Precautions	U05a, U20b, E15a, E34, D01, D02, D09a		
Bandrift Plus	Ciba Specialty	-	drift retardant
Contains	A non-ionic polyamide dispersed in oil		
Use with	A wide range of pesticides. See label for restrictions on use with wettable powders, suspension concentrates and water dispersible granules, and other usage limitations		
Protective clothing	A, C, M		
Precautions	U05a, U08, U20a, E13c, E34, D01, D02, D09a		

Product	Supplier	Adj. No.	Type
Banka	Interagro	A0245	spreader/wetter
Contains	29.2% w/w alkyl pyrrolidones		
Use with	Potato fungicides and a wide range of other pesticides on specified crops		
Protective clothing	A, C		
Precautions	R38, R41, R52, R58, U02a, U05a, U11, U19a, U20b, E15a, E34, E37, D01, D02, D05, D09a, H04		
Barclay Actol	Barclay	A0126	mineral oil
Contains	99% highly refined paraffinic oil		
Use with	Any approved pesticide for which the addition of a spraying oil is recommended		
Precautions	U19a, U20b, E13c, D09a, D10b		
Barclay Clinger	Barclay	A0198	sticker/wetter
Contains	96% w/v poly-1-p-menthene		
Use with	Barclay Gallup, Barclay Gallup Amenity and a wide range of approved fungicides and insecticides. See label for details		
Precautions	U19a, U20b, E13c, D09a, D10a		
Barramundi	Interagro	A0376	mineral oil
Contains	95% w/w mineral oil		
Use with	Sugar beet herbicides, oilseed rape herbicides, cereal graminicides and a wide range of other pesticides that have a label recommendation for use with authorised adjuvant oils on specified crops. Refer to label or contact supplier for further details		
Protective clothing	A		
Precautions	R22b, R38, U02a, U05a, U08, U20b, E15a, D01, D02, D05, D09a, D10a, M05b, H03		
Bio Syl	Intracrop	A0385	spreader/sticker/wetter
Contains	1% w/w polyoxyethylen-alpha-methyl-omega-[3-(1,3,3,3-tetramethyl-3-trimethylsiloxy)-disiloxanyl] propylether and 32.67% w/w ethylene oxide condensate		
Use with	Recommended rates of approved pesticides on a range of specified fruit and vegetable crops and on managed amenity turf. See label for details		
Protective clothing	A, C		
Precautions	R36, R38, U05a, U08, U20c, E15a, D01, D02, D10a, H04		
Bioduo	Intracrop	A0299	wetter
Contains	85% alkyl polyglycol ether and fatty acids		
Use with	A wide range of pesticides used in grassland, agriculture and horticulture and with pesticides used in non-crop situations		
Protective clothing	A, C		
Precautions	R22a, R36, R38, U05a, U08, U20c, E13c, E34, D01, D02, D09a, D10a, M03, H03, H04, H08		

SECTION 4

Product	Supplier	Adj. No.	Type
Biofilm	Intracrop	A0359	sticker/wetter
Contains	96% poly-1-p-menthene		
Use with	All approved fungicides and insecticides on edible crops and all approved formulations of glyphosate used pre-harvest on wheat, barley, oilseed rape, stubble, and in non-crop situations and grassland destruction		
Precautions	U19a, U20c, E13c, D09a, D11a		
BioPower	Bayer CropScience	A0454	wetter
Contains	6.7% w/w 3,6-dioxaeicosylsulphate sodium salt and 20.2% w/w 3,6-dioxaoctadecylsulphate sodium salt		
Use with	Atlantis and all other approved cereal herbicides		
Protective clothing	A, C		
Precautions	R36, R38, U02a, U05a, U08, U13, U19a, U20b, E13c, E34, D01, D02, D05, D09a, D10a, H04		
Biothene	Intracrop	A0360	sticker/wetter
Contains	96% poly-1-p-menthene		
Use with	All approved fungicides and insecticides on edible crops up to 30 d before harvest, and all approved formulations of glyphosate used pre-harvest on wheat, barley, oilseed rape, stubble, and in non-crop situations and grassland destruction. Must not be used in mixture with adjuvant oils or surfactants		
Precautions	U19a, U20c, E13c, D09a, D11a		
Bond	De Sangosse	A0184	drift retardant/extender/sticker/wetter
Contains	450 g/l synthetic latex and 100 g/l alkylphenylhydroxy-polyoxyethylene		
Use with	Blight sprays on potatoes and other pesticides on a wide range of crops. See label or contact supplier for details		
Protective clothing	A, C, H		
Precautions	R36, R38, U11, U14, U19a, E15a, E19b, D01, D02, D05, D10a, D12a, H04		
Byo-Flex	Greenaway	A0481	sticker/wetter
Contains	95 g/l vegetable oil		
Use with	GLY 490 (MAPP 11064)		
Precautions	U05a, U08, U20b, E37, D02, D05, D10a		
C-Cure	Interagro	A0467	mineral oil
Contains	60% w/w refined mineral oil		
Use with	Pre-emergence herbicides		
Protective clothing	A		
Precautions	R22b, R38, U02a, U05a, U08, U20b, E15a, D01, D02, D05, D09a, D10a, M05b, H03		
Celect	Fargro	A0334	extender/sticker/wetter
Contains	95% emulsified vegetable oil		
Use with	approved pesticides in horticulture (but see label for certain active ingredient exclusions)		
Protective clothing	A, C		
Precautions	U19a, U20b, D09a, D10b		

Product	Supplier	Adj. No.	Type
Celect	Fargro	A0413	extender/sticker/wetter
Contains	95% emulsified vegetable oil		
Use with	Approved pesticides in horticulture (but see label for certain active ingredient exclusions)		
Protective clothing	A, C		
Precautions	U19a, U20b, D09a, D10b		
Ceres Platinum	Interagro	A0445	penetrant/spreader
Contains	500 g/l alkoxylated fatty amine + 500 g/l polyoxyethylene monolaurate		
Use with	Cereal growth regulators, cereal herbicides, cereal fungicides, oilseed rape fungicides and a wide range of other pesticides on specified crops		
Precautions	R51, R58, U05a, E15a, E34, D01, D02, D05, D09a, D10a, H11		
Cerround	Helena	A0510	adjuvant/vegetable oil
Contains	80% w/w methylated soybean oil and 15% w/w polyalkylene oxide modified heptamethyl trisiloxane		
Use with	Approved pesticides on non-edible crops, cereals and stubbles of all edible crops, grassland (destruction), and pesticides approved for use on growing edible crops when used at half recommended dose or less. On specified crops product may be used at a maximum spray concentration of 0.5% with approved pesticides at their full approved rate up to the growth stages shown in the label		
Protective clothing	A, H		
Precautions	R38, R41, U02a, U05a, U08, U11, U19a, U20b, E13c, E34, E37, D01, D02, D05, D09a, M03, H04		
Codacide Oil	Microcide	A0011	vegetable oil
Contains	95% emulsified vegetable oil		
Use with	All approved pesticides and tank mixes. See label for details		
Protective clothing	A, C		
Precautions	U20b, D09a, D10b		
Companion PCT12	Ciba Specialty	A0482	spreader/sticker/wetter
Contains	25% w/w polyacryalmide		
Use with	Approved herbicides on cereals and any pesticide on non-food crops. See label for restrictions on use with wettable powders, suspension concentrates and water dispersible granules, and other usage limitations		
Precautions	U05a, U08, E13c, D01, D02, D09a		
Contact Plus	Interagro	A0418	mineral oil
Contains	95% w/w mineral oil		
Use with	Sugar beet herbicides, oilseed rape herbicides, cereal graminicides and a wide range of other pesticides that have a label recommendation for use with authorised adjuvant oils on specified crops. Refer to label or contact supplier for further details		
Protective clothing	A, C		
Precautions	R22b, R38, U02a, U05a, U08, U20b, E15a, D01, D02, D09a, D10a, M05b, H03		

SECTION 4

Product	Supplier	Adj. No.	Type
Designer	De Sangosse	A0322	drift retardant/extender/sticker/wetter
Contains	8.44% w/w organosilicone wetter and 50% w/w synthetic latex		
Use with	A wide range of fungicides, insecticides and trace elements for cereals and specified agricultural and horticultural crops		
Protective clothing	A, C, H		
Precautions	R36, R38, R52, U02a, U11, U14, U15, U19a, E15a, E37, D01, D02, D05, D09a, D10a, H04		
Drill	De Sangosse	A0409	adjuvant/wetter
Contains	589.4 g/l esterified rapeseed oil, 225 g/l alkylphenyl hydroxypolyoxyethylene and 45 g/l natural fatty acids		
Use with	All approved pesticides on edible and non-edible crops when used at half their approved dose or less. Also with approved pesticides on specified crops, up to specified growth stages, at up to their full approved dose		
Protective clothing	A, C, H		
Precautions	R36, R58, U05a, U11, U14, U19a, U20b, E15a, E19b, E34, D01, D02, D05, D10a, D12a, H04		
Elan	Intracrop	A0392	spreader/sticker/wetter
Contains	20% w/w polyoxyethylen-alpha-methyl-omega-[3-(1,3,3,3-tetramethyl-3-trimethylsiloxy)-disiloxanyl] propylether		
Use with	Recommended rates of approved pesticides up to the growth stage indicated for specified crops, and with half or less than the recommended rate on these crops after the stated growth stages. Also for use with recommended rates of herbicides on managed amenity turf		
Protective clothing	A, C, H		
Precautions	R36, R38, U05a, U08, U20c, E13c, E15a, E37, D01, D02, D09a, D10a, H04		
Emerald	Intracrop	A0031	anti-transpirant/extender
Contains	96% di-1-p-menthene		
Use with	Recommended rates of approved pesticides up to the growth stage indicated for specified crops, and with half or less than the recommended rate on these crops after the stated growth stages. Also for use alone on transplants, turf, fruit crops, glasshouse crops and Christmas trees. Must not be used in mixture with adjuvant oils or surfactants		
Precautions	U19a, U20c, E13c, E37, D09a, D11a		
Esterol	Interagro	A0330	vegetable oil
Contains	95% w/w methylated rapeseed oil		
Use with	Plant protection products that have a label recommendation for use with adjuvant oils. Refer to label or contact supplier for further details		
Precautions	U05a, U20b, E15a, D01, D02, D09a, D10a		
Felix	Intracrop	A0178	spreader/wetter
Contains	60% ethylene oxide condensate		
Use with	Mecoprop, 2,4-D in cereals and amenity turf, and a range of grass weedkillers in agriculture. See label for details		
Protective clothing	A, C		
Precautions	R36, R38, U05a, U08, U20a, E15a, D01, D02, D09a, D10a, H04, H08		

Product	Supplier	Adj. No.	Type
FMC Oil	Belchim	A0435	adjuvant/vegetable oil
Contains	842 g/l esterified rapeseed oil		
Use with	Spotlight 24 EC (MAPP 10702 and 11617)		
Precautions	E13c, E37		
Frigate	Unicrop	A0325	sticker/wetter
Contains	800 g/l tallow amine ethoxylate		
Use with	Roundup formulations		
Protective clothing	A, C		
Precautions	R22a, R23, R41, R51, R58, R67, U05a, U11, U19a, E38, D01, D02, D12a, M04, H02, H08, H11		
Gex 1664	Headland	A0366	spreader/wetter
Contains	900 g/l ethylene oxide concentrate		
Use with	All approved pesticides on edible and non-edible crops when used at half their approved dose or less. Also with approved pesticides on specified crops, up to specified growth stages, at up to their full approved dose		
Protective clothing	A, C		
Precautions	R36, R58, U05a, U08, U11, U12, U15, U20b, E13c, E34, E37, E38, D01, D02, D09b, D10b, H04		
Gly-Flex	Greenaway	A0370	spreader/wetter
Contains	140 g/l refined mineral oil		
Use with	All approved formulations of 360 g/l glyphosate		
Protective clothing	A, C, D, H, M		
Precautions	R36, R38, U08, U20b, E13c, E15a, E37, D02, D05, D10a, H04		
Green Gold	Intracrop	A0250	spreader/wetter
Contains	95% refined rapeseed oil		
Use with	All pesticides which have a recommendation for the addition of a wetter/spreader		
Precautions	U08, U20b, E13c, E34, D09a, D10b		
Greencrop Astra	Greencrop	A0417	adjuvant/penetrant/spreader
Contains	500 g/l alkoxylated fatty amine and 500 g/l polyoxylene monolaurate		
Use with	Approved pesticides on a range of specified arable crops, and in forestry, managed amenity turf, stubbles and non-crop areas and situations		
Precautions	U05a, E13c, E34, D01, D02, D05, D09a, D10b		
Greencrop Dolmen	Greencrop	A0419	adjuvant/spreader
Contains	64% w/w polyalkeneoxide modified heptamethyltrisiloxane		
Use with	Approved pesticides on a range of specified arable crops, and in forestry, managed amenity turf, stubbles and non-crop areas and situations		
Protective clothing	A, C, H		
Precautions	R21, R22a, R38, R41, R43, U02a, U05a, U08, U11, E13c, E34, D01, D02, D09a, D10b, M03, H03, H04		

SECTION 4

Product	Supplier	Adj. No.	Type
Greencrop Rookie	Greencrop	A0515	wetter
Contains	20.2% w/w 3,6-dioxoctadecylsulphate sodium salt and 6.7% w/w 3,6-dioxaeicosylsulphate sodium salt		
Use with	All approved herbicides in cereals		
Protective clothing	A, C		
Precautions	R36, R38, R52, U02a, U05a, U08, U11, U13, U19a, E15a, D01, D02, D05, D09a, D10b, H04		
Grip	De Sangosse	A0211	drift retardant/extender/sticker/wetter
Contains	450 g/l synthetic latex and 100 g/l alkylphenyl hydroxypolyoxyethylene		
Use with	Blight sprays on potatoes and other pesticides on a wide range of crops. See label or contact supplier for details		
Protective clothing	A, C, H		
Precautions	R36, R38, U11, U14, U19a, U20b, E15a, E19b, D01, D02, D05, D10a, D12a, M03, H04		
Grounded	Helena	A0456	mineral oil
Contains	732 g/l refined paraffinic petroleum oil		
Use with	All approved pesticides on edible and non-edible crops when used at half their approved dose or less. Also with approved pesticides on specified crops, up to specified growth stages, at up to their full approved dose		
Protective clothing	A, C		
Precautions	R53a, U02a, U05a, U08, U20b, E15a, E34, E37, D01, D02, D05, D09a, D10c, H11		
Guide	De Sangosse	A0208	acidifier/drift retardant/penetrant
Contains	350 g/l modified soya lecithin, 100 g/l alkylphenyl hydroxypolyoxyethylene and 350 g/l propionic acid		
Use with	All plant growth regulators for cereals and oilseed rape, dimethoate and chlorpyrifos for wheat bulb fly control. For other uses see label or contact supplier for details		
Protective clothing	A, C, H		
Precautions	R36, R38, U11, U14, U15, U19a, E15a, E19b, D01, D02, D05, D10a, D12a, H04		
Headland Fortune	Headland	A0277	penetrant/spreader/vegetable oil/wetter
Contains	75% w/w mixed methylated fatty acid esters of seed oil and N-butanol		
Use with	Herbicides and fungicides in a wide range of crops. See label for details		
Protective clothing	A, C		
Precautions	R43, U02a, U05a, U14, U20a, E15a, E34, E38, D01, D02, D05, D09a, D10b, H04		

Product	Supplier	Adj. No.	Type
Headland Guard 2000	Headland	A0369	extender/sticker/wetter
Contains	52% w/w synthetic latex solution		
Use with	All approved pesticides that recommend the use of a sticking, wetting or extending agent; also with all approved pesticides for use on crops not destined for human or animal consumption; also with all approved pesticides on beans, peas, edible podded peas, oilseed rape, linseed, sugar beet, cereals, maize, Brussels sprouts, potatoes, cauliflowers		
Precautions	U05a, U08, U19a, U20b, E13b, E34, E37, D01, D02, D05, D10b		
Headland Guard Pro	Headland Amenity	A0423	extender/sticker
Contains	52% synthetic latex solution		
Use with	All approved pesticides, micronutrients and sea-weed based plant growth stimulants in amenity grass, managed amenity turf and amenity vegetation		
Precautions	U05a, U08, U19a, U20b, E13b, E34, E37, D01, D02, D05, D10b		
Headland Inflo XL	Headland Amenity	A0329	wetter
Contains	85% w/w polyalkylene oxide modified heptamethyl siloxane		
Use with	Any pesticide for use on amenity grassland or managed amenity turf (except any product applied in or near water) where the use of a wetting, spreading and penetrating surfactant is recommended to improve foliar coverage, and with any approved pesticide on crops not intended for human or animal consumption		
Protective clothing	A, C		
Precautions	R21, R22a, R38, R41, R43, R51, R58, U02a, U05a, U08, U14, U19a, U20b, E15a, E34, E38, D01, D02, D05, D09a, D10b, M03, H03, H11		
Headland Intake	Headland	A0074	penetrant
Contains	450 g/l propionic acid		
Use with	All approved pesticides on any crop not intended for human or animal consumption, and with all approved pesticides on beans, peas, edible podded peas, oilseed rape, linseed, sugar beet, cereals (except triazole fungicides), maize, Brussels sprouts, potatoes, cauliflowers. See label for detailed advice on timing on these crops		
Protective clothing	A, C		
Precautions	R34, U02a, U05a, U10, U11, U14, U15, U19a, E34, D01, D02, D05, D09b, D10b, M04, H05		
Headland Rhino	Headland	A0328	wetter
Contains	85% w/w polyalkylene oxide modified heptamethyl siloxane		
Use with	Any herbicide, systemic fungicide, systemic insecticide or plant growth regulator (except any product applied in or near water) where the use of a wetting, spreading and penetrating surfactant is recommended to improve foliar coverage		
Protective clothing	A, C		
Precautions	R21, R22a, R38, R41, R43, R51, R58, U02a, U05a, U08, U11, U14, U19a, U20b, E13c, E34, D01, D02, D05, D09a, D10b, M03, H03, H11		

SECTION 4

Product	Supplier	Adj. No.	Type
Hyspray	Fine	A0020	cationic surfactant
Contains	800 g/l polyethoxylated tallow amine		
Use with	All approved formulations of glyphosate		
Protective clothing	A, C		
Precautions	R22a, R36, R38, U02a, U05a, U08, U13, U19a, U20a, E13b, E34, D01, D02, D09a, D10b, H03, H04, H08		
Intracrop BLA	Intracrop	A0453	extender/sticker
Contains	42.5% w/w styrene butadiene co-polymer and 20% w/w alkylphenol ethylene oxide condensate		
Use with	All potato blight fungicides. Also with recommended rates of approved pesticides in certain non-crop situations and up to the growth stage indicated for specified crops, and with half or less than the recommended rate on these crops after the stated growth stages. Also with pesticides in grassland at half or less their recommended rate		
Precautions	U08, U20b, E13c, E34, E37, D01, D05, D09a, D10a		
Intracrop Green Oil	Intracrop	A0262	wetter
Contains	95% refined rapeseed oil		
Use with	Recommended rates of approved pesticides up to the growth stage indicated for specified crops, and with half or less than the recommended rate on these crops after the stated growth stages. Also for use with pesticides on grassland at half their recommended rate		
Precautions	U08, U20b, E13c, E37, D09a, D10b		
Intracrop Neotex	Intracrop	A0460	extender/sticker
Contains	42.5% w/w styrene butadiene co-polymer and 20% w/w alkylphenol ethylene oxide condensate		
Use with	Recommended rates of approved pesticides up to specified growth stages of a wide range of agricultural arable and horticultural crops, and for non-crop uses. Use on edible crops beyond specified growth stages, and in grass, should only be with half recommended rates of the pesticide or less. See label for details of growth stage restrictions. In addition may be used with all potato blight fungicides at their recommended rates of use up to the latest recommended timing of the fungicide		
Precautions	U08, U20b, E13c, E34, E37, D01, D05, D09a, D10a		
Intracrop Novatex	Intracrop	A0462	extender/sticker
Contains	42.5% w/w styrene butadiene co-polymer and 20% w/w alkylphenol ethylene oxide condensate		
Use with	Recommended rates of approved pesticides up to the growth stage indicated for specified crops, and with half or less than the recommended rate on these crops after the stated growth stages. Also for use with recommended rates of pesticides on grassland and specified non-crop situations		
Precautions	U08, U20b, E13c, E34, E37, D01, D09a, D10a		

Product	Supplier	Adj. No.	Type
Intracrop Predict	Intracrop	A0503	adjuvant/vegetable oil

Contains — 91% w/w methylated rapeseed oil

Use with — All approved pesticides on non-edible crops, and all pesticides approved for use on growing edible crops when used at half recommended dose or less. On specified crops product may be used at a maximum spray concentration of 1% with approved pesticides at their full approved rate up to the growth stages shown in the label

Precautions — U19a, U20b, D05, D09a, D10b

Intracrop Questor	Intracrop	A0495	activator/non-ionic surfactant/spreader

Contains — 75% polyoxyethylene polypropoxypropanol and 15% alkyl polyglycol ether

Use with — All approved pesticides on non-edible crops and pesticides used in non-crop production, and all pesticides approved for use on growing edible crops when used at half recommended dose or less. On specified crops product may be used at a maximum spray concentration of 0.3% with approved pesticides at their full approved rate up to the growth stages shown in the label

Protective clothing — A, C

Precautions — R36, R38, U04a, U05a, U08, U19a, E13b, D01, D02, D09a, D10b, M03, M05a, H04

Intracrop Retainer	Intracrop	A0508	adjuvant/vegetable oil

Contains — 60% w/w methylated rapeseed oil

Use with — All approved pesticides on non-edible crops, and all pesticides approved for use on growing edible crops when used at half recommended dose or less. On specified crops product may be used at a maximum spray concentration of 1% with approved pesticides at their full approved rate up to the growth stages shown in the label

Precautions — U19a, U20b, D05, D09a, D10b

Intracrop Rigger	Intracrop	A0479	adjuvant/vegetable oil

Contains — 91.7% w/w methylated rapeseed oil

Use with — All approved pesticides on non-edible crops, and all pesticides approved for use on growing edible crops when used at half recommended dose or less. On specified crops product may be used at a maximum spray concentration of 1.78% with approved pesticides at their full approved rate up to the growth stages shown in the label

Precautions — U19a, U20b, D05, D09a, D10b

Intracrop Saturn	Intracrop	A0494	activator/non-ionic surfactant/spreader

Contains — 75% polyoxyethylene polypropoxypropanol and 15% alkyl polyglycol ether

Use with — All approved pesticides on non-edible crops and pesticides used in non-crop production, and all pesticides approved for use on growing edible crops when used at half recommended dose or less. On specified crops product may be used at a maximum spray concentration of 0.3% with approved pesticides at their full approved rate up to the growth stages shown in the label

Protective clothing — A, C

Precautions — R36, R38, U04a, U05a, U08, U19a, E13b, D01, D02, D09a, D10b, M03, M05a, H04

SECTION 4

Product	Supplier	Adj. No.	Type
Intracrop Sprinter	Intracrop	A0513	spreader/wetter

Contains	192 g/l primary alcohol ethoxylate
Use with	Recommended rates of approved pesticides up to the growth stage indicated for specified crops, and with half or less than the recommended rate on these crops after the stated growth stages. Also with herbicides on managed amenity turf at recommended rates, and on grassland at half or less than recommended rates
Protective clothing	A, C
Precautions	R36, R38, R41, U05a, U08, U11, U14, U15, U20c, E15a, D01, D02, D09a, D10b, M03, H04

Product	Supplier	Adj. No.	Type
Intracrop Status	Intracrop	A0506	adjuvant/vegetable oil

Contains	91% w/w methylated rapeseed oil
Use with	All approved pesticides on non-edible crops, and all pesticides approved for use on growing edible crops when used at half recommended dose or less. On specified crops product may be used at a maximum spray concentration of 1.0% with approved pesticides at their full approved rate up to the growth stages shown in the label
Precautions	U19a, U20b, D05, D09a, D10b

Product	Supplier	Adj. No.	Type
Intracrop Stay-Put	Intracrop	A0507	vegetable oil

Contains	91% w/w methylated rapeseed oil
Use with	Recommended rates of approved pesticides for non-crop uses and with recommended rates of approved pesticides up to the growth stage indicated for specified crops, and with half or less than the recommended rate on these crops after the stated growth stages
Precautions	U19a, U20b, D05, D09a, D10b

Product	Supplier	Adj. No.	Type
Intracrop Super Rapeze MSO	Intracrop	A0491	adjuvant/vegetable oil

Contains	91.7% w/w methylated rapeseed oil
Use with	All approved pesticides on non-edible crops, and all pesticides approved for use on growing edible crops when used at half recommended dose or less. On specified crops product may be used at a maximum spray concentration of 1.78% with approved pesticides at their full approved rate up to the growth stages shown in the label
Precautions	U19a, U20b, D05, D09a, D10b

Product	Supplier	Adj. No.	Type
Intracrop Warrier	Intracrop	0514	spreader/wetter

Contains	192 g/l primary alcohol ethylene
Use with	Recommended rates of approved pesticides up to the growth stage indicated for specified crops, and with half or less than the recommended rate on these crops after the stated growth stages. Also with herbicides on managed amenity turf at recommended rates, and on grassland at half or less than recommended rates
Protective clothing	A, C
Precautions	R36, R38, R41, U05a, U08, U11, U14, U15, U20c, E15a, D01, D02, D09a, D10b, M03, H04

Product	Supplier	Adj. No.	Type
Katalyst	Interagro	A0450	penetrant/water conditioner
Contains	90% w/w alkoxylated fatty amine		
Use with	Glyphosate and a wide range of other pesticides on specified crops. Refer to label or contact supplier for further details		
Protective clothing	A, C		
Precautions	R22a, R36, R38, R50, R58, U02a, U05a, U08, U19a, U20b, E15a, E34, E37, D01, D02, D05, D09a, D10a, M03, H03, H11		
Kinetic	Helena	A0252	spreader/wetter
Contains	80% w/w polyoxypropylene-polyoxyethylene glycol 20% w/w polyalkylene oxide modified heptamethyl trisiloxane		
Use with	Approved pesticides on non-edible crops and cereals and stubbles of all edible crops when used at full recommended dose, and with pesticides approved for use on growing edible crops when used at half recommended dose or less. On specified crops product may be used at a maximum spray concentration of 0.2% with approved pesticides at their full approved rate up to the growth stages shown in the label		
Protective clothing	A, H		
Precautions	R38, R41, U02a, U05a, U08, U19a, U20b, E13c, E34, E37, D01, D02, D05, D09a, D10c, M03, H04		
Klipper	Amega	A0260	spreader/wetter
Contains	600 g/l ethylene oxide condensate		
Use with	Approved salt formulations of mecoprop alone or in mixtures with 2,4-D on managed amenity turf		
Protective clothing	A, C		
Precautions	R36, R38, U05a, U08, U20c, E15a, D01, D02, D09a, D10a, H04, H08		
Landgold Minoan Two	Teliton	A0523	wetter
Contains	20.2 % w/w sodium salt of 3,6-dioxaoctadecylsulphate and 6.7% w/w sodium salt of 3,6-dioxaeicosylsulphate		
Use with	Approved cereal herbicides		
Protective clothing	A, C		
Precautions	R36, R38, R52, U02a, U05a, U08, U13, U19a, E15a, D01, D02, D05, D09a, D10b, H04		
Leaf-Koat	Helena	A0511	spreader/wetter
Contains	80% w/w polyoxypropylene-polyoxyethylene glycol 20% w/w polyalkylene oxide modified heptamethyl trisiloxane		
Use with	Approved pesticides on non-edible crops and cereals and stubbles of all edible crops when used at full recommended dose, and with pesticides approved for use on growing edible crops when used at half recommended dose or less. On specified crops product may be used at a maximum spray concentration of 0.2% with approved pesticides at their full approved rate up to the growth stages shown in the label		
Protective clothing	A, H		
Precautions	R38, R41, U02a, U05a, U08, U19a, U20b, E13c, E34, E37, D01, D02, D05, D09a, D10c, M03, H04		

SECTION 4

Product	Supplier	Adj. No.	Type
LI-700	De Sangosse	A0176	acidifier/drift retardant/penetrant
Contains	350 g/l modified soya lecithin, 100 g/l alkylphenyl hydroxypolyoxyethylene and 350 g/l propionic acid		
Use with	All plant growth regulators for cereals and oilseed rape, dimethoate and chlorpyrifos for wheat bulb fly control. For other uses see label or contact supplier for details		
Protective clothing	A, C, H		
Precautions	R36, R38, U11, U14, U15, U19a, E15a, E19b, D01, D02, D05, D10a, D12a, H04		
Libsorb	Ciba Specialty	A0438	spreader/wetter
Contains	alkyl alcohol ethoxylate		
Use with	Any spray for which additional wetter is approved and recommended		
Protective clothing	A, C		
Precautions	R36, R38, U05a, U08, U20a, E13c, D01, D02, D09a, D10a, H04		
Logic	Microcide	A0288	vegetable oil
Contains	95% emulsified vegetable oil		
Use with	All approved pesticides on edible and non-edible crops for ground or aerial application		
Protective clothing	A, C		
Precautions	U19a, U20b, D09a, D10b		
Logic Oil	Microcide	A0293	vegetable oil
Contains	95% emulsified vegetable oil		
Use with	All approved pesticides on edible and non-edible crops for ground or aerial application		
Protective clothing	A, C		
Precautions	U19a, U20b, D09a, D10b		
Low Down	Helena	A0459	mineral oil
Contains	732 g/l refined paraffinic petroleum oil		
Use with	All approved pesticides on edible and non-edible crops when used at half their approved dose or less. Also with approved pesticides on specified crops, up to specified growth stages, at up to their full approved dose		
Protective clothing	A, C		
Precautions	R53a, U02a, U05a, U08, U20b, E15a, E34, E37, D01, D02, D05, D09a, D10c, H11		
Luxan Non-Ionic Wetter	Luxan	A0139	non-ionic surfactant/spreader/wetter
Contains	900 g/l alkyl phenol ethylene oxide condensate		
Use with	Any spray for which additional wetter/spreader is approved and recommended		
Protective clothing	A, C		
Precautions	R36, R38, U05a, U08, U20c, E13c, E34, D01, D02, D09a, D10b, H04, H08		

Product	Supplier	Adj. No.	Type
Mangard	Mandops	A0132	adjuvant/anti-transpirant
Contains	96% w/w di-1-p-menthene		
Use with	Glyphosate and a wide range of other herbicides, fungicides and insecticides. See label for details		
Precautions	U20c, E15a, D09a, D10a		
Mediator Sun	Nufarm UK	A0512	wetter
Contains	655 g/l terpenic alcohols		
Use with	All approved pesticides and plant growth regulators on a wide range of specified crops at full dose up to specified growth stages, and at half dose at later growth stages		
Precautions	R36, R52, R53a, U11, U15, D01, M05a, H04		
Mixture B	Amega	A0161	non-ionic surfactant/spreader/wetter
Contains	500 g/l nonyl phenol ethylene oxide condensate and 500 g/l primary alcohol ethylene oxide condensate		
Use with	Approved pesticides applied in amenity and forestry situations and on a range of specified arable and vegetable crops		
Protective clothing	A, C		
Precautions	R21, R22a, R36, R38, U05a, U08, U20b, E13c, E34, E37, D01, D02, D09a, D10a, M03, H03, H04		
Nettle	Interagro	A0466	mineral oil
Contains	60% w/w mineral oil		
Use with	Cereal graminicides and a wide range of other pesticides that have a label recommendation for use with authorised adjuvant oils on specified crops. Refer to label or contact supplier for further details		
Protective clothing	A		
Precautions	R22b, R38, U02a, U05a, U08, U20b, E15a, D01, D02, D05, D09a, D10a, M05b, H03		
Newman Cropspray 11E	De Sangosse	A0195	mineral oil
Contains	99% highly refined mineral oil		
Use with	Approved herbicides for which the addition of an adjuvant oil is recommended		
Protective clothing	A, C		
Precautions	R22a, U10, U19a, U20a, E15a, E19b, E34, E37, D01, D02, D05, D10a, M05b, H03		
Newman's T-80	De Sangosse	A0192	spreader/wetter
Contains	78% w/w polyoxyethylene tallow amine		
Use with	Glyphosate		
Protective clothing	A, C, H		
Precautions	R22a, R37, R38, R41, R50, R58, R67, U05a, U11, U14, U19a, U20a, E15a, E19b, E34, E37, D01, D02, D05, D10a, D12a, M03, M04, H03, H08, H11		

SECTION 4

Product	Supplier	Adj. No.	Type
Nion	Amega	A0415	spreader/wetter

Contains	90% (900 g/l) ethylene oxide condensate
Use with	Recommended rates of approved pesticides up to specified growth stages of a wide range of agricultural arable and horticultural crops. Use on edible crops beyond specified growth stages should only be with half recommended rates of the pesticide or less. See label for details of growth stage restrictions.
Protective clothing	A, C
Precautions	R36, R38, U05a, U08, U19a, U20b, E13c, E37, D01, D02, D09a, D10a, H04

Product	Supplier	Adj. No.	Type
Nu Film P	Intracrop	A0039	sticker/wetter

Contains	96% poly-1-p-menthene
Use with	Glyphosate and many other pesticides and growth regulators for which a protectant is recommended. Do not use in mixture with adjuvant oils or surfactants
Precautions	U19a, U20b, E13c, D09a, D10a

Product	Supplier	Adj. No.	Type
Nufarm Cropoil	Nufarm UK	A0447	mineral oil

Contains	99% highly refined mineral oil
Use with	Approved pesticides for use on certain specified crops (cereals, combinable break crops, field and dwarf beans, peas, oilseed rape, brassicas, potatoes, carrots, parsnips, sugar beet, fodder beet, mangels, red beet, maize, sweetcorn, onions, leeks, horticultural crops, forestry, amenity, grassland)
Protective clothing	A, C
Precautions	R43, U19a, U20b, E13c, D05, D09a, D10b, H04

Product	Supplier	Adj. No.	Type
Nufarm Cropoil Gold	Nufarm UK	A0446	adjuvant/vegetable oil

Contains	97.5% w/w methylated rapeseed oil
Use with	A wide range of pesticides used in agriculture, horticulture, forestry and amenity situations. See label for details of crops and dose rates.
Protective clothing	A, H
Precautions	R43, U05a, U15, U19a, U20b, E13c, D01, D02, D05, D09a, D10b, H04

Product	Supplier	Adj. No.	Type
Output	Syngenta	A0429	mineral oil/surfactant

Contains	60% mineral oil and 40% surfactants
Use with	Grasp
Protective clothing	A, C
Precautions	R38, U05a, U08, U19a, U20a, E13c, D01, D02, D09a, D10b, H04

Product	Supplier	Adj. No.	Type
Pan Oasis	Pan Agriculture	A0411	vegetable oil

Contains	95% w/w methylated rapeseed oil
Use with	A wide range of pesticides that have a label recommendation for use with authorised adjuvant oils. Contact distributor for further details
Protective clothing	A
Precautions	R36, U02a, U05a, U08, U20b, E15a, D01, D02, D05, D09a, D10a, H04

Product	Supplier	Adj. No.	Type
Pan Panorama	Pan Agriculture A0412		mineral oil
Contains	95% w/w mineral oil		
Use with	A wide range of pesticides that have a label recommendation for use with adjuvant oils. Contact distributor for details		
Protective clothing	A		
Precautions	R22b, R38, U02a, U05a, U08, U20b, E13c, D01, D02, D05, D09a, D10a, M05b, H03		
Phase II	De Sangosse A0331		vegetable oil
Contains	95% w/w esterified rapeseed oil		
Use with	Pesticides approved for use in sugar beet, oilseed rape, cereals (for grass weed control), and other specified agricultural and horticultural crops		
Protective clothing	A, C, H		
Precautions	U19a, U20b, E15a, E19b, E34, D01, D02, D05, D10a, D12a		
Pin-o-Film	Intracrop A0209		sticker/wetter
Contains	96% di-1-p-menthene		
Use with	Recommended rates of approved pesticides up to the growth stage indicated for specified crops, and with half or less than the recommended rate on these crops after the stated growth stages. Also for use with pesticides on grassland at half or less than their recommended rates. Must not be used in mixture with adjuvant oils or surfactants		
Precautions	U19a, U20c, E13c, C02, D09a, D11a		
Planet	Intracrop A0224		non-ionic surfactant/spreader/wetter
Contains	85% alkyl polyglycol ether and fatty acid		
Use with	Any spray for which additional wetter is recommended		
Precautions	R22a, R36, U05a, U08, U19a, U20b, E13c, D01, D09a, D10a, D11a, H03, H04, H08		
Prima	De Sangosse A0310		mineral oil
Contains	99% highly refined mineral oil		
Use with	All pesticides on all crops when used up to 50% of maximum approved dose for that use (mixtures with Roundup must only be used for treatment of stubbles); all approved pesticides at full dose on listed crops (see label). For other uses see label or contact supplier for details		
Protective clothing	A, C		
Precautions	R22a, U10, U19a, E15a, E19b, E34, E37, D01, D02, D05, D10a, D12a, M05b, H03		
Profit Oil	Microcide A0294		extender/sticker/wetter
Contains	95% rape/soya oil		
Use with	All approved pesticides		
Protective clothing	A, C		
Precautions	U19a, U20b, D09a, D10b		

SECTION 4

Product	Supplier	Adj. No.	Type
Qdos Brefi	Me2	A0539	wetter
Contains	20.2 % w/w sodium salt of 3,6-dioxaoctadecylsulphate and 6.7% w/w sodium salt of 3,6-dioxaeicosylsulphate		
Use with	Approved cereal herbicides		
Protective clothing	A, C		
Precautions	R36, R38, U02a, U05a, U08, U13, U19a, U20b, E13c, E34, D01, D02, D05, D09a, D10b, H04		
Qdos Chikara	Me2	A0540	adjuvant/vegetable oil
Contains	95.2% esterified rapeseed oil		
Use with	All approved pesticides on non-edible crops. Also with approved pesticides on specified edible crops, up to specified growth stages, up to their full approved dose and on other edible crops only when used at half or less than half the approved pesticide dose		
Protective clothing	A, C, H		
Precautions	U02a, U20b, E15a, E34, D01, D09a, D10b		
QuikSilver	Interagro	A0471	non-ionic surfactant/spreader/wetter
Contains	75% w/w nonylphenol ethylene oxide condensate + fatty acids		
Use with	Pesticides that have a label recommendation for the use of an authorised non-ionic spreader or wetter		
Protective clothing	A, C		
Precautions	R22a, R36, R38, R51, R58, U02a, U05a, U08, U20a, E15a, E34, D01, D02, D05, D09a, M03, H03, H11		
Ranger	De Sangosse	A0296	acidifier/drift retardant/penetrant
Contains	350 g/l modified soya lecithin, 100 g/l alkylphenyl hydroxypolyoxyethylene and 350 g/l propionic acid		
Use with	All plant growth regulators for cereals and oilseed rape, dimethoate and chlorpyrifos for wheat bulb fly control. For other uses see label or contact supplier for details		
Protective clothing	A, C, H		
Precautions	R36, R38, U11, U14, U15, U19a, E15a, E19b, D01, D02, D05, D10a, D12a, H04		
Rapide	Intracrop	A0116	penetrant/surfactant
Contains	40% propionic acid		
Use with	A wide range of pesticides and growth regulators, especially chlormequat		
Protective clothing	A, C		
Precautions	R34, R36, R38, U02a, U05a, U08, U19a, U20a, D01, D02, D09a, D10a, H04, H05		
Reward Oil	Microcide	A0295	extender/sticker/wetter
Contains	95% rape/soya oil		
Use with	All approved pesticides		
Protective clothing	A, C		
Precautions	U19a, U20b, D09a, D10b		

Product	Supplier	Adj. No.	Type
Ryda	Interagro	A0168	cationic surfactant/wetter

Contains: 800 g/l polyethoxylated tallow amine

Use with: Glyphosate

Protective clothing: A, C

Precautions: R22a, R36, R38, R41, R51, R58, U02a, U05a, U08, U13, U19a, U20a, E15a, E34, E37, D01, D02, D05, D09a, D10b, M03, H03, H08, H11

Product	Supplier	Adj. No.	Type
Saracen	Interagro	A0368	vegetable oil

Contains: 95% w/w methylated vegetable oil

Use with: Cereal graminicides and a wide range of other pesticides that have a recommendation for use with authorised adjuvant oils on specified crops. Refer to label or contact supplier for further details

Precautions: U05a, U20b, E15a, D01, D02, D09a, D10a

Product	Supplier	Adj. No.	Type
SAS 90	Intracrop	A0311	spreader/wetter

Contains: 100% polyoxyethylen-alpha-methyl-omega-[3-(1,3,3,3-tetramethyl-3-trimethylsiloxy)-disiloxanyl] propylether

Use with: A wide range of pesticides on specified crops in agriculture and horticulture, and on amenity vegetation and land not intended for cropping

Precautions: U02a, U08, U19a, U20b, E13c, E34, D09a, D10a

Product	Supplier	Adj. No.	Type
Scout	De Sangosse	A0207	acidifier/drift retardant/non-ionic surfactant/penetrant

Contains: 350 g/l modified soya lecithin, 100 g/l alkylphenyl hydroxypolyoxyethylene and 350 g/l propionic acid

Use with: All plant growth regulators for cereals and oilseed rape, dimethoate and chlorpyrifos for wheat bulb fly control. For other uses see label or contact supplier for details

Protective clothing: A, C, H

Precautions: R36, R38, U11, U14, U15, U19a, E15a, E19b, D01, D02, D05, D10a, D12a, H04

Product	Supplier	Adj. No.	Type
Signal	Intracrop	A0308	extender/sticker

Contains: 450 g/l synthetic latex

Use with: All approved pesticides where addition of a sticker/extender is required and recommended

Protective clothing: A, C

Precautions: R36, R38, U02a, U05a, U08, U19a, U20b, E13c, D01, D02, D09a, D10a, M03, H04

Product	Supplier	Adj. No.	Type
Siltex	Intracrop	A0398	spreader/sticker/wetter

Contains: 27% w/w polyoxyethylen-alpha-methyl-omega-[3-(1,3,3,3-tetramethyl-3-trimethylsiloxy)-disiloxanyl] propylether and 7.3% w/w cocoiminodipropionate

Use with: Approved pesticides used at full dose on a range of specified fruit and vegetable crops, and on managed amenity turf

Protective clothing: A, C

Precautions: U05a, U08, U20c, E13c, E37, D01, D02, D09a, D10a

SECTION 4

Product	Supplier	Adj. No.	Type
Silwet L-77	De Sangosse	A0193	drift retardant/spreader/wetter
Contains	80% w/w polyalkylene oxide modified heptamethyltrisiloxane and a maximum of 20 % w/w allyloxypolyethylene glycol methyl ether		
Use with	All approved fungicides on winter and spring sown cereals; all approved pesticides applied at 50% or less of their full approved dose. A wide range of other uses. See label or contact supplier for details		
Protective clothing	A, C, H		
Precautions	R20, R21, R22a, R36, R43, R48, R51, R58, U11, U15, U19a, E15a, E19b, D01, D02, D05, D10a, D12a, H03, H11		
Slippa	Interagro	A0206	spreader/wetter
Contains	655 g/l polyalkyleneoxide modified heptamethyltrisiloxane		
Use with	Cereal fungicides and a wide range of other pesticides and trace elements on specified crops		
Protective clothing	A, C, H		
Precautions	R20, R38, R41, R43, R48, R51, R58, U02a, U05a, U08, U11, U19a, U20a, E15a, E34, E37, D01, D02, D05, D09a, D10a, M03, H03, H11		
SM99	De Sangosse	A0134	mineral oil
Contains	99% w/w highly refined paraffinic oil		
Use with	Approved herbicides for which the addition of an adjuvant oil is recommended		
Protective clothing	A, C		
Precautions	R22a, U10, U19a, E15a, E19b, E34, E37, D01, D02, D05, D10a, D12a, M05b, H03		
Solar	Intracrop	A0225	activator/non-ionic surfactant/spreader
Contains	75% polypropoxypropanol and 15% alkyl polyglycol ether		
Use with	Foliar applied plant growth regulators		
Protective clothing	A, C		
Precautions	R36, R38, U04a, U05a, U08, U20a, E13c, E34, D01, D02, D09a, D10b, D11a, M03, H04		
Spartan	Interagro	A0375	penetrant/water conditioner
Contains	500 g/l alkoxylated fatty amine + 400 g/l polyoxyethylene monolaurate		
Use with	Cereal growth regulators, cereal herbicides, oilseed rape fungicides and a wide range of other pesticides on specified crops. Refer to label or contact supplier for further details		
Protective clothing	A, C		
Precautions	R22a, R36, R38, R51, R58, U02a, U05a, U19a, U20b, E15a, E34, E37, D01, D02, D05, D09a, D10a, M03, H03, H11		
Sprayfast	Mandops	A0131	extender/sticker/wetter
Contains	334 g/l di-1-p-menthene and nonyl phenol ethylene oxide condensate		
Use with	Glyphosate and other pesticides, growth regulators or nutrients for which a coating agent is approved and recommended		
Precautions	U20b, E13c, D09a, D10a		

Product	Supplier	Adj. No.	Type
Sprayfix	De Sangosse	A0297	drift retardant/extender/sticker/wetter
Contains	450 g/l synthetic latex and 100 g/l alkylphenyl hydroxypolyoxyethylene		
Use with	Blight sprays on potatoes and other pesticides on a wide range of crops. See label or contact supplier for details		
Protective clothing	A, C, H		
Precautions	R36, R38, U11, U14, U19a, E15a, E19b, D01, D02, D05, D09a, D10a, D12a, H04		
Spraygard	Mandops	A0394	anti-transpirant/extender/sticker/wetter
Contains	400 g/l di-1-p-menthene		
Use with	Approved pesticides on all edible crops (except herbicides on peas) and as an anti-transpirant on vegetables before transplanting, evergreens, deciduous trees, shrubs, bushes and on turf		
Precautions	U20b, E13c, E37, D09a, D10a, M03		
Spraymac	De Sangosse	A0298	acidifier/non-ionic surfactant
Contains	350 g/l propionic acid and 100 g/l alkylphenyl hydroxypolyoxyethylene		
Use with	All plant growth regulators for cereals and oilseed rape and any spray for which the addition of a wetter/spreader is recommended. See label or contact supplier for details		
Protective clothing	A, C, H		
Precautions	R34, U04a, U10, U11, U14, U19a, U20b, E15a, E19b, E34, D01, D02, D05, D10a, D12a, M04, H05		
Sprayprover	Fine	A0238	mineral oil
Contains	95% highly refined mineral oil		
Use with	Any spray for which the addition of a mineral oil is approved and recommended		
Protective clothing	A, C		
Precautions	R36, R38, U05a, U08, U19a, U20a, E13c, D01, D02, D09a, D10b, H04		
Spread and Seal	Mandops	A0449	extender/sticker/wetter
Contains	334 g/l d-1-p-menthene		
Use with	All pesticides on all edible crops when used at up to 50% of their approved dose, and all approved forulations of glyphosate when used on cereals. Must not be used in cobination with herbicides applied to pea crops		
Precautions	U20c, E13c, E15a, D09a, D10a		
Stamina	Interagro	A0202	penetrant
Contains	100% w/w alkoxylated fatty amine		
Use with	Glyphosate and a wide range of other pesticides on specified crops		
Protective clothing	A, C		
Precautions	R22a, R38, R50, R58, U02a, U05a, U20a, E15a, E34, D01, D02, D05, D09a, D10a, M03, H03, H11		

SECTION 4

Product	Supplier	Adj. No.	Type
Standon Shiva	Standon	A0519	wetter

Contains	20.2 % w/w sodium salt of 3,6-dioxaoctadecylsulphate and 6.7% w/w sodium salt of 3,6-dioxaeicosylsulphate
Use with	Approved cereal herbicides
Protective clothing	A, C
Precautions	R36, R38, R52, U02a, U05a, U08, U13, U19a, E15a, D01, D02, D05, D09a, D10b, H04

Stika	De Sangosse	A0452	extender/sticker/wetter

Contains	225 g/l synthetic latex and 100 g/l alkylphenyl hydroxypolyoxyethylene
Use with	Approved pesticides on non-edible crops where the use of a sticking, wetting or extending agent is recommended, and pesticides approved for use on growing edible crops when used at half recommended dose or less. On specified crops product may be used at a maximum spray concentration of 0.14% with approved pesticides at their full approved rate up to the growth stages shown in the label
Protective clothing	A, C, H
Precautions	U11, U14, U19a, E15a, E19b, D01, D02, D05, D10a, D12a

Super Nova	Interagro	A0364	penetrant

Contains	500 g/l alkoxylated fatty amine + 500 g/l polyoxyethylene monolaurate
Use with	Cereal growth regulators, cereal herbicides, cereal fungicides, oilseed rape fungicides and a wide range of other pesticides on specified crops
Precautions	R51, R58, U05a, E15a, E34, D01, D02, D05, D09a, D10a, H11

Sward	Amega	A0215	spreader/wetter

Contains	15.2% w/w polyalkylene oxide modified heptomethyltrisiloxane
Use with	Specified pesticides in amenity situations
Protective clothing	A, C
Precautions	R41, R43, U05a, U08, U20c, E13c, D01, D02, D09a, D10a, H04

Talzene	Intracrop	A0233	wetter

Contains	800 g/l polyoxyethylene tallow amine
Use with	Herbicides and dessicants up to their latest recommended time of application in agriculture, horticulture, amenity and forestry
Protective clothing	A, C
Precautions	R22a, R36, R38, R41, R48, U05a, U08, U11, U13, U19a, U20b, E13c, E34, D01, D02, D09a, D10a, M03, H03

Toil	Interagro	A0248	vegetable oil

Contains	95% w/w methylated rapeseed oil
Use with	Sugar beet herbicides, oilseed rape herbicides, cereal graminicides and a wide range of other pesticides that have a label recommendation for use with authorised adjuvant oils on specified crops. Refer to label or contact supplier for further details
Precautions	U05a, U20b, E15a, D01, D02, D05, D09a, D10a

Product	Supplier	Adj. No.	Type
Torpedo II	De Sangosse	A0393	penetrant/wetter
Contains	215.5 g/litre alkoxylated tallow amine, 375 g/litre alkylphenyl hydroxypolyoxyethylene and 75 g/litre natural fatty acids		
Use with	All pesticides on edible and non-edible crops when used at half or less than the full approved dose and with products up to their full approved dose on a wide range of specified fruit and vegetable crops.		
Protective clothing	A, C, H		
Precautions	R38, R41, R52, R58, U05a, U10, U11, U19a, E15a, E19b, E34, E37, D01, D02, D05, D10a, D12a, M03, H04, H08, H11		
Transcend	Helena	A0333	adjuvant/vegetable oil
Contains	80% w/w methylated soybean oil and 15% w/w polyalkylene oxide modified heptamethyl trisiloxane		
Use with	Approved pesticides on non-edible crops, cereals and stubbles of all edible crops, grassland (destruction), and pesticides approved for use on growing edible crops when used at half recommended dose or less. On specified crops product may be used at a maximum spray concentration of 0.5% with approved pesticides at their full approved rate up to the growth stages shown in the label		
Protective clothing	A, H		
Precautions	R38, R41, U02a, U05a, U08, U11, U19a, U20b, E13c, E34, E37, D01, D02, D05, D09a, M03, H04		
Try-Flex	Greenaway	A0489	sticker/wetter
Contains	95% refined rapeseed oil		
Use with	Gly-480 (MAPP 09837) and Freeway (MAPP 11129) on natural Surfaces not intended to bear vegetation, permeable surfaces overlying soil, hard surfaces and forest		
Protective clothing	A, C, D, H, M		
Precautions	U08, U20b, E13c, E37, D02, D05, D10a		
Wetta Plus	Interagro	A0480	non-ionic surfactant/spreader/wetter
Contains	75% w/w nonylphenol ethylene oxide condensate + fatty acids		
Use with	Pesticides that recommend the use of an authorised non-ionic spreader or wetter		
Protective clothing	A, C		
Precautions	R22a, R36, R38, R51, R58, U02a, U05a, U08, U20a, E15a, E34, D01, D02, D05, D09a, M03, H03, H11		
Zarado	De Sangosse	A0516	vegetable oil
Contains	70% w/w esterified rapeseed oil		
Use with	All approved pesticides on edible and non-edible crops when used at half their approved dose or less. Also with approved pesticides on specified crops, up to specified growth stages, at up to their full approved dose		
Protective clothing	A, C, H		
Precautions	R43, R52, U05a, U13, U19a, U20b, E15a, E19b, E34, D01, D02, D05, D10a, M05a, H04		

SECTION 4

SECTION 5
USEFUL INFORMATION

Pesticide Legislation

Anyone who advertises, sells, supplies, stores or uses a pesticide is bound by legislation, including those who use pesticides in their own homes, gardens or allotments. There are numerous UK statutory controls, but the major legal instruments are outlined below.

The Food and Environment Protection Act 1985 (FEPA) and Control of Pesticides Regulations 1986 (COPR)

FEPA introduced statutory powers to control pesticides, with the aims of protecting human beings, creatures and plants, safeguarding the environment, ensuring safe, effective and humane methods of controlling pests, and making pesticide information available to the public.

Control of pesticides is achieved by COPR. These Regulations lay down the Approvals required before any pesticide may be sold, stored, supplied, advertised or used, subject to conditions which are contained in a series of Schedules. Schedule 1 relates to advertisement; Schedule 2 to sale, supply and storage; Schedule 3 to use; and Schedule 4 to aerial application of pesticides. The Schedules may be changed at any time, following Parliamentary approval. Details are given on the websites of the Pesticides Safety Directorate (PSD, www.pesticides.gov.uk) and the Health and Safety Executive (HSE, www.hse.gov.uk).

The controls currently in force include the following.

- Only approved products may be sold, supplied, stored, advertised or used.
- No advertisement may contain any claim for safety beyond that which is permitted in the approved label text.
- Only products specifically approved for the purpose may be applied from the air.
- A recognised Storeman's Certificate of Competence is required by anyone who stores for sale or supply pesticides approved for agricultural use.
- A recognised Certificate of Competence is required by anyone who gives advice when selling or supplying pesticides approved for agricultural use.
- Users of pesticides must comply with the Conditions of Approval relating to use.
- A recognised Certificate of Competence is required for all contractors and for persons born after 31 December 1964 applying pesticides approved for agricultural use (unless working under direct supervision of a certificate holder). Proposals are now in place that every user of agricultural pesticides will have to hold a Certificate of Competence regardless of age or supervision. A suitable transition period will allow this requirement to be enacted.
- Only those adjuvants authorised by PSD may be used.
- Regarding tank-mixes, 'no person shall combine or mix for use two or more pesticides which are anti-cholinesterase compounds unless such a mixture is expressly permitted by the conditions of approval given in relation to at least one of those pesticides', and 'no person shall combine or mix for use two or more pesticides unless all the conditions of Approval given in relation to each of those pesticides can be complied with'.

The Plant Protection Products Directive

European Council Directive 91/414/EEC is intended to harmonise national arrangements for the authorisation of plant protection products within the European Community. It became effective in the UK on 25 July 1993. Under the provisions of the Directive, individual Member States are responsible for authorisation within their own territory of products containing active substances that appear in a list agreed at Community level. This list, known as Annex I, is being created over a period of time by review of existing active ingredients (to ensure they meet present safety standards) and authorisation of new ones. Nearly 100 active substances had achieved Annex I listing by the time of printing this edition.

Individual Member States are required to amend their national arrangements and legislation in order to meet the requirements of Directive 91/414/EEC. In the UK this has been achieved by a series of Plant Protection Products Regulations (PPPR) under which, over a period of time, all agricultural and horticultural pesticides will come to be regulated. Meanwhile, existing product approvals are being maintained under COPR, and new ones are granted for products containing active ingredients that were already on the market by 25 July 1993. Products containing new active substances that were not on the market at this date are being granted provisional approval in advance of Annex I listing of their active ingredients. As active ingredients are placed on Annex I of the Directive, products containing only those active ingredients will be regulated solely under PPPR. Active ingredients in this *Guide* that have been included in Annex I are identified in the 'Approval' section of the profile. The Directive also provides for a system of 'mutual recognition' of products registered in other Member States. Annex I listing, and a relevant approval in the Member State on which the mutual recognition is to be based, are essential prerequisites. Authorisation may be subject to conditions set to take account of differences such as climate and agricultural practice.

The Review Programme

The process of reviewing existing active ingredients is taking considerably longer than originally anticipated. The programme is designed to ensure that all available plant protection products are supported by up-to-date information on safety and efficacy. At the start of the programme it was envisaged that all active substances that were on the market on 25 July 1993 would have been reviewed by 25 July 2003, but the process is now not scheduled for completion until December 2008. Approximately 850 substances on the EU market are being handled in four phases but, because of the cost of providing this information, many are not being supported and their approvals are being revoked. By the end of 2003 about half the original list had been lost. For the remainder, the complex packages of data have to be evaluated, and further losses are likely following scientific risk assessments. A few substances are temporarily reprieved by derogations for 'Essential Uses' granted by the European Commission (see below). However, even allowing for this temporary relief, the effect on the horticulture industry will be serious, with fewer products, especially herbicides, available and little prospect of new developments for vegetable crops coming forward.

Derogations for 'Essential Uses' Permitted until 31 December 2007

Because of the scale of the likely loss of compounds and products, and the impact this would have on some sectors, requests have been made to the European Commission by Member States to allow the use of a few active substances after 2003 for key uses where no alternative exists. UK requests were coordinated by BCPC and the Horticulture Development Council (HDC). Against this background, the Commission has made provision for certain 'Essential Uses' to be maintained until December 2007, in order to provide time for alternatives to be researched and developed. In the UK this includes 11 herbicides for vegetables, two fungicides, two insecticides, two disinfectants, one acaricide, one nematicide and one fumigant (see list below).

These 'Essential Uses' are only for the crops specified and apply only to SOLAs or products already registered for the use. Extrapolations to other crops are not permitted other than those specifically identified as permissible under the Long-Term Arrangements for Extension of Use in Table 1.1 below. It is possible that there will be further losses of products as a result of product rationalisation by companies, or failure of some active substances to achieve Annex I listing. Even where active substances are supported in EC Review rounds 2 and 3, not all uses will be supported, and growers need to be aware of these so that requests for SOLAs can be made.

Table 1.1 Active substances and products permitted, for the following 'Essential Uses' only, until 31 Dec 2007 (unless approval expiry precedes this date)

Activity	Product	Active substance	Approved Essential Use(s)	Off-label Essential Use(s)
Acaricides	Bye Bye 20EC (09346)	amitraz	pears, trees after harvest	
	Cleancrop Bye Bye 20EC (11530)	amitraz	pears, trees after harvest	
	Mitac HF (07358)	amitraz	pears, trees after harvest	
Fumigants	Certis 2-Aminobutane (11182)	2-aminobutane	potato (seed) stored	
	Hortichem 2-Aminobutane (06147)	2-aminobutane	potato (seed) stored	
Fungicides	Nectec Paste* (08510)	azaconazole + imazalil	ornamental plant production	
	Plantvax 75 (01601)	oxycarboxin	ornamentals (outdoor and protected)	
Herbicides	Aconite 50 (10642)	atrazine	sweetcorn	
	Alpha Atrazine 50 SC (04877)	atrazine	farm forestry, forest, sweetcorn	
	Alpha Atrazine 50 WP (04793)	atrazine	sweetcorn	
	Alpha Prometryne 50 WP (04871)	prometryn	carrots, celery, leeks (transplanted), parsley, parsnips (LTAEU)	outdoor and protected herbs
	Alpha Prometryne 80 WP (04795)	prometryn	carrots, celery, leeks (transplanted), parsnips (LTAEU)	
	Alpha Simazine 50 SC (04801)	simazine	broad beans, field beans, hops, ornamental plant production, strawberries	asparagus, rhubarb, runner beans
	Alpha Simazine 50 WP (04879)	simazine	asparagus, broad beans, field beans, hops, ornamental shrubs, ornamental trees, rhubarb, strawberries	
	Alpha Simazine 80 WP (04800)	simazine	asparagus, broad beans, field beans, hops, ornamental shrubs, ornamental trees, rhubarb, strawberries	

SECTION 5

Activity	Product	Active substance	Approved Essential Use(s)	Off-label Essential Use(s)
Herbicides *cont.*	Atlas Brown* (07703)	pentanochlor + chlorpropham	carrots, celeriac, celery, Chrysanthemums, herbs (LTAEU), *Narcissus*, ornamentals (other) (LTAEU), parsley, parsnips, tulips	
	Atlas Solan 40 (07726)	pentanochlor	Anemone de Caen, carnations, carrots, celeriac, celery, chrysanthemums, foxgloves, freesias, herbs (LTAEU), larkspur, nursery stock (including conifers), ornamentals (other) (LTAEU), parsley, parsnips, roses, sweet peas, sweet williams, wallflowers	
	Atlas Somon (03045 and 07727)	sodium monochloro-acetate	Brussels sprouts, cabbages, collards (LTAEU), kale, leeks, onions	raspberries, loganberries, hybrid *Rubus* spp.
	Atrazol (07598)	atrazine	coniferous forestry, sweetcorn	
	Batallion* (08305)	terbutryn + terbuthylazine	broad beans, combining peas, field beans, lupins (LTAEU), edible podded peas, vining peas	
	Bullet* (11204)	cyanazine + pendimethalin	spring field beans, garlic (LTAEU), combining peas, shallots (LTAEU)	
	CleanCrop Vectro (10475)	cyanazine	bulb onions, broad beans, combining peas, garlic (LTAEU), leeks (LTAEU), winter oilseed rape, ornamental plant production (narcissi only), salad onions, shallots (LTAEU), vining peas	
	Croptex Bronze (04087)	pentanochlor	carrots, chrysanthemums, nursery stock, ornamentals (other) (LTAEU), parsnips, sweet peas	
	Croptex Bronze (11329)	pentanochlor	carrots, celery, chrysanthemums, herbs (LTAEU), nursery stock, ornamentals (other) (LTAEU), parsley, parsnips, sweet peas	

Activity	Product	Active substance	Approved Essential Use(s)	Off-label Essential Use(s)
Herbicides *cont.*	Croptex Chrome* (02415 and 11180)	fenuron + chlorpropham	spinach	runner beans (including crops grown under covers)
	Croptex Steel (02418)	sodium monochloro-acetate	Brussels sprouts, bulb onions, cabbages, collards (LTAEU), kale, leeks	cauliflowers, broccoli, calabrese, hops, raspberries, blackberries, loganberries, hybrid *Rubus* spp
	DG90 (11200)	atrazine	sweetcorn	
	Dosaflo (09351)	metoxuron	carrots, parsnips (LTAEU)	
	Flex (08885)	fomesafen	dwarf French beans, runner beans (LTAEU)	outdoor soybeans
	Fortrol (11174)	cyanazine	combining peas, bulb onions, broad beans, garlic (LTAEU), leeks (LTAEU), ornamental plant production (propagating material) (narcissi only), winter oilseed rape, salad onions, shallots (LTAEU), vining peas	broccoli, cabbages, calabrese, cauliflowers, collards, kale, farm forestry, forestry
	Gesagard (08410)	prometryn	carrots, celery (outdoor), parsley, parsnips (LTAEU)	outdoor and protected herbs, direct drilled and transplanted leeks and onions
	Gesaprim (08411)	atrazine	sweetcorn	
	Gesatop (08412)	simazine	broad beans, field beans, hops, ornamental plant production, roses, strawberries, tic beans	asparagus, runner beans, rhubarb
	Greencrop Amaize (11973)	atrazine	farm forestry, forest, sweetcorn	
	IT Cyanazine (10179)	cyanazine	bulb onions, broad beans, combining peas, garlic (LTAEU), winter oilseed rape, ornamental plant production (propagating material) (narcissi only), salad onions, strawberries (LTAEU), vining peas	
	Opogard* (08427)	terbutryn + terbuthylazine	broad beans, field beans, lupins, peas	

Activity	Product	Active substance	Approved Essential Use(s)	Off-label Essential Use(s)
Herbicides *cont.*	Pennine 50 (10638)	prometryn	carrots, celery, leeks (transplanted), parsley, parsnips (LTAEU), herbs (LTAEU)	
	Reflex T* (08884)	fomesafen + terbutryn	broad beans (spring sown), spring field beans, lupins (LTAEU), peas (spring sown)	
	Simazine 90 WG (10641)	simazine	asparagus, broad beans, field beans, hops, ornamental plant production, strawberries	
	Sinbar (01956)	terbacil		protected and outdoor herbs
	Sipcam Simazine Flowable (07622)	simazine	asparagus, broad beans, field beans, hops, ornamental plant production, rhubarb, roses (field grown nursery stock), shrubs (field grown nursery stock), strawberries, trees (field grown nursery stock)	
	Solan 40 (11897)	pentanochlor	Anemone de Caen, carnations, carrots, celeriac, celery, chrysanthemums, foxgloves, freesias, herbs (LTAEU), larkspur, nursery stock (including conifers), ornamental (other) (LTAEU), parsley, parsnips, roses, sweet peas, sweet williams, wallflowers	
	Unicrop Atrazine 50 (02645)	atrazine	forestry (coniferous), sweetcorn	
	Unicrop Atrazine FL (08045)	atrazine	sweetcorn	
	Unicrop Flowable Atrazine (05446)	atrazine	forestry (coniferous), sweetcorn	
	Unicrop Simazine 50 (02646)	simazine	asparagus, broad beans, field beans, hops, rhubarb, roses, shrubs, strawberries, trees (field grown nursery stock)	
	Unicrop Simazine FL (08032)	simazine	broad beans, field beans, hops, strawberries	

Activity	Product	Active substance	Approved Essential Use(s)	Off-label Essential Use(s)
Insecticides	Meothrin (07206 and 10400)	fenpropathrin	blackcurrants	
	Pynosect 30 Water Miscible* (01653 and 10934)	resmethrin + pyrethrins	mushrooms	
Nematicides	Agriguard Aldicarb (10481)	aldicarb	bulb onions, carrots, parsnips, potatoes	
	Me2 New Aldee (10434)	aldicarb	bulb onions, carrots, parsnips, potatoes	
	Standon Aldocarb 10G (11024)	aldicarb	bulb onions, carrots, parsnips, potatoes	
	Temik 10 G (10021)	aldicarb	potatoes, carrots, parsnips, onions	chrysanthemums, dahlias, herbaceous plants, nursery stock, shrubs, trees, protected roses, protected stool beds, protected ornamentals, pot plants
Disinfectants	Enforcer (09288)	dichlorophen	managed amenity turf, hard surfaces, ornamentals, glasshouses, crop standing areas	
	Jeyes Fluid (04606)	tar acids	surfaces in greenhouses, containers, boxes etc	

*Product in mixture with a supported active substance.
SOLA, specific off-label approval; LTAEU, long-term arrangements for extension of use.

The Control of Substances Hazardous to Health Regulations 1988 (COSHH)

The COSHH regulations, which came into force on 1 October 1989, were made under the Health and Safety at Work Act 1974, and are also important as a means of regulating the use of pesticides. The regulations cover virtually all substances hazardous to health, including those pesticides classed as Very toxic, Toxic, Harmful, Irritant or Corrosive; other chemicals used in farming or industry; and substances with occupational exposure limits. They also cover harmful micro-organisms, dusts and any other material, mixture, or compound used at work which can harm people's health.

The original Regulations, together with all subsequent amendments, have been consolidated into a single set of regulations: The Control of Substances Hazardous to Health Regulations 1994 (COSHH 1994).

The basic principle underlying the COSHH regulations is that the risks associated with the use of any substance hazardous to health must be assessed before it is used, and the appropriate measures taken to control the risk. The emphasis is changed from that pertaining under the

SECTION 5

Poisonous Substances in Agriculture Regulations 1984 (now repealed), whereby the principal method of ensuring safety was the use of protective clothing, to preventing or controlling exposure to hazardous substances by a combination of measures. In order of preference, the measures should be:

1 substitution with a less hazardous chemical or product
2 technical or engineering controls (e.g. the use of closed handling systems, etc.)
3 operational controls (e.g. operators located in cabs fitted with air-filtration systems, etc.)
4 use of personal protective equipment, which includes protective clothing.

Consideration must be given to whether it is necessary to use a pesticide at all in a given situation, and if so, the product posing the least risk to humans, animals and the environment must be selected. Where other measures do not provide adequate control of exposure and the use of personal protective equipment is necessary, the items stipulated on the product label must be used as a minimum. It is essential that equipment is properly maintained and the correct procedures adopted. Where necessary, the exposure of workers must be monitored, health checks carried out, and employees must be instructed and trained in precautionary techniques. Adequate records of all operations involving pesticide application must be made and retained for at least 3 years.

Certificates of Competence – the roles of BASIS and the NPTC

COPR, COSHH and other legislation places certain obligations on those who handle and use pesticides. Minimum standards are laid down for the transport, storage and use of pesticides, and the law requires those who act as storekeepers, sellers and advisors to hold recognised Certificates of Competence.

BASIS is an independent registration scheme for the pesticide industry, recognised under COPR. It is responsible for organising training courses and examinations to enable staff to obtain a Certificate of Competence. In addition, BASIS undertakes annual assessments of pesticide supply stores, enabling distributors, contractors and seedsmen to meet their obligations under the Code of Practice for Suppliers of Pesticides. Further information can be obtained from BASIS (www.basis-reg.co.uk; see Appendix 2).

Certain spray operators also require Certificates of Competence under the Control of Pesticides Regulations. These certificates are awarded by NPTC to candidates who pass an assessment carried by an approved NPTC or Scottish Skills Testing Service Assessor in one or more of a series of competence modules. Holders are required to produce their certificate on demand for inspection to any person authorised to enforce COPR. Further information can be obtained from NPTC (www.nptc.org.uk; see Appendix 2).

Maximum Residue Levels

A small number of pesticides are liable to leave residues in foodstuffs, even when used correctly. Where residues can occur, statutory limits, known as Maximum Residue Levels (MRLs), have been established. MRLs provide a check that products have been used as directed; they are not safety limits. However, they do take account of consumer safety because they are set at levels that ensure normal dietary intake of residues presents no risk to health. Wide safety margins are built in, and eating food containing residues above the MRL does not automatically imply a risk to health. Nevertheless, it is an offence to put into circulation any produce where the MRL is exceeded.

The surrounding legislation is complex. MRLs may be specified by several different bodies. The UK has set statutory MRLs since 1988. The European Union intends eventually to introduce MRLs for all pesticide/commodity combinations. These are being introduced initially by a series of priority lists, but will subsequently be covered by the review programme under Directive 91/414. However, in cases where no information is available, EU Directive 2000/42/EC requires many MRLs to be set at the limit of determination (LOD). As a result of this, certain approvals are being withdrawn where such use would leave residues above the MRL set in the Directive.

MRLs apply to imported as well as to home-produced foodstuffs. Details of those that have been set have been published in *The Pesticides (Maximum Residue Levels in Crops, Food and Feeding Stuffs) Regulations 1994*, and successive amendments to these Regulations. These Statutory Instruments are available from The Stationery Office.

MRLs are set for many chemicals not currently marketed in Britain. Because of this, and the ever-changing information on MRLs, no quantitative information is given in this *Guide*. Instead, users of the sister CD-ROM publication *The e-UK Pesticide Guide 2006* may gain direct weblink access to the MRL databases on the Pesticides Safety Directorate website (https://secure.pesticides.gov.uk/MRLs). This online database sets out in table form those levels specified by UK Regulations, EC Directives, and the *Codex Alimentarius* for each commodity.

The e-UK Pesticide Guide 2006 edition may be purchased from BCPC Publication Sales, 01420 593200, or online via www.bcpc.org/bookshop

SECTION 5

Approval (On-label and Off-label)

Only officially approved pesticides may be marketed and used in the UK. Approvals are granted by UK Government Ministers in response to applications that are supported by satisfactory data on safety, efficacy and, where relevant, humaneness. The Pesticides Safety Directorate (PSD, www.pesticides.gov.uk) is the UK Government Agency for regulating agricultural pesticides and plant protection products. The Health and Safety Executive (HSE, www.hse.gov.uk) fulfils the same role for other pesticides. The main focus of the regulatory process is the protection of human health and the environment.

Approvals are normally granted only in relation to individual products and for specified uses. It is an offence to use non-approved products or to use approved products in a manner that does not comply with the statutory conditions of use, except where the crop or situation is the subject of an off-label extension of use (see below).

Statutory Conditions of Use

Statutory conditions have been laid down for the use of individual products and may include:

- field of use (e.g. agriculture, horticulture etc.)
- crop or situations for which treatment is permitted
- maximum individual dose
- maximum number of treatments or maximum total dose
- maximum area or quantity which may be treated
- latest time of application or harvest interval
- operator protection or training requirements
- any other specific restrictions relating to particular pesticides.

All products must display these statutory conditions of use in a 'statutory box' on the label.

Types of Approval

There are three categories of approval that may be granted, under the Control of Pesticides Regulations, to products containing active ingredients that were already on the market by 25 July 1993:

- **Full** (granted for an unstipulated period)
- **Provisional** (granted for a stipulated period)
- **Experimental Permit** (granted for the purposes of testing and developing new products, formulations or uses). Products with only an Experimental Permit may not be advertised or sold and do not appear in this *Guide*.

Products containing new active substances, and those containing older ingredients once they have been listed in Annex I to Directive 91/414 (see previous section), are granted approval under the Plant Protection Products Regulations. Again, there are three categories, similar to those listed above:

- **Standard approval** (granted for a period not exceeding 10 years). Only applicable to products whose active substances are listed in Annex I.
- **Provisional approval** (granted for a period not exceeding three years, but renewable). Applicable where Annex I listing of the active substance is awaited.
- **Approval for research and development** (to enable field experiments or tests to be carried out with active substances not otherwise approved).

The official lists of approved products, including all the above categories except those for experimental purposes, are shown on the websites of PSD (www.pesticides.gov.uk) and HSE (www.hse.gov.uk).
Details of new approvals, amendments to existing approvals and off-label approvals (see below) are published regularly on the above websites.

Withdrawal of Approval

Product approvals may be reviewed, amended, suspended or revoked at any time. Revocation may occur for various reasons, such as commercial withdrawal or failure by the approval holder to meet data requirements. Where there are no safety concerns, a phased approval revocation is implemented, specifying a 'wind-down' period to allow the using up of existing stocks by persons other than the approval holder.

Until April 2003 the wind-down period was normally set by PSD at 2 or 3 years, but it became apparent that the European Commission considers 18 months a reasonable period for the withdrawal or use of stocks in the supply chain.

Since 1 April 2003 phased revocations have been as follows.

- A **maximum** of six months for the approval holder to advertise, sell, supply and use the product.
- A further **maximum** of 12 months for sale, supply and use by persons other than the approval holder. This allows a total of 18 months from the date of revocation for stocks to pass through the supply chain and be used. Anyone may store the product during this 18-month period.

Shorter periods will continue to be applied where necessary.

The expiry dates shown in this *Guide* continue to be the final date of legal use.

Approval of Commodity Substances

Commodity substances are compounds that have a variety of alternative and often widespread non-pesticidal uses, but also have potential for use as a pesticide. For a commodity substance to be used as a pesticide, it requires approval under the Control of Pesticides Regulations. Approval is given only for the *use* of the substance; approval is not given for sale, storage, supply or advertisement. There is no approval holder, and no pesticide product label, but the approval and associated conditions of use are published in the Guide to Approved Pesticides on the PSD website. Twenty such substances have been approved for certain specified uses as laid down in the Guide to Approved Pesticides. Of these, ethylene, carbon dioxide, formaldehyde, methyl bromide, peroxyacetic acid, potassium bicarbonate, sodium chloride, sodium hypochlorite, strychnine hydrochloride, sulphuric acid, and urea are approved for agricultural or horticultural use, and are included in this *Guide*.

Off-label Extension of Use

Products may legally be used in a manner not covered by the printed label in several ways:

- in accordance with the 'Long-Term Arrangements for Extension of Use' (see below)
- in accordance with a specific off-label approval (SOLA). SOLAs are uses for which approval has been sought by individuals or organisations other than the manufacturer. The Notices of Approval are published by Defra and are widely available from ADAS or National Farmers' Union offices. Users of SOLAs must first obtain a copy of the relevant Notice of Approval and comply strictly with the conditions laid down therein
- in tank mixture with other approved pesticides in accordance with Consent C(i) made under FEPA. Full details of Consent C(i) are given in Annex A of the Guide to Approved Pesticides on the PSD website (currently www.pesticides.gov.uk/publications.asp?id=499), but there are two essential requirements for tank mixes. First, all the conditions of approval of all the components of a mixture must be complied with. Second, no person may mix or combine pesticides which are cholinesterase compounds unless allowed by the label of at least one of the pesticides in the mixture
- in conjunction with authorised adjuvants
- in reduced spray volume under certain conditions
- the use of certain herbicides on specified set-aside areas subject to restrictions which differ between Scotland and the rest of the UK
- by mutual recognition of a use fully approved in another Member State of the European Union and authorised by PSD.

SECTION 5

Although approved, off-label uses are not endorsed by manufacturers and such treatments are made entirely at the risk of the user.

Long-Term Arrangements for Extension of Use

Since 1 January 1990, arrangements have been in place allowing many approved products to be used for additional minor uses, without the need for Specific Off-Label Approvals (SOLAs). The arrangements – the Long-Term Arrangements for Extension of Use (LTAEU) – permit growers to use pesticides on certain named minor crops provided an approval exists for use on specified major crops. Growers of minor crops are thus provided with access to a large number of important crop protection products. There is no approval documentation associated with these uses.

The scheme is founded on the principle that the risks to the operator, the consumer, wildlife and the environment arising from the extended (off-label) use are acceptable and no greater than those which arise from the approved uses of a product, which appear on its label. All off-label approvals, including those covered by these arrangements, undergo a full safety assessment by the PSD, in consultation with the Advisory Committee on Pesticide. There is, however, no test of efficacy – **use of these extensions of use, as with any off-label use, is made at the user's choosing, and the commercial risk is entirely theirs.**

Because the LTAEU are national arrangements which have no parallel in European Community (EC) legislation they are becoming increasingly difficult to sustain as they are not compatible with the Plant Protection Products Directive and procedures for setting maximum residue levels. After consultation, the Pesticides Safety Directorate concluded in November 2004 that the best and most cost-effective way of retaining these uses is by converting them to SOLAs. The Horticultural Development Council has identified the treatments most used under the LTAEU, and PSD will produce SOLAs for these uses by March 2006. The LTAEU will not be revoked until the exercise is complete. This exercise relates only to edible uses; ornamentals and other non-edible uses will be reviewed at a later date.

The following is a **summary** of the main points and uses of the current arrangements.

Specific restrictions for extension of use

Certain restrictions are necessary to ensure that the extension of use does not increase the risk to the operator, the consumer or the environment.

- The arrangements apply to label and SOLA recommendations, but ONLY for the use of products approved for use as Agricultural and/or Horticultural pesticides.
- Safety precautions and statutory conditions relating to use (shown in the statutory box on the label) MUST be observed. If extrapolation from a SOLA is to be used, all conditions specified on the SOLA Notice of Approval MUST be observed in addition to the safety precautions and statutory conditions on the specified product's label.
- Pesticides must only be used in the same situation (i.e. outdoor or protected) as that specified on the product label or SOLA Notice. Approvals for use on tomatoes, cucumbers, lettuces, chrysanthemums and mushrooms include protected crops unless otherwise stated. Use on other protected crops may only occur when the product label specifically allows use under protection on the crop on which the extrapolation is based. Similarly, pesticides approved only for use in protected situations must not be used outdoors. **If a label or SOLA Notice does not specify a situation, then extrapolation only to an outdoor use is permitted**.
- The application method must be as stated on the product label and in accordance with relevant Codes of Practice and COSHH requirements. Where application of a pesticide under these arrangements is planned to be made by hand-held equipment users MUST ensure that hand-held use is appropriate for the current label or SOLA Notice conditions.
- **Note:** unless otherwise stated, spray applications to protected crops include hand-held uses.

In cases where hand-held application is not appropriate to the use on which extrapolation is to be made, hand-held application may nevertheless be permitted provided certain conditions specified in the arrangements are met (consult the PSD website, currently www.pesticides.gov.uk/publications.asp?id=52). When planning to apply a pesticide under these arrangements by broadcast air-assisted sprayer, only those products with a specific label or SOLA recommendation for such use may be used. Any associated buffer zone restrictions must also be observed.

- Use in or near water (including coastal waters), or by aerial application, is not permitted under the arrangements.
- Rodenticides and other vertebrate control agents are not included, nor is use on land not intended for cropping.
- Pesticides classed as harmful, dangerous, extremely dangerous or high risk to bees must not be used off-label during flowering unless otherwise permitted. Applications must also not be made when flowering weeds are present or when bees are actively foraging. If the label or SOLA Notice specifies an aquatic buffer zone, users must conduct a Local Environment Risk Assessment for Pesticides (LERAP) assessment for the extension of use.
- The user of a pesticide under these arrangements must take all reasonable precautions to safeguard wildlife and the environment.

The arrangements apply as follows.

Non-edible crops and plants

Subject to the specific restrictions set out above:

- Pesticides approved for use on any growing crop may be used on commercial agricultural and horticultural holdings and forest nurseries on: (i) ornamental crops including hardy nursery stock, plants, bulbs, flowers and seed crops where neither the seed nor any part of the plant is to be consumed by humans or animals; (ii) forest nursery crops prior to final planting out..
- In addition, pesticides approved for use on any growing edible crop (except seed treatments) may be used on non-ornamental crops grown for seed subject to the same consumption restrictions as above (but not including seed crops of potatoes, cereals, oilseeds, peas, beans and other pulses).
- Pesticides (except seed treatments) approved for use on oilseed rape may be used on hemp grown for fibre and woad (*Isatis tinctoria* and *I. indigotica*).
- Pesticides (except seed treatments) approved for use on cereals, grass and forage maize may be used on commercial agricultural and horticultural holdings on *Miscanthus* spp. (Elephant grass), but not after the crop is 1 m high. The crops and its products must not be used for food or feed.
- Herbicides approved for use on oilseed rape may be used on sunflower, quinoa, fodder radish and sweet clover grown as game cover.
- Herbicides approved for use on cereals and forage maize may be used on canary grass, tanka millet, white millet and sorghum grown as game cover.

Farm forestry and rotational cropping

Subject to the specific restrictions set out above:

- Herbicides approved for use on cereals may be used in the first 5 years of establishment in farm forestry on land previously under arable cultivation or improved grassland, and reclaimed brownfield sites.
- Herbicides approved for use on cereals, oilseed rape, sugar beet, potatoes, peas and beans may be used in the first year of regrowth after cutting in coppices established on land previously under arable cultivation or improved grassland, or reclaimed brownfield sites.

Nursery fruit crops

Subject to the specific restrictions set out above:

- Pesticides approved for use on any crop for human or animal consumption may be used on commercial agricultural and horticultural holdings on nursery fruit trees, nursery grape vines prior to final planting out, bushes, canes and non-fruiting strawberry plants provided any fruit harvested within 12 months is destroyed. Applications must NOT be made if fruit is present.

SECTION 5

Hops

Subject to the specific restrictions set out above:

- On commercial and horticultural holdings pesticides may be used on mature stock or mother plants kept specifically for propagation, hop propagules prior to final planting out, and first year 'nursery hops' in their final planting position but not taken to harvest in that year. Treated hops must not be harvested for human or animal consumption (including idling) within 12 months of treatment.

Crops used partly or wholly for consumption by humans or livestock

Subject to the specific restrictions set out above:

- Pesticides may be used on commercial agricultural and horticultural holdings on crops in the first column of Tables 1.2 and 1.3 below if they have been approved for use on the crop(s) opposite them in the second column. **However, it is the responsibility of the user to ensure that the proposed use does not result in any statutory UK MRL being exceeded. These extrapolations may NOT be used where the MRL for the crop in column 1 is lower than the MRL for the crop in column 2, or where an MRL has been established for the crop in column 1 but not for the crop in column 2.**

Table 1.2 Crops used partly or wholly for consumption by humans or livestock

Minor use	Crops on which use is approved	Additional special conditions
Arable crops		
Poppy (grown for oilseed), sesame	Sunflower	
Mustard, linseed, evening primrose, honesty	Oilseed rape	
Borage (grown for oilseed), canary flower e.g. *Echium vulgare, E. plantagineum* (grown for oilseed)	Oilseed rape	Seed treatments are not permitted
Rye, triticale	Barley	Treatments applied before first spikelet of inflorescence just visible
Rye, triticale	Wheat	
Grass seed crop	Grass for grazing or fodder	
Grass seed crop	Wheat, barley, oats, rye, triticale	Treated crops must not be grazed or cut for fodder until 90 days after treatment Seed treatments are not permitted. Use of chlormequat-containing products is not permitted
Lupins	Combining peas or field beans	

Minor use	Crops on which use is approved	Additional special conditions
Fruit crops		
Almond, chestnut, hazelnut, walnut	Apple, cherry or plum	For herbicides used on the orchard floor ONLY
Quince, crab apple	Apple or pear	
Almond, chestnut, hazelnut, walnut	Any two of the following: almond, chestnut, hazelnut, walnut	
Nectarine, apricot	Peach	
Blackberry, dewberry, *Rubus* species (e.g. tayberry, loganberry)	Raspberry	
Whitecurrant, bilberry, cranberry, blueberry, other *Vaccinium* species	Blackcurrant or redcurrant	
Gooseberry	Blackcurrant	
Redcurrant	Blackcurrant	
Vegetable crops		
Parsley root	Carrot or radish	
Fodder beet, mangel	Sugar beet	
Horseradish	Carrot or radish	
Parsnip	Carrot	
Salsify	Carrot or celeriac	
Swede	Turnip	
Turnip	Swede	
Garlic, shallot	Bulb onion	
Leek	Salad onion	Herbicides only Latest timing up to two true leaves of the crop
Cayenne pepper	Pepper	
Aubergine	Tomato	
Squash, pumpkin, marrow, watermelon,	Melon	
Broccoli	Calabrese	
Calabrese	Broccoli	

SECTION 5

Minor use	Crops on which use is approved	Additional special conditions
Roscoff cauliflower	Cauliflower	
Collards	Kale	
Chinese cabbage, pak choi, choi sum	Kale	
Lamb's lettuce, frisée/ frise, radicchio, cress, scarole	Lettuce	
Leaf herbs and edible flowers*	Lettuce or spinach or parsley or sage or mint or tarragon	
Leafy brassicas grown for baby leaf production	Lettuce	Seed treatments are not permitted Pre-crop emergence use not permitted
Spinach	Lettuce	Seed treatments are not permitted
Beet leaves, red chard, white chard, yellow chard	Spinach	
Edible podded peas (e.g. mange-tout, sugar snap)	Edible podded beans	
Runner beans	Dwarf French beans	
Broad beans	Vining pea	Applications must NOT be made during flowering or where bees are actively foraging Seed treatments are not permitted
Rhubarb, cardoon	Celery	
Edible fungi other than mushroom (e.g. oyster mushroom)	Mushroom	

*This extension of use applies to the following leaf herbs and edible flowers: *Agastache* spp., angelica, balm, basil, bay, borage, burnet (salad), caraway, catnip, camomile, chervil, chives, clary, coriander, curry plant, dill, dragonhead, fennel, fenugreek, feverfew, hyssop, land cress, lavender, lavandin, lovage, marjoram, marigold, mint, nasturtium, nettle, oregano, *Origanum heracleoticum*, parsley, rocket, rosemary, rue, sage, savory, sorrel, spike lavender, tarragon, thyme, verbena (lemon), woodruff, violet.

For applications in store on crops **partly or wholly for human or animal consumption**, the extensions of use are set out in Table 1.3.

Table 1.3 Applications in store on crops used partly or wholly for consumption by humans or animals

Minor use	Crops on which use is approved
Rye, barley, oats, buckwheat, millet, sorghum, triticale	Wheat
Dried peas	Dried beans
Dried beans	Dried peas
Mustard, sunflower, honesty, sesame, evening primrose, poppy (grown for oilseed), borage (grown for oilseed), canary flower e.g. *Echium vulgare/E. plantaginium* (grown for oilseed)	Oilseed rape

Clarifications

Under these arrangements the crops listed in Table 1.4 are considered synonymous or equivalent and, as such, uses on crops in column 1 can be read across to uses in column 2.

Table 1.4 Crops considered synonymous or equivalent

Column 1	Column 2 (equivalent)
Hazelnut	Cobnuts, filberts
French bean	Navy bean
Vining pea	Picking pea, shelling pea, non-edible podded pea
Linseed	Linola, flax
Wheat	Durum wheat

SECTION 5

Using Crop Protection Chemicals

Use of Herbicides In or Near Water

Products in this *Guide* approved for use in or near water are listed in Table 1.5. Before use of any product in or near water, the appropriate water regulatory body (Environment Agency/Local Rivers Purification Authority; or in Scotland the Scottish Environment Protection Agency) must be consulted. Guidance and definitions of the situation covered by approved labels are given in the Defra publication *Guidelines for the Use of Herbicides on Weeds in or near Watercourses and Lakes*. Always read the label before use.

Table 1.5 Products approved for use in or near water

Chemical	Product
2,4-D	Dormone
dichlobenil	Casoron G, Luxan Dichlobenil Granules, Midstream GSR
glyphosate	Asteroid, Barclay Barbarian, Barclay Gallup 360, Barclay Gallup Amenity, Barclay Gallup Biograde Amenity, Barclay Gallup Hi-Aktiv, Buggy SG, Clinic, Dow Agrosciences Glyphosate 360, Envision, Flame, Glyfos Gold, Glyfos Proactive, Glyphogan, Glyphosate 360, Greenaway Gly-490, Habitat, Kernel, Manifest, Nufosate, Romany, Roundup Amenity, Roundup Biactive, Roundup Pro Biactive, Roundup Pro-Green, Roundup ProVide, Spasor, Spasor Biactive, Tangent, Trustee Elite, Typhoon 360
maleic hydrazide	Regulox K
terbutryn	Clarosan

Use of Pesticides in Forestry

Table 1.6 Products in this *Guide* approved for use in forestry

Chemical	Product	Use
2,4-D	Dicotox Extra	Herbicide
2,4-D + dicamba + triclopyr	Broadsword, Greengard, Nu-Shot	Herbicide
4-indol-3-yl-butyric acid + 2-(1-naphthyl)acetic acid with dichlorophen	Synergol	Growth regulator/stimulator/sprout suppressant
aluminium ammonium sulphate	Curb Crop Spray Powder, Liquid Curb Crop Spray	Animal deterrent/repellent
ammonium sulphamate	Amcide, Root-Out	Herbicide, Tree killer
asulam	Asulox, Greencrop Frond, Spitfire	Herbicide
atrazine	Greencrop Amaize	Herbicide
carbosulfan	Marshal Soil Insecticide suSCon CR granules, Marshal Soil Insecticide suSCon CR Sachets	Insecticide
chlorpyrifos	Dursban WG, Equity, Greencrop Pontoon, Lorsban T	Acaricide, Insecticide
cycloxydim	Greencrop Pomeroy, Greencrop Valentia, Laser, Marnoch Clodim, Standon Cycloxydim	Herbicide
diflubenzuron	Dimilin Flo	Insecticide
diquat + paraquat	ASAP, Fernpath Pronto, PDQ	Desiccant, Herbicide
fluazifop-P-butyl	Fusilade Max	Herbicide
glufosinate-ammonium	Harvest, Kaspar	Herbicide
isoxaben	Flexidor 125	Herbicide
metazachlor	Alpha Metazachlor 50 SC, Greencrop Monogram, Landgold Metazachlor 50 SC, Me2 Booty, Me2 Booty 2, Rapsan 500 SC, Standon Metazachlor 500, Sultan 50 SC	Herbicide
paraquat	Agriguard Paraquat, Fernpath Graminite, Gramoxone 100	Herbicide
propaquizafop	Bulldog, Cleancrop GYR, Emrald Eyetort, Falcon, Greencrop Satchmo, Landgold PQF 100, Raptor, Shogun, Standon Propaquizafop	Herbicide

SECTION 5

Chemical	Product	Use
propyzamide	Agriguard Propyzamide 50 WP, Bulwark Flo, Kerb 50 W, Kerb Granules, Kerb Pro Flo, Kerb Pro Granules, Menace 80 EDF, Propose 50 W	Herbicide
simazine	Alpha Simazine 50 SC, Gesatop, Sipcam Simazine Flowable	Herbicide
thiabendazole	Storite Clear Liquid	Fungicide
triclopyr	Garlon 4, Nomix Garlon 4, Timbrel	Herbicide
urea (commodity substance)	urea	Fungicide
ziram	AAprotect	Animal deterrent/repellent

Pesticides Used as Seed Treatments

Information on the target pests for these products can be found in the relevant pesticide profile in Section 2.

Table 1.7 Products used as seed treatments (including treatments on seed potatoes)

Chemical	Product	Formulation	Crop(s)
aluminium ammonium sulphate	Guardsman STP Seed Dressing Powder	DS	Corms, Flower bulbs, Seeds
beta-cyfluthrin + clothianidin	Poncho Beta	FS	Fodder beet, Sugar beet
beta-cyfluthrin + imidacloprid	Chinook Blue	LS	Spring oilseed rape, Winter oilseed rape
	Chinook Colourless	LS	Winter oilseed rape
bitertanol + fuberidazole	Sibutol	FS	Spring oats, Spring rye, Spring wheat, Triticale, Winter oats, Winter rye, Winter wheat
bitertanol + fuberidazole + imidacloprid	Sibutol Secur	LS	Winter oats, Winter wheat
carboxin + thiram	Anchor	FS	Spring barley, Spring oats, Spring rye, Spring wheat, Triticale, Winter barley, Winter oats, Winter rye, Winter wheat
clothianidin	Deter	FS	Winter barley, Winter wheat
clothianidin + prothioconazole	Redigo Deter	FS	Winter barley, Winter wheat
cymoxanil + fludioxonil + metalaxyl-M	Wakil XL	WS	Broad beans, Carrots, Combining peas, Parsnips, Poppies, Red beet, Spring field beans, Vining peas, Winter field beans
fludioxonil	Beret Gold	FS	Spring barley, Spring oats, Spring wheat, Winter barley, Winter oats, Winter wheat
fluoxastrobin + prothioconazole	Redigo Twin	FS	Winter wheat
fluquinconazole	Galmano	FS	Winter wheat
	Jockey F	FS	Winter barley, Winter wheat

SECTION 5

Chemical	Product	Formulation	Crop(s)
fluquinconazole + prochloraz	Epona	FS	Winter wheat
	Galmano Plus	FS	Winter wheat
	Jockey	FS	Winter barley, Winter wheat
flutolanil	Rhino	FS	Potatoes
fuberidazole + imidacloprid + triadimenol	Baytan Secur	FS	Winter barley, Winter oats, Winter wheat
fuberidazole + triadimenol	Baytan Flowable	FS	Spring barley, Spring oats, Spring rye, Spring wheat, Triticale, Winter barley, Winter oats, Winter rye, Winter wheat
guazatine	Panoctine	LS	Spring barley, Spring oats, Winter barley, Winter oats
	Ravine	LS	Spring barley, Spring oats, Spring wheat, Winter barley, Winter oats, Winter wheat
guazatine + imazalil	Panoctine Plus	LS	Spring barley, Spring oats, Winter barley, Winter oats
	Ravine Plus	LS	Spring barley, Spring oats, Winter barley, Winter oats
imazalil	Fungazil 100 SL	LS	Seed potatoes, Ware potatoes
imazalil + pencycuron	Monceren IM	DS	Potatoes
imazalil + thiabendazole	Extratect Flowable	FS	Potatoes, Seed potatoes
imazalil + triticonazole	Robust	FS	Spring barley, Winter barley
imidacloprid	Gaucho	WS	Broccoli, Brussels sprouts, Cabbages, Calabrese, Cauliflowers, Collards, Kale, Lettuce, Sugar beet
imidacloprid + tebuconazole + triazoxide	Raxil Secur	LS	Winter barley
iprodione	Rovral Liquid FS	FS	Flax, Poppies, Seed potatoes, Spring oilseed rape, Winter oilseed rape
pencycuron	Agriguard Pencycuron	DS	Potatoes
	Me2 Penny	DS	Potatoes

Chemical	Product	Formulation	Crop(s)
	Monceren DS	DS	Potatoes
	Monceren Flowable	FS	Potatoes
	Standon Pencycuron DP	DS	Potatoes
prochloraz	Prelude 20LF	LS	Flax, Linseed
prochloraz + thiram	Agrichem Hy-Pro Duet	FS	Winter oilseed rape
prochloraz + triticonazole	Kinto	FS	Winter barley, Winter wheat
prothioconazole	Redigo	FS	Spring wheat, Winter barley, Winter wheat
prothioconazole + tebuconazole + triazoxide	Raxil Pro	FS	Spring barley, Winter barley
silthiofam	Latitude	FS	Spring wheat, Winter barley, Winter wheat
tebuconazole + triazoxide	Raxil S	LS	Spring barley, Winter barley
tefluthrin	Evict	CF	Spring barley, spring oats, spring wheat, winter barley, winter oats, winter wheat
	Force ST	CF	Fodder beet, sugar beet
thiabendazole	Storite Flowable	FS	Seed potatoes, Ware potatoes
thiabendazole + thiram	Hy-TL	FS	Broad beans, Bulb onions, Peas, Spring field beans, Winter field beans
thiram	Agrichem Flowable Thiram	FS	Broad beans, Cabbages, Carrots, Cauliflowers, Celery, Dwarf beans, Fodder beet, Grass seed, Leeks, Lettuce, Maize, Mangels, Onions, Parsley, Peas, Poppies, Radishes, Red beet, Runner beans, Salad onions, Spring field beans, Spring oilseed rape, Sugar beet, Turnips, Winter field beans, Winter oilseed rape

SECTION 5

Chemical	Product	Formulation	Crop(s)
	Thiraflo	FS	Broad beans, Combining peas, Dwarf beans, Grass seed, Maize, Runner beans, Spring field beans, Spring oilseed rape, Vining peas, Winter field beans, Winter oilseed rape
	Thyram Plus	FS	Broad beans, Cabbages, Carrots, Cauliflowers, Dwarf beans, Grass seed, Leeks, Lettuce, Maize, Onions, Peas, Radishes, Runner beans, Salad onions, Spring field beans, Spring oilseed rape, Turnips, Winter field beans, Winter oilseed rape
tolclofos-methyl	Me2 Cindy	FS	Potatoes
	Rizolex	DS	Potatoes
	Rizolex Flowable	FS	Potatoes

Aerial Application of Pesticides

Only those products specifically approved for aerial application may be so applied, and they may only be applied to specific crops or for specified uses. A complete list of products approved for application from the air can be accessed in Guide to Pesticides on the website of the Pesticides Safety Directorate (PSD, www.pesticides.gov.uk). The list in Table 1.8 is taken mainly from this source, with updating from issues of *The Pesticides Monitor* and product labels.

It is emphasised that the list is for guidance only – reference must be made to the product labels for detailed conditions of use, which must be complied with. The list does not include those products which have been granted restricted aerial application approval limiting the area which may be treated.

Detailed rules are imposed on aerial application regarding prior notification of the Nature Conservancy, water authorities, bee keepers, Environmental Health Officers, neighbours, hospitals, schools, etc. and the conditions under which application may be made. The full conditions are available from Defra and must be consulted before any aerial application is made.

Table 1.8 Products approved for aerial application

Chemical	Product	Crops
2-chloroethylphosphonic acid	Agriguard Cerusite	Winter barley
	Cerone	Winter barley
	Pan Stiffen	Winter barley
asulam	Asulox	Amenity grass, Forest, Permanent pasture, Rough grazing
	Greencrop Frond	Amenity grass, Forest, Permanent pasture, Rough grazing
	Spitfire	Amenity grass, Forest, Permanent pasture, Rough grazing
benalaxyl + mancozeb	Galben M	Potatoes
	Intro Plus	Potatoes
	Tairel	Potatoes
chlormequat	Agriguard Chlormequat 720	Spring oats, Spring wheat, Winter oats, Winter wheat
	Barclay Take 5	Spring oats, Spring wheat, Triticale, Winter barley, Winter oats, Winter rye, Winter wheat
	BASF 3C Chlormequat 720	Spring oats, Spring rye, Spring wheat, Triticale, Winter barley, Winter oats, Winter rye, Winter wheat
	Greencrop Carna	Spring oats, Spring wheat, Winter barley, Winter oats, Winter wheat
	Hive	Winter wheat

SECTION 5

Chemical	Product	Crops
	Mandops Chlormequat 700	Spring oats, Spring wheat, Winter oats, Winter wheat
	Mirquat	Spring oats, Spring wheat, Winter oats, Winter wheat
	New 5C Cycocel	Spring oats, Spring rye, Spring wheat, Triticale, Winter barley, Winter oats, Winter rye, Winter wheat
	New 5C Quintacel	Spring oats, Spring rye, Spring wheat, Triticale, Winter barley, Winter oats, Winter rye, Winter wheat
	Sigma PCT	Spring wheat, Winter barley, Winter wheat
chlorothalonil	Agriguard Chlorothalonil	Potatoes
	Bombardier FL	Potatoes
	Bravo 500	Potatoes
	Clortosip 500	Potatoes
	Delphi	Potatoes
	Flute	Potatoes
	Greencrop Orchid	Potatoes
	Greencrop Orchid B	Potatoes
	Repulse	Potatoes
	Sipcam Echo 75	Potatoes
chlorotoluron	Lentipur CL 500	Durum wheat, Triticale, Winter barley, Winter wheat
copper oxychloride	Cuprokylt	Potatoes
diflubenzuron	Dimilin Flo	Forest
dimethoate	BASF Dimethoate 40	Spring rye, Spring wheat, Sugar beet, Sugar beet seed crops, Triticale, Winter rye, Winter wheat
	Danadim Progress	Spring rye, Spring wheat, Sugar beet, Triticale, Winter rye, Winter wheat

Chemical	Product	Crops
	Rogor L40	Spring rye, Spring wheat, Sugar beet, Triticale, Winter rye, Winter wheat
mancozeb	Dithane 945	Potatoes
	Dithane NT	Potatoes
	Dithane NT Dry Flowable	Potatoes
	Dithane Superflo	Potatoes
	Micene 80	Potatoes
	Micene DF	Potatoes
	Penncozeb WDG	Potatoes
	Quell Flo	Potatoes
metaldehyde	Bristol Blues	All edible crops, All non-edible crops, Natural surfaces not intended to bear vegetation
	CDP Minis	All edible crops, All non-edible crops, Natural surfaces not intended to bear vegetation
	Condor	All edible crops, All non-edible crops, Natural surfaces not intended to bear vegetation
	Dixie 6	All edible crops, All non-edible crops, Natural surfaces not intended to bear vegetation
	Doff Horticultural Slug Killer Blue Mini Pellets	All edible crops, All non-edible crops, Cultivated land/soil
	Entice	All edible crops, All non-edible crops, Natural surfaces not intended to bear vegetation
	ESP	All edible crops, All non-edible crops
	Lincoln VI	All edible crops, All non-edible crops, Natural surfaces not intended to bear vegetation
	Luxan Deal	All edible crops, All non-edible crops
	Optimol	All edible crops, All non-edible crops, Natural surfaces not intended to bear vegetation

SECTION 5

Chemical	Product	Crops
	Super Six	All edible crops, All non-edible crops, Natural surfaces not intended to bear vegetation
pirimicarb	Agriguard Pirimicarb	Spring barley, Spring oats, Spring wheat, Winter barley, Winter oats, Winter wheat
	Aphox	Durum wheat, Spring barley, Spring oats, Spring rye, Spring wheat, Triticale, Winter barley, Winter oats, Winter rye, Winter wheat
	Greencrop Glenroe	Durum wheat, Spring barley, Spring oats, Spring rye, Spring wheat, Triticale, Winter barley, Winter oats, Winter rye, Winter wheat
	Phantom	Durum wheat, Spring barley, Spring oats, Spring rye, Spring wheat, Triticale, Winter barley, Winter oats, Winter rye, Winter wheat
tri-allate	Avadex Excel 15G	Combining peas, Fodder beet, Lucerne, Red beet, Red clover, Sainfoin, Spring barley, Spring field beans, Sugar beet, Triticale, Vetches, Vining peas, White clover, Winter barley, Winter field beans, Winter wheat

Resistance Management

Pest species are, by definition, adaptable organisms. The development of resistance to some crop protection chemicals is just one example of this adaptability. Repeated use of products with the same mode of action will clearly favour those individuals in the pest population able to tolerate the treatment. This leads to a situation where the tolerant (or resistant) individuals can dominate the population and the product is ineffective. In general, the more rapidly the pest species reproduces and the more mobile it is, the faster is the emergence of resistant populations, although some weeds seem able to evolve resistance more quickly than would be expected.

The speed of appearance of resistance depends on the mode of action of the crop protection chemicals, as well as the manner in which they are used. Resistance among insects and fungal diseases has been evident for much longer than weed resistance to herbicides, but examples in all three categories are now widespread and increasing. This has created a need for agreement on the advice given for the use of crop protection chemicals in order to reduce the likelihood of the development of resistance and to avoid the loss of potentially valuable products in the chemical armoury.

In the UK, key independent research organisations, chemical manufacturers and other organisations have collaborated to share knowledge and expertise on resistance issues through three action groups. Participants include ADAS, the Pesticides Safety Directorate, universities, colleges and the Home Grown Cereals Authority. The groups have a common aim of monitoring resistance in UK and devising and publishing management strategies designed to combat it where it occurs. The groups are:

The Weed Resistance Action Group (WRAG), formed in 1989 (Secretary: Dr Stephen Moss, Rothamsted Research, Harpenden, Herts AL5 2JQ *Tel: 01582 7631330 ext. 2521*)

The Fungicide Resistance Action Group (FRAG), formed in 1995 (Secretary: Mr Oliver Macdonald, Pesticides Safety Directorate, Mallard House, King's Pool, 3 Peasholme Green, York YO1 2PX *Tel: 01904 455864*)

The Insecticide Resistance Action Group (IRAG), formed in 1997 (Secretary: Dr Stephen Foster, Rothamsted Research, Harpenden, Herts AL5 2JQ *Tel: 01582 763133 ext. 2324*)

The above groups publish detailed advice on resistance management relevant for each sector and, in some cases, specific to a pest problem. This information, together with further details about the function of each group, can be obtained from the Pesticides Safety Directorate web site: www.pesticides.gov.uk/rags.asp

The general guidelines for resistance management are similar for all three problem areas.

Preparation in advance

- Be aware of the factors that favour the development of resistance, such as repeated annual use of the same product, and assess the risk.
- Plan ahead and aim to integrate all possible means of control.
- Use cultural measures such as rotations, stubble hygiene, variety selection and, for fungicides, removal of primary inoculum sources, to reduce reliance on chemical control.
- Monitor crops regularly.
- Keep aware of local resistance problems.
- Monitor effectiveness of actions taken and take professional advice, especially in cases of unexplained poor control.

Using crop protection products

- Optimise product efficacy by using it as directed, at the right time, in good conditions.
- Treat pest problems early.
- Mix or alternate chemicals with different modes of action.
- Avoid repeated applications of very low doses.
- Keep accurate field records.

SECTION 5

Label guidance depends on the appropriate strategy for the product. Most frequently it consists of a warning of the possibility of poor performance due to resistance, and a restriction on the number of treatments that should be applied in order to minimise the development of resistance. This information is summarised in the profiles in this *Guide*, but detailed guidance must always be obtained by reading the label itself before use.

International action committees

Resistance to crop protection products is an international problem. Agrochemical industry collaboration on a global scale is via three action committees whose aims are to support a coordinated industry approach to the management of resistance worldwide. In particular they produce lists of crop protection chemicals classified according to their mode of action. These lists and other information can be obtained from the respective websites:

Herbicide Resistance Action Committee (HRAC): www.plantprotection.org/hrac

Fungicide Resistance Action Committee (FRAC): www.frac.info

Insecticide Resistance Action Committee (IRAC): www.irac-online.org

Poisons and Poisoning

Chemicals Subject to the Poisons Law

Certain products in this book are subject to the provisions of the Poisons Act 1972, the Poisons List order 1982 and the Poisons Rules 1982 (copies of all these are obtainable from The Stationery Office). These rules include general and specific provisions for the storage, sale and supply of listed non-medicine poisons. Full details can be accessed in Annex C of the Guide to Approved Pesticides on the PSD website (currently www.pesticides.gov.uk/publications.asp?id=499). The nature of the formulation and the concentration of the active ingredient allow some products to be exempted from the rules (see below).

The chemicals approved for use in the UK and included in this book are specified under Parts I and II of the Poisons List as follows.

Part I Poisons

Sale restricted to registered retail pharmacists and to registered non-pharmacy businesses provided sales do not take place on retail premises:

aluminium phosphide   magnesium phosphide
chloropicrin   strychnine

Part II Poisons

Sale restricted to registered retail pharmacists and listed sellers registered with a local authority:

aldicarb   oxamyl (a)
alphachloralose   paraquat (c)
formaldehyde   sulphuric acid
nicotine (b)   zinc phosphide

(a) Granular formulations containing up to 12% w/w of this, or a combination of similarly flagged poisons, are exempt.
(b) Formulations containing not more than 7.5% of nicotine are exempt.
(c) Pellets containing not more than 5% paraquat ion are exempt.

Note:

1. The European Commission Ozone Depleting Substances Regulation (EC Regulation 2037/2000) banned the use of methyl bromide in the UK, with the exception of a few 'Critical Use Exemptions', which will be reviewed annually by the EC. Suppliers of products containing methyl bromide which had previously been listed in this Guide withdrew them in 2004.
2. All approvals for dichlorvos-containing products expired in April 2004.

Occupational Exposure Limits

A fundamental requirement of the COSHH Regulations is that exposure of employees to substances hazardous to health should be prevented or adequately controlled. Exposure by inhalation is usually the main hazard, and in order to measure the adequacy of control of exposure by this route, various substances have been assigned occupational exposure limits.

There are two types of occupational exposure limits defined under COSHH: Occupational Exposure Standards (OES) and Maximum Exposure Limits (MEL). The key difference is that an OES is set at a level at which there is no indication of risk to health; for an MEL a residual risk may exist and the level takes socio-economic factors into account. In practice, MELs have most often been allocated to carcinogens and other substances for which no threshold of effect can be identified, and for which there is no doubt about the seriousness of the effects of exposure.

OESs and MELs are set on the recommendation of the Advisory Committee on Toxic Substances. Full details are published by the Health and Safety Executive (HSE) in *Occupational Exposure Limits* 1995, EH 40/95 (ISBN 0 7176 0876 X).

As far as pesticides are concerned, OESs and MELs have been set for relatively few active ingredients. This is because pesticide products usually contain other substances in their formulation, including solvents, which may have their own OES/MEL. In practice, inhalation of solvent may be at least, or more, important than that of the active ingredient. These factors are taken into account in the approval process of the pesticide product under the Regulations. This indicates one of the reasons why a change of pesticide formulation usually necessitates a new approval assessment.

First Aid Measures

If pesticides are handled in accordance with the required safety precautions, as given on the container label, poisoning should not occur. It is difficult, however, to guard completely against the occasional accidental exposure. Thus if a person handling, or exposed to, pesticides becomes ill, it is a wise precaution to apply first aid measures appropriate to pesticide poisoning even though the cause of illness may eventually prove to have been quite different. An employer has a legal duty to make adequate first aid provision for employees. Regular pesticide users should consider appointing a trained first aider even if numbers of employees are not large, as there is a specific hazard.

The first essential in a case of suspected poisoning is for the person involved to stop work, to be moved away from any area of possible contamination, and for a doctor to be called at once. If no doctor is available the patient should be taken to hospital as quickly as possible. In either event it is most important that the name of the chemical being used should be recorded, and preferably the whole product label or leaflet should be shown to the doctor or hospital concerned.

Some pesticides which are unlikely to cause poisoning in normal use are extremely toxic if swallowed accidentally or deliberately. In such cases get the patient to hospital as quickly as possible, with all the information you have.

General measures

Measures appropriate in all cases of suspected poisoning include:

- remove any protective or other contaminated clothing (taking care to avoid personal contamination)
- wash any contaminated areas carefully with water or with soap and water if available
- in cases of eye contamination, flush with plenty of clean water for at least 15 minutes
- lay the patient down, keep at rest and under shelter; cover with one clean blanket or coat, and avoid overheating
- monitor level of consciousness, breathing and pulse rate
- if consciousness is lost, place the casualty in the recovery position (on his or her side with head down and tongue forward to prevent inhalation of vomit)
- if breathing ceases or weakens, commence mouth-to-mouth resuscitation: ensure that the mouth is clear of obstructions such as false teeth, that the breathing passages are clear, and that tight clothing around the neck, chest and waist has been loosened; if a poisonous chemical has been swallowed, it is essential that the first aider is protected by the use of a resuscitation device (several types are available on the market).

Specific measures

In case of poisoning with particular chemical groups, the following measures may be taken before transfer to hospital.

Organophosphorus and carbamate insecticides

Keep the patient at rest. The patient may suddenly stop breathing, so be ready to give artificial respiration.

Organochlorine compounds

If convulsions occur, do not interfere unless the patient is in danger of injury; if so any restraint must be gentle. When convulsions cease, place the casualty on his or her side with head down, tongue forward (recovery position).

Paraquat, diquat

Irrigate skin and eye splashes copiously with water. If any chemical has been swallowed, take to hospital for tests.

Cyanide (including sodium cyanide)

Send for medical aid. Remove casualty to fresh air, if necessary using breathing apparatus and protective clothing. Remove casualty's contaminated clothing. Gently brush solid particles from the skin, making sure you protect your own skin from contamination. Wash the skin and eyes copiously with water. Transfer casualty to nearest accident and emergency hospital by the quickest possible means together with the first aid cyanide antidote, if held on the premises.

Reporting pesticide poisoning

Any cases of poisoning by pesticides must be reported without delay to an HM Agricultural Inspector of the Health and Safety Executive. In addition, any cases of poisoning by substances named in Schedule 2 of The Reporting of Injuries, Diseases and Dangerous Occurrences Regulations 1985 must also be reported to HM Agricultural Inspectorate (this includes organophosphorus chemicals, mercury and some fumigants).

Cases of pesticide poisoning should also be reported to the manufacturer concerned.

Additional information

General advice on the safe use of pesticides is given in a range of Health and Safety Executive leaflets available from HSE Books (see Appendix 2).

A useful booklet, *Guidelines for Emergency Measures in Cases of Pesticide Poisoning*, is available from CropLife International (see Appendix 2).

The major agrochemical companies are able to provide authoritative medical advice about their own pesticide products. Detailed advice is also available to doctors from the National Poisons Information Service, New Cross Hospital, London SE14 5ER (020 7635 9191) and from regional centres (see Appendix 2).

SECTION 5

Environmental Protection

Protection of Bees

Honey bees

Honey bees are a source of income for their owners, and important to farmers and growers as pollinators of their crops. It is irresponsible and unnecessary to use pesticides in such a way that may endanger them. Pesticides vary in their toxicity to bees, but those that present a special hazard are classed as 'harmful', 'dangerous' or 'extremely dangerous' to bees. Products so classified carry a specific warning in the precautions section of the label. These are indicated in this *Guide* in the **hazard classification and safety precautions** section of the pesticide profile by the numbers E12c, E12d, E12e, E12f, depending on which phrase applies. The phrases are stated in full in Appendix 4.

In July 1996 a new classification was introduced that more accurately reflects the actual risk to honey bees when a product is used. Products assessed by PSD as posing such a risk are classified as 'High risk to bees'. Such products carry a new warning in the precautions section of the label, indicated in this *Guide* by the numbers E12a and E12b.

Product labels indicate the necessary environmental precautions to take, but where use of an insecticide on a flowering crop is contemplated, the British Beekeepers Association (see Appendix 2) has produced the following guidelines for growers:

- target insect pests with the most appropriate product
- choose a product that will cause minimal harm to beneficial species
- follow the manufacturer's instructions carefully
- inspect and monitor crops regularly
- avoid spraying crops in flower or where bees are actively foraging
- keep down flowering weeds
- spray late in the day, in still conditions
- avoid excessive spray volume and run-off
- adjust sprayer pressure to reduce production of fine droplets and drift
- give local beekeepers as much warning of your intention as possible.

Wild bees

Wild bees also play an important role. Bumblebees are useful pollinators of spring-flowering crops and fruit trees because they forage in cool, dull weather when honey bees are inactive. They play a particularly important part in pollinating field beans, red and white clover, lucerne and borage. Bumblebees nest and overwinter in field margins and woodland edges. Avoidance of direct or indirect spray contamination of these areas, in addition to the creation of hedgerows and field margins, and late cutting or grazing of meadows, all helps the survival of these valuable insects.

The Campaign Against Illegal Poisoning of Wildlife

The Campaign Against Illegal Poisoning of Wildlife, aimed at protecting some of Britain's rarest birds of prey and wildlife while also safeguarding domestic animals, was launched in March 1991 by the (then) Ministry of Agriculture, Fisheries and Food and the Department of the Environment, Transport and the Regions. The main objective is to deter those who may be considering using pesticides illegally. The Campaign is supported by a range of organisations associated with animal welfare, nature preservation, field sports and game keeping including the Royal Society for the Protection of Birds, English Nature, the Countryside Alliance and the Game Conservancy Trust.

The three objectives are to:

- advise farmers, gamekeepers and other land managers on legal ways of controlling pests
- advise the public on how to report illegal poisoning incidents and to respect the need for legal alternatives
- investigate incidents and prosecute offenders.

A freephone number (0800 321 600) is available to make it easier for the public to report incidents, and numerous leaflets, posters, postcards, coasters and stickers have been created to publicise the existence of the Campaign.

The Campaign arose from the results of the Wildlife Incident Investigation Scheme for the investigation of possible cases of illegal poisoning. Under this scheme, all reported incidents are considered and thoroughly investigated where appropriate. Enforcement action is taken wherever sufficient evidence of an offence can be obtained, and numerous prosecutions have been made since the start of the Campaign.

Further information about the Campaign is available from:

Helena Cooke
Pesticides Safety Directorate
Room 317, Mallard House
3 Peasholme Green
York YO1 7PX
e-mail: helena.cooke@psd.defra.gsi.gov.uk
website: www.pesticides.gov.uk/environment.asp?id=504

Water Quality

The Food and Environment Protection Act 1985 (FEPA) places a special obligation on users of pesticides to 'safeguard the environment and in particular avoid the pollution of water'. Under the Water Resources Act 1991 it is an offence to pollute any controlled waters (watercourses or groundwater), either deliberately or accidentally. Protection of controlled waters from pollution is the responsibility of the Environment Agency. Users of pesticides therefore have a duty to adopt responsible working practices and, unless they are applying herbicides in or near water, to prevent them getting into water. Guidance on how to achieve this is given in the Defra *Code of Good Agricultural Practice for the Protection of Water* (available from The Stationery Office; see Appendix 2). The duty of care covers not only the way in which a pesticide is sprayed, but also its storage, preparation and disposal of surplus, sprayer washings and the container. Products in this *Guide* that are a major hazard to fish, other aquatic life or aquatic higher plants carry one of several specific label precautions in their profile, depending on the assessed hazard level. Advice on any water pollution problems is available from the Environment Agency and the Crop Protection Association (see Appendix 2).

Protecting surface waters

Surface waters are particularly vulnerable to contamination. One of the best ways of preventing those pesticides that carry the greatest risk to aquatic wildlife from reaching surface waters is to prohibit their application within a boundary adjacent to the water. Such areas are known as no-spray, or buffer, zones. Certain products in this *Guide* are restricted in this way, and have a legally binding label precaution to make sure the potential exposure of aquatic organisms to pesticides that might harm them is minimised. Before 1999 the protected zones were measured from the edge of the water. The distances were 2 m for hand-held or knapsack sprayers, 6 m for ground crop sprayers, and a variable distance (but often 18 m) for broadcast air-assisted applications, such as in orchards. The introduction of LERAPs (see below) has changed the method of measuring buffer zones.

'Surface water' includes lakes, ponds, reservoirs, streams, rivers and watercourses (natural or artificial). It also includes temporarily or seasonally dry ditches, which have the potential to carry water at different times of the year. Buffer zone restrictions do not necessarily apply to all products containing the same active ingredient. Those in formulations that are not likely to contaminate surface water through spray drift do not pose the same risk to aquatic life and are not subject to the restrictions.

Local Environmental Risk Assessments for Pesticides

Local Environment Risk Assessments for Pesticides (LERAPs) were introduced in March 1999, and revised guidelines were issued in January 2002. They give users of most products currently subject to a buffer zone restriction the option of continuing to comply with the existing buffer zone restriction (using the new method of measurement), or carrying out a LERAP and possibly reducing the size of the buffer zone as a result. In either case, there is a new legal obligation for users to record his decision, including the results of the LERAP. The scheme has changed the method of measuring the buffer zone. Previously the zone was measured from the edge of the water, but it is now the distance from the top of the bank of the watercourse to the edge of the spray area.

The LERAP provides a mechanism for taking into account other factors that may reduce the risk, such as dose reduction, the use of low drift nozzles, and whether the watercourse is dry or flowing. The previous arrangements applied the same restriction regardless of whether there was actually water present. Now there is a standard zone of 1 m from the top of a dry ditch bank.

Other factors to include in a LERAP that may allow a reduction in the buffer zone are:

- size of the watercourse: the wider it is, the greater the dilution factor and the lower the risk of serious pollution
- dose applied: the lower the dose, the less is the risk
- application equipment: sprayers and nozzles are star-rated according to their ability to reduce spray drift fallout - equipment offering the greatest reductions achieves the highest rating of three stars. The scheme was initially restricted to ground crop sprayers; new, more flexible rules introduced in February 2002 included broadcast air-assisted orchard and hop sprayers.

Other changes introduced in 2002 allow the reduction of a buffer zone if there is an appropriate living windbreak between the sprayed area and a watercourse.

Not all products that had a label buffer zone restriction are included in the LERAP scheme. The option to reduce the buffer zone does not apply to organophosphorus or synthetic pyrethroid insecticides. These groups are classified as Category A products. All other products that had a label buffer zone restriction are classified as Category B. The wording of the buffer zone label precautions has been amended for all products to take account of the new method of measurement, and whether or not the particular product qualifies for inclusion in the LERAP scheme.

Products in this *Guide* that are in Category A or B are identified in the **Environmental Safety** section of the profile. Updates to the list are published regularly by PSD and details can be obtained from www.pesticides.gov.uk/psd_databases.asp?id=325

The introduction of LERAPs is an important step forward because it demonstrates a willingness to reduce the impact of regulation on users of pesticides by allowing flexibility where local conditions make it safe to do so. This places a legal responsibility on users to ensure the risk assessment is done either personally, or by the spray operator or a professional consultant or advisor. It is compulsory to record the LERAP and make it available for inspection by enforcement authorities.

Groundwater regulations

New Groundwater Regulations were introduced in 1999 to complete the implementation in the UK of the EU Groundwater Directive (Protection of Groundwater Against Pollution Caused by Certain Dangerous Substances - 80/68/EEC). These Regulations help prevent the pollution of groundwater by controlling discharges or disposal of certain substances, including all pesticides.

Groundwater is defined under the Regulations as any water contained in the ground below the water table. Pesticides must not enter groundwater unless it is deemed by the appropriate

Agency to be permanently unsuitable for other uses. The Regulations make it a criminal offence to dispose of pesticides onto land without official authorisation from the Environment Agency (in England and Wales) or the Scottish Environmental Protection Agency. Normal use of a pesticide in accordance with product approval does not require authorisation. This includes spraying the washings and rinsings back on the crop provided that, in so doing, the maximum approved dose for that product on that crop is not exceeded.

The Agencies will review all authorisations within a 4-year interval. They may also grant authorisations for a limited period. The Agencies can serve notice at any time to modify the conditions of an authorisation where necessary to prevent pollution of groundwater. In practice, the best advice to farmers and growers is to plan to use all diluted spray within the crop and to dispose of all washings via the same route, making sure they stay within the conditions of approval of the product.

SECTION 5

SECTION 6
APPENDICES

Appendix 1
Suppliers of Pesticides and Adjuvants

Ag-Chem: Ag-Chem Direct Limited
P O Box 82
Thirsk
N. Yorks
YO7 4YY
Tel: (0797) 423 5105
Fax: (0797) 113 8857

Agrichem: Agrichem (International) Ltd
Industrial Estate
Station Road
Whittlesey
Cambs.
PE7 2EY
Tel: (01733) 204019
Fax: (01733) 204162
Email: info@agrichem.co.uk
Web: www.agrichem.co.uk

AgriGuard: AgriGuard Ltd
Unit 3
Tally House
Broomfield Business Park
Malahide
Co. Dublin
Ireland
Tel: (+353) 1 846 2044
Fax: (+353) 1 846 2489
Email: info@agriguard.ie
Web: www.agriguard.ie

Agropharm: Agropharm Limited
Buckingham Place
Church Road
Penn
High Wycombe
Bucks.
HP10 8LN
Tel: (01494) 816575
Fax: (01494) 816578
Web: www.agropharm.co.uk

Amega: Amega Sciences
Lanchester Way
Royal Oak Industrial Estate
Daventry
Northants.
NN11 5PH
Tel: (01327) 704444
Fax: (01327) 71154
Email: admin@amega-sciences.com
Web: www.amega-sciences.com

Amenity Land: Amenity Land Services Limited
Long Lane
Wellington
Telford
Shropshire
TF6 6HA
Tel: (01952) 641949
Fax: (01952) 247369
Email: sales@amenity.co.uk
Web: www.amenity.co.uk

Antec Biosentry: Antec Biosentry
DuPont Animal Health Solutions
Windham Road
Chilton Industrial Estate
Sudbury
Suffolk
CO10 2XD
Tel: (01787) 377305
Fax: (01787) 310846
Email: biosecurity@antecint.com
Web: www.antecint.com

B H & B: Battle Hayward & Bower Ltd
Victoria Chemical Works
Crofton Drive
Allenby Road Industrial Estate
Lincoln
LN3 4NP
Tel: (01522) 529206
Fax: (01522) 538960

Banks Cargill: Banks Cargill Agriculture Ltd
Fleet Road Industrial Estate
Holbeach
Spalding
Lincs.
PE12 8LY
Tel: (01406) 421405
Email: les_sykes@bankscargill.co.uk
Web: www.bankscargill.co.uk

Barclay: Barclay Chemicals Manufacturing Ltd
Damastown Way
Damastown Industrial Park
Mulhuddart
Dublin 15
Ireland
Tel: (+353) 1 822 4555
Fax: (+353) 1 822 4678
Email: info@barclay.ie
Web: www.barclay.ie

Barrier: Barrier BioTech Ltd
36/37 Haverscroft Industrial Estate
New Road
Attleborough
Norfolk
NR17 1YE
Tel: (01953) 456363
Fax: (01953) 455594
Email: sales@barrier-biotech.com
Web: www.barrier-biotech.com

BASF: BASF plc.
Agricultural Divison
PO Box 4, Earl Road
Cheadle Hulme
Cheshire
SK8 6QG
Tel: (0845) 602 2553
Fax: (0161) 485 2229
Web: www.agricentre.co.uk

Bayer CropScience:
Bayer CropScience Limited
230 Cambridge Science Park
Milton Road
Cambridge
CB4 0WB
Tel: (01223) 226500
Fax: (01223) 426240
Web: www.bayercropscience.co.uk

Bayer Environ.: Bayer Environmental Science
Durkan House
214-224 High Street
Waltham Cross
Herts.
EN8 7DP
Tel: (01992) 784270
Fax: (01992) 784276
Email: john.hall1@bayercropscience.com
Web: www.bayer-escience.co.uk

Belchim: Belchim Crop Protection Ltd
Suite 2, Unit 3 Phoenix Park
Eaton Socon
St Neots
Cambs.
PE19 8EP
Tel: (01480) 403333
Fax: (01480) 403444
Email: info@belchim.com
Web: www.belchim.com

Cardel: Cardel Agro SPRL
Avenue de Tervuren 270-272
B-1150 Brussels
Belgium
Tel: (+32) 2776 7652

Certis: Certis
1b Mills Way
Boscombe Down Business Park
Amesbury
Wilts.
SP4 7RX
Tel: (01980) 676500
Fax: (01980) 626555
Email: certis@certiseurope.co.uk
Web: www.certiseurope.co.uk

Chiltern: Chiltern Farm Chemicals Ltd
East Mellwaters
Stainmore
Bowes
Barnard Castle
Co. Durham
DL12 9RH
Tel: (01833) 628282
Fax: (01833) 628020
Web: www.chilternfarm.com

Ciba Specialty: Ciba Specialty Chemicals
Water Treatments Limited
P O Box 38
Low Moor
Bradford
W. Yorks.
BD12 0JZ
Tel: (01274) 417549
Fax: (01274) 417305
Email: soil.additives@cibasc.com
Web: www.cibasc.com

Crompton: Crompton Europe Ltd
A Chemtura Company
Kennet House
4 Langley Quay
Slough
Berks.
SL3 6EH
Tel: (01753) 603000
Fax: (01753) 603077

DAPT: DAPT Agrochemicals Limited
14 Monks Walk
Southfleet
Gravesend
Kent
DA13 9NZ
Tel: (01474) 834448
Fax: (01474) 834449
Email: rkjltd@supanet.com

Dax: Dax Products Ltd
    18 Marlborough Road
    Woodthorpe
    Nottingham
    NG5 4FG
    Tel: (0115) 926 9996
    Fax: (0115) 966 1173
    Email: info@daxproducts.co.uk
    Web: www.daxproducts.co.uk

De Sangosse: De Sangosse UK
    Hillside Mill
    Quarry Lane
    Swaffham Bulbeck
    Cambridge
    CB5 0LU
    Tel: (01223) 811215
    Fax: (01223) 810020
    Email: info@desangosse.co.uk
    Web: www.desangosse.co.uk

Dewco-Lloyd: Dewco-Lloyd Ltd
    Cyder House
    Ixworth
    Suffolk
    IP31 2HT
    Tel: (01359) 230555
    Fax: (01359) 232553

Doff Portland: Doff Portland Ltd
    Aerial Way
    Hucknall
    Nottingham
    NG15 6DW
    Tel: (0115) 963 2842
    Fax: (0115) 963 8657
    Email: info@doff.co.uk
    Web: www.doff.co.uk

Dow: Dow AgroSciences
    Latchmore Court
    Brand Street
    Hitchin
    Herts.
    SG5 1NH
    Tel: (01462) 457272
    Fax: (01462) 426605
    Email: fhihotl@dow.com
    Web: www.dowagro.com/uk

DuPont: DuPont (UK) Ltd
    Crop Protection Products Department
    Wedgwood Way
    Stevenage
    Herts.
    SG1 4QN
    Tel: (01438) 734450
    Fax: (01438) 734969
    Web: www.gbr.ag.dupont.com

Elliott: Thomas Elliott Ltd
    Bencewell Granary
    Oakley Road
    Bromley Common
    Kent
    BR2 8HG
    Tel: (0208) 462 6622
    Fax: (0208) 462 5599

Fargro: Fargro Ltd
    Toddington Lane
    Littlehampton
    Sussex
    BN17 7PP
    Tel: (01903) 721591
    Fax: (01903) 730737
    Email: promos@fargro.co.uk
    Web: www.fargro.co.uk

FCC: Farmers Crop Chemicals Ltd
    P O Box 12379
    Alcester
    Warwicks.
    B49 9AA
    Tel: (01789) 774726
    Fax: (01789) 774726
    Email:
    enquiries@farmerscropchemicals.com
    Web: www.farmerscropchemicals.com

Fine: Fine Agrochemicals Ltd
    Hill End House
    Whittington
    Worcester
    WR5 2RQ
    Tel: (01905) 361800
    Fax: (01905) 361810
    Email: enquire@fine-agrochemicals.com
    Web: www.fine-agrochemicals.com

Ford Smith: Ford Smith & Co. Ltd
    Lyndean Industrial Estate
    Felixstowe Road
    Abbey Wood
    London
    SE2 9SG
    Tel: (020) 8310 8127
    Fax: (020) 8310 9563
    Email: fordsmithltd@aol.com

Gowan: Gowan International
    Onen House
    Onen
    Tal-y-Coed
    Monmouth
    NP25 5EN
    Tel: (01600) 780543
    Fax: (01600) 780543

SECTION 6

Greenaway: Greenaway Amenity Ltd
   7 Browntoft Lane
   Donington
   Spalding
   Lincs
   PE11 4TQ
   Tel: (01775) 821031
   Fax: (01775) 821034
   Email: greenawayamenity@aol.com
   Web: www.greenawaycda.com

Greencrop: Greencrop Technology Ltd
   Burren House
   2 Cowbrook Court
   Glossop
   Derbyshire
   SK13 8SL
   Tel: (01457) 856001
   Fax: (01457) 857137
   Email: mail@greencrop-technology.co.uk
   Web: www.greencrop-technology.co.uk

Headland: Headland Agrochemicals Ltd
   Norfolk House
   Gt. Chesterford Court
   Gt. Chesterford
   Saffron Walden
   Essex
   CB10 1PF
   Tel: (01799) 530146
   Fax: (01799) 530229
   Email: enquiry@headlandgroup.com
   Web: www.headland-ag.co.uk

Headland Amenity:
Headland Amenity Limited
   1010 Cambourne Business Park
   Cambourne
   Cambs.
   CB3 6DP
   Tel: (01223) 597834
   Fax: (01223) 598052
   Email: info@headlandamenity.com
   Web: www.headlandamenity.com

Helena: Helena Chemical Company
   Cambridge House
   Nottingham Road
   Stapleford
   Nottingham
   NG9 8AB
   Tel: (0115) 939 0202
   Fax: (0115) 939 8031

Hermoo: Hermoo (UK)
   21 Victoria Road
   Wargrave
   Berks.
   RG10 8AD
   Tel: (0118) 940 4264
   Fax: (0118) 940 4264
   Email: johnhudson23@aol.com
   Web: www.hermoo.com

Interagro: Interagro (UK) Ltd
   Sworders Barn
   Sworders Yard
   North Street
   Bishop's Stortford
   Herts.
   CM23 2LD
   Tel: (01279) 501995
   Fax: (01279) 501996
   Email: info@interagro.co.uk
   Web: www.interagro.co.uk

Interfarm: Interfarm (UK) Ltd
   Kinghams's Place
   36 Newgate Street
   Doddington
   Cambs.
   PE15 0SR
   Tel: (01354) 741414
   Fax: (01354) 741004
   Email: technical@interfarm.co.uk
   Web: www.interfarm.co.uk

Intracrop: Intracrop
   Little Hay
   Broadwell
   Lechlade
   Glos.
   GL7 3QS
   Tel: (01367) 860255
   Fax: (01926) 634798

Irish Drugs: Irish Drugs Ltd
   Burnfoot
   Lifford
   Co. Donegal
   Ireland
   Tel: (+353) 74 9368104
   Fax: (+353) 74 9368311
   Email: idl@eircom.net

K & S Fumigation:
K & S Fumigation Services Ltd
  Shirley Farmhouse
  Moor Lane
  Woodchurch
  Ashford
  Kent
  TN26 3SS
  Tel: (01233) 758252
  Fax: (01233) 758343
  Email: nick@kstreatments.co.uk

Killgerm: Killgerm Chemicals Ltd
  115 Wakefield Road
  Flushdyke
  Ossett
  W. Yorks.
  WF5 9AR
  Tel: (01924) 268400
  Fax: (01924) 264757
  Email: info@killgerm.com
  Web: www.killgerm.com

Koppert: Koppert (UK) Ltd
  Homefield Road
  Haverhill
  Suffolk
  CB9 8QP
  Tel: (01440) 704488
  Fax: (01440) 704487
  Email: m.cook@koppert.co.uk
  Web: www.koppert.com

Landgold: Landgold & Co. Ltd
  See Teliton Ltd

Landseer: Landseer Ltd
  Lodge Farm
  Goat Hall Lane
  Galleywood
  Chelmsford
  Essex
  CM2 8PH
  Tel: (01245) 357109
  Fax: (01245) 494165

Lanxess: Lanxess Distribution GmbH
  Business Unit CropChem
  Piccolominstrasse 2 .
  D-51063 Cologne
  Germany
  Tel: (+49) 214 30 43158
  Fax: (+49) 214 959 43158
  Web: www.lanxess-cropchem.com

Luxan: Luxan (UK) Ltd
  Crown Business Park
  Old Dalby
  Leics.
  LE14 3NQ
  Tel: (01664) 820052
  Fax: (01664) 820216
  Email: enquiry@luxan.co.uk
  Web: www.luxan.co.uk

Makhteshim: Makhteshim-Agan (UK) Ltd
  Unit 16
  Thatcham Business Village
  Colthrop Way
  Thatcham
  Berks.
  RG19 4LW
  Tel: (01635) 860555
  Fax: (01635) 861555
  Email: admin@mauk.co.uk
  Web: www.mauk.co.uk

Mandops: Mandops (UK) Ltd
  36 Leigh Road
  Eastleigh
  Hants.
  SO50 9DT
  Tel: (023) 8064 1826
  Fax: (023) 8062 9106
  Email: enquiries@mandops.co.uk
  Web: www.mandops.co.uk

Me2: Me2 Crop Protection Ltd
  Tower House
  Fishergate
  York
  YO10 4HA
  Tel: (01904) 567331
  Fax: (01904) 720094
  Email: sales@me2cpl.co.uk
  Web: www.me2cpl.co.uk

Microcide: Microcide Ltd
  Shepherds Grove
  Stanton
  Bury St. Edmunds
  Suffolk
  IP31 2AR
  Tel: (01359) 251077
  Fax: (01359) 251545
  Email: microcide@microcide co.uk
  Web: www.microcide.co.uk

SECTION 6

Monsanto: Monsanto (UK) Ltd
The Maris Centre
45 Hauxton Road
Trumpington
Cambridge
CB2 2LQ
Tel: (01223) 849200
Fax: (01223) 849414
Email:
technical.helpline.uk@monsanto.com
Web: www.monsanto-ag.co.uk

Nickerson: Nickerson UK Ltd
Rothwell
Market Rasen
Lincs.
LN7 6DT
Tel: (01472) 371471
Fax: (01472) 371602
Email: enquiries@Nickerson.co.uk
Web: www.nickerson.co.uk

Nomix Enviro: Nomix Enviro Limited
Portland Building
Portland Street
Staple Hill
Bristol
BS16 4PS
Tel: (0117) 957 4574
Fax: (0117) 956 3461
Email: info@nomix.co.uk
Web: www.nomix.co.uk

Nufarm UK: Nufarm UK Ltd
Crabtree Manorway North
Belvedere
Kent
DA17 6BQ
Tel: (020) 8319 7222
Fax: (020) 8319 7280
Email: infouk@uk.nufarm.com
Web: www.ag.nufarm.co.uk

Pan Agriculture: Pan Agriculture Ltd
8 Cromwell Mews
Station Road
St Ives
Huntingdon
Cambs.
PE27 5HJ
Tel: (01480) 467790
Fax: (01480) 467041
Email: info@panagriculture.co.uk

Rentokil: Rentokil Initial plc
Felcourt
East Grinstead
W. Sussex
RH19 2JY
Tel: (01342) 833022
Fax: (01342) 326229
Web: www.ri-research.com

Rhizopon: Rhizopon UK Ltd
Croda Rosa
12 Bixley Road
Ipswich
Suffolk
IP3 8PL
Tel: (01473) 712666
Fax: (01473) 712666

Rigby Taylor: Rigby Taylor Ltd
Rigby Taylor House
Crown Lane
Horwich
Bolton
Lancs.
BL6 5HP
Tel: (01204) 677777
Fax: (01204) 677765
Email: info@rigbytaylor.com
Web: www.rigbytaylor.com

Scotts: The Scotts Company (UK) Ltd
Paper Mill Lane
Bramford
Ipswich
Suffolk
IP8 4BZ
Tel: (01473) 830492
Fax: (01473) 830386

Sherriff Amenity: Sherriff Amenity Services
The Pines
Fordham Road
Newmarket
Suffolk
CB8 7LG
Tel: (01638) 721888
Fax: (01638) 721815
Web: www.sherriffamenity.co.

Sinclair: William Sinclair Horticulture Ltd
Firth Road
Lincoln
LN6 7AH
Tel: (01522) 537561
Fax: (01522) 513609
Web: www.william-sinclair.co.uk

Sipcam: Sipcam UK Ltd
3 The Barn
27 Kneesworth Street
Royston
Herts.
SG8 5AB
Tel: (01763) 212100
Fax: (01763) 212101

Sorex: Sorex Ltd
St Michael's Industrial Estate
Widnes
Cheshire
WA8 8TJ
Tel: (0151) 420 7151
Fax: (0151) 495 1163
Email: enquiries@sorex.com
Web: www.sorex.com

Sphere: Sphere Laboratories (London) Ltd
The Yews
Main Street
Chilton
Oxon.
OX11 0RZ
Tel: (01235) 831802
Fax: (01235) 833896

Standon: Standon Chemicals Ltd
48 Grosvenor Square
London
W1K 2HT
Tel: (020) 7493 8648
Fax: (020) 7493 4219

SumiAgro: SumiAgro (UK) Ltd
Merlin House
Falconry Court
Bakers Lane
Epping
Essex
CM16 5DQ
Tel: (01992) 563700
Fax: (01992) 563800
Email: sumiagro@sumiagro.co.uk
Web: www.sumiagro.co.uk

SumiAgro Amenity: SumiAgro Amenity, a
division of SumiAgro (UK) Ltd
See SumiAgro (UK) Ltd

Syngenta:
Syngenta Crop Protection UK Limited
Whittlesford
Cambridge
CB2 4QT
Tel: (0800) 169 6058
Fax: (01223) 493700
Web: www.syngenta-crop.co.uk

Syngenta Bioline: Syngenta Bioline
Telstar Nursery
Holland Road
Little Clacton
Essex
CO16 9QG
Tel: (01255) 863200
Fax: (01255) 863206
Email: syngenta.bioline@syngenta.com
Web: www.syngenta-bioline.co.uk

Teliton: Teliton Ltd
Brook Point
1412 High Road
London
N20 9BH
Tel: (01279) 843344

Tomen: Tomen (UK) plc
Tomen House
13 Charles II Street
London
SW1Y 4QT
Tel: (020) 7808 3200
Fax: (020) 7976 1163
Email: andy_meads@ov.tomen.com
Web: www.arvesta.com

Truchem: Truchem Ltd
The Knoll
The Cross
Horsley
Stroud
Gloucestershire
GL6 0PR
Tel: (01453) 833293
Fax: (01453) 833293
Email: truchem@btopenworld.com

Unicrop: Universal Crop Protection Ltd
Park House
Maidenhead Road
Cookham
Berks.
SL6 9DS
Tel: (01628) 526083
Fax: (01628) 810457
Email: enquiries@unicrop.com

United Phosphorus: United Phosphorus Ltd
Chadwick House
Birchwood Park
Warrington
Cheshire
WA3 6AE
Tel: (01925) 859008
Fax: (01925) 811951
Email: ahoare@uniphos.co.uk

SECTION 6

Vitax: Vitax Ltd
   Owen Street
   Coalville
   Leicester
   LE67 3DE
   Tel: (01530) 510060
   Fax: (01530) 510299
   Email: info@vitax.co.uk
   Web: www.vitax.co.uk

Whyte Agrochemicals:
Whyte Agrochemicals Ltd
   Marlborough House
   298 Regents Park Road
   Finchley
   London
   N3 2UA
   Tel: (020) 8346 5946
   Fax: (020) 8349 4589

Appendix 2
Useful Contacts

Agricultural Industries Confederation Ltd (AIC)
Confederation House
East of England Showground
Peterborough PE2 6XE
Tel: (01733) 385236
Fax: (01733) 385270
Web: www.agindustries.org.uk

BASIS Ltd
Bank Chambers
34 St John Street
Ashbourne
Derbyshire DE6 1GH
Tel: (01335) 343945/346138
Fax: (01335) 346488
Web: www.basis-reg.co.uk

British Beekeepers' Association
National Agricultural Centre
Stoneleigh
Kenilworth
Warwickshire CV8 2LZ
Tel: (024) 7669 6679
Fax: (024) 7669 0682

British Beer & Pub Association
Market Towers
1 Nine Elms Lane
London SW8 5NQ
Tel: (0207) 627 9191
Fax: (0207) 627 9123

BCPC (British Crop Production Council)
7 Omni Business Centre
Omega Park
Alton
Hampshire GU34 2QD
Tel: (01420) 593200
Fax: (01420) 593209
Web: www.bcpc.org

BCPC Publications Sales
7 Omni Business Centre
Omega Park
Alton
Hampshire GU34 2QD
Tel: (01420) 593200
Fax: (01420) 593209
Web: www.bcpc.org/bookshop
e-mail: publications@bcpc.org

British Pest Control Association
1 Gleneagles House
Vernon Gate
South Street
Derby DE1 1UP
Tel: (01332) 294288
Fax: (01332) 295904
Web: www.bpca.org.uk

Crop Protection Association Ltd
18 & 20 Evans Business Centre
Cully Court
Bakewell Road
Orton Southgate
Peterborough PE2 6XS
Tel: (01733) 367213
Fax: (01733) 367212
Web: www.cropprotection.org.uk

CropLife International
(previously the Global Crop Protection Federation)
Avenue Louise 143
B-1050 Brussels
Belgium
Tel: (+32) 2 542 0410
Fax: (+32) 2 542 0419
Web: www.croplife.org

Department of Agriculture and Rural Development (Northern Ireland)
Pesticides Section
Dundonald House
Upper Newtownards Road
Belfast BT4 3SB
Tel: (028) 9052 4704
Fax: (028) 9052 4266

Department of Environment, Food and Rural Affairs (Defra)
Nobel House
17 Smith Square
London SW1P 3JR
Tel: (020) 7238 6000
Fax: (020) 7238 6591
Web: www.defra.gov.uk

SECTION 6

Environment Agency
Rio House
Waterside Drive
Aztec West
Almondsbury
Bristol BS12 4UD
Tel: (01454) 624400
Fax: (01454) 624409
Web: www.environment-agency.gov.uk

European Crop Protection Association (ECPA)
Avenue E van Nieuwenhuyse 6
B-1160 Brussels
Belgium
Tel: (+32) 2 663 1550
Fax: (+32) 2 663 1560
Web: www.ecpa.be

Farmers' Union of Wales
Llys Amaeth
Queen's Square
Aberystwyth
Dyfed SY23 2EA
Tel: (01970) 612755
Fax: (01970) 624369

Forestry Commission
231 Corstorphine Road
Edinburgh EH12 7AT
Tel: (0131) 334 0303
Fax: (0131) 334 3047
Web: www.forestry.gov.uk

Health and Safety Executive
Information Services
Room 318, Daniel House
Stanley Precinct
Bootle
Merseyside L20 3TW
Tel: (0151) 951 3191
Fax: (0151) 951 3467

Health and Safety Executive
Biocides & Pesticides Assessment Unit
Room 123, Magdalen House
Bootle
Merseyside L20 3QZ
Tel: (0151) 951 3535
Fax: (0151) 951 3317

Health and Safety Executive – Books
PO Box 1999
Sudbury
Suffolk CO10 2WA
Tel: (01787) 881165
Fax: (01787) 313995

Horticultural Development Council
Bradbourne House
East Malling
Kent ME19 6DZ
Tel: (01732) 848383
Fax: (01732) 848498

Lantra
Lantra House
Stoneleigh Park
Nr Coventry
Warwickshire CV8 2LG
Tel: (024) 7669 6996
Fax: (024) 7669 6732
Web: www.lantra.co.uk

National Association of Agricultural Contractors (NAAC)
Samuelson House
Paxton Road
Orton Centre
Peterborough
Cambs. PE2 5LT
Tel: (01733) 362920
Fax: (01733) 362921

National Farmers' Union
Agriculture House
164 Shaftesbury Avenue
London WC2H 8HL
Tel: (020) 7331 7200
Fax: (020) 7331 7313
Web: www.nfu.org.uk

National Poisons Information Service
Guys' and St Thomas' Hospital Trust
London SE14 5ER
Tel: (020) 7635 9191

The Royal Hospitals
Belfast BT12 6BA
Tel: (028) 9024 0503

City Hospital NHS Trust
Birmingham B18 7QH
Tel: (0121) 507 5588/5589

Llandough Hospital
Penarth CF64 2XX
Tel: (029) 2070 9901

Royal Infirmary
Edinburgh
Tel: (0131) 536 2300

The General Infirmary
Leeds LS1 3EX
Tel: (0113) 243 0715

Royal Victoria Infirmary
Newcastle NE1 4LP
Tel: (0191) 232 5131

National Turfgrass Council
Hunter's Lodge
Dr Brown's Road
Minchinhampton
Gloucestershire GL6 9BT
Tel: (01453) 883588
Fax: (01453) 731449

NPTC
National Agricultural Centre
Stoneleigh
Kenilworth
Warwickshire CV8 2LG
Tel: (024) 7669 6553
Fax: (024) 7669 6128
Web: www.nptc.org.uk

Pesticides Safety Directorate
Mallard House
King's Pool
3 Peasholme Green
York YO1 2PX
Tel: (01904) 640500
Fax: (01904) 455733
Web: www.pesticides.gov.uk

Processors and Growers Research Organisation
The Research Station
Great North Road
Thornhaugh
Peterborough
Cambs. PE8 6HJ
Tel: (01780) 782585
Fax: (01780) 783993
Web: www.pgro.co.uk

Scottish Beekeepers' Association
North Trinity House
114 Trinity Road
Edinburgh EH5 3JZ
Tel: (0131) 552 5341

Scottish Environment Protection Agency (SEPA)
Erskine Court
The Castle Business Park
Stirling FK9 4TR
Tel: (01786) 457 700
Fax: (01786) 446 885
Web: www.sepa.org.uk

TSO (The Stationery Office)
Publications Centre
PO Box 276
London SW8 5DT
Tel: (020) 7873 9090 (orders)/
(020) 7873 0011 (enquiries)
Fax: (020) 7873 8200
Web: www.thestationeryoffice.com

Ulster Beekeepers' Association
57 Liberty Road
Carrickfergus
Co. Antrim BT38 9DJ
Tel: (01960) 362998

Welsh Beekeepers' Association
Trem y Clawdd
Fron Isaf
Chirk
Wrexham
Clwyd LL14 5AH
Tel/Fax: (01691) 773300

SECTION 6

Appendix 3
Keys to Crop and Weed Growth Stages

Decimal Code for the Growth Stages of Cereals
Illustrations of these growth stages can be found in The e-UK Pesticide Guide, *in the reference indicated below and in some company product manuals.*

0 Germination

- 00 Dryseed (caryopsis)
- 01 Beginning of seed imbibition
- 03 Seed imbibition complete
- 05 Radicle emerged from caryopsis
- 06 Radicle elongated, root hairs and/or side roots visible
- 07 Coleoptile emerged from caryopsis
- 09 Emergence: coleoptile penetrates soil surface (cracking stage)

1 Leaf development

- 10 First leaf through coleoptile
- 11 First leaf unfolded
- 12 2 leaves unfolded
- 13 3 leaves unfolded
- 1... Stages continuous until...
- 19 9 or more leaves unfolded

2 Tillering

- 20 No tillers
- 21 Beginning of tillering: first tiller detectable
- 22 2 tillers detectable
- 23 3 tillers detectable
- 2... Stages continuous until...
- 29 End of tillering: maximum number of tillers detectable

3 Stem elongation

- 30 Beginning of item elongation: pseudostem and tillers erect, first internode begins to elongate, top of infloresecence at least 1 cm above tillering node
- 31 First node at least 1 cm above tillering node
- 32 Node 2 at least 1 cm above node 1
- 33 Node 3 at least 1 cm above node 2
- 3... Stages continuous until...
- 37 Flag leaf just visible, still rolled
- 39 Flag leaf stage: flag leaf fully unrolled, ligule just visible

4 Booting

- 41 Early boot stage: flag leaf sheath extending
- 43 Mid boot stage: flag leaf sheath just visibly swollen
- 45 Late boot stage: flag leaf sheath swollen
- 47 Flag leaf sheath opening
- 49 First awns visible (in awned forms only)

5 Inflorescence emergente, heading

- 51 Beginning of heading: tip of inflorescence emerged from sheath, first spikelet just visible
- 52 20% of inflorescence emerged
- 53 30% of inflorescence emerged
- 54 40% of inflorescence emerged
- 55 Middle of heading: 50% of inflorescence emerged
- 56 60% of inflorescence emerged
- 57 70% of inflorescence emerged
- 58 80% of inflorescence emerged
- 59 End of heading: inflorescence fully emerged

6 Flowering, anthesis

- 61 Beginning of flowering: first anthers visible
- 65 Full flowering: 50% of anthers mature
- 69 End of flowering: all spikelets have completed flowering but some dehydrated anthers may remain

7 Development of fruit

- 71 Watery ripe: first grains have reached half their final size
- 73 Early milk
- 75 Medium milk: grain content milky, grains have reached final size, still green
- 77 Late milk

8 Ripening

- 83 Early dough
- 85 Soft dough; grain content soft but dry. Fingernail impression not held
- 87 Hard dough: grain content solid. Fingernail impresion held
- 89 Fully ripe: grain hard, difficult to divide with thumbnail

9 Senescence

- 92 Over-ripe: grain very hard, cannot be dented by thumbnail
- 93 Grains loosening in day-time
- 97 Plant dead and collapsing
- 99 Harvested product

[From Meier, U. (1997) BBCH-Monograph: *Growth stages of plants*. Blackwell Wissenschaftsverlag, Berlin and Vienna.]

Stages in Development of Oilseed Rape

Illustrations of these growth stages can be found in The e-UK Pesticide Guide, *in the reference indicated below and in some company product manuals.*

0 Germination and emergence

00	Dry seed
01	Beginning of seed imbibition
03	Seed imbibition complete
05	Radicle emerged from seed
07	Hypocotyl with cotyledons emerged from seed
08	Hypocotyl with cotyledons growing towards soil surface
09	Emergence: cotyledons emerge through soil surface

1 Leaf development

10	Cotyledons completely unfolded
11	First leaf unfolded
12	2 leaves unfolded
13	3 leaves unfolded
1...	Stages continuous until...
19	9 or more leaves unfolded

2 Formation of side shoots

20	No side shoots
21	Beginning of side shoot development: first side shoot detectable
22	2 side shoots detectable
23	3 side shoots detectable
2...	Stages continuous until...
29	End of side shoot development: 9 or more side shoots detectable

3 Stem elongation

30	Beginning of stem elongation: no internodes (rosette)
31	1 visibly extended internode
32	2 visibly extended internodes
33	3 visibly extended internodes
3...	Stages continuous until...
39	9 or more visibly extended internodes

5 Inflorescence emergence

50	Flower buds present, still enclosed by leaves
51	Flower buds visible from above (green bud)
52	Flower buds free, level with the youngest leaves
53	Flower buds raised above the youngest leaves
55	Individual flower buds (main inflorescence) visible but still closed
57	Individual flower buds (secondary inflorescences) visible but still closed
59	First petals visible, flower buds still closed (yellow bud)

6 Flowering

60	First flowers open
61	10% of flowers on main raceme open, main raceme elongating
62	20% of flowers on main raceme open
63	30% of flowers on main raceme open
64	40% of flowers on main raceme open
65	Full flowering: 50% flowers on main raceme open, older petals falling
67	Flowering declining: majority of petals fallen
69	End of flowering

7 Development of fruit

71	10% of pods have reached final size
72	20% of pods have reached final size
73	30% of pods have reached final size
74	40% of pods have reached final size
75	50% of pods have reached final size
76	60% of pods have reached final size
77	70% of pods have reached final size
78	80% of pods have reached final size
79	Nearly all pods have reached final size

8 Ripening

80	Beginning of ripening: seed green, filling pod cavity
81	10% of pods ripe, seeds dark and hard
82	20% of pods ripe, seeds dark and hard
83	30% of pods ripe, seeds dark and hard
84	40% of pods ripe, seeds dark and hard
85	50% of pods ripe, seeds dark and hard
86	60% of pods ripe, seeds dark and hard
87	70% of pods ripe, seeds dark and hard
88	80% of pods ripe, seeds dark and hard
89	Fully ripe: nearly all pods ripe, seeds dark and hard

9 Stem senescence

97	Plant dead and dry
99	Harvested product

[From Meier, U. (1997) BBCH-Monograph: *Growth stages of plants*. Blackwell Wissenschaftsverlag, Berlin and Vienna.]

SECTION 6

Stages in Development of Peas

Illustrations of these growth stages can be found in The e-UK Pesticide Guide, *in the reference indicated below and in some company product manuals.*

0   Germination

00   Dry seed
01   Beginning of seed imbibition
03   Seed imbibition complete
05   Radicle emerged from seed
07   Shoot breaking through seed coat
08   Shoot growing towards soil surface; hypocotyl arch visible
09   Emergence: shoot breaks through soil surface (cracking stage)

1   Leaf development

10   Pair of scale leaves visible
11   First true leaf (with stipules) unfolded or first tendril developed
12   2 leaves (with stipules) unfolded or 2 tendrils developed
13   3 leaves (with stipules) unfolded or 3 tendrils developed
1... Stages continuous until...
19   9 or more leaves (with stipules) unfolded or 9 or more tendrils developed

3   Stem elongation (main shoot)

30   Beginning of stem elongation
31   1 visibly extended internode1
32   2 visibly extended internodes1
33   3 visibly extended internodes1
3... Stages continuous until...
39   9 or more visibly extended internodes

5   Inflorescence emergence

51   First flower buds visible outside leaves
55   First separated flower buds visible outside leaves but still closed
59   First petals visible, flowers still closed

6   Flowering

60   First flowers open (sporadically within the population)
61   Beginning of flowering: 10% of flowers open
62   20% of flowers open
63   30% of flowers open
64   40% of flowers open
65   Full flowering: 50% of flowers open
67   Flowering declining
69   End of flowering

7   Development of fruit

71   10% of pods have reached typical length; juice exudes if pressed
72   20% of pods have reached typical length; juice exudes if pressed
73   30% of pods have reached typical length; juice exudes if pressed. Tenderometer value: 80 TE
74   40% of pods have reached typical length; juice exudes if pressed. Tenderometer value: 95 TE
75   50% of pods have reached typical length; juice exudes if pressed. Tenderometer value: 105 TE
76   60% of pods have reached typical length; juice exudes if pressed. Tenderometer value: 115 TE
77   70% of pods have reached typical length. Tenderometer value: 130 TE
79   Pods have reached typical size (green ripe); peas fully formed

8   Ripening of fruit and seed

81   10% of pods ripe, seeds final colour, dry and hard
82   20% of pods ripe, seeds final colour, dry and hard
83   30% of pods ripe, seeds final colour, dry and hard
84   40% of pods ripe, seeds final colour, dry and hard
85   50% of pods ripe, seeds final colour, dry and hard
86   60% of pods ripe, seeds final colour, dry and hard
87   70% of pods ripe, seeds final colour, dry and hard
88   80% of pods ripe, seeds final colour, dry and hard
89   Fully ripe: all pods dry and brown. Seeds dry and hard (dry ripe)

9   Senescence

97   Plants dead and dry
99   Harvested product

[From Meier, U. (1997) BBCH-Monograph: *Growth stages of plants*. Blackwell Wissenschaftsverlag, Berlin and Vienna.]

Stages in Development of Faba Beans

Illustrations of these growth stages can be found in The e-UK Pesticide Guide, *in the reference indicated below and in some company product manuals.*

0 Germination

00	Dry seed
01	Beginning of seed imbibition
03	Seed imbibition complete
05	Radicle emerged from seed
07	Shoot emerged from seed (plumule apparent)
08	Shoot growing towards soil surface
09	Emergence: shoot emerges through soil surface

1 Leaf development

10	Pair of scale leaves visible (may be eaten or lost)
11	First leaf unfolded
12	2 leaves unfolded
13	3 leaves unfolded
1...	Stages continuous until...
19	9 or more leaves unfolded

2 Formation of side shoots

20	No side shoots
21	Beginning of side shoot development: first side shoot detectable
22	2 side shoots detectable
23	3 side shoots detectable
2...	Stages continuous until...
29	End of side shoot development: 9 or more side shoots detectable

3 Stem elongation

30	Beginning of stem elongation
31	1 visibly extended internode2
32	2 visibly extended internodes
33	3 visibly extended internodes
3...	Stages continuous until...
39	9 or more visibly extended internodes

5 Inflorescence emergence

50	Flower buds present, still enclosed by leaves
51	First flower buds visible outside leaves
55	First individual flower buds visible outside leaves but still closed
59	First petals visible, many individual flower buds, still closed

6 Flowering

60	First flowers open
61	Flowers open on first raceme
63	Flowers open 3 racemes per plant
65	Full flowering: flowers open on 5 racemes per plant
67	Flowering declining
69	End of flowering

7 Development of fruit

70	First pods have reached final length (flat pod)
71	10% of pods have reached final length
72	20% of pods have reached final length
73	30% of pods have reached final length
74	40% of pods have reached final length
75	50% of pods have reached final length
76	60% of pods have reached final length
77	70% of pods have reached final length
78	80% of pods have reached final length
79	Nearly all pods have reached final length

8 Ripening

80	Beginning of ripening: seed green, filling pod cavity
81	10% of pods ripe, seeds dry and hard
82	20% of pods ripe, seeds dry and hard
83	30% of pods ripe and dark, seeds dry and hard
84	40% of pods ripe and dark, seeds dry and hard
85	50% of pods ripe and dark, seeds dry and hard
86	60% of pods ripe and dark, seeds dry and hard
87	70% of pods ripe and dark, seeds dry and hard
88	80% of pods ripe and dark, seeds dry and hard
89	Fully ripe: nearly all pods dark, seeds dry and hard

9 Senescence

93	Stems begin to darken
95	50% of stems brown or black
97	Plant dead and dry
99	Harvested product

[From Meier, U. (1997) BBCH-Monograph: *Growth stages of plants*. Blackwell Wissenschaftsverlag, Berlin and Vienna.]

SECTION 6

Stages in Development of Potato

Illustrations of these growth stages can be found in The e-UK Pesticide Guide, *in the reference indicated below and in some company product manuals.*

0   Sprouting/Germination

000   Innate or enforced dormancy: dry seed tuber not sprouted

001   Beginning of sprouting: beginning of sprouts visible (<1 mm) seed imbibition

002   Sprouts upright (<2 mm)

003   End of dormancy: sprouts 2-3 mm, seed imbibition complete

005   Beginning of root formation: radicle (root) emerged from seed

007   Beginning of stem formation: hypocotyl with cotyledons breaking

008   Stems growing towards soil surface: hypocotyl with formation of scale leaves in the axils; cotyledons growing

009   Emergence: Emergence: stems break through soil surface; cotyledons break through soil surface

(021-029 For second generation sprouts)

1   Leaf development

100   From tuber: first leaves begin to extend. From seed: cotyledons completely unfolded

101   1st leaf of main stem unfolded (>4 cm)

102   2nd leaf of main stem unfolded (>4 cm)

103   3rd leaf of main stem unfolded (>4 cm)

10...   Stages continuous until...

109   9 leaves of main stem unfolded (>4 cm)

110   10th leaf of main stem unfolded (>4 cm)

11...   Stages continuous until...

119   19th leaf of main stem unfolded (>4 cm)

121   First leaf of 2nd order branch above first inflorescence unfolded (>4 cm)

122   2nd leaf of 2nd order branch above first inflorescence unfolded (>4 cm)

12...   Stages continuous until...

131   First leaf of 3rd order branch above 2nd inflorescence unfolded (>4 cm)

132   2nd leaf of 3rd order branch above 2nd inflorescence unfolded (>4 cm)

13...   Stages continuous until...

1NX   Xth leaf of nth order branch above n-1th inflorescence unfolded (>4 cm)

2   Formation of basal side shoots below and above soil surface (main stem)

201   First basal side shoot visible (>5 cm)

202   2nd basal side shoot visible (>5 cm)

203   3rd basal side shoot visible (>5 cm)

20...   Stages continuous until...

209   9 or more basal side shoots visible (>5 cm)

3   Main stem elongation (crop cover)

301   Beginning of crop cover: 10% of plants meet between rows

302   20% off plants meet between rows

305   50% of plants meet between rows

306   60% of plants meet between rows

307   70% of plants meet between rows

308   80% of plants meet between rows

309   Crop cover complete: about 90% of plants meet between rows

4   Tuber formation

400   Tuber initiation: swelling of first stolon tips to twice the diameter of subtending stolon

401   10% of total final tuber mass reached

402   20% of total final tuber mass reached

403   30% of total final tuber mass reached

404   40% of total final tuber mass reached

405   50% of total final tuber mass reached

406   60% of total final tuber mass reached

407   70% of total final tuber mass reached

408   Maximum of total tuber mass reached, tubers detach easily from stolons, skin set not yet complete (skin easily removed)

409   Skin set complete: (skin at apical end of tuber not removable with thumb) 95% of tubers in this stage

5 Inflorescence emergence

501 First individual buds (1-2 mm) of first inflorescence visible (main stem)

505 Buds of first inflorescence extended to 5 mm

509 First flower petals of first inflorescence visible

521 Individual buds of 2nd inflorescence visible (second order branch)

525 Buds of 2nd inflorescence extended to 5 mm open (main stem)

529 First flower petals of 2nd inflorescence visible above sepals

531 Individual buds of 3rd inflorescence visible (3rd order branch)

535 Buds of 3rd inflorescence extended to 5 mm

539 First flower petals of 3rd inflorescence visible above sepals

5N   Nth inflorescence emerging

6 Flowering

600 First open flowers in population

601 Beginning of flowering: 10% of flowers in the first inflorescence open (main stem)

602 20% of flowers in the first inflorescence open

603 30% of flowers in the first inflorescence open

604 40% of flowers in the first inflorescence open

605 Full flowering: 50% of flowers in the first inflorescence open

606 60% of flowers in the first inflorescence open

607 70% of flowers in the first inflorescence open

608 80% of flowers in the first inflorescence open

609 End of flowering in the first inflorescence

621 Beginning of flowering: 10% of flowers in the 2nd inflorescence open (second order branch)

625 Full flowering: 50% of flowers in the 2nd inflorescence open

629 End of flowering in the 2nd inflorescence

631 Beginning of flowering: 10% of flowers in the 3rd inflorescence open (third order branch)

635 Full flowering: 50% of flowers in the 3rd inflorescence open

639 End of flowering in the 3rd inflorescence

6N   Nth inflorescence flowering

6N9 End of flowering

7 Development of fruit

700 First berries visible

701 10% of berries in the first fructification have reached full size (main stem)

702 20% of berries in the first fructification have reached full size

703 30% of berries in the first fructification have reached full size

70... Stages continuous until...

721 10% of berries in the 2nd fructification have reached full size (second order branch)

7N   Development of berries in nth fructification

7N9 Nearly all berries in the nth fructification have reached full size (or have been shed)

8 Ripening of fruit and seed

801 Berries in the first fructification still green, seed light-coloured (main stem)

805 Berries in the first fructification ochre-coloured or brownish

809 Berries in the first fructification shrivelled, seed dark

821 Berries in the 2nd fructification still green, seed light-coloured (second order branch)

8N   Ripening of fruit and seed in nth fructification

9 Senescence

901 Beginning of leaf yellowing

903 Most of the leaves yellowish

905 50% of the leaves brownish

907 Leaves and stem dead, stems bleached and dry

909 Harvested product

SECTION 6

[From Meier, U. (1997) BBCH-Monograph: *Growth stages of plants.* Blackwell Wissenschaftsverlag, Berlin and Vienna.]

Stages in Development of Linseed

Illustrations of these growth stages can be found in The e-UK Pesticide Guide, *in the reference indicated below and in some company product manuals.*

0   Germination and emergence

00	Dry seed
01	Imbibed seed
02	Radicle apparent
04	Hypocotyl extending
05	Emergence
06	Cotyledon unfolding from seed case
07	Cotyledons unfolded and fully expanded

1   Vegetative stage (of main stem)

10	True leaves visible
12	First pair of true leaves fully expanded
13	Third pair of true leaves fully expanded
1n	n leaf fully expanded

2   Basal branching

21	One branch
22	Two branches
23	Three branches
2n	n branches

3   Flower bud development (on main stem)

31	Enclosed bud visible in leaf axils
33	Bud extending from axil
35	Corymb formed
37	Buds enclosed but petals visible
39	First flower open

4   Flowering (whole plant)

41	10% of flowers open
43	30% of flowers open
45	50% of flowers open
49	End of flowering

5   Capsule formation (whole plant)

51	10% of capsules formed
53	30% Of capsules formed
55	50% of capsules formed
59	End of capsule formation

6   Capsule senescence (on most advanced plant)

61	Capsules expanding
63	Capsules green and full size
65	Capsules turning yellow
67	Capsules all yellow brown but soft
69	Capsules brown, dry and senesced

7   Stem senescence (whole plant)

71	Stems mostly green below panicle
73	Most stems 30% brown
75	Most stems 50% brown
77	Stems 75% brown
79	Stems completely brown

8   Stems rotting (retting)

81	Outer tissue rotting
85	Vascular tissue easily removed
89	Stems completely collapsed

9   Seed development (whole plant)

91	Seeds expanding
92	Seeds white but full size
93	Most seeds turning ivory yellow
94	Most seeds turning brown
95	All seeds brown and hard
98	Some seeds shed from capsule
99	Most seeds shed from capsule

[From Freer (1991) *Aspects of Applied Biology,* **28**, 33-40.]

Stages in Development of Annual Grass Weeds

Illustrations of these growth stages can be found in The e-UK Pesticide Guide, *in the reference indicated below and in some company product manuals.*

0   Germination and emergence

00	Dry seed
01	Start of imbibition
03	Imbibition complete
05	Radicle emerged from caryopsis
07	Coleoptile emerged from caryopsis
09	Leaf just at coleoptile tip

1   Seedling growth

10	First leaf through coleoptile
11	First leaf unfolded
12	2 leaves unfolded
13	3 leaves unfolded
14	4 leaves unfolded
15	5 leaves unfolded
16	6 leaves unfolded
17	7 leaves unfolded
18	8 leaves unfolded
19	9 or more leaves unfolded

2   Tillering

20	Main shoot only
21	Main shoot and 1 tiller
22	Main shoot and 2 tillers
23	Main shoot and 3 tillers
24	Main shoot and 4 tillers
25	Main shoot and 5 tillers
26	Main shoot and 6 tillers
27	Main shoot and 7 tillers
28	Main shoot and 8 tillers
29	Main shoot and 9 or more tillers

3   Stem elongation

31	First node detectable
32	2nd node detectable
33	3rd node detectable
34	4th node detectable
35	5th node detectable
36	6th node detectable
37	Flag leaf just visible
39	Flag leaf ligule just visible

4   Booting

41	Flag leaf sheath extending
43	Boots just visibly swollen
45	Boots swollen
47	Flag leaf sheath opening
49	First awns visible

5   Inflorescence emergence

51	First spikelet of inflorescence just visible
53	25% of inflorescence emerged
55	50% of inflorescence emerged
57	75% of inflorescence emerged
59	Emergence of inflorescence completed

6   Anthesis

61	Beginning of anthesis
65	Anthesis half-way
69	Anthesis complete

[From Lawson & Read (1992) *Annals of Applied Biology*, **12**, 211-214.]

Growth Stages of Annual Broad-leaved Weeds

Preferred Descriptive Phrases
Illustrations of these growth stages can be found in The e-UK Pesticide Guide, *in the reference indicated below and in some company product manuals*

Pre-emergence
Early cotyledons
Expanded cotyledons
One expanded true leaf
Two expanded true leaves
Four expanded true leaves
Six expanded true leaves
Plants up to 25 mm across/high

Plants up to 50 mm across/high
Plants up to 100 mm across/high
Plants up to 150 mm across/high
Plants up to 250 mm across/high
Flower buds visible
Plant flowering
Plant senescent

[From Lutman & Tucker (1987) *Annals of Applied Biology*, **110**, 683-687.]

Appendix 4
Key to Hazard Classifications and Safety Precautions

Every product label contains information to warn users of the risks from using the product, together with precautions that must be followed in order to minimise the risks. A hazard classification (if any) and symbol is shown with associated risk phrases, followed by a series of safety precautions. These are represented in the pesticide profiles in Section 2 by code letters and numbers under the heading **Hazard classification and safety precautions**.

The codes are defined below, under the same sub-headings as they appear in the pesticide profiles.

Where a product label specifies the use of personal protective equipment (PPE), the requirements are listed under the sub-heading **Operator protection**, using letter codes to denote the protective items, according to the list below. Often PPE requirements are different for specified operations, e.g. handling the concentrate, cleaning equipment etc., but it is not possible to list them separately. The lists of PPE are therefore an indication of what the user may need to have available to use the product in different ways. **When making a COSHH assessment it is therefore essential that the product label is consulted for information on the particular use that is being assessed**.

Where the generalised wording includes a phrase such as '... for *xx* days', the specific requirement for each pesticide is shown in brackets after the code.

Hazard

H01	Very toxic
H02	Toxic
H03	Harmful
H04	Irritant
H05	Corrosive
H06	Extremely flammable
H07	Highly flammable
H08	Flammable
H09	Oxidising agent
H10	Explosive
H11	Dangerous for the environment

Risk phrases

R08	Contact with combustible material may cause fire
R09	Explosive when mixed with combustible material
R16	Explosive when mixed with oxidising substances
R20	Harmful by inhalation
R21	Harmful in contact with skin
R22a	Harmful if swallowed
R22b	May cause lung damage if swallowed
R23	Toxic by inhalation
R24	Toxic in contact with skin
R25	Toxic if swallowed
R26	Very toxic by inhalation
R27	Very toxic in contact with skin
R28	Very toxic if swallowed
R34	Causes burns
R35	Causes severe burns
R36	Irritating to eyes
R37	Irritating to respiratory system

R38	Irritating to skin
R39	Danger of very serious irreversible effects
R40	Limited evidence of a carcinogenic effect
R41	Risk of serious damage to eyes
R42	May cause sensitization by inhalation
R43	May cause sensitization by skin contact
R45	May cause cancer
R46	May cause heritable genetic damage
R48	Danger of serious damage to health by prolonged exposure
R50	Very toxic to aquatic organisms
R51	Toxic to aquatic organisms
R52	Harmful to aquatic organisms
R53a	May cause long-term adverse effects in the aquatic environment
R53b	Dangerous to aquatic organisms
R54	Toxic to flora
R55	Toxic to fauna
R56	Toxic to soil organisms
R57	Toxic to bees
R58	May cause long-term adverse effects in the environment
R60	May impair fertility
R61	May cause harm to the unborn child
R62	Possible risk of impaired fertility
R63	Possible risk of harm to the unborn child
R64	May cause harm to breast-fed babies
R66	Repeated exposure may cause skin dryness or cracking
R67	Vapours may cause drowsiness and dizziness
R68	Possible risk of irreversible effects

Operator protection

A	Suitable protective gloves (the product label should be consulted for any specific requirements about the material of which the gloves should be made)
B	Rubber gauntlet gloves
C	Face-shield
D	Approved respiratory protective equipment
E	Goggles
F	Dust mask
G	Full face-piece respirator
H	Coverall
J	Hood
K	Apron/Rubber apron
L	Waterproof coat
M	Rubber boots
N	Waterproof jacket and trousers
P	Suitable protective clothing
U01	To be used only by operators instructed or trained in the use of chemical/product/type of produce and familiar with the precautionary measures to be observed
U02a	Wash all protective clothing thoroughly after use, especially the inside of gloves
U02b	Avoid excessive contamination of coveralls and launder regularly
U02c	Remove and wash contaminated gloves immediately
U03	Wash splashes off gloves immediately
U04a	Take off immediately all contaminated clothing
U04b	Take off immediately all contaminated clothing and wash underlying skin. Wash clothes before re-use
U04c	Wash clothes before re-use
U05a	When using do not eat, drink or smoke
U05b	When using do not eat, drink, smoke or use naked lights
U06	Handle with care and mix only in a closed container
U07	Open the container only as directed
U08	Wash concentrate/dust from skin or eyes immediately
U09a	Wash any contamination/splashes/dust/powder/concentrate from skin or eyes immediately
U09b	Wash any contamination/splashes/dust/powder/concentrate from eyes immediately
U10	After contact with skin or eyes wash immediately with plenty of water

U11	In case of contact with eyes rinse immediately with plenty of water and seek medical advice
U12	In case of contact with skin rinse immediately with plenty of water and seek medical advice
U13	Avoid contact by mouth
U14	Avoid contact with skin
U15	Avoid contact with eyes
U16	Ensure adequate ventilation in confined spaces
U18	Extinguish all naked flames, including pilot lights, when applying the fumigant/dust/liquid/product
U19a	Do not breathe dust/fog/fumes/gas/smoke/spray mist/vapour. Avoid working in spray mist
U19b	Do not work in confined spaces or enter spaces in which high concentrations of vapour are present. Where this precaution cannot be observed distance breathing or self-contained breathing apparatus must be worn, and the work should be done by trained oper
U20a	Wash hands and exposed skin before eating, drinking or smoking and after work
U20b	Wash hands and exposed skin before meals and after work
U20c	Wash hands before meals and after work
U21	Before entering treated crops, cover exposed skin areas, particularly arms and legs
U22a	Do not touch sachet with wet hands or gloves/Do not touch water soluble bag directly
U22b	Protect sachets from rain or water
U23a	Do not apply by knapsack sprayer/hand-held equipment
U23b	Do not apply through hand held rotary atomisers
U24	Do not handle grain unnecessarily

Environmental protection

E02a	Keep unprotected persons/animals out of treated/fumigation areas for at least xx hours/days
E02b	Prevent access by livestock, pets and other non-target mammals and birds to buildings under fumigation and ventilation
E02c	Vacate treatment areas before application
E03	Label treated seed with the appropriate precautions, using the printed sacks, labels or bag tags supplied
E05a	Do not apply directly to livestock/poultry
E05b	Keep poultry out of treated areas for at least xx days/weeks
E06a	Keep livestock out of treated areas for at least xx days/weeks after treatment
E06b	Dangerous to livestock. Keep all livestock out of treated areas/away from treated water for at least xx days/weeks. Bury or remove spillages
E06c	Harmful to livestock. Keep all livestock out of treated areas/away from treated water for at least xx days/weeks. Bury or remove spillages
E07a	Keep livestock out of treated areas for at least two weeks following treatment and until poisonous weeds, such as ragwort, have died down and become unpalatable
E07b	Dangerous to livestock. Keep livestock out of treated areas/away from treated water for at least xx weeks and until foliage of any poisonous weeds, such as ragwort, has died and become unpalatable
E07c	Harmful to livestock. Keep livestock out of treated areas/away from treated water for at least xx days/weeks and until foliage of any poisonous weeds such as ragwort has died and become unpalatable
E08	Do not feed treated straw or haulm to livestock within xx days/weeks of spraying
E09	Do not use on crops if the straw is to be used as animal feed/bedding
E10a	Dangerous to game, wild birds and animals
E10b	Harmful to game, wild birds and animals
E11	Paraquat can be harmful to hares; spray stubbles early in the day
E12a	High risk to bees. Do not apply to crops in flower or to those in which bees are actively foraging. Do not apply when flowering weeds are present
E12b	High risk to bees. Do not apply to crops in flower, or to those in which bees are actively foraging, except as directed on [crop]. Do not apply when flowering weeds are present
E12c	Extremely dangerous to bees. Do not apply to crops in flower or to those in which bees are actively foraging. Do not apply when flowering weeds are present
E12d	Dangerous to bees. Do not apply to crops in flower or to those in which bees are actively foraging. Do not apply when flowering weeds are present

SECTION 6

E12e	Dangerous to bees. Do not apply to crops in flower, or to those in which bees are actively foraging, except as directed on [crop]. Do not apply when flowering weeds are present
E12f	Harmful to bees. Do not apply to crops in flower or to those in which bees are actively foraging. Do not apply when flowering weeds are present
E12g	Apply away from bees
E13a	Extremely dangerous to fish or other aquatic life. Do not contaminate surface waters or ditches with chemical or used container
E13b	Dangerous to fish or other aquatic life. Do not contaminate surface waters or ditches with chemical or used container
E13c	Harmful to fish or other aquatic life. Do not contaminate surface waters or ditches with chemical or used container
E13d	Apply away from fish
E14a	Extremely dangerous to aquatic higher plants. Do not contaminate surface waters or ditches with chemical or used container
E14b	Dangerous to aquatic higher plants. Do not contaminate surface waters or ditches with chemical or used container
E15a	Do not contaminate surface waters or ditches with chemical or used container
E15b	Do not contaminate water with product or its container. Do not clean application equipment near surface water. Avoid contamination via drains from farmyards or roads
E16a	Do not allow direct spray from horizontal boom sprayers to fall within 5 m of the top of the bank of a static or flowing waterbody, unless a Local Environment Risk Assessment for Pesticides (LERAP) permits a narrower buffer zone, or within 1 m of the top
E16b	Do not allow direct spray from hand-held sprayers to fall within 1 m of the top of the bank of a static or flowing waterbody. Aim spray away from water
E16c	Do not allow direct spray from horizontal boom sprayers to fall within 5 m of the top of the bank of a static or flowing waterbody, or within 1m of the top of a ditch which is dry at the time of application. Aim spray away from water. This product is not
E16d	Do not allow direct spray from hand-held sprayers to fall within 1 m of the top of the bank of a static or flowing waterbody. Aim spray away from water. This product is not eligible for buffer zone reduction under the LERAP horizontal boom sprayers sche
E16e	Do not allow direct spray from horizontal boom sprayers to fall within 5 m of the top of the bank of a static or flowing water body or within 1 m from the top of any ditch which is dry at the time of application. Spray from hand held sprayers must not in
E16f	Do not allow direct spray/granule applications from vehicle mounted/drawn hydraulic sprayers/applicators to fall within 6 m of surface waters or ditches/Do not allow direct spray/granule applications from hand-held sprayers/applicators to fall within 2 m
E17a	Do not allow direct spray from broadcast air-assisted sprayers to fall within xx m of surface waters or ditches. Direct spray away from water
E17b	Do not allow direct spray from broadcast air-assisted sprayers to fall within xx m of the top of the bank of a static or flowing waterbody, unless a Local Environmental Risk Assessment for Pesticides (LERAP) permits a narrower buffer zone, or within 5 m o
E18	Do not spray from the air within 250 m horizontal distance of surface waters or ditches
E19a	Do not dump surplus herbicide in water or ditch bottoms
E19b	Do not empty into drains
E20	Prevent any surface run-off from entering storm drains
E21	Do not use treated water for irrigation purposes within xx days/weeks of treatment
E22a	High risk to non-target insects or other arthropods. Do not spray within 6 m of the field boundary
E22b	Risk to certain non-target insects or other arthropods. For advice on risk management and use in Integrated Pest Management (IPM) see directions for use
E22c	Risk to non-target insects or other arthropods
E23	Avoid damage by drift onto susceptible crops or water courses
E34	Do not re-use container for any purpose/Do not re-use container for any other purpose
E35	Do not burn this container
E36	Do not rinse out the container
E37	Do not use with any pesticide which is to be applied in or near water
E38	Use appropriate containment to avoid environmental contamination

Consumer protection

C01	Do not use on food crops
C02	Do not harvest for human or animal consumption for at least xx days/weeks after last application
C04	Do not apply to surfaces on which food/feed is stored, prepared or eaten
C05	Remove/cover all foodstuffs before application
C06	Remove exposed milk before application
C07	Collect eggs before application
C08	Protect food preparing equipment and eating utensils from contamination during application
C09	Cover water storage tanks before application
C10	Protect exposed water/feed/milk machinery/milk containers from contamination
C11	Remove all pets/livestock/fish tanks before treatment/spraying
C12	Ventilate treated areas thoroughly when smoke has cleared/Ventilate treated rooms thoroughly before occupying

Storage and disposal

D01	Keep out of reach of children
D02	Keep away from food, drink and animal feeding-stuffs
D03	Store away from seeds, fertilizers, fungicides and insecticides
D04	Store well away from corms, bulbs, tubers and seeds
D05	Protect from frost
D06a	Store away from heat
D06b	Do not store near heat or open flame
D07	Store under cool, dry conditions
D08	Store in a safe, dry, frost-free place designated as an agrochemical store
D09a	Keep in original container, tightly closed, in a safe place
D09b	Keep in original container, tightly closed, in a safe place, under lock and key
D10a	Wash out container thoroughly and dispose of safely
D10b	Wash out container thoroughly, empty washings into spray tank and dispose of safely
D10c	Rinse container thoroughly by using an integrated pressure rinsing device or manually rinsing three times. Add washings to sprayer at time of filling and dispose of container safely
D11a	Empty container completely and dispose of safely/Dispose of used generator safely
D11b	Empty container completely and dispose of it in the specified manner
D12a	This material (and its container) must be disposed of in a safe way
D12b	This material and its container must be disposed of as hazardous waste
D13	Treat used container as if it contained pesticide
D14	Return empty container as instructed by supplier

Treated Seed

S01	Do not handle treated seed unnecessarily
S02	Do not use treated seed as food or feed
S03	Keep treated seed secure from people, domestic stock/pets and wildlife at all times during storage and use
S04a	Bury or remove spillages
S04b	Harmful to birds/game and wildlife. Treated seed should not be left on the soil surface. Bury or remove spillages
S04c	Dangerous to birds/game and wildlife. Treated seed should not be left on the soil surface. Bury or remove spillages
S04d	To protect birds/wild animals, treated seed should not be left on the soil surface. Bury or remove spillages
S05	Do not reuse sacks or containers that have been used for treated seed for food or feed
S06a	Wash hands and exposed skin before meals and after work
S06b	Wash hands and exposed skin after cleaning and re-calibrating equipment
S07	Do not apply treated seed from the air
S08	Treated seed should not be broadcast

SECTION 6

Vertebrate/Rodent control products

V01a   Prevent access to baits/powder by children, birds and other animals, particularly cats, dogs, pigs and poultry

V01b   Prevent access to bait/gel/dust by children, birds and non-target animals, particularly dogs, cats, pigs, poultry

V02   Do not prepare/use/lay baits/dust/spray where food/feed/water could become contaminated

V03a   Remove all remains of bait, tracking powder or bait containers after use and burn or bury

V03b   Remove all remains of bait and bait containers/exposed dust/after treatment (except where used in sewers) and dispose of safely (e.g. burn/bury). Do not dispose of in refuse sacks or on open rubbish tips.

V04a   Search for and burn or bury all rodent bodies. Do not place in refuse bins or on rubbish tips

V04b   Search for rodent bodies (except where used in sewers) and dispose of safely (e.g. burn/bury). Do not dispose of in refuse sacks or on open rubbish tips

V04c   Dispose of safely any rodent bodies and remains of bait and bait containers that are recovered after treatment (e.g. burn/bury). Do not dispose of in refuse sacks or on open rubbish tips

V05   Use bait containers clearly marked POISON at all surface baiting points

Medical advice

M01   This product contains an anticholinesterase organophosphorus compound. DO NOT USE if under medical advice NOT to work with such compounds

M02   This product contains an anticholinesterase carbamate compound. DO NOT USE if under medical advice NOT to work with such compounds

M03   If you feel unwell, seek medical advice immediately (show the label where possible)

M04   In case of accident or if you feel unwell, seek medical advice immediately (show the label where possible)

M05a   If swallowed, seek medical advice immediately and show this container or label

M05b   If swallowed, do not induce vomiting: seek medical advice immediately and show this container or label

M05c   If swallowed induce vomiting if not already occurring and take patient to hospital immediately

M06   This product contains an anticholinesterase carbamoyl triazole compound. DO NOT USE if under medical advice NOT to work with such compounds

Appendix 5
Key to Abbreviations and Acronyms

The abbreviations of formulation types in the following list are used in Section 2 (Pesticide Profiles) and are derived from the *Catalogue of Pesticide Formulation Types and International Coding System* (CropLife International Technical Monograph 2, 4th edn, April 1999)

1 Formulation Types

AB	Grain bait
AE	Aerosol generator
AL	Other liquids to be applied undiluted
AP	Any other powder
BB	Block bait
BR	Briquette
CB	Bait concentrate
CF	Capsule suspension for seed treatment
CG	Encapsulated granule (controlled release)
CL	Contact liquid or gel (for direct application)
CP	Contact powder (for direct application)
CR	Crystals
CS	Capsule suspension
DC	Dispersible concentrate
DP	Dustable powder
DS	Powder for dry seed treatment
EC	Emulsifiable concentrate
EG	Emulsifiable granule
ES	Emulsion for seed treatment
EW	Oil in water emulsion
FG	Fine granules
FP	Smoke cartridge
FS	Flowable concentrate for seed treatment
FT	Smoke tablet
FU	Smoke generator
FW	Smoke pellets
GA	Gas
GB	Granular bait
GE	Gas-generating product
GG	Macrogranules
GL	Emulsifiable gel
GP	Flo-dust (for pneumatic application)
GR	Granules
GS	Grease
GW	Water soluble gel
HN	Hot fogging concentrate
KK	Combi-pack (solid/liquid)
KL	Combi-pack (liquid/liquid)
KN	Cold-fogging concentrate
KP	Combi-pack (solid/solid)
LA	Lacquer
LI	Liquid, unspecified
LS	Solution for seed treatment
ME	Microemulsion
MG	Microgranules
OL	Oil miscible liquid
PA	Paste
PC	Gel or paste concentrate
PS	Seed coated with a pesticide
PT	Pellet
RB	Ready-to-use bait

RH	Ready-to-use spray in hand-operated sprayer
SA	Sand
SC	Suspension concentrate (= flowable)
SE	Suspo-emulsion
SG	Water soluble granules
SL	Soluble concentrate
SP	Water soluble powder
SS	Water soluble powder for seed treatment
ST	Water soluble tablet
SU	Ultra low-volume suspension
TB	Tablets
TC	Technical material
TP	Tracking powder
UL	Ultra low-volume liquid
VP	Vapour releasing product
WB	Water soluble bags
WG	Water dispersible granules
WP	Wettable powder
WS	Water dispersible powder for slurry treatment of seed
WT	Water dispersible tablet
XX	Other formulations
ZZ	Not Applicable

2 Other Abbreviations and Acronyms

ACP	Advisory Committee on Pesticides
ACTS	Advisory Committee on Toxic Substances
ADAS	Agricultural Development and Advisory Service
a.i.	active ingredient
AIC	Agriculture Industries Confederation
BBPA	British Beer and Pub Association
CD	Acontrolled droplet application
CPA	Crop Protection Association
cm	centimetre(s)
COPR	Control of Pesticides Regulations 1986
COSHH	Control of Substances Hazardous to Health Regulations
d	day(s)
Defra	Department for Environment Food and Rural Affairs
EA	Environment Agency
EBDC	ethylene-bis-dithiocarbamate fungicide
FEPA	Food and Environment Protection Act 1985
g	gram(s)
GCPF	Global Crop Protection Federation (now CropLife International)
GS	growth stage (unless in formulation column)
h	hour(s)
ha	hectare(s)
HBN	hydroxybenzonitrile herbicide
HI	harvest interval
HSE	Health and Safety Executive
ICM	integrated crop management
IPM	integrated pest management
kg	kilogram(s)
l	litre(s)
LERAP	Local Environmental Risk Assessments for Pesticides
m	metre(s)
MAFF	Ministry of Agriculture, Fisheries and Food (now Defra)
MBC	methyl benzimidazole carbamate fungicide
MEL	maximum exposure limit
min	minute(s)
mm	millimetre(s)
MRL	maximum residue level
mth	month(s)
NA	Notice of Approval

NFU	National Farmers' Union
OES	Occupational Exposure Standard
OLA	off-label approval
PPE	personal protective equipment
PPPR	Plant Protection Products Regulations
PSD	Pesticides Safety Directorate
SOLA	specific off-label approval
ULV	ultra-low volume
VI	Voluntary Initiative
w/v	weight/volume
w/w	weight/weight
wk	week(s)
yr	year(s)

SECTION 6

Appendix 6
Definitions

The descriptions used in this *Guide* for the crops or situations in which products are approved for use are those used on the approved product labels. These are now standardised in a Crop Hierarchy published by the Pesticides Safety Directorate in which definitions are given. To assist users of this *Guide* the definitions of some of the terminology where misunderstandings can occur are reproduced below.

Rotational grass: Short-term grass crops grown on land that is likely to be growing different crops in future years (*e.g. short-term intensively managed leys for one to three years that may include clover*)

Permanent grassland: Grazed areas that are intended to be permanent in nature (*e.g. permanent pasture and moorland that can be grazed*).

Ornamental Plant Production: All ornamental plants that are grown for sale or are produced for replanting into their final growing position (*e.g. flowers, house plants, nursery stock, bulbs grown in containers or in the ground*).

Managed Amenity Turf: Areas of frequently mown, intensively managed, turf that is not intended to flower and set seed. It includes areas that may be for intensive public use (*e.g. all types of sports turf*).

Amenity Grassland: Areas of semi-natural or planted grassland subject to minimal management. It includes areas that may be accessed by the public (*e.g. railway and motorway embankments, airfields, and grassland nature reserves*). These areas may be managed for their botanical interest, and the relevant authority should be contacted before using pesticides in such locations.

Amenity Vegetation: Areas of semi-natural or ornamental vegetation, including trees, or bare soil around ornamental plants, or soil intended for ornamental planting. It includes areas to which the public have access. It does NOT include hedgerows around arable fields.

Natural surfaces not intended to bear vegetation: Areas of soil or natural outcroppings of rock that are not intended to bear vegetation, including areas such as sterile strips around fields. It may include areas to which the public have access. It does not include the land between rows of crops.

Hard surfaces: Man-made impermeable surfaces that are not intended to bear vegetation (*e.g. pavements, tennis courts, industrial areas, railway ballast*).

Permeable surfaces overlying soil: Any man-made permeable surface (excluding railway ballast) such as gravel that overlies soil and is not intended to bear vegetation

Green Cover on Land Temporarily Removed from Production: Includes fields covered by natural regeneration or by a planted green cover crop that will not be harvested (*e.g. green cover on setaside*). It does NOT include industrial crops.

Forest Nursery: Areas where young trees are raised outside for subsequent forest planting.

Forest: Groups of trees being grown in their final positions. Covers all woodland grown for whatever objective, including commercial timber production, amenity and recreation, conservation and landscaping, ancient traditional coppice and farm forestry, and trees from natural regeneration, colonisation or coppicing. Also includes restocking of established woodlands and new planting on both improved and unimproved land.

Farm forestry: Groups of trees established on arable land or improved grassland including those planted for short rotation coppicing. It includes mature hedgerows around arable fields.

Indoors (for rodenticide use): Situations where the bait is placed within a building or other enclosed structure, and where the target is living or feeding predominantly within that building or structure.

Herbs: Reference to Herbs or Protected Herbs when used in Section 2 may include any or all of the following. The particular label or SOLA Notice will indicate which species are included in the approval.

Agastache spp.	Lemon thyme
Angelica	Lemon verbena
Applemint	Lovage
Balm	Marigold
Basil	Marjoram
Bay	Mint
Borage (except when grown for oilseed)	Mother of thyme
Camomile	Nasturtium
Caraway	Nettle
Catnip	Oregano
Chervil	*Origanum heracleoticum*
Clary	Parsley root
Clary sage	Peppermint
Coriander	Pineapplemint
Curry plant	Rocket
Dill	Rosemary
Dragonhead	Rue
English chamomile	Sage
Fennel	Salad burnet
Fenugreek	Savory
Feverfew	Sorrel
French lavender	Spearmint
Gingermint	Spike lavender
Hyssop	Tarragon
Korean mint	Thyme
Land cress	*Thymus camphoratus*
Lavandin	Violet
Lavender	Winter savory
Lemon balm	Woodruff
Lemon peppermint	

Herbs for Medicinal Uses: Reference to Herbs for Medicinal Uses when used in Section 2 may include any or all of the following. The particular label or SOLA Notice will indicate which species are included in the approval

Black cohosh	Goldenseal
Burdock	Liquorice
Dandelion	Nettle
Echinacea	Valerian
Ginseng	

SECTION 6

Appendix 7
References

The information given in *The UK Pesticide Guide* provides some of the answers needed to assess health risks, including the hazard classification and the level of operator protection required. However, the *Guide* cannot provide all the details needed for a complete hazard assessment, which must be based on the product label itself and, where necessary, the Health and Safety Data Sheet and other official literature.

Detailed guidance on how to comply with the Regulations is available from several sources.

Pesticides: Code of Practice
Code of Practice for the Safe Use of Pesticides on Farms and Holdings, 1998
(ISBN 0 11 242892 4)

Known as the 'Green Code', this Defra publication (PB3528), which promotes the safe use of pesticides, covers the requirements of the Control of Pesticides Regulations 1986 (COPR) and the Control of Substances Hazardous to Health Regulations 2002 (COSHH). The principal source of information for making a COSHH assessment is the approved product label. In most cases the label provides all the necessary information but in certain circumstances other sources must be consulted, and these are listed in the Code of Practice. **Note that this Code was being revised, and amalgamated with the 'Orange Code' (which covers pesticide use in amenity and industrial areas), when this edition went to press. The new code is due for publication in 2006. Details can be found on the PSD website (www.pesticides.gov.uk).**

Other Codes of Practice
Code of Practice for Suppliers of Pesticides to Agriculture, Horticulture and Forestry (the 'Yellow Code') (Defra Booklet PB 3529)

Code of Good Agricultural Practice for the Protection of Soil (Defra Booklet PB 0617)

Code of Good Agricultural Practice for the Protection of Water (Defra Booklet PB 0587)

Code of Good Agricultural Practice for the Protection of Air (Defra Booklet PB 0618)

Control of Substances Hazardous to Health Regulations 2002. Approved Code of Practice and Guidance (ISBN 0 7176 2534 6)

Approved Code of Practice for the Control of Substances Hazardous to Health in Fumigation Operations. Health and Safety Commission (ISBN 0 7176 1195 7)

Other Guidance and Practical Advice
HSE (by mail order from HSE Books) (See Appendix 2)
Recommendations for Training Users of Non-Agricultural Pesticides. Health and Safety Commission (ISBN 0 11 885548 4)

Defra (from The Stationery Office – see Appendix 2)
Local Environment Risk Assessments for Pesticides – A Practical Guide (PB4168)

Crop Protection Association (see Appendix 2)

Every Drop Counts: Keeping Water Clean

For the Benefit of Biodiversity – A Biodiversity Strategy and Action Plan for a Better Environment

Best Practice Guides. A range of leaflets giving guidance on best practice when dealing with pesticides before, during and after application.

British Crop Production Council (see Appendix 2)

The Pesticide Manual (13th edition) (ISBN 1 90139 613 4)

The e-Pesticide Manual PC CD-ROM (Version 3.2) (ISBN 1 901396 40 1)

The Manual of Biocontrol Agents (3rd edition of *The BioPesticide Manual*) (ISBN 1 90139 635 5)

IdentiPest PC CD-ROM (ISBN 1 90139 605 3)

Garden Detective PC CD-ROM (ISBN 1 90139 632 0)

Hand-held and Amenity Sprayers Handbook (ISBN 1 90139 603 7)

Boom and Fruit Sprayers Handbook (ISBN 1 90139 602 9)

Using Pesticides: A Complete Guide to Safe Effective Spraying (ISBN 1 90139 601 0)

Safety Equipment Handbook (ISBN 1 90139 606 1)

The Environment Agency (see Appendix 2)

The Prevention of Pollution by Pesticides (Leaflet PPG9)

SECTION 6

SECTION 7
INDEX

Index of Proprietary Names of Products

The references are to entry numbers, not to pages. Adjuvant names are referred to as 'Adj' and are listed separately in Section 4

REFERENCES ARE TO ENTRY NUMBERS NOT PAGES

REFERENCES ARE TO ENTRY NUMBERS NOT PAGES

INDEX

REFERENCES ARE TO ENTRY NUMBERS NOT PAGES

REFERENCES ARE TO ENTRY NUMBERS NOT PAGES

INDEX

REFERENCES ARE TO ENTRY NUMBERS NOT PAGES

REFERENCES ARE TO ENTRY NUMBERS NOT PAGES

REFERENCES ARE TO ENTRY NUMBERS NOT PAGES

THE UK PESTICIDE GUIDE 2006

RE-ORDERS

☐ Please send me _____ more copies of *The UK Pesticide Guide 2006* at £36.95 each

☐ Please send me —— copies of *The e-UK Pesticide Guide 2006* (PC CD-ROM, single-user version) at £52.50 + VAT (£61.68 incl. VAT)

Name _____ Position_____

Institution _____ Department_____

Address _____

City _____ Region _____ Postcode_____

Country _____

　　　　Tel_____ Fax_____ E-mail_____

EU countries except UK – VAT No:_____

Payment (pre-payment is required)

☐ I enclose a cheque/draft for £_____ payable to BCPE Ltd. Please send me a receipt.

☐ I wish to pay by credit card: ☐ Visa   ☐ Mastercard   ☐ Amex   ☐ Switch

Please charge to my card £_____ and send me a receipt. Name of issuing bank_____

Card no. ☐☐☐☐ ☐☐☐☐ ☐☐☐☐ ☐☐☐☐

Expiry date ☐☐/☐☐ Security code ☐☐/☐☐ Switch cards only: Start date ☐☐/☐☐ Issue no. ☐

Signature_____ Date_____

Name and address of cardholder if different from above:_____

_____

Please photocopy and return to:
BCPC BCPC Publications Sales, 7 Omni Business Centre, Omega Park, Alton, Hants GU34 2QD, UK
Tel: 01420 593 200, Fax: 01420 593 209, Email: publications@bcpc.org, Web: www.bcpc.org

FUTURE EDITIONS

☐ I wish to take out an annual order for _____ copies of each new edition of *The UK Pesticide Guide*

☐ Please send me advance price details for the 2007 edition of *The UK Pesticide Guide* when available

Order by phone: 01420 593 200 or online: www.bcpc.org/bookshop

Bookshop orders to:	Bulk discount:	
Customer Services, CABI Publishing, CAB International,	100+ copies	30%
Nosworthy Way, Wallingford, Oxfordshire OX10 8DE, UK	50–99	25%
Tel: 01491 832111, Fax: 01491 829292,	10–49	15%
Email: publishing@cabi.org, Web: www.cabi-publishing.org	List price £36.95	

Thank you for your order